Critical Values of t

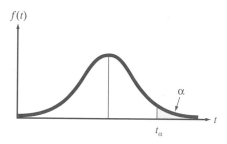

ν	$t_{.100}$	$t_{.050}$	$t_{.025}$	$t_{.010}$	$t_{.005}$	$t_{.001}$	$t_{.0005}$
1	3.078	6.314	12.706	31.821	63.657	318.31	636.62
2	1.886	2.920	4.303	6.965	9.925	22.326	31.598
3	1.638	2.353	3.182	4.541	5.841	10.213	12.924
4	1.533	2.132	2.776	3.747	4.604	7.173	8.610
5	1.476	2.015	2.571	3.365	4.032	5.893	6.869
6	1.440	1.943	2.447	3.143	3.707	5.208	5.959
7	1.415	1.895	2.365	2.998	3.499	4.785	5.408
8	1.397	1.860	2.306	2.896	3.355	4.501	5.041
9	1.383	1.833	2.262	2.821	3.250	4.297	4.781
10	1.372	1.812	2.228	2.764	3.169	4.144	4.587
11	1.363	1.796	2.201	2.718	3.106	4.025	4.437
12	1.356	1.782	2.179	2.681	3.055	3.930	4.318
13	1.350	1.771	2.160	2.650	3.012	3.852	4.221
14	1.345	1.761	2.145	2.624	2.977	3.787	4.140
15	1.341	1.753	2.131	2.602	2.947	3.733	4.073
16	1.337	1.746	2.120	2.583	2.921	3.686	4.015
17	1.333	1.740	2.110	2.567	2.898	3.646	3.965
18	1.330	1.734	2.101	2.552	2.878	3.610	3.922
19	1.328	1.729	2.093	2.539	2.861	3.579	3.883
20	1.325	1.725	2.086	2.528	2.845	3.552	3.850
21	1.323	1.721	2.080	2.518	2.831	3.527	3.819
22	1.321	1.717	2.074	2.508	2.819	3.505	3.792
23	1.319	1.714	2.069	2.500	2.807	3.485	3.767
24	1.318	1.711	2.064	2.492	2.797	3.467	3.745
25	1.316	1.708	2.060	2.485	2.787	3.450	3.725
26	1.315	1.706	2.056	2.479	2.779	3.435	3.707
27	1.314	1.703	2.052	2.473	2.771	3.421	3.690
28	1.313	1.701	2.048	2.467	2.763	3.408	3.674
29	1.311	1.699	2.045	2.462	2.756	3.396	3.659
30	1.310	1.697	2.042	2.457	2.750	3.385	3.646
40	1.303	1.684	2.021	2.423	2.704	3.307	3.551
60	1.296	1.671	2.000	2.390	2.660	3.232	3.460
120	1.289	1.658	1.980	2.358	2.617	3.160	3.373
∞	1.282	1.645	1.960	2.326	2.576	3.090	3.291

Source: This table is reproduced with the kind permission of the Trustees of Biometrika from E. S. Pearson and H.O. Hartley (eds.), *The Biometrika Tables for Statisticians*, Vol. 1, 3d ed., Biometrika, 1966.

Annotated Instructor's Edition

A FIRST COURSE IN
STATISTICS

Eighth Edition

James T. McClave
Info Tech, Inc.
University of Florida

Terry Sincich
University of South Florida

Prentice
Hall

Pearson Education, Inc.   Upper Saddle River, New Jersey 07458

Editor in Chief: Sally Yagan
Vice President/Director of Production and Manufacturing: David W. Riccardi
Executive Managing Editor: Kathleen Schiaparelli
Senior Managing Editor: Linda Mihatov Behrens
Project Management: Elm Street Publishing Services, Inc.
Manufacturing Buyer: Alan Fischer
Manufacturing Manager: Trudy Pisciotti
Marketing Assistant: Rachel Beckman
Assistant Editor: Vince Jansen
Assistant Managing Editor, Math Media Production: John Matthews
Media Production Editor: Donna Crilly
Editorial Assistant/Supplements Editor: Joanne Wendelken
Art Director: John Christiana
Interior and Cover Designer: Jonathan Boylan
Art Editor: Thomas Benfatti
Creative Director: Carole Anson
Director of Creative Services: Paul Belfanti
Photo Editor: Beth Boyd
Interior Photos: Photodisc
    Page 1: Photographer: Karl Weatherly; page 19: Scenics of America/Photolink; page 107: Photographer: Karl Weatherly;
    page 163: F. Schussler/Photolink; page 237: Scenics of America/Photolink; page 277: S. Alden/Photolink; page 321:
    Photographer: Karl Weatherly; page 397: J. Luke/Photolink; page 439: Photographer: Karl Weatherly
Cover Photo: Mike Hewitt/Getty Images Stone
Art Studio: Elm Street Publishing Services, Inc.

© 2003 by Pearson Education, Inc.
Upper Saddle River, NJ 07458

Printed in the United States of America
10   9   8   7   6   5   4   3   2   1

ISBN 0-13-067423-0   (0-13-065596-1 Student Edition)

Pearson Education LTD., *London*
Pearson Education Australia PTY, Limited, *Sydney*
Pearson Education Singapore, Pte. Ltd.
Pearson Education North Asia Ltd., *Hong Kong*
Pearson Education Canada, Ltd., *Toronto*
Pearson Educación de Mexico, S.A. de C.V.
Pearson Education—Japan, *Tokyo*
Pearson Education Malaysia, Pte. Ltd.

CONTENTS

P R E F A C E

STATISTICS IN THE NEW MILLENIUM

This eighth edition of *A First Course in Statistics* is an introductory text emphasizing inference, with extensive coverage of data collection and analysis as needed to evaluate the reported results of statistical studies and make good decisions. As in earlier editions, the text stresses the development of statistical thinking, the assessment of credibility and value of the inferences made from data, both by those who consume and those who produce them. This one-semester text covers basic statistical and probability topics through simple linear regression. It assumes a mathematical background of basic algebra.

NEW TO THE EIGHTH EDITION

- Exercise sets have been revised to provide a greater variety in level of difficulty. In addition to "Learning the Mechanics" exercises, "Applied Exercises" are categorized into "Basic," "Intermediate," and "Advanced" at the end of each section.
- All printouts from statistical software (SAS, SPSS, MINITAB, and STATISTIX) have been revised to reflect the latest versions of the software.
- The sections in Chapter 8 (Comparing Population Proportions) have been reorganized with emphasis on whether one or two qualitative variables are analyzed.

TRADITIONAL STRENGTHS

We have maintained the pedagogical features of *A First Course in Statistics* that we believe make it unique among introductory statistics texts. These features, which assist the student in achieving an overview of statistics and an understanding of its relevance in the social and life sciences, business, and everyday life, are as follows:

- **Use of Examples as a Teaching Device**   Almost all new ideas are introduced and illustrated by data-based applications and examples. We believe that students better understand definitions, generalizations, and abstractions *after* seeing an application.
- **Exploring Data with Statistical Computer Software**   Each statistical analysis method presented is demonstrated using output from four leading statistical software packages: SAS, SPSS, MINITAB, or STATISTIX. In addition, output and keystroke instructions for the TI-83 Graphing Calculator are covered in optional boxes.
- **Nonparametric Methods Integrated**   Throughout the text, optional sections on alternative distribution-free (nonparametric) procedures follow the relevant sections.
- **Statistics in Action**   Each chapter concludes with a case study on a contemporary, controversial, or high-profile issue. Data from the study are presented for analysis and questions, prompting the students to evaluate the findings and

to think through the statistical issues involved. Additional cases appear on the Web site.

- **Real Data Exercises**   The text includes more than 800 exercises illustrated by applications in almost all areas of research. All the applied exercises employ the use of current real data extracted from a wide variety of publications (e.g., newspapers, magazines, journals, and the Internet). Many students have trouble learning the mechanics of statistical techniques when all problems are couched in terms of realistic applications. For this reason, all exercise sections are divided into four parts:

    Learning the Mechanics.   Designed as straightforward applications of new concepts, these exercises allow students to test their ability to comprehend a mathematical concept or a definition.

    Applying the Concepts—Basic.   Based on applications taken from a wide variety of journals, newspapers, and other sources, these short exercises help the student begin developing the skills necessary to diagnose and analyze real-world problems.

    Applying the Concepts—Intermediate.   Based on more detailed real-world applications, these exercises require students to apply their knowledge of the technique presented in the section.

    Applying the Concepts—Advanced.   These more challenging real-data exercises require students to utilize their critical thinking skills.

- **Quick Review**   Each chapter ends with a list of key terms and formulas, with reference to the page number where they first appear.
- **Language Lab**   Following the Quick Review is a pronunciation guide to Greek letters and other special terms. Usage notes are also provided.
- **Student Projects**   Presented at the end of each chapter, these projects emphasize gathering data, analyzing data, and/or report writing.
- **Footnotes**   Although the text is designed for students with a non-calculus background, footnotes explain the role of calculus in various derivations. Footnotes are also used to inform the student about some of the theory underlying certain methods of analysis. These footnotes allow additional flexibility in the mathematical and theoretical level at which the material is presented.

ACKNOWLEDGMENTS

This book reflects the efforts of a great many people over a number of years. First, we would like to thank the following professors, whose reviews and comments on this and prior editions of *Statistics* and/or *A First Course in Statistics* have contributed to the eighth edition:

Reviewers of Previous Editions

Bill Adamson
South Dakota State University

Ibrahim Ahmad
Northern Illinois University

David Atkinson
Olivet Nazarene University

Mary Sue Beersman
Northeast Missouri State University

William H. Beyer
University of Akron

Patricia M. Buchanan
Pennsylvania State University

Dean S. Burbank
Gulf Coast Community College

Kathryn Chaloner
University of Minnesota

Hanfeng Chen
Bowling Green State University

Gerardo Chin-Leo
The Evergreen State College

Linda Brant Collins
Iowa State University

Brant Deppa
Winona State University

John Dirkse
California State University—Bakersfield

N. B. Ebrahimi
Northern Illinois University

John Egenolf
University of Alaska—Anchorage

Dale Everson
University of Idaho

Christine Franklin
University of Georgia

Rudy Gideon
University of Montana

Victoria Marie Gribshaw
Seton Hill College

Larry Griffey
Florida Community College

David Groggel
Miami University at Oxford

John E. Groves
*California Polytechnic State University—
San Luis Obispo*

Dale K. Hathaway
Olivet Nazarene University

Shu-ping Hodgson
Central Michigan University

Jean L. Holton
Virginia Commonwealth University

Soon Hong
Grand Valley State University

Ina Parks S. Howell
Florida International University

Gary Itzkowitz
Rowan College of New Jersey

John H. Kellermeier
State University College at Plattsburgh

Timothy J. Killeen
University of Connecticut

William G. Koellner
Montclair State University

James R. Lackritz
San Diego State University

Diane Lambert
AT&T/Bell Laboratories

Edwin G. Landauer
Clackamas Community College

James Lang
Valencia Junior College

Glenn Larson
University of Regina

John J. Lefante, Jr.
University of South Alabama

Pi-Erh Lin
Florida State University

R. Bruce Lind
University of Puget Sound

Rhonda Magel
North Dakota State University

Linda C. Malone
University of Central Florida

Allen E. Martin
California State University—Los Angeles

Rick Martinez
Foothill College

Brenda Masters
Oklahoma State University

Leslie Matekaitis
Cal Genetics

E. Donice McCune
Stephen F. Austin State University

Mark M. Meerschaert
University of Nevada—Reno

Greg Miller
Steven F. Austin State University

Satya Narayan Mishra
University of South Alabama

Christopher Morrell
Loyola College in Maryland

A. Mukherjea
University of South Florida

Thomas O'Gorman
Northern Illinois University

Bernard Ostle
University of Central Florida

William B. Owen
Central Washington University

Won J. Park
Wright State University

John J. Peterson
Smith Kline & French Laboratories

Ronald Pierce
Eastern Kentucky University

Betty Rehfuss
North Dakota State University—Bottineau

Andrew Rosalsky
University of Florida

C. Bradley Russell
Clemson University

Rita Schillaber
University of Alberta

James R. Schott
University of Central Florida

Susan C. Schott
University of Central Florida

George Schultz
St. Petersburg Junior College

Carl James Schwarz
University of Manitoba

Mike Seyfried
Shippensburg University

Arvind K. Shah
University of South Alabama

Lewis Shoemaker
Millersville University

Charles W. Sinclair
Portland State University

Robert K. Smidt
California Polytechnic State University—San Luis Obispo

Vasanth B. Solomon
Drake University

W. Robert Stephenson
Iowa State University

Thaddeus Tarpey
Wright State University

Kathy Taylor
Clackamas Community College

Barbara Treadwell
Western Michigan University

Dan Voss
Wright State University

Augustin Vukov
University of Toronto

Dennis D. Wackerly
University of Florida

Matthew Wood
University of Missouri—Columbia

Reviewers of Current Edition

Ann Cascarelle
St. Petersburg College

Dale K. Hathaway
Olivet Nazarene University

Gary S. Itzkowitz
Rowan College of New Jersey

Mir Mortazavi
Eastern New Mexico University

Steve Nimmo
Morningside College (Iowa)

James R. Schott
University of Central Florida

Special thanks are due to our ancillary authors, Nancy Shafer Boudreau, Mark Dummeldinger, Engin Sungur, Anne Drougas, Augustin Vukov, Singh Kelly, and to typist Kelly Barber, several of whom have worked with us for many years. Sudhir Goel has done an excellent job of accuracy checking and has helped us to ensure a highly accurate, clean text. Finally, the Prentice Hall staff of Rachel Beckman, Joanne Wendelken, Vince Jansen, Linda Behrens, and Alan Fischer and Elm Street Publishing Services' Martha Beyerlein, Melissa Morgan, and Irene Sotiroff helped greatly with all phases of the text development, production, and marketing effort.

SUPPLEMENTS FOR THE INSTRUCTOR

Annotated Instructor's Edition (AIE) (ISBN 0-13-067423-0)

Marginal notes placed next to discussions of essential teaching concepts include:

- Teaching Tips—suggest alternative presentations or point out common student errors
- Exercises—reference specific section and chapter exercises that reinforce the concept
- Short Answers—section and chapter exercise answers are provided next to the selected exercises

Instructor's Solutions Manual by Nancy S. Boudreau (ISBN 0-13-067419-2)

Solutions to all of the even-numbered exercises are given in this manual. Careful attention has been paid to ensure that all methods of solution and notation are consistent with those used in the core text. Solutions to the odd-numbered exercises are found in the *Student's Solutions Manual*.

Test Bank by Mark Dummeldinger (ISBN 0-13-067410-9)

The *Test Bank* includes more than 1,000 problems that correlate to problems presented in the text.

Instructor's Notes by Mark Dummeldinger (ISBN 0-13-066074-4)

This printed resource contains suggestions for using the questions at the end of the *Statistics in Action* cases as the basis for class discussion and a complete short answer book with letter of permission to duplicate for student use.

TestGen-EQ (0-13-066078-7)

TestGen-EQ is a computerized test bank which allows you to view and edit test bank questions, transfer them to exams, and print in a variety of formats. The program offers many options for organizing and displaying test banks and tests and a built-in random number and test generator allows instructors to create multiple versions of exams.

WebCT/BlackBoard by Engin Sungur

The *BlackBoard* and *WebCT* sites offer course-compatible content including the *Student's Solutions Manual*, technology help, quizzes, and lecture content. These sites also include Web-based statistical applets (STATLETS™) designed specifically for the text by StatPoint, LLC, which encourage students to develop their conceptual understanding of statistics by providing accessible computational and graphing capabilities matched with specific questions to reinforce understanding of key concepts from the text.

CourseCompass by Engin Sungur

CourseCompass is the perfect course management solution that combines quality Prentice Hall content with state-of-the-art *BlackBoard* technology! It is a dynamic, interactive online course management tool powered by *BlackBoard*. This exciting product allows you to teach with market-leading Prentice Hall content in an easy-to-use customizable format. This site offers course-compatible content, including the *Student's Solutions Manual*, technology help, quizzes, and lecture content. This site also includes Web-based statistical applets (STATLETS™) designed specifically for the eighth edition of the text by StatPoint, LLC, which encourage students to develop their conceptual understanding of statistics by providing accessible computational and graphing capabilities matched with specific questions to reinforce understanding of key concepts from the text.

Companion Website (0-13-066073-6)

Located at *http://www.prenhall.com/mcclave*, the *Companion Website* is an online, interactive study guide, matched to each chapter of the text, with an online syllabus builder, technology help, quizzes, objectives, Internet destinations, PowerPoint downloads, and the text's data files available for download.

SUPPLEMENTS AVAILABLE FOR THE STUDENT

Data CD

The text is accompanied by a CD which contains files for all of the data sets marked with a CD icon in the text. These include data sets for text examples, exercises, and *Statistics in Action* cases. All data files are saved in ASCII format for easy importing into statistical software (e.g., SAS, SPSS, MINITAB, or STATISTIX). A list of the data sets, with file names and variables, is provided in Appendix B.

Student's Solutions Manual by Nancy S. Boudreau (ISBN 0-13-067418-4)

Fully worked-out solutions to all of the odd-numbered exercises are provided in this manual. Careful attention has been paid to ensure that all methods of solution and notation are consistent with those used in the core text.

TI-83 Manual by Singh Kelly (0-13-066919-9)

MINITAB Manual by Augustin Vukov (0-13-066922-9)

Microsoft Excel Manual by Anne Drougas (0-13-066921-0)

Each spiral-bound companion manual works hand-in-glove with the text. Step-by-step keystroke-level instructions, with screen captures, provide detailed help for using the technology to work pertinent exercises. A cross-reference chart indicates which text examples are included and the exact page reference in both the text and technology manual. Output with brief instruction is provided for selected odd-numbered exercises to reinforce the examples.

WebCT/BlackBoard by Engin Sungur

The *BlackBoard* and *WebCT* sites offer course-compatible content including the *Student's Solutions Manual*, technology help, quizzes, and lecture content. These sites also include Web-based statistical applets (STATLETS™) designed specifically for the text by StatPoint, LLC, which encourage students to develop their conceptual understanding of statistics by providing accessible computational and graphing capabilities matched with specific questions to reinforce understanding of key concepts from the text.

CourseCompass by Engin Sungur

CourseCompass is the perfect course management solution that combines quality Prentice Hall content with state-of-the-art *BlackBoard* technology! It is a dynamic, interactive online course management tool powered by *BlackBoard*. This exciting product allows you to learn with market-leading Prentice Hall content in an easy-to-use customizable format. This site offers course-compatible content, including the *Student's Solutions Manual*, technology help, quizzes, and lecture content. This site also includes Web-based statistical applets (STATLETS™) designed specifically for the text by StatPoint, LLC, which encourage students to develop their conceptual understanding of statistics by providing accessible computational and graphing capabilities matched with specific questions to reinforce understanding of key concepts from the text.

Companion Website

Located at *http://www.prenhall.com/mcclave*, the *Companion Website* is an online, interactive study guide, matched to each chapter of the text, with technology help, quizzes, objectives, Internet destinations, PowerPoint downloads, and the text's data files available for download.

LIST OF DATA SOURCES

ABC News
Academy of Management Journal
Adapted Physical Activity Quarterly
Alachua County (Florida) Property
    Appraisers Office
American Association for Marriage and
    Family Therapy
*American Association of Nurse
    Anesthetists Journal*
American Automobile Association
American Cancer Society
American Education Research Journal
American Journal of Audiology
American Journal of Dance Therapy
American Journal of Orthopsychiatry
American Journal of Political Science
American Journal of Psychiatry
American Journal of Psychology
American Journal of Public Health
*American Journal of Speech-Language
    Pathology*
American Journal on Mental Retardation
American Psychologist
American Psychosomatic Society
American Rifleman
American Sociological Review
American Scientist
American Statistical Association
American Statistician
AmStat News
Animal Behavior
*Annals of the Association of American
    Geographers*
Annals of Internal Medicine
Annals of Tourism Research
Applied Animal Behaviour Science
Applied Psycholinguistics
Applied Spectroscopy
Aquatic Botany
Arctic and Alpine Research
Archives of Disease in Childhood
Archives of Internal Medicine
Association for the Advancement of
    Applied Sport Psychology
Astronomical Journal
Astronomy
Australian Journal of Zoology
Bayonet Point Hospital (St. Petersburg,
    Florida)
Beanie World Magazine
Benefits Quarterly
Biogen, Inc.
Brain and Behaviour Evolution
Brain and Language
British Journal of Sports Medicine
Business Week
CACI Marketing Systems
Card Fax

Car and Driver
Cell Biology
Centers for Disease Control and
    Prevention
Chance
Chemosphere
Chest
Children and Youth Services Review
Christianity Today
Chromatographia
Clinical Kinesiology
Clinical Psychology Review
Cognitive Aging Conference
College Entrance Examination Board
College Student Journal
Collegium Antropologicum
Communication Research
Community Mental Health Journal
Consumer Reports
Consumer's Research
Dermatology
Dini Miller, Virginia Tech University
Discover
Ear & Hearing
Ecological Applications
Economic Botany
Educational Technology
Elizabeth Schreiber, University of Kansas
Entrepreneur
Environmental Entomology
Environmental Protection Agency
Environmental Science & Engineering
Environmental Science & Technology
Ethology
*European Journal of Operational
    Research*
Exceptional Children
Explore (University of Florida)
Federal Trade Commission
Fisheries Science
Florida Attoney General's Office
Florida Department of Health and
    Administrative Services
Florida Employer Opinion Survey
Florida Scientist
Florida Shredder's Association
Florida Trend
Food and Drug Administration
Forbes
Fortune
Frances VanVoorhis
Genetical Research
Geoderma
Geography
Geological Magazine
Geriatric Psychiatry
Health Education Journal
Health Psychology

Herman Kelting
Hillsborough County (Florida) Property
    Appraiser's Office
Hillsborough County (Florida) Water
    Department
Howard Journal of Criminal Justice
Human Factors
*IBM Journal of Research and
    Development*
*IEEE Engineering in Medicine and
    Biology Magazine*
IEEE Transactions
*IEEE Transactions on Semiconductor
    Manufacturing*
*IEEE Transactions on Speech and Audio
    Processing*
Industrial Engineering
Info Tech, Inc.
Interfaces
*International Journal of Circuit Theory
    and Applications*
*International Journal of Radiation
    Oncology and Biological Physics*
International Rhino Federation
Johnson & Johnson, Inc.
*Journal of the Academy of Rehabilitative
    Audiology*
*Journal of Adlerian Theory, Research and
    Practice*
Journal of Advanced Nursing
*Journal of Agricultural, Biological, and
    Environmental Statistics*
*Journal of Allergy and Clinical
    Immunology*
*Journal of the American Medical
    Association*
*Journal of the American Mosquito
    Control Association*
*Journal of the American Statistical
    Association*
Journal of Applied Biology
Journal of Applied Biomechanics
Journal of Applied Business Research
Journal of Applied Ecology
Journal of Applied Phycology
Journal of Applied Physics
Journal of Applied Psychology
Journal of Behavioral Medicine
Journal of Business Communications
Journal of Business & Economic Statistics
Journal of Business Ethics
*Journal of Business Finance and
    Accounting*
*Journal of Child Psychology and
    Psychiatry*
Journal of Colloid and Interface Science
Journal of Communication Disorders
Journal of Computer Information Systems

Journal of Consulting and Clinical Psychology
Journal of Drug Issues
Journal of Economics
Journal of Educational Statistics
Journal of Ethnopharmacology
Journal of Experimental Psychology
Journal of Experimental Zoology
Journal of Family Violence
Journal of Fish Biology
Journal of Gang Research
Journal of Genetic Psychology
Journal of Geography
Journal of Hazardous Materials
Journal of Head Trauma Rehabilitation
Journal of Human Stress
Journal of Information Science
Journal of Infrastructure Systems
Journal of Intellectual Disability Research
Journal of Leisure Research
Journal of Marketing Research
Journal of Marriage and the Family
Journal of the Mosquito Control Association
Journal of the National Cancer Institute
Journal of Negro Education
Journal of Nonverbal Behavior
Journal of Nursing Education
Journal of Nutrition
Journal of Occupational Psychology
Journal of Organic Chemistry
Journal of Orthopsychiatry
Journal of Parapsychology
Journal of Peace Research
Journal of Political Economy
Journal of Portfolio Management
Journal of Psychology
Journal of Psychology and Aging
Journal of Quality Technology
Journal of Quantitative Criminology
Journal of Real Estate Research
Journal of Rehabilitation
Journal of Research in Music Education
Journal of Retailing
Journal of Shellfish Research
Journal of Socio-Economics
Journal of Soil and Water Conservation
Journal of Speech, Language, and Hearing Research
Journal of Sport and Social Issues
Journal of Sport Behavior
Journal of Sports Medicine and Physical Fitness
Journal of STAR Research
Journal of Statistics Education
Journal of Studies in Technical Careers
Journal of Testing and Evaluation
Journal of Visual Impairment & Blindness
Journal of Wildlife Management
Journal of Zoology

Kelly Uscategui, University of Connecticut
Ladies Home Journal
Lei Lei, Rutgers University
Library Acquisitions: Practice & Theory
Liquid Assets: The International Guide to Fine Wines
Major League Baseball
Management Science
Marine Technology
Mathematical Geology
Memory & Cognition
Merck Research Labs
Meteoritics
Minneapolis Star and Tribune
Minnesota Department of Transportation
Moffitt Cancer Research Center (University of South Florida)
Motivation and Emotion
National Aeronautics and Space Administration
National Center for Education Statistics
National Center for Environmental Health
National Direct Student Loan Program
National Earthquake Information Center
National Highway Traffic Safety Administration
National Institute of Environmental Health Services
National Oceanic and Atmospheric Administration
National Wildlife
Nature
Naval Research Logistics
Networks
Newark Star-Ledger
New England Journal of Medicine
New Jersey Business
New Scientist
New York Times
Newsweek
Nilson Report
Optometry and Vision Science
Organizational Behaviour and Human Decision Processes
Organizational Science
Parade Magazine
Pediatric Exercise Science
Perception & Psychophysics
Physical Therapy
Phytopathology
Planta Medica
Political Science & Politics
Principles of Operation Management
Proceedings of the National Academy of Sciences
Production and Inventory Management Journal
Professional Geographer
Progress in Natural Science

Psychological Assessment
Psychological Reports
Psychology and Aging
Psychosomatic Medicine
Quality Congress Transactions
Quality Engineering
Quality Management Journal
Quality Progress
Queuing Models
Rehabilitative Psychology
Research Review
Robinson Appraisal Co., Inc.
Science
Science Education
Science News
Scientific American
Sleep
Social Psychology Quarterly
Sociological Methods & Research
Sociology of Sport Journal
Soil Science
Sporting News
Sports Illustrated
Statistical Abstract of the United States
Statistical Bulletin
Statistics in Sport
St. Petersburg Times
Susan Farber, *Identical Twins Reared Apart*
Tampa Tribune
Teaching Psychology
Teaching Sociology
Technometrics
The Bell Curve: Intelligence and Class Structure in American Life
The Condor
The Movie Times
Time
UCLA Journal of Environmental Law & Policy
University of Florida
University of South Florida Office of Research
U.S. Army Corps of Engineers
U.S. Bureau of Labor Statistics
U.S. Chess Federation
U.S. Department of Energy
U.S. Golf Association
U.S. News & World Report
USA Today
USF Magazine
Wall Street Journal
Washington Post
Wetlands
Wine Spectator
Work and Occupations
World Archaeology
World Development
World Economy
Zoological Science

HOW TO USE THIS BOOK

To the Student

The following pages will demonstrate how to use this text in the most effective way to make studying easier and to understand the connection between statistics and your world.

| CHAPTER | 3 |

Probability

Contents

3.1 Events, Sample Spaces, and Probability
3.2 Unions and Intersections
3.3 Complementary Events
3.4 The Additive Rule and Mutually Exclusive Events
3.5 Conditional Probability
3.6 The Multiplicative Rule and Independent Events
3.7 Random Sampling

Statistics in Action

Game Show Strategy: To Switch or Not to Switch?

Where We've Been

We've identified inference, from a sample to a population, as the goal of statistics. And we've seen that to reach this goal, we must be able to describe a set of measurements. Thus, we explored the use of graphical and numerical methods for describing both quantitative and qualitative data sets and for phrasing inferences.

Where We're Going

Now that we know how to phrase an inference about a population, we turn to the problem of making the inference. What is it that permits us to make the inferential jump from sample to population and then to give a measure of reliability for the inference? As you'll see, the answer is *probability*. This chapter is devoted to a study of probability—what it is and some of the basic concepts of the theory behind it.

Page 107

Chapter Openers Show Links Between Old and New Topics

- **Where We've Been** quickly reviews how information learned previously applies to the chapter at hand.

- **Where We're Going** highlights how the chapter topics fit into your growing understanding of statistics.

Easy to Locate Boxes Highlight Important Information

- Definitions, Strategies, Key Formulas, and other important information is highlighted for quick reference.

- Helps students to prepare for quizzes and tests by reviewing the highlighted information.

4.3   THE BINOMIAL DISTRIBUTION

Many experiments result in *dichotomous* responses—i.e., responses for which there exist two possible alternatives, such as Yes–No, Pass–Fail, Defective–Nondefective, or Male–Female. A simple example of such an experiment is the coin-toss experiment. A coin is tossed a number of times, say 10. Each toss results in one of two outcomes, Head or Tail, and the probability of observing each of these two outcomes remains the same for each of the 10 tosses. Ultimately, we are interested in the probability distribution of *x*, the number of heads observed. Many other experiments are equivalent to tossing a coin (either balanced or unbalanced) a fixed number *n* of times and observing the number *x* of times that one of the two possible outcomes occurs. Random variables that possess these characteristics are called **binomial random variables**.

Public opinion and consumer preference polls (e.g., the CNN, Gallup, and Harris polls) frequently yield observations on binomial random variables. For example, suppose a sample of 100 students is selected from a large student body and each person is asked whether he or she favors (a Head) or opposes (a Tail) a certain campus issue. Suppose we are interested in *x*, the number of students in the sample who favor the issue. Sampling 100 students is analogous to tossing the coin 100 times. Thus, you can see that opinion polls that record the number of people who favor a certain issue are real-life equivalents of coin-toss experiments. We have been describing a **binomial experiment**; it is identified by the following characteristics.

Characteristics of a Binomial Random Variable

1. The experiment consists of *n* identical trials.
2. There are only two possible outcomes on each trial. We will denote one outcome by *S* (for Success) and the other by *F* (for Failure).
3. The probability of *S* remains the same from trial to trial. This probability is denoted by *p*, and the probability of *F* is denoted by *q*. Note that $q = 1 - p$.
4. The trials are independent.
5. The binomial random variable *x* is the number of *S*'s in *n* trials.

Page 177

EXAMPLE 5.5   According to *True Odds: How Risk Affects Your Everyday Life* (Walsh, 1997), the probability of being the victim of a violent crime is less than .01. Suppose that in a random sample of 200 Americans, 3 were victims of a violent crime. Estimate the true proportion of Americans who were victims of a violent crime using a 95% confidence interval.

Solution

Let *p* represent the true proportion of Americans who were victims of a violent crime. Since *p* is near 0, an "extremely large" sample is required to estimate its value using the usual large-sample method. Since we are unsure whether the sample size of 200 is large enough, we will apply Wilson's adjustment outlined in the box.

The number of "successes" (i.e., number of violent crime victims) in the sample is *x* = 3. Therefore, the adjusted sample proportion is

$$p^* = \frac{x + 2}{n + 4} = \frac{3 + 2}{200 + 4} = \frac{5}{204} = .025$$

Note that this adjusted sample proportion is obtained by adding a total of four observations—two "successes" and two "failures"—to the sample data. Substituting $p^* = .025$ into the equation for a 95% confidence interval, we obtain

$$p^* \pm 1.96\sqrt{\frac{p^*(1 - p^*)}{n + 4}} = .025 \pm 1.96\sqrt{\frac{(.025)(.975)}{204}}$$
$$= .025 \pm .021$$

or (.004, .046). Consequently, we are 95% confident that the true proportion of Americans who are victims of a violent crime falls between .004 and .046. ∎

Page 262

Learn by Example Approach

- New ideas are introduced and illustrated with data-based applications and examples. Complete solutions and explanations are also provided to help students prepare for end-of-section exercises.

- Tangible applications and examples help students to digest theory because they are always provided *before* definitions, generalizations, or abstractions.

- All examples are numbered for easy reference.

- The end of the solution is marked with a ∎ symbol.

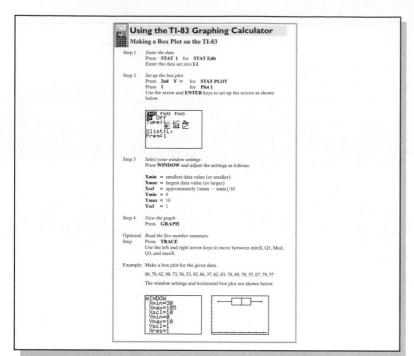

Page 82

TI-83 Graphing Calculator Key Stroke Level Instruction Contained in Boxes

- Gives students step-by-step instruction on how to use the TI-83.

- Helps students to see how technology can be used as a substitute for hand computation.

- Teaches students valuable skills they may use in later careers.

Emphasis on Real Data

- Most of the 1200+ exercises contain data or information taken from current newspaper articles, magazines, journals, and the internet. Statistics are all around you.

- Real data is also integrated into examples, chapter content, graphs, tables, etc.

Computer Output Integrated Throughout

Exploring Data with Statistical Computer Software

- Statistical software packages, such as SPSS, MINITAB, SAS, or STATISTIX, crunch data quickly so students can spend time analyzing the results. Learning how to interpret statistical output will prove helpful in future classes or on the job.

- When computer output appears in examples, the solution explains how to read and interpret the output.

- Output and instructions for the TI-83 graphing calculator are also provided.

	Sample Size	Youngest Tertile Mean Height	Middle Tertile Mean Height	Oldest Tertile Mean Height	F Value	p-Value
Boys	1439	0.33	0.33	0.16	4.57	0.01
Girls	1409	0.27	0.18	0.21	0.85	0.43

Source: Wake, M., Coghlan, D., and Hesketh, K. "Does height influence progression through primary school grades?" *The Archives of Disease in Childhood*, Vol. 82, Apr. 2000 (Table 2).

a. What is the null hypothesis for the ANOVA of the boys' data?
b. Interpret the results of the test, part **a**. Use $\alpha = .05$.
c. Repeat parts **a** and **b** for the girls' data.
d. Summarize the results of the hypothesis tests in the words of the problem.

7.82 The *Journal of Hazardous Materials* (July 1995) published the results of a study of the chemical properties of three different types of hazardous organic solvents used to clean metal parts: aromatics, chloroalkanes, and esters. One variable studied was sorption rate, measured as mole percentage. Independent samples of solvents from each type were tested and their sorption rates were recorded, as shown in the table. A MINITAB analysis of variance of the data is provided.

🌐 SORPRATE

Aromatics		Chloroalkanes		Esters		
1.06	.95	1.58	1.12	.29	.43	.06
.79	.65	1.45	.91	.06	.51	.09
.82	1.15	.57	.83	.44	.10	.17
.89	1.12	1.16	.43	.61	.34	.60
1.05				.55	.53	.17

Source: Reprinted from *Journal of Hazardous Materials*, Vol. 42, No. 2, J. D. Ortego *et al.*, "A review of polymeric geosynthetics used in hazardous waste facilities." p. 142 (Table 9), July 1995, Elsevier Science-NL, Sara Burgerhartstraat 25, 1055 KV Amsterdam, The Netherlands.

```
MINITAB Output for Exercise 7.82

Analysis of Variance for SORPRATE

Source    DF      SS       MS       F        P
SOLVENT    2   3.3054   1.6527   24.51    0.000
Error     29   1.9553   0.0674
Total     31   5.2608
```

a. Construct an ANOVA table from the MINITAB printout.
b. Is there evidence of differences among the mean sorption rates of the three organic solvent types? Test using $\alpha = .10$.

Applying the Concepts—Intermediate

7.83 Studies conducted at the University of Melbourne (Australia) indicate that there may be a difference between the pain thresholds of blondes and brunettes. Men and women of various ages were divided into four categories according to hair color: light blond, dark blond, light brunette, and dark brunette. The purpose of the experiment was to determine whether hair color is related to the amount of pain evoked by common types of mishaps and assorted types of trauma. Each person in the experiment was given a pain threshold score based on his or her performance in a pain sensitivity test (the higher the score, the higher the person's pain tolerance). SAS was used to conduct the analysis of variance of the data listed in the table on p. 385. The SAS printout provided below.

Page 384

Lots of Exercises for Practice

Every section in the book is followed by an Exercise Set divided into four parts:

- **Learning the Mechanics** Designed as straightfoward applications of new concepts, these exercises allow students to test their ability to comprehend a mathematical concept or presentation.

- **Applying the Concepts—Basic** Based on applications taken from a wide variety of journals, newspapers, and other sources, these short exercises help the student to begin developing the skills necessary to diagnose and analyze real-world problems.

 SUPPLEMENTARY EXERCISES 6.86–6.107

Note: List the assumptions necessary for the valid implementation of the statistical procedures you use in solving all these exercises. Starred (*) exercises refer to the optional section in this chapter.

Learning the Mechanics

6.86 Specify the differences between a large-sample and small-sample test of hypothesis about a population mean μ. Focus on the assumptions and test statistics.

6.87 Which of the elements of a test of hypothesis can and should be specified *prior* to analyzing the data that are to be utilized to conduct the test?

6.88 A random sample of 20 observations selected from a normal population produced $\bar{x} = 72.6$ and $s^2 = 19.4$.
  a. Form a 90% confidence interval for the population mean.
  b. Test $H_0: \mu = 80$ against $H_a: \mu < 80$. Use $\alpha = .05$.
  c. Test $H_0: \mu = 80$ against $H_a: \mu \neq 80$. Use $\alpha = .01$.
  d. Form a 99% confidence interval for μ.
  e. How large a sample would be required to estimate μ to within 1 unit with 95% confidence?

6.89 *Complete the following statement:* The smaller the *p*-value associated with a test of hypothesis, the stronger the support for the _____ hypothesis. Explain your answer.

6.90 *Complete the following statement:* The larger the *p*-value associated with a test of hypothesis, the stronger the support for the _____ hypothesis. Explain your answer.

6.91 A random sample of $n = 200$ observations from a binomial population yields $\hat{p} = .29$.
  a. Test $H_0: p = .35$ against $H_a: p < .35$. Use $\alpha = .05$.
  b. Test $H_0: p = .35$ against $H_a: p \neq .35$. Use $\alpha = .05$.
  c. Form a 95% confidence interval for p.
  d. Form a 99% confidence interval for p.
  e. How large a sample would be required to estimate p to within .05 with 99% confidence?

6.92 A random sample of 175 measurements possessed a mean $\bar{x} = 8.2$ and a standard deviation $s = .79$.
  a. Form a 95% confidence interval for μ.
  b. Test $H_0: \mu = 8.3$ against $H_a: \mu \neq 8.3$. Use $\alpha = .05$.
  c. Test $H_0: \mu = 8.4$ against $H_a: \mu \neq 8.4$. Use $\alpha = .05$.

6.93 A *t*-test is conducted for the null hypothesis $H_0: \mu = 10$ versus the alternative $H_a: \mu > 10$ for a random sample of $n = 17$ observations. The data are analyzed using MINITAB, with the results shown below.

```
MINITAB Output for Exercise 6.93

Test of mu = 10 vs mu > 10

Variable         N       Mean      StDev    SE Mean
X               17      12.50       8.78       2.13

Variable         95.0% CI                 T          P
X            (  8.00,   17.00)          1.17      .1288
```

---

  a. Interpret the *p*-value.
  b. What assumptions are necessary for the validity of this test?
  c. Calculate and interpret the *p*-value assuming the alternative hypothesis was instead $H_a: \mu \neq 10$.

Applying the Concepts—Basic

6.94 An article in the *Annals of the Association of American Geographers* (June 1992) revealed that only 133 of 337 randomly selected residences in Los Angeles County were protected by earthquake insurance.
  a. What are the appropriate null and alternative hypotheses to test the research hypothesis that less than 40% of the residents of Los Angeles County were protected by earthquake insurance?
  b. Do the data provide sufficient evidence to support the research hypothesis? Use $\alpha = .10$.
  c. Calculate and interpret the *p*-value for the test.

6.95 When a new drug is formulated, the pharmaceutical company must subject it to lengthy and involved testing before receiving the necessary permission from the Food and Drug Administration (FDA) to market the drug. The FDA requires the pharmaceutical company to provide substantial evidence that the new drug is safe for potential consumers.
  a. If the new drug testing were to be placed in a test of hypothesis framework, would the null hypothesis be that the drug is safe or unsafe? The alternative hypothesis?
  b. Given the choice of null and alternative hypotheses in part **a**, describe Type I and Type II errors in terms of this application. Define α and β in terms of this application.
  c. If the FDA wants to be very confident that the drug is safe before permitting it to be marketed, is it more important that α or β be small? Explain.

*6.96 The American Cancer Society funds medical research focused on finding treatments for various forms of cancer. One new treatment is being tested to determine whether the average remission time for a particularly aggressive form of cancer is extended by the treatment. Assume that current treatments have been shown to provide a median remission time of 4.5 years. Seven patients with this cancer are given the new treatment, and their remission times (in years) are as follows:

⊘ REMISSION

5.3	7.3	3.6	5.2	6.1	4.8	8.4

  Use the sign test to determine if the median remission time is increased by the new treatment. Test at $\alpha = .05$.

6.97 In a British study, 12 healthy college students, deprived of one night's sleep, received an array of tests intended to measure thinking time, fluency, flexibility, and originality of thought. The overall test scores of the sleep-deprived students were compared to the average score expected from students who received their accustomed sleep (*Sleep*, Jan. 1989). Suppose the overall scores of the 12 sleep-deprived students had a mean of $\bar{x} = 63$ and a standard deviation of 17. (Lower scores are associated with a decreased ability to think creatively.)
  a. Test the hypothesis that the true mean score of sleep-deprived subjects is less than 80, the mean score of subjects who received sleep prior to taking the test. Use $\alpha = .05$.
  b. What assumption is required for the hypothesis test of part **a** to be valid?

6.98 "Take the Pepsi Challenge" was a marketing campaign used by the Pepsi-Cola Company. Coca-Cola drinkers participated in a blind taste test in which they tasted unmarked cups of Pepsi and Coke and were asked to select their favorite. Pepsi claimed that "in recent blind taste tests, more than half the Diet Coke drinkers surveyed said they preferred the taste of Diet Pepsi" (*Consumer's Research*, May 1993). Suppose 100 Diet Coke drinkers took the Pepsi Challenge and 56 preferred the taste of Diet Pepsi. Test the hypothesis that more than half of all Diet Coke drinkers will select Diet Pepsi in a blind taste test. Use $\alpha = .05$.

Applying the Concepts—Intermediate

6.99 In order to be effective, a certain mechanical component used in a spacecraft must have a mean length of life greater than 1,100 hours. Owing to the prohibitive cost of this component, only three were tested under simulated space conditions. The lifetimes (hours) of the components were recorded and the computed statistics were $\bar{x} = 1,173.6$ and, $s = 36.3$. These data were analyzed using MINITAB, as shown below.

```
MINITAB Output for Exercise 6.99

Test of mu = 1100 vs mu > 1100

Variable         N       Mean      StDev    SE Mean
X                3     1173.6       36.3      20.96

Variable         95.0% CI                 T          P
X            ( 1083.5, 1263.7)          3.51      .0362
```

  a. Verify that the software has correctly calculated the *t* statistic, and use Table IV of Appendix A to determine whether the *p*-value is in the appropriate range.
  b. Interpret the *p*-value.
  c. What assumptions are necessary for the validity of this test?
  d. Which type of error, I or II, is of greater concern for this test? Explain.
  e. Would you recommend that this component be passed as meeting specifications?

- **Applying the Concepts—Intermediate** Based on more detailed real-world applications, these exercises require students to apply their knowledge of the technique presented in the section.

- **Applying the Concepts—Advanced** These more challenging real-data exercises require students to utilize their critical thinking skills.

Statistics In Action Case Studies Explore High Interest Issues

- Each chapter concludes with a data-based case study that showcases controversial, contemporary, or high-profile issues.

- **Focus** questions are provided to help you evaluate the findings.

- Additional cases are available on the companion website located at **www.prenhall.com/mcclave**

STATISTICS IN ACTION:
The "Eye Cue" Test: Does Experience Improve Performance?

In 1948, famous child psychologist Jean Piaget devised a test of basic perceptual and conceptual skills dubbed the "water-level task." Subjects were shown a drawing of a glass being held perfectly still (at a 45° angle) by an invisible hand so that any water in it had to be at rest (see Figure 2.39).

FIGURE 2.39

Drawing of the Water-Level Task

The task for the subject was to draw a line representing the surface of the water—a line that would touch the black dot pictured on the right side of the glass. Piaget found that young children typically failed the test. Fifty years later, research psychologists still use the water-level task to test the perception of both adults and children. Surprisingly, about 40% of the adult population also fail. In addition, males tend to do better than females, and younger adults tend to do better than older adults.

Will people with experience handling liquid-filled containers perform better on this "eye cue" test? This question was the focus of research conducted by psychologists Heiko Hecht (NASA) and Dennis R. Proffitt (University of Virginia) and published in *Psychological Science* (Mar. 1995). The researchers presented the task to each of six different groups: (1) male college students, (2) female college students, (3) professional waitresses, (4) housewives, (5) male bartenders, and (6) male bus drivers. A total of 120 subjects (20 per group) participated in the study. Two of the groups, waitresses and bartenders, were assumed to have considerable experience handling liquid-filled glasses.

After each subject completed the drawing task, the researchers recorded the deviation* (in angle degrees) of the judged line from the true line. If the deviation was within 5° of the true water surface angle, the answer was considered correct. Deviations of more than 5° in either direction were considered incorrect answers.

The data for the water-level task (simulated based on summary results presented in the journal article) are provided in the file EYECUE on the data disk that accompanies this text. For each of the 120 subjects in the experiment, the following variables (in the order they appear on the data file) were measured:

⊘ EYECUE

Variables

GENDER (F or M)
GROUP (Student, Waitres, Wife, Bartend, or Busdriv)
DEVIATION (angle, in degrees, of the judged line from the true line)
JUDGE (Within5, More5Above, More5Below)

Focus

a. Use a graphical method to describe the qualitative responses for all 120 subjects.
b. Construct graphs that researchers could use to describe the effect of gender on the qualitative measure of task performance. Do the results tend to support or refute the prevailing theory that males do better than females?
c. Construct graphs that the researchers could use to describe the effect of age on the qualitative measure of task performance. [*Note:* The college students in the study were much younger than the subjects in the other four groups.] Do the results tend to support or refute the prevailing theory that younger adults do better than older adults?
d. Construct graphs that the researchers could use to describe the effect of experience in handling liquid-filled containers on the qualitative measure of task performance. Comment on the theory that experience improves task performance.
e. Refer to parts **a–d**. What are the drawbacks to using graphical displays of sample data to infer the nature of the population?
f. Repeat parts **a–e**, but use the quantitative measure of task performance, i.e., the actual deviations in water-line angles (in degrees) for each subject.
g. Discuss which graphs—those of parts **a–d** or those of part **f**—are more informative.
h. For each group of subjects, find and interpret numerical descriptive measures of the data for the quantitative measure of task performance. What do you conclude from this analysis?

---

*The true surface line is perfectly parallel to the table top.

Robust End of Chapter Review Material

- Each chapter ends with information designed to help students check their understanding of the material, study for tests, and expand their knowledge of statistics.
- **Quick Review** provides a list of key terms and formulas with page number references.
- **Language Lab** helps students learn the language of statistics through pronunciation guides, descriptions of symbols, names, etc.
- **Supplementary Exercises** review all of the important topics covered in the chapter.

Data sets for use with problems are available on the free data CD packaged free with every new text.

QUICK REVIEW

Key Terms

Note: Starred (*) items are from the optional sections to this chapter

Bar graph 22	Measures of central tendency 44	Quartiles* 76
Bivariate relationship* 86	Measures of relative standing 70	Range 56
Box plots* 76	Measures of variation or spread 55	Rare-event approach* 81
Central tendency 44		Sample standard deviation 57
Chebyshev's Rule 61	Median 46	Sample variance 57
Class 20	Middle quartile* 76	Scattergram* 86
Class frequency 20	Modal class 48	Scatterplot* 86
Class relative frequency 20	Mode 48	Skewness 48
Dot plot 30	Mound-shaped distribution 61	Standard deviation 57
Empirical Rule 62	Numerical descriptive measures 44	Stem-and-leaf display 30
Hinges* 76		Symmetric distribution 48
Histogram 31	Outer fences* 77	Upper quartile* 76
Inner fences* 77	Outlier* 75	Variability 44
Interquartile range* 76	Percentile 71	Whiskers* 77
Lower quartile* 76	Pie chart 23	z-score 72
Mean 44		
Measurement classes 32		

SUPPLEMENTARY EXERCISES 2.116–2.140

Note: Starred (*) exercises refer to the optional sections in this chapter.

Learning the Mechanics

2.116 Construct a relative frequency histogram for the data summarized in the table below.

Measurement Class	Relative Frequency	Measurement Class	Relative Frequency
.00–.75	.02	5.25–6.00	.15
.75–1.50	.01	6.00–6.75	.12
1.50–2.25	.03	6.75–7.50	.09
2.25–3.00	.05	7.50–8.25	.05
3.00–3.75	.10	8.25–9.00	.04
3.75–4.50	.14	9.00–9.75	.01
4.50–5.25	.19		

2.117 Consider the following three measurements: 50, 70, 80. Find the z-score for each measurement if they are from a population with a mean and standard deviation equal to
  a. $\mu = 60, \sigma = 10$
  b. $\mu = 60, \sigma = 5$
  c. $\mu = 40, \sigma = 10$
  d. $\mu = 40, \sigma = 100$

2.122 For each of the following data sets, compute $\bar{x}$, s^2, and s:
  a. 13, 1, 10, 3, 3
  b. 13, 6, 6, 0
  c. 1, 0, 1, 10, 11, 11, 15
  d. 3, 3, 3, 3
  e. For each data set in parts **a–d**, form the interval $\bar{x} \pm 2s$ and calculate the percentage of the measurements that fall in the interval.

*2.123 Construct a scatterplot for the data listed here. Do you detect any trends?

Variable #1:	174	268	345	119	400	520	190	448	307	252
Variable #2:	8	10	15	7	22	31	15	20	11	9

2.124 The following data sets have been invented to demonstrate that the lower bounds given by Chebyshev's Rule are appropriate. Notice that the data are contrived and would not be encountered in a real-life problem.
  a. Consider a data set that contains ten 0s, two 1s, and ten 2s. Calculate $\bar{x}$, s^2, and s. What percentage of the measurements are in the interval $\bar{x} \pm s$? Compare this result to Chebyshev's Rule.
  b. Consider a data set that contains five 0s, thirty-two

STUDENT PROJECT

We list here several sources of real-life data sets (many in the list have been obtained from Wasserman and Bernero's *Statistics Sources* and many can be accessed via the Internet). This index of data sources is very complete and is a useful reference for anyone interested in finding almost any type of data. First we list some almanacs:

CBS News Almanac
Information Please Almanac
World Almanac and Book of Facts

United States Government publications are also rich sources of data:

Agricultural Statistics
Digest of Educational Statistics
Handbook of Labor Statistics
Housing and Urban Development Yearbook
Social Indicators
Uniform Crime Reports for the United States
Vital Statistics of the United States
Business Conditions Digest
Economic Indicators
Monthly Labor Review
Survey of Current Business
Bureau of the Census Catalog
Statistical Abstract of the United States

Main data sources are published on an annual basis:

Commodity Yearbook
Facts and Figures on Government Finance
Municipal Yearbook
Standard and Poor's Corporation, Trade and Securities: Statistics

Some sources contain data that are international in scope:

Compendium of Social Statistics
Demographic Yearbook
United Nations Statistical Yearbook
World Handbook of Political and Social Indicators

Utilizing the data sources listed, sources suggested by your instructor, or your own resourcefulness, find one real-life quantitative data set that stems from an area of particular interest to you.

a. Describe the data set by using a relative frequency histogram, stem-and-leaf display, or dot plot.
b. Find the mean, median, variance, standard deviation, and range of the data set.
c. Use Chebyshev's Rule and the Empirical Rule to describe the distribution of this data set. Count the actual number of observations that fall within 1, 2, and 3 standard deviations of the mean of the data set and compare these counts with the description of the data set you developed in part **b**.

Additional End of Chapter Resources

Student Projects provide challenging projects for further exploration individually or with a group of students. These projects provide good practice in gathering and analyzing data and report writing—skills that will be important in future classes and in the workplace.

CHAPTER 1

Statistics, Data, and Statistical Thinking

Contents

Statistics in Action

A "20/20" View of Survey Results: Fact or Fiction?

☛ Where We're Going

Statistics? Is it a field of study, or is it a group of numbers that summarizes the state of our national economy, the performance of a football team, or the social conditions in a particular locale? Or, as one popular book (Tanur *et al.*, 1989) suggests, is it "a guide to the unknown"? We'll see in Chapter 1 that each of these descriptions is applicable in understanding what statistics is. We'll also see that there are two areas of statistics: *descriptive statistics*, which focuses on developing graphical and numerical summaries that describe some phenomenon, and *inferential statistics*, which uses these numerical summaries to assist in making decisions. The primary theme of this text is inferential statistics. Thus, we'll concentrate on showing how you can use statistics to interpret data and use them to make decisions. Many jobs in industry, government, medicine, and other fields require you to make data-driven decisions so understanding these methods offers you important practical benefits.

1.1 THE SCIENCE OF STATISTICS

What does statistics mean to you? Does it bring to mind batting averages, Gallup polls, unemployment figures, numerical distortions of facts (lying with statistics!)? Or is it simply a college requirement you have to complete? We hope to persuade you that statistics is a meaningful, useful science whose broad scope of applications to business, government, and the physical and social sciences is almost limitless. We also want to show that statistics can lie only when they are misapplied. Finally, we wish to demonstrate the key role statistics plays in critical thinking—whether in the classroom, on the job, or in everyday life. Our objective is to leave you with the impression that the time you spend studying this subject will repay you in many ways.

The *Random House College Dictionary* defines *statistics* as "the science that deals with the collection, classification, analysis, and interpretation of information or data." Thus, a statistician isn't just someone who calculates batting averages at baseball games or tabulates the results of a Gallup Poll. Professional statisticians are trained in *statistical science*. That is, they are trained in collecting numerical information in the form of **data**, evaluating it, and drawing conclusions from it. Furthermore, statisticians determine what information is relevant in a given problem and whether the conclusions drawn from a study are to be trusted.

> **DEFINITION 1.1**
> **Statistics** is the science of data. This involves collecting, classifying, summarizing, organizing, analyzing, and interpreting numerical information.

In the next section, you'll see several real-life examples of statistical applications that involve making decisions and drawing conclusions.

1.2 TYPES OF STATISTICAL APPLICATIONS

Statistics means "numerical descriptions" to most people. Monthly housing starts, the failure rate of liver transplants, and the proportion of African-Americans who feel brutalized by local police all represent statistical descriptions of large sets of data collected on some phenomenon. Often the data are selected from some larger set of data whose characteristics we wish to estimate. We call this selection process *sampling*. For example, you might collect the ages of a sample of customers at a video store to estimate the average age of *all* customers of the store. Then you could use your estimate to target the store's advertisements to the appropriate age group. Notice that statistics involves two different processes: (1) describing sets of data and (2) drawing conclusions (making estimates, decisions, predictions, etc.) about the sets of data based on sampling. So, the applications of statistics can be divided into two broad areas: *descriptive statistics* and *inferential statistics*.

TEACHING TIP
Descriptive statistics summarizes the data set collected.

> **DEFINITION 1.2**
> **Descriptive statistics** utilizes numerical and graphical methods to look for patterns in a data set, to summarize the information revealed in a data set, and to present that information in a convenient form.

TEACHING TIP
Inferences make statements
about populations (the data for
which information is desired).

DEFINITION 1.3

Inferential statistics utilizes sample data to make estimates, decisions, predictions, or other generalizations about a larger set of data.

Although we'll discuss both descriptive and inferential statistics in the following chapters, the primary theme of the text is **inference**.

Let's begin by examining some studies that illustrate applications of statistics.

Study 1 "1999–2000 Financial Report, H. Lee Moffitt Cancer Center & Research Institute" (*Today's Tomorrow*, Winter 2000/2001)

TEACHING TIP
Use Studies 1–3 to point out the
difference between descriptive
and inferential statistics. Focus on
the data that statements are being
made about (i.e., the sample or
population).

The Moffitt Cancer Center at the University of South Florida treated a total of 28,196 patients during the 2000 fiscal year. The center classified the type of cancer treated for each patient. The results, extracted from the center's annual financial report, are summarized in the graphic, Figure 1.1. From the graph, you can clearly see that 19% of the patients were treated for breast cancer and 13% for prostate cancer. Since Figure 1.1 describes the type of cancer treated in all Moffitt patients in fiscal year 2000, the graphic is an example of descriptive statistics.

FIGURE 1.1

Distribution of Moffitt Cancer Patients

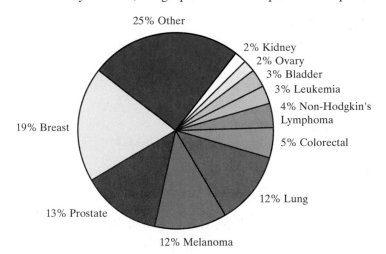

Study 2 "Does Height Influence Progression through Primary School Grades?" (*The Archives of Disease in Childhood*, April 2000)

Researchers from the University of Melbourne (Australia) analyzed the heights of over 2,800 students in primary school (grades 1 through 6). After dividing the students into three equal groups based on age (youngest third, middle third, and oldest third) within each grade level, the researchers found that the oldest group of students were the shortest in height, on average—and that this phenomenon was mostly due to the 133 children (mostly boys) who had been held back a grade level. From the analysis, the researchers inferred "that older boys within grades were relatively shorter than their younger peers" and that "when boys experienced school difficulties, height [is] one factor influencing the final decision to [hold back the student a grade level]." Thus, inferential statistics was applied to arrive at this conclusion.

Study 3 "The Significance of Belief and Expectancy Within the Spiritual Healing Encounter" (*Social Science & Medicine*, July 1995)

"Spiritual healing techniques have been a fundamental component of the healing rituals of virtually all societies since the advent of man," says researcher Daniel P. Wirth. What effect, if any, do spiritual healing encounters have on the mental and physical health of patients in today's society? And do the effects depend on the patient's expectations? To answer these questions, Wirth conducted a scientific study of 48 individuals residing in a northern California suburb. The mental and physical health of each subject was determined via a battery of standardized tests. Each individual then underwent a spiritual healing encounter that involved a "magnetic laying-on-of-hands approach." One week following the healing session, subjects were retested.

An analysis of the data revealed a significant difference between the pretest and posttest health scores for the subjects. Further analysis led Wirth to conclude that "high healer and patient expectancy [are] important ... predictors as well as facilitators of the healing process."

Like Study 2, this study is an example of the use of inferential statistics. Wirth used data from 48 individuals to make inferences about the beliefs and health improvements of all patients who experience a spiritual healing encounter.

These studies provide three real-life examples of the uses of statistics. Notice that each involves an analysis of data, either for the purpose of describing the data set (Study 1) or for making inferences about a data set (Studies 2 and 3).

1.3   FUNDAMENTAL ELEMENTS OF STATISTICS

Statistical methods are particularly useful for studying, analyzing, and learning about *populations*.

> **DEFINITION 1.4**
> A **population** is a set of units (usually people, objects, transactions, or events) that we are interested in studying.

For example, populations may include (1) *all* employed workers in the United States, (2) *all* registered voters in California, (3) *everyone* who is afflicted with AIDS, (4) *all* the cars produced last year by a particular assembly line, (5) the entire stock of spare parts available at United Airlines' maintenance facility, (6) *all* sales made at the drive-in window of a McDonald's restaurant during a given year, and (7) the set of *all* accidents occurring on a particular stretch of interstate highway during a holiday period. Notice that the first three population examples (1-3) are sets (groups) of people, the next two (4-5) are sets of objects, the next (6) is a set of transactions, and the last (7) is a set of events. Also notice that each set includes all the units in the population.

In studying a population, we focus on one or more characteristics or properties of the units in the population. We call such characteristics *variables*. For example, we may be interested in the variables age, gender, and/or the number of years of education of the people currently unemployed in the United States.

> **DEFINITION 1.5**
> A **variable** is a characteristic or property of an individual population unit.

The name "variable" is derived from the fact that any particular characteristic may vary among the units in a population.

In studying a particular variable it is helpful to be able to obtain a numerical representation for it. Often, however, numerical representations are not readily available, so the process of measurement plays an important supporting role in statistical studies. **Measurement** is the process we use to assign numbers to variables of individual population units. We might, for instance, measure the performance of the President by asking a registered voter to rate it on a scale from 1 to 10. Or we might measure workforce age simply by asking each worker, "How old are you?" In other cases, measurement involves the use of instruments such as stop watches, scales, and calipers.

TEACHING TIP
We very rarely collect all the data contained in the population.

If the population you wish to study is small, it is possible to measure a variable for every unit in the population. For example, if you are measuring the GPA for all incoming first-year students at your university, it is at least feasible to obtain every GPA. When we measure a variable for every unit of a population, it is called a **census** of the population. Typically, however, the populations of interest in most applications are much larger, involving perhaps many thousands or even an infinite number of units. Examples of large populations are those following Definition 1.4, as well as all graduates of your university or college, all potential buyers of a new fax machine, and all pieces of first-class mail handled by the U.S. Post Office. For such populations conducting a census would be prohibitively time-consuming and/or costly. A reasonable alternative would be to select and study a *subset* (or portion) of the units in the population.

TEACHING TIP
Emphasize the word *subset*.

> **DEFINITION 1.6**
> A **sample** is a subset of the units of a population.

TEACHING TIP
Use an example and define the population and the sample to illustrate the difference.

Suggested Exercise 1.22.

For example, instead of polling all 120 million registered voters in the United States during a presidential election year, a pollster might select and question a sample of just 1,500 voters (see Figure 1.2). If he is interested in the variable "presidential preference," he would record (measure) the preference of each sampled vote.

After the variable(s) of interest for every unit in the sample (or population) are measured, the data are analyzed, either by descriptive or inferential statistical methods. The pollster, for example, may be interested only in *describing* the voting patterns of the sample of 1,500 voters. More likely, however, he will want to use the information in the sample to make inferences about the population of all 120 million voters.

> **DEFINITION 1.7**
> A **statistical inference** is an estimate, prediction, or some other generalization about a population based on information contained in a sample.

That is, *we use the information contained in the smaller sample to learn about the larger population.** Thus, from the sample of 1,500 voters, the pollster may estimate

---

*The terms *population* and *sample* are often used to refer to the sets of measurements themselves, as well as to the units on which the measurements are made. When a single variable of interest is being measured, this usage causes little confusion. But when the terminology is ambiguous, we'll refer to the measurements as *population data sets* and *sample data sets*, respectively.

FIGURE 1.2

A Sample of Voter Registration Cards for All Registered Voters

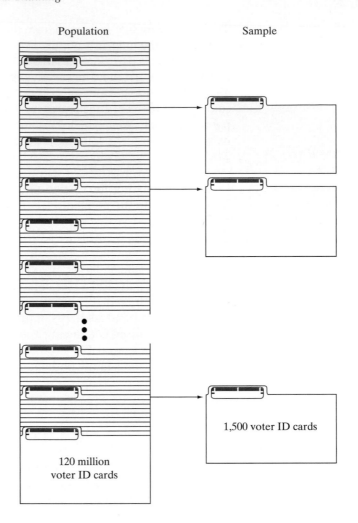

the percentage of all the voters who would vote for each presidential candidate if the election were held on the day the poll was conducted, or he might use the results to predict the outcome on election day.

EXAMPLE 1.1

According to *USA Today* (Dec. 30, 1999), the average age of viewers of MSNBC cable television news programming is 50 years. Suppose a rival network executive hypothesizes that the average age of MSNBC news viewers is less than 50. To test her hypothesis, she samples 500 MSNBC news viewers and determines the age of each.

a. Describe the population.
b. Describe the variable of interest.
c. Describe the sample.
d. Describe the inference.

Solution

a. The population is the set of units of interest to the cable TV executive, which is the set of all MSNBC news viewers.
b. The age (in years) of each viewer is the variable of interest.
c. The sample must be a subset of the population. In this case, it is the 500 MSNBC news viewers selected by the executive.

d. The inference of interest involves the *generalization* of the information contained in the sample of 500 viewers to the population of all MSNBC viewers. In particular, the executive wants to estimate the average age of the viewers in order to determine whether it is less than 50 years. She might accomplish this by calculating the average age in the sample and using the sample average to estimate the population average.

E X A M P L E   1 . 2   "Cola wars" is the popular term for the intense competition between Coca-Cola and Pepsi displayed in their marketing campaigns. Their campaigns have featured movie and television stars, rock videos, athletic endorsements, and claims of consumer preference based on taste tests. Suppose, as part of a Pepsi marketing campaign, 1,000 cola consumers are given a blind taste test (i.e., a taste test in which the two brand names are disguised). Each consumer is asked to state a preference for brand A or brand B.

a. Describe the population.
b. Describe the variable of interest.
c. Describe the sample.
d. Describe the inference.

Solution

a. The population of interest is the collection or set of all consumers.
b. The characteristic that Pepsi wants to measure is the consumer's cola preference as revealed under the conditions of a blind taste test, so *cola preference* is the variable of interest.
c. The sample is the 1,000 cola consumers selected from the population of all cola consumers.
d. The inference of interest is the *generalization* of the cola preferences of the 1,000 sampled consumers to the population of all cola consumers. In particular, the preferences of the consumers in the sample can be used to *estimate* the percentage of all cola consumers who prefer each brand.

The preceding definitions and examples identify four of the five elements of an inferential statistical problem: a population, one or more variables of interest, a sample, and an inference. But making the inference is only part of the story. We also need to know its **reliability**—that is, how good the inference is. The only way we can be certain that an inference about a population is correct is to include the entire population in our sample. However, because of *resource constraints* (i.e., insufficient time and/or money), we usually can't work with whole populations, so we base our inferences on just a portion of the population (a sample). Thus, we introduce an element of *uncertainty* into our inferences. Consequently, whenever possible, it is important to determine and report the reliability of each inference made. Reliability, then, is the fifth element of inferential statistical problems.

The measure of reliability that accompanies an inference separates the science of statistics from the art of fortune-telling. A palm reader, like a statistician, may examine a sample (your hand) and make inferences about the population (your life). However, unlike statistical inferences, the palm reader's inferences include no measure of reliability.

Suppose, like the cable TV executive in Example 1.1, we are interested in estimating the average age of a population of cable TV viewers from the average age of a sample of viewers. Using statistical methods, we can determine a *bound on the estimation error.* This bound is simply a number that our estimation error (the difference between the average age of the sample and the average age of the population of viewers) is not likely to exceed. We'll see in later chapters that this bound is a measure of the uncertainty of our inference. The reliability of statistical inferences is discussed throughout this text. For now, we simply want you to realize that an inference is incomplete without a measure of its reliability.

> **DEFINITION 1.8**
> A **measure of reliability** is a statement (usually quantified) about the degree of uncertainty associated with a statistical inference.

Let's conclude this section with a summary of the elements of both descriptive and inferential statistical problems and an example to illustrate a measure of reliability.

Four Elements of Descriptive Statistical Problems

1. The population or sample of interest
2. One or more variables (characteristics of the population or sample units) that are to be investigated
3. Tables, graphs, or numerical summary tools
4. Identification of patterns in the data

Five Elements of Inferential Statistical Problems

1. The population of interest
2. One or more variables that are to be investigated
3. The sample of population units
4. The inference about the population based on information contained in the sample
5. A measure of reliability for the inference

EXAMPLE 1.3

Refer to Example 1.2, in which the cola preferences of 1,000 consumers were indicated in a taste test. Describe how the reliability of an inference concerning the preferences of all cola consumers in the Pepsi bottler's marketing region could be measured.

Solution

When the preferences of 1,000 consumers are used to estimate those of all consumers in the region, the estimate will not exactly mirror the preferences of the population. For example, if the taste test shows that 56% of the 1,000 consumers preferred Pepsi, it does not follow (nor is it likely) that exactly 56% of all cola drinkers in the region prefer Pepsi. Nevertheless, we can use sound statistical reasoning (which we'll explore later in the text) to ensure that the sampling procedure will generate estimates that are almost certainly within a specified limit of the true percentage of all consumers who prefer Pepsi. For example, such reasoning might assure us that the estimate of the preference for Pepsi is almost certainly within 5% of the actual population preference. The implication is that the actual preference for Pepsi is between 51% [i.e., $(56 - 5)$%] and 61% [i.e., $(56 + 5)$%]—that is, (56 ± 5)%. This interval represents a measure of reliability for the inference.   ∎

1.4   TYPES OF DATA

You have learned that statistics is the science of data and that data are obtained by measuring the values of one or more variables on the units in the sample (or population). All data (and hence the variables we measure) can be classified as one of two general types: *quantitative data* and *qualitative data*.

Quantitative data are data that are measured on a naturally occurring numerical scale.* The following are examples of quantitative data:

1. The temperature (in degrees Celsius) at which each in a sample of 20 pieces of heat-resistant plastic begins to melt
2. The current unemployment rate (measured as a percentage) for each of the 50 states
3. The scores of a sample of 150 law school applicants on the LSAT, a standardized law school entrance exam administered nationwide
4. The number of convicted murderers who receive the death penalty each year over a 10-year period

> **DEFINITION 1.9**
> **Quantitative data** are measurements that are recorded on a naturally occurring numerical scale.

TEACHING TIP
Use data collected from students to illustrate quantitative versus qualitative variables. Quantitative examples: age of student, grade point average, cost of books for a semester. Qualitative examples: year in school, live on/off campus, type of transportation to school.

Suggested Exercise 1.19.

In contrast, qualitative data cannot be measured on a natural numerical scale; they can only be classified into categories.† Examples of qualitative data include:

1. The political party affiliation (Democrat, Republican, or Independent) in a sample of 50 voters
2. The defective status (defective or not) of each of 100 computer chips manufactured by Intel
3. The size of a car (subcompact, compact, mid-size, or full-size) rented by each of a sample of 30 business travelers
4. A taste-tester's ranking (best, worst, etc.) of four brands of barbecue sauce for a panel of 10 testers

Often, we assign arbitrary numerical values to qualitative data for ease of computer entry and analysis. But these assigned numerical values are simply codes: They cannot be meaningfully added, subtracted, multiplied, or divided. For example, we might code Democrat = 1, Republican = 2, and Independent = 3. Similarly, a taste tester might rank the barbecue sauces from 1 (best) to 4 (worst). These are simply arbitrarily selected numerical codes for the categories and have no utility beyond that.

---

*Quantitative data can be subclassified as either *interval data* or *ratio data*. For ratio data, the origin (i.e., the value 0) is a meaningful number. But the origin has no meaning with interval data. Consequently, we can add and subtract interval data, but we can't multiply and divide them. Of the four quantitative data sets listed above, (1) and (3) are interval data, while (2) and (4) are ratio data.

†Qualitative data can be subclassified as either *nominal data* or *ordinal data*. The categories of an ordinal data set can be ranked or meaningfully ordered, but the categories of a nominal data set can't be ordered. Of the four qualitative data sets listed above, (1) and (2) are nominal and (3) and (4) are ordinal.

> **DEFINITION 1.10**
> **Qualitative data** are measurements that cannot be measured on a natural numerical scale; they can only be classified into one of a group of categories.

EXAMPLE 1.4

1. Qualitative
2. Qualitative
3. Quantitative
4. Quantitative
5. Quantitative

Chemical and manufacturing plants often discharge toxic-waste materials such as DDT into nearby rivers and streams. These toxins can adversely affect the plants and animals inhabiting the river and the river bank. The U.S. Army Corps of Engineers conducted a study of fish in the Tennessee River (in Alabama) and its three tributary creeks: Flint Creek, Limestone Creek, and Spring Creek. A total of 144 fish were captured, and the following variables were measured for each:

1. River/creek where each fish was captured
2. Species (channel catfish, largemouth bass, or smallmouth buffalofish)
3. Length (centimeters)
4. Weight (grams)
5. DDT concentration (parts per million)

Classify each of the five variables measured as quantitative or qualitative.

Solution

The variables length, weight, and DDT concentration are quantitative because each is measured on a numerical scale: length in centimeters, weight in grams, and DDT in parts per million. In contrast, river/creek and species cannot be measured quantitatively: They can only be classified into categories (e.g., channel catfish, largemouth bass, and smallmouth buffalofish for species). Consequently, data on river/creek and species are qualitative.

As you would expect, the statistical methods for describing, reporting, and analyzing data depend on the type (quantitative or qualitative) of data measured. We demonstrate many useful methods in the remaining chapters of the text. But first we discuss some important ideas on data collection.

1.5   COLLECTING DATA

Once you decide on the type of data—quantitative or qualitative—appropriate for the problem at hand, you'll need to collect the data. Generally, you can obtain data in four different ways:

1. Data from a *published source*
2. Data from a *designed experiment*
3. Data from a *survey*
4. Data from an *observational study*

TEACHING TIP
Bring examples from the library reference section to class to illustrate published sources of data.

Sometimes, the data set of interest has already been collected for you and is available in a **published source**, such as a book, journal, or newspaper. For example, you may want to examine and summarize the divorce rates (i.e., number of divorces per 1,000 population) in the 50 states of the United States. You can find this data set (as well as numerous other data sets) at your library in the *Statistical Abstract of the United States*, published annually by the U.S. government. Similarly, someone who is interested in monthly mortgage applications for new home construction would find this data set in the *Survey of Current Business*, another government publication. Other examples of published data sources include *The Wall Street Journal* (financial data) and *The Sporting News* (sports information). The Internet (World Wide Web) now provides a medium by which data from published sources are readily obtained.*

TEACHING TIP
Illustrate using a scientific or medical study that controls the units in the study.

Suggested Exercise 1.24.

A second method of collecting data involves conducting a **designed experiment**, in which the researcher exerts strict control over the units (people, objects, or things) in the study. For example, a recent medical study investigated the potential of aspirin in preventing heart attacks. Volunteer physicians were divided into two groups—the *treatment* group and the *control* group. In the treatment group, each physician took one aspirin tablet a day for one year, while each physician in the control group took an aspirin-free placebo made to look like an aspirin tablet. The researchers, not the physicians under study, controlled who received the aspirin (the treatment) and who received the placebo. As you'll learn in Chapter 10, a properly designed experiment allows you to extract more information from the data than is possible with an uncontrolled study.

TEACHING TIP
Use consumer confidence of presidential approval rating as examples of survey information.

Suggested Exercise 1.15.

Surveys are a third source of data. With a **survey**, the researcher samples a group of people, asks one or more questions, and records the responses. Probably the most familiar type of survey is the political poll conducted by any one of a number of organizations (e.g., Harris, Gallup, Roper, and CNN) and designed to predict the outcome of a political election. Another familiar survey is the Nielsen survey, which provides the major networks with information on the most watched programs on television. Surveys can be conducted through the mail, with telephone interviews, or with in-person interviews. Although in-person surveys are more expensive than mail or telephone surveys, they may be necessary when complex information is to be collected.

Suggested Exercise 1.20.

Finally, observational studies can be employed to collect data. In an **observational study**, the researcher observes the experimental units in their natural setting and records the variable(s) of interest. For example, a child psychologist might observe and record the level of aggressive behavior of a sample of fifth graders playing on a school playground. Similarly, a zoologist may observe and measure the weights of newborn elephants born in captivity. Unlike a designed experiment, an observational study is one in which the researcher makes no attempt to control any aspect of the experimental units.

Regardless of which data collection method is employed, it is likely that the data will be a sample from some population. And if we wish to apply inferential statistics, we must obtain a *representative sample*.

DEFINITION 1.11
A **representative sample** exhibits characteristics typical of those possessed by the target population.

TEACHING TIP
Explain the importance of a representative sample when using a sample to make an inference about the population it was sampled from.

For example, consider a political poll conducted during a presidential election year. Assume the pollster wants to estimate the percentage of all 120 million registered voters in the United States who favor the incumbent President. The pollster would be unwise to base the estimate on survey data collected for a sample

---

*With published data, we often make a distinction between the *primary source* and a *secondary source*. If the publisher is the original collector of the data, the source is primary. Otherwise, the data is secondary source data.

of voters from the incumbent's own state. Such an estimate would almost certainly be *biased* high; consequently, it would not be very reliable.

TEACHING TIP
Random sampling is the easiest sampling procedure that will ensure a representative sample.

The most common way to satisfy the representative sample requirement is to select a random sample. A **random sample** ensures that every subset of fixed size in the population has the same chance of being included in the sample. If the pollster samples 1,500 of the 120 million voters in the population so that every subset of 1,500 voters has an equal chance of being selected, he has devised a random sample. The procedure for selecting a random sample is discussed in Chapter 3. Here, however, let's look at two examples involving actual sampling studies.

EXAMPLE 1.5

"Have you ever seen anything that you believe was a spacecraft from another planet?" This was the question put to 1,500 American adults in a national poll conducted by ABC News and *The Washington Post*. The pollsters used random-digit telephone dialing to contact adult Americans, until 1,500 responded. Ten percent (i.e., 150) of the respondents answered that they had, in fact, seen an alien spacecraft (*Chance*, Summer 1997). No information was provided on how many adults were called and, for one reason or another, did not answer the question.

a. Survey
b. Unknown

a. Identify the data collection method.
b. Identify the target population.
c. Are the sample data representative of the population?

Solution

a. The data collection method is a telephone survey: 1,500 adults were contacted via random-digit dialing of telephone numbers.
b. Since the pollsters conducted a *national* survey, presumably the target population is all American adults.
c. Because the 1,500 respondents clearly make up a subset of the target population, they do form a sample. Whether or not the sample is representative is unclear, since we are given no information on the 1,500 respondents. However, telephone surveys (and surveys in general) often suffer from **nonresponse bias**. It is possible that many adults who chose not to respond would have answered the question "no," leading to a percentage of affirmative answers less than 10%. ▪

EXAMPLE 1.6

Psychologists at the University of Tennessee carried out a study of the susceptibility of people to hypnosis (*Psychological Assessment*, Mar. 1995). In a random sample of 130 undergraduate psychology students at the university, each experienced both traditional hypnosis and computer-assisted hypnosis. Approximately half were randomly assigned to undergo the traditional procedure first, followed by the computer-assisted procedure. The other half were randomly assigned to experience computer-assisted hypnosis first, then traditional hypnosis. Following the hypnosis episodes, all students filled out questionnaires designed to measure a student's susceptibility to hypnosis. The susceptibility scores of the two groups of students were compared.

a. Designed Experiment
b. Yes

a. Identify the data collection method.
b. Is the sample data representative of the target population?

Solution

a. The researchers controlled which type of hypnosis—traditional or computer-assisted—the students experienced first. Consequently, a designed experiment was used to collect the data for the two groups.

b. The sample of 130 psychology students was randomly selected from all psychology students at the University of Tennessee. If the target population is *all University of Tennessee psychology students*, it is likely that the sample is representative. However, the researchers warn that the sample data should not be used to make inferences about other, more general, populations. ⎯⎯⎯⎯⎯ ■

1.6 THE ROLE OF STATISTICS IN CRITICAL THINKING

According to H. G. Wells, author of such science-fiction classics as *The War of the Worlds* and *The Time Machine,* "*Statistical thinking* will one day be as necessary for efficient citizenship as the ability to read and write." Written more than a hundred years ago, Wells' prediction is proving true today.

The growth in data collection associated with scientific phenomena, business operations, and government activities (quality control, statistical auditing, forecasting, etc.) has been remarkable in the past several decades. Every day the media present us with published results of political, economic, and social surveys. In increasing government emphasis on drug and product testing, for example, we see vivid evidence of the need to be able to evaluate data sets intelligently. Consequently, each of us has to develop a discerning sense—an ability to use rational thought to interpret the meaning of data. This ability can help you make intelligent decisions, inferences, and generalizations; that is, it helps you *think critically* using statistics.

TEACHING TIP
The concept of statistical thinking is fundamental to using statistics in any discipline.

DEFINITION 1.12
Statistical thinking involves applying rational thought to assess data and the inferences made from them critically.

To gain some insight into the role statistics plays in critical thinking, let's look at a *New York Times* (June 17, 1995) article titled "The Case for No Helmets." The article, written by the editor of a magazine for Harley-Davidson bikers, lists several reasons why motorcyclists should not be required by law to wear helmets. First, he argued that helmets may actually kill, since in collisions at speeds greater than 15 miles an hour, the heavy helmet may protect the head but snap the spine. Second, the editor cited a "study" that claimed "nine states without helmet laws had a lower fatality rate (3.05 deaths per 10,000 motorcycles) than those that mandated helmets (3.38)." Finally, he reported that "in a survey of 2,500 [at a rally], 98% of the respondents opposed such laws."

You can use "statistical thinking" to help you critically evaluate the arguments presented in this article. For example, before you accept the 98% estimate, you would want to know how the data were collected for the study cited by the editor of the biker magazine. If a survey was conducted, it's possible that the 2,500 bikers in the sample were not selected at random from the target population of all bikers, but rather were "self-selected." (Remember, they were all attending a rally—maybe even a rally for bikers who oppose the law.) If the respondents were likely to have strong opinions regarding the helmet law (e.g., strongly oppose the law), the resulting estimate is probably biased high.

You'd also want more information about the study comparing the motorcycle fatality rate of the nine states without a helmet law to those states that mandate helmets: Were the data obtained from a published source? Were all 50 states included in the study? That is, are you seeing sample data or population data? Furthermore, do the helmet laws vary among states? If so, can you really compare the fatality rates?

These questions actually led to the discovery of two scientific and statistically sound studies on helmets. The first, a UCLA study of nonfatal injuries, disputed the charge that helmets shift injuries to the spine. The second study reported a dramatic *decline* in motorcycle crash deaths after California passed its helmet law.

In the remaining chapters of the text, you'll become familiar with the tools essential for building a firm foundation in statistics and statistical thinking.

TEACHING TIP
Suggestions for class discussion of the case on page 15 can be found in the *Instructor's Guide.*

QUICK REVIEW

Key Terms

Census   5
Data   2
Descriptive statistics   2
Designed experiment   11
Inference   3
Inferential statistics   3
Measure of reliability   8
Measurement   5

Nonresponse bias   12
Observational study   11
Population   4
Published source   10
Qualitative data   10
Quantitative data   9
Random sample   12
Reliability   7

Representative sample   11
Sample   5
Statistical inference   5
Statistical thinking   13
Statistics   2
Survey   11
Variable   4

 EXERCISES 1.1–1.27

Learning the Mechanics

1.1   What is statistics?
1.2   Explain the difference between descriptive and inferential statistics.
1.3   List and define the five elements of an inferential statistical analysis.
1.4   List the four major methods of collecting data and explain their differences.
1.5   Explain the difference between quantitative and qualitative data.
1.6   Explain how populations and variables differ.
1.7   Explain how populations and samples differ.
1.8   What is a representative sample? What is its value?
1.9   Why would a statistician consider an inference incomplete without an accompanying measure of its reliability?
1.10   Define statistical thinking.
1.11   Suppose you're given a data set that classifies each sample unit into one of four categories: A, B, C, or D. You plan to create a computer database consisting of these data, and you decide to code the data as A = 1, B = 2, C = 3, and D = 4. Are the data consisting of the classifications A, B, C, and D qualitative or quantitative? After the data are input as 1, 2, 3, or 4, are they qualitative or quantitative? Explain your answers.

Applying the Concepts—Basic

1.12   Colleges and universities are requiring an increasing amount of information about applicants before making acceptance and financial aid decisions. Classify each of the following types of data required on a college application as quantitative or qualitative.
   a. High school GPA   Quantitative
   b. High school class rank   Quantitative
   c. Applicant's score on the SAT or ACT   Quantitative
   d. Gender of applicant   Qualitative
   e. Parents' income   Quantitative
   f. Age of applicant   Quantitative
1.13   Classify the following examples of data as either qualitative or quantitative:
   a. The bacteria count in the water at each of 30 city swimming pools   Quantitative
   b. The occupation of each of 200 shoppers at a supermarket   Qualitative
   c. The marital status of each person living on a city block   Qualitative
   d. The time (in months) between auto maintenance for each of 100 used cars   Quantitative

Did you ever notice that, no matter where you stand on popular issues of the day, you can always find statistics or surveys to back up your point of view—whether to take vitamins, whether day care harms kids, or what foods can hurt you or save you? There is an endless flow of information to help you make decisions, but is this information accurate, unbiased? John Stossel decided to check that out, and you may be surprised to learn if the picture you're getting doesn't seem quite right, maybe it isn't.

Barbara Walters gave this introduction to a March 31, 1995, segment of the popular prime-time ABC television program, "20/20." The story is titled "Facts or Fiction?—Exposés of So-Called Surveys." One of the surveys investigated by ABC correspondent John Stossel compared the discipline problems experienced by teachers in the 1940s and those experienced today. The results: In the 1940s, teachers worried most about students talking in class, chewing gum, and running in the halls. Today, they worry most about being assaulted! This information was highly publicized in the print media—daily newspapers, weekly magazines, Ann Landers' column, *The Congressional Quarterly*, and *The Wall Street Journal*, among others—and referenced in speeches by a variety of public figures, including former first lady Barbara Bush and former Education Secretary William Bennett.

"Hearing this made me yearn for the old days when life was so much simpler and gentler, but was life that simple then?", asks Stossel. "Wasn't there juvenile delinquency [in the 1940s]? Is the survey true?" With the help of a Yale School of Management professor, Stossel found the original source of the teacher survey—Texas oilman T. Colin Davis—and discovered it wasn't a survey at all. Davis had simply identified certain disciplinary problems encountered by teachers and published his list in a conservative newsletter—a list he admitted was not obtained from a statistical survey, but from Davis' personal knowledge of the problems in the 1940s ("I was in school then") and his understanding of the problems today ("I read the papers").

Stossel's commonsense questions of the teacher "survey" is a clear example of the critical thinking process in action. Several more misleading surveys were presented on the ABC program. Listed here, most of these surveys were conducted by businesses or special interest groups with specific objectives in mind.

The "20/20" segment ended with an interview of Cynthia Crossen, author of *Tainted Truth*, an exposé of misleading and biased surveys. Crossen warns:

> If everybody is misusing numbers and scaring us with numbers to get us to do something, however good [that something] is, we've lost the power of numbers. Now, we know certain things from research. For example, we know that smoking cigarettes is hard on your lungs and heart, and because we know that, many people's lives have been extended or saved. We don't want to lose the power of information to help us make decisions, and that's what I worry about.

Focus

a. Consider the false March of Dimes report on domestic violence and birth defects. Discuss the type of data required to investigate the impact of domestic violence on birth defects. What data collection method would you recommend?

b. Refer to the American Association of University Women (AAUW) study of self-esteem of high school girls. Explain why the results of the AAUW study are likely to be misleading. What data might be appropriate for assessing the self-esteem of high school girls?

c. Refer to the Food Research and Action Center study of hunger in America. Explain why the results of the study are likely to be misleading. What data would provide insight into the proportion of hungry American children?

Reported Information (Source)	Actual Study Information
• Eating oat bran is a cheap and easy way to reduce your cholesterol count. (Quaker Oats)	• Diet must consist of nothing but oat bran to achieve a slightly lower cholesterol count.
• 150,000 women a year die from anorexia. (Feminist group)	• Approximately 1,000 women a year die from problems that were likely caused by anorexia.
• Domestic violence causes more birth defects than all medical issues combined. (March of Dimes)	• No study—false report.
• Only 29% of high school girls are happy with themselves, compared to 66% of elementary school girls. (American Association of University Women)	• Of 3,000 high school girls, 29% responded to the statement, "I am happy the way I am." Most answered "Sort of true" and "Sometimes true."
• One in four American children under age 12 is hungry or at risk of hunger. (Food Research and Action Center)	• Based on responses to questions: "Do you ever cut the size of meals?", "Do you ever eat less than you feel you should?", "Did you ever rely on limited numbers of foods to feed your children because you were running out of money to buy food for a meal?"

1.14 A food-products company is considering marketing a new snack food. To see how consumers react to the product, the company conducted a taste test using a sample of 100 randomly selected shoppers at a suburban shopping mall. The shoppers were asked to taste the snack food and then fill out a short questionnaire that requested the following information:
  1. What is your age?   Quantitative
  2. Are you the person who typically does the food shopping for your household?   Qualitative
  3. How many people are in your family?   Quantitative
  4. How would you rate the taste of the snack food on a scale of 1 to 10, where 1 is least tasty?   Quantitative
  5. Would you purchase this snack food if it were available on the market?   Qualitative
  6. If you answered yes to part **5**, how often would you purchase the product?   Quantitative
  a. Identify the data collection method.   Survey
  b. Classify the data generated for each question as quantitative or qualitative. Justify your classifications.

1.15 A poll is to be conducted in which 2,000 individuals are asked whether the 1999 presidential impeachment trial of Bill Clinton was conducted fairly. The 2,000 individuals are selected by random-digit telephone dialing and asked the question over the phone.
  a. What is the relevant population?
  b. What is the variable of interest? Is it quantitative or qualitative?   Qualitative
  c. What is the sample?
  d. What is the inference of interest to the pollster?
  e. What method of data collection is employed?   Survey
  f. How likely is the sample to be representative?

1.16 Consider the set of all students enrolled in your statistics course this term. Suppose you're interested in learning about the current grade point averages (GPAs) of this group.
  a. Define the population and variable of interest.   GPA
  b. Is the variable qualitative or quantitative?
  c. Suppose you determine the GPA of every member of the class. Would this represent a census or a sample?
  d. Suppose you determine the GPA of 10 members of the class. Would this represent a census or a sample?
  e. If you determine the GPA of every member of the class and then calculate the average, how much reliability does this have as an "estimate" of the class average GPA?   100%
  f. If you determine the GPA of 10 members of the class and then calculate the average, will the number you get necessarily be the same as the average GPA for the whole class? On what factors would you expect the reliability of the estimate to depend?
  g. What must be true in order for the sample of 10 students you select from your class to be considered a random sample?

1.17 The Cutter Consortium surveyed 154 U.S. companies to determine the extent of their involvement in electronic commerce (called *e-commerce*). Four of the questions they asked follow (*Internet Week*, Sept. 6, 1999).
  1. Do you have an overall e-commerce strategy?
  2. If you don't already have an e-commerce plan, when will you implement one?
  3. Are you delivering products over the Internet?
  4. What was your company's total revenue in the last fiscal year?
  a. For each question, determine the variable of interest and classify it as quantitative or qualitative.
  b. Do the data collected for the 154 companies represent a sample or a population? Explain.   Sample

Applying the Concepts—Intermediate

1.18 *The American Association of Nurse Anesthetists Journal* (Feb. 2000) published the results of a study on the use of herbal medicines before surgery. Of 500 surgical patients that were randomly selected for the study, 51% used herbal or alternative medicines (e.g., garlic, ginkgo, kava, fish oil) against their doctor's advice prior to surgery.
  a. Do the 500 surgical patients represent a population or a sample? Explain.
  b. If your answer was sample in part **a**, is the sample representative of the population? If you answered population in part **a**, explain how to obtain a representative sample from the population.
  c. For each patient, what variable is measured? Is the data collected quantitative or qualitative?

1.19 All highway bridges in the United States are inspected periodically for structural deficiency by the Federal Highway Administration (FHWA). Data from the FHWA inspections are compiled into the National Bridge Inventory (NBI). Several of the nearly 100 variables maintained by the NBI are listed below. Classify each variable as quantitative or qualitative.
  a. Length of maximum span (feet)   Quantitative
  b. Number of vehicle lanes   Quantitative
  c. Toll bridge (yes or no)   Qualitative
  d. Average daily traffic   Quantitative
  e. Condition of deck (good, fair, or poor)   Qualitative
  f. Bypass or detour length (miles)   Quantitative
  g. Route type (interstate, U.S., state, county, or city)

1.20 Refer to Exercise 1.19. The most recent NBI data were analyzed, and the results published in the *Journal of Infrastructure Systems* (June 1995). Using the FHWA inspection ratings, each of the 470,515 highway bridges in the United States was categorized as structurally deficient, functionally obsolete, or safe. About 26% of the bridges were found to be structurally deficient, while 19% were functionally obsolete.
  a. What is the variable of interest to the researchers?
  b. Is the variable of part **a** quantitative or qualitative?
  c. Is the data set analyzed a population or a sample? Explain.   Population
  d. How did the researchers obtain the data for their study?   Observational study

1.21 Does a massage enable the muscles of tired athletes to recover from exertion faster than usual? To answer this question, researchers recruited eight amateur boxers to participate in an experiment (*British Journal of Sports Medicine*, April 2000). After a 10-minute workout in which each boxer threw 400 punches, half the boxers were given a 20-minute massage and half just rested for 20 minutes. Before returning to the ring for a second workout, the heart rate (beats per minute) and blood lactate level (micromoles) were recorded for each boxer. The researchers found no difference in the means of the two groups of boxers for either variable.

  a. Identify the data collection method used by the researchers.   Designed experiment
  b. Identify the experimental units of the study.   A boxer
  c. Identify the variables measured and their type (quantitative or qualitative).
  d. What is the inference drawn from the analysis?
  e. Comment on whether this inference can be made about all athletes.

1.22 A Gallup Youth Poll was conducted to determine the topics that teenagers most want to discuss with their parents. The findings show that 46% would like more discussion about the family's financial situation, 37% would like to talk about school, and 30% would like to talk about religion. The survey was based on a national sampling of 505 teenagers, selected at random from all US. teenagers.

  a. Describe the sample.
  b. Describe the population from which the sample was selected.
  c. Is the sample representative of the population?
  d. What is the variable of interest?
  e. How is the inference expressed?
  f. Newspaper accounts of most polls usually give a *margin of error* (i.e., plus or minus 3%) for the survey result. What is the purpose of the margin of error and what is its interpretation?

1.23 A group of University of South Florida surgery researchers believe that the drug aprotinin is effective in reducing the blood loss of burn patients who undergo skin replacement surgery (*USF Office of Research Annual Report, 1997–1998*). In a clinical trial of 14 burn patients, half were randomly assigned to receive the drug and half a placebo (no drug). A preliminary analysis of the patients' blood loss revealed that the group with the drug had a significant reduction of bleeding.

  a. Identify the data collection method used for this study.   Designed experiment
  b. Does the study involve descriptive or inferential statistics? Explain.   Inferential statistics
  c. What is the population (or sample) of interest to the researchers?

1.24 A first-year chemistry student conducts an experiment to determine the amount of hydrochloric acid necessary to neutralize 2 milliliters (ml) of a basic solution. The student prepares five 2-ml portions of the solution and adds a known concentration of hydrochloric acid to each. The amount of acid necessary to achieve neutrality of the solution is recorded for each of the five portions.

  a. Describe the population of interest to the student.
  b. What is the variable of interest? Is it quantitative or qualitative?   Quantitative
  c. Describe the sample.
  d. Describe the data collection method.

Applying the Concepts—Advanced

1.25 *USA Today* (Aug.14,1995) reported on a study that suggests "frequently 'heading' the ball in soccer lowers players' IQs." A psychologist tested 60 male soccer players, ages 14–29, who played up to five times a week. Players who averaged 10 or more headers a game had an average IQ of 103, while players who headed one or fewer times per game had an average IQ of 112.

  a. Describe the population of interest to the psychologist.
  b. Identify the variables of interest.
  c. Identify the type (qualitative or quantitative) of the variables, part **b**.   Quantitative
  d. Describe the sample.
  e. What is the inference made by the psychologist?
  f. Discuss possible reasons why the inference, part **e**, may be misleading.

1.26 Are men or women more adept at remembering where they leave misplaced items (like car keys)? According to University of Florida psychology professor Robin West, women show greater competence in actually finding these objects (*Explore*, Fall 1998). Approximately 300 men and women from Gainesville, Florida, participated in a study in which each person placed 20 common objects in a 12-room "virtual" house represented on a computer screen. Thirty minutes later, the subjects were asked to recall where they put each of the objects. For each object, a recall variable was measured as "yes" or "no."

  a. Identify the population of interest to the psychology professor.
  b. Identify the sample.
  c. Does the study involve descriptive or inferential statistics? Explain.   Inferential
  d. Are the variables measured in the study quantitative or qualitative?   Qualitative

1.27 The Holocaust—the Nazi Germany extermination of the Jews during World War II—has been well documented through film, books, and interviews with the concentration camp survivors. But a recent Roper poll found that one in five U.S. residents doubted the Holocaust really occurred. A national sample of more than 1,000 adults and high school students were asked, "Does it seem possible or does it seem impossible to you that the Nazi extermination of the Jews never happened?" Twenty-two percent of the adults and 20% of the high school students responded "Yes" (*New York Times*, May 16, 1994).

a. Describe the population of interest to the pollsters.

b. What is the variable of interest? Is it quantitative or qualitative?   Qualitative

c. Describe the sample.

d. Identify the data collection method employed.

e. What inference was made by the pollsters?

f. Comment on the reliability of the inference. [*Hint:* Examine, closely, the survey question.]

STUDENT PROJECTS

Scan your daily newspaper or weekly news magazine or search the Internet for articles that contain numerical data. The data might be a summary of the results of a public opinion poll, the results of a vote by the U.S. Senate, or a list of crime rates, birth or death rates, etc. For each article you find, answer the following questions:

a. Do the data constitute a sample or an entire population? If a sample has been taken, clearly identify both the sample and the population; otherwise, identify the population.

b. What type of data (quantitative or qualitative) has been collected?

c. What is the data source?

d. If a sample has been observed, is it likely to be representative of the population?

e. If a sample has been observed, does the article present an explicit (or implied) inference about the population of interest? If so, state the inference made in the article.

f. If an inference has been made, has a measure of reliability been included? What is it?

g. Use your answers to questions d–f to critically evaluate the article.

REFERENCES

Careers in Statistics. American Statistical Association, Biometric Society, Institute of Mathematical Statistics and Statistical Society of Canada, 1995.

Cochran, W.G. *Sampling Techniques*, 3rd ed. New York: Wiley, 1977.

Deming, W. E. *Sample Design in Business Research.* New York: Wiley, 1963.

Ethical Guidelines for Statistical Practice. American Statistical Association, 1995.

Hansen, M. H., Hurwitz, W. N., and Madow, W. G. *Sample Survey Methods and Theory*, Vol. 1. New York: Wiley & Sons, Inc., 1953.

Kirk, R. E. ed. *Statistical Issues: A Reader for the Behavioral Sciences.* Monterey, Ca.: Brooks/Cole, 1972.

Kish, L. *Survey Sampling.* New York: Wiley & Sons, Inc., 1965.

Scheaffer, R., Mendenhall, W., and Ott, R. L. *Elementary Survey Sampling*, 2nd ed. Boston: Duxbury, 1979.

Tanur, J. M., Mosteller, F. Kruskal, W. H., Link, R. E, Pieters, R. S., and Rising, G. R. *Statistics: A Guide to the Unknown* (E. L. Lehmann, special editor). San Francisco: Holden-Day, 1989.

What Is a Survey? Section on Survey Research Methods, American Statistical Association, 1995.

Yamane, T. *Elementary Sampling Theory,* 3rd ed. Englewood Cliffs, N.J.: Prentice-Hall, 1967.

Methods for Describing Sets of Data

Contents

Statistics in Action

The "Eye Cue" Test: Does Experience Improve Performance?

✊ Where We've Been

In Chapter 1 we looked at some typical examples of the use of statistics and we discussed the role that statistical thinking plays in supporting decision making. We examined the difference between descriptive and inferential statistics and described the five elements of inferential statistics: a population, one or more variables, a sample, an inference, and a measure of reliability for the inference. We also learned that data can be one of two types—quantitative or qualitative.

☛ Where We're Going

Before we make an inference, we must be able to describe a data set. We can do this by using graphical and/or numerical methods, which we discuss in this chapter. As we'll see in Chapter 5, we use some sample numerical descriptive measures to estimate the values of corresponding population descriptive measures. Therefore, our efforts in this chapter will ultimately lead us to statistical inference.

TEACHING TIP
Explain to the students that descriptive techniques will also be useful in inferential statistics for generating the sample statistics necessary to make inferences and also in generating the graphs necessary to check assumptions that will be made.

TEACHING TIP
Use data collected in the class to illustrate the techniques for describing qualitative data. Collect data such as year in school, major discipline, and state of residency. Use this data to illustrate class frequency and class relative frequency.

Suppose you wish to evaluate the mathematical capabilities of a class of 1,000 first-year college students based on their quantitative Scholastic Aptitude Test (SAT) scores. How would you describe these 1,000 measurements? Characteristics of interest include the typical or most frequent SAT score, the variability in the scores, the highest and lowest scores, the "shape" of the data, and whether or not the data set contains any unusual scores. Extracting this information isn't easy. The 1,000 scores provide too many bits of information for our minds to comprehend. Clearly we need some method for summarizing and characterizing the information in such a data set. Methods for describing data sets are also essential for statistical inference. Most populations are large data sets. Consequently, we need methods for describing a data set that let us make descriptive statements (inferences) about a population based on information contained in a sample.

Two methods for describing data are presented in this chapter, one *graphical* and the other *numerical*. Both play an important role in statistics. Section 2.1 presents both graphical and numerical methods for describing qualitative data. Graphical methods for describing quantitative data are illustrated in Sections 2.2, 2.8, and 2.9; numerical descriptive methods for quantitative data are presented in Sections 2.3–2.7. We end this chapter with a section on the *misuse* of descriptive techniques.

2.1   DESCRIBING QUALITATIVE DATA

Consider a study of aphasia published in the *Journal of Communication Disorders* (Mar. 1995). Aphasia is the "impairment or loss of the faculty of using or understanding spoken or written language." Three types of aphasia have been identified by researchers: Broca's, conduction, and anomic. They wanted to determine whether one type of aphasia occurs more often than any other, and, if so, how often. Consequently, they measured aphasia type for a sample of 22 adult aphasiacs. Table 2.1 (p. 21) gives the type of aphasia diagnosed for each aphasiac in the sample.

For this study, the variable of interest, aphasia type, is qualitative in nature. Qualitative data are nonnumerical in nature; thus, the value of a qualitative variable can only be classified into categories called *classes*. The possible aphasia types—Broca's, conduction, and anomic—represent the classes for this qualitative variable. We can summarize such data numerically in two ways: (1) by computing the *class frequency*—the number of observations in the data set that fall into each class; or (2) by computing the *class relative frequency*—the proportion of the total number of observations falling into each class.

DEFINITION 2.1
A **class** is one of the categories into which qualitative data can be classified.

DEFINITION 2.2
The **class frequency** is the number of observations in the data set falling in a particular class.

DEFINITION 2.3
The **class relative frequency** is the class frequency divided by the total number of observations in the data set, i.e.,

$$\text{class relative frequency} = \frac{\text{class frequency}}{n}$$

● APHASIA

TABLE 2.1   Data on 22 Adult Aphasiacs

Subject	Type of Aphasia
1	Broca's
2	Anomic
3	Anomic
4	Conduction
5	Broca's
6	Conduction
7	Conduction
8	Anomic
9	Conduction
10	Anomic
11	Conduction
12	Broca's
13	Anomic
14	Broca's
15	Anomic
16	Anomic
17	Anomic
18	Conduction
19	Broca's
20	Anomic
21	Conduction
22	Anomic

Source: Li, E. C., Williams, S. E., and Volpe, R. D., "The effects of topic and listener familiarity of discourse variables in procedural and narrative discourse tasks." *Journal of Communication Disorders*, Vol. 28, No. 1, Mar. 1995, p. 44 (Table 1).

Examining Table 2.1, we observe that 5 aphasiacs in the study were diagnosed as suffering from Broca's aphasia, 7 from conduction aphasia, and 10 from anomic aphasia. These numbers—5, 7, and 10—represent the class frequencies for the three classes and are shown in the summary table, Table 2.2.

Table 2.2 also gives the relative frequency of each of the three aphasia classes. From Definition 2.3, we know that we calculate the relative frequency by dividing the class frequency by the total number of observations in the data set. Thus, the relative frequencies for the three types of aphasia are

$$\text{Broca's:} \quad \frac{5}{22} = .227$$

$$\text{Conduction:} \quad \frac{7}{22} = .318$$

$$\text{Anomic:} \quad \frac{10}{22} = .455$$

TABLE 2.2   Summary Table for Data on 22 Adult Aphasiacs

Class	Frequency	Relative Frequency
Type of Aphasia	Number of Subjects	Proportion
Broca's	5	.227
Conduction	7	.318
Anomic	10	.455
Totals	22	1.000

From these relative frequencies we observe that nearly half (45.5%) of the 22 subjects in the study are suffering from anomic aphasia.

Although the summary table of Table 2.2 adequately describes the data of Table 2.1, we often want a graphical presentation as well. Figures 2.1 and 2.2 show two of the most widely used graphical methods for describing qualitative data—bar graphs and pie charts. Figure 2.1 shows the frequencies of aphasia types in a **bar graph** produced with SAS. Note that the height of the rectangle, or "bar," over each class is equal to the class frequency. (Optionally, the bar heights can be proportional to class relative frequencies.)

Suggested Exercise 2.1

FIGURE 2.1

SAS Bar Graph for Data on 22 Aphasiacs

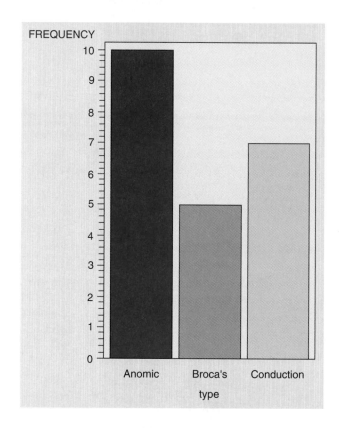

FIGURE 2.2

SPSS Pie Chart for Data on 22 Aphasiacs

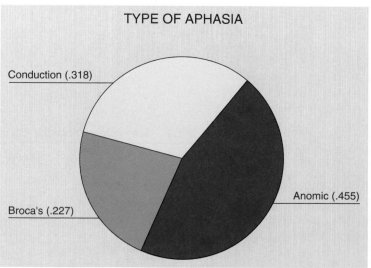

Suggested Exercise 2.7

In contrast, Figure 2.2 (p. 22) shows the relative frequencies of the three types of aphasia in a **pie chart** generated with SPSS. Note that the pie is a circle (spanning 360°) and the size (angle) of the "pie slice" assigned to each class is proportional to the class relative frequency. For example, the slice assigned to anomic aphasia is 45.5% of 360°, or $(.455)(360°) = 163.8°$.

Let's look at a practical example that requires interpretation of the graphical results.

■ **E X A M P L E   2 . 1**

Suggested Exercise 2.11

⚫ BLOODLOSS

A group of cardiac physicians in southwest Florida have been studying a new drug designed to reduce blood loss in coronary bypass operations. Blood loss data for 114 coronary bypass patients (some who received a dosage of the drug and others who did not) are saved in the **BLOODLOSS** file. Although the drug shows promise in reducing blood loss, the physicians are concerned about possible side effects and complications. So their data set includes not only the qualitative variable DRUG, which indicates whether or not the patient received the drug, but also the qualitative variable COMP, which specifies the type (if any) of complication experienced by the patient. The four values of COMP recorded in Appendix B are (1) redo surgery, (2) post-op infection, (3) both, or (4) none.

a. Figure 2.3, generated using SAS, shows summary tables for the two qualitative variables, DRUG and COMP. Interpret the results.
b. Interpret the SAS output shown in Figures 2.4 and 2.5.

Solution

a. The top table in Figure 2.3 is a summary frequency table for DRUG. Note that exactly half (57) of the 114 coronary bypass patients received the drug and half did not. The bottom table in Figure 2.3 is a summary frequency table for COMP. The class relative frequencies are given in the Percent column. We see that almost 75% of the 114 patients had no complications, leaving about 25% who experienced either a redo surgery, a post-op infection, or both.
b. Figure 2.4 displays bar graphs for the data. The four bars in the top graph represent the frequencies of COMP for the 57 patients who did not receive the drug; the four bars in the bottom graph represent the frequencies of COMP for the 57 patients who did receive a dosage of the drug. The graph clearly shows that patients who did not receive the drug suffered fewer complications. The exact percentages are displayed in the SAS summary tables of Figure 2.5 (p. 25). Over 82% of the patients who got the drug had no complications, compared to about 68% for the patients who got no drug.

FIGURE 2.3

SAS Summary Tables for DRUG and COMP

The FREQ Procedure

DRUG	Frequency	Percent	Cumulative Frequency	Cumulative Percent
NO	57	50.00	57	50.00
YES	57	50.00	114	100.00

COMP	Frequency	Percent	Cumulative Frequency	Cumulative Percent
BOTH	4	3.51	4	3.51
INFECT	12	10.53	16	14.04
NONE	86	75.44	102	89.47
REDO	12	10.53	114	100.00

FIGURE 2.4

SAS Bar Graphs for
COMP by Value of DRUG

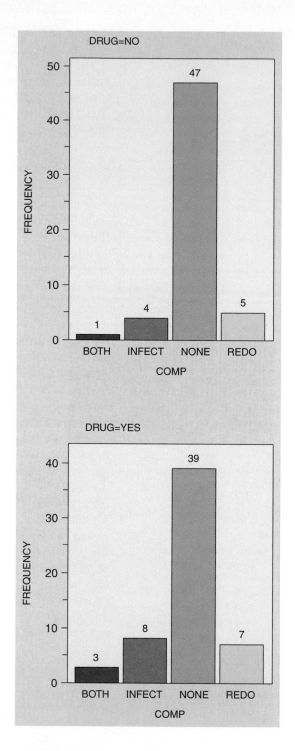

Although these results show that the drug may be effective in reducing blood loss, they also imply that patients on the drug may have a higher risk of complications. But before using this information to make a decision about the drug, the physicians will need to provide a measure of reliability for the inference. That is, the physicians will want to know whether the difference between the percentages of patients with complications observed in this sample of 114 patients is generalizable to the population of all coronary bypass patients.

FIGURE 2.5

SAS Summary Tables for COMP by Value of DRUG

--- DRUG=NO --

COMP	Frequency	Percent	Cumulative Frequency	Cumulative Percent
BOTH	1	1.75	1	1.75
INFECT	4	7.02	5	8.77
NONE	47	82.46	52	91.23
REDO	5	8.77	57	100.00

--- DRUG=YES --

COMP	Frequency	Percent	Cumulative Frequency	Cumulative Percent
BOTH	3	5.26	3	5.26
INFECT	8	14.04	11	19.30
NONE	39	68.42	50	87.72
REDO	7	12.28	57	100.00

 EXERCISES 2.1–2.15

Learning the Mechanics

2.1 Complete the following table.

Grade on Statistics Exam	Frequency	Relative Frequency
A: 90–100		.08
B: 80–89	36	
C: 65–79	90	
D: 50–64	30	
F: Below 50	28	
Total	200	1.00

2.2 A qualitative variable with three classes (X, Y, and Z) is measured for each of 20 units randomly sampled from a target population. The data (observed class for each unit) are listed below.

Y X X Z X Y Y Y Y X X Z X
Y Y X Z Y Y Y X

a. Compute the frequency for each of the three classes.
b. Compute the relative frequency for each of the three classes.
c. Display the results, part **a**, in a frequency bar graph.
d. Display the results, part **b**, in a pie chart.

Applying the Concepts—Basic

2.3 The International Rhino Federation estimates that there are 13,585 rhinoceroses living in the wild in Africa and Asia. A breakdown of the number of rhinos of each species is reported in the accompanying table.

Rhino Species	Population Estimate
African Black	2,600
African White	8,465
(Asian) Sumatran	400
(Asian) Javan	70
(Asian) Indian	2,050
Total	13,585

Source: International Rhino Federation, July 1998.

a. Construct a relative frequency table for the data.
b. Display the relative frequencies in a bar graph.
c. What proportion of the 13,585 rhinos are African rhinos? Asian?   .8145; .1855

2.4 *USA Today* (Dec. 31, 1999) reported on a Gallup Survey which asked: "Do you think cigarette smoking is one of the causes of lung cancer?" The accompanying table shows the results of the 1999 survey compared to the results of a similar survey conducted 45 years earlier.

Is Smoking a Cause of Lung Cancer?	1954 Survey	1999 Survey
Yes	41%	92%
No	31%	6%
No opinion	28%	2%

a. Construct a pie chart for the 1954 survey results.
b. Construct a pie chart for the 1999 survey results.

c. Compare the two graphs. Do the graphs show a dramatic shift in people's opinion on whether smoking causes lung cancer?

2.5   Driver-side and passenger-side air bags are installed in all new cars to prevent serious or fatal injury in an automobile crash. However, air bags have been found to cause deaths in children and small people or people with handicaps in low-speed crashes. Consequently, in 1998 the federal government began allowing vehicle owners to request installation of an on-off switch for air bags. The table describes the reasons for requesting the installation of passenger-side on-off switches given by car owners in 1998 and 1999.

Reason	Number of Requests
Infant	1,852
Child	17,148
Medical	8,377
Infant & Medical	44
Child & Medical	903
Infant & Child	1,878
Infant & Child & Medical	135
Total	30,337

Source: National Highway Transportation Safety Administration, Sept. 2000.

a. What type of variable, quantitative or qualitative, is summarized in the table? Give the values that the variable could assume. Qualitative

b. Calculate the relative frequencies for each reason.

c. Display the information in the table in an appropriate graph.

d. What proportion of the car owners who requested on-off air bag switches gave Medical as one of the reasons? .3118

2.6   In *Psychology and Aging* (Dec. 2000), University of Michigan School of Public Health researchers studied the roles that elderly people feel are the most important to them in late life. The accompanying table summarizes the most salient roles identified by each in a national sample of 1,102 adults, 65 years or older.

Most Salient Role	Number
Spouse	424
Parent	269
Grandparent	148
Other relative	59
Friend	73
Homemaker	59
Provider	34
Volunteer, club, church member	36
Total	1,102

Source: Krause, N., and Shaw, B. A. "Role-specific feelings of control and mortality," *Psychology and Aging*, Vol. 15, No. 4, Dec. 2000 (Table 2).

a. Describe the qualitative variable summarized in the table. Give the categories associated with the variable.

b. Are the numbers in the table frequencies or relative frequencies?   Frequencies

c. Display the information in the table in a bar graph.

d. Which role is identified by the highest percentage of elderly adults? Interpret the relative frequency associated with this role.   Spouse

2.7   The Moffitt Cancer Center at the University of South Florida treats over 25,000 patients a year. The graphic (reproduced from Figure 1.1) describes the types of cancer treated in Moffitt's patients during fiscal year 2000.

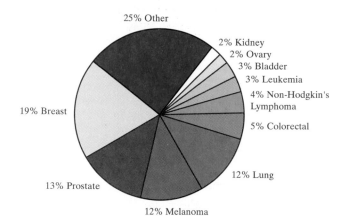

a. What type of graph is portrayed?   Pie chart

b. Which type of cancer is treated most often at Moffitt?

c. What percentage of Moffitt's patients are treated for melanoma, lymphoma, or leukemia?   19%

Applying the Concepts—Intermediate

2.8   Archaeologists excavating the ancient Greek settlement at Phylakopi classified the pottery found in trenches (*Chance*, Fall 2000). The accompanying table describes the collection of 837 pottery pieces uncovered in a particular layer at the excavation site. Construct and interpret a graph that will aid the archaeologists in understanding the distribution of the pottery types found at the site.

Pot Category	Number Found
Burnished	133
Monochrome	460
Slipped	55
Painted in curvilinear decoration	14
Painted in geometric decoration	165
Painted in naturalistic decoration	4
Cycladic white clay	4
Conical cup clay	2
Total	837

Source: Berg, I., and Bliedon, S. "The Pots of Phylakopi: Applying Statistical Techniques to Archaeology," Vol. 13, No. 4, Fall 2000.

2.9 *Forbes* magazine periodically conducts a salary survey of chief executive officers. In addition to salary information, *Forbes* collects and reports personal data on the CEOs, including level of education. Do most CEOs have advanced degrees, such as masters degrees or doctorates? The data in the table represent the highest degree obtained for each of the top 25 best-paid CEOs of 1999. Use a graphical method to summarize the highest degree obtained for these CEOs. What is your opinion about whether most CEOs have advanced degrees?

⊚ CEODEGREES

CEO	Company	Degree
1. Michael D. Eisner	Walt Disney	Bachelors
2. Mel Karmazin	CBS	Bachelors
3. Stephen M. Case	America Online	Bachelors
4. Stephen C. Hilbert	Conseco	none
5. Craig R. Barrett	Intel	Doctorate
6. Millard Drexler	GAP	Masters
7. John F. Welch, Jr.	General Electric	Doctorate
8. Thomas G. Stemberg	Staples	Masters
9. Henry R. Silverman	Cendant	JD
10. Reuben Mark	Colgate-Palmolive	Masters
11. Philip J. Purcell	Morgan Stanley Dean Witter	Masters
12. Scott G. McNealy	Sun Microsystems	Masters
13. Margaret C. Whitman	eBay	Masters
14. Louis V. Gerstner, Jr.	IBM	Masters
15. John F. Gifford	Maxim Integrated Products	Bachelors
16. Robert L. Waltrip	Service Corp. International	Bachelors
17. M. Douglas Ivester	Coca-Cola	Bachelors
18. Gordon M. Binder	Amgen	Masters
19. Charles R. Schwab	Charles Schwab	Masters
20. William R. Steere, Jr.	Pfizer	Bachelors
21. Nolan D. Archibald	Black & Decker	Masters
22. Charles A. Heimbold, Jr.	Bristol-Myers Squibb	LLB (law)
23. William L. Larson	Network Associates	JD
24. Maurice R. Greenberg	American International Group	LLB (law)
25. Richard Jay Kogan	Schering-Plough	Masters

Source: Forbes, May 17, 1999.

2.10 Audiologists have recently developed a rehabilitation program for hearing-impaired patients in a Canadian home for senior citizens (*Journal of the Academy of Rehabilitative Audiology*, 1994). Each of the 30 residents of the home were diagnosed for degree and type of sensorineural hearing loss, coded as follows: 1 = hear within normal limits, 2 = high-frequency hearing loss, 3 = mild loss, 4 = mild-to-moderate loss, 5 = moderate loss, 6 = moderate-to-severe loss, and 7 = severe-to-profound loss. The data are listed in the accompanying table. Use a graph to portray the results. Which type of hearing loss appears to be the most prevalent among nursing home residents?

⊚ HEARLOSS

6	7	1	1	2	6	4	6	4	2	5	2	5
1	5	4	6	6	5	5	5	2	5	3	6	4
6	6	4	2									

Source: Jennings, M. B., and Head, B. G. "Development of an ecological audiologic rehabilitation program in a home-for-the-aged." *Journal of the Academy of Rehabilitative Audiology*, Vol. 27, 1994, p. 77 (Table 1).

2.11 *Choice* magazine provides new-book reviews each issue. A random sample of 375 *Choice* book reviews in American history, geography, and area studies was selected and the "overall opinion" of the book stated in each review was ascertained (*Library Acquisitions: Practice and Theory*, Vol. 19, 1995). Overall opinion was coded as follows: 1 = would not recommend, 2 = cautious or very little recommendation, 3 = little or no preference, 4 = favorable/recommended, 5 = outstanding/significant contribution. A summary of the data is provided in the accompanying bar graph.

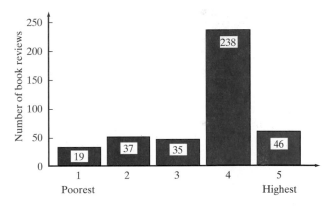

Source: Carlo, P.W., and Natowitz, A. "*Choice* book reviews in American history, geography, and area studies: An analysis for 1988–1993." *Library Acquisitions: Practice & Theory*, Vol. 19, No. 2, 1995, p. 159 (Figure 1).

a. Find the opinion that occurred most often. What proportion of the books reviewed had this opinion?

b. Do you agree with the following statement extracted from the study: "A majority (more than 75%) of books reviewed are evaluated favorably and recommended for purchase."?   Yes

Applying the Concepts—Advanced

2.12 Drowsy driving leads to thousands of automobile crashes each year. A report on driver fatigue and drowsiness, sponsored by the National Center on Sleep Disorders Research (NCSDR) and the National Highway Traffic Safety Administration (NHTSA), found that the time of day when the accident occurs depends on the age of the drowsy driver. Based on data collected from police reports for accidents attributed to drowsy driving or the driver falling asleep, bar graphs showing the distribution

a) 25 and under

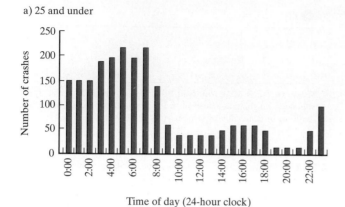

b) 26–45 years

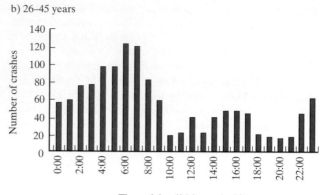

c) 46–65 years

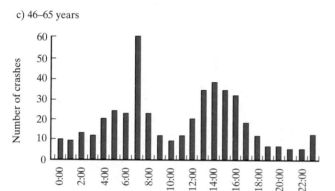

d) 65 and over

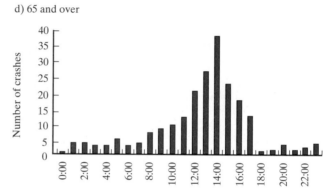

Source: National Highway Traffic Safety Administration, 1996.

of the time of day in which the crash occurred were constructed. These graphs, one for each of four driver age groups (25 and under, 26 to 45, 46 to 65, and over 65), are displayed above. Compare the time-of-day distributions. What patterns do the graphs reveal?

2.13 Marine scientists who study dolphin communication have discovered that bottlenose dolphins exhibit an individualized whistle contour known as the *signature whistle*. A study was conducted to categorize the signature whistles of ten captive adult bottlenose dolphins in socially interactive contexts (*Ethology*, July 1995). A total of 185 whistles were recorded during the study period; each whistle contour was analyzed and assigned to a category using a contour similarity (CS) technique. The results are reported in the accompanying table. Use a graphical method to summarize the results. Do you detect any patterns in the data that might be helpful to marine scientists?

 DOLPHIN

Whistle Category	Number of Whistles
Type a	97
Type b	15
Type c	9
Type d	7
Type e	7
Type f	2
Type g	2
Type h	2
Type i	2
Type j	4
Type k	13
Other types	25

Source: McCowan, B., and Reiss, D. "Quantitative comparison of whistle repertoires from captive adult bottlenose dolphins (Delphiniae, *Tursiops truncatus*): A re-evaluation of the signature whistle hypothesis." *Ethology*, Vol. 100, No. 3, July 1995, p. 200 (Table 2).

2.14 According to *Benford's Law*, certain digits $(1, 2, 3, \ldots, 9)$ are more likely to occur as the first significant digit in a randomly selected number than other digits. For example, the law predicts that the number "1" is the most likely to occur (30% of the time) as the first digit. In a study reported in the *American Scientist* (July–Aug. 1998) to test Benford's Law, 743 first-year college students were asked to write down a six-digit number at random. The first significant digit of each number was recorded and its distribution summarized in the table on p. 29.

⦿ DIGITS First Digit	Number of Occurrences
1	109
2	75
3	77
4	99
5	72
6	117
7	89
8	62
9	43
Total	743

Source: Hill, T. P "The First Digit Phenomenon." *American Scientist*, Vol. 86, No. 4, July–Aug. 1998, p. 363 (Figure 5).

a. Describe the first digit of the "random guess" data with an appropriate graph.

b. Does the graph support Benford's Law? Explain.

2.15 Owing to several major ocean oil spills by tank vessels, Congress passed the 1990 Oil Pollution Act, which requires all tankers to be designed with thicker hulls. Further improvements in the structural design of a tank vessel have been proposed since then, each with the objective of reducing the likelihood of an oil spill and decreasing the amount of outflow in the event of a hull puncture. To aid in this development, *Marine Technology* (Jan. 1995) reported on the spillage amount and cause of puncture for 50 recent major oil spills from tankers and carriers. The data are reproduced in the table below.

a. Use a graphical method to describe the cause of oil spillage for the 50 tankers.

b. Does the graph, part **a**, suggest that any one cause is more likely to occur than any other? How is this information of value to the design engineers?

⦿ OILSPILL Tanker	Spillage (metric tons, thousands)	Cause
Atlantic Empress	257	Collision (C)
Castillo De Bellver	239	Fire/Explosion (FE)
Amoco Cadiz	221	Hull Failure (HF)
Odyssey	132	FE
Torrey Canyon	124	Grounding (G)

Tanker	Spillage	Cause
Sea Star	123	C
Hawaiian Patriot	101	HF
Independenta	95	C
Urquiola	91	G
Irenes Serenade	82	FE
Khark 5	76	FE
Nova	68	C
Wafra	62	G
Epic Colocotronis	58	G
Sinclair Petrolore	57	FE
Yuyo Maru No 10	42	C
Assimi	50	FE
Andros Patria	48	FE
World Glory	46	HF
British Ambassador	46	HF
Metula	45	G
Pericles G. C.	44	FE
Mandoil II	41	C
Jakob Maersk	41	G
Burmah Agate	41	C
J. Antonio Lavalleja	38	G
Napier	37	G
Exxon Valdez	36	G
Corinthos	36	C
Trader	36	HF
St. Peter	33	FE
Gino	32	C
Golden Drake	32	FE
Lonnis Angelicoussis	32	FE
Chryssi	32	HF
Ionnis Challenge	31	HF
Argo Merchant	28	G
Heimvard	31	C
Pegasus	25	Unknown (U)
Pacocean	31	HF
Texaco Oklahoma	29	HF
Scorpio	31	G
Ellen Conway	31	G
Caribbean Sea	30	HF
Cretan Star	26	U
Grand Zenith	26	HF
Athenian Venture	26	FE
Venoil	26	C
Aragon	24	HF
Ocean Eagle	21	G

Source: Daidola, J. C. "Tanker structure behavior during collision and grounding." *Marine Technology*, Vol. 32, No. 1, Jan. 1995, p. 22 (Table 1).

2.2 GRAPHICAL METHODS FOR DESCRIBING QUANTITATIVE DATA

Recall from Section 1.4 that quantitative data sets consist of data that are recorded on a meaningful numerical scale. For describing, summarizing, and detecting patterns in such data, we can use three graphical methods: *dot plots, stem-and-leaf displays,* and *histograms*. Since most statistical software packages can be used to construct these displays, we'll focus here on their interpretation rather than their construction.

⊘ EPAGAS

TABLE 2.3   EPA Mileage Ratings on 100 Cars

36.3	41.0	36.9	37.1	44.9	36.8	30.0	37.2	42.1	36.7
32.7	37.3	41.2	36.6	32.9	36.5	33.2	37.4	37.5	33.6
40.5	36.5	37.6	33.9	40.2	36.4	37.7	37.7	40.0	34.2
36.2	37.9	36.0	37.9	35.9	38.2	38.3	35.7	35.6	35.1
38.5	39.0	35.5	34.8	38.6	39.4	35.3	34.4	38.8	39.7
36.3	36.8	32.5	36.4	40.5	36.6	36.1	38.2	38.4	39.3
41.0	31.8	37.3	33.1	37.0	37.6	37.0	38.7	39.0	35.8
37.0	37.2	40.7	37.4	37.1	37.8	35.9	35.6	36.7	34.5
37.1	40.3	36.7	37.0	33.9	40.1	38.0	35.2	34.8	39.5
39.9	36.9	32.9	33.8	39.8	34.0	36.8	35.0	38.1	36.9

For example, the Environmental Protection Agency (EPA) performs extensive tests on all new car models to determine their mileage ratings. Suppose that the 100 measurements in Table 2.3 represent the results of such tests on a certain new car model. How can we summarize the information in this rather large sample?

Suggested Exercise 2.28

A visual inspection of the data indicates some obvious facts. For example, most of the mileages are in the 30s, with a smaller fraction in the 40s. But it is difficult to provide much additional information on the 100 mileage ratings without resorting to some method of summarizing the data. One such method is a dot plot.

Dot Plots

TEACHING TIP
The dot plot condenses the data by grouping all values that are the same together in the plot.

A MINITAB **dot plot** for the 100 EPA mileage ratings is shown in Figure 2.6. The horizontal axis of Figure 2.6 is a scale for the quantitative variable in miles per gallon (mpg). The numerical value of each measurement in the data set is located on the horizontal scale by a dot. When data values repeat, the dots are placed above one another, forming a pile at that particular numerical location. As you can see, this dot plot verifies that almost all of the mileage ratings are in the 30s, with most falling between 35 and 40 miles per gallon.

FIGURE 2.6

A MINITAB Dot Plot for the EPA Mileage Ratings on 100 Cars

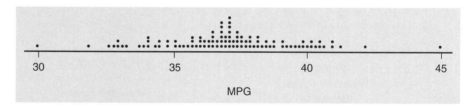

Stem-and-Leaf Display

TEACHING TIP
The stem-and-leaf display condenses the data by grouping all data with the same stem together in the graph.

Suggested Exercise 2.24

Another graphical representation of these same data, a MINITAB **stem-and-leaf display**, is shown in Figure 2.7. In this display the *stem* is the portion of the measurement (mpg) to the left of the decimal point, while the remaining portion to the right of the decimal point is the *leaf*.

TEACHING TIP
Choices for the stems and the leaves are critical to producing the most meaningful stem-and-leaf display. Encourage students to try different options until they produce the display that they think best characterizes the data.

In Figure 2.7, the stems for the data set are listed in the second column from the smallest (30) to the largest (44). Then the leaf for each observation is listed to the right in the row of the display corresponding to the observation's stem. For example, the leaf 3 of the first observation (36.3) in Table 2.3 appears in the row corresponding to the stem 36. Similarly, the leaf 7 for the second observation (32.7) in Table 2.3 appears in the row corresponding to the stem 32, while the leaf 5 for the third observation (40.5) appears in the row corresponding to the stem 40. (The stems and leaves for these first three observations are highlighted in Figure 2.7.)

FIGURE 2.7
MINITAB Stem-and-Leaf Display for the EPA Mileage Ratings on 100 Cars

```
Stem-and-leaf of MPG          N = 100
Leaf Unit = 0.10

   1   30  0
   2   31  8
   6   32  5799
  12   33  126899
  18   34  024588
  29   35  01235667899
  49   36  01233445566777888999
 (21)  37  000011122334456677899
  30   38  0122345678
  20   39  00345789
  12   40  0123557
   5   41  002
   2   42  1
   1   43
   1   44  9
```

Typically, the leaves in each row are ordered as shown in the MINITAB stem-and-leaf display.

The stem-and-leaf display presents another compact picture of the data set. You can see at a glance that the 100 mileage readings were distributed between 30.0 and 44.9, with most of them falling in stem rows 35 to 39. The six leaves in stem row 34 indicate that six of the 100 readings were at least 34.0 but less than 35.0. Similarly, the eleven leaves in stem row 35 indicate that eleven of the 100 readings were at least 35.0 but less than 36.0. Only five cars had readings equal to 41 or larger, and only one was as low as 30.

The definitions of the stem and leaf for a data set can be modified to alter the graphical description. For example, suppose we had defined the stem as the tens digit for the gas mileage data, rather than the ones and tens digits. With this definition, the stems and leaves corresponding to the measurements 36.3 and 32.7 would be as follows:

Stem	Leaf
3	6

Stem	Leaf
3	2

Note that the decimal portion of the numbers has been dropped. Generally, only one digit is displayed in the leaf.

If you look at the data, you'll see why we didn't define the stem this way. All the mileage measurements fall in the 30s and 40s, so all the leaves would fall into just two stem rows in this display. The resulting picture would not be nearly as informative as Figure 2.7.

Histograms

TEACHING TIP
The histogram condenses the data by grouping similar data values in the same class in the graph.

A STATISTIX **histogram** for these 100 EPA mileage readings is shown in Figure 2.8. The horizontal axis of Figure 2.8, which gives the miles per gallon for a given automobile, is divided into *intervals* commencing with the interval from 30.0–31.5 and proceeding in intervals of equal size to 43.5–45.0 mpg. The vertical axis gives the number (or *frequency*) of the 100 readings that fall in each interval.

FIGURE 2.8

STATISTIX Histogram for EPA Mileage Data

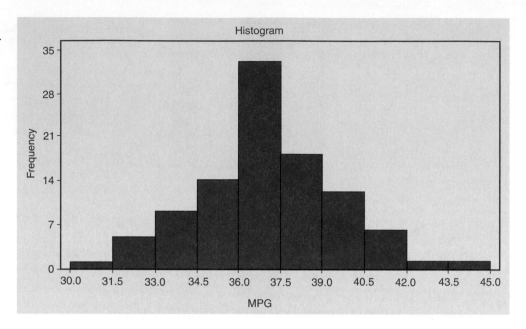

Suggested Exercise 2.26

It appears that about 33 of the 100 cars, or 33%, obtained a mileage between 36.0 and 37.5. This interval contains the highest frequency, and the intervals tend to contain a smaller number of the measurements as the mileages get smaller or larger.

Histograms can be used to display either the frequency or relative frequency of the measurements falling into specified intervals known as **measurement classes**. The measurement classes, frequencies, and relative frequencies for the EPA car mileage data are shown in the summary table, Table 2.4.

TABLE 2.4   Measurement Classes, Frequencies, and Relative Frequencies for the Car Mileage Data

Measurement Class	Frequency	Relative Frequency
30.0–31.5	1	.01
31.5–33.0	5	.05
33.0–34.5	9	.09
34.5–36.0	14	.14
36.0–37.5	33	.33
37.5–39.0	18	.18
39.0–40.5	12	.12
40.5–42.0	6	.06
42.0–43.5	1	.01
43.5–45.0	1	.01
Totals	100	1.00

TEACHING TIP
Classes of equal width should be used when generating a histogram.

TEACHING TIP
Use the data from exercise 2.30 and have different students use different graphical techniques to summarize the data. Use the students' work to compare the techniques.

By summing the relative frequencies in the intervals 34.5–36.0, 36.0–37.5, and 37.5–39.0, you can see that 65% of the mileages are between 34.5 and 39.0. Similarly, only 2% of the cars obtained a mileage rating over 42.0. Many other summary statements can be made by further study of the histogram and accompanying summary table.

When interpreting a histogram (say, the histogram in Figure 2.8), consider two important facts. First, the proportion of the total area under the histogram that falls above a particular interval on the *x*-axis is equal to the relative frequency of measurements falling in the interval. For example, the relative frequency for the class interval 36.0–37.5 is .33. Consequently, the rectangle above the interval contains .33 of the total area under the histogram.

Second, imagine the appearance of the relative frequency histogram for a very large set of data (say, a population). As the number of measurements in a data set is increased, you can obtain a better description of the data by decreasing the width of the class intervals. When the class intervals become small enough, a relative frequency histogram will (for all practical purposes) appear as a smooth curve (see Figure 2.9).

TEACHING TIP
When constructing histograms, use more classes as the number of values in the data set gets larger.

FIGURE 2.9

The Effect of the Size of a Data Set on the Outline of a Histogram

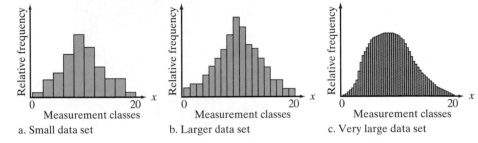

a. Small data set   b. Larger data set   c. Very large data set

While histograms provide good visual descriptions of data sets—particularly very large ones—they do not let us identify individual measurements. In contrast, each of the original measurements is visible to some extent in a dot plot and clearly visible in a stem-and-leaf display. The stem-and-leaf display arranges the data in ascending order, so it's easy to locate the individual measurements. For example, in Figure 2.7 we can easily see that two of the gas mileage measurements are equal to 36.3, but can't see that fact by inspecting the histogram in Figure 2.8. However, stem-and-leaf displays can become unwieldy for very large data sets. A very large number of stems and leaves causes the vertical and horizontal dimensions of the display to become cumbersome, diminishing the usefulness of the visual display.

EXAMPLE 2.2

The data in Table 2.5 give, by state, the percentages of the total number of college or university student loans that are in default.
a. Use a statistical software package to create a histogram for these data.
b. Use a statistical software package to create a stem-and-leaf display for these data.
c. Compare and interpret the two graphic displays of these data.

⊘ LOANDEFAULT

TABLE 2.5   Percentage of Student Loans (per State) in Default

State	%	State	%	State	%	State	%
Ala.	12.0	Ill.	9.3	Mont.	6.4	R.I.	8.8
Alaska	19.7	Ind.	6.7	Nebr.	4.9	S.C.	14.1
Ariz.	12.1	Iowa	6.2	Nev.	10.1	S. Dak.	5.5
Ark.	12.9	Kans.	5.7	N.H.	7.9	Tenn.	12.3
Calif.	11.4	Ky.	10.3	N.J.	12.0	Tex.	15.2
Colo.	9.5	La.	13.5	N. Mex.	7.5	Utah	6.0
Conn.	8.8	Maine	9.7	N.Y.	11.3	Vt.	8.3
Del.	10.9	Md.	16.6	N.C.	15.5	Va.	14.4
D.C.	14.7	Mass.	8.3	N. Dak.	4.8	Wash.	8.4
Fla.	11.8	Mich.	11.4	Ohio	10.4	W Va.	9.5
Ga.	14.8	Minn.	6.6	Okla.	11.2	Wis.	9.0
Hawaii	12.8	Miss.	15.6	Oreg.	7.9	Wyo.	2.7
Idaho	7.1	Mo.	8.8	Pa.	8.7		

Source: National Direct Student Loan Program.

FIGURE 2.10

SPSS Frequency Histogram for Student Loan Default Rate

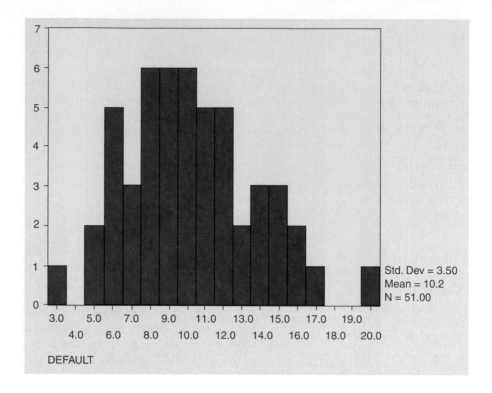

Std. Dev = 3.50
Mean = 10.2
N = 51.00

DEFAULT

Solution

a. We used SPSS to generate the frequency histogram in Figure 2.10. Note that 18 classes were formed by the SPSS program. The classes are identified by their *midpoints* rather than their endpoints. Thus, the first interval has a midpoint of 3, the second of 4, and so on. The corresponding measurement classes based on these midpoints are therefore 2.5–3.5, 3.5–4.5, etc. Note that the measurement classes with midpoints of 8, 9, and 10 (ranging from 7.5 to 10.5) contain 18 of the 51 default measurements.

b. We also used SPSS to generate the stem-and-leaf display in Figure 2.11. Note that the stem, which is the *second* column in the printout, has been defined as

FIGURE 2.11

SPSS Stem-and-Leaf Display for Defaulted Student Loan Data

```
DEFAULT Stem-and-Leaf Plot

 Frequency     Stem  &  Leaf

     1.00        0  .  2
     4.00        0  .  4455
     9.00        0  .  666667777
    12.00        0  .  888888899999
     9.00        1  .  000011111
     7.00        1  .  2222223
     7.00        1  .  4444555
     1.00        1  .  6
     1.00   Extremes      (>=20)


Stem width:      10.0
Each leaf:            1 case(s)
```

the number *two* places to the left of the decimal. The leaf is the third column in the printout, and is the number one place to the left of the decimal. (The digit to the right of the decimal is not shown.) SPSS also indicates the frequency of measurements (first column of the printout) in each stem.

c. As is usually the case for data sets that are not too large (say, fewer than 100 measurements), the stem-and-leaf display provides more detail than the histogram without being unwieldy. For instance, the stem-and-leaf display in Figure 2.11 clearly indicates that most states' rates are relatively evenly distributed from 6 to 15 (note that the five stem rows in the middle of the display have between 7 and 12 measurements "attached" to each stem). The stem-and-leaf display also clearly indicates the values of the individual measurements in the data set. For example, the lowest default rate (representing the measurement 2.7) is shown in the first stem row.

The histogram in Figure 2.10 also indicates a relatively even distribution of rates over the ten measurement classes with midpoints from 6 to 15. However, it's harder to determine the boundaries since we can't "see" the individual measurements in the display. Similarly, the two small bars over the outside classes with midpoints of 3 and 20 indicate the presence of some extreme measurements, but we can't see the exact number and location of these measurements. However, histograms are most useful for displaying very large data sets when the overall shape of the distribution of measurements is more important than the identification of individual measurements. ⬛

Almost all statistical software packages can be used to generate histograms, stem-and-leaf displays, and dot plots. All three are useful tools for graphically describing data sets. We recommend that you generate and compare the displays whenever you can. You'll find that histograms are generally more useful for very large data sets, while stem-and-leaf displays and dot plots provide useful detail for smaller data sets.

Using the TI-83 Graphing Calculator

Making a Histogram on the TI-83

Step 1   *Enter the data*
Press   **STAT   1**   for   **STAT edit**
Use the arrow and **ENTER** keys to enter the data set into **L1**.

Step 2   *Set up the histogram plot*
Press   **2nd   Y=**   for   **STAT PLOT**
Press   **1**   for   **Plot 1**

Use the arrow and **ENTER** keys to set up the screen as shown below.

USING THE TI-83 GRAPHING CALCULATOR (*continued*)

Step 3   *Select your window settings*
           Press WINDOW and adjust the settings as follows:

           Xmin = lowest class boundary
           Xmax = highest class boundary
           Xscl = class width
           Ymin = 0
           Ymax = greatest class frequency
           Yscl = approximately Ymax/10
           Xres = 1

Step 4   *View the graph*
           Press **GRAPH**

Optional   *Read class frequencies and class boundaries*

Step   You can press **TRACE** to read the class frequencies and class
           boundaries.
           Use the arrow keys to move between bars.

Example   The figures below show TI-83 window settings and histogram for
           the following sample data:

           86, 70, 62, 98, 73, 56, 53, 92, 86, 37, 62, 83, 78, 49, 78, 37, 67, 79, 57

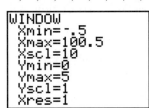 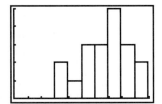

Making a Histogram from a Frequency Table

Step 1   *Enter the data*
           Enter a value from the middle of each class in **L1**
           Enter the class frequencies or relative frequencies in **L2**

Step 2   *Set up the histogram plot*
           Press   **2nd   Y=**   for   **STAT PLOT**
           Press   **1**   for   **Plot1**

           Use the arrow and **ENTER** keys to set up the screen as shown
           below.

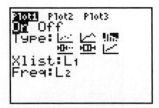

Steps 3-4   Follow steps 3-4 given above.

EXERCISES 2.16–2.32

Learning the Mechanics

2.16 Graph the relative frequency histogram for the 500 measurements summarized in the accompanying relative frequency table.

Measurement Class	Relative Frequency
.5–2.5	.10
2.5–4.5	.15
4.5–6.5	.25
6.5–8.5	.20
8.5–10.5	.05
10.5–12.5	.10
12.5–14.5	.10
14.5–16.5	.05

2.17 Refer to Exercise 2.16. Calculate the number of the 500 measurements falling into each of the measurement classes. Then graph a frequency histogram for these data.

2.18 STATISTIX was used to generate the stem-and-leaf display shown below.

STATISTIX Output for Exercise 2.18

```
STEM AND LEAF PLOT OF X

LEAF DIGIT UNIT = 1        MINIMUM  0.0000
5  1   REPRESENTS 51.      MEDIAN  25.000
                          MAXIMUM  51.000

        STEM   LEAVES
    3     0    012
    3     0
    6     1    224
    7     1    8
   11     2    1134
   (3)    2    599
    9     3    0003
    5     3    6
    4     4    4
    3     4    57
    1     5    1

23 CASES INCLUDED    0 MISSING CASES
```

a. How many observations were in the original data set?

b. In the top row of the stem-and-leaf display, identify the stem, the leaves, and the numbers in the original data set represented by this stem and its leaves.

c. Re-create all the numbers in the data set and construct a dot plot.

2.19 MINITAB was used to generate the following histogram:

MINITAB Output for Exercise 2.19

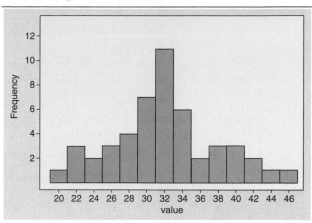

a. Is this a frequency histogram or a relative frequency histogram? Explain.   Frequency

b. How many measurement classes were used in the construction of this histogram?   14

c. How many measurements are there in the data set described by this histogram?   49

Applying the Concepts—Basic

2.20 The graph below summarizes the scores obtained by 100 students on a questionnaire designed to measure aggressiveness. (Scores are integer values that range from 0 to 20. A high score indicates a high level of aggression.)

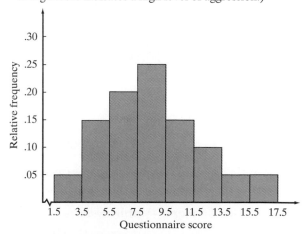

a. Which measurement class contains the highest proportion of test scores?   7.5–9.5

b. What is the proportion of scores that lie between 3.5 and 5.5?   .15

c. What is the proportion of scores that are higher than 11.5?   .20

d. How many students scored less than 5.5?   20

2.21 The National Earthquake Information Center located 16,287 earthquakes worldwide in 1998. The magnitudes (on the Richter scale) of the earthquakes are summarized in the table. Display the data in a graph.

Magnitude	Number of Earthquakes
0	1,774
0.1–0.9	6
1.0–1.9	580
2.0–2.9	3,080
3.0–3.9	4,518
4.0–4.9	5,548
5.0–5.9	674
6.0–6.9	97
7.0–7.9	9
8.0–8.9	1

Source: National Earthquake Information Center, Department of Interior, U.S. Geological Survey.

2.22 The United States Chess Federation (USCF) establishes a numerical rating for each competitive chess player. The USCF rating is a number between 0 and 4,000 that changes over time depending on the outcome of tournament games. The higher the rating, the better (more successful) the player. The graph describes the rating distribution of 27,563 players who were active competitors in 1997.

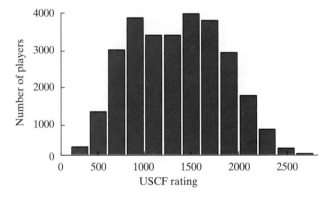

Source: United States Chess Federation, Jan. 1998.

a. What type of graph is displayed?

b. Is the variable displayed on the graph quantitative or qualitative?   Quantitative

c. What percentage of players has a USCF rating above 1,000?   ≈ 71%

Applying the Concepts—Intermediate

2.23 The United States Golf Association (USGA) Handicap System is designed to allow golfers of differing abilities to enjoy fair competition. The handicap index is a measure of a player's potential scoring ability on an 18-hole golf course of standard difficulty. For example, on a par-72 course, a golfer with a handicap of 7 will typically have a score of 79 (seven strokes over par). Over 4.5 million golfers have an official USGA Handicap index. The handicap indexes for both male and female golfers were obtained from the USGA and are summarized in the two MINITAB histograms shown below.

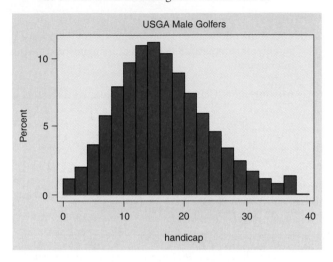

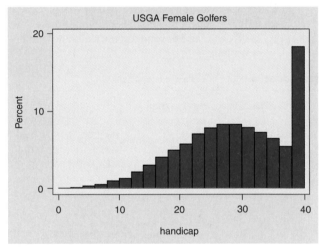

a. What percentage of male USGA golfers have a handicap greater than 20?   ≈ 28.5%

b. What percentage of female USGA golfers have a handicap greater than 20?   ≈ 71%

c. Identify the type of skewness (if any) present in the male handicap distribution.   Skewed to the right

d. Identify the type of skewness (if any) present in the female handicap distribution.   Skewed to the left

2.24 A group of University of Virginia biologists studied nectivory (nectar drinking) in crab spiders to determine if adult males were feeding on nectar to prevent fluid loss (*Animal Behavior*, June 1995). Nine male spiders were weighed and then placed on the flowers of Queen Anne's lace. One hour later, the spiders were removed and reweighed. The evaporative fluid loss (in milligrams) of each of the nine male spiders is given in the next table.

⊘ SPIDERS

Male Spider	Fluid Loss
A	.018
B	.020
C	.017
D	.024
E	.020
F	.024
G	.003
H	.001
I	.009

Source: Pollard, S. D., *et al.* "Why do male crab spiders drink nectar?" *Animal Behavior*, Vol. 49, No. 6, June 1995, p. 1445 (Table II).

a. Summarize the fluid losses of male crab spiders with a stem-and-leaf display.

b. Of the nine spiders, only three drank any nectar from the flowers of Queen Anne's lace. These three spiders are identified as G, H, and I in the table. Locate and circle these three fluid losses on the stem-and-leaf display. Does the pattern depicted in the graph give you any insight into whether feeding on flower nectar reduces evaporative fluid loss for male crab spiders? Explain.

2.25 *Postmortem interval* (PMI) is defined as the elapsed time between death and an autopsy. Knowledge of PMI is considered essential when conducting medical research on human cadavers. The data in the table are the PMIs of 22 human brain specimens obtained at autopsy in a recent study (*Brain and Language*, June 1995). Graphically describe the PMI data with a dot plot. Based on the plot, make a summary statement about the PMI of the 22 human brain specimens.

⊘ BRAINPMI
Postmortem Intervals for 22 Human Brain Specimens

5.5	14.5	6.0	5.5	5.3	5.8	11.0	6.1
7.0	14.5	10.4	4.6	4.3	7.2	10.5	6.5
3.3	7.0	4.1	6.2	10.4	4.9		

Source: Hayes. T. L., and Lewis, D. A. "Anatomical specialization of the anterior motor speech area: Hemispheric differences in magnopyramidal neurons." *Brain and Language*, Vol. 49, No. 3, June 1995, p. 292 (Table 1).

2.26 It's not uncommon for hearing aids to cancel the desired signal. *IEEE Transactions on Speech and Audio Processing* (May 1995) reported on an audio processing system designed to limit the amount of signal cancellation that may occur. The system utilizes a mathematical equation that involves a variable, V, called a *sufficient norm constraint*. A histogram for realizations of V, produced using simulation, is shown here.

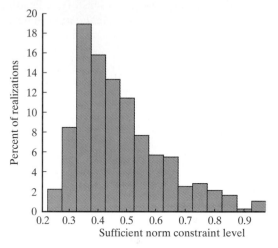

Source: Hoffman, M. W., and Buckley, K. M. "Robust time-domain processing of broadband microphone array data." *IEEE Transactions on Speech and Audio Processing*, Vol. 3, No. 3, May 1995, p. 199 (Figure 4).

a. Estimate the percentage of realizations of V with values ranging from .425 to .675.   .45

b. Cancellation of the desired signal is limited by selecting a norm constraint V so that only 10% of the realizations have values below the selected level. Find this value of V.   .325

2.27 Data from a psychology experiment were reported and analyzed in *The American Statistician* (May 2001). Two samples of female students participated in the experiment. One sample consisted of 11 students known to suffer from the eating disorder bulimia; the other sample consisted of 14 students with normal eating habits. Each student completed a questionnaire from which a "fear of negative evaluation" (FNE) score was produced. (The higher the score, the greater the fear of negative evaluation.) The data are displayed in the table.

a. Construct a dot plot or stem-and-leaf display for the FNE scores of all 25 female students.

b. Highlight the bulimic students on the graph, part **a**. Does it appear that bulimics tend to have a greater fear of negative evaluation? Explain.

c. Why is it important to attach a measure of reliability to the inference made in part **b**?

⊘ BULIMIA

Bulimic students:	21	13	10	20	25	19	16	21	24	13	14			
Normal students:	13	6	16	13	8	19	23	18	11	19	7	10	15	20

Source: Randles, R. H. "On Neutral Responses (Zeros) in the Sign Test and Ties in the Wilcoxon-Mann-Whitney Test." *The American Statistician*, Vol. 55, No. 2, May 2001 (Figure 3).

2.28 Neurological single-photon-emission computed tomography (neuroSPECT) is a relatively new method for detecting abnormalities of the brain that are not apparent on CT or MRI scans. Fourteen mild head injury patients with normal CT or MRI scans were examined using neuroSPECT compared to five age-matched hospital employees (controls) who had no history of head injury (*Journal of Head Trauma Rehabilitation*, June 1995). Density ratios for the right temporal lobes of the subjects are shown below.

⊘NEUROSPECT

Patients		Controls
.852	1.004	1.012
.901	.942	1.111
.849	.977	1.094
.779	.891	1.216
.842	.898	.989
.960	.866	
.782	1.004	

Source: Varney, N. R., *et al.* "NeuroSPECT correlates of disabling mild head injury: Preliminary findings." *Journal of Head Trauma Rehabilitation*, Vol. 10, No. 3, June 1995, p. 24 (Table 2), Aspen Publishers, Inc.

a. Construct a dot plot for the neuroSPECT ratios of the 14 head injury patients.
b. Add the neuroSPECT ratios for the five controls to the dot plot, part **a**. Use a different-colored dot for these five measurements.
c. Do you detect a pattern in the dot plot, part **b**? Explain.

2.29 Saturn has five satellites that rotate around the planet. *Astronomy* (Aug. 1995) lists 19 different events involving eclipses or occults of Saturnian satellites during the month of August. For each event, the percent of light lost by the eclipsed or occulted satellite at midevent is recorded in the table.

⊘SATURN

Date	Event	Light Loss (%)
Aug. 2	Eclipse	65
4	Eclipse	61
5	Occult	1
6	Eclipse	56
8	Eclipse	46
8	Occult	2
9	Occult	9
11	Occult	5
12	Occult	39
14	Occult	1
14	Eclipse	100
15	Occult	5
15	Occult	4
16	Occult	13
20	Occult	11
23	Occult	3
23	Occult	20
25	Occult	20
28	Occult	12

Source: Astronomy magazine, Aug. 1995, p. 60.

a. Construct a stem-and-leaf display for light loss percentage of the 19 events.
b. Locate on the stem-and-leaf plot, part **a**, the light losses associated with eclipses of Saturnian satellites. (Circle the light losses on the plot.)
c. Based on the marked stem-and-leaf display, part **b**, make an inference about which event type (eclipse or occult) is more likely to lead to a greater light loss.

Applying the Concepts—Advanced

2.30 Educators are constantly evaluating the efficacy of public schools in the education and training of American students. One quantitative assessment of change over time is the difference in scores on the SAT, which has been used for decades by colleges and universities as one criterion for admission. The following table shows the average SAT scores for each of the 50 states and District of Columbia for 1990 and 2000:

⊘SAT

State	1990	2000
Alabama	1079	1114
Alaska	1015	1034
Arizona	1041	1044
Arkansas	1077	1117
California	1002	1015
Colorado	1067	1071
Connecticut	1002	1017
Delaware	1006	998
D.C.	950	980
Florida	988	998
Georgia	951	974
Hawaii	985	1007
Idaho	1066	1081
Illinois	1089	1154
Indiana	972	999
Iowa	1172	1189
Kansas	1129	1154
Kentucky	1089	1098
Louisiana	1088	1120
Maine	991	1004
Maryland	1008	1016
Massachusetts	1001	1024
Michigan	1063	1126
Minnesota	1110	1175
Mississippi	1090	1111
Missouri	1089	1149
Montana	1082	1089
Nebraska	1121	1131

Nevada	1022	1027
N. Hampshire	1028	1039
N. Jersey	993	1011
N. Mexico	1100	1092
New York	985	1000
N. Carolina	948	988
N. Dakota	1157	1197
Ohio	1048	1072
Oklahoma	1095	1123
Oregon	1024	1054
Pennsylvania	987	995
R. Island	986	1005
S. Carolina	942	966
S. Dakota	1150	1175
Tennessee	1102	1116
Texas	979	993
Utah	1121	1139
Vermont	1000	1021
Virginia	997	1009
Washington	1024	1054
W. Virginia	1034	1037
Wisconsin	1111	1181
Wyoming	1072	1090

Source: College Entrance Examination Board, 2001.

a. Use graphs to display the two SAT score distributions. How have the distributions of average state scores changed over the last decade?

b. As another method of comparing the 1990 and 2000 average SAT scores, compute the **paired difference** by subtracting the 1990 score from the 2000 score for each state. Summarize these differences with a graph.

c. Interpret the graph, part **b**. How do your conclusions compare to those you reached when comparing the two graphs in part **a**?

d. What is the largest improvement in the average score between 1990 and 2000 as indicated on the graph, part **b**? With which state is this improvement associated in the table of data?   70; Wisconsin

2.31 The role that listener knowledge plays in the perception of imperfectly articulated speech was investigated in the *American Journal of Speech–Language Pathology* (Feb. 1995). Thirty female college students, randomly divided into three groups of ten, participated as listeners in the study. All subjects were required to listen to a 48-sentence audiotape of a Korean woman with cerebral palsy and asked to transcribe her entire speech. For the first group of students (the *control group*), the speaker used her normal manner of speaking. For the second group (the *treatment group*), the speaker employed a learned breathing pattern (called breath-group strategy) to improve speech efficiency. The third group (the *familiarity group*) also listened to the tape with the breath-group strategy, but only after they had practiced listening twice to another tape in which they were told exactly what the speaker was saying. At the end of the listening/transcribing session, two quantitative variables were measured for each listener: (1) the total number of words transcribed (called the rate of response) and (2) the percentage of words correctly transcribed (called the accuracy score). The data for all 30 subjects are provided in the table below.

a. MINITAB dot plots of the response rates (RESPRATE) for each listener group are displayed on p. 42. Use the information in the plots to describe the differences in the distributions of response rates among the three groups.

b. Use a graphical method to describe the distribution of accuracy scores for the three listener groups. Construct one graph for each group. Interpret the results.

2.32 Mark McGwire of the St. Louis Cardinals and Sammy Sosa of the Chicago Cubs hit 70 and 66 home runs, respectively, during the 1998 Major League Baseball season, breaking a record held by Roger Maris (61 home runs) since 1961. J. S. Simonoff of New York University collected data on the number of runs scored by their respective teams in games in which McGwire and Sosa hit home runs (*Journal of Statistics Education*, Vol. 6, 1998).

⦿ LISTEN

Control Group		Treatment Group		Familiarization Group	
Rate of Response	**Percent Correct**	**Rate of Response**	**Percent Correct**	**Rate of Response**	**Percent Correct**
250	23.6	254	26.0	193	36.0
230	26.0	178	32.6	223	41.0
197	26.0	139	32.6	232	43.0
238	26.7	249	33.0	214	44.0
174	29.2	236	34.4	269	44.0
263	29.5	231	36.5	256	46.0
275	31.3	161	38.5	224	46.0
193	32.9	255	40.0	225	48.0
204	35.4	275	41.7	288	49.0
168	29.2	181	44.8	244	52.0

Source: Tjaden, K., and Liss, J. M. "The influence of familiarity on judgments of treated speech." *American Journal of Speech–Language Pathology*, Vol. 4, No. 1, Feb. 1995, p. 43 (Table 1).

MINITAB Output for Exercise 2.31

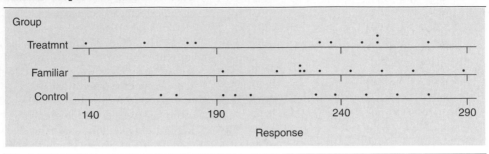

The data is reproduced in the accompanyng table. (An asterisk indicates a game in which McGwire or Sosa hit multiple home runs.)

a. Construct a stem-and-leaf display for the number of runs scored by St. Louis during games when McGwire hit a home run.

b. Repeat part a for Chicago and Sosa.

c. Compare and contrast the two distributions.

d. On the stem-and-leaf display, circle the games in which McGwire or Sosa hit multiple home runs. Do you detect any patterns?

⊘ STLRUNS

St. Louis Cardinals

6	6	3	11	13
8	1	10	6	7
5	8	9	6*	3
8	2	3	6*	6
15*	2	8*	5	6
8	6	2	3	5
5	8	4	10*	
8	9	4	4	
3	5	3	11*	
5	7	6	1	
2	3	8	4	
6	2	7*	6*	
3	7	14*	4*	

⊘ CUBSRUNS

Chicago Cubs

3	6*	8*	6	7*
4	5	2	7	2
1	8*	6	13	
1	9*	6	8	
4	2	3*	10	
3	6	9	6*	
5	4	10	5	
5	4	4	9	
2	9	5	7	
5*	5	4	8	
5*	6	5	15	
10*	9	8	11*	
5	3	11	6	

2.3 SUMMATION NOTATION

Now that we've examined some graphical techniques for summarizing and describing quantitative data sets, we turn to numerical methods for accomplishing this objective. Before giving the formulas for calculating numerical descriptive measures, let's look at some shorthand notation that will simplify our calculation instructions. Remember that such notation is used for one reason only—to avoid repeating the same verbal descriptions over and over. If you mentally substitute the verbal definition of a symbol each time you read it, you'll soon get used to it.

We denote the measurements of a quantitative data set as

$$x_1, x_2, x_3, \ldots, x_n$$

where x_1 is the first measurement in the data set, x_2 is the second measurement in the data set, x_3 is the third measurement in the data set, . . . , and x_n is the nth (and last) measurement in the data set. Thus, if we have five measurements in a set of data, we will write x_1, x_2, x_3, x_4, x_5 to represent the measurements. If the actual numbers are 5, 3, 8, 5, and 4, we have $x_1 = 5, x_2 = 3, x_3 = 8, x_4 = 5$, and $x_5 = 4$.

Most of the formulas we use require a summation of numbers. For example, one sum we'll need to obtain is the sum of all the measurements in the data set, or $x_1 + x_2 + x_3 + \cdots + x_n$. To shorten the notation, we use the symbol $\sum$ for the summation. That is, $x_1 + x_2 + x_3 + \cdots + x_n = \sum_{i=1}^{n} x_i$. Verbally translate $\sum_{i=1}^{n} x_i$ as follows: "The sum of the measurements, whose typical member is x_i, beginning with the member x_1 and ending with the member x_n."

Suppose, as in our earlier example, $x_1 = 5$, $x_2 = 3$, $x_3 = 8$, $x_4 = 5$, and $x_5 = 4$. Then the sum of the five measurements, denoted $\sum_{i=1}^{5} x_i$ is obtained as follows:

$$\sum_{i=1}^{5} x_i = x_1 + x_2 + x_3 + x_4 + x_5$$
$$= 5 + 3 + 8 + 5 + 4 = 25$$

Another important calculation requires that we square each measurement and then sum the squares. The notation for this sum is $\sum_{i=1}^{n} x_i^2$. For the five measurements, we have

$$\sum_{i=1}^{5} x_i^2 = x_1^2 + x_2^2 + x_3^2 + x_4^2 + x_5^2$$
$$= 5^2 + 3^2 + 8^2 + 5^2 + 4^2$$
$$= 25 + 9 + 64 + 25 + 16 = 139$$

In general, the symbol following the summation sign $\sum$ represents the variable (or function of the variable) that is to be summed.

The Meaning of Summation Notation $\sum_{i=1}^{n} x_i$

Sum the measurements on the variable that appears to the right of the summation symbol, beginning with the 1st measurement and ending with the nth measurement.

 EXERCISES 2.33–2.36

Learning the Mechanics

Note: In all exercises, $\sum$ represents $\sum_{i=1}^{n}$.

2.33 A data set contains the observations 5, 1, 3, 2, 1. Find:

  a. $\sum x$ 12

  b. $\sum x^2$ 40

  c. $\sum (x - 1)$ 7

  d. $\sum (x - 1)^2$ 21

  e. $\left(\sum x \right)^2$ 144

2.34 Suppose a data set contains the observations 3, 8, 4, 5, 3, 4, 6. Find:

  a. $\sum x$ 33

  b. $\sum x^2$ 175

  c. $\sum (x - 5)^2$ 20

  d. $\sum (x - 2)^2$ 71

  e. $\left(\sum x \right)^2$ 1,089

2.35 Refer to Exercise 2.33. Find:

  a. $\sum x^2 - \dfrac{\left(\sum x \right)^2}{5}$ **b.** $\sum (x - 2)^2$ **c.** $\sum x^2 - 10$

2.36 A data set contains the observations 6, 0, −2, −1, 3. Find:

  a. $\sum x$ **b.** $\sum x^2$ **c.** $\sum x^2 - \dfrac{\left(\sum \right)^2}{5}$

2.4   NUMERICAL MEASURES OF CENTRAL TENDENCY

When we speak of a data set, we refer to either a sample or to a population. If statistical inference is our goal, we'll wish ultimately to use sample **numerical descriptive measures** to make inferences about the corresponding measures for a population.

As you'll see, a large number of numerical methods are available to describe quantitative data sets. Most of these methods measure one of two data characteristics:

1. The **central tendency** of the set of measurements—that is, the tendency of the data to cluster, or center, about certain numerical values (see Figure 2.12a).
2. The **variability** of the set of measurements—that is, the spread of the data (see Figure 2.12b).

FIGURE 2.12

Numerical Descriptive Measures

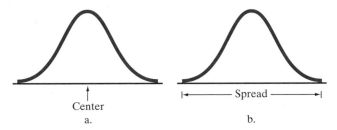

In this section we concentrate on **measures of central tendency**. In the next section, we discuss measures of variability.

The most popular and best understood measure of central tendency for a quantitative data set is the *arithmetic mean* (or simply the mean) of a data set.

DEFINITION 2.4

The **mean** of a set of quantitative data is the sum of the measurements divided by the number of measurements contained in the data set.

In everyday terms, the mean is the average value of the data set and is often used to represent a "typical" value. We denote the **mean of a sample** of measurements by $\bar{x}$ (read "x-bar"), and represent the formula for its calculation as shown in the box:

Calculating a Sample Mean

$$\bar{x} = \frac{\sum_{i=1}^{n} x_i}{n}$$

E X A M P L E   2 . 3   Calculate the mean of the following five sample measurements: 5, 3, 8, 5, 6.

$\bar{x} = 5.4$

Solution

Using the definition of sample mean and the summation notation, we find

$$\bar{x} = \frac{\sum_{i=1}^{5} x_i}{5} = \frac{5 + 3 + 8 + 5 + 6}{5} = \frac{27}{5} = 5.4$$

Thus, the mean of this sample is 5.4.*

EXAMPLE 2.4

$\bar{x} = 36,994$

Calculate the sample mean for the 100 EPA mileages given in Table 2.3.

Solution

The mean gas mileage for the 100 cars is denoted

$$\bar{x} = \frac{\sum_{i=1}^{100} x_i}{100}$$

TEACHING TIP
When calculating a population mean, the denominator is the population size, N.

Rather than compute $\bar{x}$ by hand (or calculator), we entered the data of Table 2.3 into a computer and employed SPSS statistical software to compute the mean. The SPSS printout is shown in Figure 2.13. The sample mean, highlighted on the printout, is $\bar{x} = 36.9940$.

FIGURE 2.13

SPSS Printout of Numerical Descriptive Measures for 100 EPA Gas Mileages

Descriptives			Statistic	Std. Error
MPG	Mean		36.9940	.2418
	95% Confidence Interval for Mean	Lower Bound	36.5142	
		Upper Bound	37.4738	
	5% Trimmed Mean		36.9922	
	Median		37.0000	
	Variance		5.846	
	Std. Deviation		2.4179	
	Minimum		30.00	
	Maximum		44.90	
	Range		14.90	
	Interquartile Range		2.7500	
	Skewness		.051	.241
	Kurtosis		.770	.478

Given this information, you can visualize a distribution of gas mileage readings centered in the vicinity of $\bar{x} \approx 37$. An examination of the relative frequency histogram (Figure 2.8) confirms that $\bar{x}$ does in fact fall near the center of the distribution.

---

*In the examples given here, $\bar{x}$ is sometimes rounded to the nearest tenth, sometimes to the nearest hundredth, sometimes to the nearest thousandth, and so on. There is no specific rule for rounding when calculating $\bar{x}$ because $\bar{x}$ is specifically defined to be the sum of all measurements divided by n; that is, it is a specific fraction. When $\bar{x}$ is used for descriptive purposes, it is often convenient to round the calculated value of $\bar{x}$ to the number of significant figures used for the original measurements. When $\bar{x}$ is to be used in other calculations, however, it may be necessary to retain more significant figures.

The sample mean $\bar{x}$ will play an important role in accomplishing our objective of making inferences about populations based on sample information. For this reason we need to use a different symbol for the *mean of a population*—the mean of the set of measurements on every unit in the population. We use the Greek letter μ (mu) for the population mean.

TEACHING TIP
Explain that Greek letters are used to represent population values throughout the text.

Symbols for the Sample Mean and the Population Mean

In this text, we adopt a general policy of using Greek letters to represent numerical descriptive measures for the population and Roman letters to represent corresponding descriptive measures for the sample. The symbols for the mean are

$\bar{x}$ = Sample mean μ = Population mean

TEACHING TIP
Average, mean, and expected value are all terms that are used to represent the same descriptive measure.

We'll often use the sample mean, $\bar{x}$, to estimate (make an inference about) the population mean, μ. For example, the EPA mileages for the population consisting of *all* cars has a mean equal to some value, μ. Our sample of 100 cars yielded mileages with a mean of $\bar{x} = 36.9940$. If, as is usually the case, we don't have access to the measurements for the entire population, we could use $\bar{x}$ as an estimator or approximator for μ. Then we'd need to know something about the reliability of our inference. That is, we'd need to know how accurately we might expect $\bar{x}$ to estimate μ. In Chapter 5, we'll find that this accuracy depends on two factors:

TEACHING TIP
Look ahead to sampling distributions to plant the idea that measures of center and spread will be used together to generate estimates of population values.

1. The *size of the sample*. The larger the sample, the more accurate the estimate will tend to be.
2. The *variability, or spread, of the data*. All other factors remaining constant, the more variable the data, the less accurate the estimate.

Another important measure of central tendency is the *median*.

DEFINITION 2.5
The **median** of a quantitative data set is the middle number when the measurements are arranged in ascending (or descending) order.

TEACHING TIP
Remind students to order the data before calculating a value for the median.

The median is of most value in describing large data sets. If the data set is characterized by a relative frequency histogram (Figure 2.14), the median is the point on the *x*-axis such that half the area under the histogram lies above the median and half lies below. [*Note*: In Section 2.2 we observed that the relative frequency associated with a particular interval on the *x*-axis is proportional to the amount of area under the histogram that lies above the interval.] We denote the *median* of a *sample* by *M*.

FIGURE 2.14

Location of the Median

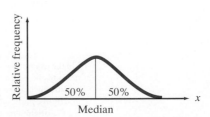

Calculating a Sample Median, *M*

Arrange the *n* measurements from the smallest to the largest.

1. If *n* is odd, *M* is the middle number.
2. If *n* is even, *M* is the mean of the middle two numbers.

E X A M P L E   2 . 5

a. 5
b. 5.5

Consider the following sample of *n* = 7 measurements: 5, 7, 4, 5, 20, 6, 2

a. Calculate the median *M* of this sample.
b. Eliminate the last measurement (the 2) and calculate the median of the remaining *n* = 6 measurements.

Solution

a. The seven measurements in the sample are ranked in ascending order: 2, 4, 5, 5, 6, 7, 20. Because the number of measurements is odd, the median is the middle measurement. Thus, the median of this sample is *M* = 5.
b. After removing the 2 from the set of measurements, we rank the sample measurements in ascending order as follows: 4, 5, 5, 6, 7, 20. Now the number of measurements is even, so we average the middle two measurements. The median is $M = (5 + 6)/2 = 5.5$.

In certain situations, the median may be a better measure of central tendency than the mean. In particular, the median is less sensitive than the mean to extremely large or small measurements. Note, for instance, that all but one of the measurements in part **a** of Example 2.5 center about *x* = 5. The single relatively large measurement, *x* = 20, does not affect the value of the median, 5, but it causes the mean, $\bar{x}$ = 7, to lie to the right of most of the measurements.

As another example of data for which the central tendency is better described by the median than the mean, consider the household incomes of a community being studied by a sociologist. The presence of just a few households with very high incomes will affect the mean more than the median. Thus, the median will provide a more accurate picture of the typical income for the community. The mean could exceed the vast majority of the sample measurements (household incomes), making it a misleading measure of central tendency.

E X A M P L E   2 . 6

M = 37

Calculate the median for the 100 EPA mileages given in Table 2.3. Compare the median to the mean computed in Example 2.4.

Solution

For this large data set, we again resort to a computer analysis. An SAS printout is displayed in Figure 2.15, with the median highlighted. You can see that the median

FIGURE 2.15

SAS Printout of Numerical Descriptive Measures for 100 EPA Mileages

Basic Statistical Measures			
Location		Variability	
Mean	36.99400	Std Deviation	2.41790
Median	37.00000	Variance	5.84623
Mode	37.00000	Range	14.90000
		Interquartile Range	2.70000

is 37.0. This value implies that half of the 100 mileages in the data set fall below 37.0 and half lie above 37.0. Note that the median, 37.0, and the mean, 36.9940, are almost equal. This fact indicates that the data form an approximately **symmetric distribution**. As indicated in the box on p. 49, a comparison of the mean and median gives an indication of the **skewness** (i.e, the tendency of the distribution to have elongated tails) of a data set. ∎

A third measure of central tendency is the *mode* of a set of measurements.

> ### DEFINITION 2.6
> The **mode** is the measurement that occurs most frequently in the data set.

Therefore, the mode shows where the data tend to concentrate.

E X A M P L E   2 . 7

mode = 9

Each of ten taste testers rated a new brand of barbecue sauce on a 10-point scale, where 1 = awful and 10 = excellent. Find the mode for the ten ratings shown below.

$$8 \quad 7 \quad 9 \quad 6 \quad 8 \quad 10 \quad 9 \quad 9 \quad 5 \quad 7$$

Solution

Since 9 occurs most often (three times), the mode of the ten taste ratings is 9. ∎

Note that the data in Example 2.7 are actually qualitative in nature (e.g., "awful," "excellent"). The mode is particularly useful for describing qualitative data. The modal category is simply the category (or class) that occurs most often. Because it emphasizes data concentration, the mode is also used with quantitative data sets to locate the region in which much of the data is concentrated. A retailer of men's clothing would be interested in the modal neck size and sleeve length of potential customers. The modal income class of the laborers in the United States is of interest to the Labor Department.

For some quantitative data sets, the mode may not be very meaningful. For example, consider the EPA mileage ratings in Table 2.3. A reexamination of the data reveals that the gas mileage of 37.0 occurs most often (four times). However, the mode of 37.0 (shown on the SAS printout, Figure 2.15) is not particularly useful as a measure of central tendency.

A more meaningful measure can be obtained from a relative frequency histogram for quantitative data. The measurement class containing the largest relative frequency is called the **modal class**. Several definitions exist for locating the position of the mode within a modal class, but the simplest is to define the mode as the midpoint of the modal class. For example, examine the frequency histogram for the EPA mileage ratings, in Figure 2.8 (p. 32). You can see that the modal class is the interval 36.0–37.5. The mode (the midpoint) is 36.75. This modal class (and the mode itself) identifies the area in which the data are most concentrated, and in that sense it is a measure of central tendency. However, for most applications involving quantitative data, the mean and median provide more descriptive information than the mode.

Comparing the Mean and the Median

If the data set is skewed to the right, the mean is greater than (to the right of) the median.

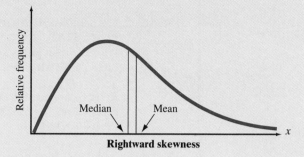

If the data set is symmetric, the mean equals the median.

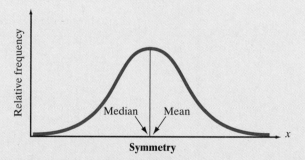

If the data set is skewed to the left, the mean is less than (to the left of) the median.

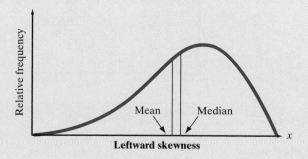

Using the TI-83 Graphing Calculator

Finding One-Variable Descriptive Statistics on the TI-83

Step 1   *Enter the data*
Press **STAT 1** for **STAT Edit**
Enter the data into one of the lists.

Step 2   *Calculate descriptive statistics*
Press **STAT**
Press the right arrow key to highlight **CALC**
Press **ENTER** for **1-Var Stats**

Enter the name of the list containing your data.
Press **2nd 1** for **L1** (or **2nd 2** for **L2** etc.)
Press **ENTER**

You should see the statistics on your screen. Some of the statistics are off the bottom of the screen. Use the down arrow to scroll through to see the remaining statistics. Use the up arrow to scroll back up.

Example   The descriptive statistics for the sample data set

86, 70, 62, 98, 73, 56, 53, 92, 86, 37, 62, 83, 78, 49, 78, 37, 67, 79, 57

are shown below, as calculated by the TI-83.

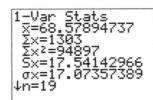 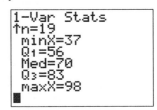

Sorting   The descriptive statistics do not include the mode. To find the mode,
data   sort your data as follows:

Press **STAT**
Press **2** for **SORTA(**
Enter the name of the list your data is in. If your data is in **L1**,
press **2nd 1**

Press **ENTER**
The screen will say: **DONE**
To see the sorted data, press **STAT 1** for **STAT Edit**

 EXERCISES 2.37–2.53

Learning the Mechanics

2.37 Calculate the mode, mean, and median of the following data:

18   10   15   13   17   15   12   15   18   16   11

2.38 Calculate the mean and median of the following grade point averages:

3.2   2.5   2.1   3.7   2.8   2.0

2.39 Explain the difference between the calculation of the median for an odd and an even number of measurements.

Construct one data set consisting of five measurements and another consisting of six measurements for which the medians are equal.

2.40 Explain how the relationship between the mean and median provides information about the symmetry or skewness of the data's distribution.

2.41 Calculate the mean for samples where

   a. $n = 10$, $\sum x = 85$ 8.5

   b. $n = 16$, $\sum x = 400$ 25

   c. $n = 45$, $\sum x = 35$ .778

   d. $n = 18$, $\sum x = 242$ 13.44

2.42 Calculate the mean, median, and mode for each of the following samples:
   a. $7, -2, 3, 3, 0, 4$ 2.5, 3, 3

   b. $2, 3, 5, 3, 2, 3, 4, 3, 5, 1, 2, 3, 4$ 3.08, 3, 3
   c. $51, 50, 47, 50, 48, 41, 59, 68, 45, 37$ 49.6, 49, 50

2.43 Describe how the mean compares to the median for a distribution as follows:
   a. Skewed to the left Mean $<$ median
   b. Skewed to the right Mean $>$ median
   c. Symmetric Mean $=$ median

Applying the Concepts—Basic

2.44 *Fortune* (Oct. 25, 1999) published a list of the 50 most powerful women in America. The data on age (in years) and title of each of these 50 women are stored in the WOMENPOWER file.
   a. Find the mean, median, and modal age of these 50 women.
   b. What do the mean and median indicate about the skewness of the age distribution?
   c. Construct a relative frequency histogram for the age data. What is the modal age class?

WOMENPOWER

Rank	Name	Age	Company	Title
1	Carly Fiorina	45	Hewlett-Packard	CEO
2	Heidi Miller	46	Citigroup	CFO
3	Mary Meeker	40	Morgan Stanley	Managing Director
4	Shelly Lazarus	52	Ogilvy & Mather	CEO
5	Meg Whitman	43	eBay	CEO
6	Debby Hopkins	44	Boeing	CFO
7	Marjorie Scardino	52	Pearson	CEO
8	Martha Stewart	58	Omnimedia	CEO
9	Nancy Peretsman	45	Allen & Co.	Ex.V.P.
10	Pat Russo	47	Lucent Technologies	Ex. V.P.
11	Patricia Dunn	46	Barclays Global Investors	Chairman
12	Abby Joseph Cohen	47	Goldman Sachs	Managing Director
13	Ann Livermore	41	Hewlett-Packard	CEO
14	Andrea Jung	41	Avon Products	COO
15	Sherry Lansing	55	Paramount Pictures	Chairman
16	Karen Katen	50	Pfizer	Ex.V.P.
17	Marilyn Carlson Nelson	60	Carlson Cos.	CEO
18	Judy McGrath	47	MTV & M2	President
19	Lois Juliber	50	Colgate-Palmolive	COO
20	Gerry Laybourne	52	Oxygen Media	CEO
21	Judith Estrin	44	Cisco Systems	Sr. V.P.
22	Cathleen Black	55	Hearst Magazines	President
23	Linda Sandford	46	IBM	General Manager
24	Ann Moore	49	Time Inc.	President
25	Jill Barad	48	Mattel	CEO
26	Oprah Winfrey	45	Harpo Entertainment	Chairman
27	Judy Lewent	50	Merck	Sr. V.P.
28	Joy Covey	36	Amazon.com	COO
29	Rebecca Mark	45	Azurix	CEO
30	Deborah Willingham	43	Microsoft	V.P.
31	Dina Dubion	46	Chase Manhattan	Ex.V.P.

(continues on next page)

WOMENPOWER *(continued)*

Rank	Name	Age	Company	Title
32	Patricia Woertz	46	Chevron	President
33	Lawton Fitt	46	Goldman Sachs	Man. Dir.
34	Ann Fudge	48	Kraft Foods	Ex. V.P.
35	Carolyn Ticknor	52	Hewlett-Packard	CEO
36	Dawn Lepore	45	Charles Schwab	CIO
37	Jeannine Rivet	51	United Healthcare	CEO
38	Jamie Gorelick	49	Fannie Mae	Vice Chairman
39	Jan Brandt	48	America Online	Mar. President
40	Bridget Macaskill	51	Oppenheimer Funds	CEO
41	Jeanne Jackson	48	Banana Republic	CEO
42	Cynthia Trudell	46	General Motors	V.P.
43	Nina DiSesa	53	McCann-Erickson	Chairman
44	Linda Wachner	53	Warnaco	Chairman
45	Darla Moore	45	Rainwater Inc.	President
46	Marion Sandler	68	Golden West	Co-CEO
47	Michelle Anthony	42	Sony Music	Ex. V.P.
48	Orit Gadlesh	48	Bain & Co.	Chairman
49	Charlotte Beers	64	J. Walter Thompson	Chairman
50	Abigail Johnson	37	Fidelity Investments	V.P.

Source: Fortune, Oct. 25, 1999.

2.45 Three-way catalytic converters have been installed in new vehicles in order to reduce pollutants from motor vehicle exhaust emissions. However, these converters unintentionally increase the level of ammonia in the air. *Environmental Science & Technology* (Sept. 1, 2000) published a study on the ammonia levels near the exit ramp of a San Francisco highway tunnel. The data in the table represent daily ammonia concentrations (parts per million) on eight randomly selected days during afternoon drive-time in the summer of 1999.

⊘ AMMONIA

1.53	1.50	1.37	1.51	1.55	1.42	1.41	1.48

   a. Find the mean daily ammonia level in air in the tunnel.   1.4713
   b. Find the median ammonia level.   1.49
   c. Interpret the values obtained in parts **a** and **b**.

2.46 *The Condor* (May 1995) published a study of competition for nest holes among collared flycatchers, a bird species. The authors collected the data for the study by periodically inspecting nest boxes located on the island of Gotland in Sweden. The nest boxes were grouped into 14 discrete locations (called plots). The accompanying table gives the number of flycatchers killed and the number of flycatchers breeding at each plot.

⊘ CONDOR

Plot Number	Number of Breeders	Number Killed
1	5	30
2	4	28
3	3	38
4	2	34
5	2	26
6	1	124
7	1	68
8	1	86
9	1	32
10	0	30
11	0	46
12	0	132
13	0	100
14	0	6

Source: Merilä, J., and Wiggins, D. A. "Interspecific competition for nest holes causes adult mortality in the collared flycatcher." *The Condor,* Vol. 97, No. 2, May 1995, p. 447 (Table 4). Cooper Ornithological Society.

   a. Calculate the mean, median, and mode for the number of flycatchers killed at the 14 study plots.
   b. Interpret the measures of central tendency, part **a**.
   c. A MINITAB printout of descriptive statistics for the number of breeders at each plot is shown on p. 53. Locate the measures of central tendency on the printout and interpret these values.

MINITAB Output for Exercise 2.46

Variable	N	Mean	Median	TrMean	StDev	SE Mean
Breeders	14	55.7	36.0	53.5	39.6	10.6

Variable	Minimum	Maximum	Q1	Q3
Breeders	6.0	132.0	29.5	89.5

2.47 Applicants for an academic position (e.g., assistant professor) at a college or university are usually required to submit at least three letters of recommendation. A study of 148 applicants for an entry-level position in experimental psychology at the University of Alaska Anchorage revealed that many did not meet the three-letter requirement (*American Psychologist*, July 1995). Summary statistics for the number of recommendation letters in each application are given below. Interpret these summary measures.

Mean = 2.28   Median = 3   Mode = 3

Applying the Concepts—Intermediate

2.48 Platelet-activating factor (PAF) is a potent chemical that occurs in patients suffering from shock, inflammation, hypotension, respiratory, and cardiovascular disorders. A bioassay was undertaken to investigate the potential of 17 traditional Chinese herbal drugs in PAF inhibition (*Progress in Natural Science*, June 1995). The prevention of the PAF binding process, measured as a percentage, for each drug is provided in the accompanying table.

a. Construct a stem-and-leaf display for the data.

PAF

Drug	PAF Inhibition (%)
Hai-feng-teng (Fuji)	77
Hai-feng-teng (Japan)	33
Shan-ju	75
Zhang-yiz-hu-jiao	62
Shi-nan-teng	70
Huang-hua-hu-jiao	12
Hua-nan-hu-jiao	0
Xiao-yie-pa-ai-xiang	0
Mao-ju	0
Jia-ju	15
Xie-yie-ju	25
Da-yie-ju	0
Bian-yie-hu-jiao	9
Bi-bo	24
Duo-mai-hu-jiao	40
Yan-sen	0
Jiao-guo-hu-jiao	31

Source: Guiqiu, H. "PAF receptor antagonistic principles from Chinese traditional drugs." *Progress in Natural Science*, Vol. 5, No. 3, June 1995, p. 301 (Table 1).

b. Compute the median inhibition percentage for the 17 herbal drugs. Interpret the result.   24
c. Compute the mean inhibition percentage for the 17 herbal drugs. Interpret the result.   27.82
d. Compute the mode of the 17 inhibition percentages. Interpret the result.   0
e. Locate the median, mean, and mode on the stem-and-leaf display, part **a**. Do these measures of central tendency appear to locate the center of the data?

2.49 "The Training Game" is an activity used in psychology in which one person shapes an arbitrary behavior by selectively reinforcing the movements of another person. A group of 15 psychology students at Georgia Institute of Technology played "The Training Game" at Zoo Atlanta while participating in an experimental psychology laboratory in which they assisted in the training of animals (*Teaching of Psychology*, May 1998). At the end of the session, each student was asked to rate the statement: "'The Training Game' is a great way for students to understand the animal's perspective during training." Responses were recorded on a 7-point scale ranging from 1 (strongly disagree) to 7 (strongly agree). The 15 responses were summarized as follows:

mean = 5.87, mode = 6.

a. Interpret the measures of central tendency in the words of the problem.
b. What type of skewness (if any) is likely to be present in the distribution of student responses? Explain.

2.50 Would you expect the data sets described below to possess relative frequency distributions that are symmetric, skewed to the right, or skewed to the left? Explain.

a. The salaries of all persons employed by a large university   Skewed right
b. The grades on an easy test   Skewed left
c. The grades on a difficult test   Skewed right
d. The amounts of time students in your class studied last week   Symmetric
e. The ages of automobiles on a used-car lot
f. The amounts of time spent by students on a difficult examination (maximum time is 50 minutes)

2.51 Clinical observations suggest that specifically language-impaired (SLI) children have great difficulty with the proper use of pronouns. This phenomenon was investigated and reported in the *Journal of Communication Disorders* (Mar. 1995). Thirty children, all from low-income families, participated in the study. Ten were 5-year-old SLI children, ten were younger (3-year-old) normally developing (YND) children, and ten were older (5-year-old) normally developing (OND) children. The table contains the gender, deviation intelligence quotient (DIQ), and percentage of pronoun errors observed for each of the 30 subjects.

⊘ SLI

Subject	Gender	Group	DIQ	Pronoun Errors (%)
1	F	YND	110	94.40
2	F	YND	92	19.05
3	F	YND	92	62.50
4	M	YND	100	18.75
5	F	YND	86	0
6	F	YND	105	55.00
7	F	YND	90	100.00
8	M	YND	96	86.67
9	M	YND	90	32.43
10	F	YND	92	0
11	F	SLI	86	60.00
12	M	SLI	86	40.00
13	M	SLI	94	31.58
14	M	SLI	98	66.67
15	F	SLI	89	42.86
16	F	SLI	84	27.27
17	M	SLI	110	33.33
18	F	SLI	107	0
19	F	SLI	87	0
20	M	SLI	95	0
21	M	OND	110	0
22	M	OND	113	0
23	M	OND	113	0
24	F	OND	109	0
25	M	OND	92	0
26	F	OND	108	0
27	M	OND	95	0
28	F	OND	87	0
29	F	OND	94	0
30	F	OND	98	0

Source: Moore, M. E. "Error analysis of pronouns by normal and language-impaired children." *Journal of Communication Disorders,* Vol. 28, No. 1, Mar. 1995, p. 62 (Table 2), p. 67 (Table 5).

a. Identify the variables in the data set as quantitative or qualitative.

b. Why is it nonsensical to compute numerical descriptive measures for qualitative variables?

c. Compute measures of central tendency for DIQ for the ten SLI children.

d. Compute measures of central tendency for DIQ for the ten YND children.

e. Compute measures of central tendency for DIQ for the ten OND children.

f. Use the results, parts **c-e,** to compare the DIQ central tendencies of the three groups of children.

g. An SPSS printout of descriptive statistics for the percentage of pronoun errors, by group, is shown at the bottom of the page. Locate the measures of central tendency on the printout and interpret their values.

h. Is it reasonable to use a single number (e.g., mean or median) to describe the center of the percentage of pronoun error distribution? Or should three "centers" be calculated, one for each of the three groups of children? Explain.

Applying the Concepts—Advanced

2.52 The conventional method of measuring the refractive status of an eye involves three quantities: (1) sphere power, (2) cylinder power, and (3) axis. Optometric researchers studied the variation in these three measures of refraction (*Optometry and Vision Science,* June 1995). Twenty-five successive refractive measurements were obtained on the eyes of over 100 university students. The cylinder power measurements for the left eye of one particular student (ID #11) are listed in the table. [*Note:* All measurements are negative values.] Numerical descriptive measures for the data set are provided in the SAS printout on p. 55.

⊘ LEFTEYE

.08	.08	1.07	.09	.16	.04	.07	.17	.11
.06	.12	.17	.20	.12	.17	.09	.07	.16
.15	.16	.09	.06	.10	.21	.06		

Source: Rubin, A., and Harris, W. F. "Refractive variation during autorefraction: Multivariate distribution of refractive status." *Optometry and Vision Science,* Vol. 72, No. 6, June 1995, p. 409 (Table 4).

a. Locate the measures of central tendency on the printout and interpret their values.

b. Note that the data contains one unusually large (negative) cylinder power measurement relative to the other measurements in the data set. Find this

SPSS Output for Exercise 2.51

PCTERROR							
GROUP	Mean	N	Std.Deviation	Median	Variance	Minimum	Maximum
OND	.00	10	.00	.00	.000	0	0
SLI	30.17	10	24.11	32.46	581.202	0	67
YND	46.88	10	38.21	43.72	1460.347	0	100
Total	25.68	30	31.98	9.38	10022.919	0	100

SAS Output for Exercise 2.52

The UNIVARIATE Procedure
Variable: CYLPOWER

Moments

N	25	Sum Weights	25
Mean	0.1544	Sum Observations	3.86
Std Deviation	0.19676721	Variance	0.03871733
Skewness	4.52208392	Kurtosis	21.7019614
Uncorrected SS	1.5252	Corrected SS	0.929216
Coeff Variation	127.4399	Std Error Mean	0.03935344

Basic Statistical Measures

Location		Variability	
Mean	0.154400	Std Deviation	0.19677
Median	0.110000	Variance	0.03872
Mode	0.060000	Range	1.03000
		Interquartile Range	0.08000

NOTE: The mode displayed is the smallest of 4 modes with a count of 3.

Extreme Observations

-------Lowest------		-------Highest-------	
Value	Obs	Value	Obs
0.04	6	0.17	12
0.06	25	0.17	15
0.06	22	0.20	13
0.06	10	0.21	24
0.07	17	1.07	3

measurement. (In Section 2.8, we call this value an **outlier**).   1.07

c. Delete the outlier, part **b**, from the data set and recalculate the measures of central tendency. Which measure is most affected by the deletion of the outlier?

2.53 The salaries of superstar professional athletes receive much attention in the media. The multimillion-dollar long-term contract is now commonplace among this elite group. Nevertheless, rarely does a season pass without negotiations between one or more of the players' associations and team owners for additional salary and fringe benefits for *all* players in their particular sports.

a. If a players' association wanted to support its argument for higher "average" salaries, which measure of central tendency do you think it should use? Why?

b. To refute the argument, which measure of central tendency should the owners apply to the players' salaries? Why?   Mean

2.5 NUMERICAL MEASURES OF VARIABILITY

Measures of central tendency provide only a partial description of a quantitative data set. The description is incomplete without a **measure of the variability**, or **spread**, of the data set. Knowledge of the data set's variability along with its center can help us visualize the shape of a data set as well as its extreme values.

If you examine the two histograms in Figure 2.16, you'll notice that both hypothetical data sets are symmetric with equal modes, medians, and means. However, data set 1 (Figure 2.16a) has measurements spread with almost equal relative frequency over the measurement classes, while data set 2 (Figure 2.16b) has most of its measurements clustered about its center. Thus, data set 2 is *less variable* than data set 1. Consequently, you can see that we need a measure of variability as well as a measure of central tendency to describe a data set.

FIGURE 2.16

Hypothetical Data Sets

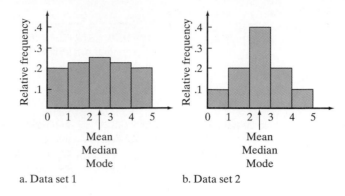

a. Data set 1   b. Data set 2

Perhaps the simplest measure of the variability of a quantitative data set is its *range*.

> **DEFINITION 2.7**
>
> The **range** of a quantitative data set is equal to the largest measurement minus the smallest measurement.

The range is easy to compute and easy to understand, but it is a rather insensitive measure of data variation when the data sets are large. This is because two data sets can have the same range and be vastly different with respect to data variation. This phenomenon is demonstrated in Figure 2.16. Both distributions of data shown in the figure have the same range, but most of the measurements in data set 2 tend to concentrate near the center of the distribution. Consequently, the data are much less variable than the data in set 1. Thus, you can see that the range does not always detect differences in data variation for large data sets.

Let's see if we can find a measure of data variation that is more sensitive than the range. Consider the two samples in Table 2.6: Each has five measurements. (We have ordered the numbers for convenience.) Note that both samples have a mean of 3 and that we have also calculated the distance between each measurement and the mean. What information do these distances contain? If they tend to be large in magnitude, as in sample 1, the data are spread out, or highly variable. If the distances are mostly small, as in sample 2, the data are clustered around the mean, $\bar{x}$, and therefore do not exhibit much variability. You can see that these distances, displayed graphically in Figure 2.17, provide information about the variability of the sample measurements.

The next step is to condense the information in these distances into a single numerical measure of variability. Averaging the distances from $\bar{x}$ won't help because the negative and positive distances cancel; that is, the sum of the deviations (and thus the average deviation) is always equal to zero.

TABLE 2.6   Two Hypothetical Data Sets

	Sample 1	**Sample 2**
Measurements	$1, 2, 3, 4, 5$	$2, 3, 3, 3, 4$
Mean	$\bar{x} = \dfrac{1 + 2 + 3 + 4 + 5}{5} = \dfrac{15}{5} = 3$	$\bar{x} = \dfrac{2 + 3 + 3 + 3 + 4}{5} = \dfrac{15}{5} = 3$
Distances of measurement values from $\bar{x}$	$(1 - 3), (2 - 3), (4 - 3),$ $(5 - 3)$ or $-2, -1, 0, 1, 2$	$(2 - 3), (3 - 3), (3 - 3), (3 - 3),$ $(4 - 3)$ or $-1, 0, 0, 0, 1$

FIGURE 2.17

Dot Plots for Two Data Sets

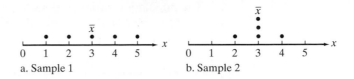

a. Sample 1   b. Sample 2

Two methods come to mind for dealing with the fact that positive and negative distances from the mean cancel. The first is to treat all the distances as though they were positive, ignoring the sign of the negative distances. We won't pursue this line of thought because the resulting measure of variability (the mean of the absolute values of the distances) presents analytical difficulties beyond the scope of this text. A second method of eliminating the minus signs associated with the distances is to square them. The quantity we can calculate from the squared distances will provide a meaningful description of the variability of a data set and presents fewer analytical difficulties in inference making.

To use the squared distances calculated from a data set, we first calculate the *sample variance*.

DEFINITION 2.8

The **sample variance** for a sample of n measurements is equal to the sum of the squared distances from the mean divided by $(n - 1)$. In symbols, using s^2 to represent the sample variance,

$$s^2 = \frac{\sum_{i=1}^{n} (x_i - \bar{x})^2}{n - 1}$$

Note: A shortcut formula for calculating s^2 is

$$s^2 = \frac{\sum_{i=1}^{n} x_i^2 - \frac{\left(\sum_{i=1}^{n} x_i\right)^2}{n}}{n - 1}$$

Referring to the two samples in Table 2.6, you can calculate the variance for sample 1 as follows:

$$s^2 = \frac{(1 - 3)^2 + (2 - 3)^2 + (3 - 3)^2 + (4 - 3)^2 + (5 - 3)^2}{5 - 1}$$

$$= \frac{4 + 1 + 0 + 1 + 4}{4} = 2.5$$

The second step in finding a meaningful measure of data variability is to calculate the *standard deviation* of the data set:

DEFINITION 2.9

The **sample standard deviation**, s, is defined as the positive square root of the sample variance, s^2. Thus,

$$s = \sqrt{s^2}$$

The population variance, denoted by the symbol σ^2 (sigma squared), is the average of the squared distances of the measurements on *all* units in the population from the mean, μ, and σ (sigma) is the square root of this quantity.

Symbols for Variance and Standard Deviation

s^2 = Sample variance
s = Sample standard deviation
σ^2 = Population variance
σ = Population standard deviation

Notice that, unlike the variance, the standard deviation is expressed in the original units of measurement. For example, if the original measurements are in dollars, the variance is expressed in the peculiar units "dollar squared," but the standard deviation is expressed in dollars.

TEACHING TIP
Let the student know that the divisor question will become clearer when they learn more about estimating parameters with sampling distributions.

You may wonder why we use the divisor $(n - 1)$ instead of n when calculating the sample variance. Wouldn't using n seem more logical, so that the sample variance would be the average squared distance from the mean? The trouble is, using n tends to produce an underestimate of the population variance, σ^2. So we use $(n - 1)$ in the denominator to provide the appropriate correction for this tendency.* Since sample statistics like s^2 are primarily used to estimate population parameters like σ^2, $(n - 1)$ is preferred to n when defining the sample variance.

E X A M P L E 2 . 8

$s^2 = .5$
$s = .71$

Calculate the variance and standard deviation of the following sample: 2, 3, 3, 3, 4.

Solution

As the number of measurements increases, calculating s^2 and s becomes very tedious. Fortunately, as we show in Example 2.9, we can use a statistical software package (or calculator) to find these values. If you must calculate these quantities by hand, it is advantageous to use the shortcut formula provided in Definition 2.8.

To do this, we need two summations: $\sum x$ and $\sum x^2$. These can easily be obtained from the following type of tabulation:

x	x^2
2	4
3	9
3	9
3	9
4	16
$\sum x = 15$	$\sum x^2 = 47$

---

*"Appropriate" here means that s^2 with a divisor of $(n - 1)$ is an *unbiased estimator* of σ^2. We define and discuss unbiasedness of estimators in Chapter 4.

FIGURE 2.18

SAS Printout of Numerical Descriptive Measures for 100 EPA Mileages

			The MEANS Procedure		
			Analysis Variable : MPG		
N	Minimum	Maximum	Mean	Variance	Std Dev
100	30.0000000	44.9000000	36.9940000	5.8462263	2.4178971

Then we use*

$$s^2 = \frac{\sum_{i=1}^{n} x_i^2 - \frac{\left(\sum_{i=1}^{n} x_i\right)^2}{n}}{n-1} = \frac{47 - \frac{(15)^2}{5}}{5-1} = \frac{2}{4} = .5$$

$$s = \sqrt{.5} = .71$$

EXAMPLE 2.9   Use the computer to find the sample variance s^2 and the sample standard deviation s for the 100 gas mileage readings given in Table 2.3.

Solution

TEACHING TIP
To illustrate the mechanics of calculating the measures of variability, use Exercise 2.62 as an example in class. For a practical comparison of different values of the standard deviation, use Exercise 2.64 as an example in class.

The SAS printout describing the gas mileage data is reproduced in Figure 2.18. The variance and standard deviation, highlighted on the printout, are (rounded) $s^2 = 5.84623$ and $s = 2.418$.

You now know that the standard deviation measures the variability of a set of data and how to calculate it. The larger the standard deviation, the more variable the data are. The smaller the standard deviation, the less variation in the data. But how can we practically interpret the standard deviation and use it to make inferences? This is the topic of Section 2.6.

---

*When calculating s^2, how many decimal places should you carry? Although there are no rules for the rounding procedure, it is reasonable to retain twice as many decimal places in s^2 as you ultimately wish to have in s. If you wish to calculate s to the nearest hundredth (two decimal places), for example, you should calculate s^2 to the nearest ten-thousandth (four decimal places).

EXERCISES 2.54–2.65

Learning the Mechanics

2.54 Answer the following questions about variability of data sets:

    a. What is the primary disadvantage of using the range to compare the variability of data sets?

    b. Describe the sample variance using words rather than a formula. Do the same with the population variance.

    c. Can the variance of a data set ever be negative? Explain. Can the variance ever be smaller than the standard deviation? Explain.   No, yes

2.55 Calculate the variance and standard deviation for samples where

    a. $n = 10, \sum x^2 = 84, \sum x = 20$ 4.89, 2.21

    b. $n = 40, \sum x^2 = 380, \sum x = 100$ 3.33, 1.83

    c. $n = 20, \sum x^2 = 18, \sum x = 17$ .187, .432

2.56 Calculate the range, variance, and standard deviation for the following samples:

    a. 4, 2, 1, 0, 1   4, 2.3, 1.52

    b. 1, 6, 2, 2, 3, 0, 3   6, 3.62, 1.90

    c. 8, −2, 1, 3, 5, 4, 4, 1, 3   10, 7.11, 2.67

    d. 0, 2, 0, 0, −1, 1, −2, 1, 0, −1, 1, −1, 0, −3, −2, −1, 0, 1

2.57 Calculate the range, variance, and standard deviation for the following samples:

    a. 39, 42, 40, 37, 41   5, 3.7, 1.92

    b. 100, 4, 7, 96, 80, 3, 1, 10, 2   99, 1949.25, 44.15

    c. 100, 4, 7, 30, 80, 30, 42, 2   98, 1307.84, 36.16

2.58 Compute $\bar{x}$, s^2, and s for each of the following data sets. If appropriate, specify the units in which your answer is expressed.
  a. 3, 1, 10, 10, 4   5.6, 17.3, 4.16
  b. 8 feet, 10 feet, 32 feet, 5 feet   13.75, 152.25, 12.34
  c. −1, −4, −3, 1, −4, −4   −2.5, 4.3, 2.07
  d. 1/5 ounce, 1/5 ounce, 1/5 ounce, 2/5 ounce, 1/5 ounce, 4/5 ounce   .333, .059, .242

2.59 Using only integers between 0 and 10, construct two data sets with at least 10 observations each so that the two sets have the same mean but different variances. Construct dot plots for each of your data sets, and mark the mean of each data set on its dot plot.

2.60 Using only integers between 0 and 10, construct two data sets with at least 10 observations each that have the same range but different means. Construct a dot plot for each of your data sets, and mark the mean of each data set on its dot plot.

2.61 Consider the following sample of five measurements: 2, 1, 1, 0, 3.
  a. Calculate the range, s^2, and s.   3, 1.3, 1.14
  b. Add 3 to each measurement and repeat part **a**.
  c. Subtract 4 from each measurement and repeat part **a**.
  d. Considering your answers to parts **a**, **b**, and **c**, what seems to be the effect on the variability of a data set by adding the same number to or subtracting the same number from each measurement?

2.62 Consider the following two samples:

  Sample 1:   10, 0, 1, 9, 10, 0, 8, 1, 1, 9
  Sample 2:   0, 5, 10, 5, 5, 5, 6, 5, 6, 5

  a. Examine both samples and identify the one that you believe has the greater variability.   Sample 1
  b. Calculate the range for each sample. Does the result agree with your answer to part **a**? Explain.
  c. Calculate the standard deviation for each sample. Does the result agree with your answer to part **a**? Explain.   4.58; 2.39
  d. Which of the two, the range or the standard deviation, provides a better measure of variability?

Applying the Concepts—Basic

2.63 Refer to the *Environmental Science & Technology* (Sept. 1, 2000) study on the ammonia levels near the exit ramp of a San Francisco highway tunnel, Exercise 2.45 (p. 52). The data (in parts per million) for 8 days during afternoon drive-time are reproduced in the table.

  ● AMMONIA

| 1.53 | 1.50 | 1.37 | 1.51 | 1.55 | 1.42 | 1.41 | 1.48 |

  a. Find the range of the ammonia levels.   .18
  b. Find the variance of the ammonia levels.   .0041
  c. Find the standard deviation of the ammonia levels.
  d. Suppose the standard deviation of the daily ammonia levels during morning drive-time at the exit ramp is

1.45 ppm. Which time, morning or afternoon drive-time, has more variable ammonia levels?   Mornings

2.64 Refer to *The Condor* (May 1995) study of collared flycatchers, Exercise 2.46 (p. 52). The data for the study is reproduced here.

● CONDOR

Plot Number	Number Killed	Number of Breeders
1	5	30
2	4	28
3	3	38
4	2	34
5	2	26
6	1	124
7	1	68
8	1	86
9	1	32
10	0	30
11	0	46
12	0	132
13	0	100
14	0	6

  a. Find the range, variance, and standard deviation of the number of flycatchers killed at the 14 plots.
  b. Specify the units in which each of your answers to part **a** is expressed.
  c. Repeat parts **a** and **b** for the number of breeders.

Applying the Concepts—Intermediate

2.65 Refer to the *Fortune* (Oct. 25, 1999) ranking of the 50 most powerful women in America, Exercise 2.44 (p. 51). The distribution of the ages of the 50 women is summarized in the STATISTIX printout below.

STATISTIX Output for Exercise 2.65

```
          DESCRIPTIVE STATISTICS

                              AGE
          N                    50
          MEAN             48.160
          SD               6.0148
          VARIANCE         36.178
          MINIMUM          36.000
          MEDIAN           47.000
          MAXIMUM          68.000
```

  a. Locate the measures of variation on the printout.
  b. Give a value of the standard deviation that would make the age distribution more variable.
  c. Give a value of the standard deviation that would make the age distribution less variable. $s < 6.0148$
  d. If the largest age in the data set (68 years) is omitted, would the standard deviation increase or decrease? Explain.   Decrease

2.6 INTERPRETING THE STANDARD DEVIATION

We've seen that if we are comparing the variability of two samples selected from a population, the sample with the larger standard deviation is the more variable of the two. Thus, we know how to interpret the standard deviation on a relative or comparative basis, but we haven't explained how it provides a measure of variability for a single sample.

To understand how the standard deviation provides a measure of variability of a data set, consider a specific data set and answer the following questions: How many measurements are within 1 standard deviation of the mean? How many measurements are within 2 standard deviations? For example, look at the 100 mileage per gallon readings given in Table 2.3. Recall that $\bar{x} = 36.99$ and $s = 2.42$. Then

$$\bar{x} - s = 34.57 \qquad \bar{x} + s = 39.41$$
$$\bar{x} - 2s = 32.15 \qquad \bar{x} + 2s = 41.83$$

If we examine the data, we find that 68 of the 100 measurements, or 68%, are in the interval

$$\bar{x} - s \text{ to } \bar{x} + s$$

Similarly, we find that 96, or 96%, of the 100 measurements are in the interval

$$\bar{x} - 2s \text{ to } \bar{x} + 2s$$

We usually write these intervals as

$$(\bar{x} - s, \bar{x} + s) \text{ and } (\bar{x} - 2s, \bar{x} + 2s)$$

Such observations identify criteria for interpreting a standard deviation that apply to *any* set of data, whether a population or a sample. The criteria, expressed as a mathematical theorem and as a rule of thumb, are presented in Tables 2.7 and 2.8. In these tables we give two sets of answers to the questions of how many measurements fall within 1, 2, and 3 standard deviations of the mean. The first, which applies to *any* set of data, is derived from a theorem proved by the Russian mathematician P L. Chebyshev (1821–1894). The second, which applies to symmetric, **mound-shaped distributions** of data (where the mean, median, and mode are all about the same), is based upon empirical evidence that has accumulated over the years. However, the percentages given for the intervals in Table 2.8 provide remarkably good approximations even when the distribution of the data is slightly skewed or asymmetric.

TABLE 2.7 Interpreting the Standard Deviation: Chebyshev's Rule

Chebyshev's Rule applies to any data set, regardless of the shape of the frequency distribution of the data.

a. No useful information is provided on the fraction of measurements that fall within 1 standard deviation of the mean, i.e., within the interval $(\bar{x} - s, \bar{x} + s)$ for samples and $(\mu - \sigma, \mu + \sigma)$ for populations.

b. At least $\frac{3}{4}$ of the measurements will fall within 2 standard deviations of the mean, i.e., within the interval $(\bar{x} - 2s, \bar{x} + 2s)$ for samples and $(\mu - 2\sigma, \mu + 2\sigma)$ for populations.

c. At least $\frac{8}{9}$ of the measurements will fall within 3 standard deviations of the mean, i.e., within the interval $(\bar{x} - 3s, \bar{x} + 3s)$ for samples and $(\mu - 3\sigma, \mu + 3\sigma)$ for populations.

d. Generally, for any number k greater than 1, at least $(1 - 1/k^2)$ of the measurements will fall within k standard deviations of the mean, i.e., within the interval $(\bar{x} - ks, \bar{x} + ks)$ for samples and $(\mu - k\sigma, \mu + k\sigma)$ for populations.

TABLE 2.8   Interpreting the Standard Deviation: The Empirical Rule

The **Empirical Rule** is a rule of thumb that applies to data sets with frequency distributions that are mound-shaped and symmetric, as shown below.

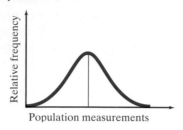

a. Approximately 68% of the measurements will fall within 1 standard deviation of the mean, i.e., within the interval $(\bar{x} - s, \bar{x} + s)$ for samples and $(\mu - \sigma, \mu + \sigma)$ for populations.

b. Approximately 95% of the measurements will fall within 2 standard deviations of the mean, i.e., within the interval $(\bar{x} - 2s, \bar{x} + 2s)$ for samples and $(\mu - 2\sigma, \mu + 2\sigma)$ for populations.

c. Approximately 99.7% (essentially all) of the measurements will fall within 3 standard deviations of the mean, i.e., within the interval $(\bar{x} - 3s, \bar{x} + 3s)$ for samples and $(\mu - 3\sigma, \mu + 3\sigma)$ for populations.

E X A M P L E   2 . 1 0

77%, 93%, 100%

Thirty students in an experimental psychology class use various techniques to train a rat to move through a maze. At the end of the course, each student's rat is timed through the maze. The results (in minutes) are listed in Table 2.9. Determine the fraction of the 30 measurements in the intervals $\bar{x} \pm s$, $\bar{x} \pm 2s$, and $\bar{x} \pm 3s$, and compare the results with those predicted in Tables 2.7 and 2.8.

⊘ RATMAZE
TABLE 2.9   Times (in Minutes) of 30 Rats Running Through a Maze

1.97	.60	4.02	3.20	1.15	6.06	4.44	2.02	3.37	3.65
1.74	2.75	3.81	9.70	8.29	5.63	5.21	4.55	7.60	3.16
3.77	5.36	1.06	1.71	2.47	4.25	1.93	5.15	2.06	1.65

Solution

First, we entered the data into the computer and used STATISTIX to produce summary statistics. The mean and standard deviation of the sample data, highlighted on the printout shown in Figure 2.19, are (rounded)

$$\bar{x} = 3.74 \text{ minutes} \quad s = 2.20 \text{ minutes}$$

FIGURE 2.19

STATISTIX Printout for Example 2.10

```
DESCRIPTIVE STATISTICS

                      RUNTIME
N                          30
MEAN                   3.7443
SD                     2.1982
VARIANCE               4.8323
MINIMUM                0.6000
MEDIAN                 3.5100
MAXIMUM                9.7000
```

Now, we form the interval

$$(\bar{x} - s, \bar{x} + s) = (3.74 - 2.20, 3.74 + 2.20) = (1.54, 5.94)$$

A check of the measurements shows that 23 of the times are within this 1 standard deviation interval around the mean. This number represents 23/30, or ≈ 77% of the sample measurements.

The next interval of interest is

$$(\bar{x} - 2s, \bar{x} + 2s) = (3.74 - 4.40, 3.74 + 4.40) = (-.66, 8.14)$$

All but two of the times are within this interval, so 28/30, or approximately 93%, are within 2 standard deviations of $\bar{x}$.

Finally, the 3-standard-deviation interval around $\bar{x}$ is

$$(\bar{x} - 3s, \bar{x} + 3s) = (3.74 - 6.60, 3.74 + 6.60) = (-2.86, 10.34)$$

All of the times fall within 3 standard deviations of the mean.

These 1-, 2-, and 3-standard-deviation percentages (77%, 93%, and 100%) agree fairly well with the approximations of 68%, 95%, and 100% given by the Empirical Rule (Table 2.8) for mound-shaped distributions. If you look at the STATISTIX frequency histogram for this data set in Figure 2.20, you'll note that the distribution is not really mound-shaped, nor is it extremely skewed. Thus, we get reasonably good results from the mound-shaped approximations. Of course, we know from Chebyshev's Rule (Table 2.7) that no matter what the shape of the distribution, we would expect at least 75% and 89% of the measurements to lie within 2 and 3 standard deviations of $\bar{x}$, respectively.

FIGURE 2.20

STATISTIX Histogram for Times for Rats to Move Through Maze

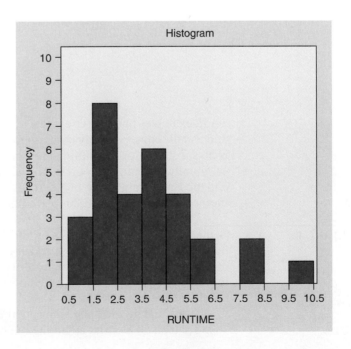

EXAMPLE 2.11

Chebyshev's Rule and the Empirical Rule (Tables 2.7 and 2.8) are useful as a check on the calculation of the standard deviation. For example, suppose we calculated the standard deviation for the gas mileage data (Table 2.3) to be 5.85. Are there any "clues" in the data that enable us to judge whether this number is reasonable?

Solution

The range of the mileage data in Table 2.3 is $44.0 - 30.0 = 14.9$. From Chebyshev's Rule and the Empirical Rule we know that most of the measurements (approximately 95% if the distribution is mound-shaped) will be within 2 standard deviations of the mean. And, regardless of the shape of the distribution and the number of measurements, almost all of them will fall within 3 standard deviations of the mean. Consequently, we would expect the range of the measurements to be between 4 (i.e., $\pm 2s$) and 6 (i.e., $\pm 3s$) standard deviations in length (see Figure 2.21). For the car mileage data, this means that s should fall between

$$\frac{\text{Range}}{6} = \frac{14.9}{6} = 2.48 \quad \text{and} \quad \frac{\text{Range}}{4} = \frac{14.9}{4} = 3.73$$

In particular, the standard deviation should not be much larger than 1/4 of the range, particularly for the data set with 100 measurements. Thus, we have reason to believe that the calculation of 5.85 is too large. A check of our work reveals that 5.85 is the variance s^2, not the standard deviation s (see Example 2.9). We "forgot" to take the square root (a common error); the correct value is $s = 2.42$. Note that this value is slightly smaller than the range divided by 6 (2.48). The larger the data set, the greater the tendency for very large or very small measurements (extreme values) to appear, and when they do, the range may exceed 6 standard deviations. ∎

FIGURE 2.21

The Relation Between the Range and the Standard Deviation

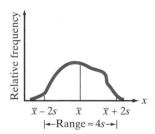

In examples and exercises we'll sometimes use $s \approx \text{range}/4$ to obtain a crude, and usually conservatively large, approximation for s. However, we stress that this is no substitute for calculating the exact value of s when possible.

Finally, and most importantly, we will use the concepts in Chebyshev's Rule and the Empirical Rule to build the foundation for statistical inference-making. The method is illustrated in Example 2.12.

EXAMPLE 2.12

A manufacturer of automobile batteries claims that the average length of life for its grade A battery is 60 months. However, the guarantee on this brand is for just 36 months. Suppose the standard deviation of the life length is known to be 10 months, and the frequency distribution of the life-length data is known to be mound-shaped.

a. Approximately what percentage of the manufacturer's grade A batteries will last more than 50 months, assuming the manufacturer's claim is true?

b. 2.5%

b. Approximately what percentage of the manufacturer's batteries will last less than 40 months, assuming the manufacturer's claim is true?

c. Suppose your battery lasts 37 months. What could you infer about the manufacturer's claim?

Solution

If the distribution of life length is assumed to be mound-shaped with a mean of 60 months and a standard deviation of 10 months, it would appear as shown in Figure 2.22. Note that we can take advantage of the fact that mound-shaped distributions are (approximately) symmetric about the mean, so that the percentages given by the Empirical Rule can be split equally between the halves of the distribution on each side of the mean. The approximations given in Figure 2.22 are more dependent on the assumption of a mound-shaped distribution than those given by the Empirical Rule (Table 2.8), because the approximations in Figure 2.22 depend on the (approximate) symmetry of the mound-shaped distribution. We saw in Example 2.10 that the Empirical Rule can yield good approximations even for skewed distributions. This will *not* be true of the approximations in Figure 2.22; the distribution *must* be mound-shaped and (approximately) symmetric.

FIGURE 2.22

Battery Life-Length Distribution: Manufacturer's Claim Assumed True

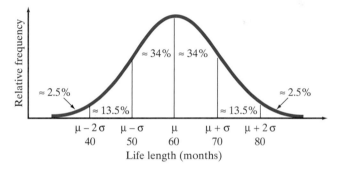

For example, since approximately 68% of the measurements will fall within 1 standard deviation of the mean, the distribution's symmetry implies that approximately $(1/2)(68\%) = 34\%$ of the measurements will fall between the mean and 1 standard deviation on each side. This concept is illustrated in Figure 2.22. The figure also shows that 2.5% of the measurements lie beyond 2 standard deviations in each direction from the mean. This result follows from the fact that if approximately 95% of the measurements fall within 2 standard deviations of the mean, then about 5% fall outside 2 standard deviations; if the distribution is approximately symmetric, then about 2.5% of the measurements fall beyond 2 standard deviations on each side of the mean.

a. It is easy to see in Figure 2.22 that the percentage of batteries lasting more than 50 months is approximately 34% (between 50 and 60 months) plus 50% (greater than 60 months). Thus, approximately 84% of the batteries should have life length exceeding 50 months.

b. The percentage of batteries that last less than 40 months can also be easily determined from Figure 2.22. Approximately 2.5% of the batteries should fail prior to 40 months, assuming the manufacturer's claim is true.

c. If you are so unfortunate that your grade A battery fails at 37 months, you can make one of two inferences: Either your battery was one of the approximately 2.5% that fail prior to 40 months, or something about the manufacturer's claim is not true. Because the chances are so small that a battery fails before 40 months, you would have good reason to have serious doubts about the manufacturer's claim. A mean smaller than 60 months and/or a standard deviation longer than 10 months would both increase the likelihood of failure prior to 40 months.*   ■

Example 2.12 is our initial demonstration of the statistical inference-making process. At this point you should realize that we'll use sample information (in Example 2.12, your battery's failure at 37 months) to make inferences about the population (in Example 2.12, the manufacturer's claim about the life length for the population of all batteries). We'll build on this foundation as we proceed.

 EXERCISES 2.66–2.81

Learning the Mechanics

2.66 To what kind of data sets can Chebyshev's Rule be applied? The Empirical Rule?

2.67 The output from a statistical computer program indicates that the mean and standard deviation of a data set consisting of 200 measurements are $1,500 and $300, respectively.
  a. What are the units of measurement of the variable of interest? Based on the units, what type of data is this: quantitative or qualitative?   Quantitative
  b. What can be said about the number of measurements between $900 and $2,100? Between $600 and $2,400? Between $1,200 and $1,800? Between $1,500 and $2,100?

2.68 For any set of data, what can be said about the percentage of the measurements contained in each of the following intervals?
  a. $\bar{x} - s$ to $\bar{x} + s$ At least 0%
  b. $\bar{x} - 2s$ to $\bar{x} + 2s$ At least 3/4
  c. $\bar{x} - 3s$ to $\bar{x} + 3s$ At least 8/9

2.69 For a set of data with a mound-shaped relative frequency distribution, what can be said about the percentage of the measurements contained in each of the intervals specified in Exercise 2.68?

2.70 The following is a sample of 25 measurements:

| 7 | 6 | 6 | 11 | 8 | 9 | 11 | 9 | 10 | 8 | 7 | 7 | 5 |
| 9 | 10 | 7 | 7 | 7 | 7 | 9 | 12 | 10 | 10 | 8 | 6 |

  a. Compute $\bar{x}$, s^2, and s for this sample.   8.24, 3.36, 1.83
  b. Count the number of measurements in the intervals $\bar{x} \pm s$, $\bar{x} \pm 2s$, and $\bar{x} \pm 3s$. Express each count as a percentage of the total number of measurements.

  c. Compare the percentages found in part **b** to the percentages given by the Empirical Rule and Chebyshev's Rule.
  d. Calculate the range and use it to obtain a rough approximation for s. Does the result compare favorably with the actual value for s found in part **a**?

2.71 Given a data set with a largest value of 760 and a smallest value of 135, what would you estimate the standard deviation to be? Explain the logic behind the procedure you used to estimate the standard deviation. Suppose the standard deviation is reported to be 25. Is this feasible? Explain. $s \approx 104.17$, no

Applying the Concepts—Basic

2.72 To minimize the potential for gastrointestinal disease outbreaks, all passenger cruise ships arriving at U.S. ports are subject to unannounced sanitation inspections. Ships are rated on a 100-point scale by the Centers for Disease Control and Prevention. A score of 86 or higher indicates that the ship is providing an accepted standard of sanitation. The May 2001 sanitation scores for 151 cruise ships are listed in the accompanying table, followed by a MINITAB printout of descriptive statistics on p. 68.
  a. Locate the mean and standard deviation on the MINITAB printout.   93.113; 5.184
  b. Calculate the intervals $\bar{x} \pm s$, $\bar{x} \pm 2s$, $\bar{x} \pm 3s$.
  c. Find the percentage of measurements in the data set that fall within each of the intervals, part **b**. Do these percentages agree with either Chebyshev's Theorem or the Empirical Rule?

---

*The assumption that the distribution is mound-shaped and symmetric may also be incorrect. However, if the distribution were skewed to the right, as life-length distributions often tend to be, the percentage of measurements more than 2 standard deviations below the mean would be even less than 2.5%.

⬤ SHIPSANIT

Ship Name	Score	Ship Name	Score	Ship Name	Score
Grande Mariner	95	Crystal Symphony	93	Scotia Prince	92
Seabourn Sun	90	Statendam	94	Splendour of the Seas	90
Olympic Voyager	95	Texas Treasure	96	Silver Wind	96
Grandeur of the Seas	94	Elation	95	Clipper Adventurer	92
Jubilee	95	Delphin	99	Sea Bird	96
Club Med 2	94	Sea Princess	97	Grande Caribe	93
Costa Victoria	96	Melody	98	Spirit of Columbia	95
Wind Spirit	98	Crystal Harmony	96	Yorktown Clipper	90
Norwegian Majesty	94	Millennium	93	Hanseatic	97
Galaxy	98	Zenith	95	Pacific Sky	91
Costa Atlantica	100	Legacy	93	Big Red Boat II	94
Volendam	98	Vision of the Seas	92	Pacific Princess	88
Norwegian Dream	100	Westerdam	92	Big Red Boat III	92
Maasdam	94	C. Columbus	98	Oceanic	91
Mercury	96	Queen Elizabeth 2	92	Niagra Prince	88
Orient Venus	87	Fuji Maru	91	Sea Lion	92
Seven Seas Navigator	100	Pacific Venus	82	Regal Princess	93
Seabourn Pride	93	Universe Explorer	95	Dolphin IV	88
Infinity	97	The Emerald	91	Asuka	94
Seabourn Legend	97	Celebration	91	Enchanted Capri	93
Radisson Diamond	96	Inspiration	98	Horizon	92
Ocean Princess	97	Grand Princess	94	Rembrandt	94
Nordic Empress	98	Regal Empress	89	Enchanted Isle	89
Vistamar	83	Ocean Breeze	95	Maxim Gorky	90
Norwegian Wind	92	Palm Beach Princess	92	Enchanted Sun	89
Bolero	96	Regal Voyager	87	The Topaz	77
Oriana	100	Rotterdam	97	Costa Romantica	90
Nippon Maru	99	Dawn Princess	96	Royal Princess	94
Paradise	95	Noordam	95	Flamenco	95
Albatross	93	Enchantment of the Seas	99	Stella Solaris	93
Viking Serenade	92	Fascination	95	Contessa 1	93
Holiday	97	Caronia	95	Aegean I	52
Island Adventure	97	Arcadia	96	Silver Cloud	98
Carnival Victory	97	Seabourn Goddess II	94	Deutschland	91
Norway	93	Explorer of the Seas	91	Spirit of Ninety Eight	91
Imagination	94	Voyager of the Sea	92	Legend of the Seas	95
Majesty of the Seas	90	Europa	96	Astor	91
Crown Princess	96	Discovery Sun	95	Paul Gauguin	86
Arkona	90	Norwegian Sea	90	Norwegian Star	78
Sun Princess	98	Tropicale	86	Black Watch	86
Monarch of the Seas	93	Ryndam	96	Spirit of Alaska	93
Carnival Destiny	96	Amsterdam	93	Aida	90
Seabourn Goddess I	92	Fantasy	93	Spirit of Discovery	96
Zaandam	97	Silver Shadow	97	Spirit of Glacier Bay	95
Aurora	98	Century	93	Seabourn Spirit	92
Disney Wonder	99	Ecstasy	94	Triton	93
Carnival Triumph	98	Norwegian Sky	97	Costa Allegra	88
Sensation	90	Nantucket Clipper	88	Bremen	90
Victoria	94	Rhapsody of the Seas	94	Costa Classica	92
Sovereign of the Seas	98	Veendam	86		
Disney Magic	96	Le Levant	91		

Source: National Center for Environmental Health, Centers for Disease Control and Prevention, May 8, 2001.

MINITAB Output for Exercise 2.72

```
Descriptive Statistics: Score

Variable          N        Mean    Median    TrMean    StDev    SE Mean
Score           151      93.113    94.000    93.585    5.184      0.422

Variable    Minimum     Maximum        Q1        Q3
Score        52.000     100.000    91.000    96.000
```

2.73 Refer to the *Marine Technology* (Jan. 1995) data on spillage amounts (in thousands of metric tons) for 50 major oil spills, Exercise 2.15 (p. 29). An SPSS histogram for the 50 spillage amounts is shown below.
   a. Interpret the histogram.
   b. Descriptive statistics for the 50 spillage amounts are also shown on the SPSS histogram. Use this information to form an interval that can be used to predict the spillage amount for the next major oil spill.

2.74 Refer to the *American Psychologist* (July 1995) study of 148 applicants for a position in experimental psychology, Exercise 2.47 (p. 53). Recall that the mean number of recommendation letters included in each application packet was $\bar{x} = 2.28$. The standard deviation was also reported in the article; its value was $s = 1.48$.
   a. Sketch the relative frequency distribution for the number of recommendation letters included in each application for the experimental psychology position. (Assume the distribution is mound-shaped and relatively symmetric.)

   b. Locate an interval on the distribution, part **a**, that captures approximately 95% of the measurements in the sample.
   c. Locate an interval on the distribution, part **a**, that captures almost all the sample measurements.

2.75 For each day of last year, the number of vehicles passing through a certain intersection was recorded by a city engineer. One objective of this study was to determine the percentage of days that more than 425 vehicles used the intersection. Suppose the mean for the data was 375 vehicles per day and the standard deviation was 25 vehicles.
   a. What can you say about the percentage of days that more than 425 vehicles used the intersection? Assume you know nothing about the shape of the relative frequency distribution for the data.   At most 25%
   b. What is your answer to part **a** if you know that the relative frequency distribution for the data is mound-shaped? $\approx 2.5\%$

SPSS Output for Exercise 2.73

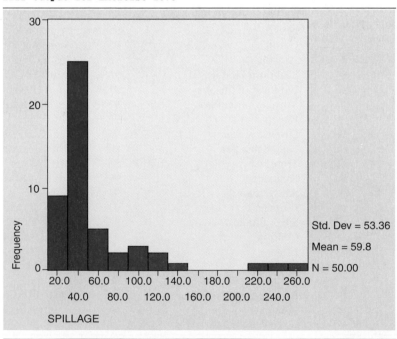

Applying the Concepts—Intermediate

2.76 A study published in *Applied Psycholinguistics* (June 1998) compared the language skills of young children (16–30 months old) from low income and middle income families. A total of 260 children—65 in the low income and 195 in the middle income group—completed the Communicative Development Inventory (CDI) exam. One of the variables measured on each child was sentence complexity score. Summary statistics for the scores of the two groups are reproduced in the table. Use this information to sketch a graph of the sentence complexity score distribution for each income group. (Assume the distributions are mound-shaped and symmetric.) Compare the distributions. What can you infer?

	Low Income	Middle Income
Sample Size	65	195
Mean	7.62	15.55
Median	4	14
Standard Deviation	8.91	12.24
Minimum	0	0
Maximum	36	37

Source: Arriaga, R. I. *et al.* "Scores on the MacArthur Communicative Development Inventory of children from low-income and middle-income families." *Applied Psycholinguistics*, Vol. 19, No. 2, June 1998, p. 217 (Table 7).

2.77 The *American Rifleman* (June 1993) reported on the velocity of ammunition fired from the FEG P9R pistol, a 9mm gun manufactured in Hungary. Field tests revealed that Winchester bullets fired from the pistol had a mean velocity (at 15 feet) of 936 feet per second and a standard deviation of 10 feet per second. Tests were also conducted with Uzi and Black Hills ammunition.
   a. Describe the velocity distribution of Winchester bullets fired from the FEG P9R pistol.
   b. A bullet, brand unknown, is fired from the FEG P9R pistol. Suppose the velocity (at 15 feet) of the bullet is 1,000 feet per second. Is the bullet likely to be manufactured by Winchester? Explain.

2.78 Astronomers theorize that cold dark matter (CDM) caused the formation of galaxies and clusters of galaxies in the universe. The theoretical CDM model requires an estimate of the velocity of light emitted from the galaxy cluster. *The Astronomical Journal* (July 1995) published a study of observed velocities for galaxies in four different galaxy clusters. Galaxy velocity was measured in kilometers per second (km/s) using a spectrograph and high-power telescope.
   a. The observed velocities of 103 galaxies located in the cluster named A2142 are summarized in the accompanying histogram. Comment on whether the Empirical Rule is applicable for describing the velocity distribution for this cluster.

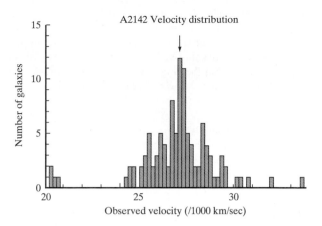

Source: Oegerle, W. R., Hill, J. M., and Fitchett, M. J. "Observations of high dispersion clusters of galaxies: Constraints on cold dark matter." *The Astronomical Journal*, Vol. 110, No. 1, July 1995, p. 37 (Figure 1).

   b. The mean and standard deviation of the 103 velocities observed in galaxy cluster A2142 were reported as $\bar{x} = 27{,}117$ km/s and $s = 1{,}280$ km/s, respectively. Use this information to construct an interval that captures approximately 95% of the galaxy velocities in the cluster.
   c. Recommend a single velocity value to be used in the CDM model for galaxy cluster A2142. Explain your reasoning   Use $\bar{x} = 27.117$

2.79 The Locus of Control (LOC) is a measure of one's perception of control over factors affecting one's life. In one study, the LOC was measured for two groups of individuals undergoing weight-reduction for treatment of obesity (*Journal of Psychology*, Mar. 1991). The LOC mean and standard deviation for a sample of 46 adults were 6.45 and 2.89, respectively, while the LOC mean and standard deviation for a sample of 19 adolescents were 10.89 and 2.48, respectively. A lower score on the LOC scale indicates a perception that internal factors are in control, while a higher score indicates a perception that external factors are in control.
   a. Calculate the 1- and 2-standard-deviation intervals around the means for each group. Plot these intervals on a line graph using different colors or symbols to represent each group.
   b. Assuming that the distributions of LOC scores are approximately mound-shaped, estimate the numbers of individuals within each interval.
   c. Based on your answers to parts **a** and **b**, do you think an inference can be made that *all* adults and adolescents undergoing weight-reduction treatment differ with respect to LOC? What factors did you consider in making this inference? (In Chapter 7 we show how to measure the reliability of this inference.)   No

Applying the Concepts—Advanced

2.80 A buyer for a lumber company must decide whether to buy a piece of land containing 5,000 pine trees. If 1,000 of the trees are at least 40 feet tall, the buyer will purchase the land; otherwise, he won't. The owner of the land reports that the height of the trees has a mean of 30 feet and a standard deviation of 3 feet. Based on this information, what is the buyer's decision?

2.81 The National Education Longitudinal Survey (NELS) tracks a nationally representative sample of U.S. students from eighth grade through high school and college. Research published in *Chance* (Winter 2001) examined the Standardized Admission Test (SAT) scores of 265 NELS students who paid a private tutor to help them improve their scores. The table summarizes the changes in both the SAT-Mathematics and SAT-Verbal scores for these students.

	SAT-Math	SAT-Verbal
Mean change in score	19	7
Standard deviation of score changes	65	49

a. Suppose one of the 265 students who paid a private tutor is selected at random. Give an interval that is likely to contain this student's change in the SAT-Math score. $(-176, 214)$

b. Repeat part **a** for the SAT-Verbal score.

c. Suppose the selected student's score increased on one of the SAT tests by 140 points. Which test, the SAT-Math or SAT-Verbal, is the one most likely to have the 140-point increase? Explain.   SAT-Math

2.7   NUMERICAL MEASURES OF RELATIVE STANDING

We've seen that numerical measures of central tendency and variability describe the general nature of a quantitative data set (either a sample or a population). In addition, we may also be interested in describing the *relative* quantitative location of a particular measurement within a data set. Descriptive measures of the relationship of a measurement to the rest of the data are called **measures of relative standing**.

One measure of the relative standing of a measurement is its *percentile ranking*. For example, suppose you scored an 80 on a test and you want to know how you fared in comparison with others in your class. If the instructor tells you that you scored at the 90th percentile, it means that 90% of the grades were lower than yours and 10% were higher. Thus, if the scores were described by the relative frequency histogram in Figure 2.23, the 90th percentile would be located at a point such that 90% of the total area under the relative frequency histogram lies below the 90th percentile and 10% lies above. If the instructor tells you that you scored in the 50th percentile (the median of the data set), 50% of the test grades would be lower than yours and 50% would be higher.

TEACHING TIP
Use the SAT and ACT college entrance examinations as an example of the need for measures of relative standing. Note that all test scores reported contain a percentile measurement for use in comparison.

FIGURE 2.23

Location of 90th Percentile for Test Grades

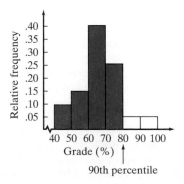

Percentile rankings are of practical value only for large data sets. Finding them involves a process similar to the one used in finding a median. The measurements are ranked in order and a rule is selected to define the location of each percentile. Since we are primarily interested in interpreting the percentile rankings of measurements (rather than finding particular percentiles for a data set), we define the *p*th *percentile* of a data set as shown in Definition 2.10.

TEACHING TIP
Use the median to illustrate that students have already been exposed to the 50th percentile of a data set.

DEFINITION 2.10

For any set of n measurements (arranged in ascending or descending order), the **pth percentile** is a number such that $p\%$ of the measurements fall below the pth percentile and $(100 - p)\%$ fall above it.

EXAMPLE 2.13

7.9, 15.6

Refer to the student default rates of the 50 states (and District of Columbia) in Table 2.5. An SAS printout describing the data is shown in Figure 2.24. Locate the 25th percentile and 95th percentile on the printout and interpret these values.

FIGURE 2.24

SAS Descriptive Statistics for Student Default Rate Data

The UNIVARIATE Procedure
Variable: DEFAULT

Moments

N	51	Sum Weights	51
Mean	10.1960784	Sum Observations	520
Std Deviation	3.504395	Variance	12.2807843
Skewness	0.32759095	Kurtosis	-0.0733636
Uncorrected SS	5916	Corrected SS	614.039216
Coeff Variation	34.3700279	Std Error Mean	0.49071345

Basic Statistical Measures

Location		Variability	
Mean	10.19608	Std Deviation	3.50439
Median	9.70000	Variance	12.28078
Mode	8.80000	Range	17.00000
		Interquartile Range	4.40000

Quantiles (Definition 5)

Quantile	Estimate
100% Max	19.7
99%	19.7
95%	15.6
90%	14.8
75% Q3	12.3
50% Median	9.7
25% Q1	7.9
10%	6.0
5%	4.9
1%	2.7
0% Min	2.7

Extreme Observations

----Lowest----		----Highest---	
Value	Obs	Value	Obs
2.7	51	15.2	44
4.8	35	15.5	34
4.9	28	15.6	25
5.5	42	16.6	21
5.7	17	19.7	2

Solution

Suggested Exercise 2.83

Both the 25th percentile and 95th percentile are highlighted on the SAS printout, Figure 2.24. These values are 7.9 and 15.6, respectively. Our interpretations are as follows: 25% of the 51 default rates fall below 7.9 and 95% of the default rates fall below 15.6.

■

Another measure of relative standing in popular use is the *z-score*. As you can see in Definition 2.11, the *z*-score makes use of the mean and standard deviation of the data set in order to specify the relative location of the measurement.

> **DEFINITION 2.11**
> The **sample z-score** for a measurement *x* is
>
> $$z = \frac{x - \bar{x}}{s}$$
>
> The **population z-score** for a measurement *x* is
>
> $$z = \frac{x - \mu}{\sigma}$$

Note that the *z*-score is calculated by subtracting $\bar{x}$ (or μ) from the measurement *x* and then dividing the result by *s* (or σ). The final result, the z-score, represents the distance between a given measurement *x* and the mean, expressed in standard deviations.

■

EXAMPLE 2.14

$z = -1.0$

Suppose a sample of 2,000 high school seniors' verbal SAT scores is selected. The mean and standard deviation are

$$\bar{x} = 550 \quad s = 75$$

Suppose Joe Smith's score is 475. What is his sample *z*-score?

Solution

You can see that Joe Smith's score lies below the mean score of the 2,000 seniors:

325	475	550	775
$\bar{x} - 3s$	Joe Smith's score	$\bar{x}$	$\bar{x} + 3s$

We compute

$$z = \frac{x - \bar{x}}{s} = \frac{475 - 550}{75} = -1.0$$

which tells us that Joe Smith's score is 1.0 standard deviation *below* the sample mean; in short, his sample *z*-score is −1.0.

■

The numerical value of the *z*-score reflects the relative standing of the measurement. A large positive *z*-score implies that the measurement is larger than almost all other measurements, whereas a large negative *z*-score indicates that the measurement is smaller than almost every other measurement. If a z-score is 0 or near 0, the measurement is located at or near the mean of the sample or population.

We can be more specific if we know that the frequency distribution of the measurements is mound-shaped. In this case, the following interpretation of the z-score can be given:

Interpretation of z-Scores for Mound-Shaped Distributions of Data

1. Approximately 68% of the measurements will have a z-score between -1 and 1.
2. Approximately 95% of the measurements will have a z-score between -2 and 2.
3. Approximately 99.7% (almost all) of the measurements will have a z-score between -3 and 3.

TEACHING TIP
Draw a picture of a mound-shaped distribution and locate the z-scores $-3, -2, -1, 0, 1, 2$, and 3 on it to help students understand what the z-score measures.

Note that this interpretation of z-scores is identical to that given by the Empirical Rule for mound-shaped distributions (Table 2.8). The statement that a measurement falls in the interval $(\mu - \sigma)$ to $(\mu + \sigma)$ is equivalent to the statement that a measurement has a population z-score between -1 and 1, since all measurements between $(\mu - \sigma)$ and $(\mu + \sigma)$ are within 1 standard deviation of μ. These z-scores are displayed in Figure 2.25.

FIGURE 2.25

Population z-Scores for a Mound-Shaped Distribution

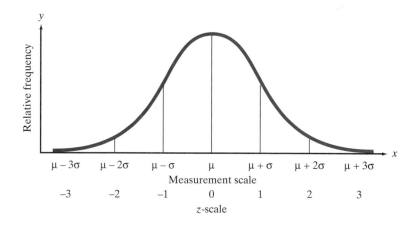

 EXERCISES 2.82–2.93

Learning the Mechanics

2.82 Compute the z-score corresponding to each of the following values of x:

   a. $x = 40$, $s = 5$, $\bar{x} = 30$ 2
   b. $x = 90$, $\mu = 89$, $\sigma = 2$ .05
   c. $\mu = 50$, $\sigma = 5$, $x = 50$ 0
   d. $s = 4$, $x = 20$, $\bar{x} = 30$ -2.5

   e. In parts **a–d**, state whether the z-score locates x within a sample or a population.
   f. In parts **a–d**, state whether each value of x lies above or below the mean and by how many standard deviations.

2.83 Give the percentage of measurements in a data set that are above and below each of the following percentiles:

a. 75th percentile   25%, 75%
b. 50th percentile   50%, 50%
c. 20th percentile   80%, 20%
d. 84th percentile   16%, 84%

2.84 What is the 50th percentile of a quantitative data set called?   Median

2.85 Compare the z-scores to decide which of the following x values lie the greatest distance above the mean and the greatest distance below the mean.
   a. $x = 100$, $\mu = 50$, $\sigma = 25$ 2
   b. $x = 1$, $\mu = 4$, $\sigma = 1$ -3
   c. $x = 0$, $\mu = 200$, $\sigma = 100$ -2
   d. $x = 10$, $\mu = 5$, $\sigma = 3$ 1.67

2.86 Suppose that 40 and 90 are two elements of a population data set and that their z-scores are -2 and 3, respectively. Using only this information, is it possible to determine the population's mean and standard deviation? If so, find them. If not, explain why it's not possible.

Applying the Concepts—Basic

2.87 The distribution of scores on a nationally administered college achievement test has a median of 520 and a mean of 540.
   a. Explain why it is possible for the mean to exceed the median for this distribution of measurements.
   b. Suppose you are told that the 90th percentile is 660. What does this mean?
   c. Suppose you are told that you scored at the 94th percentile. Interpret this statement.

2.88 According to the National Center for Education Statistics (2000), scores on a mathematics assessment test for United States eighth graders have a mean of 500, a 5th percentile of 356, a 25th percentile of 435, a 75th percentile of 563, and a 95th percentile of 653. Interpret each of these numerical descriptive measures.

2.89 Refer to the sanitation levels of cruise ships, Exercise 2.72 (p. 66). The MINITAB descriptive statistics printout is reproduced below.
   a. Give a measure of relative standing for the Norwegian Star's score of 78. Interpret the result.
   b. Give a measure of relative standing for the Rotterdam's score of 97. Interpret the result.

Applying the Concepts—Intermediate

2.90 The U.S. Environmental Protection Agency (EPA) sets a limit on the amount of lead permitted in drinking water. The EPA *Action Level* for lead is .015 milligrams

per liter (mg/L) of water. Under EPA guidelines, if 90% of a water system's study samples have a lead concentration less than .015 mg/L, the water is considered safe for drinking. I (co-author Sincich) received a report on a study of lead levels in the drinking water of homes in my subdivision. The 90th percentile of the study sample had a lead concentration of .00372 mg/L. Are water customers in my subdivision at risk of drinking water with unhealthy lead levels? Explain.

2.91 Refer to the *Optometry and Vision Science* (June 1995) study of refractive variation in eyes, Exercise 2.52 (p. 54). A portion of the SAS descriptive statistics printout for the data set consisting of 25 cylinder power measurements is displayed on p. 75.
   a. Find the 10th percentile of cylinder power measurements on the printout. Interpret the result.   .06
   b. Find the 95th percentile of cylinder power measurements on the printout. Interpret the result.
   c. Use the information on the SAS printout to calculate the z-score for the cylinder power measurement of 1.07. Interpret the result. $z = 4.65$

2.92 In a study of how external clues influence performance, psychology professors at the University of Alberta and Pennsylvania State University gave two different forms of a midterm examination to a large group of introductory psychology students. The questions on the exam were identical and in the same order, but one exam was printed on blue paper and the other on red paper (*Teaching Psychology*, May 1998). Grading only the difficult questions on the exam, the researchers found that scores on the blue exam had a distribution with a mean of 53% and a standard deviation of 15%, while scores on the red exam had a distribution with a mean of 39% and a standard deviation of 12%. (Assume that both distributions are approximately mound-shaped and symmetric.)
   a. Give an interpretation of the standard deviation for the students who took the blue exam.
   b. Give an interpretation of the standard deviation for the students who took the red exam.
   c. Suppose a student is selected at random from the group of students who participated in the study and the student's score on the difficult questions is 20%. Which exam form is the student more likely to have taken, the blue or the red exam? Explain.

2.93 At one university, the students are given z-scores at the end of each semester rather than the traditional GPAs. The mean and standard deviation of all students'

MINITAB Output for Exercise 2.89

```
Descriptive Statistics: Score

Variable          N        Mean      Median      TrMean      StDev     SE Mean
Score           151      93.113      94.000      93.585      5.184       0.422

Variable    Minimum     Maximum          Q1          Q3
Score        52.000     100.000      91.000      96.000
```

cumulative GPAs, on which the z-scores are based, are 2.7 and .5, respectively.

a. Translate each of the following z-scores to corresponding GPA scores: $z = 2.0$, $z = -1.0$, $z = .5$, $z = -2.5$. 3.7, 2.2, 2.95, 1.45

b. Students with z-scores below -1.6 are put on probation. What is the corresponding probationary GPA?

c. The president of the university wishes to graduate the top 16% of the students with *cum laude* honors and the top 2.5% with *summa cum laude* honors. Where (approximately) should the limits be set in terms of z-scores? In terms of GPAs? What assumption, if any, did you make about the distribution of the GPAs at the university?

SAS Output for Exercise 2.91

The UNIVARIATE Procedure
Variable: CYLPOWER

Basic Statistical Measures

Location		Variability	
Mean	0.154400	Std Deviation	0.19677
Median	0.110000	Variance	0.03872
Mode	0.060000	Range	1.03000
		Interquartile Range	0.08000

NOTE: The mode displayed is the smallest of 4 modes with a count of 3.

Quantiles (Definition 5)

Quantile	Estimate
100% Max	1.07
99%	1.07
95%	0.21
90%	0.20
75% Q3	0.16
50% Median	0.11
25% Q1	0.08
10%	0.06
5%	0.06
1%	0.04
0% Min	0.04

2.8 METHODS FOR DETECTING OUTLIERS (OPTIONAL)

Sometimes it is important to identify inconsistent or unusual measurements in a data set. An observation that is unusually large or small relative to the data values we want to describe is called an *outlier*.

Outliers are often attributable to one of several causes. First, the measurement associated with the outlier may be invalid. For example, the experimental procedure used to generate the measurement may have malfunctioned, the experimenter may have misrecorded the measurement, or the data might have been coded incorrectly in the computer. Second, the outlier may be the result of a misclassified measurement. That is, the measurement belongs to a population different from that from which the rest of the sample was drawn. Finally, the measurement associated with the outlier may be recorded correctly and from the same population as the rest of the sample, but represents a rare (chance) event. Such outliers occur most often when the relative frequency distribution of the sample data is extremely skewed, because such a distribution has a tendency to include extremely large or small observations relative to the others in the data set.

DEFINITION 2.12

An observation (or measurement) that is unusually large or small relative to the other values in a data set is called an **outlier**. Outliers typically are attributable to one of the following causes:

1. The measurement is observed, recorded, or entered into the computer incorrectly.
2. The measurement comes from a different population.
3. The measurement is correct, but represents a rare (chance) event.

Two useful methods for detecting outliers, one graphical and one numerical, are **box plots** and *z*-scores. The box plot is based on the *quartiles* of a data set. **Quartiles** are values that partition the data set into four groups, each containing 25% of the measurements. The *lower quartile* Q_L is the 25th percentile, the *middle quartile* is the median *M* (the 50th percentile), and the *upper quartile* Q_U is the 75th percentile (see Figure 2.26).

FIGURE 2.26

The Quartiles for a Data Set

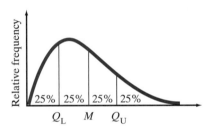

DEFINITION 2.13

The **lower quartile Q_L** is the 25th percentile of a data set. The **middle quartile M** is the median. The **upper quartile Q_U** is the 75th percentile.

A box plot is based on the *interquartile range (IQR)*, the distance between the lower and upper quartiles:

$$IQR = Q_U - Q_L$$

DEFINITION 2.14

The **interquartile range (IQR)** is the distance between the lower and upper quartiles:

$$IQR = Q_U - Q_L$$

A vertical MINITAB box plot for the gas mileage data (Table 2.3) is shown in Figure 2.27.* Note that a rectangle (the *box*) is drawn, with the bottom and top of the rectangle (the **hinges**) drawn at the quartiles Q_L and Q_U, respectively. By

---

*Although box plots can be generated by hand, the amount of detail required makes them particularly well suited for computer generation. We use computer software to generate the box plots in this section.

FIGURE 2.27

MINITAB Box Plot for Gas Mileage Data

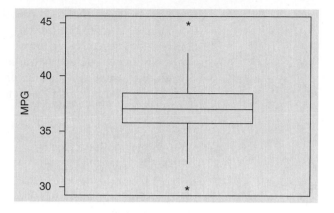

definition, then, the "middle" 50% of the observations—those between Q_L and Q_U—fall inside the box. For the gas mileage data, these quartiles appear to be at (approximately) 35.5 and 38.5. Thus,

$$IQR = 38.5 - 35.5 = 3.0 \text{ (approximately)}$$

The median is shown at about 37 by a horizontal line within the box.

To guide the construction of the "tails" of the box plot, two sets of limits, called **inner fences** and **outer fences**, are used. Neither set of fences actually appears on the box plot. Inner fences are located at a distance of 1.5(IQR) from the hinges. Emanating from the hinges of the box are vertical lines called the **whiskers**. The two whiskers extend to the most extreme observation inside the inner fences. For example, the inner fence on the lower side of the gas mileage box plot is (approximately)

$$\text{Lower inner fence} = \text{Lower hinge} - 1.5(\text{IQR})$$
$$\approx 35.5 - 1.5(3.0)$$
$$= 35.5 - 4.5 = 31.0$$

The smallest measurement *inside* this fence is the second smallest measurement, 31.8. Thus, the lower whisker extends to 31.8. Similarly, the upper whisker extends to 42.1, the largest measurement inside the upper inner fence at about $38.5 + 4.5 = 43.0$.

Values that are beyond the inner fences are deemed *potential outliers* because they are extreme values that represent relatively rare occurrences. In fact, for mound-shaped distributions, fewer than 1% of the observations are expected to fall outside the inner fences. Two of the 100 gas mileage measurements, 30.0 and 44.9, fall beyond the inner fences, one on each end of the distribution. Each of these potential outliers is represented by the symbol * (an asterisk).

The other two imaginary fences, the outer fences, are defined at a distance 3(IQR) from each end of the box. Measurements that fall beyond the outer fences are represented by 0s (zeros) and are very extreme measurements that require special analysis. Since less than one-hundredth of 1% (.01% or .0001) of the measurements from mound-shaped distributions are expected to fall beyond the outer fences, these measurements are considered to be *outliers*.

Recall that outliers may be incorrectly recorded observations, members of a population different from the rest of the sample, or, at the least, very unusual measurements from the same population. The box plot of Figure 2.27 detected two potential outliers—the two gas mileage measurements beyond the inner fences. When we analyze these measurements, we find that they are correctly recorded. Perhaps they

TEACHING TIP

Use data collected in class to generate the values of Q_L and Q_U. Using these values, construct a box plot for the data. Pay particular attention to the extreme values in the data set. Discuss whether they are outliers or not. Calculate z-scores for these observations and discuss the results.

represent mileages that correspond to exceptional models of the car being tested or to unusual gas mixtures. Outlier analysis often reveals useful information of this kind and therefore plays an important role in the statistical inference-making process.

In addition to detecting outliers, box plots provide useful information on the variation in a data set. The elements (and nomenclature) of box plots are summarized in the next box. Some aids to the interpretation of box plots are also given.

Suggested Exercise 2.97

Elements of a Box Plot

1. A rectangle (the **box**) is drawn with the ends (the **hinges**) drawn at the lower and upper quartiles (Q_L and Q_U). The median of the data is shown in the box, usually by a line.
2. The points at distances 1.5(IQR) from each hinge mark the **inner fences** of the data set. Lines (the **whiskers**) are drawn from each hinge to the most extreme measurement inside the inner fence.
3. A second pair of fences, the **outer fences**, appear at a distance of 3 interquartile ranges, 3(IQR), from the hinges. One symbol (usually "*") is used to represent measurements falling between the inner and outer fences, and another (usually "0") is used to represent measurements beyond the outer fences. Thus, outer fences are not shown unless one or more measurements lie beyond them.
4. The symbols used to represent the median and the extreme data points (those beyond the fences) will vary depending on the software you use to construct the box plot. (You may use your own symbols if you are constructing a box plot by hand.) You should consult the program's documentation to determine exactly which symbols are used.

Aids to the Interpretation of Box Plots

1. Examine the length of the box. The IQR is a measure of the sample's variability and is especially useful for the comparison of two samples (see Example 2.16).
2. Visually compare the lengths of the whiskers. If one is clearly longer, the distribution of the data is probably skewed in the direction of the longer whisker.
3. Analyze any measurements that lie beyond the fences. Fewer than 5% should fall beyond the inner fences, even for very skewed distributions. Measurements beyond the outer fences are probably outliers, with one of the following explanations:

    a. The measurement is incorrect. It may have been observed, recorded, or entered into the computer incorrectly.

    b. The measurement belongs to a population different from the population that the rest of the sample was drawn from (see Example 2.16).

    c. The measurement is correct *and* from the same population as the rest. Generally, we accept this explanation only after carefully ruling out all others.

EXAMPLE 2.15

Use a statistical software package to draw a box plot for the student loan default data, Table 2.5. Identify any outliers in the data set.

Solution

The MINITAB box plot for the student loan default rates is shown in Figure 2.28. Note that the median appears to be about 9.5, and, with the exception of a single extreme observation, the distribution appears to be symmetrically distributed between approximately 3% and 17%. The single outlier is beyond the inner fence

FIGURE 2.28

MINITAB Box Plot for Student Default Rates

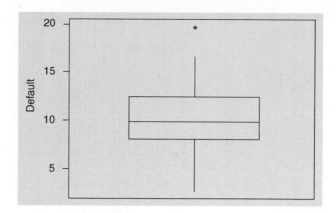

but inside the outer fence. Examination of the data reveals that this observation corresponds to Alaska's default rate of 19.7%.

EXAMPLE 2.16

A Ph.D. student in psychology conducted a stimulus reaction experiment as a part of her dissertation research. She subjected 50 subjects to a threatening stimulus and 50 to a nonthreatening stimulus. The reaction times of all 100 students, recorded to the nearest tenth of a second, are listed in Table 2.10. Box plots of the two resulting samples of reaction times, generated using SAS, are shown in Figure 2.29. Interpret the box plots.

REACTION

TABLE 2.10   Reaction Times of Students

Nonthreatening Stimulus									
2.0	1.8	2.3	2.1	2.0	2.2	2.1	2.2	2.1	2.1
2.0	2.0	1.8	1.9	2.2	2.0	2.2	2.4	2.1	2.0
2.2	2.1	2.2	1.9	1.7	2.0	2.0	2.3	2.1	1.9
2.0	2.2	1.6	2.1	2.3	2.0	2.0	2.0	2.2	2.6
2.0	2.0	1.9	1.9	2.2	2.3	1.8	1.7	1.7	1.8

Threatening Stimulus									
1.8	1.7	1.4	2.1	1.3	1.5	1.6	1.8	1.5	1.4
1.4	2.0	1.5	1.8	1.4	1.7	1.7	1.7	1.4	1.9
1.9	1.7	1.6	2.5	1.6	1.6	1.8	1.7	1.9	1.9
1.5	1.8	1.6	1.9	1.3	1.5	1.6	1.5	1.6	1.5
1.3	1.7	1.3	1.7	1.7	1.8	1.6	1.7	1.7	1.7

Solution

In SAS, the median is represented by the horizontal line through the box, while the asterisk (*) symbol represents the mean. Analysis of the box plots on the same numerical scale reveals that the distribution of times corresponding to the threatening stimulus lies below that of the nonthreatening stimulus. The implication is that the reaction times tend to be faster to the threatening stimulus. Note, too, that the upper whiskers of both samples are longer than the lower whiskers, indicating that the reaction times are positively skewed.

No observations in the two samples fall between the inner and outer fences. However, there is one outlier—the observation of 2.5 seconds corresponding to the threatening stimulus that is beyond the outer fence (denoted by the square symbol in SAS). When the researcher examined her notes from the experiments, she found that the

FIGURE 2.29

SAS Box Plots for Reaction Time Data

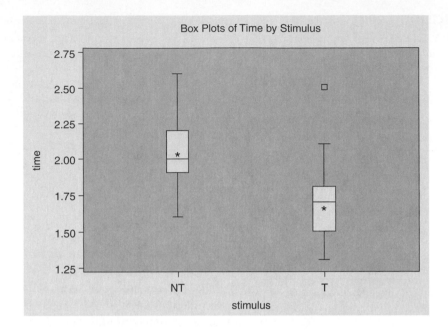

subject whose time was beyond the outer fence had mistakenly been given the nonthreatening stimulus. You can see in Figure 2.29 that his time would have been within the upper whisker if moved to the box plot corresponding to the nonthreatening stimulus. The box plots should be reconstructed since they will both change slightly when this misclassified reaction time is moved from one sample to the other.

The researcher concluded that the reactions to the threatening stimulus were faster than those to the nonthreatening stimulus. However, she was asked by her Ph.D. committee whether the results were *statistically significant.* Their question addresses the issue of whether the observed difference between the samples might be attributable to chance or sampling variation rather than to real differences between the populations. To answer their question, the researcher must use inferential statistics rather than graphical descriptions. We discuss how to compare two samples using inferential statistics in Chapter 7. ■

The following example illustrates how *z*-scores can be used to detect outliers and make inferences.

EXAMPLE 2.17

Suppose a female bank employee believes that her salary is low as a result of sex discrimination. To substantiate her belief, she collects information on the salaries of her male counterparts in the banking business. She finds that their salaries have a mean of $44,000 and a standard deviation of $2,000. Her salary is $37,000. Does this information support her claim of sex discrimination?

Solution

The analysis might proceed as follows: First, we calculate the *z*-score for the woman's salary with respect to those of her male counterparts. Thus,

$$z = \frac{\$37{,}000 - \$44{,}000}{\$2{,}000} = -3.5$$

The implication is that the woman's salary is 3.5 standard deviations *below* the mean of the male salary distribution. Furthermore, if a check of the male salary data shows that the frequency distribution is mound-shaped, we can infer that very few salaries in this distribution should have a *z*-score less than −3, as shown

FIGURE 2.30

Male Salary Distribution

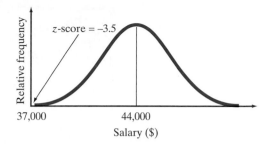

in Figure 2.30. Clearly, a *z*-score of −3.5 represents an outlier—a measurement from a distribution different from the male salary distribution or a very unusual (highly improbable) measurement for the male salary distribution.

Which of the two situations do you think prevails? Do you think the woman's salary is simply unusually low in the distribution of salaries, or do you think her claim of sex discrimination is justified? Most people would probably conclude that her salary does not come from the male salary distribution. However, the careful investigator should require more information before inferring sex discrimination as the cause. We would want to know more about the data collection technique the woman used and more about her competence at her job. Also, perhaps other factors such as length of employment should be considered in the analysis.   ■

Examples 2.16 and 2.17 exemplify an approach to statistical inference that might be called the **rare-event approach**. An experimenter hypothesizes a specific frequency distribution to describe a population of measurements. Then a sample of measurements is drawn from the population. If the experimenter finds it unlikely that the sample came from the hypothesized distribution, the hypothesis is concluded to be false. Thus, in Example 2.17 the woman believes her salary reflects discrimination. She hypothesizes that her salary should be just another measurement in the distribution of her male counterparts' salaries if no discrimination exists. However, it is so unlikely that the sample (in this case, her salary) came from the male frequency distribution that she rejects that hypothesis, concluding that the distribution from which her salary was drawn is different from the distribution for the men.

This rare-event approach to inference-making is discussed further in later chapters. Proper application of the approach requires a knowledge of probability, the subject of our next chapter.

We conclude this section with some rules of thumb for detecting outliers.

Rules of Thumb for Detecting Outliers*

Box plots: Observations falling between the inner and outer fences are deemed **suspect outliers**. Observations falling beyond the outer fence are deemed **highly suspect outliers**.

z-scores: Observations with *z*-scores greater than 3 in absolute value are considered outliers. (For some highly skewed data sets, observations with *z*-scores greater than 1 in absolute value may be outliers.)

---

*The *z*-score and box plot methods both establish rule-of-thumb limits outside of which a measurement is deemed to be an outlier. Usually, the two methods produce similar results. However, the presence of one or more outliers in a data set can inflate the computed value of *s*. Consequently, it will be less likely that an errant observation would have a *z*-score larger than 3 in absolute value. In contrast, the values of the quartiles used to calculate the intervals for a box plot are not affected by the presence of outliers.

Using the TI-83 Graphing Calculator

Making a Box Plot on the TI-83

Step 1   *Enter the data*
Press   **STAT 1**   for   **STAT Edit**
Enter the data set into **L1**.

Step 2   *Set up the box plot*
Press   **2nd Y =**   for   **STAT PLOT**
Press   **1**   for   **Plot 1**
Use the arrow and **ENTER** keys to set up the screen as shown below.

Step 3   *Select your window settings*
Press **WINDOW** and adjust the settings as follows:

Xmin = smallest data value (or smaller)
Xmax = largest data value (or larger)
Xscl = approximately (xmax − xmin)/10
Ymin = 0
Ymax = 10
Yscl = 1

Step 4   *View the graph*
Press   **GRAPH**

Optional   *Read the five number summary*
Step   Press   **TRACE**
Use the left and right arrow keys to move between minX, Q1, Med, Q3, and maxX.

Example: Make a box plot for the given data.

86, 70, 62, 98, 73, 56, 53, 92, 86, 37, 62, 83, 78, 49, 78, 37, 67, 79, 57

The window settings and horizontal box plot are shown below.

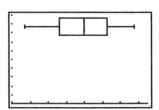

EXERCISES 2.94–2.107

Learning the Mechanics

2.94 A sample data set has a mean of 57 and a standard deviation of 11. Determine whether each of the following sample measurements are outliers.
  a. 65   No
  b. 21   Yes
  c. 72   No
  d. 98   Yes

2.95 Define the 25th, 50th, and 75th percentiles of a data set. Explain how they provide a description of the data.

2.96 Suppose a data set consisting of exam scores has a lower quartile $Q_L = 60$, a median $M = 75$, and an upper quartile $Q_U = 85$. The scores on the exam range from 18 to 100. Without having the actual scores available to you, construct as much of the box plot as possible.

2.97 Consider the box plot at right:
  a. What is the median of the data set (approximately)?
  b. What are the upper and lower quartiles of the data set (approximately)?   5.5, 3
  c. What is the interquartile range of the data set (approximately)?   2.5
  d. Is the data set skewed to the left, skewed to the right, or symmetric?   Skewed right
  e. What percentage of the measurements in the data set lie to the right of the median? To the left of the upper quartile?   50%, 75%
  f. Identify any outliers in the data.

2.98 Consider the following two sample data sets:

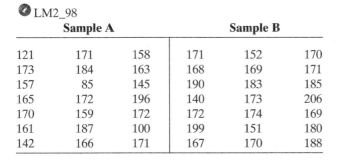

● LM2_98

	Sample A			Sample B	
121	171	158	171	152	170
173	184	163	168	169	171
157	85	145	190	183	185
165	172	196	140	173	206
170	159	172	172	174	169
161	187	100	199	151	180
142	166	171	167	170	188

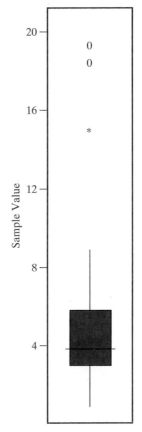

  a. Use a statistical software package to construct a box plot for each data set.
  b. Using information reflected in your box plots, describe the similarities and differences in the two data sets.
  c. Identify any outliers that may exist in the two data sets.

Applying the Concepts—Basic

2.99 Refer to the *Brain and Language* data on postmortem intervals (PMIs) of 22 human brain specimens, Exercise 2.25 (p. 39). An SAS descriptive statistics printout for the PMI variable is shown below.

SAS Output for Exercise 2.99

	The MEANS Procedure		
	Analysis Variable : PMI		
N	Mean	Median	Variance	Std Dev
22	7.3000000	6.1500000	10.1438095	3.1849348

a. Find the *z*-score for the PMI value of 3.3.   −1.26

b. Is the PMI value of 3.3 considered an outlier? Explain.   No

2.100 The table contains the top salary offer (in thousands of dollars) received by each member of a sample of 50 MBA students who recently graduated from the Graduate School of Management at Rutgers University.

⊘ MBASAL

61.1	48.5	47.0	49.1	43.5
50.8	62.3	50.0	65.4	58.0
53.2	39.9	49.1	75.0	51.2
41.7	40.0	53.0	39.6	49.6
55.2	54.9	62.5	35.0	50.3
41.5	56.0	55.5	70.0	59.2
39.2	47.0	58.2	59.0	60.8
72.3	55.0	41.4	51.5	63.0
48.4	61.7	45.3	63.2	41.5
47.0	43.2	44.6	47.7	58.6

Source: Career Services Office, Graduate School of Management, Rutgers University.

a. The mean and standard deviation are 52.33 and 9.22, respectively. Find and interpret the *z*-score associated with the highest salary offer, the lowest salary offer, and the mean salary offer.

b. Would you consider the highest offer to be unusually high? Why or why not?

2.101 Refer to Exercise 2.30 (p. 40) in which we compared states' average SAT scores in 1990 and 2000. SPSS was used to generate the following box plots for the states' SAT scores in 1990 and 2000.

SPSS Output for Exercise 2.101

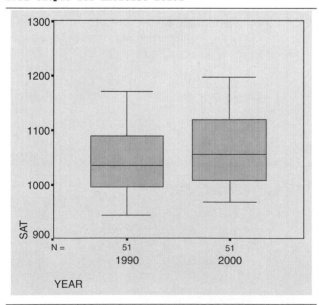

a. Compare the central tendency of the SAT scores for the 2 years. [*Note:* The median is represented by the horizontal line within each SPSS box plot.]

b. Compare the variability of the SAT scores for the 2 years.

c. Are any states' SAT scores outliers in either year? If so, identify them.   No

Applying the Concepts—Intermediate

2.102 Refer to the *American Journal of Speech–Language Pathology* (Feb. 1995) study, Exercise 2.31 (p. 41). Recall that three groups of college students listened to an audio tape of a woman with imperfect speech. The groups were called *control, treatment,* and *familiarity.* At the end of the session, each student transcribed the spoken words. Box plots, constructed using STATISTIX, for the percentage of words correctly transcribed (i.e., accuracy rate) by each group are shown here.

STATISTIX Output for Exercise 2.102

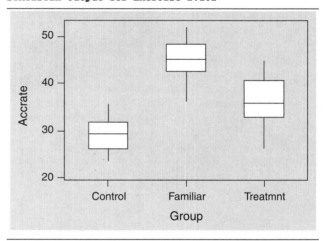

a. How do the median accuracy rates compare for the three groups?

b. How do the variabilities of the accuracy rates compare for the three groups?

c. The standard deviations of the accuracy rates are 3.56 for the control group, 5.45 for the treatment group, and 4.46 for the familiarity group. Do the standard deviations agree with the interquartile ranges (part **b**) with regard to the comparison of the variabilities of the accuracy rates?   Yes

d. Is there evidence of outliers in any of the three distributions?   No

2.103 A manufacturer of minicomputers is investigating the down time of its systems. The 40 most recent customers were surveyed to determine the amount of down time (in hours) they had experienced during the previous month. These data are listed in the table on p. 85.

STATISTIX Output for Exercise 2.105

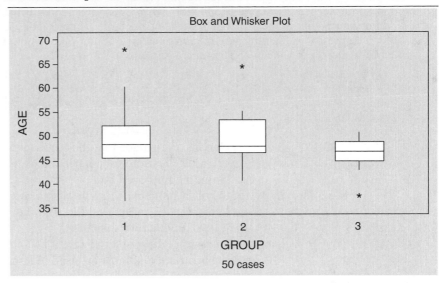

⊘ DOWNTIME

Customer Number	Down Time	Customer Number	Down Time	Customer Number	Down Time
230	12	244	2	258	28
231	16	245	11	259	19
232	5	246	22	260	34
233	16	247	17	261	26
234	21	248	31	262	17
235	29	249	10	263	11
236	38	250	4	264	64
237	14	251	10	265	19
238	47	252	15	266	18
239	0	253	7	267	24
240	24	254	20	268	49
241	15	255	9	269	50
242	13	256	22		
243	8	257	18		

a. Use a statistical software package to construct a box plot for these data. Use the information reflected in the box plot to describe the frequency distribution of the data set. Your description should address central tendency, variation, and skewness.

b. Use your box plot to determine which customers are having unusually lengthy down times.

c. Find and interpret the z-scores associated with customers you identified in part **b**.

2.104 Refer to the data on sanitation levels of cruise ships, Exercises 2.72 (p. 66) and 2.89 (p. 74).

a. Use the box plot method to detect any outliers in the data.

b. Use the z-score method to detect any outliers in the data.

c. Do the two methods agree? If not, explain why.

2.105 Refer to the *Fortune* (Oct. 25, 1999) ranking of the 50 most powerful women in America, Exercise 2.44 (p. 51). Use the STATISTIX box plots above to compare the ages of the women in three groups based on their position within the firm: Group 1 (CEO, CFO, CIO, or COO); Group 2 (Chairman, President, or Director); and Group 3 (Vice President, Vice Chairman, or General Manager).

Applying the Concepts—Advanced

2.106 A city librarian claims that books have been checked out an average of seven (or more) times in the last year. You suspect he has exaggerated the checkout rate (book usage) and that the mean number of checkouts per book per year is, in fact, less than seven. Using the computerized card catalog, you randomly select one book and find that it has been checked out four times in the last year. Assume that the standard deviation of the number of checkouts per book per year is approximately 1.

a. If the mean number of checkouts per book per year really is seven, what is the z-score corresponding to four? $z = -3$

b. Considering your answer to part **a**, do you have reason to believe that the librarian's claim is incorrect?

c. If you knew that the distribution of the number of checkouts were mound-shaped, would your answer to part **b** change? Explain. No

d. If the standard deviation of the number of checkouts per book per year were 2 (instead of 1), would your answers to parts **b** and **c** change? Explain.

2.107 A chemical company produces a substance composed of 98% cracked corn particles and 2% zinc phosphide for use in controlling rat populations in sugarcane fields. Production must be carefully controlled to maintain the 2% zinc phosphide because too much zinc phosphide will cause damage to the sugarcane and too little will be ineffective in controlling the rat population. Records from past production indicate that the distribution of the actual percentage of zinc phosphide present in the substance is approximately mound-shaped, with a mean of 2.0% and a standard deviation of .08%. Suppose one batch chosen randomly actually contains 1.80% zinc phosphide. Does this indicate that there is too little zinc phosphide in today's production? Explain your reasoning.

2.9 GRAPHING BIVARIATE RELATIONSHIPS (OPTIONAL)

The claim is often made that the crime rate and the unemployment rate are "highly correlated." Another popular belief is that smoking and lung cancer are "related." Some people even believe that the Dow Jones Industrial Average and the lengths of fashionable skirts are "associated." The words "correlated," "related," and "associated" imply a relationship between two variables—in the examples above, two *quantitative* variables.

One way to describe the relationship between two quantitative variables—called a **bivariate relationship**—is to plot the data in a **scattergram** (or **scatterplot**). A scattergram is a two-dimensional plot, with one variable's values plotted along the vertical axis and the other along the horizontal axis. For example, Figure 2.31 is a scattergram relating (1) the cost of mechanical work (heating, ventilating, and plumbing) to (2) the floor area of the building for a sample of 26 factory and warehouse buildings. Note that the scattergram suggests a general tendency for mechanical cost to increase as building floor area increases.

When an increase in one variable is generally associated with an increase in the second variable, we say that the two variables are "positively related" or "positively correlated."* Figure 2.31 implies that mechanical cost and floor area are positively correlated. Alternatively, if one variable has a tendency to decrease as the other increases, we say the variables are "negatively correlated." Figure 2.32 shows several hypothetical scattergrams that portray a positive bivariate relationship (Figure 2.32a), a negative bivariate relationship (Figure 2.32b), and a situation where the two variables are unrelated (Figure 2.32c).

FIGURE 2.31

Scattergram of Cost vs. Floor Area

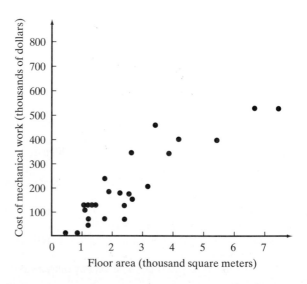

---

*A formal definition of correlation is given in Chapter 9.

FIGURE 2.32

*Hypothetical Bivariate
Relationship*

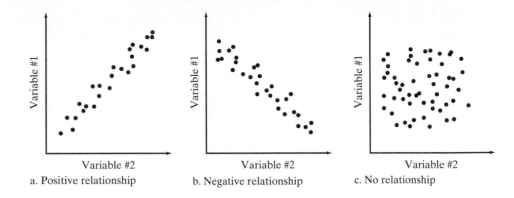

a. Positive relationship   b. Negative relationship   c. No relationship

EXAMPLE 2.18

A medical item used to administer to a hospital patient is called a *factor*. For example, factors can be intravenous (IV) tubing, IV fluid, needles, shave kits, bedpans, diapers, dressings, medications, and even code carts. The coronary care unit at Bayonet Point Hospital (St. Petersburg, Florida) recently investigated the relationship between the number of factors administered per patient and the patient's length of stay (in days). Data on these two variables for a sample of 50 coronary care patients are given in Table 2.11. Use a scattergram to describe the relationship between the two variables of interest, number of factors and length of stay.

⊘ MEDFACTORS

TABLE 2.11   Data on Patient's Factors and Length of Stay

Number of Factors	Length of Stay (days)	Number of Factors	Length of Stay (days)
231	9	354	11
323	7	142	7
113	8	286	9
208	5	341	10
162	4	201	5
117	4	158	11
159	6	243	6
169	9	156	6
55	6	184	7
77	3	115	4
103	4	202	6
147	6	206	5
230	6	360	6
78	3	84	3
525	9	331	9
121	7	302	7
248	5	60	2
233	8	110	2
260	4	131	5
224	7	364	4
472	12	180	7
220	8	134	6
383	6	401	15
301	9	155	4
262	7	338	8

Source: Bayonet Point Hospital, Coronary Care Unit.

FIGURE 2.33

MINITAB Scatterplot of Data in Table 2.11

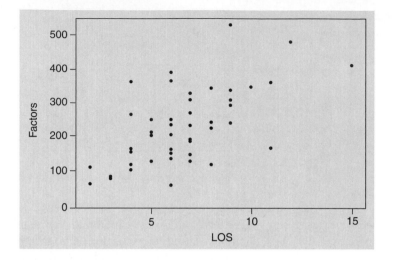

Solution

Suggested Exercise 2.11

Rather than construct the plot by hand, we resort to a statistical software package. The MINITAB plot of the data in Table 2.11, with length of stay (LOS) on the horizontal axis and number of factors (FACTORS) on the vertical axis, is shown in Figure 2.33.

Although the plotted points exhibit a fair amount of variation, the scattergram clearly shows an increasing trend. It appears that a patient's length of stay is positively correlated with the number of factors administered to the patient. Hospital administrators may use this information to improve their forecasts of lengths of stay for future patients. ▬▬▬▬▬ ■

The scattergram is a simple but powerful tool for describing a bivariate relationship. However, keep in mind that it is only a graph. No measure of reliability can be attached to inferences made about bivariate populations based on scattergrams of sample data. The statistical tools that enable us to make inferences about bivariate relationships are presented in Chapter 9.

 ## Using the TI-83 Graphing Calculator

Making Scatterplots

To make a scatterplot for paired data,

Step 1   *Enter the data*
             Press   **STAT 1**   for   **STAT Edit**
             Enter your x-data in **L1** and your y-data in **L2**.

Step 2   *Set up the scatterplot*
             Press   **2nd Y =**   for   **STAT PLOT**
             Press   **1**   for   **Plot1**

USING THE TI-83 GRAPHING CALCULATOR *(continued)*

Use the arrow and **ENTER** keys to set up the screen as shown below.

Step 3   *View the scatterplot*
Press   **ZOOM 9**   for   **ZoomStat**

Example   The figures below show a table of data entered on the TI-83 and the scatterplot of the data obtained using the steps given above.

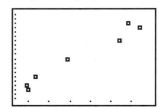

  EXERCISES 2.108–2.115

Learning the Mechanics

2.108 Construct a scatterplot for the data in the following table. Do you detect a trend?

Variable #1:   .5   1   1.5   2   2.5   3   3.5   4   4.5   5
Variable #2:   2   1   3   4   6   10   9   12   17   17

2.109 Construct a scatterplot for the data in the following table. Do you detect a trend?

Variable #1:   5   3   −1   2   7   6   4   0   8
Variable #2:   14   3   10   1   8   5   3   2   12

Applying the Concepts—Basic

2.110 Zoologists at the University of Western Australia conducted a study of the feeding behavior of blackbream fish spawned in aquariums (*Brain and Behavior Evolution*, April 2000). In one experiment, the zoologists recorded the number of aggressive strikes of two blackbream fish feeding at the bottom of the aquarium in the 10-minute period following the addition of food. The number of strikes and age of the fish (in days) was recorded approximately each week for nine weeks, as shown in the table.

⊘ BLACKBREAM

Week	Number of Strikes	Age of Fish (days)
1	85	120
2	63	136
3	34	150
4	39	155
5	58	162
6	35	169
7	57	178
8	12	184
9	15	190

Source: J. Shand, *et al.* "Variability in the location of the retinal ganglion cell area centralis is correlated with ontogenetic changes in feeding behavior in the Blackbream, Acanthopagrus 'butcher'," *Brain and Behavior Evolution*, Vol. 55, No. 4, April 2000 (Figure H).

a. Construct a scattergram for the data, with number of strikes on the *y*-axis and age of the fish on the *x*-axis.
b. Examine the scattergram of part **a**. Do you detect a trend?

2.111 Baseball wisdom says if you can't hit, you can't win. But is the number of games won by a major league baseball team in a season related to the team's batting average?

The accompanying table shows the number of games won and the batting averages for the 14 teams in the American League for the 2000 season. Construct a two-dimensional plot for the data. Do you observe a trend?

⊘ ALWINS

Team	Games Won	Team Batting Average
Cleveland	90	.288
New York	87	.277
Boston	85	.267
Toronto	83	.275
Texas	71	.283
Detroit	79	.275
Minnesota	69	.270
Baltimore	74	.272
Anaheim	82	.280
Tampa Bay	69	.257
Seattle	91	.269
Kansas City	77	.288
Oakland	91	.270
Chicago	95	.286

Source: Compiled by the Major League Baseball (MLB) Baseball Information System, 2001.

2.112 *The Condor* data on number of flycatchers killed and number of flycatchers breeding at 14 nest sites, Exercise 2.46 (p. 52), are reproduced in the table. Graphically examine the relationship between the two variables. Do you detect a trend?

⊘ CONDOR

Plot Number	Number Killed	Number of Breeders
1	5	30
2	4	28
3	3	38
4	2	34
5	2	26
6	1	124
7	1	68
8	1	86
9	1	32
10	0	30
11	0	46
12	0	132
13	0	100
14	0	6

Source: Merilä, J., and Wiggins, D. A. "Interspecific competition for nest holes causes adult mortality in the collared flycatcher." *The Condor*, Vol. 97, No. 2, May 1995, p. 447 (Table 4) Cooper Ornithological Society.

2.113 A "self-avoiding walk" describes a path in which you never retrace your steps or cross your own path. An "unrooted walk" is a path in which it is impossible to distinguish between the starting point and ending point of the path. The *American Scientist* (July–Aug. 1998) investigated the relationship between self-avoiding and unrooted walks. The table gives the number of unrooted walks and possible number of self-avoiding walks of various lengths, where length is measured as number of steps.

⊘ WALKS

Walk Length (Number of Steps)	Unrooted Walks	Self-Avoiding Walks
1	1	4
2	2	12
3	4	36
4	9	100
5	22	284
6	56	780
7	147	2,172
8	388	5,916

Source: Hayes, B. "How to avoid yourself." *American Scientist*, Vol. 86, No. 4, July–Aug.1998, p. 317 (Figure 5).

a. Construct a plot to investigate the relationship between total possible number of self-avoiding walks and walk length. What pattern (if any) do you observe?

b. Repeat part **a** for unrooted walks.

Applying the Concepts—Intermediate

2.114 Research published in the *American Journal of Psychiatry* (July 1995) attempted to establish a link between hippocampal (brain) volume and short-term verbal memory of patients with combat-related post-traumatic stress disorder (PTSD). A sample of 21 Vietnam veterans with a history of combat-related PTSD participated in the study. Magnetic resonance imaging was used to measure the volume of the right hippocampus (in cubic millimeters) of each subject. Verbal memory retention of each subject was measured by the percent retention subscale of the Wechsler Memory Scale. The data for the 21 patients is plotted in the scattergram. The researchers "hypothesized that smaller hippocampal volume would be associated with deficits in short-term verbal memory in patients with PTSD." Does the scattergram provide visual evidence to support this theory?

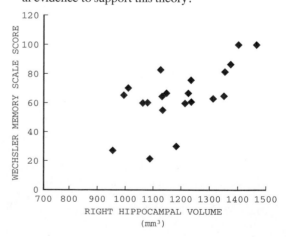

2.115 The *Journal of Communication Disorders* data on deviation intelligence quotient (DIQ) and percent pronoun errors for 30 children, Exercise 2.51 (p. 53), are reproduced below.

a. Plot all the data to investigate a possible trend between DIQ and proper use of pronouns. What do you observe?

b. Plot the data for the ten SLI children only. Is there a trend between DIQ and proper use of pronouns?

⊘ SLI

Subject	Gender	Group	DIQ	Pronoun Errors (%)
1	F	YND	110	94.40
2	F	YND	92	19.05
3	F	YND	92	62.50
4	M	YND	100	18.75
5	F	YND	86	0
6	F	YND	105	55.00
7	F	YND	90	100.00
8	M	YND	96	86.67
9	M	YND	90	32.43
10	F	YND	92	0
11	F	SLI	86	60.00
12	M	SLI	86	40.00
13	M	SLI	94	31.58
14	M	SLI	98	66.67
15	F	SLI	89	42.86
16	F	SLI	84	27.27
17	M	SLI	110	33.33
18	F	SLI	107	0
19	F	SLI	87	0
20	M	SLI	95	0
21	M	OND	110	0
22	M	OND	113	0
23	M	OND	113	0
24	F	OND	109	0
25	M	OND	92	0
26	F	OND	108	0
27	M	OND	95	0
28	F	OND	87	0
29	F	OND	94	0
30	F	OND	98	0

Source: Moore, M. E. "Error analysis of pronouns by normal and language-impaired children." *Journal of Communication Disorders*, Vol. 28, No. 1, Mar. 1995, p. 62 (Table 2), p. 67 (Table 5).

2.10 DISTORTING THE TRUTH WITH DESCRIPTIVE TECHNIQUES

A picture may be "worth a thousand words," but pictures can also color messages or distort them. In fact, the pictures in statistics—histograms, bar charts, and other graphical descriptions—are susceptible to distortion, so we have to examine each of them with care. We'll mention a few of the pitfalls to watch for when interpreting a chart or graph.

One common way to change the impression conveyed by a graph is to change the scale on the vertical axis, the horizontal axis, or both. For example, Figure 2.34 is a bar graph that shows the market share of sales for a company for each of the years 1997 to 2002. If you want to show that the change in firm A's market share over time is moderate, you should pack in a large number of units per inch on the vertical axis. That is, make the distance between successive units on the vertical scale small, as shown in Figure 2.34. You can see that a change in the firm's market share over time is barely apparent.

TEACHING TIP
Use this section to emphasize the importance of looking past the picture to the information it is trying to convey. If a student can successfully interpret the graph, she will be able to see through the deception.

FIGURE 2.34

Firm A's Market Share from 1997 to 2002: Packed Vertical Axis

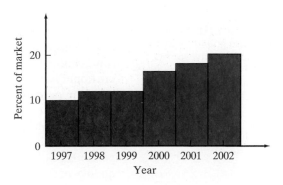

FIGURE 2.35

Firm A's Market Share from 1997 to 2002: Stretched Vertical Axis

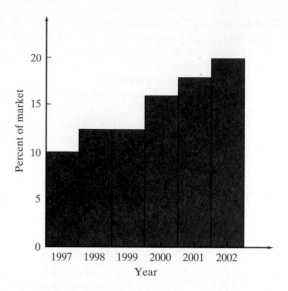

If you want to use the same data to make the changes in firm A's market share appear large, you should increase the distance between successive units on the vertical axis. That is, stretch the vertical axis by graphing only a few units per inch as in Figure 2.35. A telltale sign of stretching is a long vertical axis, but this is often hidden by starting the vertical axis at some point above 0. The same effect can be achieved by using a broken line—called a *scale break*—for the vertical axis, as shown in Figure 2.36.

FIGURE 2.36

Firm A's Market Share from 1997 to 2002: Scale Break

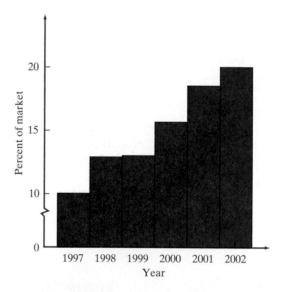

The changes in categories indicated by a bar graph can also be emphasized or deemphasized by stretching or shrinking the vertical axis. Another method of achieving visual distortion with bar graphs is by making the width of the bars proportional to height. For example, look at the bar chart in Figure 2.37a, which depicts the percentage of the total number of motor vehicle deaths in a year that occurred on each of four major highways. Now suppose we make both the width and the height grow as the percentage of fatal accidents grows. This change is shown in Figure 2.37b. The reader may tend to equate the *area* of the bars with the percentage of deaths occurring at each highway. But in fact, the true relative frequency of fatal accidents is proportional only to the *height* of the bars.

FIGURE 2.37

*Relative Frequency of Fatal
Motor Vehicle Accidents on
Each of Four Major Highways*

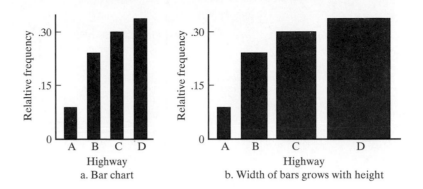

a. Bar chart

b. Width of bars grows with height

Although we've discussed only a few of the ways that graphs can be used to convey misleading pictures of phenomena, the lesson is clear. Look at all graphical descriptions of data with a critical eye. Particularly, check the axes and the size of the units on each axis. Ignore the visual changes and concentrate on the actual numerical changes indicated by the graph or chart.

The information in a data set can also be distorted by using numerical descriptive measures, as Example 2.19 indicates.

EXAMPLE 2.19

Suppose you're considering working for a small law firm—one that currently has a senior member and three junior members. You inquire about the salary you could expect to earn if you join the firm. Unfortunately, you receive two answers:

Answer A: The senior member tells you that an "average employee" earns $57,500

Answer B: One of the junior members later tells you that an "average employee" earns $45,000

Which answer can you believe?

Solution

The confusion exists because the phrase "average employee" has not been clearly defined. Suppose the four salaries paid are $45,000 for each of the three junior members and $95,000 for the senior member. Thus,

$$\bar{x} = \frac{3(\$45,000) + \$95,000}{4} = \frac{\$230,000}{4} = \$57,500$$

Median = $45,000

You can now see how the two answers were obtained. The senior member reported the mean of the four salaries, and the junior member reported the median. The information you received was distorted because neither person stated which measure of central tendency was being used. ∎

Another distortion of information in a sample occurs when *only* a measure of central tendency is reported. Both a measure of central tendency and a measure of variability are needed to obtain an accurate mental image of a data set.

Suppose you want to buy a new car and are trying to decide which of two models to purchase. Since energy and economy are both important issues, you decide to purchase model A because its EPA mileage rating is 32 miles per gallon in the city, whereas the mileage rating for model B is only 30 miles per gallon in the city.

FIGURE 2.38

Mileage Distributions for Two Car Models

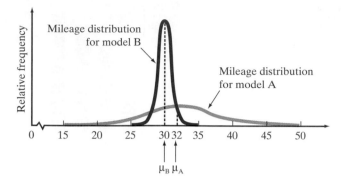

However, you may have acted too quickly. How much variability is associated with the ratings? As an extreme example, suppose that further investigation reveals that the standard deviation for model A mileages is 5 miles per gallon, whereas that for model B is only 1 mile per gallon. If the mileages form a mound-shaped distribution, they might appear as shown in Figure 2.38. Note that the larger amount of variability associated with model A implies that more risk is involved in purchasing model A. That is, the particular car you purchase is more likely to have a mileage rating that will greatly differ from the EPA rating of 32 miles per gallon if you purchase model A, while a model B car is not likely to vary from the 30 miles per gallon rating by more than 2 miles per gallon.

We conclude this section with another example on distorting the truth with numerical descriptive measures.

EXAMPLE 2.20

Children Out of School in America is a report on delinquency of school-age children by the Children's Defense Fund (CDF). Consider the following three reported results of the CDF survey.

• Reported result: 25% of the 16- and 17-year-olds in the Portland, Maine, Bayside East Housing Project were out of school. Actual data: *Only eight children were surveyed; two were found to be out of school.*

• Reported result: Of all the secondary school students who had been suspended more than once in census tract 22 in Columbia, South Carolina, 33% had been suspended two times and 67% had been suspended three or more times. Actual data: *CDF found only three children in that entire census tract who had been suspended; one child was suspended twice and the other two children, three or more times.*

• Reported result: In the Portland Bayside East Housing Project, 50% of all the secondary school children who had been suspended more than once had been suspended three or more times. Actual data: *The survey found two secondary school children had been suspended in that area; one of them had been suspended three or more times.*

• Identify the potential distortions in the results reported by the CDF.

Solution

In each of these examples the reporting of percentages (i.e., relative frequencies) instead of the numbers themselves is misleading. No inference we might draw from the cited examples would be reliable. (We'll see how to measure the reliability of estimated percentages in Chapter 5). In short, either the report should state the numbers alone instead of percentages, or, better yet, it should state that the numbers were too small to report by region. If several regions were combined, the numbers (and percentages) would be more meaningful.

STATISTICS IN ACTION:
The "Eye Cue" Test: Does Experience Improve Performance?

In 1948, famous child psychologist Jean Piaget devised a test of basic perceptual and conceptual skills dubbed the "water-level task." Subjects were shown a drawing of a glass being held perfectly still (at a 45° angle) by an invisible hand so that any water in it had to be at rest (see Figure 2.39).

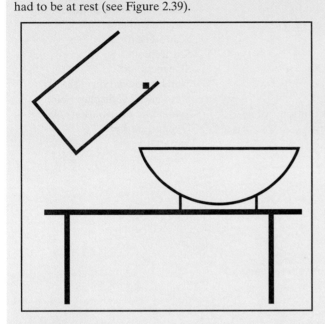

FIGURE 2.39

Drawing of the Water-Level Task

The task for the subject was to draw a line representing the surface of the water—a line that would touch the black dot pictured on the right side of the glass. Piaget found that young children typically failed the test. Fifty years later, research psychologists still use the water-level task to test the perception of both adults and children. Surprisingly, about 40% of the adult population also fail. In addition, males tend to do better than females, and younger adults tend to do better than older adults.

Will people with experience handling liquid-filled containers perform better on this "eye cue" test? This question was the focus of research conducted by psychologists Heiko Hecht (NASA) and Dennis R. Proffitt (University of Virginia) and published in *Psychological Science* (Mar. 1995). The researchers presented the task to each of six different groups: (1) male college students, (2) female college students, (3) professional waitresses, (4) housewives, (5) male bartenders, and (6) male bus drivers. A total of 120 subjects (20 per group) participated in the study. Two of the groups, waitresses and bartenders, were assumed to have considerable experience handling liquid-filled glasses.

After each subject completed the drawing task, the researchers recorded the deviation* (in angle degrees) of the

judged line from the true line. If the deviation was within 5° of the true water surface angle, the answer was considered correct. Deviations of more than 5° in either direction were considered incorrect answers.

The data for the water-level task (simulated based on summary results presented in the journal article) are provided in the file EYECUE on the data disk that accompanies this text. For each of the 120 subjects in the experiment, the following variables (in the order they appear on the data file) were measured:

⊘ EYECUE

Variables

GENDER (F or M)
GROUP (Student, Waitres, Wife, Bartend, or Busdriv)
DEVIATION (angle, in degrees, of the judged line from the true line)
JUDGE (Within5, More5Above, More5Below)

Focus

a. Use a graphical method to describe the qualitative responses for all 120 subjects.

b. Construct graphs that researchers could use to describe the effect of gender on the qualitative measure of task performance. Do the results tend to support or refute the prevailing theory that males do better than females?

c. Construct graphs that the researchers could use to describe the effect of age on the qualitative measure of task performance. [*Note*: The college students in the study were much younger than the subjects in the other four groups.] Do the results tend to support or refute the prevailing theory that younger adults do better than older adults?

d. Construct graphs that the researchers could use to describe the effect of experience in handling liquid-filled containers on the qualitative measure of task performance. Comment on the theory that experience improves task performance.

e. Refer to parts **a–d**. What are the drawbacks to using graphical displays of sample data to infer the nature of the population?

f. Repeat parts **a–e**, but use the quantitative measure of task performance, i.e., the actual deviations in water-line angles (in degrees) for each subject.

g. Discuss which graphs—those of parts **a–d** or those of part **f**—are more informative.

h. For each group of subjects, find and interpret numerical descriptive measures of the data for the quantitative measure of task performance. What do you conclude from this analysis?

---

*The true surface line is perfectly parallel to the table top.

>>>>QUICK REVIEW

Key Terms
Note: Starred (*) items are from the optional sections to this chapter

Bar graph   22
Bivariate relationship*   86
Box plots*   76
Central tendency   44
Chebyshev's Rule   61
Class   20
Class frequency   20
Class relative frequency   20
Dot plot   30
Empirical Rule   62
Hinges*   76
Histogram   31
Inner fences*   77
Interquartile range*   76
Lower quartile*   76
Mean   44
Measurement classes   32

Measures of central
   tendency   44
Measures of relative
   standing   70
Measures of variation or
   spread   55
Median   46
Middle quartile*   76
Modal class   48
Mode   48
Mound-shaped distribution   61
Numerical descriptive
   measures   44
Outer fences*   77
Outlier*   75
Percentile   71
Pie chart   23

Quartiles*   76
Range   56
Rare-event approach*   81
Sample standard deviation   57
Sample variance   57
Scattergram*   86
Scatterplot*   86
Skewness   48
Standard deviation   57
Stem-and-leaf display   30
Symmetric distribution   48
Upper quartile*   76
Variability   44
Whiskers*   77
z-score   72

Key Formulas

$\dfrac{\text{Class frequency}}{n}$	Class relative frequency 20
$\bar{x} = \dfrac{\sum\limits_{i=1}^{n} x_i}{n}$	Sample mean 44
$s^2 = \dfrac{\sum\limits_{i=1}^{n} (x_i - \bar{x})^2}{n-1} = \dfrac{\sum\limits_{i=1}^{n} x_i^2 - \dfrac{\left(\sum\limits_{i=1}^{n} x_i\right)^2}{n}}{n-1}$	Sample variance 57
$s = \sqrt{s^2}$	Sample standard deviation 57
$z = \dfrac{x - \bar{x}}{s}$	Sample z-score 72
$z = \dfrac{x - \mu}{\sigma}$	Population z-score 72
$\text{IQR} = Q_U - Q_L$*	Interquartile range 76

Language Lab

Symbol	Pronunciation	Description
$\sum$	sum of	Summation notation; $\sum\limits_{i=1}^{n} x_i$ represents the sum of the measurements $x_1, x_2, \ldots, x_n$
μ	mu	Population mean
$\bar{x}$	x-bar	Sample mean
σ^2	sigma-squared	Population variance
σ	sigma	Population standard deviation
s^2		Sample variance
s		Sample standard deviation

Symbol	Pronunciation	Description
z		z-score for a measurement
M		Median (middle quartile) of a sample data set
Q_L		Lower quartile (25th percentile)
Q_U		Upper quartile (75th percentile)
IQR		Interquartile range

 SUPPLEMENTARY EXERCISES 2.116–2.140

Note: Starred (*) exercises refer to the optional sections in this chapter.

Learning the Mechanics

2.116 Construct a relative frequency histogram for the data summarized in the table below.

Measurement Class	Relative Frequency	Measurement Class	Relative Frequency
.00–.75	.02	5.25–6.00	.15
.75–1.50	.01	6.00–6.75	.12
1.50–2.25	.03	6.75–7.50	.09
2.25–3.00	.05	7.50–8.25	.05
3.00–3.75	.10	8.25–9.00	.04
3.75–4.50	.14	9.00–9.75	.01
4.50–5.25	.19		

2.117 Consider the following three measurements: 50, 70, 80. Find the z-score for each measurement if they are from a population with a mean and standard deviation equal to
  a. $\mu = 60, \sigma = 10$ $-1, 1, 2$
  b. $\mu = 60, \sigma = 5$ $-2, 2, 4$
  c. $\mu = 40, \sigma = 10$ $1, 3, 4$
  d. $\mu = 40, \sigma = 100$ $.1, .3, .4$

*2.118 Refer to Exercise 2.117. For parts **a–d**, determine whether the three measurements 50, 70, and 80 are outliers.

2.119 Compute s^2 for data sets with the following characteristics:
  a. $\sum_{i=1}^{n} x_i^2 = 246, \sum_{i=1}^{n} x_i = 63, n = 22$ 3.1234
  b. $\sum_{i=1}^{n} x_i^2 = 666, \sum_{i=1}^{n} x_i = 106, n = 25$ 9.0233
  c. $\sum_{i=1}^{n} x_i^2 = 76, \sum_{i=1}^{n} x_i = 11, n = 7$ 9.7857

2.120 If the range of a set of data is 20, find a rough approximation to the standard deviation of the data set.

2.121 For each of the following data sets, compute $\bar{x}, s^2$, and s. If appropriate, specify the units in which your answers are expressed.
  a. 4, 6, 6, 5, 6, 7   5.67, 1.0667, 1.03
  b. −$1, $4, −$3, $0, −$3, −$6   −$1.5, 11.5, $3.39
  c. 3/5%, 4/5%, 2/5%, 1/5%, 1/16%
  d. Calculate the range of each data set in parts **a–c**.

2.122 For each of the following data sets, compute $\bar{x}, s^2$, and s:
  a. 13, 1, 10, 3, 3   6, 27, 5.196
  b. 13, 6, 6, 0   6.25, 28.25, 5.135
  c. 1, 0, 1, 10, 11, 11, 15   7, 37.667, 6.137
  d. 3, 3, 3, 3   3, 0, 0
  e. For each data set in parts **a–d**, form the interval $\bar{x} \pm 2s$ and calculate the percentage of the measurements that fall in the interval.

*2.123 Construct a scatterplot for the data listed here. Do you detect any trends?

Variable #1:   174   268   345   119   400   520   190   448   307   252
Variable #2:   8   10   15   7   22   31   15   20   11   9

2.124 The following data sets have been invented to demonstrate that the lower bounds given by Chebyshev's Rule are appropriate. Notice that the data are contrived and would not be encountered in a real-life problem.
  a. Consider a data set that contains ten 0s, two 1s, and ten 2s. Calculate $\bar{x}, s^2$, and s. What percentage of the measurements are in the interval $\bar{x} \pm s$? Compare this result to Chebyshev's Rule.   1, .9524, .976
  b. Consider a data set that contains five 0s, thirty-two 1s, and five 2s. Calculate $\bar{x}, s^2$, and s. What percentage of the measurements are in the interval $\bar{x} \pm 2s$? Compare this result to Chebyshev's Rule.
  c. Consider a data set that contains three 0s, fifty 1s, and three 2s. Calculate $\bar{x}, s^2$, and s. What percentage of the measurements are in the interval $\bar{x} \pm 3s$? Compare this result to Chebyshev's Rule.   1, .1091, .330
  d. Draw a histogram for each of the data sets in parts **a, b**, and **c**. What do you conclude from these graphs and the answers to parts **a, b**, and **c**?

Applying the Concepts—Basic

2.125 Each year, the National Highway Traffic Safety Administration (NHTSA) crash tests new car models to determine how well they protect the driver and front-seat passenger in a head-on collision. The NHTSA has developed a "star" scoring system for the frontal crash test, with results ranging from one star (*) to five stars (*****). The more stars in the rating, the better the level

CRASH

of crash protection in a head-on collision. The NHTSA crash test results for 98 cars (model year 1997) are stored in the data file named CRASH. The driver-side star ratings for the 98 cars are summarized in the accompanying MINITAB printout. Use the information in the printout to form a pie chart. Interpret the graph.

MINITAB Output for Exercise 2.125

```
Tally for Discrete Variables: DrivStar

DrivStar   Count   Percent
       2       4      4.08
       3      17     17.35
       4      59     60.20
       5      18     18.37
     N=       98
```

2.126 Refer to Exercise 2.125 and the National Highway Traffic Safety Administration (NHTSA) crash test data. One quantitative variable recorded by the NHTSA is driver's severity of head injury (measured on a scale from 0 to 1,500). The mean and standard deviation for the 98 driver head-injury ratings in the CRASH file are displayed in the MINITAB printout below. Use these values to find the z-score for a driver head-injury rating of 408. Interpret the result. $z = -1.06$

2.127 Beanie Babies are toy stuffed animals that have become valuable collector's items. *Beanie World Magazine* provided the information on 50 Beanie Babies shown in the table on p. 99.

    a. Summarize the retired/current status of the 50 Beanie Babies with an appropriate graph. Interpret the graph.

    b. Summarize the values of the 50 Beanie Babies with an appropriate graph. Interpret the graph.

    *****c.** Use a graph to portray the relationship between a Beanie Baby's value and its age. Do you detect a trend?

2.128 A survey was conducted to investigate the impact of the mass media on the public's perception of mental illness (*Health Education Journal,* Sept. 1994). The media coverage for each of 562 items related to mental health in Scotland was classified into one of five categories: violence to others, sympathetic coverage, harm to self, comic images, and criticism of accepted defini-

tions of mental illness. A summary of the results is provided in the accompanying table.

Media Coverage	Number of Items
Violence to others	373
Sympathetic	102
Harm to self	71
Comic images	12
Criticism of definitions	4
Total	562

Source: Philo, G., *et al.* "The impact of the mass media on public images of mental illness: Media content and audience belief." *Health Education Journal,* Vol. 53, No. 3, Sept. 1994, p.274 (Table 1).

    a. Construct a relative frequency table for the data.

    b. Display the relative frequencies in a graph.

    c. Discuss the findings.

2.129 The extent of computer anxiety among secondary technical education teachers in West Virginia was examined in the *Journal of Studies in Technical Careers* (Vol. 15, 1995). Level of computer anxiety was measured by the Computer Anxiety Scale (COMPAS), with scores ranging from 10 to 50. (Lower scores reflect a higher degree of anxiety.) One objective was to compare the computer anxiety levels of male and female secondary technical education teachers in West Virginia. The data for the 116 teachers who participated in the survey are summarized in the following table.

	Male Teachers	Female Teachers
n	68	48
$\bar{x}$	26.4	24.5
s	10.6	11.2

Source: Gordon, H. R. D. "Analysis of the computer anxiety levels of secondary technical education teachers in West Virginia." *Journal of Studies in Technical Careers*, Vol. 15, No. 2, 1995, pp. 26–27 (Table 1).

    a. Interpret the mean COMPAS score for male teachers.

    b. Repeat part **a** for female teachers.

    c. Give an interval that contains about 95% of the COMPAS scores for male teachers. (Assume the distribution of scores for males is symmetric and mound-shaped.)

    d. Repeat part **c** for female teachers.

MINITAB Output for Exercise 2.126

```
Descriptive Statistics: DrivHead

Variable        N       Mean    Median    TrMean    StDev   SE Mean
DrivHead       98      603.7     605.0     600.3    185.4      18.7

Variable   Minimum   Maximum        Q1        Q3
DrivHead     216.0    1240.0     475.0     724.3
```

● BEANIE

Name	Age (Months) as of Sept. 1998	Retired (R) Current (C)	Value ($)
1. Ally the Alligator	52	R	55.00
2. Batty the Bat	12	C	12.00
3. Bongo the Brown Monkey	28	R	40.00
4. Blackie the Bear	52	C	10.00
5. Bucky the Beaver	40	R	45.00
6. Bumble the Bee	28	R	600.00
7. Crunch the Shark	21	C	10.00
8. Congo the Gorilla	28	C	10.00
9. Derby the Coarse Mane Horse	28	R	30.00
10. Digger the Red Crab	40	R	150.00
11. Echo the Dolphin	17	R	20.00
12. Fetch the Golden Retriever	5	C	15.00
13. Early the Robin	5	C	20.00
14. Flip the White Cat	28	R	40.00
15. Garcia the Teddy	28	R	200.00
16. Happy the Hippo	52	R	20.00
17. Grunt the Razorback	28	R	175.00
18. Gigi the Poodle	5	C	15.00
19. Goldie the Goldfish	52	R	45.00
20. Iggy the Iguana	10	C	10.00
21. Inch the Inchworm	28	R	20.00
22. Jake the Mallard Duck	5	C	20.00
23. Kiwi the Toucan	40	R	165.00
24. Kuku the Cockatoo	5	C	20.00
25. Mystic the Unicorn	11	R	45.00
26. Mel the Koala Bear	21	C	10.00
27. Nanook the Husky	17	C	15.00
28. Nuts the Squirrel	21	C	10.00
29. Peace the Tie Dyed Teddy	17	C	25.00
30. Patty the Platypus	64	R	800.00
31. Quacker the Duck	40	R	15.00
32. Puffer the Penguin	10	C	15.00
33. Princess the Bear	12	C	65.00
34. Scottie the Scottie	28	R	28.00
35. Rover the Dog	28	R	15.00
36. Rex the Tyrannosaurus	40	R	825.00
37. Sly the Fox	28	C	10.00
38. Slither the Snake	52	R	1,900.00
39. Skip the Siamese Cat	21	C	10.00
40. Splash the Orca Whale	52	R	150.00
41. Spooky the Ghost	28	R	40.00
42. Snowball the Snowman	12	R	40.00
43. Stinger the Scorpion	5	C	15.00
44. Spot the Dog	52	R	65.00
45. Tank the Armadillo	28	R	85.00
46. Stripes the Tiger (Gold/Black)	40	R	400.00
47. Teddy the 1997 Holiday Bear	12	R	50.00
48. Tuffy the Terrier	17	C	10.00
49. Tracker the Basset Hound	5	C	15.00
50. Zip the Black Cat	28	R	40.00

Source: Beanie World Magazine, Sept. 1998.

TORNADO

State	Number of Tornadoes per Year	Month of Peak Occurrence	State	Number of Tornadoes per Year	Month of Peak Occurrence
AL	23	April	MT	5	June
AK	0	—	NE	36	June
AZ	4	August	NV	1	June
AR	21	April	NH	2	July
CA	4	February	NJ	3	July
CO	24	June	NM	9	June
CT	1	July	NY	5	July
DE	1	July	NC	14	April
FL	52	June	ND	20	July
GA	21	April	OH	16	May
HI	1	January	OK	47	May
ID	2	July	OR	1	May
IL	27	May	PA	10	June
IN	23	May	RI	0.23	August
IA	35	May	SC	10	April
KS	36	May	SD	28	June
KY	10	May	TN	12	April
LA	27	April	TX	137	May
ME	2	July	UT	2	June
MD	3	May	VT	1	July
MA	3	August	VA	6	May
MI	21	June	WA	1	May
MN	19	June	WV	2	May
MS	26	April	WI	18	July
MO	27	May	WY	11	June

Source: "Tornadoes ... nature's most violent storms." U.S. Dept. of Commerce, National Oceanic and Atmospheric Administration, Sept. 1992.

2.130 A tornado is defined as a violently rotating column of air extending from a thunderstorm to the ground. The most violent tornadoes, with wind speeds of 250 mph or more, are capable of tremendous destruction. The accompanying table lists, for each state, the reported number of tornadoes per year and the month of peak tornado occurrence.

a. Identify the variables measured, number of tornadoes per year and month of peak occurrence, as quantitative or qualitative.

b. Use a graphical technique to summarize the month of peak tornado occurrence in the 50 states. Interpret the graph.

2.131 Refer to Exercise 2.130. The descriptive statistics and stem-and-leaf display shown below were obtained by

SPSS Output for Exercise 2.131

Descriptives

			Statistic	Std. Error
TORNUM	Mean		16.2000	3.0862
	95% Confidence Interval for Mean	Lower Bound	9.9980	
		Upper Bound	22.4020	
	5% Trimmed Mean		13.2667	
	Median		10.0000	
	Variance		476.245	
	Std. Deviation		21.8230	
	Minimum		.00	
	Maximum		137.00	
	Range		137.00	
	Interquartile Range		21.2500	
	Skewness		3.692	.337
	Kurtosis		18.867	.662

```
Frequency     Stem &  Leaf

   19.00       0 .  0011111112222233344
    4.00       0 .  5569
    6.00       1 .  000124
    3.00       1 .  689
    7.00       2 .  0111334
    5.00       2 .  67778
     .00       3 .
    3.00       3 .  566
     .00       4 .
    1.00       4 .  7
    1.00       5 .  2
    1.00   Extremes    (>=137)

Stem width:      10.00
Each leaf:        1 case(s)
```

using SPSS to analyze the reported number of tornadoes per year of the 50 states.

a. Use the stem-and-leaf display to give a verbal description of the data set. Do any of the measurements appear to be outliers?   Yes

b. Examine the computer output to determine the value of $\bar{x}$, the median, s^2, and s for the number of tornadoes per year.   16.2, 10, 476.245, 21.82

c. Based on the stem-and-leaf display and the relationship of the mean and median, is the distribution of these data approximately mound-shaped, or is it skewed to the right or left?   Skewed right

d. Based on your answer to part **c**, what percentage of the measurements do you expect to find in the intervals $\bar{x} \pm s$, $\bar{x} \pm 2s$, and $\bar{x} \pm 3s$?

e. Count the number of measurements that actually fall in each interval of part **d** and express each interval count as a percentage of the total number of measurements. Compare these results to your estimates from part **d**.   94%, 98%, 98%

Applying the Concepts—Intermediate

2.132 Archaeologists gain insight into the social life of ancient tribes by measuring the distance (in centimeters) of each artifact found at a site from the "central hearth" (i.e., the middle of an artifact scatter). A graphical summary of these distances, in the form of a histogram, is called a *ring diagram* (*World Archaeology*, Oct. 1997). Ring diagrams for two archaeological sites (A and G) in Europe are shown at right.

a. Identify the type of skewness (if any) present in the data collected at the two sites.   None, skewed left

b. Archaeologists have associated unimodal ring diagrams with open air hearths and multimodal ring diagrams with hearths inside dwellings. Can you identify the type of hearth (*open air or inside dwelling*) that was most likely present at each site?

2.133 In 1961, Russian cosmonaut Yuri Gagarin was the first human to fly into space. Since then, more than 770 astronauts and cosmonauts have made space flights. The graphic at the top of p. 102, produced by the National Aeronautics and Space Administration (*NASA*), summarizes the annual number of people who flew into space between 1961 and 1998.

a. What type of graph is used to describe the number of humans in space each year?   Frequency histogram

b. In 1986, the space shuttle *Challenger* exploded, killing 7 American astronauts. Does the graph depict any trends that resulted from the disaster? Explain.

2.134 One method used to control, trap, and kill predator species by the US. Fish and Wildlife Service is the steel-jaw leghold trap. Unfortunately, the traps often end up capturing and killing animals other than the intended predators (*UCLA Journal of Environmental Law & Policy*, Vol. 13, 1994/95). A study revealed that in one year, 44,982 animals were killed by steel traps intended for coyotes, but only 25,026 of these animals were

Histogram for Exercise 2.132

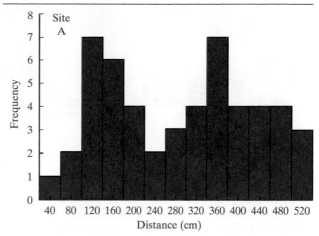

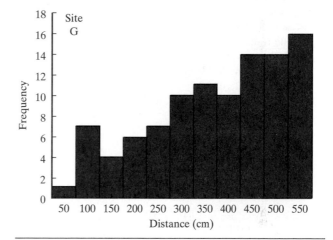

actually coyotes. The accompanying table lists the species of all animals captured by steel traps during that year and the number of each killed. Use a graphic method to summarize the data. Interpret the graph.

⊘ TRAPS

Species	Number Killed by Steel Traps
Coyotes	25,026
Opossums	2,698
Porcupines	1,367
Racoons	3,345
Skunks	6,348
Beavers	682
Dogs	273
Cats	73
Goats	49
Foxes	114
Rabbits	98
Deer	20
Muskrats	52
Other	4,837

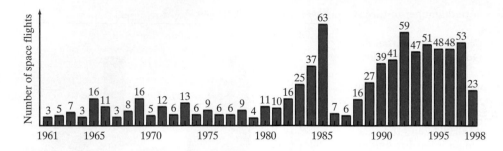

2.135 The *Journal of Agricultural, Biological, and Environmental Statistics* (Sept. 2000) published a study on the impact of the *Exxon Valdez* tanker oil spill on the seabird population in Prince William Sound, Alaska. A subset of the data analyzed is stored in the EVOS file. Data were collected on 96 shoreline locations (called transects) of constant width but variable length. For each transect, the number of seabirds found is recorded as well as the length (in kilometers) of the transect and whether or not the transect was in an oiled area. (The first five and last five observations in the EVOS file are listed in the table.)

a. Identify the variables measured as quantitative or qualitative.

b. Identify the experimental unit.

⊘EVOS

Transect	Seabirds	Length	Oil
1	0	4.06	No
2	0	6.51	No
3	54	6.76	No
4	0	4.26	No
5	14	3.59	No
⋮	⋮	⋮	⋮
92	7	3.40	Yes
93	4	6.67	Yes
94	0	3.29	Yes
95	0	6.22	Yes
96	27	8.94	Yes

Source: McDonald, T. L., Erickson, W. P., and McDonald, L. L. "Analysis of count data from before–after control-impact studies," *Journal of Agricultural, Biological, and Environmental Statistics*, Vol. 5, No. 3, Sept. 2000, pp. 277–8 (Table A.1).

c. Use a pie chart to describe the percentage of transects in oiled and unoiled areas.

d. Use a graphical method to examine the relationship between observed number of seabirds and transect length.

e. Observed seabird density is defined as the observed count divided by the length of the transect. MINITAB descriptive statistics for seabird densities in unoiled and oiled transects are displayed in the printout shown below, followed by histograms for each type of transect on p. 103. Assess whether the distribution of seabird densities differs for transects in oiled and unoiled areas.

f. For unoiled transects, give an interval of values that is likely to contain at least 75% of the seabird densities.

g. For oiled transects, give an interval of values that is likely to contain at least 75% of the seabird densities.

h. Which type of transect, an oiled or unoiled one, is more likely to have a seabird density of 16? Explain.

2.136 A study in the *Journal of Leisure Research* (Vol. 24, 1992) investigated the relationship between academic performance and leisure activities. Each in a sample of 159 high school students was asked to state how many leisure activities (sports, fishing, music, drama, photography, writing, watching TV, etc.) they participated in weekly. From this list, activities that involved reading, writing, or arithmetic were labeled "academic leisure activities." Some of the results of the study are presented below:

	$\bar{x}$	s
GPA	2.96	.71
Number of leisure activities	12.38	5.07
Number of academic leisure activities	2.77	1.97

MINITAB Output for Exercise 2.135

```
Descriptive Statistics: DENSITY by OIL

Variable   OIL              N      Mean    Median    TrMean    StDev
DENSITY    no              36      3.27      0.89      2.05     6.70
           yes             60     3.495     0.700     2.542    5.968

Variable   OIL         SE Mean   Minimum   Maximum        Q1       Q3
DENSITY    no             1.12      0.00     36.23      0.00     3.87
           yes           0.770     0.000    32.836     0.000    5.233
```

MINITAB Histograms for Exercise 2.135

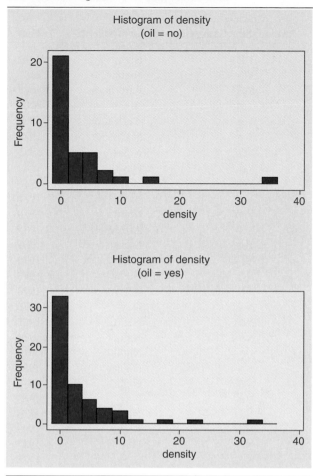

Applying the Concepts—Advanced

2.137 Refer to *The Astronomical Journal* study of galaxy velocities, Exercise 2.78 (p. 69). A second cluster of galaxies, named A1775, is thought to be a *double cluster*; that is, two clusters of galaxies in close proximity. Fifty-one velocity observations (in kilometers per second, km/s) from cluster A1775 are listed in the table.

🧭 GALAXY2

22922	20210	21911	19225	18792	21993	23059
20785	22781	23303	22192	19462	19057	23017
20186	23292	19408	24909	19866	22891	23121
19673	23261	22796	22355	19807	23432	22625
22744	22426	19111	18933	22417	19595	23408
22809	19619	22738	18499	19130	23220	22647
22718	22779	19026	22513	19740	22682	19179
19404	22193					

Source: Oegerle, W. R., Hill, J. M., and Fitchett, M. J. "Observations of high dispersion clusters of galaxies: Constraints on cold dark matter." *The Astronomical Journal*, Vol. 110, No. 1, July 1995, p. 34 (Table 1), p. 37 (Figure 1).

a. Use a graphical method to describe the velocity distribution of galaxy cluster A1775.

b. Examine the graph, part **a**. Is there evidence to support the double cluster theory? Explain.

c. Calculate numerical descriptive measures (e.g., mean and standard deviation) for galaxy velocities in cluster A1775. Depending on your answer to part **b**, you may need to calculate two sets of numerical descriptive measures, one for each of the clusters (say, A1775A and A1775B) within the double cluster.

d. Suppose you observe a galaxy velocity of 20,000 km/s. Is this galaxy likely to belong to cluster A1775A or A1775B? Explain.

2.138 *US News & World Report* (Mar. 5, 1990) reported on many factors contributing to the breakdown of public education. One study mentioned in the article found that over 90% of the nation's school districts reported that their students were scoring "above the national average" on standardized tests. Using your knowledge of measures of central tendency, explain why the schools' reports are incorrect. Does your analysis change if the term "average" refers to the mean? To the median? Explain what effect this misinformation might have on the perception of the nation's schools.

2.139 Suicide is the leading cause of death of Americans incarcerated in correctional facilities. To determine what factors increase the risk of suicide in urban jails, a group of researchers collected data on all 37 suicides that occurred from 1967 to 1992 in the Wayne County Jail, Detroit, Michigan (*American Journal of Psychiatry*, July 1995). The data for each suicide victim is listed in the table shown on p. 104.

a. Identify the type (quantitative or qualitative) of each variable measured.

a. For GPA, calculate the intervals $\bar{x} \pm s, \bar{x} \pm 2s$, and $\bar{x} \pm 3s$. Based on these intervals, comment on the skewness or symmetry you would expect in these data. (Remember, the range of GPAs is 0 to 4.) Approximately what percentage of the students would you expect to find in each interval?

b. For number of leisure activities, calculate the intervals $\bar{x} \pm s, \bar{x} \pm 2s$, and $\bar{x} \pm 3s$. Based on these intervals, comment on the skewness or symmetry you would expect in these data. (Remember, the number of leisure activities cannot be negative.) Approximately what percentage of the students would you expect to find in each interval?

c. For number of academic leisure activities, calculate the intervals $\bar{x} \pm s, \bar{x} \pm 2s$, and $\bar{x} \pm 3s$. Based on these intervals, comment on the skewness or symmetry you would expect in these data. Approximately what percentage of the students would you expect to find in each interval?

d. Based on your answers, which of the variables' distributions would you expect to be most skewed? Why?

SUICIDE

Victim	Days in Jail Before Suicide	Marital Status	Race	Murder/ Manslaughter Charge	Time of Suicide	Year
1	3	Married	W	Yes	Night	1972
2	4	Single	W	Yes	Night	1987
3	5	Single	NW	Yes	Afternoon	1975
4	7	Widowed	NW	Yes	Night	1981
5	10	Single	NW	Yes	Afternoon	1982
6	15	Married	NW	Yes	Night	1970
7	15	Divorced	NW	Yes	Night	1985
8	19	Married	NW	Yes	Day	1980
9	22	Single	NW	Yes	Night	1983
10	29	Married	W	Yes	Night	1981
11	31	Single	NW	Yes	Night	1970
12	85	Single	NW	Yes	Afternoon	1985
13	126	Married	NW	Yes	Night	1981
14	221	Divorced	W	Yes	Night	1970
15	14	Single	NW	No	Night	1980
16	22	Single	W	No	Afternoon	1970
17	42	Single	W	No	Day	1976
18	122	Single	NW	No	Night	1983
19	309	Married	NW	No	Night	1970
20	69	Married	NW	No	Night	1979
21	143	Single	NW	No	Night	1985
22	6	Married	NW	No	Night	1971
23	1	Single	NW	No	Day	1968
24	1	Single	NW	No	Night	1973
25	1	Single	W	No	Night	1989
26	2	Single	W	No	Day	1987
27	3	Single	NW	No	Night	1985
28	4	Married	W	No	Afternoon	1969
29	4	Single	W	No	Afternoon	1974
30	4	Single	NW	No	Night	1980
31	6	Single	W	No	Day	1968
32	6	Married	NW	No	Night	1975
33	10	Single	W	No	Night	1967
34	18	Single	NW	No	Night	1979
35	26	Married	W	No	Night	1976
36	41	Single	NW	No	Night	1985
37	86	Married	W	No	Night	1968

Source: DuRand, C. J., *et al.* "A quarter century of suicide in a major urban jail: Implications for community psychiatry." *American Journal of Psychiatry*, Vol. 152, No. 7, July 1995, p. 1078 (Table 1).

b. Are suicides at the jail more likely to be committed by inmates charged with murder/manslaughter or with lesser crimes? Illustrate with a graph.

c. Are suicides at the jail more likely to be committed at night? Illustrate with a graph.

d. What is the mean length of time an inmate is in jail before committing suicide? What is the median? Interpret these two numbers.

e. Is it likely that a future suicide at the jail will occur after 200 days? Explain.

*f. Have suicides at the jail declined over the years? Support your answer with a graph.

2.140 The final grades given by two professors in introductory statistics courses have been carefully examined. The students in the first professor's class had a grade point average of 3.0 and a standard deviation of .2. Those in the second professor's class had grade points with an average of 3.0 and a standard deviation of 1.0. If you had a choice, which professor would you take for this course? Explain.   First professor

STUDENT PROJECT

We list here several sources of real-life data sets (many in the list have been obtained from Wasserman and Bernero's *Statistics Sources* and many can be accessed via the Internet). This index of data sources is very complete and is a useful reference for anyone interested in finding almost any type of data. First we list some almanacs:

CBS News Almanac
Information Please Almanac
World Almanac and Book of Facts

United States Government publications are also rich sources of data:

Agricultural Statistics
Digest of Educational Statistics
Handbook of Labor Statistics
Housing and Urban Development Yearbook
Social Indicators
Uniform Crime Reports for the United States
Vital Statistics of the United States
Business Conditions Digest
Economic Indicators
Monthly Labor Review
Survey of Current Business
Bureau of the Census Catalog
Statistical Abstract of the United States

Main data sources are published on an annual basis:

Commodity Yearbook
Facts and Figures on Government Finance
Municipal Yearbook
Standard and Poor's Corporation, Trade and Securities: Statistics

Some sources contain data that are international in scope:

Compendium of Social Statistics
Demographic Yearbook
United Nations Statistical Yearbook
World Handbook of Political and Social Indicators

Utilizing the data sources listed, sources suggested by your instructor, or your own resourcefulness, find one real-life quantitative data set that stems from an area of particular interest to you.

a. Describe the data set by using a relative frequency histogram, stem-and-leaf display, or dot plot.
b. Find the mean, median, variance, standard deviation, and range of the data set.
c. Use Chebyshev's Rule and the Empirical Rule to describe the distribution of this data set. Count the actual number of observations that fall within 1, 2, and 3 standard deviations of the mean of the data set and compare these counts with the description of the data set you developed in part **b**.

REFERENCES

Huff, D. *How to Lie with Statistics*. New York: Norton, 1954.

Mendenhall, W., Beaver, R. J., and Beaver, B. M. *Introduction to Probability and Statistics*, 10th ed. North Scituate, Mass.: Duxbury, 1999.

Tufte, E. R. *Envisioning Information*. Cheshire, Conn.: Graphics Press, 1990.

Tufte, E.R. *Visual Explanations*. Cheshire, Conn.: Graphics Press, 1997.

Tufte, E. R. *Visual Display of Quantitative Information*. Cheshire, Conn.: Graphics Press, 1983.

Tukey, J. *Exploratory Data Analysis*. Reading, Mass.: Addison-Wesley, 1977.

Probability

Contents

Statistics in Action

Game Show Strategy: To Switch or Not to Switch?

🖑 *Where We've Been*

We've identified inference, from a sample to a population, as the goal of statistics. And we've seen that to reach this goal, we must be able to describe a set of measurements. Thus, we explored the use of graphical and numerical methods for describing both quantitative and qualitative data sets and for phrasing inferences.

☞ *Where We're Going*

Now that we know how to phrase an inference about a population, we turn to the problem of making the inference. What is it that permits us to make the inferential jump from sample to population and then to give a measure of reliability for the inference? As you'll see, the answer is *probability*. This chapter is devoted to a study of probability—what it is and some of the basic concepts of the theory behind it.

TEACHING TIP
Use these simple examples of
probability to show how knowing
probabilities of past performance
provides valuable information
that can be used to make
decisions concerning future
outcomes.

Recall that one branch of statistics is concerned with decisions about a population based on sample information. You can see how this is accomplished more easily if you understand the relationship between population and sample—a relationship that becomes clearer if we reverse the statistical procedure of making inferences from sample to population. In this chapter we assume the population is *known* and calculate the chances of obtaining various samples from the population. Thus, we show that probability is the reverse of statistics: In probability, we use the population information to infer the probable nature of the sample.

Probability plays an important role in inference making. Suppose, for example, you have an opportunity to invest in an oil exploration company. Past records show that out of ten previous oil drillings (a sample of the company's experiences), all ten came up dry. What do you conclude? Do you think the chances are better than 50:50 that the company will hit a gusher? Should you invest in this company? Chances are, your answer to these questions will be an emphatic No. If the company's exploratory prowess is sufficient to hit a producing well 50% of the time, a record of ten dry wells out of ten drilled is an event that is just too improbable.

Or suppose you're playing poker with what your opponents assure you is a well-shuffled deck of cards. In three consecutive five-card hands, the person on your right is dealt four aces. Based on this sample of three deals, do you think the cards are being adequately shuffled? Again, your answer is likely to be No because dealing three hands of four aces is just too improbable if the cards were properly shuffled.

Note that the decisions concerning the potential success of the oil drilling company and the adequacy of card shuffling both involve knowing the chance—or probability—of a certain sample result. Both situations were contrived so that you could easily conclude that the probabilities of the sample results were small. Unfortunately, the probabilities of many observed sample results aren't so easy to evaluate intuitively. For these cases we need the assistance of a theory of probability.

3.1   EVENTS, SAMPLE SPACES, AND PROBABILITY

TEACHING TIP
Use examples of other outcomes
or measurements and ask the
students to identify the
experiment needed to generate
the result.

Let's begin our treatment of probability with straightforward examples that are easily described. With the aid of simple examples, we can introduce important definitions that will help us develop the notion of probability more easily.

Suppose a coin is tossed once and the up face is recorded. The result we see is called an *observation*, or *measurement*, and the process of making an observation is called an *experiment*. Notice that our definition of experiment is broader than the one used in the physical sciences, where you tend to picture test tubes, microscopes, and other laboratory equipment. Among other things, statistical experiments may include recording an Internet user's preference for a Web browser, recording a voter's opinion on an important political issue, measuring the amount of dissolved oxygen in a polluted river, observing the level of anxiety of a test taker, counting the number of errors in an inventory, and observing the fraction of insects killed by a new insecticide. The point is that a statistical experiment can be almost any act of observation as long as the outcome is uncertain.

DEFINITION 3.1
An **experiment** is an act or process of observation that leads to a single outcome that cannot be predicted with certainty.

Consider another simple experiment consisting of tossing a die and observing the number on the up face. The six possible outcomes to this experiment are:

1. Observe a 1
2. Observe a 2
3. Observe a 3
4. Observe a 4
5. Observe a 5
6. Observe a 6

Note that if this experiment is conducted once, *you can observe one and only one of these six basic outcomes, and the outcome cannot be predicted with certainty.* Also, these possibilities cannot be decomposed into more basic outcomes. Because observing the outcome of an experiment is similar to selecting a sample from a population, the basic possible outcomes to an experiment are called *sample points.**

> **DEFINITION 3.2**
>
> A **sample point** is the most basic outcome of an experiment.

EXAMPLE 3.1

HH, HT, TH, TT

Two coins are tossed, and their up faces are recorded. List all the sample points for this experiment.

Solution

Even for a seemingly trivial experiment, we must be careful when listing the sample points. At first glance, we might expect three basic outcomes: Observe two heads; Observe two tails; or Observe one head and one tail. However, further reflection reveals that the last of these, Observe one head and one tail, can be decomposed into two outcomes: Head on coin 1, Tail on coin 2; and Tail on coin 1, Head on coin 2.[†] Thus, we have four sample points:

1. Observe *HH*
2. Observe *HT*
3. Observe *TH*
4. Observe *TT*

where *H* in the first position means "Head on coin 1," *H* in the second position means "Head on coin 2," and so on.

We often wish to refer to the collection of all the sample points of an experiment. This collection is called the *sample space* of the experiment. For example, there are six sample points in the sample space associated with the die-toss experiment. The sample spaces for the experiments discussed thus far are shown in Table 3.1.

> **DEFINITION 3.3**
>
> The **sample space** of an experiment is the collection of all its sample points.

---

*Alternatively, the term "simple event" can be used.

[†]Even if the coins are identical in appearance, there are, in fact, two distinct coins. Thus, the designation of one coin as coin 1 and the other as coin 2 is legitimate in any case.

TABLE 3.1   Experiments and Their Sample Spaces

Experiment: Observe the up face on a coin.
Sample Space: 1. Observe a head
　　　　　　　　2. Observe a tail.

This sample space can be represented in set notation as a set containing two sample points:

$$S: \quad \{H, T\}$$

where H represents the sample point Observe a head and T represents the sample point Observe a tail.

Experiment: Observe the up face on a die.
Sample Space: 1. Observe a 1
　　　　　　　　2. Observe a 2
　　　　　　　　3. Observe a 3
　　　　　　　　4. Observe a 4
　　　　　　　　5. Observe a 5
　　　　　　　　6. Observe a 6

This sample space can be represented in set notation as a set of six sample points:

$$S: \quad \{1, 2, 3, 4, 5, 6\}$$

Experiment: Observe the up faces on two coins.
Sample Space: 1. Observe HH
　　　　　　　　2. Observe HT
　　　　　　　　3. Observe TH
　　　　　　　　4. Observe TT

This sample space can be represented in set notation as a set of four sample points:

$$S: \quad \{HH, HT, TH, TT\}$$

TEACHING TIP

Explain that sample spaces can be described in several ways. For example, the results of the die experiment could be considered as odd or even outcomes. This leads into the definition of an event very nicely.

TEACHING TIP

Venn diagrams are useful as both a teaching and learning tool in probability. Note that the Venn diagram contains all possible outcomes of the sample space.

Just as graphs are useful in describing sets of data, a pictorial method for presenting the sample space will often be useful. Figure 3.1 shows such a representation for each of the experiments in Table 3.1. In each case, the sample space is shown as a closed figure, labeled **S**, containing all possible sample points. Each sample point is represented by a solid dot (i.e., a "point") and labeled accordingly. Such graphical representations are called **Venn diagrams**.

Now that we know that an experiment will result in *only one* basic outcome—called a sample point—and that the sample space is the collection of all possible sample points, we're ready to discuss the probabilities of the sample points. You have undoubtedly used the term *probability* and have some intuitive idea about its meaning. Probability is generally used synonymously with "chance," "odds," and similar concepts. For example, if a fair coin is tossed, we might reason that both the sample points, Observe a head and Observe a tail, have the same *chance* of occurring. Thus, we might state that "the probability of observing a head is 50%" or "the odds of seeing a head are 50:50." Both these statements are based on an informal knowledge of probability. We'll begin our treatment of probability by using such informal concepts and then solidify what we mean later.

The probability of a sample point is a number between 0 and 1 that measures the likelihood that the outcome will occur when the experiment is performed. This number is usually taken to be the relative frequency of the occurrence of a sample point in a very long series of repetitions of an experiment.* For example, if we are

---

*The result derives from an axiom in probability theory called the *Law of Large Numbers*. Phrased informally, the law states that the relative frequency of the number of times that an outcome occurs when an experiment is replicated over and over again (i.e., a large number of times) approaches the true (or theoretical) probability of the outcome.

FIGURE 3.1

Venn Diagrams for the Three Experiments from Table 3.1

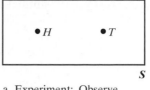

a. Experiment: Observe the up face on a coin

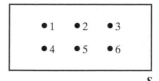

b. Experiment: Observe the up face on a die

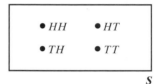

c. Experiment: Observe the up faces on two coins

assigning probabilities to the two sample points in the coin-toss experiment (Observe a head and Observe a tail), we might reason that if we toss a balanced coin a very large number of times, the sample points Observe a head and Observe a tail will occur with the same relative frequency of .5.

Our reasoning is supported by Figure 3.2. The figure, generated using SAS, plots the relative frequency of the number of times that a head occurs when

FIGURE 3.2

SAS Scatterplot Showing the Proportion of Heads in N Tosses of a Coin

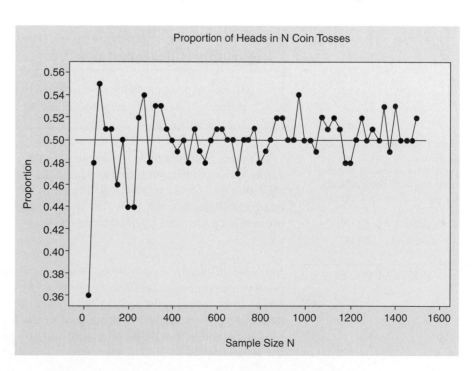

simulating (by computer) the toss of a coin *N* times, where *N* ranges from as few as 25 tosses to as many as 1,500 tosses of the coin. You can see that when *N* is large (i.e., $N = 1,500$), the relative frequency is converging to .5. Thus, the probability of each sample point in the coin tossing experiment is .5.

For some experiments, we may have little or no information on the relative frequency of occurrence of the sample points; consequently, we must assign probabilities to the sample points based on general information about the experiment. For example, if the experiment is to invest in a business venture and to observe whether it succeeds or fails, the sample space would appear as in Figure 3.3.

FIGURE 3.3

*Experiment: Invest in a
Business Venture and
Observe Whether It
Succeeds (S) or Fails (F)*

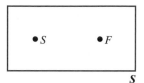

We are unlikely to be able to assign probabilities to the sample points of this experiment based on a long series of repetitions since unique factors govern each performance of this kind of experiment. Instead, we may consider factors such as the personnel managing the venture, the general state of the economy at the time, the rate of success of similar ventures, and any other information deemed pertinent. If we finally decide that the venture has an 80% chance of succeeding, we assign a probability of .8 to the sample point Success. This probability can be interpreted as a measure of our degree of belief in the outcome of the business venture; that is, it is a subjective probability. Notice, however, that such probabilities should be based on expert information that is carefully assessed. If not, we may be misled on any decisions based on these probabilities or based on any calculations in which they appear. [*Note:* For a text that deals in detail with the subjective evaluation of probabilities, see Winkler (1972) or Lindley (1985).]

No matter how you assign the probabilities to sample points, the probabilities assigned must obey two rules:

Probability Rules for Sample Points

1. All sample point probabilities *must* lie between 0 and 1.
2. The probabilities of all the sample points within a sample space *must* sum to 1.

TEACHING TIP
Ask the class to assign
probabilities to the six possible
sample points of the single die
experiment. Show how these two
probability rules apply to the
die example.

Assigning probabilities to sample points is easy for some experiments. For example, if the experiment is to toss a fair coin and observe the face, we would probably all agree to assign a probability of $\frac{1}{2}$ to the two sample points, Observe a head and Observe a tail. However, many experiments have sample points whose probabilities are more difficult to assign.

EXAMPLE 3.2   Many American hotels offer complimentary shampoo in their guest rooms. Suppose you randomly select one hotel from a registry of all hotels in the United States and check whether or not the hotel offers complimentary shampoo. Show how this problem might be formulated in the framework of an experiment with sample points and a sample space. Indicate how probabilities might be assigned to the sample points.

Solution

The experiment can be defined as the selection of an American hotel and the observation of whether or not complimentary shampoo is offered in the hotel's guest rooms. There are two sample points in the sample space corresponding to this experiment:

  S: {The hotel offers complimentary shampoo}
  N: {No complimentary shampoo is offered by the hotel}

The difference between this and the coin-toss experiment becomes apparent when we attempt to assign probabilities to the two sample points. What probability should we assign to the sample point S? If you answer .5, you are assuming that the events S and N should occur with equal likelihood, just like the sample points Heads and Tails in the coin-toss experiment. But assignment of sample point probabilities for the hotel-shampoo experiment is not so easy. In fact, a recent survey of American hotels found that 80% now offer complimentary shampoo to guests. Then it might be reasonable to approximate the probability of the sample point S as .8 and that of the sample point N as .2. Here we see that sample points are not always equally likely, so assigning probabilities to them can be complicated—particularly for experiments that represent real applications (as opposed to coin- and die-toss experiments). ∎

Although the probabilities of sample points are often of interest in their own right, it is usually probabilities of collections of sample points that are important. Example 3.3 demonstrates this point.

EXAMPLE 3.3

$P(\text{win}) = .5$

A fair die is tossed, and the up face is observed. If the face is even, you win $1. Otherwise, you lose $1. What is the probability that you win?

Solution

Recall that the sample space for this experiment contains six sample points:

$$S: \{1, 2, 3, 4, 5, 6\}$$

Since the die is balanced, we assign a probability of $\frac{1}{6}$ to each of the sample points in this sample space. An even number will occur if one of the sample points, Observe a 2, Observe a 4, or Observe a 6, occurs. A collection of sample points such as this is called an *event*, which we denote by the letter A. Since the event A contains three sample points—all with probability $\frac{1}{6}$—and since no sample points can occur simultaneously, we reason that the probability of A is the sum of the probabilities of the sample points in A. Thus, the probability of A is $\frac{1}{6} + \frac{1}{6} + \frac{1}{6} = \frac{1}{2}$. This implies that, *in the long run*, you will win $1 half the time and lose $1 half the time. ∎

Figure 3.4 is a Venn diagram depicting the sample space associated with a die-toss experiment and the event A, Observe an even number. The event A is represented by the closed figure inside the sample space S. This closed figure A contains all the sample points that comprise it.

To decide which sample points belong to the set associated with an event A, test each sample point in the sample space S. If event A occurs, then that sample point is in the event A. For example, the event A, Observe an even number, in the die-toss experiment will occur if the sample point Observe a 2 occurs. By the same reasoning, the sample points Observe a 4 and Observe a 6 are also in event A.

TEACHING TIP
An even number is the event that represents the collection of the sample points 2, 4, and 6 in the single die experiment. An odd number represents the collection of the sample points 1, 3, and 5.

FIGURE 3.4

*Die-Toss Experiment with
Event A: Observe an Even
Number*

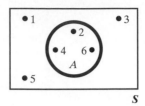

To summarize, we have demonstrated that an event can be defined in words or it can be defined as a specific set of sample points. This leads us to the following general definition of an *event*:

DEFINITION 3.4

An **event** is a specific collection of sample points.

E X A M P L E   3 . 4

Consider the experiment of tossing two *unbalanced* coins. Because the coins are *not* balanced, their outcomes (*H* or *T*) are not equiprobable. Suppose the correct probabilities associated with the sample points are given in the accompanying table. [*Note:* The necessary properties for assigning probabilities to sample points are satisfied.]

Consider the events

$$A: \{\text{Observe exactly one head}\}$$
$$B: \{\text{Observe at least one head}\}$$

$P(A) = \frac{4}{9}$
$P(B) = \frac{8}{9}$

Calculate the probability of *A* and the probability of *B*.

Sample Point	Probability
HH	$\frac{4}{9}$
HT	$\frac{2}{9}$
TH	$\frac{2}{9}$
TT	$\frac{1}{9}$

Solution

Event *A* contains the sample points *HT* and *TH*. Since two or more sample points cannot occur at the same time, we can easily calculate the probability of event *A* by summing the probabilities of the two sample points. Thus, the probability of observing exactly one head (event *A*), denoted by the symbol $P(A)$, is

$$P(A) = P(\text{Observe } HT) + P(\text{Observe } TH) = \frac{2}{9} + \frac{2}{9} = \frac{4}{9}$$

Similarly, since *B* contains the sample points *HH, HT,* and *TH,*

$$P(B) = \frac{4}{9} + \frac{2}{9} + \frac{2}{9} = \frac{8}{9}$$

The preceding example leads us to a general procedure for finding the probability of an event A:

> ## Probability of an Event
>
> The probability of an event A is calculated by summing the probabilities of the sample points in the sample space for A.

Thus, we can summarize the steps for calculating the probability of any event, as indicated in the next box.

> ## Steps for Calculating Probabilities of Events
>
> 1. Define the experiment, that is, describe the process used to make an observation and the type of observation that will be recorded.
> 2. List the sample points.
> 3. Assign probabilities to the sample points.
> 4. Determine the collection of sample points contained in the event of interest.
> 5. Sum the sample point probabilities to get the event probability.

TEACHING TIP
Point out that this procedure works well with sample points but does not always work with events of an experiment because more than one event can occur at the same time (i.e., an even number and a number greater than 3 for the single die experiment).

EXAMPLE 3.5

The American Association for Marriage and Family Therapy (AAMFT) is a group of professional therapists and family practitioners that treats many of the nation's couples and families. The AAMFT released the findings of a study that tracked the post-divorce history of 100 pairs of former spouses with children. Each divorced couple was classified into one of four groups, nicknamed "perfect pals (PP)," "cooperative colleagues (CC)," "angry associates (AA)," and "fiery foes (FF)." The proportions classified into each group are shown in Table 3.2.

Suppose one of the 100 couples is selected at random.

a. Define the experiment that generated the data in Table 3.2, and list the sample points.
b. Assign probabilities to the sample points.
c. What is the probability that the former spouses are "fiery foes"?
d. What is the probability that the former spouses have at least some conflict in their relationship?

c. .25
d. .88

TABLE 3.2   Results of AAMFT Study of Divorced Couples

Group	Proportion
Perfect Pals (PP) (Joint-custody parents who get along well)	.12
Cooperative Colleagues (CC) (Occasional conflict, likely to be remarried)	.38
Angry Associates (AA) (Cooperate on children-related issues only, conflicting otherwise)	.25
Fiery Foes (FF) (Communicate only through children, hostile toward each other)	.25

FIGURE 3.5

Venn Diagram for AAMFT Survey

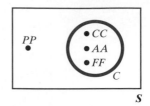

Solution

Suggested Exercise 3.21

Suggested Exercise 3.24

a. The experiment is the act of classifying the randomly selected couple. The sample points, the simplest outcomes of the experiment, are the four groups (categories) listed in Table 3.2. They are shown in the Venn diagram in Figure 3.5.
b. If, as in Example 3.1, we were to assign equal probabilities in this case, each of the response categories would have a probability of one-fourth (1/4), or .25. But, by examining Table 3.2 you can see that equal probabilities are not reasonable in this case because the response percentages are not all the same in the four categories. It is more reasonable to assign a probability equal to the response proportion in each class, as shown in Table 3.3.*

TEACHING TIP
Illustrate the need for counting rules by discussing the possible outcomes when ten coins are simultaneously flipped. There are 1,024 sample points in this example.

TABLE 3.3   Sample Point Probabilities for AAMFT Survey

Sample Point	Probability
PP	.12
CC	.38
AA	.25
FF	.25

c. The event that the former spouses are "fiery foes" corresponds to the sample point *FF*. Consequently, the probability of the event is the probability of the sample point. From Table 3.3, we find $P(FF) = .25$. Therefore, there is a .25 probability (or one-fourth chance) that the couple we select are "fiery foes."
d. The event that the former spouses have at least some conflict in their relationship, call it event *C*, is not a sample point because it consists of more than one of the response classifications (the sample points). In fact, as shown in Figure 3.5, the event *C* consists of three sample points: *CC, AA*, and *FF*. The probability of *C* is defined to be the sum of the probabilities of the sample points in *C*:

$$P(C) = P(CC) + P(AA) + P(FF) = .38 + .25 + .25 = .88$$

Thus, the chance that we observe a divorced couple with some degree of conflict in their relationship is .88—a fairly high probability. _____ ∎

For the experiments discussed so far, listing the sample points has been easy. For more complex experiments, the number of sample points may be so large that listing them is impractical. In solving probability problems for experiments with many sample points, we employ the same principles as for experiments with few sample points. The only difference is that we need *counting rules* for determining the number of sample points without actually enumerating all of them.

---

*Since the response percentages were based on a sample of divorced comples, these assigned probabilities are estimates of the true population response percentages. You will learn how to measure the reliability of probability estimates in Chapter 5.

EXERCISES 3.1–3.24

Learning the Mechanics

3.1 An experiment results in one of the following sample points: E_1, E_2, E_3, E_4, or E_5.
 a. Find $P(E_3)$ if $P(E_1) = .1, P(E_2) = .2, P(E_4) = .1$, and $P(E_5) = .1$ .5
 b. Find $P(E_3)$ if $P(E_1) = P(E_3)$, $P(E_2) = .1$, $P(E_4) = .2$, and $P(E_5) = .1$ .3
 c. Find $P(E_3)$ if $P(E_1) = P(E_2) = P(E_4) = P(E_5) = .1$ .6

3.2 The accompanying diagram describes the sample space of a particular experiment and events A and B.

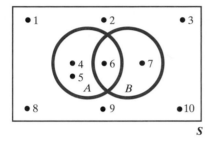

 a. What is this type of diagram called?
 b. Suppose the sample points are equally likely. Find $P(A)$ and $P(B)$ .3, .2
 c. Suppose $P(1) = P(2) = P(3) = P(4) = P(5) = \frac{1}{20}$ and $P(6) = P(7) = P(8) = P(9) = P(10) = \frac{3}{20}$. Find $P(A)$ and $P(B)$.   .25, .3

3.3 The sample space for an experiment contains five sample points with probabilities as shown in the table. Find the probability of each of the following events:

A: {Either 1, 2, or 3 occurs}   .55
B: {Either 1, 3, or 5 occurs}   .50
C: {4 does not occur}   .70

Sample Points	Probabilities
1	.05
2	.20
3	.30
4	.30
5	.15

3.4 Consider the experiment of tossing a die and observing the up face.
 a. Draw a Venn diagram for the experiment. On your diagram indicate the event Observe a number greater than 4. Call this event A. Also indicate the event Observe an even number. Call this event B.
 b. We would all agree that the probability of observing a 3 on the toss of a "fair" die is $\frac{1}{6}$. Explain what it means for a die to be fair (or balanced) and explain how knowing that a die is fair leads us to $P(3) = \frac{1}{6}$.
 c. For your Venn diagram of part **a**, assume the die is fair and find $P(A)$ and $P(B)$. $\frac{2}{6}, \frac{3}{6}$
 d. If you knew that a particular die were unfair (i.e., "loaded"), how would you determine the probability of observing a 3?

3.5 The Venn diagram depicts an experiment with six sample points. The events A and B are also shown. The probabilities of the sample points are:

$$P(1) = P(2) = P(4) = \frac{2}{9}$$
$$P(3) = P(5) = P(6) = \frac{1}{9}$$

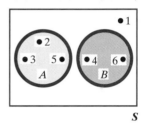

 a. Find $P(A)$. $\frac{4}{9}$
 b. Find $P(B)$. $\frac{1}{3}$
 c. Find the probability that the events A and B occur *simultaneously*.   0

3.6 Two fair dice are tossed, and the up face on each die is recorded.
 a. List the 36 sample points contained in the sample space.
 b. Find the probability of observing each of the following events:

A: {A 3 appears on each of the two dice} $\frac{1}{36}$
B: {The sum of the numbers is even} $\frac{1}{2}$
C: {The sum of the numbers is equal to 7} $\frac{1}{6}$
D: {A 5 appears on at least one of the dice} $\frac{11}{36}$
E: {The sum of the numbers is 10 or more} $\frac{1}{6}$

3.7 Consider the experiment composed of one roll of a fair die followed by one toss of a fair coin. List the sample points. Assign a probability to each sample point. Determine the probability of observing each of the following events:

A: {6 on the die; H on the coin} $\frac{1}{12}$
B: {Even number on the die; T on the coin} $\frac{1}{4}$
C: {Even number on the die} $\frac{1}{2}$
D: {T on the coin} $\frac{1}{2}$

3.8 Two marbles are drawn at random and without replacement from a box containing two blue marbles and three red marbles. Determine the probability of observing each of the following events:

A: {Two blue marbles are drawn} $\frac{1}{10}$
B: {A red and a blue marble are drawn} $\frac{6}{10}$
C: {Two red marbles are drawn} $\frac{3}{10}$

3.9 Simulate the experiment described in Exercise 3.8 using any five identically shaped objects, two of which are one color and three, another. Mix the objects, draw two, record the results, and then replace the objects. Repeat the experiment a large number of times (at least 100). Calculate the proportion of times events *A, B,* and *C* occur. How do these proportions compare with the probabilities you calculated in Exercise 3.8? Should these proportions equal the probabilities? Explain.

Applying the Concepts—Basic

3.10 According to *USA Today* (Sept. 19, 2000), there are 650 members of the International Nanny Association (INA). Of these, only three are men. Find the probability that a randomly selected member of the INA is a man.

3.11 *The Wall Street Journal* (Sept. 1, 2000) reported on an independent study of postal workers and violence at post offices. In a sample of 12,000 postal workers, 600 of them were physically assaulted on the job in the past year. Use this information to estimate the probability that a randomly selected postal worker will be physically assaulted on the job during the year.   .05

3.12 Two species of rhinoceros native to Africa are black rhinos and white rhinos. The International Rhino Federation estimates that the African rhinoceros population consists of 2,600 white rhinos and 8,400 black rhinos. Suppose one rhino is selected at random from the African rhino population and its species (black or white) is observed.

　a. List the sample points for this experiment.

　b. Assign probabilities to the sample points based on the estimates made by the International Rhino Federation.

3.13 The United States Department of Agriculture (USDA) reports that, under its standard inspection system, one in every 100 slaughtered chickens pass inspection with fecal contamination (*Tampa Tribune*, Mar. 31, 2000).

　a. If a slaughtered chicken is selected at random, what is the probability that it passes inspection with fecal contamination? $1/100 = .01$

　b. The probability of part **a** was based on a USDA study that found that 306 of 32,075 chicken carcasses passed inspection with fecal contamination. Do you agree with the USDA's statement about the likelihood of a slaughtered chicken passing inspection with fecal contamination?   Yes

3.14 Answer the following question posed in the *Atlanta Journal-Constitution* (Feb. 7, 2000): When a meteorologist says that "the probability of rain this afternoon is .4," does it mean that 40% of the time during the afternoon, it will be raining?   No

Applying the Concepts—Intermediate

3.15 Refer to the *American Scientist* (July–Aug.1998) study of *Benford's Law*, Exercise 2.14 (p. 28). Recall that *Benford's Law* states that the integer 1 is more likely to occur as the first significant digit in a randomly selected number than the other integers. The table summarizes

the results of a study where 743 college freshmen selected a six-digit number at random. Assume that one of these 743 numbers is selected and the first significant digit is observed.

⬤ DIGITS

First Digit	Number of Occurrences
1	109
2	75
3	77
4	99
5	72
6	117
7	89
8	62
9	43
Total	**743**

Source: Hill, T. P "The First Digit Phenomenon." *American Scientist,* Vol. 86, No. 4, July-Aug. 1998, p. 363 (Figure 5).

　a. List the sample points for this experiment.

　b. Explain why the sample points are not equally likely to occur.

　c. Assign reasonable probabilities to the sample points. Verify that the probabilities sum to 1.

　d. What is the probability that the first significant digit is 1 or 2? $184/743$

　e. What is the probability that the first significant digit is greater than 5? $311/743$

3.16 The maximum score possible on the Standardized Admission Test (SAT) is 1600. According to the test developers, the chance that a student scores a perfect 1600 on the SAT is 5 in 10,000.

　a. Find the probability that a randomly selected student scores a 1600 on the SAT. $5/10,000$

　b. In a recent year, 545 students scored a perfect 1600 on the SAT. How is this possible given the probability calculated in part **a**?

3.17 "Urban" and "rural" describe geographic areas upon which land zoning regulations, school district policy, and public service policy are often set. However, the characteristics of urban/rural areas are not clearly defined. Researchers at the University of Nevada (Reno) asked a sample of county commissioners to give their perception of the single most important factor in identifying urban counties (*Professional Geographer*, Feb. 2000). In all, five factors were mentioned by the commissioners: total population, agricultural change, presence of industry, growth, and population concentration. The survey results are displayed in the SPSS pie chart (p. 119). Suppose one of the commissioners is selected at random and the most important factor specified by the commissioner is recorded.

　a. List the sample points for this experiment.

　b. Assign reasonable probabilities to the sample points.

　c. Find the probability that the most important factor specified by the commissioner is population-related.

SPSS Output for Exercise 3.17

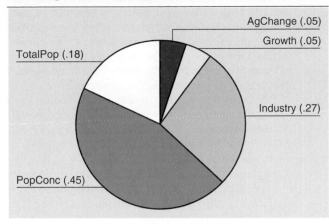

Most Salient Role	Number
Spouse	424
Parent	269
Grandparent	148
Other relative	59
Friend	73
Homemaker	59
Provider	34
Volunteer, club, church member	36
Total	**1,102**

Source: Krause, N., and Shaw, B. A. "Role-specific feelings of control and mortality." *Psychology and Aging,* Vol. 15, No. 4, Dec. 2000 (Table 2).

3.18 Refer to the National Highway Transportation Safety Administration (NHTSA) study of requests for installation of on-off switches for passenger-side air bags, Exercise 2.5 (p. 26). The table summarizing the reasons for requesting the on-off switches for 1998–1999 is reproduced here. Suppose we select one of the 30,337 car owners who made such a request over this time period and record the reason for the request.

Reason	Number of requests
Infant	1,852
Child	17,148
Medical	8,377
Infant & medical	44
Child & medical	903
Infant & child	1,878
Infant & child & medical	135
Total	**30,337**

Source: National Highway Transportation Safety Administration, Sept. 2000.

  a. List the sample points for this experiment.
  b. Assign reasonable probabilities to the sample points.
  c. Find the probability that the car owner uses medical as a reason.   .312
  d. Find the probability that the car owner does not use an infant as a reason.   .871

3.19 Refer to the *Psychology and Aging* (Dec. 2000) study of elderly people and their roles, Exercise 2.6 (p. 26). The table summarizing the most salient roles identified by each in a national sample of 1,102 adults, 65 years or older, is reproduced in the next column.
  a. What is the probability that a randomly selected elderly person feels his/her most salient role is a spouse?
  b. What is the probability that a randomly selected elderly person feels his/her most salient role is a parent or grandparent?   .378
  c. What is the probability that a randomly selected elderly person feels his/her most salient role does not involve a spouse or relative of any kind?   .183

3.20 Scientists have already discovered two genes (on chromosomes 19 and 21) that mutate to cause the early onset of Alzheimer's disease. *Science News* (July 8,1995) reported on a search for a third gene that causes Alzheimer's. An international team of scientists gathered genetic information from 21 families afflicted by the early-onset form of the disease. In six of these families, mutations in the S182 gene on chromosome 14 accounted for the early onset of Alzheimer's.
  a. Consider a family afflicted by the early onset of Alzheimer's disease. Find the approximate probability that the researchers will find mutations in the S182 gene on chromosome 14 for this family   .286
  b. Discuss the reliability of the probability, part **a**. How could a more accurate estimate of the probability be obtained?

Applying the Concepts—Advanced

3.21 Three people play a game called "Odd Man Out." In this game, each player flips a fair coin until the outcome (heads or tails) for one of the players is not the same as the other two players. This player is then "the odd man out" and loses the game. Find the probability that the game ends (i.e., either exactly one of the coins will fall heads or exactly one of the coins will fall tails) after only one toss by each player. Suppose one of the players, hoping to reduce his chances of being the odd man, uses a two-headed coin. Will this ploy be successful? Solve by listing the sample points in the sample space.

3.22 Consider the following question posed to Marilyn vos Savant in her weekly newspaper column "Ask Marilyn":

> I have two pairs of argyle socks, and they look nearly identical—one navy blue and the other black. [When doing the laundry] my wife matches the socks incorrectly much more often than she does correctly. ... If all four socks are in front of her, it seems to me that her chances are 50% for a wrong match and 50% for a right match. What do you think? *Source: Parade Magazine,* Feb. 27, 1994.

Use your knowledge of probability to answer this question. [*Hint:* List the sample points in the experiment.]

3.23 Handicappers for horse races express their belief about the probabilities of each horse winning a race in terms of **odds**. If the probability of event E is $P(E)$, then the *odds in favor of E are $P(E)$ to $1 - P(E)$*. Thus, if a handicapper assesses a probability of .25 that Snow Chief will win the Belmont Stakes, the odds in favor of Snow Chief are $^{25}/_{100}$ to $^{75}/_{100}$, or 1 to 3. It follows that the *odds against E are $1 - P(E)$ to $P(E)$*, or 3 to 1 against a win by Snow Chief. In general, if the odds in favor of event E are a to b, then $P(E) = a/(a + b)$.

a. A second handicapper assesses the probability of a win by Snow Chief to be $^1/_3$. According to the second handicapper, what are the odds in favor of a Snow Chief win?   1 to 2

b. A third handicapper assesses the odds in favor of Snow Chief to be 1 to 1. According to the third handicapper, what is the probability of a Snow Chief win?   .5

c. A fourth handicapper assesses the odds against Snow Chief's winning to be 3 to 2. Find this handicapper's assessment of the probability that Snow Chief will win.

3.24 An individual's genetic makeup is determined by the genes obtained from each parent. For every genetic trait, each parent possesses a gene pair; and each contributes one-half of this gene pair, with equal probability, to their offspring, forming a new gene pair. The offspring's traits (eye color, baldness, etc.) come from this new gene pair, where each gene in this pair possesses some characteristic.

For the gene pair that determines eye color, each gene trait may be one of two types: dominant brown (B) or recessive blue (b). A person possessing the gene pair BB or Bb has brown eyes, whereas the gene pair bb produces blue eyes.

a. Suppose both parents of an individual are brown-eyed, each with a gene pair of the type Bb. What is the probability that a randomly selected child of this couple will have blue eyes?

b. If one parent has brown eyes, type Bb, and the other has blue eyes, what is the probability that a randomly selected child of this couple will have blue eyes?

c. Suppose one parent is brown-eyed, type BB. What is the probability that a child has blue eyes?   0

3.2   UNIONS AND INTERSECTIONS

An event can often be viewed as a composition of two or more other events. Such events, which are called **compound events**, can be formed (composed) in two ways, as defined and illustrated here.

DEFINITION 3.5

The **union** of two events A and B is the event that occurs if either A or B or both occur on a single performance of the experiment. We denote the union of events A and B by the symbol $A \cup B$. $A \cup B$ consists of all the sample points that belong to *A or B or both*. (See Figure 3.6a.)

DEFINITION 3.6

The **intersection** of two events A and B is the event that occurs if both A and B occur on a single performance of the experiment. We write $A \cap B$ for the intersection of A and B. $A \cap B$ consists of all the sample points belonging to *both A and B*. (See Figure 3.6b.)

FIGURE 3.6

Venn Diagrams for Union and Intersection

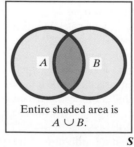

Entire shaded area is
$A \cup B$.

a. Union

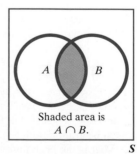

Shaded area is
$A \cap B$.

b. Intersection

EXAMPLE 3.6

a. $\{1, 2, 3, 4, 6\}$
b. $\{2\}$
c. $\frac{5}{6}, \frac{1}{6}$

Consider the die-toss experiment. Define the following events:

A: {Toss an even number}
B: {Toss a number less than or equal to 3}

a. Describe $A \cup B$ for this experiment.
b. Describe $A \cap B$ for this experiment.
c. Calculate $P(A \cup B)$ and $P(A \cap B)$ assuming the die is fair.

Solution

Draw the Venn diagram as shown in Figure 3.7.

FIGURE 3.7

Venn Diagram for Die Toss

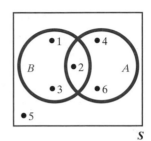

a. The union of A and B is the event that occurs if we observe either an even number, a number less than or equal to 3, or both on a single throw of the die. Consequently, the sample points in the event $A \cup B$ are those for which A occurs, B occurs, or both A and B occur. Checking the sample points in the entire sample space, we find that the collection of sample points in the union of A and B is

$$A \cup B = \{1, 2, 3, 4, 6\}$$

b. The intersection of A and B is the event that occurs if we observe *both* an even number and a number less than or equal to 3 on a single throw of the die. Checking the sample points to see which imply the occurrence of *both* events A and B, we see that the intersection contains only one sample point:

$$A \cap B = \{2\}$$

In other words, the intersection of A and B is the sample point Observe a 2.

c. Recalling that the probability of an event is the sum of the probabilities of the sample points of which the event is composed, we have

$$P(A \cup B) = P(1) + P(2) + P(3) + P(4) + P(6)$$
$$= \frac{1}{6} + \frac{1}{6} + \frac{1}{6} + \frac{1}{6} + \frac{1}{6} = \frac{5}{6}$$

and

$$P(A \cap B) = P(2) = \frac{1}{6}$$

TEACHING TIP
Use Venn diagrams to illustrate what the union and intersection look like in a picture. The union Venn diagram can be used to illustrate the additive rule later.

Unions and intersections can be defined for more than two events. For example, the event $A \cup B \cup C$ represents the union of three events—A, B, and C. This event, which includes the set of sample points in A, B, or C, will occur if any one or more of the events A, B, or C occurs. Similarly, the intersection $A \cap B \cap C$ is the event that all three of the events A, B, and C occur. Therefore, $A \cap B \cap C$ is the set of sample points that are in all three of the events A, B, and C.

EXAMPLE 3.7

a. $\{1, 2, 3, 4, 5, 6\}$
b. $\{2\}$

Refer to Example 3.6 and define the event

   C: {Toss a number greater than 1}

   Find the sample points in:

a. $A \cup B \cup C$
b. $A \cap B \cap C$

where

   A: {Toss an even number}
   B: {Toss a number less than or equal to 3}

Solution

a. Event C contains the sample points corresponding to tossing a 2, 3, 4, 5, or 6; event A contains the sample points 2, 4, and 6; and event B contains the sample points 1, 2, and 3. Therefore, the event that either A, B, or C occurs contains all six sample points in S—that is, those corresponding to tossing a 1, 2, 3, 4, 5, or 6.
b. You will observe all of the events A, B, and C only if you observe a 2. Therefore, the intersection $A \cap B \cap C$ contains the single sample point Toss a 2.

EXAMPLE 3.8

a. .79, .09
b. .83
c. .05

Family Planning Perspectives (Nov./Dec. 1997) reported on a study of over 200,000 births in New Jersey over a recent two-year period. The study investigated the link between the mother's race and the age at which she gave birth (called maternal age). The percentages of the total number of births in New Jersey are given by the maternal age and race classifications in Table 3.4. Define the following events:

   A: {A New Jersey birth mother is white}
   B: {A New Jersey mother was a teenager when giving birth}

a. Find $P(A)$ and $P(B)$
b. Find $P(A \cup B)$
c. Find $P(A \cap B)$

TABLE 3.4   Percentage of New Jersey Birth Mothers in Age–Race Classes

Maternal Age (years)	Race	
	White	**Black**
≤17	2%	2%
18–19	3%	2%
20–29	41%	12%
≥30	33%	5%

Source: Reichman, N. E., and Pagnini, D. L. "Maternal age and birth outcomes: Data from New Jersey." *Family Planning Perspectives*, Vol. 29, No. 6, Nov./Dec. 1997, p. 269 (adapted from Table 1).

Solution

TEACHING TIP
When working with tables like Example 3.8, it is helpful to circle the values in the table that are in the event of interest.

Following the steps for calculating probabilities of events, we first note that the objective is to characterize the race and maternal age distribution of New Jersey birth mothers. To accomplish this, we define the experiment to consist of selecting a birth mother from the collection of all New Jersey birth mothers during the two-year period and observing her race and maternal age class. The sample points are the eight different age-race classifications:

E_1:	{≤17 yrs., white}	E_5:	{≤17 yrs., black}
E_2:	{18–19 yrs., white}	E_6:	{18–19 yrs., black}
E_3:	{20–29 yrs., white}	E_7:	{20–29 yrs., black}
E_4:	{≥30 yrs., white}	E_8:	{≥30 yrs., black}

Next, we assign probabilities to the sample points. If we blindly select one of the birth mothers, the probability that she will occupy a particular age–race classification is just the proportion, or relative frequency, of birth mothers in the classification. These proportions (as percentages) are given in Table 3.4. Thus,

$P(E_1)$ = Relative frequency of birth mothers in age–race class $\{\leq 17$ yrs., white$\}$ = .02
$P(E_2) = .03$
$P(E_3) = .41$
$P(E_4) = .33$
$P(E_5) = .02$
$P(E_6) = .02$
$P(E_7) = .12$
$P(E_8) = .05$

You may verify that the sample point probabilities add to 1.

a. To find $P(A)$, we first determine the collection of sample points contained in event A. Since A is defined as $\{$white$\}$, we see from Table 3.4 that A contains the four sample points represented by the first column of the table. In words, the event A consists of the race classification $\{$white$\}$ in all four age classifications. The probability of A is the sum of the probabilities of the sample points in A:

$$P(A) = P(E_1) + P(E_2) + P(E_3) + P(E_4) = .02 + .03 + .41 + .33 = .79$$

Similarly, $B = \{$teenage mother, age ≤ 19 years$\}$ consists of the four sample points in the first and second rows of Table 3.4:

$$P(B) = P(E_1) + P(E_2) + P(E_5) + P(E_6) = .02 + .03 + .02 + .02 = .09$$

b. The union of events A and B, $A \cup B$, consists of all sample points in *either A or B or both*. That is, the union of A and B consists of all birth mothers who are white *or* who gave birth as a teenager. In Table 3.4 this is any sample point found in the first column *or* the first two rows. Thus,

$$P(A \cup B) = .02 + .03 + .41 + .33 + .02 + .02 = .83$$

c. The intersection of events A and B, $A \cap B$, consists of all sample points in *both A and B*. That is, the intersection of A and B consists of all birth mothers who are white *and* who gave birth as a teenager. In Table 3.4 this is any sample point found in the first column *and* the first two rows. Thus,

$$P(A \cap B) = .02 + .03 = .05.$$

■

3.3   COMPLEMENTARY EVENTS

TEACHING TIP
Emphasize the word *not* when working with complementary events. The complement of event A is the event that A does not occur, or *not A*.

Suggested Exercise 3.32

A very useful concept in the calculation of event probabilities is the notion of **complementary events**:

> **DEFINITION 3.7**
> The **complement** of an event A is the event that A does *not* occur—that is, the event consisting of all sample points that are not in event A. We denote the complement of A by A^c.

FIGURE 3.8

Venn Diagram of
Complementary Events

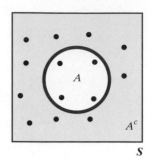

An event A is a collection of sample points, and the sample points included in A^c are those not in A. Figure 3.8 demonstrates this idea. Note from the figure that all sample points in S are included in *either A or A^c* and that *no* sample point is in both A and A^c. This leads us to conclude that the probabilities of an event and its complement *must sum to 1:*

Summing Probabilities of Complementary Events

The sum of the probabilities of complementary events equals 1; that is,

$$P(A) + P(A^c) = 1.$$

In many probability problems calculating the probability of the complement of the event of interest is easier than calculating the event itself. Then, because

$$P(A) + P(A^c) = 1$$

we can calculate $P(A)$ by using the relationship

$$P(A) = 1 - P(A^c).$$

EXAMPLE 3.9

$P(A)\ ^3\!/_4$

Consider the experiment of tossing two fair coins. (The sample space for this experiment is shown in Figure 3.9.) Calculate the probability of event A: {Observing at least one head}.

Solution

We know that the event A: {Observing at least one head} consists of the sample points

A: $\{HH, HT, TH\}$

The complement of A is defined as the event that occurs when A does not occur.

FIGURE 3.9

Complementary Events in
the Toss of Two Coins

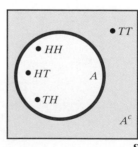

Therefore,

$$A^c: \{\text{Observe no heads}\} = \{TT\}$$

This complementary relationship is shown in Figure 3.9. Assuming the coins are balanced,

$$P(A^c) = P(TT) = \frac{1}{4}$$

and

$$P(A) = 1 - P(A^c) = 1 - \frac{1}{4} = \frac{3}{4}.$$

■

EXAMPLE 3.10

$P(A) = .999$

A fair coin is tossed 10 times, and the up face is recorded after each toss. What is the probability of event *A*: {Observe at least one head}?

Solution

We solve this problem by following the five steps for calculating probabilities of events (see Section 3.1).

TEACHING TIP
Use Example 3.10 or a similar example to illustrate the point that the complement approach makes solving this problem very simple.

Step 1 Define the experiment. The experiment is to record the results of the 10 tosses of the coin.

Step 2 List the sample points. A sample point consists of a particular sequence of ten heads and tails. Thus, one sample point is *HHTTTHTHTT*, which denotes head on first toss, head on second toss, tail on third toss, etc. Others are *HTHHHTTTTT* and *THHTHTHTTH*. Obviously, the number of sample points is very large—too many to list. It can be shown that there are $2^{10} = 1{,}024$ sample points for this experiment.

Step 3 Assign probabilities. Since the coin is fair, each sequence of heads and tails has the same chance of occurring, and therefore all the sample points are equally likely. Then

$$P(\text{Each sample point}) = \frac{1}{1{,}024}$$

Step 4 Determine the sample points in event *A*. A sample point is in *A* if at least one *H* appears in the sequence of 10 tosses. However, if we consider the complement of *A*, we find that

$$A^c = \{\text{No heads are observed in 10 tosses}\}$$

Thus, A^c contains only one sample point:

$$A^c: \quad \{TTTTTTTTTT\}$$

and $P(A^c) = \dfrac{1}{1{,}024}$

Step 5 Now we use the relationship of complementary events to find $P(A)$:

$$P(A) = 1 - P(A^c) = 1 - \frac{1}{1{,}024} = \frac{1{,}023}{1{,}024} = .999$$

That is, we are virtually certain of observing at least one head in 10 tosses of the coin.

■

3.4   THE ADDITIVE RULE AND MUTUALLY EXCLUSIVE EVENTS

TEACHING TIP
Stress that the additive rule is useful only when we are working with *or* probabilities, i.e., $P(A\ or\ B)$.

In Section 3.2 we saw how to determine which sample points are contained in a union and how to calculate the probability of the union by adding the probabilities of the sample points in the union. It is also possible to obtain the probability of the union of two events by using the *additive rule of probability*.

By studying the Venn diagram in Figure 3.10, you can see that the probability of the union of two events, A and B, can be obtained by summing $P(A)$ and $P(B)$ and subtracting $P(A \cap B)$. We must subtract $P(A \cap B)$ because the sample point probabilities in $A \cap B$ have been included twice—once in $P(A)$ and once in $P(B)$.

FIGURE 3.10

Venn Diagram of Union

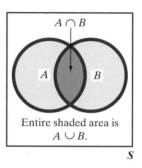

Entire shaded area is $A \cup B$.

TEACHING TIP
Use a table similar to the one used in Example 3.8 to give another illustration of how the additive rule works. It reinforces the Venn diagram.

The formula for calculating the probability of the union of two events is given in the next box.

TEACHING TIP
Point out that the additive rule uses the *intersection* of events A and B to find the probability of the *union* of events A and B.

Additive Rule of Probability

The probability of the union of events A and B is the sum of the probability of events A and B minus the probability of the intersection of events A and B, that is

$$P(A \cup B) = P(A) + P(B) - P(A \cap B)$$

EXAMPLE 3.11

.26

Hospital records show that 12% of all patients are admitted for surgical treatment, 16% are admitted for obstetrics, and 2% receive both obstetrics and surgical treatment. If a new patient is admitted to the hospital, what is the probability that the patient will be admitted either for surgery, obstetrics, or both?

Solution

Consider the following events:

    A: {A patient admitted to the hospital receives surgical treatment}
    B: {A patient admitted to the hospital receives obstetrics treatment}

Suggested Exercise 3.33

Then, from the given information

$$P(A) = .12$$
$$P(B) = .16$$

and the probability of the event that a patient receives both obstetrics and surgical treatment is

$$P(A \cap B) = .02$$

The event that a patient admitted to the hospital receives either surgical treatment, obstetrics treatment, or both is the union $A \cup B$. The probability of $A \cup B$ is given by the additive rule of probability:

$$P(A \cup B) = P(A) + P(B) - P(A \cap B)$$
$$= .12 + .16 - .02 = .26$$

Thus, 26% of all patients admitted to the hospital receive either surgical treatment, obstetrics treatment, or both.   ■

A very special relationship exists between events A and B when $A \cap B$ contains no sample points. In this case we call the events A and B *mutually exclusive events*.

DEFINITION 3.8
Events A and B are **mutually exclusive** if $A \cap B$ contains no sample points, that is, if A and B have no sample points in common.

Figure 3.11 shows a Venn diagram of two mutually exclusive events. The events A and B have no sample points in common, that is, A and B cannot occur simultaneously and $P(A \cap B) = 0$. Thus, we have the important relationship given in the box.

Probability of Union of Two Mutually Exclusive Events

If two events A and B are *mutually exclusive*, the probability of the union of A and B equals the sum of the probabilities of A and B; that is, $P(A \cup B) = P(A) + P(B)$.

Caution: The formula shown above is *false* if the events are *not* mutually exclusive. In this case (i.e., two nonmutually exclusive events), you must apply the general additive rule of probability.

FIGURE 3.11
Venn Diagram of Mutually Exclusive Events

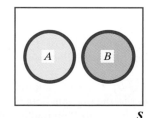

S

■
EXAMPLE 3.12   Consider the experiment of tossing two balanced coins. Find the probability of observing *at least* one head.

Solution

Define the events

    A: {Observe at least one head}
    B: {Observe exactly one head}
    C: {Observe exactly two heads}

FIGURE 3.12

Venn Diagram for Coin Toss Experiment

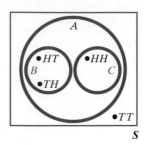

Note that

$$A = B \cup C$$

and that $B \cap C$ contains no sample points (see Figure 3.12). Thus, B and C are mutually exclusive, so that

$$P(A) = P(B \cup C) = P(B) + P(C) = \frac{1}{2} + \frac{1}{4} = \frac{3}{4}$$

∎

Although Example 3.12 is very simple, it shows us that writing events with verbal descriptions that include the phrases "at least" or "at most" as unions of mutually exclusive events is very useful. This practice enables us to find the probability of the event by adding the probabilities of the mutually exclusive events.

 EXERCISES 3.25–3.39

Learning the Mechanics

3.25 A fair coin is tossed three times and the events A and B are defined as follows:

A: {At least one head is observed}
B: {The number of heads observed is odd}

a. Identify the sample points in the events $A, B, A \cup B$, A^c, and $A \cap B$.
b. Find $P(A), P(B), P(A \cup B), P(A^c)$, and $P(A \cap B)$ by summing the probabilities of the appropriate sample points. $\frac{7}{8}, \frac{1}{2}, \frac{7}{8}, \frac{1}{8}, \frac{1}{2}$
c. Find $P(A \cup B)$ using the additive rule. Compare your answer to the one you obtained in part **b**. $\frac{7}{8}$
d. Are the events A and B mutually exclusive? Why?

3.26 A pair of fair dice is tossed. Define the following events:

A: {You will roll a 7 (i.e., the sum of the numbers of dots on the upper faces of the two dice is equal to 7)}
B: {At least one of the two dice is showing a 4}

a. Identify the sample points in the events $A, B, A \cap B$, $A \cup B$, and A^c.
b. Find $P(A), P(B), P(A \cap B), P(A \cup B)$, and $P(A^c)$ by summing the probabilities of the appropriate sample points.

c. Find $P(A \cup B)$ using the additive rule. Compare your answer to that for the same event in part **b**.
d. Are A and B mutually exclusive? Why?

3.27 Consider the Venn diagram, where $P(E_1) = P(E_2) = P(E_3) = \frac{1}{5}$, $P(E_4) = P(E_5) = \frac{1}{20}$, $P(E_6) = \frac{1}{10}$, and $P(E_7) = \frac{1}{5}$. Find each of the following probabilities.

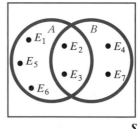

a. $P(A)$ $\frac{3}{4}$
b. $P(B)$ $\frac{13}{20}$
c. $P(A \cup B)$ 1
d. $P(A \cap B)$ $\frac{2}{5}$
e. $P(A^c)$ $\frac{1}{4}$
f. $P(B^c)$ $\frac{7}{20}$
g. $P(A \cup A^c)$ 1
h. $P(A^c \cap B)$ $\frac{1}{4}$

3.28 Consider the accompanying Venn diagram, where $P(E_1) = .10$, $P(E_2) = .05$, $P(E_3) = P(E_4) = .2$, $P(E_5) = .06$, $P(E_6) = .3$, $P(E_7) = .06$, and $P(E_8) = .03$. Find the following probabilities:

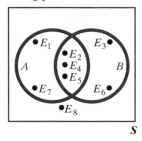

S

a. $P(A^c)$ .53
b. $P(B^c)$ .19
c. $P(A^c \cap B)$ .5
d. $P(A \cup B)$ .97
e. $P(A \cap B)$ .31
f. $P(A^c \cup B^c)$ .69
g. Are events A and B mutually exclusive? Why?   No

3.29 The outcomes of two variables are (Low, Medium, High) and (On, Off), respectively. An experiment is conducted in which the outcomes of each of the two variables are observed. The probabilities associated with each of the six possible outcome pairs are given in the accompanying table.

	Low	**Medium**	**High**
On	.50	.10	.05
Off	.25	.07	.03

Consider the following events:

A: {On}
B: {Medium or On}
C: {Off and Low}

a. Find $P(A)$.   .65
b. Find $P(B)$.   .72
c. Find $P(C)$.   .25
d. Find $P(A^c)$.   .35
e. Find $P(A \cup B)$.   .72
f. Find $P(A \cap B)$.   .10
g. Consider each pair of events (A and B, A and C, B and C). List the pairs of events that are mutually exclusive. Justify your choices. A and C, B and C

3.30 Three fair coins are tossed. We wish to find the probability of the event A: {Observe at least one head}.

a. Express A as the union of three mutually exclusive events. Find the probability of A using this expression. $\frac{7}{8}$
b. Express A as the complement of an event. Find the probability of A using this expression. $\frac{7}{8}$

Applying the Concepts—Basic

3.31 According to the National Center for Education Statistics (2000), only 5% of United States eighth graders score above 653 on a mathematics assessment test. Make a probability statement about the event that a randomly selected eighth grader has a score of 653 or below on the mathematics assessment test.   .95

3.32 A study of binge alcohol drinking by college students was published in the *American Journal of Public Health* (July 1995). Suppose an experiment consists of randomly selecting one of the undergraduate students who participated in the study. Consider the following events:

A: {The student is a binge drinker}
B: {The student is a male}
C: {The student lives in a coed dorm}

Describe each of the following events in terms of unions, intersections, and complements ($A \cup B$, $A \cap B$, A^c, etc.):
a. The student is male and a binge drinker. $A \cap B$
b. The student is not a binge drinker. A^c
c. The student is male or lives in a coed dorm. $B \cup C$
d. The student is female and not a binge drinker.

3.33 *Roulette* is a very popular game in many American casinos. In Roulette, a ball spins on a circular wheel that is divided into 38 arcs of equal length, bearing the numbers $00, 0, 1, 2, \ldots, 35, 36$. The number of the arc on which the ball stops is the outcome of one play of the game. The numbers are also colored in the manner shown in the table.

Red:	1, 3, 5, 7, 9, 12, 14, 16, 18, 19, 21, 23, 25, 27, 30, 32, 34, 36
Black:	2, 4, 6, 8, 10, 11, 13, 15, 17, 20, 22, 24, 26, 28, 29, 31, 33, 35
Green:	00, 0

Players may place bets on the table in a variety of ways, including bets on odd, even, red, black, high, low, etc. Consider the following events:

A: {Outcome is an odd number (00 and 0 are considered neither odd nor even)}
B: {Outcome is a black number}
C: {Outcome is a low number (1–18)}

a. Define the event $A \cap B$ as a specific set of sample points.
b. Define the event $A \cup B$ as a specific set of sample points.
c. Find $P(A)$, $P(B)$, $P(A \cap B)$, $P(A \cup B)$, and $P(C)$ by summing the probabilities of the appropriate sample points.
d. Define the event $A \cap B \cap C$ as a specific set of sample points.
e. Find $P(A \cup B)$ using the additive rule. Are events A and B mutually exclusive? Why?
f. Find $P(A \cap B \cap C)$ by summing the probabilities of the sample points given in part **d**.
g. Define the event $(A \cup B \cup C)$ as a specific set of sample points.
h. Find $P(A \cup B \cup C)$ by summing the probabilities of the sample points given in part **g**.

Applying the Concepts—Intermediate

3.34 *Ear and Hearing* (Apr. 1995) reported on a study of elderly hearing-impaired subjects who wear conventional analog hearing aids. Two of the variables measured for each hearing aid wearer were duration of experience with hearing aids and daily hearing aid use. The results are summarized in the table, where the numbers in the table represent the percentages of hearing aid wearers who fall into the respective categories. (Note that the percentages add to 100%.) Consider a randomly selected elderly hearing-impaired subject who wears a conventional analog hearing aid.

| Hearing Aid Experience | Daily Use (Hours) | | | | |
	Less than 1	1–4	4–8	8–16	Total
Less than 1 year	4	6	7	15	32
1–10 years	4	9	6	28	47
More than 10 years	0	3	6	12	21
Total	8	18	19	55	100

Source: Cox, R. M., and Alexander, G. C. "The abbreviated profile of hearing aid benefit." *Ear & Hearing*, Vol. 16, No. 2, Apr. 1995, p. 178 (Table 2).

a. List all the sample points for this experiment.
b. What is the set of all sample points called?
c. Let C be the event that an elderly hearing-impaired subject has more than 10 years of experience wearing hearing aids. Find $P(C)$ by summing the probabilities of the sample points in C.   .21
d. Let G be the event that an elderly hearing-impaired subject uses hearing aids 8–16 hours per day. Find $P(G)$.   .55
e. Let A be the event that an elderly hearing-impaired subject has less than 1 year of experience wearing hearing aids. Find $P(A)$.   .32
f. Let D be the event that an elderly hearing-impaired subject uses hearing aids less than 1 hour per day. Find $P(D)$.   .08
g. Let E be the event that an elderly hearing-impaired subject uses hearing aids between 1 and 4 hours per day. Find $P(E)$.   .18

3.35 Refer to Exercise 3.34. Define the characteristics of an elderly hearing-impaired person portrayed by the following events, then find the probability of each. For each union, use the additive rule to find the probability. Also, determine whether the events are mutually exclusive.
a. $A \cap G$ .15, no
b. $C \cup E$ .36, no
c. $C \cap D$ 0, yes
d. $A \cup G$ .72, no
e. $A \cup D$ .36, no

3.36 In professional sports, "stacking" is a term used to describe the practice of African-American players being excluded from certain positions because of race. To illustrate the stacking phenomenon, the *Sociology of Sport Journal* (Vol. 14, 1997) presented the table shown here. The table summarizes the race and positions of 368 National Basketball Association (NBA) players in 1993. Suppose an NBA player is selected at random from that year's player pool.

| | Position | | | |
	Guard	Forward	Center	Totals
White	26	30	28	84
Black	128	122	34	284
Totals	154	152	62	368

a. What is the probability that the player is white?   .228
b. What is the probability that the player is a center?
c. What is the probability that the player is African-American and plays guard?   .348
d. What is the probability that the player is not a guard?
e. What is the probability that the player is white or a center?   .320

3.37 A Victoria University psychologist investigated whether Asian immigrants to Australia differed from Anglo-Australians in their attitudes toward mental illness (*Community Mental Health Journal*, Feb. 1999). Each in a sample of 139 Australian students was classified according to ethnic group and degree of contact with mentally ill people. The number in each category is displayed in the table at the top of p. 131. Suppose we randomly select one of the 139 students.

a. Find the probability that the student is an Anglo-Australian who has had little or no contact with mentally ill people.   .216
b. Find the probability that the student has had close contact with mentally ill people.   .151
c. Find the probability that the student is not an Anglo-Australian.   .547
d. Find the probability that the student has had close contact with mentally ill people or is an Asian immigrant.   .475

3.38 The National Gang Crime Research Center (NGCRC) has developed a six-level gang classification system for both adults and juveniles. The NGCRC collected data on approximately 7,500 confined offenders and assigned each a score using the gang classification system (*Journal of Gang Research*, Winter 1997). One of several other variables measured by the NGCRC was whether or not the offender has ever carried a homemade weapon (e.g., knife) while in custody. The second table on p. 131 gives the number of confined offenders in each of the gang score and homemade weapon categories. Assume one of the confined offenders is randomly selected.

a. Find the probability that the offender has a gang score of 5.   .175
b. Find the probability that the offender has carried a homemade weapon.   .192
c. Find the probability that the offender has a gang score below 3.   .569

Mental Illness	Anglo-Australian	Short-Term Asian	Long-Term Asian	Other	Totals
Little or no contact	30	17	20	15	82
Some contact	17	5	3	11	36
Close contact	16	0	1	4	21
Totals	63	22	24	30	139

Source: Fan, C. "A comparison of attitudes towards mental illness and knowledge of mental health services between Australian and Asian students." *Community Mental Health Journal,* Vol. 35, No. 1, Feb. 1999, p. 54 (Table 2).

d. Find the probability that the offender has a gang score of 5 and has carried a homemade weapon.

e. Find the probability that the offender has a gang score of 0 or has never carried a homemade weapon.

f. Are the events described in parts **a** and **b** mutually exclusive? Explain.   No

g. Are the events described in parts **a** and **c** mutually exclusive? Explain.   Yes

Applying the Concepts—Advanced

3.39 A buyer for a large metropolitan department store must choose two firms from the four available to supply the store's fall line of men's slacks. The buyer has not dealt with any of the four firms before and considers their products equally attractive. Unknown to the buyer, two of the four firms are having serious financial problems that may result in their not being able to deliver the fall line of slacks as soon as promised. The four firms are identified as G_1 and G_2 (firms in good financial condition) and P_1 and P_2 (firms in poor financial condition). If the buyer selects the firms at random, find

a. The probability that firm P_1 is selected.

b. The probability that at least one of the selected firms is in good financial condition.

| | Weapon | | |
Gang Classification Score	Yes	No	Totals
0 (Never joined a gang, no close friends in a gang)	255	2,551	2,806
1 (Never joined a gang, 1–4 close friends in a gang)	110	560	670
2 (Never joined a gang, 5 or more friends in a gang)	151	636	787
3 (Inactive gang member)	271	959	1,230
4 (Active gang member, no position of rank)	175	513	688
5 (Active gang member, holds position of rank)	476	831	1,307
Totals	1,438	6,050	7,488

Source: Knox, G.W., *et al.* "A gang classification system for corrections." *Journal of Gang Research,* Vol. 4, No. 2, Winter 1997, p. 54 (Table 4).

3.5   CONDITIONAL PROBABILITY

The event probabilities we've been discussing give the relative frequencies of the occurrences of the events when the experiment is repeated a very large number of times. Such probabilities are often called **unconditional probabilities** because no special conditions are assumed, other than those that define the experiment.

Often, however, we have additional knowledge that might affect the outcome of an experiment, so we may need to alter the probability of an event of interest. A probability that reflects such additional knowledge is called the **conditional probability** of the event. For example, we've seen that the probability of observing an even number (event A) on a toss of a fair die is $1/2$. But suppose we're given the information that on a particular throw of the die the result was a number less than or equal to 3 (event B). Would the probability of observing an even number

FIGURE 3.13

Reduced Sample Space for the Die Toss Experiment: Given that Event B Has Occurred

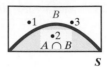

on that throw of the die still be equal to $\frac{1}{2}$? It can't be, because making the assumption that B has occurred reduces the sample space from six sample points to three sample points (namely, those contained in event B). This reduced sample space is as shown in Figure 3.13.

Because the sample points for the die-toss experiment are equally likely, each of the three sample points in the reduced sample space is assigned an equal *conditional probability* of $\frac{1}{3}$. Since the only even number of the three in the reduced sample space B is the number 2 and the die is fair, we conclude that the probability that A occurs *given that B occurs* is $\frac{1}{3}$. We use the symbol $P(A|B)$ to represent the probability of event A given that event B occurs. For the die-toss example

$$P(A|B) = \frac{1}{3}$$

To get the probability of event A given that event B occurs, we proceed as follows. We divide the probability of the part of A that falls within the reduced sample space B, namely $P(A \cap B)$, by the total probability of the reduced sample space, namely, $P(B)$. Thus, for the die-toss example with event A: {Observe an even number} and event B: {Observe a number less than or equal to 3}, we find

$$P(A|B) = \frac{P(A \cap B)}{P(B)} = \frac{P(2)}{P(1) + P(2) + P(3)} = \frac{\frac{1}{6}}{\frac{3}{6}} = \frac{1}{3}$$

The formula for $P(A|B)$ is true in general:

Formula for $P(A|B)$

To find the *conditional probability that event A occurs given that event B occurs*, divide the probability that *both A* and *B* occur by the probability that B occurs, that is,

$$P(A|B) = \frac{P(A \cap B)}{P(B)} \qquad \text{[We assume that } P(B) \neq 0.\text{]}$$

This formula adjusts the probability of $A \cap B$ from its original value in the complete sample space $\boldsymbol{S}$ to a conditional probability in the reduced sample space B. If the sample points in the complete sample space are equally likely, then the formula will assign equal probabilities to the sample points in the reduced sample space, as in the die-toss experiment. If, on the other hand, the sample points have unequal probabilities, the formula will assign conditional probabilities proportional to the probabilities in the complete sample space. This is illustrated by the following examples.

EXAMPLE 3.13

Many medical researchers have conducted experiments to examine the relationship between cigarette smoking and cancer. Consider an individual randomly selected from the adult male population. Let A represent the event that the individual smokes, and let A^c denote the complement of A (the event that the individual does

FIGURE 3.14

Sample Space for Example 3.13

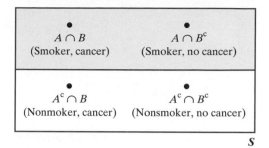

$$
\begin{array}{|cc|}
\hline
\bullet & \bullet \\
A \cap B & A \cap B^c \\
\text{(Smoker, cancer)} & \text{(Smoker, no cancer)} \\
& \\
\bullet & \bullet \\
A^c \cap B & A^c \cap B^c \\
\text{(Nonmoker, cancer)} & \text{(Nonsmoker, no cancer)} \\
\hline
\end{array}
$$

S

not smoke). Similarly, let B represent the event that the individual develops cancer, and let B^c be the complement of that event. Then the four sample points associated with the experiment are shown in Figure 3.14, and their probabilities for a certain section of the United States are given in Table 3.5. Use these sample point probabilities to examine the relationship between smoking and cancer.

TABLE 3.5   Probabilities of Smoking and Developing Cancer

	Develops Cancer	
Smoker	**Yes, B**	**No, B^c**
Yes, A	.05	.20
No, A^c	.03	.72

Solution

One method of determining whether these probabilities indicate that smoking and cancer are related is to compare the *conditional probability* that an adult male acquires cancer given that he smokes with the conditional probability that an adult male acquires cancer given that he does not smoke, i.e., compare $P(B|A)$ to $P(B|A^c)$.

First, we consider the reduced sample space A corresponding to adult male smokers. This reduced sample space is highlighted in Figure 3.14. The two sample points $A \cap B$ and $A \cap B^c$ are contained in this reduced sample space, and the adjusted probabilities of these two sample points are the two conditional probabilities:

$$
P(B|A) = \frac{P(A \cap B)}{P(A)} \quad \text{and} \quad P(B^c|A) = \frac{P(A \cap B^c)}{P(A)}
$$

The probability of event A is the sum of the probabilities of the sample points in A:

$$
P(A) = P(A \cap B) + P(A \cap B^c) = .05 + .20 = .25
$$

Then the values of the two conditional probabilities in the reduced sample space A are:

$$
P(B|A) = \frac{.05}{.25} = .20 \quad \text{and} \quad P(B^c|A) = \frac{.20}{.25} = .80
$$

These two numbers represent the probabilities that an adult male smoker develops cancer and does not develop cancer, respectively. Notice that the conditional probabilities .80 and .20 are in the same 4 to 1 ratio as the original unconditional probabilities, .20 and .05. The conditional probability formula simply adjusts the unconditional probabilities so that they add to 1 in the reduced sample space, A, of adult male smokers.

Suggested Exercise 3.42

In a like manner, the conditional probabilities of an adult male nonsmoker developing cancer and not developing cancer are

$$P(B|A^c) = \frac{P(A^c \cap B)}{P(A^c)} = \frac{.03}{.75} = .04$$

$$P(B^c|A^c) = \frac{P(A^c \cap B^c)}{P(A^c)} = \frac{.72}{.75} = .96$$

Observe that the conditional probabilities .96 and .04 are in the same 24 to 1 ratio as the unconditional probabilities .72 and .03.

Two of the conditional probabilities give some insight into the relationship between cancer and smoking: the probability of developing cancer given that the adult male is a smoker, and the probability of developing cancer given that the adult male is not a smoker. The conditional probability that an adult male smoker develops cancer (.20) is five times the probability that a nonsmoker develops cancer (.04). This does not imply that smoking *causes* cancer, but it does suggest a pronounced link between smoking and cancer.   ■

E X A M P L E   3 . 1 4

The investigation of consumer product complaints by the Federal Trade Commission (FTC) has generated much interest by manufacturers in the quality of their products. A manufacturer of an electromechanical kitchen utensil conducted an analysis of a large number of consumer complaints and found that they fell into the six categories shown in Table 3.6. If a consumer complaint is received, what is the probability that the cause of the complaint was product appearance given that the complaint originated during the guarantee period?

Suggested Exercise 3.53

TABLE 3.6   Distribution of Product Complaints

	Reason For Complaint			
	Electrical	**Mechanical**	**Appearance**	**Totals**
During Guarantee Period	18%	13%	32%	63%
After Guarantee Period	12%	22%	3%	37%
Totals	30%	35%	35%	100%

Solution

Let A represent the event that the cause of a particular complaint is product appearance, and let B represent the event that the complaint occurred during the guarantee period. Checking Table 3.6, you can see that $(18 + 13 + 32)\% = 63\%$ of the complaints occur during the guarantee period. Hence, $P(B) = .63$. The percentage of complaints that were caused by appearance and occurred during the guarantee period (the event $A \cap B$) is 32%. Therefore, $P(A \cap B) = .32$.

Using these probability values, we can calculate the conditional probability $P(A|B)$ that the cause of a complaint is appearance given that the complaint occurred during the guarantee time:

$$P(A|B) = \frac{P(A \cap B)}{P(B)} = \frac{.32}{.63} = .51$$

Consequently, we can see that slightly more than half the complaints that occurred during the guarantee period were due to scratches, dents, or other imperfections in the surface of the kitchen devices.   ■

![checkmark] EXERCISES 3.40–3.56

Learning the Mechanics

3.40 Consider the experiment defined by the accompanying Venn diagram, with the sample space **S** containing five sample points. The sample points are assigned the following probabilities: $P(E_1) = .1$, $P(E_2) = .1$, $P(E_3) = .2$, $P(E_4) = .5$, $P(E_5) = .1$.

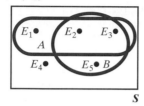

a. Calculate $P(A)$, $P(B)$, and $P(A \cap B)$.   .4, .4, .3

b. Suppose we know event A has occurred, so the reduced sample space consists of the three sample points in A: E_1, E_2, and E_3. Use the formula for conditional probability to determine the probabilities of these three sample points given that A has occurred. Verify that the conditional probabilities are in the same ratio to one another as the original sample point probabilities.   .25, .25, .5

c. Calculate the conditional probability $P(B|A)$, in two ways: First, add the adjusted (conditional) probabilities of the sample points in the intersection $A \cap B$, since these represent the event that B occurs given that A has occurred. Second, use the formula for conditional probability:

$$P(B|A) = \frac{P(A \cap B)}{P(A)}$$

Verify that the two methods yield the same result.   .75

3.41 Given that $P(A) = .3$, $P(B) = .6$, and $P(A \cap B) = .15$, find $P(A|B)$ and $P(B|A)$.   .25, .5

3.42 A sample space contains six sample points and events A, B, and C as shown in the Venn diagram. The probabilities of the sample points are $P(1) = .20$, $P(2) = .05$, $P(3) = .25$, $P(4) = .10$, $P(5) = .15$, $P(6) = .25$. Use the Venn diagram and the probabilities of the sample points to find:

a. $P(A)$, $P(B)$, and $P(C)$.   .5, .35, .40
b. $P(A \cap B)$, $P(A \cap C)$, and $P(B \cap C)$.   .25, 0, 0

c. Suppose you know that event A has occurred. Assign conditional probabilities to the three sample points contained in A. Verify that they add to 1 and are in the same ratio as the original (unconditional) probabilities.

d. Use the conditional probabilities from part **c** to calculate $P(B|A)$. Use the formula for $P(B|A)$ to verify your answer.   .5

e. Use the formula for conditional probability to calculate $P(C|A)$ and $P(C|A^c)$. Verify the results by inspection of the Venn diagram, remembering that the "given event" is a reduced sample space for a conditional probability.

3.43 Two fair coins are tossed and the events A and B are defined as follows:

A: {At least one head appears}
B: {Exactly one head appears}

a. Draw a Venn diagram for the experiment, labeling each sample point and showing events A and B. Assign probabilities to the sample points.

b. Find $P(A)$, $P(B)$, and $P(A \cap B)$. $\frac{3}{4}, \frac{1}{2}, \frac{1}{2}$

c. Use the formula for conditional probability to find $P(A|B)$ and $P(B|A)$. Verify your answers by inspecting the Venn diagram and using the concept of reduced sample spaces. $1, \frac{2}{3}$

3.44 A box contains two white, two red, and two blue poker chips. Two chips are randomly chosen without replacement and their colors are noted. Define the following events:

A: {Both chips are of the same color}
B: {Both chips are red}
C: {At least one chip is red or white}

Find $P(B|A)$, $P(B|A^c)$, $P(B|C)$, $P(A|C)$, and $P(C|A^c)$.

Applying the Concepts—Basic

3.45 To develop programs for business travelers staying at convention hotels, Hyatt Hotels Corp. commissioned a study of executives who play golf. The study revealed that 55% of the executives admitted they had cheated at golf. Also, 20% of the executives admitted they had cheated at golf and had lied in business. Given an executive had cheated at golf, what is the probability that the executive also had lied in business?   .364

3.46 *Forbes* (July 26, 1999) conducted a survey of the 20 largest non-domestic public companies in the world. Of these 20 companies, 6 were trading companies based in Japan. A total of 11 Japanese companies were on the top 20 list. Suppose we select one of these 20 companies at random. Given the company is based in Japan, what is the probability that it is a trading company?   .545

3.47 The New York Yankees, a member of the Eastern Division of the American League in Major League Baseball (MLB), recently won three consecutive World Series. The table summarizes the ten MLB World Series winners from 1990 to 2000 by division and league. (There was no World Series in 1994 due to a players' strike.) One of these ten World Series winners is to be chosen at random.

a. Given that the winner is a member of the American League, what is the probability that the winner plays in the Eastern Division?   .857

b. If the winner plays in the Central Division, what is the probability that the winner is a member of the National League?   .5

c. If the winner is a member of the National League, what is the probability that the winner plays in either the Central or Western Division?   .333

		League	
		National	**American**
Division	**Eastern**	2	6
	Central	1	1
	Western	0	0

Source: Major League Baseball.

3.48 Refer to the *Chance* (Fall 2000) study of ancient pottery found at the Greek settlement of Phylakopi, Exercise 2.8 (p. 26). Of the 837 pottery pieces uncovered at the excavation site, 183 were painted. These painted pieces included 14 painted in a curvilinear decoration, 165 painted in a geometric decoration, and 4 painted in a naturalistic decoration. Suppose one of the 837 pottery pieces is selected and examined.

a. What is the probability that the pottery piece is painted?   .219

b. Given that the pottery piece is painted, what is the probability that it is painted in a curvilinear decoration?   .077

3.49 Refer to the *Forbes* (May 17, 1999) salary survey of chief executive officers, Exercise 2.9 (p. 27). The data for the top 25 best-paid CEOs are reproduced in the table. Suppose you randomly select 5 of the CEOs (without replacement) and record the highest degree obtained by each.

a. What is the probability that the highest degree obtained by the first CEO you select is a bachelor's degree?   .28

b. Suppose the highest degree obtained by each of the first four CEOs you select is a bachelor's degree. What is the probability that the highest degree obtained by the fifth CEO you select is a bachelor's degree?   .143

CEODEGREES

CEO Name	Company	Degree
1. Michael D. Eisner	Walt Disney	Bachelors
2. Mel Karmazin	CBS	Bachelors
3. Stephen M. Case	America Online	Bachelors
4. Stephen C. Hilbert	Conseco	none
5. Craig R. Barrett	Intel	Doctorate
6. Millard Drexler	GAP	Masters
7. John F. Welch, Jr.	General Electric	Doctorate
8. Thomas G. Stemberg	Staples	Masters
9. Henry R. Silverman	Cendant	JD
10. Reuben Mark	Colgate-Palmolive	Masters
11. Philip J. Purcell	Morgan Stanley Dean Witter	Masters
12. Scott G. McNealy	Sun Microsystems	Masters
13. Margaret C. Whitman	eBay	Masters
14. Louis V. Gerstner, Jr.	IBM	Masters
15. John F. Gifford	Maxim Integrated Products	Bachelors
16. Robert L. Waltrip	Service Corp. International	Bachelors
17. M. Douglas Ivester	Coca-Cola	Bachelors
18. Gordon M. Binder	Amgen	Masters
19. Charles R. Schwab	Charles Schwab	Masters
20. William R. Steere, Jr.	Pfizer	Bachelors
21. Nolan D. Archibald	Black & Decker	Masters
22. Charles A. Heimbold, Jr.	Bristol-Myers Squibb	LLB (law)
23. William L. Larson	Network Associates	JD
24. Maurice R. Greenberg	American International Group	LLB (law)
25. Richard Jay Kogan	Schering-Plough	Masters

Source: Forbes, May 17, 1999.

Applying the Concepts—Intermediate

3.50 The *Journal of the National Cancer Institute* (Feb. 16, 2000) published the results of a study that investigated the association between cigar smoking and death from tobacco-related cancers. Data were obtained for a national sample of 137,243 American men. The results are summarized in the table below. Each male in the study was classified according to his cigar-smoking status and whether or not he died from a tobacco-related cancer.

| | **Died from Cancer** | | |
Cigars	**Yes**	**No**	**Totals**
Never Smoked	782	120,747	121,529
Former Smoker	91	7,757	7,848
Current Smoker	141	7,725	7,866
Totals	1,014	136,229	137,243

Source: Shapiro, J.A., Jacobs, E.J., and Thun, M.J. "Cigar Smoking in Men and Risk of Death from Tobacco-related Cancers." *Journal of the National Cancer Institute*, Vol. 92, No. 4, Feb. 16, 2000 (Table 2).

a. Find the probability that a randomly selected man never smoked cigars and died from cancer. .006

b. Find the probability that a randomly selected man was a former cigar smoker and died from cancer. .0007

c. Find the probability that a randomly selected man was a current cigar smoker and died from cancer. .001

d. Given that a male was a current cigar smoker, find the probability that he died from cancer. .018

e. Given that a male never smoked cigars, find the probability that he died from cancer. .006

3.51 Refer to the *Sociology of Sport Journal* (Vol. 14, 1997) study of "stacking" in the National Basketball Association (NBA), Exercise 3.36 (p. 130). Reconsider the table below, which summarizes the race and positions of 368 NBA players in 1993. Suppose an NBA player is selected at random from that year's player pool.

| | **Position** | | | |
	Guard	**Forward**	**Center**	**Totals**
White	26	30	28	84
Black	128	122	34	284
Totals	154	152	62	368

a. Given the player is white, what is the probability that he is a center? .333

b. Given the player is African-American, what is the probability that he is a center? .120

c. Are the events {White player} and {Center} independent? No

d. Recall that "stacking" refers to the practice of African-American players being excluded from certain positions because of race. Use your answers to parts **a**–**c** to make an inference about stacking in the NBA.

3.52 Refer to the *Community Mental Health Journal* study of 139 Australian students' contact with mentally ill patients, Exercise 3.37 (p. 130). The results are reproduced in the table below. Suppose one of the 139 students is selected at random.

| | **Contact** | | | |
Mental Illness	**Little/None**	**Some**	**Close**	**Totals**
Anglo-Australian	30	17	16	63
Short-Term Asian	17	5	0	22
Long-Term Asian	20	3	1	24
Other	15	11	4	30
Totals	82	36	21	139

Source: Fan, C. "A comparison of attitudes towards mental illness and knowledge of mental health services between Australian and Asian students." *Community Mental Health Journal,* Vol. 35, No. 1, Feb. 1999, p. 54 (Table 2).

a. Given the student is an Asian immigrant, what is the probability that he or she has had little or no contact with mentally ill patients? .804

b. Given the student is Anglo-Australian, what is the probability that he or she has had little or no contact with mentally ill patients? .476

c. Which ethnic group, Anglo-Australians or Asian immigrants, is most likely to have had little or no contact with mentally ill patients? Asian

3.53 A British crime survey examined the relationship between the race of the attacker, the race of the victim, and the degree of the injury sustained in reported crimes (*The Howard Journal of Criminal Justice*, Aug. 1992). The table on p. 138 is cited in the article.

a. What is the probability that a randomly selected reported crime involved a white attacker and a white victim? .557

b. What is the probability that a randomly selected reported crime involved serious injuries? .075

c. Given that both the attacker and victim were white, what is the probability that a randomly selected reported crime involved fatalities? .028

Degree of Injury	White/ White	White/ Nonwhite	Nonwhite/ Nonwhite	Nonwhite/ White	Totals
			Attacker/Victim		
Fatal	183	18	18	18	237
Serious	580	49	111	141	881
Slight	4,336	440	594	1,656	7,026
None	1,422	136	214	1,801	3,573
Totals	6,521	643	937	3,616	11,717

Source: Shah, R., and Pease, K. "Crime, race and reporting to the police." *The Howard Journal of Criminal Justice*, Volume 31, No. 3, Aug. 1992.

   d. If no injury was reported, what is the probability that a randomly selected reported crime involved a nonwhite attacker and a white victim?   .504

3.54 There are several methods of typing, or classifying, human blood. A method that is not as well known examines phosphoglucomutase (PGM) and classifies the blood into one of three main categories, 1-1, 2-1, or 2-2. Suppose a certain geographic region of the United States has the PGM percentages shown in the accompanying table. A person is to be chosen at random from this region.

Race	1-1	2-1	2-2
White	46.3%	39.2%	4.0%
Black	6.7%	3.4%	.4%

   a. What is the probability that a black person is chosen?

   b. Given that a black person is chosen, what is the probability he or she is PGM type 1-1?   .638

   c. Given that a white person is chosen, what is the probability he or she is PGM type 1-1?   .517

Applying the Concepts—Advanced

3.55 "Go" is one of the oldest and most popular strategic board games in the world, especially in Japan and Korea. The two-player game is played on a flat surface marked with 19 vertical and 19 horizontal lines. The objective is to control territory by placing pieces called "stones" on vacant points on the board. Players alternate placing their stones. The player using black stones goes first, followed by the player using white stones. [*Note:* The University of Virginia requires MBA students to learn Go to understand how the Japanese conduct business.] *Chance* (Summer 1995) published an article that investigated the advantage of playing first (i.e., using the black stones) in Go. The results of 577 games recently played by professional Go players were analyzed.

   a. In the 577 games, the player with the black stones won 319 times and the player with the white stones won 258 times. Use this information to estimate the probability of winning when you play first in Go.   .553

   b. Professional Go players are classified by level. Group C includes the top-level players, followed by Group B (middle-level) and Group A (low-level) players.

The table describes the number of games won by the player with the black stones, categorized by level of the black player and level of the opponent. Estimate the probability of winning when you play first in Go for each combination of player and opponent level.

   c. If the player with the black stones is ranked higher than the player with the white stones, what is the probability that black wins?   .856

   d. Given that the players are of the same level, what is the probability that the player with the black stones wins?

Black Player Level	Opponent Level	Number of Wins	Number of Games
C	A	34	34
C	B	69	79
C	C	66	118
B	A	40	54
B	B	52	95
B	C	27	79
A	A	15	28
A	B	11	51
A	C	5	39
	Totals	319	577

Source: J. Kim, and H. J. Kim, "The advantage of playing first in Go." *Chance*, Vol. 8, No. 3, Summer 1995, p. 26 (Table 3).

3.56 Physicians and pharmacists sometimes fail to inform patients adequately about the proper application of prescription drugs and the precautions to take in order to avoid potential side effects. One method of increasing patients' awareness of the problem is for physicians to provide Patient Medication Instruction (PMI) sheets. The American Medical Association, however, has found that only 20% of the doctors who prescribe drugs frequently distribute PMI sheets to their patients. Assume that 20% of all patients receive the PMI sheet with their prescriptions and that 12% receive the PMI sheet and are hospitalized because of a drug-related problem. What is the probability that a person will be hospitalized for a drug-related problem given that the person has received the PMI sheet?   .60

3.6   THE MULTIPLICATIVE RULE AND INDEPENDENT EVENTS

The probability of an intersection of two events can be calculated using the *multiplicative rule*, which employs the conditional probabilities we defined in the previous section. Actually, we've already developed the formula in another context. Recall that the conditional probability of B given A is

$$P(B|A) = \frac{P(A \cap B)}{P(A)}$$

Multiplying both sides of this equation by $P(A)$, we obtain a formula for the probability of the intersection of events A and B. This is often called the **multiplicative rule of probability**.

TEACHING TIP
Stress that the multiplicative rule is useful when solving *and* probabilities, i.e., $P(A \text{ and } B)$.

Multiplicative Rule of Probability

$P(A \cap B) = P(A)P(B|A)$,   or equivalently, $P(A \cap B) = P(B)P(A|B)$

EXAMPLE 3.15

.0005

An agriculturist, who is interested in planting wheat next year, is concerned with the following events:

B: {The production of wheat will be profitable}

A: {A serious drought will occur}

Based on available information, the agriculturist believes that the probability is .01 that production of wheat will be profitable *assuming* a serious drought will occur in the same year and that the probability is .05 that a serious drought will occur. That is,

$$P(B|A) = .01 \quad \text{and} \quad P(A) = .05$$

Based on the information provided, what is the probability that a serious drought will occur *and* that a profit will be made? That is, find $P(A \cap B)$, the probability of the intersection of events A and B.

Solution

We want to calculate $P(A \cap B)$. Using the formula for the multiplicative rule, we obtain

$$P(A \cap B) = P(A)P(B|A) = (.05)(.01) = .0005$$

The probability that a serious drought occurs *and* the production of wheat is profitable is only .0005. As we might expect, this intersection is a very rare event. ∎

Suggested Exercise 3.57

Intersections often contain only a few sample points. In this case, the probability of an intersection is easy to calculate by summing the appropriate sample point probabilities. However, the formula for calculating intersection probabilities is invaluable when the intersection contains numerous sample points, as the next example illustrates.*

---

*The multiplicative rule of probability also plays an important role in an area of statistics known as *Bayesian statistics*. Consult the references at the end of the chapter for detailed discussions of Bayesian statistics.

EXAMPLE 3.16

$\frac{1}{15}$

A county welfare agency employs 10 welfare workers who interview prospective food stamp recipients. Periodically the supervisor selects, at random, the forms completed by two workers to audit for illegal deductions. Unknown to the supervisor, three of the workers have regularly been giving illegal deductions to applicants. What is the probability that both of the two workers chosen have been giving illegal deductions?

Solution

Define the following two events:

    A: {First worker selected gives illegal deductions}
    B: {Second worker selected gives illegal deductions}

We want to find the probability of the event that both selected workers have been giving illegal deductions. This event can be restated as: {First worker gives illegal deductions *and* second worker gives illegal deductions}. Thus, we want to find the probability of the intersection, $A \cap B$. Applying the multiplicative rule, we have

$$P(A \cap B) = P(A)P(B|A)$$

To find $P(A)$ it is helpful to consider the experiment as selecting one worker from the 10. Then the sample space for the experiment contains 10 sample points (representing the 10 welfare workers), where the three workers giving illegal deductions are denoted by the symbol I (I_1, I_2, I_3), and the seven workers not giving illegal deductions are denoted by the symbol N $(N_1, \ldots, N_7)$. The resulting Venn diagram is illustrated in Figure 3.15.

FIGURE 3.15

Venn Diagram for Finding
$P(A)$

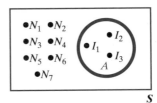

Suggested Exercise 3.73

Since the first worker is selected at random from the 10, it is reasonable to assign equal probabilities to the 10 sample points. Thus, each sample point has a probability of $\frac{1}{10}$. The sample points in event A are $\{I_1, I_2, I_3\}$—the three workers who are giving illegal deductions. Thus,

$$P(A) = P(I_1) + P(I_2) + P(I_3) = \frac{1}{10} + \frac{1}{10} + \frac{1}{10} = \frac{3}{10}$$

To find the conditional probability, $P(B|A)$, we need to alter the sample space S. Since we know A has occurred—i.e., the first worker selected is giving illegal deductions—only two of the nine remaining workers in the sample space are giving illegal deductions. The Venn diagram for this new sample space is shown in Figure 3.16.

FIGURE 3.16

Venn Diagram for Finding
$P(B|A)$

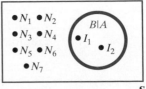

Each of these nine sample points is equally likely, so each is assigned a probability of $\frac{1}{9}$. Since the event $B|A$ contains the sample points $\{I_1, I_2\}$, we have

$$P(B|A) = P(I_1) + P(I_2) = \frac{1}{9} + \frac{1}{9} = \frac{2}{9}$$

Substituting $P(A) = \frac{3}{10}$ and $P(B|A) = \frac{2}{9}$ into the formula for the multiplicative rule, we find

$$P(A \cap B) = P(A)P(B|A) = \left(\frac{3}{10}\right)\left(\frac{2}{9}\right) = \frac{6}{90} = \frac{1}{15}$$

Thus, there is a 1 in 15 chance that both workers chosen by the supervisor have been giving illegal deductions to food stamp recipients. ∎

The sample space approach is only one way to solve the problem posed in Example 3.16. An alternative method employs the concept of a **tree diagram**. Tree diagrams are helpful for calculating the probability of an intersection.

To illustrate, a tree diagram for Example 3.16 is displayed in Figure 3.17. The tree begins at the far left with two branches. These branches represent the two possible outcomes N (no illegal deductions) and I (illegal deductions) for the first worker selected. The unconditional probability of each outcome is given (in parentheses) on the appropriate branch. That is, for the first worker selected, $P(N) = \frac{7}{10}$ and $P(I) = \frac{3}{10}$. (These can be obtained by summing sample point probabilities as in Example 3.16.)

The next level of the tree diagram (moving to the right) represents the outcomes for the second worker selected. The probabilities shown here are conditional probabilities since the outcome for the first worker is assumed to be known. For example, if the first worker is giving illegal deductions (I), the probability that the second worker is also giving illegal deductions (I) is $\frac{2}{9}$ since of the nine workers left to be selected, only two remain who are giving illegal deductions. This conditional probability, $\frac{2}{9}$, is shown in parentheses on the bottom branch of Figure 3.17.

TEACHING TIP
Point out that there are sometimes a couple of options to solving probability questions. The students should be free to choose that option that is the most comfortable for them. Example 3.14 in the last section and Example 3.15 are good illustrations of this.

FIGURE 3.17

Tree Diagram for Example 3.16

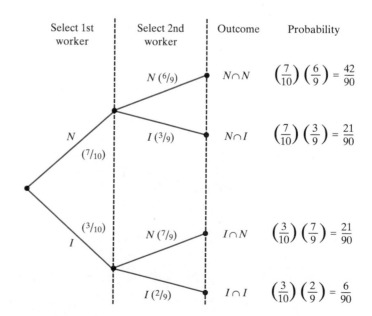

Finally, the four possible outcomes of the experiment are shown at the end of each of the four tree branches. These events are intersections of two events (outcome of first worker *and* outcome of second worker). Consequently, the multiplicative rule is applied to calculate each probability, as shown in Figure 3.17. You can see that the intersection $\{I \cap I\}$—the event that both workers selected are giving illegal deductions—has probability $^{6}\!/_{90} = {}^{1}\!/_{15}$, which is the same value obtained in Example 3.16.

In Section 3.5 we showed that the probability of event A may be substantially altered by the knowledge that an event B has occurred. However, this will not always be the case. In some instances the assumption that event B has occurred will *not* alter the probability of event A at all. When this occurs, we say that the two events A and B are *independent events*.

DEFINITION 3.9

Events A and B are **independent events** if the occurrence of B does not alter the probability that A has occurred; that is, events A and B are independent if

$$P(A|B) = P(A)$$

When events A and B are independent, it is also true that

$$P(B|A) = P(B)$$

Events that are not independent are said to be **dependent**.

EXAMPLE 3.17

Yes

Consider the experiment of tossing a fair die and let

$A = \{$Observe an even number$\}$

$B = \{$Observe a number less than or equal to 4$\}$

Are events A and B independent?

Solution

The Venn diagram for this experiment is shown in Figure 3.18. We first calculate

$$P(A) = P(2) + P(4) + P(6) = \frac{1}{2}$$

$$P(B) = P(1) + P(2) + P(3) + P(4) = \frac{4}{6} = \frac{2}{3}$$

$$P(A \cap B) = P(2) + P(4) = \frac{2}{6} = \frac{1}{3}$$

FIGURE 3.18

Venn Diagram for Die-Toss Experiment

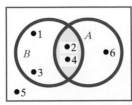

Now assuming B has occurred, the conditional probability of A given B is

$$P(A|B) = \frac{P(A \cap B)}{P(B)} = \frac{\frac{1}{3}}{\frac{2}{3}} = \frac{1}{2} = P(A)$$

Thus, assuming that event B occurs does not alter the probability of observing an even number—it remains $\frac{1}{2}$. Therefore, the events A and B are independent. Note that if we calculate the conditional probability of B given A, our conclusion is the same:

$$P(B|A) = \frac{P(A \cap B)}{P(A)} = \frac{\frac{1}{3}}{\frac{1}{2}} = \frac{2}{3} = P(B)$$

EXAMPLE 3.18

No

Refer to the consumer product complaint study in Example 3.14. The percentages of complaints of various types during and after the guarantee period are shown in Table 3.6 (p. 134). Define the following events:

    A: {Cause of complaint is product appearance}

    B: {Complaint occurred during the guarantee term}

Are A and B independent events?

Solution

Events A and B are independent if $P(A|B) = P(A)$. We calculated $P(A|B)$ in Example 3.14 to be .51, and from Table 3.6 we see that

$$P(A) = .32 + .03 = .35$$

Therefore, $P(A|B)$ is not equal to $P(A)$, and A and B are dependent events.

    To gain an intuitive understanding of independence, think of situations in which the occurrence of one event does not alter the probability that a second event will occur. For example, new medical procedures are often tested on laboratory animals. The scientists conducting the tests generally try to perform the procedures on the animals so the results for one animal do not affect the results for the others. That is, the event that the procedure is successful on one animal is *independent* of the result for another. In this way, the scientists can get a more accurate idea of the efficacy of the procedure than if the results were dependent, with the success or failure for one animal affecting the results for other animals.

    As a second example, consider an election poll in which 1,000 registered voters are asked their preference between two candidates. Pollsters try to use procedures for selecting a sample of voters so that the responses are independent. That is, the objective of the pollster is to select the sample so the event that one polled voter prefers candidate A does not alter the probability that a second polled voter prefers candidate A.

    Now consider the world of sports. Do you think the results of a batter's successive trips to the plate in baseball, or of a basketball player's successive shots at the basket, are independent? If a basketball player makes two successive shots, is the probability of making the next shot altered from its value if the result of the first shot is not known? If a player makes two shots in a row, the probability of a third successful shot is likely to be different from what we would assign if we knew nothing about the first two shots. Why should this be so? Research has shown that many such results in sports tend to be *dependent* because players (and even teams) tend to get on "hot" and "cold" streaks, during which their probabilities of success may increase or decrease significantly.

We will make three final points about independence. The first is that the property of independence, unlike the mutually exclusive property, cannot be shown on or gleaned from a Venn diagram. This means *you can't trust your intuition*. In general, the only way to check for independence is by performing the calculations of the probabilities in the definition.

The second point concerns the relationship between the mutually exclusive and independence properties. Suppose that events *A* and *B* are mutually exclusive, as shown in Figure 3.19, and both events have nonzero probabilities. Are these events independent or dependent? That is, does the assumption that *B* occurs alter the probability of the occurrence of *A*? It certainly does, because if we assume that *B* has occurred, it is impossible for *A* to have occurred simultaneously. That is, $P(A|B) = 0$. Thus, *mutually exclusive events are dependent events* since $P(A) \neq P(A|B)$.

FIGURE 3.19

Mutually Exclusive Events are Dependent Events

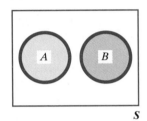

S

The third point is that the probability of the intersection of independent events is very easy to calculate. Referring to the formula for calculating the probability of an intersection, we find

$$P(A \cap B) = P(A)P(B|A)$$

Thus, since $P(B|A) = P(B)$ when *A* and *B* are independent, we have the following useful rule:

Probability of Intersection of Two Independent Events

If events A and B are independent, the probability of the intersection of *A* and *B* equals the product of the probabilities of *A* and *B*; that is

$$P(A \cap B) = P(A)P(B)$$

The converse is also true: If $P(A \cap B) = P(A)P(B)$, then events *A* and *B* are independent.

In the die-toss experiment, we showed in Example 3.17 that the two events *A*: {Observe an even number} and *B*: {Observe a number less than or equal to 4} are independent if the die is fair. Thus,

$$P(A \cap B) = P(A)P(B) = \left(\frac{1}{2}\right)\left(\frac{2}{3}\right) = \frac{1}{3}$$

This agrees with the result that we obtained in the example:

$$P(A \cap B) = P(2) + P(4) = \frac{2}{6} = \frac{1}{3}$$

EXAMPLE 3.19

a. 0625

b. .000001

Refer to Example 3.5 (p. 115). Recall that the American Association for Marriage and Family Therapy (AAMFT) found that 25% of divorced couples are classified as "fiery foes," i.e., they communicate through their children and are hostile towards each other.

a. What is the probability that in a sample of two divorced couples, both are classified as "fiery foes"?
b. What is the probability that in a sample of ten divorced couples, all ten are classified as "fiery foes"?

Solution

a. Let F_1 represent the event that divorced couple 1 is classified as "fiery foes" and F_2 represent the event that divorced couple 2 are also "fiery foes." The event that *both* couples are "fiery foes" is the intersection of the two events, $F_1 \cap F_2$. Based on the AAMFT survey that found that 25% of divorced couples are "fiery foes," we could reasonably conclude that $P(F_1) = .25$ and $P(F_2) = .25$. However, in order to compute the probability of $F_1 \cap F_2$ from the multiplicative rule, we must make the assumption that the two events are independent. Since the classification of any divorced couple is not likely to affect the classification of another divorced couple, this assumption is reasonable. Assuming independence, we have

Suggested Exercise 3.66

$$P(F_1 \cap F_2) = P(F_1)P(F_2) = (.25)(.25) = .0625$$

b. To see how to compute the probability that ten of ten divorced couples will all be classified as "fiery foes," first consider the event that three of three couples are "fiery foes." If F_3 represents the event that the third divorced couple are "fiery foes," then we want to compute the probability of the intersection $F_1 \cap F_2 \cap F_3$. Again assuming independence of the classifications, we have

$$P(F_1 \cap F_2 \cap F_3) = P(F_1)P(F_2)P(F_3) = (.25)(.25)(.25) = .015625$$

Similar reasoning leads us to the conclusion that the intersection of ten such events can be calculated as follows:

$$P(F_1 \cap F_2 \cap F_3 \cap \ldots \cap F_{10}) = P(F_1)P(F_2)\cdots P(F_{10}) = (.25)^{10} = .000001$$

Thus, the probability that ten of ten sampled divorced couples are all classified as "fiery foes" is 1 in 1 million, assuming the probability of each couple being classified as "fiery foes" is .25 and the classification decisions are independent. ■

We conclude this section with an example that uses probability and statistics to make an inference.

EXAMPLE 3.20

.004096

Suppose a research psychologist is studying the hypothesis that trained rats will pass on at least part of their training to their offspring. To test the hypothesis, three offspring (no two with the same parents) of trained rats are randomly selected and subjected to a training test. From many previous experiments the psychologist knows that the relative frequency distribution of the scores for untrained rats is mound-shaped and symmetric with a mean (μ) of 60 and a standard deviation (σ) of 10. Suppose all three of the trained rats' offspring score more than 70 on the test. Find the probability of this event. What can the research psychologist conclude?

Solution

The relative frequency distribution of the scores for untrained rats is shown in Figure 3.20. If the distribution is mound-shaped and approximately symmetric about the mean, we can conclude that approximately 16% of untrained rats will score more than 70 on the test (see Table 2.8).

FIGURE 3.20

Relative Frequency Distribution of Training Test Scores, Example 3.20

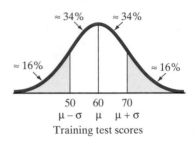

Now define the events

A_1: {Offspring 1 scores more than 70}

A_2: {Offspring 2 scores more than 70}

A_3: {Offspring 3 scores more than 70}

We want to find $P(A_1 \cap A_2 \cap A_3)$, the probability that all three offspring score more than 70 on the training test. Since the offspring are selected so that they have different parents, it may be plausible to assume that the events A_1, A_2, and A_3 are independent. That is,

$$P(A_2|A_1) = P(A_2)$$

In words, knowing that the first offspring scores more than 70 on the test does not affect the probability that the second offspring scores more than 70. With the assumption of independence, we can calculate the probability of the intersection by multiplying the individual probabilities:

$$P(A_1 \cap A_2 \cap A_3) = P(A_1)P(A_2)P(A_3)$$
$$\approx (.16)(.16)(.16) = .004096$$

Thus, the probability that the research psychologist will observe all three offspring scoring more than 70 is only about .004 *if the offspring are untrained*. If this event were to occur, the psychologist might conclude that it lends credence to the theory that the offspring inherit some of the parents' training *since it is so unlikely to occur if they are untrained*. Such a conclusion would be an application of the rare-event approach to statistical inference.

 EXERCISES 3.57–3.75

Learning the Mechanics

3.57 Three fair coins are tossed and the following events are defined:

A: {Observe at least one head}

B: {Observe exactly two heads}

C: {Observe exactly two tails}

D: {Observe at most one head}

a. Sum the probabilities of the appropriate sample points to find: $P(A)$, $P(B)$, $P(C)$, $P(D)$, $P(A \cap B)$, $P(A \cap D)$, $P(B \cap C)$, and $P(B \cap D)$.

b. Use your answers to part **a** to calculate $P(B|A)$, $P(A|D)$, and $P(C|B)$.

c. Which pairs of events, if any, are independent? Why?

3.58 An experiment results in one of five sample points with the following probabilities: $P(E_1) = .22$, $P(E_2) = .31$, $P(E_3) = .15$, $P(E_4) = .22$, and $P(E_5) = .1$. The following events have been defined:

A: $\{E_1, E_3\}$
B: $\{E_2, E_3, E_4\}$
C: $\{E_1, E_5\}$

Find each of the following probabilities:

a. $P(A)$ .37
b. $P(B)$ .68
c. $P(A \cap B)$ .15
d. $P(A|B)$ .221
e. $P(B \cap C)$ 0
f. $P(C|B)$ 0
g. Consider each pair of events: A and B, A and C, and B and C. Are any of the pairs of events independent? Why?   No

3.59 Two fair dice are tossed, and the following events are defined:

A: $\{$Sum of the numbers showing is odd$\}$
B: $\{$Sum of the numbers showing is 9, 11, or 12$\}$

Are events A and B independent? Why?   No

3.60 A sample space contains six sample points and events A, B, and C as shown in the accompanying Venn diagram. The probabilities of the sample points are $P(1) = .20$, $P(2) = .05$, $P(3) = .30$, $P(4) = .10$, $P(5) = .10$, $P(6) = .25$.

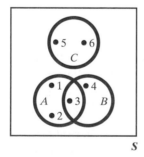

S

a. Which pairs of events, if any, are mutually exclusive? Why? $(A, C), (B, C)$
b. Which pairs of events, if any, are independent? Why?
c. Find $P(A \cup B)$ by adding the probabilities of the sample points and then by using the additive rule. Verify that the answers agree. Repeat for $P(A \cup C)$.

3.61 Defend or refute each of the following statements:

a. Dependent events are always mutually exclusive.
b. Mutually exclusive events are always dependent.
c. Independent events are always mutually exclusive.

3.62 For two events, A and B, $P(A) = .4$ and $P(B) = .2$.

a. If A and B are independent, find $P(A \cap B)$, $P(A|B)$, and $P(A \cup B)$.   .08, .4, .52
b. If A and B are dependent, with $P(A|B) = .6$, find $P(A \cap B)$ and $P(B|A)$.   .12, .30

Applying the Concepts—Basic

3.63 The National Center for Education Statistics (NCES) conducted a 1999 survey on the condition of America's public school facilities. The survey revealed the following information. The probability that a public school building has inadequate plumbing is .25. Of the buildings with inadequate plumbing, the probability that the school has plans for repairing the building is .38. Find the probability that a public school building has inadequate plumbing and will be repaired.   .095

3.64 According to the Children's Oncology Group Research Data Center at the University of Florida, 20% of children with neuroblastoma (a form of brain cancer) undergo surgery rather than the traditional treatment of chemotherapy or radiation. The surgery is successful in curing the disease 95% of the time (*Explore*, Spring 2001). Consider a child diagnosed with neuroblastoma. What are the chances that the child undergoes surgery and is cured?   .19

3.65 Consider an experiment in which 10 identical small boxes are placed side-by-side on a table. A crystal is placed, at random, inside one of the boxes. A self-professed "psychic" is asked to pick the box that contains the crystal.

a. If the "psychic" simply guesses, what is the probability that she picks the box with the crystal? $^1/_{10}$
b. If the experiment is repeated seven times, what is the probability that the "psychic" guesses correctly at least once?   .522
c. A group called the Tampa Bay Skeptics recently tested a self-proclaimed "psychic" using the above test. The "psychic" failed to pick the correct box all seven times (*Tampa Tribune*, Sept. 20, 1998). What would you infer about this person's psychic ability?

3.66 Refer to *The American Association of Nurse Anesthetists Journal* (Feb. 2000) study on the use of herbal medicines before surgery, Exercise 1.18 (p. 16). Recall that 51% of surgical patients use herbal medicines against their doctor's advice prior to surgery.

a. What is the probability that a randomly selected surgical patient will use herbal medicines against his or her doctor's advice?   .51
b. What is the probability that in a sample of two independently selected surgical patients, both will use herbal medicines against their doctor's advice?   .2601
c. What is the probability that in a sample of five independently selected surgical patients, all five will use herbal medicines against their doctor's advice?   .0345
d. Would you expect the event in part **c** to occur? Explain.   No

3.67 "Channel One" is an education television network for which participating secondary schools are equipped with TV sets in every classroom. According to *Educational Technology* (May–June 1995), 40% of all U.S. secondary schools subscribe to the Channel One Communications Network (CCN). Of these subscribers, 5% never use the CCN broadcasts, while 20% use CCN more than five times per week.

a. Find the probability that a randomly selected U.S. secondary school subscribes to CCN and never uses the CCN broadcasts.   .02

b. Find the probability that a randomly selected U.S. secondary school subscribes to CCN and uses the broadcasts more than five times per week.   .08

Applying the Concepts—Intermediate

3.68 In Italy, all high school students must take a High School Diploma (HSD) exam and write a paper. In *Organizational Behavior and Human Decision Processes* (July 2000), University of Milan researcher L. Macchi provided the following information to a group of college undergraduates. *Fact 1*: In Italy, 360 out of every 1,000 students fail their HSD exam. *Fact 2*: Of those who fail the HSD, 75% also fail the written paper. *Fact 3*: Of those who pass the HSD, only 20% fail the written paper. Define events A and B as follows:

A = {Student fails the HSD exam}
B = {Student fails the written paper}

a. Write Fact 1 as a probability statement involving events A and/or B. $P(A)$

b. Write Fact 2 as a probability statement involving events A and/or B. $P(B|A)$

c. Write Fact 3 as a probability statement involving events A and/or B. $P(B|A^c)$

d. State $P(A \cap B)$ in the words of the problem.

e. Find $P(A \cap B)$.   .27

3.69 A certain stretch of highway on the New Jersey Turnpike has a posted speed limit of 60 mph. A survey of drivers on this portion of the turnpike revealed the following facts (reported in *The Washington Post*, Aug. 16, 1998):

(1) 14% of the drivers on this stretch of highway were African Americans

(2) 98% of the drivers were exceeding the speed limit by at least 5 mph (and, thus, subject to being stopped by the state police)

(3) Of these violators, 15% of the drivers were African Americans

(4) Of the drivers stopped for speeding by the New Jersey state police, 35% were African Americans

a. Convert each of the reported percentages above into a probability statement.

b. Find the probability that a driver on this stretch of highway is African American and exceeding the speed limit.   .147

c. Given a driver on this stretch of highway is exceeding the speed limit, what is the probability the driver is not African American?   .85

d. Given that a driver on this stretch of highway is stopped by the New Jersey state police, what is the probability that the driver is not African American?   .65

e. For this stretch of highway, are the events {African American driver} and {Driver exceeding the speed limit} independent (to a close approximation)? Explain.   No

f. Use the probabilities in *The Washington Post* report to make a statement about whether African Americans are stopped more often for speeding than expected on this stretch of turnpike.

3.70 Enterococci are bacteria that cause blood infections in hospitalized patients. One antibiotic used to battle enterococci is vancomycin. A study by the Federal Centers for Disease Control and Prevention revealed that 8% of all enterococci isolated in hospitals nationwide were resistant to vancomycin (*New York Times*, Sept. 12, 1995). Consider a random sample of three patients with blood infections caused by the enterococci bacteria. Assume that all three patients are treated with the antibiotic vancomycin.

a. What is the probability that all three patients are successfully treated? What assumption did you make concerning the patients?   .7787

b. What is the probability that the bacteria resist the antibiotic for at least one patient?   .9995

3.71 The genetic origin and properties of maize (modern-day corn) was investigated in *Economic Botany* (Jan.–Mar. 1995). Seeds from maize ears carry either single spikelets or paired spikelets, but not both. Progeny tests on approximately 600 maize ears revealed the following information. Forty percent of all seeds carry single spikelets, while 60% carry paired spikelets. A seed with single spikelets will produce maize ears with single spikelets 29% of the time and paired spikelets 71% of the time. A seed with paired spikelets will produce maize ears with single spikelets 26% of the time and paired spikelets 74% of the time.

a. Find the probability that a randomly selected maize ear seed carries a single spikelet and produces ears with single spikelets.   .116

b. Find the probability that a randomly selected maize ear seed produces ears with paired spikelets.   .728

3.72 A new type of lie detector—called the Computerized Voice Stress Analyzer (CVSA)—has been developed. The manufacturer claims that the CVSA is 98% accurate, and, unlike a polygraph machine, will not be thrown off by drugs and medical factors. However, laboratory studies by the U.S. Defense Department found that the CVSA had an accuracy rate of 49.8%—slightly less than pure chance (*Tampa Tribune*, Jan. 10, 1999). Suppose the CVSA is used to test the veracity of four suspects. Assume the suspects' responses are independent.

a. If the manufacturer's claim is true, what is the probability that the CVSA will correctly determine the veracity of all four suspects?   .9224

b. If the manufacturer's claim is true, what is the probability that the CVSA will yield an incorrect result for at least one the four suspects?   .0776

c. Suppose that in a laboratory experiment conducted by the U.S. Defense Department on four suspects, the CVSA yielded incorrect results for two of the suspects. Use this result to make an inference about the true accuracy rate of the new lie detector.

Applying the Concepts—Advanced

3.73 One of the problems encountered in organ transplants is the body's rejection of the transplanted tissue. If the antigens attached to the tissue cells of the donor and receiver match, the body will accept the transplanted tissue. While the antigens in identical twins always match, the probability of a match in other siblings is .25 and that of a match in two people from the population at large is .001. Suppose you need a kidney, and you have two brothers and a sister.

   a. If one of your three siblings offers a kidney, what is the probability that the antigens will match? .25

   b. If all three siblings offer a kidney, what is the probability that all three antigens will match? .0156

   c. If all three siblings offer a kidney, what is the probability that none of the antigens will match? .4219

   a. Repeat parts **b** and **c**, this time assuming that the three donors were obtained from the population at large.

3.74 In October 1994, a flaw was discovered in the Pentium microchip installed in personal computers. The chip produced an incorrect result when dividing two numbers. Intel, the manufacturer of the Pentium chip, initially announced that such an error would occur once in 9 billion divides, or "once in every 27,000 years" for a typical user; consequently, it did not immediately offer to replace the chip. Assume the probability of a divide error with the Pentium chip is, in fact, 1/9,000,000,000.

   a. For a division performed using the flawed Pentium chip, what is the probability that no error will occur?

   b. Consider two successive divisions performed using the flawed chip. What is the probability that neither result will be in error? (Assume that any one division has no impact on the result of any other division performed by the chip.) .9999999998

   c. Depending on the procedure, statistical software packages may perform an extremely large number of divisions to produce the required output. For heavy users of the software, 1 billion divisions over a short time frame is not unusual. Calculate the probability that 1 billion divisions performed using the flawed Pentium chip will result in no errors. .8948

   d. Use the result, part **c**, to compute the probability of at least one error in the 1 billion divisions. [*Note:* Two months after the flaw was discovered, Intel agreed to replace all Pentium chips free of charge.] .1052

3.75 In *Parade Magazine*'s (Nov. 26, 2000) column "Ask Marilyn," the following question was posed: "I have just tossed a [balanced] coin 10 times, and I ask you to guess which of the following three sequences was the result. One (and only one) of the sequences is genuine."

(1) H H H H H H H H H H
(2) H H T T H T T H H H
(3) T T T T T T T T T T

   a. Demonstrate that prior to actually tossing the coins, the three sequences are equally likely to occur.

   b. Find the probability that the 10 coin tosses result in all heads or all tails.

   c. Find the probability that the 10 coin tosses result in a mix of heads and tails.

   d. Marilyn's answer to the question posed was: "Though the chances of the three specific sequences occurring randomly are equal . . . it's reasonable for us to choose sequence (2) as the most likely genuine result." If you know that only one of the three sequences actually occurred, explain why Marilyn's answer is correct. [*Hint:* Compare the probabilities in parts **b** and **c**.]

3.7 RANDOM SAMPLING

How a sample is selected from a population is of vital importance in statistical inference because the probability of an observed sample will be used to infer the characteristics of the sampled population. To illustrate, suppose you deal yourself four cards from a deck of 52 cards and all four cards are aces. Do you conclude that your deck is an ordinary bridge deck, containing only four aces, or do you conclude that the deck is stacked with more than four aces? It depends on how the cards were drawn. If the four aces were always placed at the top of a standard bridge deck, drawing four aces is not unusual—it is certain. On the other hand, if the cards are thoroughly mixed, drawing four aces in a sample of four cards is highly improbable. The point, of course, is that in order to use the observed sample of four cards to draw inferences about the population (the deck of 52 cards), you need to know how the sample was selected from the deck.

One of the simplest and most frequently employed sampling procedures is implied in the previous examples and exercises. It produces what is known as a *random sample*. We learned in Section 1.5 (p. 10) that a random sample is likely to be *representative* of the population that it is selected from.

> **DEFINITION 3.10**
>
> If *n* elements are selected from a population in such a way that every set of *n* elements in the population has an equal probability of being selected, the *n* elements are said to be a **random sample**.*

EXAMPLE 3.21

Suppose a lottery consists of 10 tickets. (This number is small to simplify our example.) One ticket stub is to be chosen, and the corresponding ticket holder will receive a generous prize. How would you select this ticket stub so that the prize will be awarded fairly?

Solution

If the prize is to be awarded fairly, it seems reasonable to require that each ticket stub have the same probability of being drawn. That is, each stub should have a probability of $1/10$ of being selected. A method to achieve the objective of equal selection probabilities is to *mix* the 10 stubs thoroughly and *blindly* pick one of the stubs. If this procedure were repeatedly used, each time replacing the selected stub, a particular stub should be chosen approximately $1/10$ of the time in a long series of draws. This method of sampling is known as **random sampling**.

If a population is not too large and the elements can be numbered on slips of paper, poker chips, etc., you can physically mix the slips of paper or chips and remove *n* elements from the total. The numbers that appear on the chips selected would indicate the population elements to be included in the sample. Since it is often difficult to achieve a thorough mix, such a procedure only provides an approximation to random sampling. Most researchers rely on **random number generators** to automatically generate the random sample. These random number generators are available in table form and they are built into most statistical software packages.

TEACHING TIP
Point out that the method of choosing the random sample is irrelevant as long as every possible subset of *n* has an equal chance of being selected.

You can think of sampling as an experiment consisting of drawing *n* elements from a population of *N* elements, with each different sample representing a sample point of the experiment. Thus, in Example 3.21, in which a lottery consisted of drawing one of 10 tickets, there are 10 different sample points. If the drawing is held so that each sample point is equally likely with probability $1/10$, then the result of the experiment is a *random sample*.

Of course, for most applications the population will consist of more than $N = 10$ elements, and the sample will consist of more than $n = 1$ element. Often, the total number of possible samples will not be easy to visualize, so a method for counting the number of samples is needed. For example, suppose the lottery consists of drawing two tickets from ten. We can list the possible samples as in Table 3.7, where T_1 represents ticket 1, T_2 ticket 2, ..., and T_{10} ticket 10. The systematic listing in Table 3.7 shows the 45 possible sample points of the experiment of sampling two elements from 10. However, the listing is tedious and only gets more so as the values of the population size *N* and the sample size *n* are increased.

A second method of determining the number of samples is to use **combinatorial mathematics**. The combinatorial symbol for the number of different ways of selecting *n* elements from *N* elements is $\binom{N}{n}$, which is read "the number of combinations of *N* elements taken *n* at a time." The formula for calculating the number is

---

*Strictly speaking, this is a *simple random sample*. There are many different types of random samples. The simple random sample is the most common.

TEACHING TIP
This counting rule will be seen again with the binomial distribution and can be introduced now to save confusion later.

TABLE 3.7 Listing of All Possible Samples of Two Tickets Drawn from 10 Tickets

T_1, T_2	T_2, T_3	T_3, T_4	T_4, T_5	T_5, T_6	T_6, T_7	T_7, T_8	T_8, T_9	T_9, T_{10}
T_1, T_3	T_2, T_4	T_3, T_5	T_4, T_6	T_5, T_7	T_6, T_8	T_7, T_9	T_8, T_{10}	
T_1, T_4	T_2, T_5	T_3, T_6	T_4, T_7	T_5, T_8	T_6, T_9	T_7, T_{10}		
T_1, T_5	T_2, T_6	T_3, T_7	T_4, T_8	T_5, T_9	T_6, T_{10}			
T_1, T_6	T_2, T_7	T_3, T_8	T_4, T_9	T_5, T_{10}				
T_1, T_7	T_2, T_8	T_3, T_9	T_4, T_{10}					
T_1, T_8	T_2, T_9	T_3, T_{10}						
T_1, T_9	T_2, T_{10}							
T_1, T_{10}								

$$\binom{N}{n} = \frac{N!}{n!(N-n)!}$$

where "!" is the factorial symbol and is a shorthand for the following multiplication:

$$n! = n(n-1)(n-2)\cdots(3)(2)(1)$$

Thus, for example, $5! = 5 \cdot 4 \cdot 3 \cdot 2 \cdot 1 = 120$. (The quantity $0!$ is defined to be 1.)

EXAMPLE 3.22

a. 10
b. 45

a. Use the combinatorial formula to count the number of different ways of drawing one lottery ticket from a total of 10 tickets.
b. Use the combinatorial formula to count the number of ways of drawing two lottery tickets from a total of 10 tickets.

Solution

a. Substituting $n = 1$ and $N = 10$ into the formula, we find

$$\binom{N}{n} = \binom{10}{1} = \frac{10!}{1!(10-1)!} = \frac{10!}{1!9!}$$
$$= \frac{10 \cdot 9 \cdot 8 \cdot 7 \cdot 6 \cdot 5 \cdot 4 \cdot 3 \cdot 2 \cdot 1}{(1)(9 \cdot 8 \cdot 7 \cdot 6 \cdot 5 \cdot 4 \cdot 3 \cdot 2 \cdot 1)} = 10$$

Thus, as we know intuitively, there are 10 different samples that can be selected when drawing one ticket from 10.

b. Substituting $n = 2$ and $N = 10$ into the formula, we find

$$\binom{N}{n} = \binom{10}{2} = \frac{10!}{2!(10-2)!} = \frac{10!}{2!8!}$$
$$= \frac{10 \cdot 9 \cdot 8 \cdot 7 \cdot 6 \cdot 5 \cdot 4 \cdot 3 \cdot 2 \cdot 1}{(2 \cdot 1)(8 \cdot 7 \cdot 6 \cdot 5 \cdot 4 \cdot 3 \cdot 2 \cdot 1)} = \frac{10 \cdot 9}{2 \cdot 1} = 45$$

This agrees with the number obtained by listing all possible samples in Table 3.7 but requires much less effort. And, as the next example illustrates, the combinatorial formula works long after listing has ceased to be practical.

EXAMPLE 3.23

a. 8.33×10^{22}

Suppose you wish to randomly sample five households from a population of 100,000 households.

a. How many different samples can be selected?
b. Use a random number generator to select a random sample.

Solution

a. Using the combinatorial rule, we find

$$
\binom{100,000}{5} = \frac{100,000!}{5!99,995!}
$$

$$
= \frac{100,000 \cdot 99,999 \cdot 99,998 \cdot 99,997 \cdot 99,996}{5 \cdot 4 \cdot 3 \cdot 2 \cdot 1}
$$

$$
= 8.33 \times 10^{22}
$$

Thus, there are 83.3 billion trillion different samples of five households that can be selected from 100,000.

b. To ensure that each of the possible samples has an equal chance of being selected, as required for random sampling, we can employ a **random number table**, as provided in Table I of Appendix A. Random number tables are constructed in such a way that every number occurs with (approximately) equal probability. Furthermore, the occurrence of any one number in a position is independent of any of the other numbers that appear in the table. To use a table of random numbers, number the N elements in the population from 1 to N.

Then turn to Table I and select a starting number in the table. Proceeding from this number either across the row or down the column, remove and record n numbers from the table.

To illustrate, first we number the households in the population from 1 to 100,000. Then, we turn to a page of Table I, say the first page. (A partial reproduction of the first page of Table I is shown in Table 3.8.) Now, we arbitrarily select a starting number, say the random number appearing in the third row, second column. This number is 48,360. Then we proceed down the second column to obtain the remaining four random numbers. In this case we have selected five random numbers, which are highlighted in Table 3.8. Using the first five digits to represent households from 1 to 99,999 and the number 00000 to represent household 100,000, we can see that the households numbered

| 48,360 | 93,093 | 39,975 | 6,907 | 72,905 |

should be included in our sample.

TEACHING TIP
Show that using $N = 10$ and $n = 2$ gives the same result as using $N = 10$ and $n = 8$. Also, write 10! as $10 \times 9 \times 8!$ and cancel with the 8! in the denominator to reduce the calculations necessary.

Suggested Exercise 3.79

TABLE 3.8   Partial Reproduction of Table I in Appendix A

Row \ Column	1	2	3	4	5	6
1	10480	15011	01536	02011	81647	91646
2	22368	46573	25595	85393	30995	89198
3	24130	48360	22527	97265	76393	64809
4	42167	93093	06243	61680	07856	16376
5	37670	39975	81837	16656	06121	91782
6	77921	06907	11008	42751	27756	53498
7	99562	72905	56420	69994	98872	31016
8	96301	91977	05463	07972	18876	20922
9	89579	14342	63661	10281	17453	18103
10	85475	36857	53342	53988	53060	59533
11	28918	69578	88231	33276	70997	79936
12	63553	40961	48235	03427	49626	69445
13	09429	93969	52636	92737	88974	33488

Note: Use only the necessary number of digits in each random number to identify the element to be included in the sample. If, in the course of recording the n numbers from the table, you select a number that has already been selected, simply discard the duplicate and select a replacement at the end of the sequence. Thus, you may have to record more than n numbers from the table to obtain a sample of n unique numbers.

Can we be sure that all 83.3 billion trillion samples have an equal chance of being selected? We can't; but to the extent that the random number table contains truly random sequences of digits, the sample should be very close to random. ∎

Table I in Appendix A is just one example of a random number generator. For most scientific studies that require a large random sample, computers are used to generate the random sample. The SAS, MINITAB, STATISTIX, and SPSS statistical software packages all have easy-to-use random number generators.

For example, suppose we required a random sample of $n = 50$ households from the population of 100,000 households in Example 3.23. Here, we might employ the SAS random number generator. Figure 3.21 shows an SAS printout listing 50 random numbers (from a population of 100,000). The households with these identification numbers would be included in the random sample.

FIGURE 3.21

SAS-Generated Random Sample of 50 Households

OBS	HOUSENUM	OBS	HOUSENUM	OBS	HOUSENUM	OBS	HOUSENUM
1	47122	14	47271	27	17098	40	4260
2	94231	15	3642	28	23259	41	58140
3	95531	16	7611	29	30512	42	22903
4	41445	17	81646	30	91548	43	65959
5	80287	18	92158	31	7673	44	13962
6	11731	19	36667	32	68549	45	25819
7	47523	20	71811	33	85433	46	66497
8	84847	21	78988	34	5231	47	79559
9	69822	22	3819	35	13455	48	87017
10	18270	23	21873	36	71666	49	28483
11	52636	24	74938	37	66280	50	91806
12	21750	25	23635	38	66210		
13	63363	26	35807	39	21998		

 EXERCISES 3.76–3.82

Learning the Mechanics

3.76 Suppose you wish to sample $n = 3$ elements from a total of $N = 6$ elements.
  a. Count the number of different samples that can be drawn, first by listing them, and then by using combinatorial mathematics. $n = 20$
  b. If random sampling is to be employed, what is the probability that any particular sample will be selected?
  c. Show how to use the random number table, Table I in Appendix A, to select a random sample of three elements from a population of six elements. Perform the sampling procedure 20 times. Do any two of the samples contain the same three elements? Given your answer to part **b**, did you expect repeated samples?

3.77 Suppose you wish to sample $n = 3$ elements from a total of $N = 600$ elements.
  a. Count the number of different samples by using combinatorial mathematics. $35,820,200$
  b. If random sampling is to be employed, what is the probability that any particular sample will be selected?
  c. Show how to use the random number table, Table I in Appendix A, to select a random sample of three elements from a population of 600 elements. Perform the sampling procedure 20 times. Do any two of the samples contain the same three elements? Given your answer to part **b**, did you expect repeated samples?
  d. Use the computer to generate a random sample of three from the population of 600 elements.

3.78 Suppose that a population contains $N = 200,000$ elements. Use a computer or Table I of Appendix A to select a random sample of $n = 10$ elements from the population. Explain how you selected your sample.

Applying the Concepts—Basic

3.79 To ascertain the effectiveness of their advertising campaigns, firms frequently conduct telephone interviews with consumers using *random-digit dialing*. With this method, a random number generator mechanically creates the sample of phone numbers to be called.
  a. Explain how the random number table (Table I of Appendix A) or a computer could be used to generate a sample of 7-digit telephone numbers.
  b. Use the procedure you described in part **a** to generate a sample of ten 7-digit telephone numbers.
  c. Use the procedure you described in part **a** to generate five 7-digit telephone numbers whose first three digits are 373.

3.80 In addition to its decennial enumeration of the population, the U.S. Bureau of the Census regularly samples the population for demographic information such as income, family size, employment, and marital status. Suppose the Bureau plans to sample 1,000 households in a city that has a total of 534,322 households. Show how the Bureau could use the random number table in Appendix A or a computer to generate the sample. Select the first 10 households to be included in the sample.

Applying the Concepts—Intermediate

3.81 In auditing a firm's financial statements, an auditor is required to assess the operational effectiveness of the accounting system. In performing the assessment, the auditor frequently relies on a random sample of actual transactions (Stickney and Weil, *Financial Accounting: An Introduction to Concepts, Methods, and Uses*, 1994).

A particular firm has 5,382 customer accounts that are numbered from 0001 to 5382.
  a. One account is to be selected at random for audit. What is the probability that account number 3,241 is selected? .000186
  b. Draw a random sample of 10 accounts and explain in detail the procedure you used.
  c. Refer to part **b**. The following are two possible random samples of size 10. Is one more likely to be selected than the other? Explain. No

Sample Number 1				
5011	0082	0963	0772	3415
2663	1126	0008	0026	4189

Sample Number 2				
0001	0003	0005	0007	0009
0002	0004	0006	0008	0010

Applying the Concepts—Advanced

3.82 Archaeologists plan to perform test digs at a location they believe was inhabited several thousand years ago. The site is approximately 10,000 meters long and 5,000 meters wide. They first draw rectangular grids over the area, consisting of lines every 100 meters, creating a total of $100 \cdot 50 = 5,000$ intersections (not counting one of the outer boundaries). The plan is to randomly sample 50 intersection points and dig at the sampled intersections. Explain how you could use a random number generator to obtain a random sample of 50 intersections. Develop at least two plans: one that numbers the intersections from 1 to 5,000 prior to selection and another that selects the row and column of each sampled intersection (from the total of 100 rows and 50 columns).

STATISTICS IN ACTION
Game Show Strategy: To Switch or Not to Switch?

Marilyn vos Savant, who is listed in *Guinness Book of World Records Hall of Fame* for "Highest IQ," writes a weekly column in the Sunday newspaper supplement, *Parade Magazine*. Her column, "Ask Marilyn," is devoted to games of skill, puzzles, and mind-bending riddles. In one issue, vos Savant posed the following question:

Suppose you're on a game show, and you're given a choice of three doors. Behind one door is a car; behind the others, goats. You pick a door—say, #1—and the host, who knows what's behind the doors, opens another door—say #3—which has a goat. He then says to you, "Do you want to pick door #2?" Is it to your advantage to switch your choice?

Marilyn's answer: "Yes, you should switch. The first door has a $\frac{1}{3}$ chance of winning [the car], but the second has a $\frac{2}{3}$ chance [of winning the car]." Predictably, vos Savant's surprising answer elicited thousands of critical letters, many of them from Ph.D. mathematicians, who disagreed with her. Some of the more interesting and critical letters, which were printed in her next column (*Parade Magazine*, Feb. 24, 1991) are condensed below:

"May I suggest you obtain and refer to a standard textbook on probability before you try to answer a question of this type again?" (University of Florida)

"Your logic is in error, and I am sure you will receive many letters on this topic from high school and college students. Perhaps you should keep a few addresses for help with future columns." (Georgia State University)

STATISTICS IN ACTION (*continued*)

"You are utterly incorrect about the game-show question, and I hope this controversy will call some public attention to the serious national crisis in mathematical education. If you can admit your error you will have contributed constructively toward the solution of a deplorable situation. How many irate mathematicians are needed to get you to change your mind?" (Georgetown University)

"I am in shock that after being corrected by at least three mathematicians, you still do not see your mistake." (Dickinson State University)

"You are the goat!" (Western State University)

"You're wrong, but look on the positive side. If all the Ph.D.'s were wrong, the country would be in serious trouble." (U.S. Army Research Institute)

The logic employed by those who disagree with vos Savant is as follows: Once the host shows you door #3 (a goat), only two doors remain. The probability of the car being behind door #1 (your door) is $\frac{1}{2}$; similarly, the probability is $\frac{1}{2}$ for door #2. Therefore, in the long run (i.e., over a long series of trials) it doesn't matter whether you switch to door #2 or keep door #1. Approximately 50% of the time you'll win a car, and 50% of the time you'll get the goat.

Who is correct, the Ph.D.'s or Marilyn? By answering the following series of questions, you'll arrive at the correct solution.

Focus

a. Before taping of the game show, the host randomly decides the door behind which to put the car; then the goats go behind the remaining two doors. List the sample points for this experiment.

b. Suppose you choose at random door #1. Now, form a new sample space for this conditional event as follows: For each sample point in part **a**, eliminate one of the remaining two doors that hides a goat. (This is the door that the host shows the contestant—always a goat.)

c. Refer to the altered sample points in part **b**. Assume your strategy is to keep door #1. Count the number of sample points for which this is a "winning" strategy (i.e., you win the car). Assuming equally likely sample points, what is the probability that you win the car?

d. Repeat part **c**, but assume your strategy is to always switch doors.

e. Based on the probabilities of parts **c** and **d**, is it to your advantage to switch your choice?

f. Repeat parts **b–e**, but assume you select door #2 at random.

g. Repeat parts **b–e**, but assume you select door #3 at random.

h. Demonstrate that your choice of doors does not impact your *long-run* strategy.

TEACHING TIP

Suggestions for class discussion of the case above can be found in the *Instructor's Guide*.

▸▸▸▸ QUICK REVIEW

Key Terms

Additive rule of probability   126
Combinatorial mathematics   150
Complementary event   123
Compound event   120
Conditional probability   131
Dependent events   142
Event   114
Experiment   108
Independent events   142

Intersection   120
Multiplicative rule
   of probability   139
Mutually exclusive events   127
Odds   120
Probability rules for sample
   points   112
Random number
   generator   150

Random number
   table   152
Random sample   150
Sample point   109
Sample space   109
Tree diagram   141
Unconditional probabilities   131
Union   120
Venn diagram   110

Key Formulas

$P(A) + P(A^c) = 1$	Complementary events 124
$P(A \cup B) = P(A) + P(B) - P(A \cap B)$	Additive rule 126
$P(A \cap B) = 0$	Mutually exclusive events 127
$P(A \cup B) = P(A) + P(B)$	Additive rule for mutually exclusive events 127

$P(A|B) = \dfrac{P(A \cap B)}{P(B)}$	Conditional probability 132
$P(A \cap B) = P(A)P(B|A) = P(B)P(A|B)$	Multiplicative rule 139
$P(A|B) = P(A)$	Independent events 142
$P(A \cap B) = P(A)P(B)$	Multiplicative rule for independent events 144
$\dbinom{N}{n} = \dfrac{N!}{n!(N-n)!}$ where $N! = N(N-1)(N-2)\cdots(2)(1)$	Combinatorial rule 151

Language Lab

Symbol	Pronunciation	Description
S		Sample space
$S: \{1,2,3,4,5\}$		Set of sample points, 1,2,3,4,5, in sample space
$A: \{1,2\}$		Set of sample points, 1,2, in event A
$P(A)$	Probability of A	Probability that event A occurs
$A \cup B$	A union B	Union of events A and B (either A or B or both occur)
$A \cap B$	A intersect B	Intersection of events A and B (both A and B occur)
A^c	A complement	Complement of event A (the event that A does not occur)
$P(A|B)$	Probability of A given B	Conditional probability that event A occurs given that event B occurs
$\dbinom{N}{n}$	N choose n	Number of combinations of N elements taken n at a time
$N!$	N factorial	Multiply $N(N-1)(N-2)\cdots(2)(1)$

 SUPPLEMENTARY EXERCISES 3.83–3.111

Learning the Mechanics

3.83 A fair die is tossed and the up face is noted. If the number is even, the die is tossed again; if the number is odd, a fair coin is tossed. Define the events:

A: {A head appears on the coin}
B: {The die is tossed only one time}

a. List the sample points in the sample space.
b. Give the probability for each of the sample points.
c. Find $P(A)$ and $P(B)$. $\frac{1}{4}, \frac{1}{2}$
d. Identify the sample points in A^c, B^c, $A \cap B$, and $A \cup B$.
e. Find $P(A^c)$, $P(B^c)$, $P(A \cap B)$, $P(A \cup B)$, $P(A|B)$, and $P(B|A)$. $\frac{3}{4}, \frac{1}{2}, \frac{1}{4}, \frac{1}{2}, \frac{1}{2}, 1$
f. Are A and B mutually exclusive events? Independent events? Why?   No, no

3.84 The accompanying Venn diagram illustrates a sample space containing six sample points and three events, A, B, and C. The probabilities of the sample points are: $P(1) = .3$, $P(2) = .2$, $P(3) = .1$, $P(4) = .1$, $P(5) = .1$, and $P(6) = .2$.

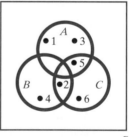

S

a. Find $P(A \cap B)$, $P(B \cap C)$, $P(A \cup C)$, $P(A \cup B \cup C)$, $P(B^c)$, $P(A^c \cap B)$, $P(B|C)$, and $P(B|A)$.
b. Are A and B independent? Mutually exclusive? Why?   No, yes
c. Are B and C independent? Mutually exclusive? Why?   No, no

3.85 Which of the following pairs of events are mutually exclusive? Justify your response.

a. St. Louis Cardinals win the World Series. Mark McGwire, Cardinals first baseman, hits 70 home runs.

b. An IBM notebook computer is purchased. An Apple notebook computer is purchased.   Yes

c. Subject A in a psychology experiment responds to a stimulus within 5 seconds. Subject A in a psychology experiment records the fastest response to a stimulus (2.3 seconds).   No

3.86 A balanced die is thrown once. If a 4 appears, a ball is drawn from urn 1; otherwise, a ball is drawn from urn 2. Urn 1 contains four red, three white, and three black balls. Urn 2 contains six red and four white balls.

a. Find the probability that a red ball is drawn.   .567

b. Find the probability that urn 1 was used given that a red ball was drawn.   .118

3.87 Two events, A and B, are independent, with $P(A) = .3$ and $P(B) = .1$.

a. Are A and B mutually exclusive? Why?   No

b. Find $P(A|B)$ and $P(B|A)$.   .3, .1

c. Find $P(A \cup B)$.   .37

3.88 A random sample of five students is to be selected from 50 sociology majors for participation in a special program.

a. In how many different ways can the sample be drawn?

b. Show how the random number table, Table I of Appendix A, can be used to select the sample of students.

Applying the Concepts—Basic

3.89 Refer to the National Highway Traffic Safety Administration (NHTSA) crash tests of new car models, Exercise 2.125 (p. 97). Recall that the NHTSA has developed a "star" scoring system, with results ranging from one star (*) to five stars (*****). The more stars in the rating, the better the level of crash protection in a head-on collision. A summary of the driver-side star ratings for 98 cars is reproduced in the accompanying MINITAB printout. Assume that one of the 98 cars is selected at random. State whether each of the following is true or false.

CRASH

MINITAB Output for Exercise 3.89

```
Tally for Discrete Variables: DrivStar
DrivStar   Count   Percent
       2       4      4.08
       3      17     17.35
       4      59     60.20
       5      18     18.37
      N=      98
```

a. The probability that the car has a rating of two stars is 4.

b. The probability that the car has a rating of four or five stars is .7857.   True

c. The probability that the car has a rating of one star is 0.

d. The car has a better chance of having a 2-star rating than of having a 5-star rating.   False

3.90 Refer to the *Journal of Agricultural, Biological, and Environmental Statistics* (Sept. 2000) study of the impact of the *Exxon Valdez* tanker oil spill on the seabird population in Prince William Sound, Alaska, Exercise 2.135

EVOS

(p. 102). Recall that data were collected on 96 shoreline locations (called transects), and it was determined whether or not the transect was in an oiled area. The data are stored in the file called EVOS. From the data, estimate the probability that a randomly selected transect in Prince William Sound is contaminated with oil.

3.91 According to a national survey conducted for CACI Marketing Systems, 25% of American adults smoke cigarettes. Of these smokers, 13% attempted (but failed) to quit smoking during the past year. Define the following events:

A: {An American adult smokes}
B: {A smoker attempted to quit smoking last year}

a. Find $P(A)$.   .25

b. Find $P(B|A)$.   .13

c. Find $P(A^c)$. State this probability in the words of the problem.   .75

d. Find $P(A \cap B)$. State this probability in the words of the problem.   .0325

3.92 The *American Journal of Public Health* (July 1995) published a study on unintentional carbon monoxide (CO) poisoning of Colorado residents. A total of 981 cases of CO poisoning were reported during a six-year period. Each case was classified as fatal or nonfatal and by source of exposure. The number of cases occurring in each of the categories is shown in the accompanying table. Assume that one of the 981 cases of unintentional CO poisoning is randomly selected.

Source of Exposure	Fatal	Nonfatal	Total
Fire	63	53	116
Auto exhaust	60	178	238
Furnace	18	345	363
Kerosene or spaceheater	9	18	27
Appliance	9	63	72
Other gas-powered motor	3	73	76
Fireplace	0	16	16
Other	3	19	22
Unknown	9	42	51
Total	**174**	**807**	**981**

Source: Cook, M. C., Simon P. A., and Hoffman, R. E., "Unintentional carbon monoxide poisoning in Colorado, 1986 through 1991." *American Journal of Public Health*, Vol. 85, No. 7, July 1995, p. 989 (Table 1).

a. List all sample points for this experiment.

b. What is the set of all sample points called?

c. Let A be the event that the CO poisoning is caused by fire. Find $P(A)$.   .118

d. Let B be the event that the CO poisoning is fatal. Find $P(B)$.   .177

e. Let C be the event that the CO poisoning is caused by auto exhaust. Find $P(C)$.   .243

f. Let D be the event that the CO poisoning is caused by auto exhaust and is fatal. Find $P(D)$.   .061

g. Let E be the event that the CO poisoning is caused by fire but is nonfatal. Find $P(E)$.   .054

h. Given that the source of the poisoning is fire, what is the probability that the case is fatal?   .543

i. Given that the case is nonfatal, what is the probability that it is caused by auto exhaust?   .221

j. If the case is fatal, what is the probability that the source is unknown?   .052

k. If the case is nonfatal, what is the probability that the source is not fire or a fireplace?   .914

3.93 Refer to the *Health Education Journal* (Sept. 1994) study of media coverage of stories involving mental illness in Scotland, Exercise 2.128 (p. 98). The media coverage of each of 562 stories was classified by type, with the results shown in the table. Assume that one of the 562 stories on mental illness is selected and the type of media coverage is noted.

Media Coverage	Number of Items
Violence to others	373
Sympathetic	102
Harm to self	71
Comic images	12
Criticism of definitions	4
Total	**562**

Source: Philo, G. *et al.* "The impact of the mass media on public images of mental illness: Media content and audience belief." *Health Education Journal*, Vol. 53, No. 3, Sept. 1994, p. 274 (Table 1).

a. List the sample points for this experiment.

b. Explain why the sample points, part **a**, are not equally likely to occur.

c. Assign reasonable probabilities to the sample points.

d. Find the probability that a randomly selected story on mental health is portrayed sympathetically or comically by the Scottish media.   .2028

3.94 The characteristics of families with young children were examined in *Children and Youth Services Review* (Vol. 17, 1995). Using data obtained from the National Child Care Survey, the income distribution and employment status of these families was summarized as shown in the table.

Income Characteristic	Percentage
No parent	1
Below poverty line; not employed	7
Below poverty line; employed	7
Above poverty line, but less than $25,000; not employed	2
Above poverty line, but less than $25,000; employed	22
$25,000 or more	61
Total	**100**

a. Find the probability that a randomly selected family with young children has an income above the poverty line, but less than $25,000.   .24

b. Find the probability that a randomly selected family with young children has unemployed parents or no parents.   .1

c. Find the probability that a randomly selected family with young children has an income below the poverty line.   .14

3.95 An entomologist is studying the effect of a chemical sex attractant (pheromone) on insects. Several insects are released at a site equidistant from the pheromone under study and a control substance. If the pheromone has an effect, more insects will travel toward it rather than toward the control. Otherwise, the insects are equally likely to travel in either direction. Suppose the pheromone under study has no effect, so that it is equally likely that an insect will move toward either the pheromone or the control. If five insects are released, what is the probability that

a. All five travel toward the pheromone? $\frac{1}{32}$

b. Exactly four? $\frac{5}{32}$

c. What inference would you make if the event in part **b** actually occurs? Explain.

Applying the Concepts—Intermediate

3.96 Psychologists tend to believe that there is a relationship between aggressiveness and order of birth. To test this belief, a psychologist chose 500 elementary school students at random and administered each a test designed to measure the student's aggressiveness. Each student was classified according to one of four categories. The percentages of students falling in the four categories are shown here.

	Firstborn	Not Firstborn
Aggressive	15%	15%
Not Aggressive	25%	45%

a. If one student is chosen at random from the 500, what is the probability that the student is firstborn?

b. What is the probability that the student is aggressive?

c. What is the probability that the student is aggressive, given the student was firstborn?   .375

d. If

A: {Student chosen is aggressive}
B: {Student chosen is firstborn}

are A and B independent? Explain.   No

3.97 In college basketball games a player may be afforded the opportunity to shoot two consecutive foul shots (free throws).

a. Suppose a player who makes (i.e., scores on) 80% of his foul shots has been awarded two free throws. If the two throws are considered independent, what is the probability that the player makes both shots? Exactly one? Neither shot?   .64, .32, .04

b. Suppose a player who makes 80% of his first attempted foul shots has been awarded two free throws, and the outcome on the second shot is dependent on the outcome of the first shot. In fact, if this player makes the first shot, he makes 90% of the second shots; and if he misses the first shot, he makes

70% of the second shots. In this case, what is the probability that the player makes both shots? Exactly one? Neither shot? .72, .22, .06

c. In parts **a** and **b**, we considered two ways of *modeling* the probability a basketball player makes two consecutive foul shots. Which model do you think is a more realistic attempt to explain the outcome of shooting foul shots; that is, do you think two consecutive foul shots are independent or dependent? Explain.

3.98 The probability that an Avon salesperson sells beauty products to a prospective customer on the first visit to the customer is .4. If the salesperson fails to make the sale on the first visit, the probability that the sale will be made on the second visit is .65. The salesperson never visits a prospective customer more than twice. What is the probability that the salesperson will make a sale to a particular customer? .79

3.99 Sandoz, a pharmaceutical firm, reports that kidney transplant patients who receive the drug cyclosporine have an 80% chance of surviving the first year. Suppose a hospital performs four kidney transplants, and all patients receive the drug cyclosporine. Assuming one patient's survival is independent of another's, what is the probability that

a. All four patients are alive at the end of 1 year?

b. None of the four patients is alive at the end of 1 year? .0016

c. At least one of the patients is alive at the end of 1 year? .9984

3.100 The system shown in the schematic below operates properly only if all three components operate properly. (The three components are said to operate *in series*.) The probability of failure for each component is listed in the accompanying table. Assume the components operate independently of each other.

Figure for Exercise 3.100

A System Comprised of Three Components in Series

Component	Probability of Failure
1	.12
2	.09
3	.11

a. Find the probability that the system operates properly.

b. What is the probability that at least one of the components will fail and therefore that the system will fail? .2873

3.101 The schematic below is a representation of a system that comprises two subsystems said to operate *in parallel*. Each subsystem has two components that operate in series (refer to Exercise 3.100). The system will operate properly as long as at least one of the subsystems functions properly. The probability of failure for each component in the system is .1. Assume the components operate independently of each other.

Figure for Exercise 3.101

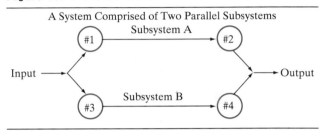

a. Find the probability that the system operates properly.

b. Find the probability that exactly one subsystem fails.

c. Find the probability that the system fails to operate properly. .0361

d. How many parallel subsystems like the two shown here would be required to guarantee that the system would operate properly at least 99% of the time? 3

3.102 A small brewery has two bottling machines. Machine A produces 75% of the bottles and machine B produces 25%. One out of every 20 bottles filled by A is rejected for some reason, while one out of every 30 bottles from B is rejected.

a. What proportion of bottles is rejected?

b. What is the probability that a randomly selected bottle comes from machine A, given that it is accepted?

3.103 Refer to *The Howard Journal of Criminal Justice* study of the relationship between the race of the attacker, the race of the victim, and the degree of injury sustained in reported crimes, Exercise 3.53 (p. 137). The information is reproduced below.

a. What is the probability that a randomly selected reported crime involved a nonwhite attacker? .3886

	Attacker/Victim				
Degree of Injury	White/White	White/Nonwhite	Nonwhite/Nonwhite	Nonwhite/White	Totals
Fatal	183	18	18	18	237
Serious	580	49	111	141	881
Slight	4,336	440	594	1,656	7,026
None	1,422	136	214	1,801	3,573
Totals	6,521	643	937	3,616	11,717

Source: Shah, R., and Pease, K. "Crime, race and reporting to the police." *The Howard Journal of Criminal Justice*, Volume 31, No. 3, Aug. 1992.

b. What is the probability that a randomly selected reported crime involved no injury?   .3049

c. What is the probability that a randomly selected reported crime involved both a nonwhite attacker and no injury?   .1720

d. Are the race of the attacker and the degree of injury independent?

3.104 A clinical psychologist is asked to view tapes in which each of six experimental subjects is discussing his or her recent dreams. Three of the six subjects have previously been classified as "high-anxiety" individuals and the other three as "low-anxiety." The psychologist is told only that there are three of each type and is asked to select the three high-anxiety subjects.

a. List all possible outcomes (sample points) for this experiment.

b. Assuming that the psychologist guesses at the classifications of the subjects, assign probabilities to the sample points.   All $\frac{1}{20}$

c. Find the probability that the psychologist guesses all classifications correctly. $\frac{1}{20}$

d. Find the probability that the psychologist guesses at least two of the three high-anxiety subjects correctly. $\frac{1}{2}$

Applying the Concepts—Advanced

3.105 In the game of Parcheesi each player rolls a pair of dice on each turn. In order to begin the game, you must throw a 5 on at least one of the dice, or a total of 5 on the two dice. What is the probability that you can begin the game on your first turn? The second turn? The third turn? The nth turn?

3.106 Pneumovax is an antipneumonia vaccine designed especially for elderly or debilitated patients who are usually the most vulnerable to bacterial pneumonia. The vaccine is 90% effective in stimulating the production of antibodies to pneumonia-producing bacteria (i.e., it is 90% successful in preventing a person exposed to pneumonia-producing bacteria from acquiring the disease). Suppose the probability that an elderly or debilitated person is exposed to these bacteria is .40 (whether inoculated or not) and, after being exposed, the probability that the person will contract bacterial pneumonia if not inoculated with the vaccine is .95. Find the probability that an elderly or debilitated person inoculated with this new vaccine acquires pneumonia. What is the probability if this person has not been inoculated?   .04, .38

3.107 Seventy-five percent of all women who submit to pregnancy tests are really pregnant. A certain pregnancy test gives a *false positive* result with probability .02 and a *valid positive result* with probability .99. If a particular woman's test is positive, what is the probability that she really is pregnant? [*Hint:* If A is the event that a woman is pregnant and B is the event that the pregnancy test is positive, then B is the union of the two mutually exclusive events $A \cap B$ and $A^c \cap B$. Also, the probability of a false positive result may be written as $P(B|A^c) = .02$.]   .993

3.108 Blackjack, a favorite game of gamblers, is played by a dealer and at least one opponent. At the outset of the game, two cards of a 52-card bridge deck are dealt to the player and two cards to the dealer. Drawing an ace and a face card is called *blackjack*. If the dealer does not draw a blackjack and the player does, the player wins. If both the dealer and player draw blackjack, a "push" (i.e., a tie) occurs.

a. What is the probability that the dealer will draw a blackjack?   .036

b. What is the probability that the player wins with a blackjack?   .035

3.109 Children who develop unexpected difficulties with acquisition of spoken language are often diagnosed with specific language impairment (SLI). A study published in the *Journal of Speech, Language, and Hearing Research* (Dec. 1997) investigated the incidence of SLI in kindergarten children. As an initial screen, each in a national sample of over 7,000 children was given a test for language performance. The percentages of children who passed and failed the screen were 73.8% and 26.2%, respectively. All children who failed the screen were tested clinically for SLI. About one-third of those who passed the screen were randomly selected and also tested for SLI. The percentage of children diagnosed with SLI in the "failed screen" group was 20.5%; the percentage diagnosed with SLI in the "pass screen" group was 2.8%.

a. For this problem, let "pass" represent a child who passed the language performance screen, "fail" represent a child who failed the screen, and "SLI" represent a child diagnosed with SLI. Now find each of the following probabilities: $P(\text{Pass})$, $P(\text{Fail})$, $P(\text{SLI}|\text{Pass})$, and $P(\text{SLI}|\text{Fail})$.   .738; .262; .028; .205

b. Use the probabilities, part **a**, to find $P(\text{Pass} \cap \text{SLI})$ and $P(\text{Fail} \cap \text{SLI})$. What probability law did you use to calculate these probabilities?   .0207; .0537

c. Use the probabilities, part **b**, to find $P(\text{SLI})$. What probability law did you use to calculate this probability?   .0744

3.110 The National Center for Health Statistics compiles data on suicide rates by race and gender. These data were used to estimate the probability of committing suicide during a lifetime in *Geriatric Psychiatry* (Vol. 27, 1994). The probabilities of lifetime suicide for four race/gender cohorts are given in the next table. For example, a black female selected at random has a $\frac{1}{535}$ chance of committing suicide during her lifetime.

Cohort	Probability of Suicide
Black female	$\frac{1}{535}$
White female	$\frac{1}{196}$
Black male	$\frac{1}{133}$
White male	$\frac{1}{66}$

a. Are the probabilities listed in the table conditional or unconditional probabilities? Explain.

b. Consider the following statement: "The probability that a randomly selected male commits suicide is obtained by summing the probabilities for black male and white male: $\frac{1}{133} + \frac{1}{66}$." Defend or refute this argument.

3.111 A version of the dice game "craps" is played in the following manner: A player starts by rolling two dice. If the result is a 7 or 11, the player wins. For most other sums appearing on the dice, the player contin-

ues to roll the dice until that sums recurs (in which case the player loses). But if on any roll the outcome is 2 or 3 (called "craps"), the game is over, and the player loses.

a. What is the probability that a player wins the game on the first roll of the dice? (Assume the dice are balanced.) $\frac{8}{36}$

b. What is the probability that a player loses the game on the first roll of the dice? $\frac{3}{36}$

c. If the player throws a total of 4 on the first roll, what is the probability that the game ends on the next roll?

STUDENT PROJECTS

Obtain a standard deck of 52 playing cards (the kind commonly used for bridge, poker, or solitaire). An experiment will consist of drawing one card at random from the deck of cards and recording which card was observed. This random drawing will be simulated by shuffling the deck thoroughly and observing the top card. Consider the following two events:

A: {Card observed is a heart}
B: {Card observed is an ace, king, queen, or jack}

a. Find $P(A)$, $P(B)$, $P(A \cap B)$, and $P(A \cup B)$.

b. Conduct the experiment 10 times and record the observed card each time. Be sure to return the observed card each time and thoroughly shuffle the deck before making the draw. After you've observed 10 cards, calculate the proportion of observations that satisfy event A,

event B, event $A \cap B$, and event $A \cup B$. Compare the observed proportions with the true probabilities calculated in part **a**.

c. Conduct the experiment 40 more times to obtain a total of 50 observed cards. Now calculate the proportion of observations that satisfy event A, event B, event $A \cap B$, and event $A \cup B$. Compare these proportions with those found in part **b** and the true probabilities found in part **a**.

d. Conduct the experiment 50 more times to obtain a total of 100 observations. Compare the observed proportions for the 100 trials with those found previously. What comments do you have concerning the different proportions found in parts **b**, **c**, and **d** as compared to the true probabilities found in part **a**? How do you think the observed proportions and true probabilities would compare if the experiment were conducted 1,000 times? 1 million times?

REFERENCES

Epstein, R. A. *The Theory of Gambling and Statistical Logic*, rev. ed. New York: Academic Press, 1977.

Feller, W. *An Introduction to Probability Theory and Its Applications*, 3d ed., Vol. 1. New York: Wiley, 1968.

Lindley, D. V. *Making Decisions*, 2d ed. London: Wiley, 1985.

Mosteller, F., Rourke, R., and Thomas, G. *Probability with Statistical Applications*, 2d ed. Reading, Mass.: Addison-Wesley, 1970.

Parzen, E. *Modern Probability Theory and Its Applications*. New York: Wiley, 1960.

Wackerly, D., Mendenhall, W., and Scheaffer, R. *Mathematical Statistics with Applications*, 6th ed. Boston: Duxbury, 2002.

Williams, B. *A Sampler on Sampling*. New York: Wiley, 1978.

Winkler, R. L. *An Introduction to Bayesian Inference and Decision*. New York: Holt, Rinehart and Winston, 1972.

Wright, G., and Ayton, P., eds. *Subjective Probability*. New York: Wiley, 1994.

CHAPTER 4

Random Variables and Probability Distributions

Contents

Statistics in Action

The Insomnia Pill

👊 Where We've Been

We saw by illustration in Chapter 3 how probability would be used to make an inference about a population from data contained in an observed sample. We also noted that probability would be used to measure the reliability of the inference.

☞ Where We're Going

Most of the experimental events we encountered in Chapter 3 were events described in words and denoted by capital letters. In real life, most sample observations are numerical—in other words, they are numerical data. In this chapter, we learn that data are observed values of random variables. We study two important random variables and learn how to find the probabilities of specific numerical outcomes. One of these, the *normal* random variable, has applications to many of the inferential methods encountered in this text.

You may have noticed that many of the examples of experiments in Chapter 3 generated quantitative (numerical) observations. The unemployment rate, the percentage of voters favoring a particular candidate, the cost of textbooks for a school term, and the amount of pesticide in the discharge waters of a chemical plant are all examples of numerical measurements of some phenomenon. Thus, most experiments have sample points that correspond to values of some numerical variable.

FIGURE 4.1
Venn Diagram for Coin Tossing Experiment

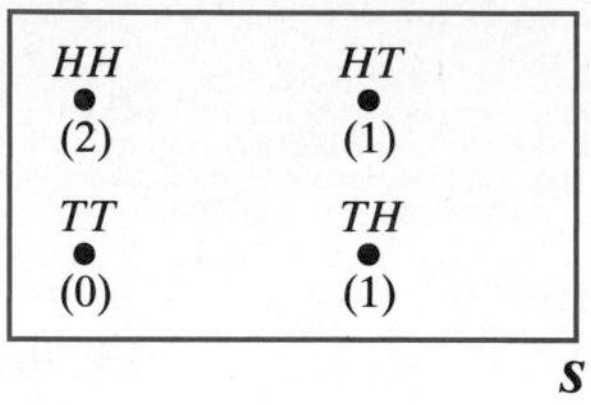

To illustrate, consider the coin-tossing experiment of Chapter 3. Figure 4.1 is a Venn diagram showing the sample points when two coins are tossed and the up faces (heads or tails) of the coins are observed. One possible numerical outcome is the total number of heads observed. These values (0, 1, or 2) are shown in parentheses on the Venn diagram, one numerical value associated with each sample point. In the jargon of probability, the variable "total number of heads observed in two tosses of a coin" is called a *random variable*.

DEFINITION 4.1

A **random variable** is a variable that assumes numerical values associated with the random outcomes of an experiment, where one (and only one) numerical value is assigned to each sample point.

TEACHING TIP
Use examples from probability (such as tossing coins) to illustrate the random variable assigning values to the outcomes of the experiment (i.e., x is the number of heads).

The term *random variable* is more meaningful than the term *variable* alone because the adjective *random* indicates that the coin-tossing experiment may result in one of the several possible values of the variable—0, 1, and 2—according to the *random* outcome of the experiment, *HH*, *HT*, *TH*, and *TT*. Similarly, if the experiment is to count the number of customers who use the drive-up window of a bank each day, the random variable (the number of customers) will vary from day to day, partly because of the random phenomena that influence whether customers use the drive-up window. Thus, the possible values of this random variable range from 0 to the maximum number of customers the window could possibly serve in a day.

We define two different types of random variables, *discrete* and *continuous*, in Section 4.1. Then we spend the remainder of this chapter discussing one specific type of discrete random variable and one specific type of continuous random variable.

4.1 TWO TYPES OF RANDOM VARIABLES

Recall that the sample point probabilities corresponding to an experiment must sum to 1. Dividing one unit of probability among the sample points in a sample space and consequently assigning probabilities to the values of a random variable is not always as easy as the examples in Chapter 3 might lead you to believe. If the number of sample points can be completely listed, the job is straightforward. But if the experiment results in an infinite number of sample points that are impossible to list, the task of assigning probabilities to the sample points is impossible without the aid of a probability model. The next three examples demonstrate the need for different probability models depending on the number of values that a random variable can assume.

EXAMPLE 4.1

31 values

A panel of 10 experts for the *Wine Spectator* (a national publication) is asked to taste a new white wine and assign a rating of 0, 1, 2, or 3. A score is then obtained by adding together the ratings of the 10 experts. How many values can this random variable assume?

Solution

A sample point is a sequence of 10 numbers associated with the rating of each expert. For example, one sample point is

$$\{1, 0, 0, 1, 2, 0, 0, 3, 1, 0\}.$$

TEACHING TIP
Explain that discrete random variables jump from one possible value to the next. For example, the number of heads is either 0 or 1 or 2 or 3 (for three coins). It could never be the value 1.5.

The random variable assigns a score to each one of these sample points by adding the 10 numbers together. Thus, the smallest score is 0 (if all 10 ratings are 0) and the largest score is 30 (if all 10 ratings are 3). Since every integer between 0 and 30 is a possible score, the random variable denoted by the symbol x can assume 31 values. Note that the value of the random variable for the sample point above is $x = 8$.*

This is an example of a *discrete random variable*, since there is a finite number of distinct possible values. Whenever all the possible values a random variable can assume can be listed (or *counted*), the random variable is *discrete*.

EXAMPLE 4.2

Suppose the Environmental Protection Agency (EPA) takes readings once a month on the amount of pesticide in the discharge water of a chemical company. If the amount of pesticide exceeds the maximum level set by the EPA, the company is forced to take corrective action and may be subject to penalty. Consider the following random variable:

Number, x, of months before the company's discharge exceeds the EPA's maximum level

$x = 1, 2, 3, 4\ldots$

What values can x assume?

Solution

The company's discharge of pesticide may exceed the maximum allowable level on the first month of testing, the second month of testing, etc. It is possible that the company's discharge will *never* exceed the maximum level. Thus, the set of possible values for the number of months until the level is first exceeded is the set of all positive integers 1, 2, 3, 4, ...

If we can list the values of a random variable x, even though the list is never-ending, we call the list *countable* and the corresponding random variable *discrete*. Thus, the number of months until the company's discharge first exceeds the limit is a *discrete random variable*.

---

*The standard mathematical convention is to use a capital letter (e.g., X) to denote the theoretical random variable. The possible values (or realizations) of the random variable are typically denoted with a lowercase letter (e.g., x). Thus, in Example 4.1, the random variable X can take on the values $x = 0, 1, 2, \ldots, 30$. Since this notation can be confusing for introductory statistics students, we simplify the notation by using the lowercase x to represent the random variable throughout.

EXAMPLE 4.3

Suggested Exercise 4.4

Refer to Example 4.2. A second random variable of interest is the amount x of pesticide (in milligrams per liter) found in the monthly sample of discharge waters from the chemical company. What values can this random variable assume?

Solution

Unlike the *number* of months before the company's discharge exceeds the EPA's maximum level, the set of all possible values for the *amount* of discharge *cannot* be listed—i.e., is not countable. The possible values for the amount x of pesticide would correspond to the points on the interval between 0 and the largest possible value the amount of the discharge could attain, the maximum number of milligrams that could occupy 1 liter of volume. (Practically, the interval would be much smaller, say, between 0 and 500 milligrams per liter.) When the values of a random variable are not countable but instead correspond to the points on some interval, we call it a *continuous random variable*. Thus, the *amount* of pesticide in the chemical plant's discharge waters is a *continuous random variable*. ∎

TEACHING TIP
For any two values of a continuous random variable, there are an infinite number of other possible values in between.

TEACHING TIP
Use many examples to illustrate the difference between discrete and continuous random variables. Try to use a similar situation to illustrate both types. For example, the number of checkout lanes open at a grocery store is a discrete random variable, while the amount of time standing in line is a continuous random variable.

DEFINITION 4.2
Random variables that can assume a *countable* number of values are called **discrete**.

DEFINITION 4.3
Random variables that can assume values corresponding to any of the points contained in one or more intervals are called **continuous**.

The following are examples of discrete random variables:

1. The number of seizures an epileptic patient has in a given week: $x = 0, 1, 2, \ldots$
2. The number of voters in a sample of 500 who favor impeachment of the president: $x = 0, 1, 2, \ldots, 500$
3. The number of students applying to medical schools this year: $x = 0, 1, 2, \ldots$
4. The number of errors on a page of an accountant's ledger: $x = 0, 1, 2, \ldots$
5. The number of customers waiting to be served in a restaurant at a particular time: $x = 0, 1, 2, \ldots$

Note that each of the examples of discrete random variables begins with the words "The number of ... ." This wording is very common, since the discrete random variables most frequently observed are counts. The following are examples of continuous random variables:

1. The length of time (in seconds) between arrivals at a hospital clinic: $0 \leq x \leq \infty$ (infinity)
2. The length of time (in minutes) it takes a student to complete a one-hour exam: $0 \leq x \leq 60$
3. The amount (in ounces) of carbonated beverage loaded into a 12-ounce can in a can-filling operation: $0 \leq x \leq 12$
4. The depth (in feet) at which a successful oil drilling venture first strikes oil: $0 \leq x \leq c$, where c is the maximum depth obtainable
5. The weight (in pounds) of a food item bought in a supermarket: $0 \leq x \leq 500$ [*Note:* Theoretically, there is no upper limit on x, but it is unlikely that it would exceed 500 pounds.]

In the succeeding sections, we will explain how to construct probability models for both discrete and continuous random variables. Then we will describe the properties of two random variables often encountered in the real world, and see how we can apply their probability distributions to solve practical problems.

EXERCISES 4.1–4.5

Applying the Concepts—Basic

4.1 What is a random variable?

4.2 How do discrete and continuous random variables differ?

4.3 Classify the following random variables according to whether they are discrete or continuous:
   a. The number of words spelled correctly by a student on a spelling test   Discrete
   b. The amount of water flowing through the Hoover Dam in a day   Continuous
   c. The length of time an employee is late for work
   d. The number of bacteria in a particular cubic centimeter of drinking water   Discrete
   e. The amount of carbon monoxide produced per gallon of unleaded gas   Continuous
   f. Your weight   Continuous

4.4 Identify the following random variables as discrete or continuous:
   a. The amount of flu vaccine in a syringe   Continuous

b. The heart rate (number of beats per minute) of an American male   Discrete
   c. The time it takes a student to complete an examination   Continuous
   d. The barometric pressure at a given location
   e. The number of registered voters who vote in a national election   Discrete
   f. Your score on the Scholastic Assessment Test (SAT)

4.5 Identify the following variables as discrete or continuous:
   a. The reaction time difference to the same stimulus before and after training   Continuous
   b. The number of violent crimes committed per month in your community   Discrete
   c. The number of commercial aircraft near-misses per month   Discrete
   d. The number of winners each week in a state lottery
   e. The number of free throws made per game by a basketball team   Discrete
   f. The distance traveled by a school bus each day

TEACHING TIP
Probability distributions can be presented as tables, graphs, or formulas. The key idea is that all forms give the possible values of the random variable, x, and the corresponding probability of observing those values of x.

4.2   PROBABILITY DISTRIBUTIONS FOR DISCRETE RANDOM VARIABLES

A complete description of a discrete random variable requires that we *specify the possible values the random variable can assume* and *the probability associated with each value*. To illustrate, consider Example 4.4.

EXAMPLE 4.4

Recall the experiment of tossing two coins (Section 4.1), and let x be the number of heads observed. Find the probability associated with each value of the random variable x, assuming the two coins are fair.

Solution

The sample space and sample points for this experiment are reproduced in Figure 4.2. Note that the random variable x can assume values $0, 1, 2$. Recall (from Chapter 3) that the probability associated with each of the four sample points is $\frac{1}{4}$. Then, identifying the probabilities of the sample points associated with each of these values of x, we have

$P(0) = \frac{1}{4}$
$P(1) = \frac{1}{2}$
$P(2) = \frac{1}{4}$

$$P(x = 0) = P(TT) = \frac{1}{4}$$

$$P(x = 1) = P(TH) + P(HT) = \frac{1}{4} + \frac{1}{4} = \frac{1}{2}$$

$$P(x = 2) = P(HH) = \frac{1}{4}$$

Thus, we now know the values the random variable can assume $(0, 1, 2)$ and how the probability is *distributed over* these values $\left(\frac{1}{4}, \frac{1}{2}, \frac{1}{4}\right)$. This completely describes

FIGURE 4.2

Venn Diagram for the Two-
Coin-Toss Experiment

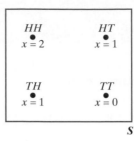

the random variable and is referred to as the *probability distribution*, denoted by the symbol $p(x)$.* The probability distribution for the coin-toss example is shown in tabular form in Table 4.1 and in graphic form in Figure 4.3. Since the probability distribution for a discrete random variable is concentrated at specific points (values of x), the graph in Figure 4.3a represents the probabilities as the heights of vertical lines over the corresponding values of x. Although the representation of the probability distribution as a histogram, as in Figure 4.3b, is less precise (since the probability is spread over a unit interval), the histogram representation will prove useful when we approximate probabilities of a discrete random variable in Section 4.7.

Suggested Exercise 4.25

Table 4.1   Probability Distribution for Coin-Toss Experiment: Tabular Form

x	$p(x)$
0	$1/4$
1	$1/2$
2	$1/4$

FIGURE 4.3

Probability Distribution for Coin-Toss Experiment: Graphical Form

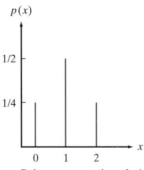

a. Point representation of $p(x)$

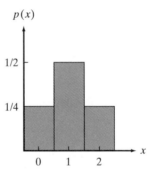

b. Histogram representation of $p(x)$

We could also present the probability distribution for x as a formula, but this would unnecessarily complicate a very simple example. We give the formulas for the probability distributions of some common discrete random variables later in this chapter. ∎

DEFINITION 4.4

The **probability distribution of a discrete random variable** is a graph, table, or formula that specifies the probability associated with each possible value the random variable can assume.

---

*In standard mathematical notation, the probability that a random variable X takes on a value x is denoted $P(X = x) = p(x)$. Thus, $P(X = 0) = p(0)$, $P(X = 1) = p(1)$, etc. In this text, we adopt the simpler $p(x)$ notation.

TEACHING TIP
Let the student view the probability distribution as a way of organizing the outcomes that were discussed in the probability chapter. For example, organize the eight outcomes of the three-coin toss example by grouping together all outcomes that result in the same number of heads.

Two requirements must be satisfied by all probability distributions for discrete random variables.

Requirements for the Probability Distribution of a Discrete Random Variable x

(1) $p(x) \geq 0$ for all values of x
(2) $\Sigma p(x) = 1$

where the summation of $p(x)$ is over all possible values of x.*

Example 4.4 illustrates how the probability distribution for a discrete random variable can be derived, but for many practical situations the task is much more difficult. Fortunately, many experiments and associated discrete random variables observed in nature possess identical characteristics. Thus, you might observe a random variable in a psychology experiment that would possess the same probability distribution as a random variable observed in an engineering experiment or a social sample survey. We classify random variables according to type of experiment, derive the probability distribution for each of the different types, and then use the appropriate probability distribution when a particular type of random variable is observed in a practical situation. The probability distributions for most commonly occurring discrete random variables have already been derived. (We describe one of these in Section 4.3.) This fact simplifies the problem of finding the probability distributions for random variables.

TEACHING TIP
Expected value is another name for mean, which measures the center of the probability distribution.

TEACHING TIP
Use any of the probability distributions from earlier in this section to illustrate the calculations of the expected value and variance (below). Discuss the interpretation of these values as they relate to the values of the random variable.

Since probability distributions are analogous to the relative frequency distributions of Chapter 2, it should be no surprise that the mean and standard deviation are useful descriptive measures. For example, if a discrete random variable x were observed a very large number of times and the data generated were arranged in a relative frequency distribution, the relative frequency distribution would be indistinguishable from the probability distribution for the random variable. Thus, the probability distribution for a random variable is a theoretical model for the relative frequency distribution of a population. To the extent that the two distributions are equivalent (and we will assume they are), the probability distribution for x possesses a mean μ and a variance σ^2 that are identical to the corresponding descriptive measures for the population.

Examine the probability distribution for x (the number of heads observed in the toss of two fair coins) in Figure 4.4. Try to locate the mean of the distribution intuitively. We may reason that the mean μ of this distribution is equal to 1 as

FIGURE 4.4

Probability Distribution for a Two-Coin Toss

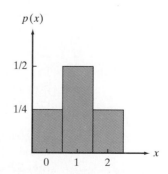

---

*Unless otherwise indicated, summations will always be over all possible values of x.

follows: In a large number of experiments, $\frac{1}{4}$ should result in $x = 0$, $\frac{1}{2}$ in $x = 1$, and $\frac{1}{4}$ in $x = 2$ heads. Therefore, the average number of heads is

$$\mu = 0\left(\tfrac{1}{4}\right) + 1\left(\tfrac{1}{2}\right) + 2\left(\tfrac{1}{4}\right) = 0 + \tfrac{1}{2} + \tfrac{1}{2} = 1$$

Note that to get the population mean of the random variable x, we multiply each possible value of x by its probability $p(x)$, and then we sum this product over all possible values of x. The *mean of* x is also referred to as the *expected value of x*, denoted $E(x)$.

DEFINITION 4.5
The **mean**, or **expected value**, of a discrete random variable x is

$$\mu = E(x) = \Sigma x p(x)$$

The term *expected* is a mathematical term and should not be interpreted as it is typically used. Specifically, a random variable might never be equal to its "expected value." Rather, the expected value is the mean of the probability distribution, or a measure of its central tendency. You can think of μ as the mean value of x in a *very large* (actually, *infinite*) number of repetitions of the experiment, where the values of x occur in proportions equivalent to the probabilities of x.

EXAMPLE   4.5

$\mu = \$280$

Suppose you work for an insurance company, and you sell a $10,000 1-year term insurance policy at an annual premium of $290. Actuarial tables show that the probability of death during the next year for a person of your customer's age, sex, health, etc., is .001. What is the expected gain (amount of money made by the company) for a policy of this type?

Solution

The experiment is to observe whether the customer survives the upcoming year. The probabilities associated with the two sample points, Live and Die, are .999 and .001, respectively. The random variable you are interested in is the gain x, which can assume the values shown in the following table.

Gain x	Sample Point	Probability
$290	Customer lives	.999
$-\$9,710$	Customer dies	.001

TEACHING TIP

Point out to the student that the gain associated with an insurance policy can never be $280. This value refers to the average gain per policy if repeated policies were sold.

If the customer lives, the company gains the $290 premium as profit. If the customer dies, the gain is negative because the company must pay $10,000, for a net "gain" of $ (290 - 10,000) = -\$9,710$. The expected gain is therefore

$$\mu = E(x) = \Sigma x p(x)$$
$$= (290)(.999) + (-9,710)(.001) = \$280$$

In other words, if the company were to sell a very large number of 1-year $10,000 policies to customers possessing the characteristics described above, it would (on the average) net $280 per sale in the next year.

Example 4.5 illustrates that the expected value of a random variable x need not equal a possible value of x. That is, the expected value is $280, but x will equal either $290 or $-\$9,710$ each time the experiment is performed (a policy is sold and a year elapses). The expected value is a measure of central tendency—and in this case represents the average over a very large number of 1-year policies—but is not a possible value of x.

We learned in Chapter 2 that the mean and other measures of central tendency tell only part of the story about a set of data. The same is true about probability distributions. We need to measure variability as well. Since a probability distribution can be viewed as a representation of a population, we will use the population variance to measure its variability.

The *population variance* σ^2 is defined as the average of the squared distance of x from the population mean μ. Since x is a random variable, the squared distance, $(x - \mu)^2$, is also a random variable. Using the same logic used to find the mean value of x, we find the mean value of $(x - \mu)^2$ by multiplying all possible values of $(x - \mu)^2$ by $p(x)$ and then summing over all possible x values.* This quantity

$$E[(x - \mu)^2] = \sum_{\text{all } x} (x - \mu)^2 p(x)$$

is also called the *expected value of the squared distance from the mean;* that is, $\sigma^2 = E[(x - \mu)^2]$. The standard deviation of x is defined as the square root of the variance σ^2.

DEFINITION 4.6

The **variance** of a random variable x is

$$\sigma^2 = E[(x - \mu)^2] = \Sigma(x - \mu)^2 p(x)$$

DEFINITION 4.7

The **standard deviation** of a discrete random variable is equal to the square root of the variance, i.e., $\sigma = \sqrt{\sigma^2}$.

Knowing the mean μ and standard deviation σ of the probability distribution of x, in conjunction with Chebyshev's Rule (Table 2.7) and the Empirical Rule (Table 2.8), we can make statements about the likelihood of values of x falling within the intervals $\mu \pm \sigma, \mu \pm 2\sigma$, and $\mu \pm 3\sigma$. These probabilities are given in the box.

Chebyshev's Rule and Empirical Rule for a Discrete Random Variable

Let x be a discrete random variable with probability distribution $p(x)$, mean μ, and standard deviation σ. Then, depending on the shape of $p(x)$, the following probability statements can be made:

	Chebyshev's Rule	Empirical Rule
	Applies to any probability distribution (see Figure 4.5a)	Applies to probability distributions that are mound-shaped and symmetric (see Figure 4.5b)
$P(\mu - \sigma < x < \mu + \sigma)$	≥ 0	$\approx .68$
$P(\mu - 2\sigma < x < \mu + 2\sigma)$	$\geq 3/4$	$\approx .95$
$P(\mu - 3\sigma < x < \mu + 3\sigma)$	$\geq 8/9$	≈ 1.00

---

*It can be shown that $E[(x - \mu)^2] = E(x^2) - \mu^2$, where $E(x^2) = \Sigma x^2 p(x)$. Note the similarity between this expression and the shortcut formula $\Sigma(x - \bar{x})^2 = \Sigma x^2 - (\Sigma x)^2/n$ given in Chapter 2.

FIGURE 4.5

Shapes of Two Probability Distributions for a Discrete Random Variable x

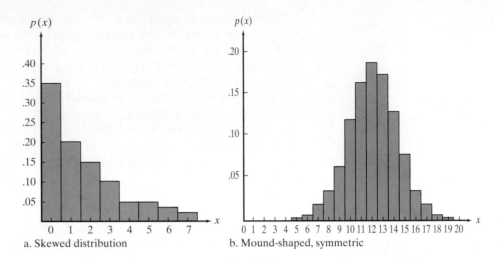

a. Skewed distribution

b. Mound-shaped, symmetric

EXAMPLE 4.6

Medical research has shown that a certain type of chemotherapy is successful 70% of the time when used to treat skin cancer. Suppose five skin cancer patients are treated with this type of chemotherapy and let x equal the number of successful cures out of the five. The probability distribution for the number x of successful cures out of five is given in the table:

x	0	1	2	3	4	5
$p(x)$	.002	.029	.132	.309	.360	.168

a. $\mu = 3.50$

b. $\sigma = 1.02$

a. Find $\mu = E(x)$. Interpret the result.
b. Find $\sigma = \sqrt{E[(x - \mu)^2]}$. Interpret the result.
c. Graph $p(x)$. Locate μ and the interval $\mu \pm 2\sigma$ on the graph. Use either Chebyshev's Rule or the Empirical Rule to approximate the probability that x falls in this interval. Compare this result with the actual probability.
d. Would you expect to observe fewer than two successful cures out of five?

No, $p(x \leq 1) = .031$

Solution

a. Applying the formula,

$$\mu = E(x) = \Sigma x p(x)$$
$$= 0(.002) + 1(.029) + 2(.132) + 3(.309) + 4(.360) + 5(.168) = 3.50$$

On average, the number of successful cures out of five skin cancer patients treated with chemotherapy will equal 3.5. Remember that this expected value only has meaning when the experiment—treating five skin cancer patients with chemotherapy—is repeated a large number of times.

Suggested Exercise 4.15

b. Now we calculate the variance of x:

$$\sigma^2 = E[(x - \mu)^2] = \Sigma(x - \mu)^2 p(x)$$
$$= (0 - 3.5)^2(.002) + (1 - 3.5)^2(.029) + (2 - 3.5)^2(.132)$$
$$+ (3 - 3.5)^2(.309) + (4 - 3.5)^2(.360) + (5 - 3.5)^2(.168)$$
$$= 1.05$$

Thus, the standard deviation is

$$\sigma = \sqrt{\sigma^2} = \sqrt{1.05} = 1.02$$

This value measures the spread of the probability distribution of x, the number of successful cures out of five. A more useful interpretation is obtained by answering parts **c** and **d**.

FIGURE 4.6

Graph of p(x) for Example 4.6

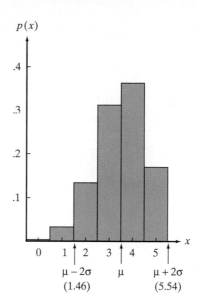

c. The graph of $p(x)$ is shown in Figure 4.6 with the mean μ and the interval $\mu \pm 2\sigma = 3.50 \pm 2(1.02) = 3.50 \pm 2.04 = (1.46, 5.54)$ shown on the graph. Note particularly that $\mu = 3.5$ locates the center of the probability distribution. Since this distribution is a theoretical relative frequency distribution that is moderately mound-shaped (see Figure 4.6), we expect (from Chebyshev's Rule) at least 75% and, more likely (from the Empirical Rule), approximately 95% of observed x values to fall between 1.46 and 5.54. You can see from Figure 4.6 that the actual probability that x falls in the interval $\mu \pm 2\sigma$ includes the sum of $p(x)$ for the values $x = 2$, $x = 3$, $x = 4$, and $x = 5$. This probability is $p(2) + p(3) + p(4) + p(5) = .132 + .309 + .360 + .168 = .969$. Therefore, 96.9% of the probability distribution lies within 2 standard deviations of the mean. This percentage is consistent with both Chebyshev's Rule and the Empirical Rule.

d. Fewer than two successful cures out of five implies that $x = 0$ or $x = 1$. Since both these values of x lie outside the interval $\mu \pm 2\sigma$, we know from the Empirical Rule that such a result is unlikely (approximate probability of .05). The exact probability, $P(x \le 1)$, is $p(0) + p(1) = .002 + .029 = .031$. Consequently, in a single experiment where five skin cancer patients are treated with chemotherapy, we would not expect to observe fewer than two successful cures. ∎

 EXERCISES 4.6–4.25

Learning the Mechanics

4.6 Consider the following probability distribution:

x	−4	0	1	3
$p(x)$	.1	.2	.4	.3

a. List the values that x may assume. $-4, 0, 1, 3$
b. What value of x is most probable? 1
c. What is the probability that x is greater than 0? .7
d. What is the probability that $x = -2$? 0

4.7 A discrete random variable x can assume five possible values: 2, 3, 5, 8, and 10. Its probability distribution is shown here:

x	2	3	5	8	10
$p(x)$	.15	.10		.25	.25

a. What is $p(5)$? .25
b. What is the probability that x equals 2 or 10? .40
c. What is $P(x \le 8)$? .75

4.8 Explain why each of the following is or is not a valid probability distribution for a discrete random variable x:

a.
x	0	1	2	3
$p(x)$	.2	.3	.3	.2

b.
x	−2	−1	0
$p(x)$	.25	.50	.20

c.
x	4	9	20
$p(x)$	−.3	1.0	.3

d.
x	2	3	5	6
$p(x)$	.15	.20	.40	.35

4.9 The random variable x has the following discrete probability distribution:

x	10	11	12	13	14
$p(x)$	.2	.3	.2	.1	.2

Since the values that x can assume are mutually exclusive events, the event $\{x \le 12\}$ is the union of three mutually exclusive events:

$$\{x = 10\} \cup \{x = 11\} \cup \{x = 12\}$$

a. Find $P(x \le 12)$.   .7
b. Find $P(x > 12)$.   .3
c. Find $P(x \le 14)$.   1
d. Find $P(x = 14)$.   .2
e. Find $P(x \le 11 \text{ or } x > 12)$.   .8

4.10 Toss three fair coins and let x equal the number of heads observed.
   a. Identify the sample points associated with this experiment and assign a value of x to each sample point.
   b. Calculate $p(x)$ for each value of x.
   c. Construct a probability histogram for $p(x)$.
   d. What is $P(x = 2 \text{ or } x = 3)$? $\frac{1}{2}$

4.11 Consider the probability distribution shown here.

x	1	2	4	10
$p(x)$	.2	.4	.2	.2

a. Find $\mu = E(x)$.   3.8
b. Find $\sigma^2 = E[(x - \mu)^2]$.   10.56
c. Find σ.   3.2496
d. Interpret the value you obtained for μ.
e. In this case, can the random variable x ever assume the value μ? Explain.   No
f. In general, can a random variable ever assume a value equal to its expected value? Explain.   Yes

4.12 Consider the probability distribution shown here.

x	−4	−3	−2	−1	0	1	2	3	4
$p(x)$	.02	.07	.10	.15	.30	.18	.10	.06	.02

a. Calculate μ, σ^2, and σ.   0; 2.94; 1.715
b. Graph $p(x)$. Locate $\mu, \mu - 2\sigma$, and $\mu + 2\sigma$ on the graph.
c. What is the probability that x will fall in the interval $\mu \pm 2\sigma$?   .96

4.13 Consider the probability distributions shown here.

x	0	1	2
$p(x)$	.3	.4	.3

y	0	1	2
$p(y)$	.1	.8	.1

a. Use your intuition to find the mean for each distribution. How did you arrive at your choice?
b. Which distribution appears to be more variable? Why?
c. Calculate μ and σ^2 for each distribution. Compare these answers to your answers in parts **a** and **b**.

Applying the Concepts—Basic

4.14 According to the American Dental Association, 60% of all dentists use nitrous oxide ("laughing gas") in their practice (*New York Times*, June 20, 1995). If x equals the number of dentists in a random sample of five dentists who use laughing gas in practice, then the probability distribution of x is

x	0	1	2	3	4	5
$p(x)$	.0102	.0768	.2304	.3456	.2592	.0778

a. Find $P(x = 4)$.
b. Find $P(x < 2)$.
c. Find $P(x \ge 3)$.
d. Find the mean number of dentists out of the five who use "laughing gas." Interpret this value. $\mu = 3.0002$

4.15 A dust mite allergen level that exceeds 2 micrograms per gram ($\mu g/g$) of dust has been associated with the development of allergies. Consider a random sample of four homes and let x be the number of homes with a dust mite level that exceeds $2 \mu g/g$. The probability distribution for x, based on a May 2000 study by the National Institute of Environmental Health Sciences, is shown in the following table.

x	0	1	2	3	4
$p(x)$	.09	.30	.37	.20	.04

a. Verify that the sum of the probabilities for x in the table sum to 1.
b. Find the probability that three or four of the homes in the sample have a dust mite level that exceeds $2 \mu g/g$.

x	5	6	7	8	9	10	11	12	13	14	15
$p(x)$	.01	.02	.03	.05	.08	.09	.11	.13	.12	.10	.08

x	16	17	18	19	20	21
$p(x)$	.06	.05	.03	.02	.01	.01

Source: Ford, R., Roberts, D., and Saxton, P. *Queuing Models.* Graduate School of Management, Rutgers University, 1992.

c. Find the probability that fewer than two homes in the sample have a dust mite level that exceeds 2 µg/g.   .39

d. Find $E(x)$. Give a meaningful interpretation of the result. $\mu = 1.8$

e. Find σ.   .9899

f. Find the exact probability that x is in the interval $\mu \pm 2\sigma$. Compare to Chebyshev's Rule and the Empirical Rule.   .96

4.16 The probability distribution of x, the number of customer arrivals per 15-minute period, at a Wendy's Restaurant in New Jersey is shown above.

a. Does this distribution meet the two requirements for the probability distribution of a discrete random variable? Justify your answer.   Yes

b. What is the probability that exactly 16 customers enter the restaurant in the next 15 minutes?   .06

c. Find $P(x \le 10)$.   .28

d. Find $P(5 \le x \le 15)$.   .82

Applying the Concepts—Intermediate

4.17 In Exercise 3.16 (p. 118) you learned that five in every 10,000 students who take the Scholastic Assessment Test (SAT) score a perfect 1600. Consider a random sample of three students who take the SAT. Let x equal the number who score a perfect 1600.

a. Find $p(x)$ for $x = 0, 1, 2, 3$.

b. Graph $p(x)$.

c. Find $P(x \le 1)$.   1.000

4.18 A weapons manufacturer uses a liquid propellant to produce gun cartridges. During the manufacturing process, the propellant can get mixed with another liquid to produce a contaminated cartridge. A University of South Florida statistician, hired by the company to investigate the level of contamination in the stored cartridges, found that 23% of the cartridges in a particular lot were contaminated. Suppose you randomly sample (without replacement) gun cartridges from this lot until you find a contaminated one. Let x be the number of cartridges sampled until a contaminated one is found. It is known that the probability distribution for x is given by the formula:

$$p(x) = (.23)(.77)^{x-1}, \qquad x = 1, 2, 3, \dots$$

a. Find $p(1)$. Interpret this result.   .23

b. Find $p(5)$. Interpret this result.   .081

c. Find $P(x \ge 2)$. Interpret this result.   .77

4.19 In his article "American Football" (*Statistics in Sport*, 1998), Iowa State University statistician Hal Stern evaluates winning strategies of teams in the National Football League (NFL). In a section on estimating the probability of winning a game, Stern used actual NFL play-by-play data to approximate the probabilities associated with certain outcomes (e.g., running plays, short pass plays, and long pass plays). The table below gives the probability distribution for the yardage gained, x, on a running play. (A negative gain represents a loss of yards on the play.)

x, Yards	Probability	x, Yards	Probability
−4	.020	6	.090
−2	.060	8	.060
−1	.070	10	.050
0	.150	15	.085
1	.130	30	.010
2	.110	50	.004
3	.090	99	.001
4	.070		

a. Find the probability of gaining 10 yards or more on a running play.   .150

b. Find the probability of losing yardage on a running play.

4.20 A team of Chinese university professors investigated the reliability of several capacitated-flow networks in the journal *Networks* (May 1995). One network examined in the article and illustrated on p. 176 is a bridge network with arcs $a_1, a_2, a_3, a_4, a_5,$ and a_6. The probability distribution of the capacity x for each of the six arcs is provided in the table.

Arc	Capacity x	$p(x)$	Arc	Capacity x	$p(x)$
a_1	3	.60	a_4	1	.90
	2	.25		0	.10
	1	.10			
	0	.05			
a_2	2	.60	a_5	1	.90
	1	.30		0	.10
	0	.10			
a_3	1	.90	a_6	2	.70
	0	.10		1	.25
				0	.05

Source: Lin, J., *et al.* "On reliability evaluation of capacitated-flow network in terms of minimal pathsets." *Networks*, Vol. 25, No. 3, May 1995, p. 135 (Table 1).

a. Verify that the properties of discrete probability distributions are satisfied for each arc capacity distribution.

b. Find the probability that the capacity for arc a_1 will exceed 1.   .85

c. Repeat part **b** for each of the remaining five arcs.

d. Compute the mean capacity of each arc and interpret its value.

e. Compute σ for each arc and interpret its value.

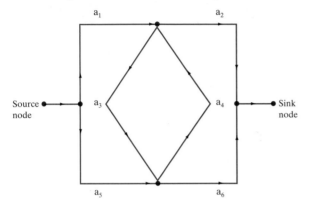

Applying the Concepts—Intermediate

4.21 The chance of winning Florida's Pick-6 Lotto game is 1 in approximately 23 million. Suppose you buy a $1 Lotto ticket in anticipation of winning the $7 million grand prize. Calculate your expected net winnings. Interpret the result.

4.22 On the popular television game show, *The Price is Right*, contestants can play "The Showcase Showdown." The game involves a large wheel with twenty nickel values, 5, 10, 15, 20, . . . , 95, 100, marked on it. Contestants spin the wheel once or twice, with the objective of obtaining the highest total score *without going over a dollar (100)*. [According to the *American Statistician* (Aug. 1995), the optimal strategy for the first spinner in a three-player game is to spin a second time only if the value of the initial spin is 65 or less.] Let x represent a single contestant playing "The Showcase Showdown." Assume a "fair" wheel, i.e., a wheel with equally likely outcomes. If the total of the player's spins exceeds 100, the total score is set to 0.

a. If the player is permitted only one spin of the wheel, find the probability distribution for x.

b. Refer to part **a**. Find $E(x)$ and interpret this value.

c. Refer to part **a**. Give a range of values within which x is likely to fall. $(-5.16, 110.16)$

d. Suppose the player will spin the wheel twice, no matter what the outcome of the first spin. Find the probability distribution for x.

e. What assumption did you make to obtain the probability distribution, part **d**? Is it a reasonable assumption?

f. Find μ and σ for the probability distribution, part **d**, and interpret the results.   33.25; 38.3577

g. Refer to part **d**. What is the probability that in two spins the player's total score exceeds a dollar (i.e., is set to 0)?   .525

h. Suppose the player obtains a 20 on the first spin and decides to spin again. Find the probability distribution for x.

i. Refer to part **h**. What is the probability that the player's total score exceeds a dollar?   .20

j. Given the player obtains a 65 on the first spin and decides to spin again, find the probability that the player's total score exceeds a dollar.   .65

k. Repeat part **j** for different first-spin outcomes. Use this information to suggest a strategy for the one-player game.

Applying the Concepts—Advanced

4.23 Odds makers try to predict which professional and college football teams will win and by how much (the *spread*). If the odds makers do this accurately, adding the spread to the underdog's score should make the final score a tie. Suppose a bookie will give you $6 for every $1 you risk if you pick the winners in three ballgames (adjusted by the spread) on a "parlay" card. What is the bookie's expected earnings per dollar wagered? Interpret this value. $.25

4.24 Geneticists use a grid—called a *Punnett square*—to display all possible gene combinations in genetic crosses. (The grid is named for Reginald Punnett, a British geneticist who developed the method in the early 1900s.) The accompanying figure is a Punnett square for a cross involving human earlobes. In humans, free earlobes (E) are dominant over attached earlobes (e). Consequently, the gene pairs EE and Ee will result in free earlobes, while the gene pair ee results in attached earlobes. Consider a couple with genes as shown in the Punnett square. Suppose the couple has seven children. Let x represent the number of children with attached earlobes (i.e., with the gene pair ee). Find the probability distribution of x.

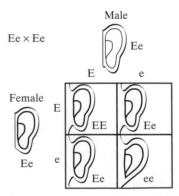

4.25 Every human possesses two sex chromosomes. A copy of one or the other (equally likely) is contributed to an offspring. Males have one X chromosome and one Y chromosome. Females have two X chromosomes. If a couple has three children, what is the probability that they have at least one boy? $7/8$

4.3 THE BINOMIAL DISTRIBUTION

Many experiments result in *dichotomous* responses—i.e., responses for which there exist two possible alternatives, such as Yes–No, Pass–Fail, Defective–Nondefective, or Male–Female. A simple example of such an experiment is the coin-toss experiment. A coin is tossed a number of times, say 10. Each toss results in one of two outcomes, Head or Tail, and the probability of observing each of these two outcomes remains the same for each of the 10 tosses. Ultimately, we are interested in the probability distribution of x, the number of heads observed. Many other experiments are equivalent to tossing a coin (either balanced or unbalanced) a fixed number n of times and observing the number x of times that one of the two possible outcomes occurs. Random variables that possess these characteristics are called **binomial random variables**.

Public opinion and consumer preference polls (e.g., the CNN, Gallup, and Harris polls) frequently yield observations on binomial random variables. For example, suppose a sample of 100 students is selected from a large student body and each person is asked whether he or she favors (a Head) or opposes (a Tail) a certain campus issue. Suppose we are interested in x, the number of students in the sample who favor the issue. Sampling 100 students is analogous to tossing the coin 100 times. Thus, you can see that opinion polls that record the number of people who favor a certain issue are real-life equivalents of coin-toss experiments. We have been describing a **binomial experiment**; it is identified by the following characteristics.

Characteristics of a Binomial Random Variable

1. The experiment consists of n identical trials.
2. There are only two possible outcomes on each trial. We will denote one outcome by S (for Success) and the other by F (for Failure).
3. The probability of S remains the same from trial to trial. This probability is denoted by p, and the probability of F is denoted by q. Note that $q = 1 - p$.
4. The trials are independent.
5. The binomial random variable x is the number of S's in n trials.

 EXAMPLE 4.7

a. Not a binomial

b. Binomial

c. Not a binomial

For the following examples, decide whether x is a binomial random variable.

a. A university scholarship committee must select two students to receive a scholarship for the next academic year. The committee receives 10 applications for the scholarships—six from male students and four from female students. Suppose the applicants are all equally qualified, so that the selections are randomly made. Let x be the number of female students who receive a scholarship.

b. Before marketing a new product on a large scale, many companies will conduct a consumer-preference survey to determine whether the product is likely to be successful. Suppose a company develops a new diet soda and then conducts a taste-preference survey in which 100 randomly chosen consumers state their preferences among the new soda and the two leading sellers. Let x be the number of the 100 who choose the new brand over the two others.

c. Some surveys are conducted by using a method of sampling other than simple random sampling (defined in Chapter 3). For example, suppose a television cable company plans to conduct a survey to determine the fraction of households in the city that would use the cable television service. The sampling method is to choose a city block at random and then survey every household on that block. This sampling technique is called *cluster sampling*. Suppose 10 blocks are so sampled, producing a total of 124 household responses. Let x be the number of the 124 households that would use the television cable service.

Solution

a. In checking the binomial characteristics, a problem arises with independence (characteristic 4 in the preceding box). Given that the first student selected is female, the probability that the second chosen is female is $\frac{3}{9}$. On the other hand, given that the first selection is a male student, the probability that the second is female is $\frac{4}{9}$. Thus, the conditional probability of a Success (choosing a female student to receive a scholarship) on the second trial (selection) depends on the outcome of the first trial, and the trials are therefore dependent. Since the trials are *not independent*, this is not a binomial random variable.

b. Surveys that produce dichotomous responses and use random sampling techniques are classic examples of binomial experiments. In our example, each randomly selected consumer either states a preference for the new diet soda or does not. The sample of 100 consumers is a very small proportion of the totality of potential consumers, so the response of one would be, for all practical purposes, independent of another.* Thus, x is a binomial random variable.

c. This example is a survey with dichotomous responses (Yes or No to the cable service), but the sampling method is not simple random sampling. Again, the binomial characteristic of independent trials would probably not be satisfied. The responses of households within a particular block would be dependent, since households within a block tend to be similar with respect to income, level of education, and general interests. Thus, the binomial model would not be satisfactory for x if the cluster sampling technique were employed. ————— ■

EXAMPLE 4.8

a. .6561

b. .0036

d. $p(x) = \binom{n}{x} p^x (1 - p)^{n-x}$

The Heart Association claims that only 10% of U.S. adults over 30 can pass the President's Physical Fitness Commission's minimum requirements. Suppose four adults are randomly selected, and each is given the fitness test.

a. Use the steps given in Chapter 3 (box on p. 115) to find the probability that none of the four adults passes the test.

b. Find the probability that three of the four adults pass the test.

c. Let x represent the number of the four adults who pass the fitness test. Explain why x is a binomial random variable.

d. Use the answers to parts **a** and **b** to derive a formula for $p(x)$, the probability distribution of the binomial random variable x.

Solution

a. 1. The first step is to define the experiment. Here we are interested in observing the fitness test results of each of the four adults: pass (S) or fail (F).

   2. Next, we list the sample points associated with the experiment. Each sample point consists of the test results of the four adults. For example, $SSSS$ represents the sample point that all four adults pass, while $FSSS$ represents the sample point that adult 1 fails, while adults 2, 3, and 4 pass the test. The 16 sample points are listed in Table 4.2.

   3. We now assign probabilities to the sample points. Note that each sample point can be viewed as the intersection of four adults' test results and,

---

*In most real-life applications of the binomial distribution, the population of interest has a finite number of elements (trials), denoted N. When N is large and the sample size n is small relative to N, say $n/N \leq .05$, the sampling procedure, for all practical purposes, satisfies the conditions of a binomial experiment.

Table 4.2   Sample Points for Fitness Test of Example 4.8

SSSS	*FSSS*	*FFSS*	*SFFF*	*FFFF*
	SFSS	*FSFS*	*FSFF*	
	SSFS	*FSSF*	*FFSF*	
	SSSF	*SFFS*	*FFFS*	
		SFSF		
		SSFF		

assuming the results are independent, the probability of each sample point can be obtained using the multiplicative rule, as follows:

$$P(SSSS) = P[(\text{adult 1 passes}) \cap (\text{adult 2 passes})$$
$$\cap (\text{adult 3 passes}) \cap (\text{adult 4 passes})]$$
$$= P(\text{adult 1 passes}) \times P(\text{adult 2 passes})$$
$$\times P(\text{adult 3 passes}) \times P(\text{adult 4 passes})$$
$$= (.1)(.1)(.1)(.1) = (.1)^4 = .0001$$

All other sample point probabilities are calculated using similar reasoning. For example,

$$P(FSSS) = (.9)(.1)(.1)(.1) = .0009$$

You can check that this reasoning results in sample point probabilities that add to 1 over the 16 points in the sample space.

4. Finally, we add the appropriate sample point probabilities to obtain the desired event probability. The event of interest is that all four adults fail the fitness test. In Table 4.2 we find only one sample point, *FFFF*, contained in this event. All other sample points imply that at least one adult passes. Thus,

$$P(\text{All four adults fail}) = P(FFFF) = (.9)^4 = .6561$$

b. The event that three of the four adults pass the fitness test consists of the four sample points in the second column of Table 4.2: *FSSS*, *SFSS*, *SSFS*, and *SSSF*. To obtain the event probability we add the sample point probabilities:

$$P(3 \text{ of } 4 \text{ adults pass}) = P(FSSS) + P(SFSS) + P(SSFS) + P(SSSF)$$
$$= (.1)^3(.9) + (.1)^3(.9) + (.1)^3(.9) + (.1)^3(.9)$$
$$= 4(.1)^3(.9) = .0036$$

Note that each of the four sample point probabilities is the same, because each sample point consists of three *S*'s and one *F*; the order does not affect the probability because the adults' test results are (assumed) independent.

c. We can characterize the experiment as consisting of four identical trials—the four test results. There are two possible outcomes to each trial, *S* or *F*, and the probability of passing, $p = .1$, is the same for each trial. Finally, we are assuming that each adult's test result is independent of all others, so that the four trials are independent. Then it follows that x, the number of the four adults who pass the fitness test, is a binomial random variable.

d. The event probabilities in parts **a** and **b** provide insight into the formula for the probability distribution $p(x)$. First, consider the event that three adults pass (part **b**). We found that

$$P(x = 3) = (\text{Number of sample points for which } x = 3) \times (.1)^{\text{Number of successes}} \times (.9)^{\text{Number of failures}}$$
$$= 4(.1)^3(.9)^1$$

In general, we can use combinatorial mathematics to count the number of sample points. For example,

Number of sample points for which $x = 3$
= Number of different ways of selecting 3 successes of the 4 trials
$$= \binom{4}{3} = \frac{4!}{3!(4-3)!} = \frac{4 \cdot 3 \cdot 2 \cdot 1}{(3 \cdot 2 \cdot 1) \cdot 1} = 4$$

The formula that works for any value of x can be deduced as follows:

$$P(x = 3) = \binom{4}{3}(.1)^3(.9)^1 = \binom{4}{x}(.1)^x(.9)^{4-x}$$

The component $\binom{4}{x}$ counts the number of sample points with x successes and the component $(.1)^x(.9)^{4-x}$ is the probability associated with each sample point having x successes.

For the general binomial experiment, with n trials and probability of Success p on each trial, the probability of x successes is

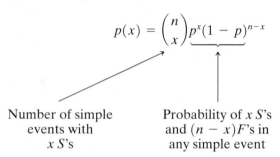

$$p(x) = \binom{n}{x}\underbrace{p^x(1-p)^{n-x}}$$

Number of simple events with x S's

Probability of x S's and $(n-x)$ F's in any simple event

In theory, you could always resort to the principles developed in Example 4.8 to calculate binomial probabilities; list the sample points and sum their probabilities. However, as the number of trials (n) increases, the number of sample points grows very rapidly (the number of sample points is 2^n). Thus, we prefer the formula for calculating binomial probabilities, since its use avoids listing sample points.

The binomial distribution* is summarized in the box.

TEACHING TIP
Using an example from the probability chapter, show the students that the binomial formula is an efficient method of finding probabilities. The formula is comprised of two parts: the first part counts the number of outcomes with x successes and $(n-x)$ failures, and the second part gives the probability of each outcome that has x successes and $(n-x)$ failures.

The Binomial Probability Distribution

$$p(x) = \binom{n}{x}p^x q^{n-x} \qquad (x = 0, 1, 2, \ldots, n)$$

where

p = Probability of a success on a single trial
$q = 1 - p$
n = Number of trials
x = Number of successes in n trials
$$\binom{n}{x} = \frac{n!}{x!(n-x)!}$$

---

*The binomial distribution is so named because the probabilities, $p(x)$, $x = 0, 1, \ldots, n$, are terms of the binomial expansion, $(q + p)^n$.

As noted in Chapter 3, the symbol 5! means $5 \cdot 4 \cdot 3 \cdot 2 \cdot 1 = 120$. Similarly, $n! = n(n-1)(n-2)\cdots 3 \cdot 2 \cdot 1$; remember, $0! = 1$.

EXAMPLE 4.9

Refer to Example 4.8. Use the formula for a binomial random variable to find the probability distribution of x, where x is the number of adults who pass the fitness test. Graph the distribution.

Solution

For this application, we have $n = 4$ trials. Since a success S is defined as an adult who passes the test, $p = P(S) = .1$ and $q = 1 - p = .9$. Substituting $n = 4$, $p = .1$, and $q = .9$ into the formula for $p(x)$, we obtain

$$p(0) = \frac{4!}{0!(4-0)!}(.1)^0(.9)^{4-0} = \frac{4 \cdot 3 \cdot 2 \cdot 1}{(1)(4 \cdot 3 \cdot 2 \cdot 1)}(.1)^0(.9)^4 = 1(1)^0(.9)^4 = .6561$$

$$p(1) = \frac{4!}{1!(4-1)!}(.1)^1(.9)^{4-1} = \frac{4 \cdot 3 \cdot 2 \cdot 1}{(1)(3 \cdot 2 \cdot 1)}(.1)^1(.9)^3 = 4(.1)(.9)^3 = .2916$$

$$p(2) = \frac{4!}{2!(4-2)!}(.1)^2(.9)^{4-2} = \frac{4 \cdot 3 \cdot 2 \cdot 1}{(2 \cdot 1)(2 \cdot 1)}(.1)^2(.9)^2 = 6(.1)^2(.9)^2 = .0486$$

$$p(3) = \frac{4!}{3!(4-3)!}(.1)^3(.9)^{4-3} = \frac{4 \cdot 3 \cdot 2 \cdot 1}{(3 \cdot 2 \cdot 1)(1)}(.1)^3(.9)^1 = 4(.1)^3(.9) = .0036$$

$$p(4) = \frac{4!}{4!(4-4)!}(.1)^4(.9)^{4-4} = \frac{4 \cdot 3 \cdot 2 \cdot 1}{(4 \cdot 3 \cdot 2 \cdot 1)(1)}(.1)^4(.9)^0 = 1(.1)^4(.9) = .0001$$

Note that these probabilities, listed in Table 4.3, sum to 1. A graph of this probability distribution is shown in Figure 4.7. ∎

Table 4.3 Probability Distribution for Physical Fitness Example: Tabular Form

x	$p(x)$
0	.6561
1	.2916
2	.0486
3	.0036
4	.0001

FIGURE 4.7

Probability Distribution for Physical Fitness Example: Graphical Form

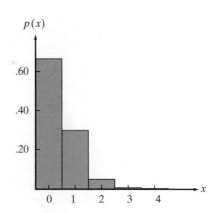

E X A M P L E   4 . 1 0

Refer to Examples 4.8 and 4.9. Calculate μ and σ, the mean and standard deviation, respectively, of the number of the four adults who pass the test. Interpret the results.

Solution

From Section 4.2 we know that the mean of a discrete probability distribution is

$$\mu = \Sigma x p(x)$$

Referring to Table 4.3, the probability distribution for the number x who pass the fitness test, we find

$$\mu = 0(.6561) + 1(.2916) + 2(.0486) + 3(.0036) + 4(.0001) = .4$$
$$= 4(.1) = np$$

Thus, in the long run, the average number of adults (out of four) who pass the test is only .4. [*Note: The relationship $\mu = np$ holds in general for a binomial random variable.*]

The variance is

$$\sigma^2 = \Sigma(x - \mu)^2 p(x) = \Sigma(x - .4)^2 p(x)$$
$$= (0 - .4)^2(.6561) + (1 - .4)^2(.2916) + (2 - .4)^2(.0486)$$
$$\quad + (3 - .4)^2(.0036) + (4 - .4)^2(.0001)$$
$$= .104976 + .104976 + .124416 + .024336 + .001296$$
$$= .36 = 4(.1)(.9) = npq$$

[*Note: The relationship $\sigma^2 = npq$ holds in general for a binomial random variable.*]

Finally, the standard deviation of the number who pass the fitness test is

$$\sigma = \sqrt{\sigma^2} = \sqrt{.36} = .6$$

Applying the Empirical Rule, we know that approximately 95% of the x values will fall in the interval $\mu \pm 2\sigma = .4 \pm 2(.6) = (-.8, 1.6)$. Since x cannot be negative, we expect (i.e., in the long run) the number of adults out of four who pass the fitness test to be less than 1.6.   ■

We emphasize that you need not use the expectation summation rules to calculate μ and σ^2 for a binomial random variable. You can find them easily using the formulas $\mu = np$ and $\sigma^2 = npq$.

Mean, Variance, and Standard Deviation for a Binomial Random Variable

Mean: $\mu = np$
Variance: $\sigma^2 = npq$
Standard deviation: $\sigma = \sqrt{npq}$

Using Binomial Tables

Calculating binomial probabilities becomes tedious when n is large. For some values of n and p the binomial probabilities have been tabulated in Table II of Appendix A. Part of Table II is shown in Table 4.4; a graph of the binomial probability distribution for $n = 10$ and $p = .10$ is shown in Figure 4.8.

Table 4.4   Reproduction of Part of Table II of Appendix A: Binomial Probabilities for $n = 10$

k \ p	.01	.05	.10	.20	.30	.40	.50	.60	.70	.80	.90	.95	.99
0	.904	.599	.349	.107	.028	.006	.001	.000	.000	.000	.000	.000	.000
1	.996	.914	.736	.376	.149	.046	.011	.002	.000	.000	.000	.000	.000
2	1.000	.988	.930	.678	.383	.167	.055	.012	.002	.000	.000	.000	.000
3	1.000	.999	.987	.879	.650	.382	.172	.055	.011	.001	.000	.000	.000
4	1.000	1.000	.998	.967	.850	.633	.377	.166	.047	.006	.000	.000	.000
5	1.000	1.000	1.000	.994	.953	.834	.623	.367	.150	.033	.002	.000	.000
6	1.000	1.000	1.000	.999	.989	.945	.828	.618	.350	.121	.013	.001	.000
7	1.000	1.000	1.000	1.000	.998	.988	.945	.833	.617	.322	.070	.012	.000
8	1.000	1.000	1.000	1.000	1.000	.988	.989	.954	.851	.624	.264	.086	.004
9	1.000	1.000	1.000	1.000	1.000	1.000	.999	.994	.972	.893	.651	.401	.096

Table II actually contains a total of nine tables, labeled (**a**) through (**i**), one each corresponding to $n = 5, 6, 7, 8, 9, 10, 15, 20$, and 25, respectively. In each of these tables the columns correspond to values of p, and the rows correspond to values of the random variable x. The entries in the table represent **cumulative binomial probabilities**. Thus, for example, the entry in the column corresponding to $p = .10$ and the row corresponding to $x = 2$ is .930 (highlighted), and its interpretation is

$$P(x \le 2) = P(x = 0) + P(x = 1) + P(x = 2) = .930$$

This probability is also highlighted in the graphical representation of the binomial distribution with $n = 10$ and $p = .10$ in Figure 4.8.

You can also use Table II to find the probability that x equals a specific value. For example, suppose you want to find the probability that $x = 2$ in the binomial distribution with $n = 10$ and $p = .10$. This is found by subtraction as follows:

$$P(x = 2) = [P(x = 0) + P(x = 1) + P(x = 2)] - [P(x = 0) + P(x = 1)]$$
$$= P(x \le 2) - P(x \le 1) = .930 - .736 = .194$$

The probability that a binomial random variable exceeds a specified value can be found using Table II and the notion of complementary events. For example, to find the probability that x exceeds 2 when $n = 10$ and $p = .10$, we use

$$P(x > 2) = 1 - P(x \le 2) = 1 - .930 = .070$$

FIGURE 4.8

Binomial Probability Distribution for n = 10 and p = .10; P(x ≤ 2) Highlighted

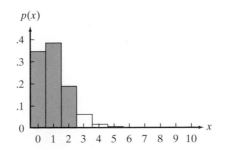

Note that this probability is represented by the unhighlighted portion of the graph in Figure 4.8.

All probabilities in Table II are rounded to three decimal places. Thus, although none of the binomial probabilities in the table is exactly zero, some are small enough (less than .0005) to round to .000. For example, using the formula to find $P(x = 0)$ when $n = 10$ and $p = .6$, we obtain

$$P(x = 0) = \binom{10}{0}(.6)^0(.4)^{10-0} = .4^{10} = .00010486$$

but this is rounded to .000 in Table II of Appendix A (see Table 4.4).

Similarly, none of the table entries is exactly 1.0, but when the cumulative probabilities exceed .9995, they are rounded to 1.000. The row corresponding to the largest possible value for x, $x = n$, is omitted, because all the cumulative probabilities in that row are equal to 1.0 (exactly). For example, in Table 4.4 with $n = 10$, $P(x \le 10) = 1.0$, no matter what the value of p.

The following example further illustrates the use of Table II.

E X A M P L E   4 . 1 1

Suppose a poll of 20 voters is taken in a large city. The purpose is to determine x, the number who favor a certain candidate for mayor. Suppose that 60% of all the city's voters favor the candidate.

a. $\mu = 12; \sigma = 2.19$

b. .245

c. .416

d. .159

a. Find the mean and standard deviation of x.
b. Use Table II of Appendix A to find the probability that $x \le 10$.
c. Use Table II to find the probability that $x > 12$.
d. Use Table II to find the probability that $x = 11$.
e. Graph the probability distribution of x and locate the interval $\mu \pm 2\sigma$ on the graph.

Solution

a. The number of voters polled is presumably small compared with the total number of eligible voters in the city. Thus, we may treat x, the number of the 20 who favor the mayoral candidate, as a binomial random variable. The value of p is the fraction of the total voters who favor the candidate; i.e., $p = .6$. Therefore, we calculate the mean and variance:

$$\mu = np = 20(.6) = 12$$
$$\sigma^2 = npq = 20(.6)(.4) = 4.8$$
$$\sigma = \sqrt{4.8} = 2.19$$

b. Looking in the $k = 10$ row and the $p = .6$ column of Table II (Appendix A) for $n = 20$, we find the value of .245. Thus,

$$P(x \le 10) = .245$$

c. To find the probability

$$P(x > 12) = \sum_{x=13}^{20} p(x)$$

Suggested Exercise 4.39

we use the fact that for all probability distributions,

$$\sum_{\text{all } x} p(x) = 1.$$

Therefore,

$$P(x > 12) = 1 - P(x \le 12) = 1 - \sum_{x=0}^{12} p(x)$$

Consulting Table II, we find the entry in row $k = 12$, column $p = .6$ to be .584. Thus,

$$P(x > 12) = 1 - .584 = .416$$

d. To find the probability that exactly 11 voters favor the candidate, recall that the entries in Table II are cumulative probabilities and use the relationship

$$P(x = 11) = [p(0) + p(1) + \cdots + p(11)] - [p(0) + p(1) + \cdots + p(10)]$$
$$= P(x \le 11) - P(x \le 10)$$

Then

$$P(x = 11) = .404 - .245 = .159$$

e. The probability distribution for x is shown in Figure 4.9. Note that

$$\mu - 2\sigma = 12 - 2(2.2) = 7.6 \qquad \mu + 2\sigma = 12 + 2(2.2) = 16.4$$

The interval $\mu - 2\sigma$ to $\mu + 2\sigma$ is shown in Figure 4.9. The probability that x falls in the interval $\mu + 2\sigma$ is $P(x = 8, 9, 10, \ldots, 16) = P(x \le 16) - P(x \le 7) = .984 - .021 = .963$. Note that this probability is very close to the .95 given by the Empirical Rule. Thus, we expect the number of voters in the sample of 20 who favor the mayoral candidate to be between 8 and 16.   ■

FIGURE 4.9

The Binomial Probability Distribution for x in Example 4.11: n = 20 and p = .6

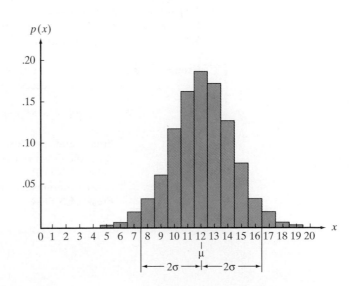

Using the TI-83 Graphing Calculator

Binomial Probabilities on the TI-83

$P(x = k)$

To compute $P(x = k)$, the probability of **k** successes in **n** trials, where the **p** is probability of success for each trial, use the **binompdf(** command. Binom**p**df stands for "binomial probability density function." This command is under the **DISTR**ibution menu and has the format **binompdf(n, p, k)**.

Example: Compute the probability of 5 successes in 8 trials where the probability of success for a single trial is 40%.

In this example, $n = 8$, $p = .4$, and $k = 5$

Press   **2nd   VARS**   for   **DISTR**
Press the down arrow key until   **0:binompdf**   is highlighted
Press   **ENTER**
After   **binompdf(**, type   **8, .4, 5)**
Press   **ENTER**

You should see

```
binompdf(8,.4,5)
        .12386304
```

Thus, $P(x = 5)$ is about 12.4%.

$P(x \leq k)$

To compute $P(x \leq k)$, the probability of **k** or fewer successes in **n** trials, where the **p** is probability of success for each trial, use the **binomcdf(** command. Binom**c**df stands for "binomial *cumulative* probability density function." This command is under the **DISTR**ibution menu and has the format **binomcdf(n, p, k)**.

Example: Compute the probability of less than 5 successes in 8 trials where the probability of success for a single trial is 40%.

In this example, $n = 8$, $p = .4$, and $k = 5$

Press   **2nd   VARS**   for   **DISTR**
Press the down arrow key until   **A:binomcdf**   is highlighted
Press   **ENTER**
After   **binomcdf(**, type   **8, .4, 5)**
Press   **ENTER**

USING THE TI-83 GRAPHING CALCULATOR (*continued*)

You should see

```
binomcdf(8,.4,5)
        .95019264
```

Thus, $P(x \le 5)$ is about 95%.

$P(x < k), P(x > k), P(x \ge k)$

To find the probability of less than k successes $P(x < k)$, more than k successes $P(x > k)$, or at least k successes $P(x \ge k)$, variations of the **binomcdf(** command must be used as shown below.

$P(x < k)$	use	**binomcdf($n, p, k - 1$)**
$P(x > k)$	use	**$1 -$ binomcdf(n, p, k)**
$P(x \ge k)$	use	**$1 -$ binomcdf($n, p, k - 1$)**

where $n =$ number of trials and $p =$ probability of success in each trial.

 EXERCISES 4.26–4.46

Learning the Mechanics

4.26 Compute the following:

a. $\dfrac{6!}{2!(6 - 2)!}$ 15

b. $\dbinom{5}{2}$ 10

c. $\dbinom{7}{0}$ 1

d. $\dbinom{6}{6}$ 1

e. $\dbinom{4}{3}$ 4

4.27 Consider the following probability distribution:

$$p(x) = \binom{5}{x}(.7)^x(.3)^{5-x} \quad (x = 0, 1, 2, \ldots, 5)$$

a. Is x a discrete or a continuous random variable? Explain.   Discrete
b. What is the name of this probability distribution?
c. Graph the probability distribution.
d. Find the mean and standard deviation of x.
e. Show the mean and the 2-standard-deviation interval on each side of the mean on the graph you drew in part **c**.

4.28 If x is a binomial random variable, compute $p(x)$ for each of the following cases:

a. $n = 5, x = 1, p = .2$ .4096
b. $n = 4, x = 2, q = .4$ .3456
c. $n = 3, x = 0, p = .7$ .027
d. $n = 5, x = 3, p = .1$ .0081
e. $n = 4, x = 2, q = .6$ .3456
f. $n = 3, x = 1, p = .9$ .027

4.29 If x is a binomial random variable, calculate μ, σ^2, and σ for each of the following:

a. $n = 25, p = .5$ 12.5; 6.25; 2.5
b. $n = 80, p = .2$ 16; 12.8; 3.578
c. $n = 100, p = .6$ 60; 24; 4.899
d. $n = 70, p = .9$ 63; 6.3; 2.510
e. $n = 60, p = .8$ 48; 9.6; 3.098
f. $n = 1,000, p = .04$ 40; 38.4; 6.197

4.30 If x is a binomial random variable, use Table II in Appendix A to find the following probabilities:

a. $P(x = 2)$ for $n = 10, p = .4$
b. $P(x \le 5)$ for $n = 15, p = .6$ .034
c. $P(x > 1)$ for $n = 5, p = .1$ .081
d. $P(x < 10)$ for $n = 25, p = .7$ 0
e. $P(x \ge 10)$ for $n = 15, p = .9$ .998
f. $P(x = 2)$ for $n = 20, p = .2$ .137

4.31 Suppose x is a binomial random variable with $n = 5$ and $p = .5$. Compute $p(x)$ for $x = 0, 1, 2, 3, 4,$ and 5 using the following two methods:

    a. List the sample points (using S for Success and F for Failure on each trial) corresponding to each value of x, assign probabilities to each sample point, and obtain $p(x)$ by adding sample point probabilities.

    b. Use the formula for the binomial probability distribution to obtain $p(x)$.

4.32 The binomial probability distribution is a family of probability distributions with each single distribution depending on the values of n and p. Assume that x is a binomial random variable with $n = 4$.

    a. Determine a value of p such that the probability distribution of x is symmetric. $p = .5$

    b. Determine a value of p such that the probability distribution of x is skewed to the right. $p < .5$

    c. Determine a value of p such that the probability distribution of x is skewed to the left. $p > .5$

    d. Graph each of the binomial distributions you obtained in parts **a**, **b**, and **c**. Locate the mean for each distribution on its graph.

    e. In general, for what values of p will a binomial distribution be symmetric? Skewed to the right? Skewed to the left?

Applying the Concepts—Basic

4.33 According to the University of South Florida's Tobacco Research and Intervention Program, only 5% of the nation's cigarette smokers ever enter into a treatment program to help them quit smoking (*USF Magazine*, Spring 2000). In a random sample of 200 smokers, let x be the number who enter into a treatment program.

    a. Explain why x is a binomial random variable (to a reasonable degree of approximation).

    b. What is the value of p? Interpret this value.   .05

    c. What is the expected value of x? Interpret this value.

4.34 The American College of Obstetricians and Gynecologists report that 22% of all births in the U.S. take place by Caesarian section each year (*USA Today*, Sept. 19, 2000).

    a. In a random sample of 1,000 births, how many, on average will take place by Caesarian section?   220

    b. What is the standard deviation of the number of Caesarian section births in a sample of 1,000 births?

    c. Use your answers to parts **a** and **b** to form an interval that is likely to contain the number of Caesarian section births in a sample of 1,000 births.   (194, 246)

4.35 A national poll conducted by *The New York Times* (May 7, 2000) revealed that 80% of Americans believe that after you die, some part of you lives on, either in a next life on earth or in heaven. Consider a random sample of 10 Americans and count x, the number who believe in life after death.

    a. Find $P(x = 3)$.   .001

    b. Find $P(x \leq 7)$.   .322

    c. Find $P(x > 4)$.   .994

4.36 According to a nationwide survey, 60% of parents with young children condone spanking their child as a regular form of punishment (*Tampa Tribune*, Oct. 5, 2000). Consider a random sample of three people, each of whom is a parent with young children. Assume that x, the number in the sample that condone spanking, is a binomial random variable.

    a. What is the probability that none of the three parents condones spanking as a regular form of punishment for their children?   .064

    b. What is the probability that at least one condones spanking as a regular form of punishment?   .936

    c. Give the mean and standard deviation of x. Interpret the results.   1.8; .8485

4.37 A Federal Trade Commission (FTC) study of the pricing accuracy of electronic checkout scanners at stores found that one of every 30 items is priced incorrectly (*Price Check II: A Follow-Up Report on the Accuracy of Checkout Scanner Prices*, Dec. 16, 1998). Suppose the FTC randomly selects five items at a retail store and checks the accuracy of the scanner price of each. Let x represent the number of the five items that is priced incorrectly.

    a. Show that x is (approximately) a binomial random variable.

    b. Use the information in the FTC study to estimate p for the binomial experiment. $\frac{1}{30}$

    c. What is the probability that exactly one of the five items is priced incorrectly by the scanner?   .1455

    d. What is the probability that at least one of the five items is priced incorrectly by the scanner?   .1559

Applying the Concepts—Intermediate

4.38 The degree to which democratic and non-democratic countries attempt to control the news media was examined in the *Journal of Peace Research* (Nov. 1997). Between 1948 and 1996, 80% of all democratic regimes allowed a free press. In contrast, over the same time period, 10% of all non-democratic regimes allowed a free press.

    a. In a random sample of 50 democratic regimes, how many would you expect to allow a free press? Give a range that is highly likely to include the number of democratic regimes with a free press.   35 to 45

    b. In a random sample of 50 non-democratic regimes, how many would you expect to allow a free press? Give a range that is highly likely to include the number of non-democratic regimes with a free press.   1 to 9

4.39 The *American Journal of Public Health* (July 1995) published a study of the relationship between passive smoking and nasal allergies in Japanese female students. The study revealed that 80% of students from heavy-smoking families showed signs of nasal allergies on physical examinations. Consider a sample of 25 Japanese female students exposed daily to heavy smoking.

a. What is the probability that fewer than 20 of the students will have nasal allergies? .383
b. What is the probability that more than 15 of the students will have nasal allergies? .983
c. On average, how many of the 25 students would you expect to show signs of nasal allergies? 20

4.40 The United States Department of Agriculture (USDA) reports that, under its standard inspection system, one in every 100 slaughtered chickens passes inspection with fecal contamination (*Tampa Tribune*, Mar. 31, 2000). In Exercise 3.13 (p. 118), you found the probability that a randomly selected slaughtered chicken passes inspection with fecal contamination. Now find the probability that, in a random sample of five slaughtered chickens, at least one passes inspection with fecal contamination. .049

4.41 According to researchers at Johns Hopkins University School of Medicine, one in every three women has been a victim of domestic abuse (*Annals of Internal Medicine*, Nov. 1995). This probability was obtained from a survey of nearly 2,000 adult women residing in Baltimore, Maryland. Suppose we randomly sample 15 women and find that four have been abused.
a. What is the probability of observing four or more abused women in a sample of 15 if the proportion p of women who are victims of domestic abuse is really $p = \frac{1}{3}$? .7908
b. Many experts on domestic violence believe that the proportion of women who are domestically abused is closer to $p = .10$. Calculate the probability of observing four or more abused women in a sample of 15 if $p = .10$. .0555
c. Why might your answers to parts **a** and **b** lead you to believe that $p = \frac{1}{3}$?

4.42 Refer to Exercise 3.65 (p. 147) and the experiment conducted by the Tampa Bay Skeptics to see whether an acclaimed psychic has extrasensory perception (ESP). Recall that a crystal is placed, at random, inside one of 10 identical boxes lying side-by-side on a table. The experiment was repeated seven times, and x, the number of correct decisions, was recorded. (Assume that the seven trials are independent.)
a. If the psychic is guessing—i.e., if the psychic does *not* possess ESP—what is the value of p, the probability of a correct decision on each trial? .1
b. If the psychic is guessing, what is the expected number of correct decisions in seven trials? .7
c. If the psychic is guessing, what is the probability of no correct decisions in seven trials? .4783
d. Now suppose the psychic has ESP and $p = .5$. What is the probability that the psychic guesses incorrectly in all seven trials? .0078
e. Refer to part **d**. Recall that the psychic failed to select the box with the crystal on all seven trials. Is this evidence against the psychic having ESP? Explain. Yes

Applying the Concepts—Advanced

4.43 According to the U.S. Golf Association (USGA), "The weight of the [golf] ball shall not be greater than 1.620 ounces avoirdupois (45.93 grams).... The diameter of the ball shall not be less than 1.680 inches.... The velocity of the ball shall not be greater than 250 feet per second" (USGA, 2001). The USGA periodically checks the specifications of golf balls sold in the United States by randomly sampling balls from pro shops around the country. Two dozen of each kind are sampled, and if more than three do not meet size and/or velocity requirements, that kind of ball is removed from the USGA's approved-ball list.
a. What assumptions must be made and what information must be known in order to use the binomial probability distribution to calculate the probability that the USGA will remove a particular kind of golf ball from its approved-ball list?
b. Suppose 10% of all balls produced by a particular manufacturer are less than 1.680 inches in diameter, and assume that the number of such balls, x, in a sample of two dozen balls can be adequately characterized by a binomial probability distribution. Find the mean and standard deviation of the binomial distribution.
c. Refer to part **b**. If x has a binomial distribution, then so does the number, y, of balls in the sample that meet the USGA's minimum diameter. [*Note:* $x + y = 24$.] Describe the distribution of y. In particular, what are p, q, and n? Also, find $E(y)$ and the standard deviation of y.

4.44 Ataxia-telangiectasia (A-T) is a neurological disorder that weakens immune systems and causes premature aging. According to *Science News* (June 24, 1995), when both members of a couple carry the A-T gene, their children have a one in five chance of developing the disease.
a. Consider 15 couples in which both members of each couple carry the A-T gene. What is the probability that more than 8 of 15 couples have children that develop the neurological disorder? .001
b. Consider 10,000 couples in which both members of each couple carry the A-T gene. Is it likely that fewer than 3,000 will have children that develop the disease? [*Hint:* Calculate the mean and standard deviation of the distribution.] Yes

4.45 Suppose you are a purchasing officer for a large company. You have purchased 5 million electrical switches and your supplier has guaranteed that the shipment will contain no more than .1% defectives. To check the shipment, you randomly sample 500 switches, test them, and find that four are defective. Based on this evidence, do you think the supplier has complied with the guarantee? Explain. No, $z = 4.95$

4.46 A literature professor decides to give a 20-question true-false quiz to determine who has read an assigned novel. She wants to choose the passing grade such that the probability of passing a student who guesses on every question is less than .05. What score should she set as the lowest passing grade? 15

4.4 PROBABILITY DISTRIBUTIONS FOR CONTINUOUS RANDOM VARIABLES

TEACHING TIP
The boxes that were drawn to illustrate the probability distributions of discrete random variables are now replaced with curves for continuous distributions.

The graphical form of the probability distribution for a continuous random variable x is a smooth curve that might appear as shown in Figure 4.10. This curve, a function of x, is denoted by the symbol $f(x)$ and is variously called a **probability density function**, a **frequency function**, or a **probability distribution**.

The areas under a probability distribution correspond to probabilities for x. For example, the area A beneath the curve between the two points a and b, as shown in Figure 4.10, is the probability that x assumes a value between a and $b (a < x < b)$. Because there is no area over a point, say $x = a$, it follows that (according to our model) the probability associated with a particular value of x is equal to 0; that is, $P(x = a) = 0$ and hence $P(a < x < b) = P(a \leq x \leq b)$. In other words, the probability is the same regardless of whether or not you include the endpoints of the interval. Also, because areas over intervals represent probabilities, it follows that the total area under a probability distribution, the probability assigned to all values of x, should equal 1. Note that probability distributions for continuous random variables possess different shapes depending on the relative frequency distributions of real data that the probability distributions are supposed to model.

TEACHING TIP
Point out the difference between continuous and discrete distributions with respect to the equal sign in the probability statements. Ask: "When is $P(x > 4) = P(x \geq 4)$?"

FIGURE 4.10

A Probability Distribution $f(x)$ for a Continuous Random Variable x

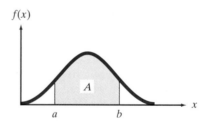

The areas under most probability distributions are obtained by using calculus or numerical methods.* Because these methods often involve difficult procedures, we will give the areas for some of the most common probability distributions in tabular form in Appendix A. Then, to find the area between two values of x, say $x = a$ and $x = b$, you simply have to consult the appropriate table.

TEACHING TIP
The area under the curve is equivalent to probabilities in continuous distributions. Therefore, the area under the curve of any probability distribution must equal one.

For the continuous random variable presented in the next section, we will give the formula for the probability distribution along with its mean μ and standard deviation σ. These two numbers will enable you to make some approximate probability statements about a random variable even when you do not have access to a table of areas under the probability distribution.

4.5 THE NORMAL DISTRIBUTION

TEACHING TIP
Let the student know the importance of the normal distribution throughout statistics.

One of the most commonly observed continuous random variables has a **bell-shaped** probability distribution as shown in Figure 4.11. It is known as a **normal random variable** and its probability distribution is called a **normal distribution**.

---

*Students with knowledge of calculus should note that the probability that x assumes a value in the interval $a < x < b$ is $P(a < x < b) = \int_a^b f(x)\, dx$, assuming the integral exists. Similar to the requirements for a discrete probability distribution, we require $f(x) \geq 0$ and $\int_{-\infty}^{\infty} f(x)\, dx = 1$.

FIGURE 4.11

A Normal Probability Distribution

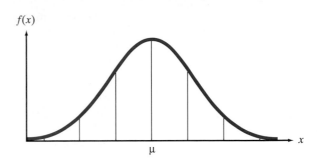

The normal distribution plays a very important role in the science of statistical inference. Moreover, many phenomena generate random variables with probability distributions that are very well approximated by a normal distribution. For example, the error made in measuring a person's blood pressure may be a normal random variable, and the probability distribution for the yearly rainfall in a certain region might be approximated by a normal probability distribution. You can determine the adequacy of the normal approximation to an existing population of data by comparing the relative frequency distribution of a large sample of the data to the normal probability distribution. Methods to detect disagreement between a set of data and the assumption of normality are presented in Section 4.6.

The normal distribution is perfectly symmetric about its mean μ, as can be seen in the examples in Figure 4.12. Its spread is determined by the value of its standard deviation σ.

FIGURE 4.12

Several Normal Distributions with Different Means and Standard Deviations

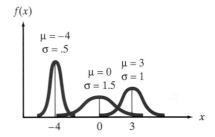

The formula for the normal probability distribution is shown in the box. When plotted, this formula yields a curve like that shown in Figure 4.11.

Probability Distribution for a Normal Random Variable x

$$f(x) = \frac{1}{\sigma\sqrt{2\pi}}e^{-(1/2)[(x-\mu)/\sigma]^2}$$

where

μ = Mean of the normal random variable x
σ = Standard deviation
π = 3.1416...
e = 2.71828...

Note that the mean μ and standard deviation σ appear in this formula, so that no separate formulas for μ and σ are necessary. To graph the normal curve we have to know the numerical values of μ and σ.

Computing the area over intervals under the normal probability distribution is a difficult task.* Consequently, we will use the computed areas listed in Table III of Appendix A. Although there are an infinitely large number of normal curves—one for each pair of values for μ and σ—we have formed a single table that will apply to any normal curve.

Table III is based on a normal distribution with mean $\mu = 0$ and standard deviation $\sigma = 1$, called a *standard normal distribution*. A random variable with a standard normal distribution is typically denoted by the symbol z. The formula for the probability distribution of z is given by

$$f(z) = \frac{1}{\sqrt{2\pi}} e^{-(1/2)z^2}$$

Figure 4.13 shows the graph of a standard normal distribution.

TEACHING TIP
Very few normal distributions in reality possess a standard normal distribution. The real benefit of the standard normal distribution is that all other normal distributions can be made to look like the standard normal and solved.

FIGURE 4.13

Standard Normal Distribution: $\mu = 0, \sigma = 1$

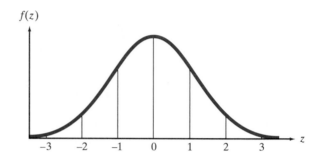

Table 4.5   Reproduction of Part of Table III in Appendix A

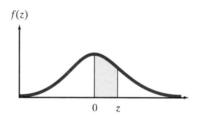

TEACHING TIP
It is often helpful to use a transparency of the normal table that can be seen while examples are being discussed in class.

z	.00	.01	.02	.03	.04	.05	.06	.07	.08	.09
.0	.0000	.0040	.0080	.0120	.0160	.0199	.0239	.0279	.0319	.0359
.1	.0398	.0438	.0478	.0517	.0557	.0596	.0636	.0675	.0714	.0753
.2	.0793	.0832	.0871	.0910	.0948	.0987	.1026	.1064	.1103	.1141
.3	.1179	.1217	.1255	.1293	.1331	.1368	.1406	.1443	.1480	.1517
.4	.1554	.1591	.1628	.1664	.1700	.1736	.1772	.1808	.1844	.1879
.5	.1915	.1950	.1985	.2019	.2054	.2088	.2123	.2157	.2190	.2224
.6	.2257	.2291	.2324	.2357	.2389	.2422	.2454	.2486	.2517	.2549
.7	.2580	.2611	.2642	.2673	.2704	.2734	.2764	.2794	.2823	.2852
.8	.2881	.2910	.2939	.2967	.2995	.3023	.3051	.3078	.3106	.3133
.9	.3159	.3186	.3212	.3238	.3264	.3289	.3315	.3340	.3365	.3389
1.0	.3413	.3438	.3461	.3485	.3508	.3531	.3554	.3577	.3599	.3621
1.1	.3643	.3665	.3686	.3708	.3729	.3749	.3770	.3790	.3810	.3830
1.2	.3849	.3869	.3888	.3907	.3925	.3944	.3962	.3980	.3997	.4015
1.3	.4032	.4049	.4066	.4082	.4099	.4115	.4131	.4147	.4162	.4177
1.4	.4192	.4207	.4222	.4236	.4251	.4265	.4279	.4292	.4306	.4319
1.5	.4332	.4345	.4357	.4370	.4382	.4394	.4406	.4418	.4429	.4441

*The student with knowledge of calculus should note that there is not a closed-form expression for $P(a < x < b) = \int_a^b f(x)\,dx$ for the normal probability distribution. The value of this definite integral can be obtained to any desired degree of accuracy by numerical approximation procedures. For this reason, it is tabulated for the user.

> **DEFINITION 4.8**
> The **standard normal distribution** is a normal distribution with $\mu = 0$ and $\sigma = 1$. A random variable with a standard normal distribution, denoted by the symbol z, is called a **standard normal random variable**.

Since we will ultimately convert all normal random variables to standard normal in order to use Table III to find probabilities, it is important that you learn to use Table III well. A partial reproduction of Table III is shown in Table 4.5. Note that the values of the standard normal random variable z are listed in the left-hand column. The entries in the body of the table give the area (probability) between 0 and z. Examples 4.12–4.15 illustrate the use of the table.

■ **E X A M P L E 4 . 1 2**

$P = .8164$

Find the probability that the standard normal random variable z falls between -1.33 and $+1.33$.

Solution

The standard normal distribution is shown again in Figure 4.14. Since all probabilities associated with standard normal random variables can be depicted as areas under the standard normal curve, you should always draw the curve and then equate the desired probability to an area.

Suggested Exercise 4.48

TEACHING TIP
Use another example such as $P(1.00 < z < 2.50)$ in which the student must subtract the two tabled probabilities. Discuss this probability together with Example 4.12.

FIGURE 4.14

Areas Under the Standard Normal Curve for Example 4.12

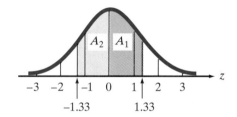

In this example we want to find the probability that z falls between -1.33 and $+1.33$, which is equivalent to the area between -1.33 and $+1.33$, shown highlighted in Figure 4.14. Table III provides the area between $z = 0$ and any value of z, so that if we look up $z = 1.33$ (the value in the 1.3 row and .03 column), we find that the area between $z = 0$ and $z = 1.33$ is .4082. This is the area labeled A_1 in Figure 4.14. To find the area A_2 located between $z = 0$ and $z = -1.33$, we note that the symmetry of the normal distribution implies that the area between $z = 0$ and any point to the left is equal to the area between $z = 0$ and the point equidistant to the right. Thus, in this example the area between $z = 0$ and $z = -1.33$ is equal to the area between $z = 0$ and $z = +1.33$. That is,

$$A_1 = A_2 = .4082$$

The probability that z falls between -1.33 and $+1.33$ is the sum of the areas of A_1 and A_2. We summarize in probabilistic notation:

$$P(-1.33 < z < +1.33) = P(-1.33 < z < 0) + P(0 < z \leq 1.33)$$
$$= A_1 + A_2 = .4082 + .4082 = .8164$$

Remember that " $<$ " and " $\leq$ " are equivalent in events involving z, because the inclusion (or exclusion) of a single point does not alter the probability of an event involving a continuous random variable. ■

EXAMPLE 4.13

$P = .0505$

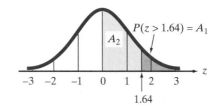

Find the probability that a standard normal random variable exceeds 1.64; that is, find $P(z > 1.64)$.

Solution

The area under the standard normal distribution to the right of 1.64 is the highlighted area labeled A_1 in Figure 4.15. This area represents the desired probability that z exceeds 1.64. However, when we look up $z = 1.64$ in Table III, we must remember that the probability given in the table corresponds to the area between $z = 0$ and $z = 1.64$ (the area labeled A_2 in Figure 4.15). From Table III we find that $A_2 = .4495$. To find the area A_1 to the right of 1.64, we make use of two facts:

1. The standard normal distribution is symmetric about its mean $z = 0$.
2. The total area under the standard normal probability distribution equals 1.

FIGURE 4.15

Areas Under the Standard Normal Curve for Example 4.13

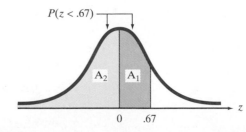

(Figure: standard normal curve with A_2 and $P(z > 1.64) = A_1$ labeled, z-axis from -3 to 3, with 1.64 marked)

Taken together, these two facts imply that the areas on either side of the mean $z = 0$ equal .5; thus, the area to the right of $z = 0$ in Figure 4.15 is $A_1 + A_2 = .5$. Then

$$P(z > 1.64) = A_1 = .5 - A_2 = .5 - .4495 = .0505$$

To attach some practical significance to this probability, note that the implication is that the chance of a standard normal random variable exceeding 1.64 is only about .05.

EXAMPLE 4.14

$P = .7486$

Find the probability that a standard normal random variable lies to the left of .67.

Solution

The event is shown as the highlighted area in Figure 4.16. We want to find $P(z < .67)$. We divide the highlighted area into two parts: the area A_1 between $z = 0$ and $z = .67$, and the area A_2 to the left of $z = 0$. We must always make such a division when the desired area lies on both sides of the mean ($z = 0$) because Table III contains areas between $z = 0$ and the point you look up. We look up $z = .67$ in Table III to find that $A_1 = .2486$. The symmetry of the standard normal distribution also implies that half the distribution lies on each side of the mean, so the area A_2 to the left of $z = 0$ is .5. Then,

$$P(z < .67) = A_1 + A_2 = .2486 + .5 = .7486$$

FIGURE 4.16

Areas Under the Standard Normal Curve for Example 4.14

(Figure: standard normal curve with $P(z < .67)$ highlighted, areas A_2 and A_1 labeled, z-axis with 0 and .67 marked)

Note that this probability is approximately .75. Thus, about 75% of the time the standard normal random variable z will fall below .67. This implies that $z = .67$ represents the approximate 75th percentile for the distribution.

E X A M P L E   4 . 1 5
P = .05

Find the probability that a standard normal random variable exceeds 1.96 in absolute value.

Solution

The event is shown highlighted in Figure 4.17. We want to find

$$P(|z| > 1.96) = P(z < -1.96 \text{ or } z > 1.96)$$

FIGURE 4.17

Areas Under the Standard Normal Curve for Example 4.15

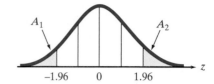

A_1 A_2

-1.96 0   1.96 z

Note that the total highlighted area is the sum of two areas, A_1 and A_2—areas that are equal because of the symmetry of the normal distribution.

We look up $z = 1.96$ and find the area between $z = 0$ and $z = 1.96$ to be .4750. Then the area to the right of 1.96, A_2, is $.5 - .4750 = .0250$ so that

$$P(|z| > 1.96) = A_1 + A_2 = .0250 + .0250 = .05$$

TEACHING TIP
Relate the normal curve to the probabilities discussed in the Empirical Rule. Explain how the normal table gives more information than the Empirical Rule and should be used whenever the normal probability distribution is appropriate.

To apply Table III to a normal random variable x with any mean μ and any standard deviation σ, we must first convert the value of x to a z-score. The population z-score for a measurement was defined (in Section 2.8) as the *distance* between the measurement and the population mean, divided by the population standard deviation. Thus, the z-score gives the distance between a measurement and the mean in units equal to the standard deviation. In symbolic form, the z-score for the measurement x is

$$z = \frac{x - \mu}{\sigma}$$

TEACHING TIP
This z-score transformation allows *all* normal distributions to be solved using the standard normal distribution.

Note that when $x = \mu$, we obtain $z = 0$.

An important property of the normal distribution is that if x is normally distributed with any mean and any standard deviation, z is *always* normally distributed with mean 0 and standard deviation 1. That is, z is a standard normal random variable.

Property of Normal Distributions

If x is a normal random variable with mean μ and standard deviation σ, then the random variable z defined by the formula

$$z = \frac{x - \mu}{\sigma}$$

has a standard normal distribution. The value z describes the number of standard deviations between x and μ.

Recall from Example 4.15 that $P(|z| > 1.96) = .05$. This probability coupled with our interpretation of z implies that any normal random variable lies more than 1.96 standard deviations from its mean only 5% of the time. Compare this to the Empirical Rule (Table 2.8, p. 62), which tells us that about 5% of the measurements in mound-shaped distributions will lie beyond 2 standard deviations from the mean. The normal distribution actually provides the model on which the Empirical Rule is based, along with much "empirical" experience with real data that often approximately obey the rule, whether drawn from a normal distribution or not.

E X A M P L E   4 . 1 6

$P = .8164$

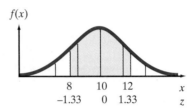

Assume that the length of time, x, between charges of a cellular phone is normally distributed with a mean of 10 hours and a standard deviation of 1.5 hours. Find the probability that the cell phone will last between 8 and 12 hours between charges.

Solution

The normal distribution with mean $\mu = 10$ and $\sigma = 1.5$ is shown in Figure 4.18. The desired probability that the cell phone lasts between 8 and 12 hours is highlighted. In order to find the probability, we must first convert the distribution to standard normal, which we do by calculating the z-score:

$$z = \frac{x - \mu}{\sigma}$$

FIGURE 4.18

Areas Under the Normal Curve for Example 4.16

$f(x)$

	8	10	12		x
	−1.33	0	1.33		z

The z-scores corresponding to the important values of x are shown beneath the x values on the horizontal axis in Figure 4.18. Note that $z = 0$ corresponds to the mean of $\mu = 10$ hours, whereas the x values 8 and 12 yield z-scores of −1.33 and +1.33, respectively. Thus, the event that the cell phone lasts between 8 and 12 hours is equivalent to the event that a standard normal random variable lies between −1.33 and +1.33. We found this probability in Example 4.12 (see Figure 4.14) by doubling the area corresponding to $z = 1.33$ in Table III. That is,

$$P(8 \le x \le 12) = P(-1.33 \le z \le 1.33) = 2(.4082) = .8164$$

The steps to follow when calculating a probability corresponding to a normal random variable are shown in the box.

Steps for Finding a Probability Corresponding to a Normal Random Variable

1. Sketch the normal distribution and indicate the mean of the random variable x. Then shade the area corresponding to the probability you want to find.

2. Convert the boundaries of the shaded area from x values to standard normal random variable z values using the formula

   $$z = \frac{x - \mu}{\sigma}$$

   Show the z values under the corresponding x values on your sketch.

3. Use Table III in Appendix A to find the areas corresponding to the z values. If necessary, use the symmetry of the normal distribution to find areas corresponding to negative z values and the fact that the total area on each side of the mean equals .5 to convert the areas from Table III to the probabilities of the event you have shaded.

Using the TI-83 Graphing Calculator

How to Graph the Area under the Standard Normal

Start from the home screen.

Step 1 Set the viewing window. (*Recall Standard Normal uses z-values, not the actual data.*)

Step 2 Access the Distributions menu.
Press **WINDOW**
Set Xmin = −5
Xmax = 5
Xscl = 1
Ymin = 0
Ymax = .5
Yscl = 0
Xres = 1

> Recall the negative sign is the gray key, not the blue key.

Step 3 *View Graph*
The graph will be displayed along with the area, lower limit, and upper limit.
Press **2ⁿᵈ VARS**
Arrow right to **DRAW**
Press **Enter**
Enter your lower limit
Press **comma**
Enter your upper limit
Press **)** *Press* **Enter**

Example What is the probability that *z* is less than 1.5 under the Standard Normal curve?

Step 1 Set your window. (See screen below.)

```
WINDOW
 Xmin=-5
 Xmax=5
 Xscl=1
 Ymin=0
 Ymax=.5
 Yscl=0
 Xres=1
```

Step 2 Enter your values (see lower left screen). Press **ENTER**.

Step 3 You will see display (see lower right screen).

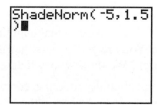

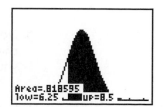

Step 4 Clear the screen for the next problem. Return to the home screen.
Press **2ⁿᵈ PRGM ENTER CLEAR CLEAR**.

EXAMPLE 4.17

Suppose an automobile manufacturer introduces a new model that has an advertised mean in-city mileage of 27 miles per gallon. Although such advertisements seldom report any measure of variability, suppose you write the manufacturer for the details of the tests, and you find that the standard deviation is 3 miles per gallon. This information leads you to formulate a probability model for the random variable x, the in-city mileage for this car model. You believe that the probability distribution of x can be approximated by a normal distribution with a mean of 27 and a standard deviation of 3.

a. .0099

a. If you were to buy this model of automobile, what is the probability that you would purchase one that averages less than 20 miles per gallon for in-city driving? In other words, find $P(x < 20)$.

b. Yes

b. Suppose you purchase one of these new models and it does get less than 20 miles per gallon for in-city driving. Should you conclude that your probability model is incorrect?

Solution

a. The probability model proposed for x, the in-city mileage, is shown in Figure 4.19. We are interested in finding the area A to the left of 20 since this area corresponds to the probability that a measurement chosen from this distribution falls below 20. In other words, if this model is correct, the area A represents the fraction of cars that can be expected to get less than 20 miles per gallon for in-city driving. To find A, we first calculate the z value corresponding to $x = 20$. That is,

$$z = \frac{x - \mu}{\sigma} = \frac{20 - 27}{3} = -\frac{7}{3} = -2.33$$

Then

$$P(x < 20) = P(z < -2.33)$$

Suggested Exercise 4.54

as indicated by the highlighted area in Figure 4.19. Since Table III gives only areas to the right of the mean (and because the normal distribution is symmetric about its mean), we look up 2.33 in Table III and find that the corresponding area is .4901. This is equal to the area between $z = 0$ and $z = -2.33$, so we find

$$P(x < 20) = A = .5 - .4901 = .0099 \approx .01$$

According to this probability model, you should have only about a 1% chance of purchasing a car of this make with an in-city mileage under 20 miles per gallon.

FIGURE 4.19

Area Under the Normal Curve for Example 4.17

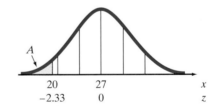

b. Now you are asked to make an inference based on a sample—the car you purchased. You are getting less than 20 miles per gallon for in-city driving. What do you infer? We think you will agree that one of two possibilities is true:

1. The probability model is correct. You simply were unfortunate to have purchased one of the cars in the 1% that get less than 20 miles per gallon in the city.
2. The probability model is incorrect. Perhaps the assumption of a normal distribution is unwarranted, or the mean of 27 is an overestimate, or the standard deviation of 3 is an underestimate, or some combination of these errors was made. At any rate, the form of the actual probability model certainly merits further investigation.

You have no way of knowing with certainty which possibility is correct, but the evidence points to the second one. We are again relying on the rare-event approach to statistical inference that we introduced earlier. The sample (one measurement in this case) was so unlikely to have been drawn from the proposed probability model that it casts serious doubt on the model. We would be inclined to believe that the model is somehow in error. ∎

Occasionally you will be given a probability and will want to find the values of the normal random variable that correspond to the probability. For example, suppose the scores on a college entrance examination are known to be normally distributed, and a certain prestigious university will consider for admission only those applicants whose scores exceed the 90th percentile of the test score distribution. To determine the minimum score for admission consideration, you will need to be able to use Table III in reverse, as demonstrated in the following example.

EXAMPLE 4.18

$z_0 = 1.28$

Find the value of z, call it z_0, in the standard normal distribution that will be exceeded only 10% of the time. That is, find z_0 such that $P(z \geq z_0) = .10$.

Solution

In this case we are given a probability, or an area, and asked to find the value of the standard normal random variable that corresponds to the area. Specifically, we want to find the value z_0 such that only 10% of the standard normal distribution exceeds z_0 (see Figure 4.20).

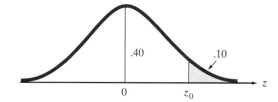

FIGURE 4.20

Areas Under the Standard Normal Curve for Example 4.18

We know that the total area to the right of the mean $z = 0$ is .5, which implies that z_0 must lie to the right of (above) 0. To pinpoint the value, we use the fact that the area to the right of z_0 is .10, which implies that the area between $z = 0$ and z_0 is $.5 - .1 = .4$. But areas between $z = 0$ and some other z value are exactly the types given in Table III. Therefore, we look up the area .4000 in the body of Table III and find that the corresponding z value is (to the closest approximation) $z_0 = 1.28$. The implication is that the point 1.28 standard deviations above the mean is the 90th percentile of a normal distribution. ∎

EXAMPLE 4.19

$z_0 = 1.96$

Find the value of z_0 such that 95% of the standard normal z values lie between $-z_0$ and $+z_0$—that is, $P(-z_0 \leq z \leq z_0) = .95$.

Solution

Here we wish to move an equal distance z_0 in the positive and negative directions from the mean $z = 0$ until 95% of the standard normal distribution is enclosed. This means that the area on each side of the mean will be equal to $\frac{1}{2}(.95) = .475$, as shown in Figure 4.21. Since the area between $z = 0$ and z_0 is .475, we look up

FIGURE 4.21

Areas Under the Standard Normal Curve for Example 4.19

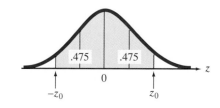

.475 in the body of Table III to find the value $z_0 = 1.96$. Thus, as we found in the reverse order in Example 4.15, 95% of a normal distribution lies between $+1.96$ and -1.96 standard deviations of the mean. ∎

Now that you have learned to use Table III to find a standard normal z value that corresponds to a specified probability, we demonstrate a practical application in Example 4.20.

■ **EXAMPLE 4.20**

$x_0 = 678$

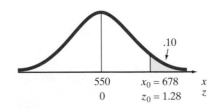

Suppose the scores, x, on a college entrance examination are normally distributed with a mean of 550 and a standard deviation of 100. A certain prestigious university will consider for admission only those applicants whose scores exceed the 90th percentile of the distribution. Find the minimum score an applicant must achieve in order to receive consideration for admission to the university.

Solution

In this example, we want to find a score, x_0, such that 90% of the scores (x values) in the distribution fall below x_0 and only 10% fall above x_0. That is,

$$P(x \le x_0) = .90$$

Converting x to a standard normal random variable, where $\mu = 550$ and $\sigma = 100$, we have

$$P(x \le x_0) = P\left(z \le \frac{x_0 - \mu}{\sigma}\right)$$
$$= P\left(z \le \frac{x_0 - 550}{100}\right) = .90$$

In Example 4.18 (see Figure 4.20) we found the 90th percentile of the standard normal distribution to be $z_0 = 1.28$. That is, we found $P(z \le 1.28) = .90$. Consequently, we know the minimum test score x_0 corresponds to a z-score of 1.28; that is,

$$\frac{x_0 - 550}{100} = 1.28$$

If we solve this equation for x_0, we find

$$x_0 = 550 + 1.28(100) = 550 + 128 = 678$$

This x value is shown in Figure 4.22. Thus, the 90th percentile of the test score distribution is 678. An applicant must score at least 678 on the entrance exam to receive consideration for admission by the university. ∎

FIGURE 4.22

Area Under the Normal Curve for Example 4.20

EXERCISES 4.47–4.68

Learning the Mechanics

4.47 Find the area under the standard normal probability distribution between the following pairs of z-scores:
  a. $z = 0$ and $z = 2.00$ .4772
  b. $z = 0$ and $z = 1.00$ .3413
  c. $z = 0$ and $z = 3$ .4987
  d. $z = 0$ and $z = .58$ .2190
  e. $z = -2.00$ and $z = 0$ .4772
  f. $z = -1.00$ and $z = 0$ .3413
  g. $z = -1.69$ and $z = 0$ .4545
  h. $z = -.58$ and $z = 0$ .2190

4.48 Find the following probabilities for the standard normal random variable z:
  a. $P(z > 1.46)$ .0721
  b. $P(z < -1.56)$ .0594
  c. $P(.67 \leq z \leq 2.41)$ .2434
  d. $P(-1.96 \leq z < -.33)$ .3457
  e. $P(z \geq 0)$ .5
  f. $P(-2.33 < z < 1.50)$ .9233
  g. $P(z \geq -2.33)$ .9901
  h. $P(z < 2.33)$ .9901

4.49 Find a value of the standard normal random variable z, call it z_0, such that
  a. $P(z \geq z_0) = .05$ 1.645
  b. $P(z \geq z_0) = .025$ 1.96
  c. $P(z \leq z_0) = .025$ -1.96
  d. $P(z \geq z_0) = .10$ 1.28
  e. $P(z > z_0) = .10$ 1.28

4.50 Suppose the random variable x is best described by a normal distribution with $\mu = 25$ and $\sigma = 5$. Find the z-score that corresponds to each of the following x values:
  a. $x = 25$ 0
  b. $x = 30$ 1
  c. $x = 37.5$ 2.5
  d. $x = 10$ -3
  e. $x = 50$ 5
  f. $x = 32$ 1.4

4.51 Give the z-score for a measurement from a normal distribution for the following:
  a. 1 standard deviation above the mean   1
  b. 1 standard deviation below the mean -1
  c. Equal to the mean   0
  d. 2.5 standard deviations below the mean -2.5
  e. 3 standard deviations above the mean   3

4.52 Suppose x is a normally distributed random variable with $\mu = 11$ and $\sigma = 2$. Find each of the following:
  a. $P(10 \leq x \leq 12)$ .3830
  b. $P(6 \leq x \leq 10)$ .3023
  c. $P(13 \leq x \leq 16)$ .1525
  d. $P(7.8 \leq x \leq 12.6)$ .7333
  e. $P(x \geq 13.24)$ .1314
  f. $P(x \geq 7.62)$ .9545

4.53 Suppose x is a normally distributed random variable with $\mu = 30$ and $\sigma = 8$. Find a value of the random variable, call it x_0, such that
  a. $P(x \geq x_0) = .5$ 30
  b. $P(x < x_0) = .025$ 14.32
  c. $P(x > x_0) = .10$ 40.24
  d. $P(x > x_0) = .95$ 16.84
  e. 10% of the values of x are less than x_0 19.76
  f. 80% of the values of x are less than x_0 36.72
  g. 1% of the values of x are greater than x_0 48.64

Applying the Concepts—Basic

4.54 Psychology students at Wittenberg University completed the Dental Anxiety Scale questionnaire (*Psychological Reports*, Aug. 1997). Scores on the scale range from 0 (no anxiety) to 20 (extreme anxiety). The mean score was 11 and the standard deviation was 3.5. Assume that the distribution of all scores on the Dental Anxiety Scale is normal with $\mu = 11$ and $\sigma = 3.5$.
  a. Suppose you score a 16 on the Dental Anxiety Scale. Find the z value for this score.   1.43
  b. Find the probability that someone scores between a 10 and a 15 on the Dental Anxiety Scale.   .4870
  c. Find the probability that someone scores above a 17 on the Dental Anxiety Scale.   .0436

4.55 The alkalinity level of water specimens collected from the Han River in Seoul, Korea, has a mean of 50 milligrams per liter and a standard deviation of 3.2 milligrams per liter (*Environmental Science & Engineering*, Sept. 1, 2000). Assume the distribution of alkalinity levels is approximately normal and find the probability that a water specimen collected from the river has an alkalinity level
  a. exceeding 45 milligrams per liter.   .9406
  b. below 55 milligrams per liter.   .9406
  c. between 51 and 52 milligrams per liter.   .1140

4.56 Refer to the *Chance* (Winter 2001) study of students who paid a private tutor to help them improve their Standardized Admission Test (SAT) scores, Exercise 2.81 (p. 70). The table summarizing the changes in both the SAT-Mathematics and SAT-Verbal scores for these students is reproduced here. Assume that both distributions of SAT score changes are approximately normal.

	SAT-Math	SAT-Verbal
Mean change in score	19	7
Standard deviation		
of score changes	65	49

  a. What is the probability that a student increases his or her score on the SAT-Math test by at least 50 points?   .3156
  b. What is the probability that a student increases his or her score on the SAT-Verbal test by at least 50 points?   .1894

4.57 A baseball player's batting average is determined by dividing the player's total number of hits by the total number of official batting attempts. The batting averages of all professional baseball players are compiled each year by Major League Baseball. The distribution of batting averages for players in the American League (which uses a designated hitter) and National League (which does not use a designated hitter) are summarized in the table. Both distributions are approximately normal.

American League	National League
$\mu = .268$	$\mu = .262$
$\sigma = .031$	$\sigma = .039$

Source: Major League Baseball, 1998.

a. Find the probability that an American League player has a batting average of .300 or higher.   .1515
b. Find the probability that a National League player has a batting average of .300 or higher.   .1660

4.58 A physical-fitness association is including the mile run in its secondary-school fitness test for students. The time for this event for students in secondary school is approximately normally distributed with a mean of 450 seconds and a standard deviation of 40 seconds. If the association wants to designate the fastest 10% as "excellent," what time should the association set for this criterion?   501.2 seconds

4.59 *Fisheries Science* (Feb. 1995) published a study of the length distributions of sardines inhabiting Japanese waters. At two years of age, fish have a length distribution that is approximately normal with $\mu = 20.20$ centimeters (cm) and $\sigma = .65$ cm.
a. Find the probability that a two-year-old sardine inhabiting Japanese waters is between 20 and 21 cm long.   .5124
b. Find the probability that a two-year-old sardine in Japanese waters is less than 19.84 cm long.   .2912
c. Find the probability that a two-year-old sardine in Japanese waters is greater than 22.01 cm long.   .0027

Applying the Concepts—Intermediate

4.60 The physical fitness of a patient is often measured by the patient's maximum oxygen uptake (recorded in milliliters per kilogram, ml/kg). The mean maximum oxygen uptake for cardiac patients who regularly participate in sports or exercise programs was found to be 24.1 with a standard deviation of 6.30 (*Adapted Physical Activity Quarterly*, Oct. 1997). Assume this distribution is approximately normal.
a. What is the probability that a cardiac patient who regularly participates in sports has a maximum oxygen uptake of at least 20 ml/kg?   .7422
b. What is the probability that a cardiac patient who regularly exercises has a maximum oxygen uptake of 10.5 ml/kg or lower?   .0154

c. Consider a cardiac patient with a maximum oxygen uptake of 10.5. Is it likely that this patient participates regularly in sports or exercise programs? Explain.

4.61 The *Journal of Visual Impairment & Blindness* (May–June 1997) published a study of the lifestyles of visually impaired students. Using diaries, the students kept track of several variables, including number of hours of sleep obtained in a typical day. These visually impaired students had a mean of 9.06 hours and a standard deviation of 2.11 hours. Assume that the distribution of the number of hours of sleep for this group of students is approximately normal.
a. Find the probability that a visually impaired student obtains less than 6 hours of sleep on a typical day.
b. Find the probability that a visually impaired student gets between 8 and 10 hours of sleep on a typical day.
c. Twenty percent of all visually impaired students obtain less than how many hours of sleep on a typical day?

4.62 A group of Florida State University psychologists examined the effects of alcohol on the reactions of people to a threat (*Journal of Abnormal Psychology*, Vol. 107, 1998). After obtaining a specified blood alcohol level, experimental subjects were placed in a room and threatened with electric shocks. Using sophisticated equipment to monitor the subjects' eye movements, the startle response (measured in milliseconds) was recorded for each subject. The mean and standard deviation of the startle responses were 37.9 and 12.4, respectively. Assume that the startle response x for a person with the specified blood alcohol level is approximately normally distributed.
a. Find the probability that x is between 40 and 50 milliseconds.   .2690
b. Find the probability that x is less than 30 milliseconds.
c. Give an interval for x, centered around 37.9 milliseconds, so that the probability that x falls in the interval is .95.   (13.6, 62.2)
d. Ten percent of the experimental subjects have startle responses above what value?   53.8

4.63 *Ecological Applications* (May 1995) published a study on the development of forests following wildfires in the Pacific Northwest. One variable of interest to the researcher was tree diameter at breast height 110 years after the fire. The population of Douglas fir trees was shown to have an approximately normal diameter distribution, with $\mu = 50$ centimeters (cm) and $\sigma = 12$ cm. Find the diameter, d, such that 30% of the Douglas fir trees in the population have diameters that exceed d.

4.64 Based on data from the National Center for Health Statistics, N. Wetzel used the normal distribution to model the length of gestation for pregnant U.S. women (*Chance*, Spring 2001). Gestation length has a mean of 280 days with a standard deviation of 20 days.
a. Find the probability that gestation length is between 275.5 and 276.5 days. (This estimates the probability that a women has her baby 4 days earlier than the "average" due date.)   .0196

b. Find the probability that gestation length is between 258.5 and 259.5 days. (This estimates the probability that a women has her baby 21 days earlier than the "average" due date.)   .0114

c. Find the probability that gestation length is between 254.5 and 255.5 days. (This estimates the probability that a women has her baby 25 days earlier than the "average" due date.)   .0090

d. The *Chance* article referenced a newspaper story about three sisters who all gave birth on the same day (March 11, 1998). Karralee had her baby 4 days early; Marrianne had her baby 21 days early; and Jennifer had her baby 25 days early. Use the results, parts **a–c**, to estimate the probability that three women have their babies 4, 21, and 25 days early, respectively. Assume the births are independent events.

4.65 A study of serum cholesterol levels in psychiatric patients of a maximum-security forensic hospital revealed that cholesterol level is approximately normally distributed with a mean of 208 milligrams per deciliter (mg/dL) and a standard deviation of 25 mg/dL (*Journal of Behavioral Medicine*, Feb. 1995). Prior research has shown that patients who exhibit violent behavior have cholesterol levels below 200 mg/dL.

a. What is the probability of observing a psychiatric patient with a cholesterol level below 200 mg/dL?

b. For three randomly selected patients, what is the probability that at least one will have a cholesterol level below 200 mg/dL?   .7553

Applying the Concepts—Advanced

4.66 The characteristics of an industrial filling process in which an expensive liquid is injected into a container were investigated in *Journal of Quality Technology* (July 1999). The quantity injected per container is approximately normally distributed with mean 10 units and standard deviation .2 units. Each unit of fill costs $20. If a container contains less than 10 units (i.e., is underfilled), it must be reprocessed at a cost of $10. A properly filled container sells for $230.

a. Find the probability that a container is underfilled.

b. A container is initially underfilled and must be reprocessed. Upon refilling it contains 10.6 units. How much profit will the company make on this container?

c. The operations manager adjusts the mean of the filling process upward to 10.5 units in order to make the probability of underfilling approximately zero. Under these conditions, what is the expected profit per container? $20

4.67 What relationship exists between the standard normal distribution and the box-plot methodology (optional Section 2.8) for describing distributions of data using quartiles? The answer depends on the true underlying probability distribution of the data. Assume for the remainder of this exercise that the distribution is normal.

a. Calculate the values of the standard normal random variable z, call them z_L and z_U, that correspond to the hinges of the box plot—i.e., the lower and upper quartiles, Q_L and Q_U—of the probability distribution.

b. Calculate the z values that correspond to the inner fences of the box plot for a normal probability distribution. $-2.68; 2.68$

c. Calculate the z values that correspond to the outer fences of the box plot for a normal probability distribution. $-4.69; 4.69$

d. What is the probability that an observation lies beyond the inner fences of a normal probability distribution? The outer fences?   .0074; 0

e. Can you better understand why the inner and outer fences of a box plot are used to detect outliers in a distribution? Explain.

4.68 A machine used to regulate the amount of dye dispensed for mixing shades of paint can be set so that it discharges an average of μ milliliters (mL) of dye per can of paint. The amount of dye discharged is known to have a normal distribution with a standard deviation of .4 mL. If more than 6 mL of dye are discharged when making a certain shade of blue paint, the shade is unacceptable. Determine the setting for μ so that only 1% of the cans of paint will be unacceptable.   5.068

4.6 DESCRIPTIVE METHODS FOR ASSESSING NORMALITY

In the chapters that follow, we learn how to make inferences about the population based on information in the sample. Several of these techniques are based on the assumption that the population is approximately normally distributed. Consequently, it will be important to determine whether the sample data come from a normal population before we can properly apply these techniques.

Several descriptive methods can be used to check for normality. In this section, we consider the three methods summarized in the box.

Determining Whether the Data Are From an Approximately Normal Distribution

1. Construct either a histogram or stem-and-leaf display for the data and note the shape of the graph. If the data are approximately normal, the shape of the histogram or stem-and-leaf display will be similar to the normal curve, Figure 4.11 (i.e., mound-shaped and symmetric about the mean).
2. Compute the intervals $\bar{x} \pm s$, $\bar{x} \pm 2s$, and $\bar{x} \pm 3s$, and determine the percentage of measurements falling in each. If the data are approximately normal, the percentages will be approximately equal to 68%, 95%, and 100%, respectively.
3. Find the interquartile range, IQR, and standard deviation, s, for the sample, then calculate the ratio IQR/s. If the data are approximately normal, then IQR/$s \approx 1.3$.
4. Construct a normal probability plot for the data. If the data are approximately normal, the points will fall (approximately) on a straight line.

TEACHING TIP
Use a statistical computer package to generate examples of the various techniques presented here. Use them in class to discuss the normality of a variety of data sets.

The first two methods come directly from the properties of a normal distribution established in Section 4.5. Method 3 is based on the fact that for normal distributions, the z values corresponding to the 25th and 75th percentiles are $-.67$ and $.67$, respectively (see Example 4.14). Since $\sigma = 1$ for a standard normal distribution,

$$\frac{\text{IQR}}{\sigma} = \frac{Q_{\text{U}} - Q_{\text{L}}}{\sigma} = \frac{.67 - (-.67)}{1} = 1.34$$

The final descriptive method for checking normality is based on a *normal probability plot*. In such a plot, the observations in a data set are ordered from smallest to largest and then plotted against the expected z-scores of observations calculated under the assumption that the data come from a normal distribution. When the data are, in fact, normally distributed, a linear (straight-line) trend will result. A nonlinear trend in the plot suggests that the data are nonnormal.

TEACHING TIP
When working with normal probability plots, point out that straight lines indicate normality while curved lines indicate non-normality.

DEFINITION 4.9

A **normal probability plot** for a data set is a scatterplot with the ranked data values on one axis and their corresponding expected z-scores from a standard normal distribution on the other axis. [*Note:* Computation of the expected standard normal z-scores are beyond the scope of this text. Therefore, we will rely on available statistical software packages to generate a normal probability plot.]

EXAMPLE 4.21
Data are normal

The EPA mileage ratings on 100 cars, first presented in Chapter 2 (p. 30), are reproduced in Table 4.6. Numerical and graphical descriptive measures for the data are shown on the MINITAB and SPSS printouts, Figure 4.23a–c. Determine whether the EPA mileage ratings are from an approximate normal distribution.

EPAGAS

Table 4.6   EPA Gas Mileage Ratings for 100 Cars (miles per gallon)

36.3	41.0	36.9	37.1	44.9	36.8	30.0	37.2	42.1	36.7
32.7	37.3	41.2	36.6	32.9	36.5	33.2	37.4	37.5	33.6
40.5	36.5	37.6	33.9	40.2	36.4	37.7	37.7	40.0	34.2
36.2	37.9	36.0	37.9	35.9	38.2	38.3	35.7	35.6	35.1
38.5	39.0	35.5	34.8	38.6	39.4	35.3	34.4	38.8	39.7
36.3	36.8	32.5	36.4	40.5	36.6	36.1	38.2	38.4	39.3
41.0	31.8	37.3	33.1	37.0	37.6	37.0	38.7	39.0	35.8
37.0	37.2	40.7	37.4	37.1	37.8	35.9	35.6	36.7	34.5
37.1	40.3	36.7	37.0	33.9	40.1	38.0	35.2	34.8	39.5
39.9	36.9	32.9	33.8	39.8	34.0	36.8	35.0	38.1	36.9

FIGURE 4.23a

MINITAB Histogram for Mileage Data

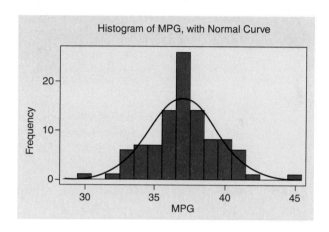

FIGURE 4.23b

MINITAB Descriptive Statistics for Mileage Data

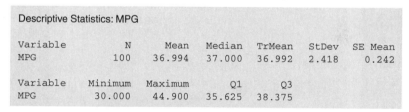

FIGURE 4.23c

SPSS Normal Probability Plot for Mileage Data

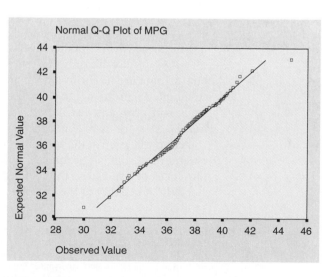

Solution

As a first check, we examine the MINITAB histogram of the data shown in Figure 4.23a. Clearly, the mileages fall in an approximately mound-shaped, symmetric distribution centered around the mean of approximately 37 mpg. Note that a normal curve is superimposed on the figure. Therefore, using check #1 in the box, the data appear to be approximately normal.

To apply check #2, we obtain $\bar{x} = 37$ and $s = 2.4$ from the MINITAB printout, Figure 4.23b. The intervals $\bar{x} \pm s, \bar{x} \pm 2s$, and $\bar{x} \pm 3s$, are shown in Table 4.7, as well as the percentage of mileage ratings that fall in each interval. (We obtained these results in Section 2.6, p. 61.) These percentages agree almost exactly with those from a normal distribution.

Table 4.7 Describing the 100 EPA Mileage Ratings

Interval	Percentage in Interval
$\bar{x} \pm s = (34.6, 39.4)$	68
$\bar{x} \pm 2s = (32.2, 41.8)$	96
$\bar{x} \pm 3s = (29.8, 44.2)$	99

Check #3 in the box requires that we find the ratio IQR/s. From Figure 4.23b, the 25th percentile (labeled Q_1 by MINITAB) is $Q_L = 35.625$ and the 75th percentile (labeled Q_3 by MINITAB) is $Q_U = 38.375$. Then, $IQR = Q_U - Q_L = 2.75$ and the ratio is

$$\frac{IRQ}{s} = \frac{2.75}{2.4} = 1.15$$

Since this value is approximately equal to 1.3, we have further confirmation that the data are approximately normal.

A fourth descriptive method is to interpret a normal probability plot. An SPSS normal probability plot for the mileage data is shown in Figure 4.23c. Notice that the ordered mileage values (shown on the horizontal axis) fall reasonably close to a straight line when plotted against the expected values from a normal distribution. Thus, check #4 also suggests that the EPA mileage data are likely to be approximately normally distributed. ∎

Suggested Exercise 4.73

The checks for normality given in the box are simple, yet powerful, techniques to apply, but they are only descriptive in nature. It is possible (although unlikely) that the data are nonnormal even when the checks are reasonably satisfied. Thus, we should be careful not to claim that the 100 EPA mileage ratings of Example 4.21 are, in fact, normally distributed. We can only state that it is reasonable to believe that the data are from a normal distribution.*

As we will learn in the remaining chapters, several inferential methods of analysis require the data to be approximately normal. If the data are clearly nonnormal, inferences derived from the method may be invalid. Therefore, it is advisable to check the normality of the data prior to conducting the analysis.

---

*Statistical tests of normality that provide a measure of reliability for the inference are available. However, these tests tend to be very sensitive to slight departures from normality, i.e., they tend to reject the hypothesis of normality for any distribution that is not perfectly symmetrical and mound shaped. Consult the references (see Ramsey & Ramsey, 1990) if you want to learn more about these tests.

Using the TI-83 Graphing Calculator

How to Graph a Normal Probability Plot

Start from the home screen.

Step 1   Enter your data into **List 1** (*Recall the Lists are accessed by pressing* **STAT ENTER**. *Always begin from a "clear" List*).

Step 2   Access the "Stat Plot" menu.
            Press   **2ⁿᵈ ENTER**
            Press   **ENTER**
            Press   **ENTER**
            Arrow down and right to last graph.
            Press   **ENTER**
            Set   Data List: **L1**
                          Data Axis: **X**

Step 3   *Press* **Zoom 9**
            Your data will be displayed against the residuals. If you see a "generally" linear relationship your data are near normal.

Example   Using a Normal Probability Plot test whether or not the data are normally distributed.

            9.7   93.1   33.0   21.2   81.4   51.1
            43.5   10.6   12.8   7.8   18.1   12.7

Step 1   Enter Data into List1 (see screen below).

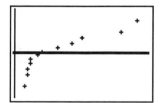

Step 2   Access the STAT PLOT Menu. ***Press 2ⁿᵈ = ENTER ENTER***.
            (You will see the screen below left after you change your setting to match.)

Step 3   View Display. ***Press* Zoom 9** (see screen above right).
            There is a noticeable curve. The data are not normally distributed.

Step 4   Clear the screen for the next problem. Return to the home screen.
            Press **2ⁿᵈY = ENTER**. Arrow right. ***Press* ENTER CLEAR**.

 EXERCISES 4.69–4.79

Learning the Mechanics

4.69 If a population data set is normally distributed, what is the proportion of measurements you would expect to fall within the following intervals?
  a. $\mu \pm \sigma$ .68
  b. $\mu \pm 2\sigma$ .95
  c. $\mu \pm 3\sigma$ 1.00

4.70 Consider a sample data set with the following summary statistics: $s = 95, Q_L = 72, Q_U = 195$.
  a. Calculate IQR.   123
  b. Calculate IQR/s.   1.295
  c. Is the value of IQR/s approximately equal to 1.3? What does this imply?   Yes

4.71 Normal probability plots for three data sets are shown below. Which plot indicates that the data are approximately normally distributed?

4.72 Examine the sample data in the table.

  ● LM4_72

5.9	5.3	1.6	7.4	8.6	1.2	2.1
4.0	7.3	8.4	8.9	6.7	4.5	6.3
7.6	9.7	3.5	1.1	4.3	3.3	8.4
1.6	8.2	6.5	1.1	5.0	9.4	6.4

  a. Construct a stem-and-leaf plot to assess whether the data are from an approximately normal distribution.
  b. Compute s for the sample data.   2.765
  c. Find the values of Q_L and Q_U and the value of s from part **b** to assess whether the data come from an approximately normal distribution.
  d. Generate a normal probability plot for the data and use it to assess whether the data are approximately normal.

Applying the Concepts—Basic

  ● WOMENPOWER

4.73 Refer to Exercise 2.44 (p. 51) and the ages of the 50 most powerful women in corporate America by *Fortune* (Oct. 25, 1999). A MINITAB printout with summary statistics for the age distribution is reproduced at the bottom of the page.
  a. Use the relevant statistics on the printout to determine whether the age distribution is approximately normal.
  b. In Exercise 2.44**d** you constructed a relative frequency histogram for the age data. Use this graph to support your conclusion in part a.

4.74 Refer to the *Astronomical Journal* (July 1995) study of galaxy velocities, Exercise 2.78 (p. 69). A histogram of the velocity for 103 galaxies located in a particular cluster named A2142 is reproduced here. Comment on whether or not the galaxy velocities are approximately normally distributed.

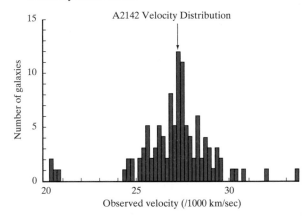

Source: Oegerle, W. R., Hill, J. M., and Fitchett, M. J. "Observations of high dispersion clusters of galaxies: Constraints on cold dark matter." *The Astronomical Journal*, Vol. 110, No. 1, July 1995, p. 37 (Figure 1).

Plots for Exercise 4.71

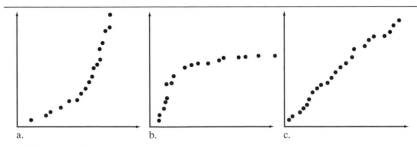

a.   b.   c.

MINITAB Output for Exercise 4.73

Variable	N	Mean	Median	Tr Mean	StDev	SE Mean
Age	50	48.160	47.000	47.795	6.015	0.851

Variable	Min	Max	Q1	Q3
Age	36.000	68.000	45.000	51.250

4.75 How accurate are you at the game of darts? Researchers at Iowa State University attempted to develop a probability model for dart throws (*Chance*, Summer 1997). For each of 590 throws made at certain targets on a dart board, the distance from the dart to the target point was measured (to the nearest millimeter). The error distribution for the dart throws is described by the frequency table shown below.

a. Construct a histogram for the data. Is the error distribution for the dart throws approximately normal?

b. Descriptive statistics for the distances from the target for the 590 throws are given below. Use this information to decide whether the error distribution is approximately normal.

$$\bar{x} = 24.4 \, \text{mm}$$
$$s = 12.8 \, \text{mm}$$
$$Q_L = 14 \, \text{mm}$$
$$Q_U = 34 \, \text{mm}$$

c. Construct a normal probability plot for the data and use it to assess whether the error distribution is approximately normal.

Applying the Concepts—Intermediate

4.76 Refer to the distribution of batting averages for Major League Baseball players, Exercise 4.57 (p. 202). The batting averages for American League players with at least 100 official at bats are stored in the data file ALBATAVG, while the batting averages for National League players with at least 100 official at bats are stored in the data file NL-BATAVG. Determine whether each distribution is approximately normal.

4.77 In Exercise 2.32 (p. 41) you read about a *Journal of Statistics Education* study of team performance on games in which Mark McGwire (of the St. Louis Cardinals) and Sammy Sosa (of the Chicago Cubs) hit home runs during their record-breaking 1998 Major League Baseball season. The data on number of runs scored by their respective teams in these games are reproduced in the table at the top of p. 210.

a. Determine whether the data on number of runs scored by the St. Louis Cardinals are approximately normal.

b. Repeat part **a** for the Chicago Cubs.

4.78 Refer to the data on May 2001 sanitation scores for 151 cruise ships, first presented in Exercise 2.72 (p. 67). Assess whether the sanitation scores (saved in the SHIP-SANIT file) are approximately normally distributed.

4.79 Refer to the *Applied Psycholinguistics* (June 1998) study of language skills in young children, Exercise 2.76 (p. 69). Recall that the mean sentence complexity score (measured on a 0 to 48 point scale) of low income children was 7.62 with a standard deviation of 8.91. Demonstrate why the distribution of sentence complexity scores for low income children is unlikely to be normally distributed.

DARTS

Distance from Target (mm)	Frequency	Distance from Target (mm)	Frequency	Distance from Target (mm)	Frequency
0	3	21	15	41	7
2	2	22	19	42	7
3	5	23	13	43	11
4	10	24	11	44	7
5	7	25	9	45	5
6	3	26	21	46	3
7	12	27	16	47	1
8	14	28	18	48	3
9	9	29	9	49	1
10	11	30	14	50	4
11	14	31	13	51	4
12	13	32	11	52	3
13	32	33	11	53	4
14	19	34	11	54	1
15	20	35	16	55	1
16	9	36	13	56	4
17	18	37	5	57	3
18	23	38	9	58	2
19	25	39	6	61	1
20	21	40	5	62	2
				66	1

Source: Stern, H. S., and Wilcox, W. "Shooting Darts." *Chance*, Vol. 10, No. 3, Summer 1997, p. 17 (adapted from Figure 2).

STLRUNS				CUBSRUNS		
St. Louis Cardinals				**Chicago Cubs**		
6	6	3	11	3	6	8
8	1	10	6	4	5	2
5	8	9	6	1	8	6
8	2	3	6	1	9	6
15	2	8		4	2	3
8	6	2		3	6	9
5	8	4		5	4	10
8	9	4		5	4	4
3	5	3		2	9	5
5	7	6		5	5	4
2	3	8		5	6	5
6	2	7		10	9	8
3	7	14		5	3	11
5	1	7		6	6	15
3	4	3		7	5	11
10	6	6		13	9	6
4	4	6		8	7	7
11	13	5		10	8	2

4.7 APPROXIMATING A BINOMIAL DISTRIBUTION WITH A NORMAL DISTRIBUTION (OPTIONAL)

When the discrete binomial random variable (Section 4.3) can assume a large number of values, the calculation of its probabilities may become very tedious. To contend with this problem, we provide tables in Appendix A to give the probabilities for some values of n and p, but these tables are by necessity incomplete. Recall that the binomial probability table (Table II) can be used only for $n = 5, 6, 7, 8, 9, 10, 15, 20,$ or 25. To deal with this limitation, we seek approximation procedures for calculating the probabilities associated with a binomial probability distribution.

When n is large, a normal probability distribution may be used to provide a good approximation to the probability distribution of a binomial random variable. To show how this approximation works, we refer to Example 4.11, in which we used the binomial distribution to model the number x of 20 eligible voters who favor a candidate. We assumed that 60% of all the eligible voters favored the candidate. The mean and standard deviation of x were found to be $\mu = 12$ and $\sigma = 2.2$. The binomial distribution for $n = 20$ and $p = .6$ is shown in Figure 4.24,

FIGURE 4.24

Binomial Distribution for $n = 20$, $p = .6$ and Normal Distribution with $\mu = 12$, $\sigma = 2.2$

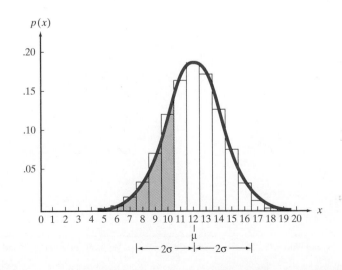

and the approximating normal distribution with mean $\mu = 12$ and standard deviation $\sigma = 2.2$ is superimposed.

As part of Example 4.11, we used Table II to find the probability that $x \leq 10$. This probability, which is equal to the sum of the areas contained in the rectangles (shown in Figure 4.24) that correspond to $p(0), p(1), p(2), \ldots, p(10)$, was found to equal .245. The portion of the approximating normal curve that would be used to approximate the area $p(0) + p(1) + p(2) + \cdots + p(10)$ is highlighted in Figure 4.24. Note that this highlighted area lies to the left of 10.5 (not 10), so we may include all of the probability in the rectangle corresponding to $p(10)$. Because we are approximating a discrete distribution (the binomial) with a continuous distribution (the normal), we call the use of 10.5 (instead of 10 or 11) a **correction for continuity**. That is, we are correcting the discrete distribution so that it can be approximated by the continuous one. The use of the correction for continuity leads to the calculation of the following standard normal z value:

$$z = \frac{x - \mu}{\sigma} = \frac{10.5 - 12}{2.2} = -.68$$

Using Table III, we find the area between $z = 0$ and $z = .68$ to be .2517. Then the probability that x is less than or equal to 10 is approximated by the area under the normal distribution to the left of 10.5, shown highlighted in Figure 4.24. That is,

$$P(x \leq 10) \approx P(z \leq -.68) = .5 - P(-.68 < z \leq 0) = .5 - .2517 = .2483$$

The approximation differs only slightly from the exact binomial probability, .245. Of course, when tables of exact binomial probabilities are available, we will use the exact value rather than a normal approximation.

Use of the normal distribution will not always provide a good approximation for binomial probabilities. The following is a useful rule of thumb to determine when n is large enough for the approximation to be effective: *The interval $\mu \pm 3\sigma$ should lie within the range of the binomial random variable x (i.e., 0 to n) in order for the normal approximation to be adequate.* The rule works well because almost all of the normal distribution falls within 3 standard deviations of the mean, so if this interval is contained within the range of x values, there is "room" for the normal approximation to work.

As shown in Figure 4.25a for the preceding example with $n = 20$ and $p = .6$, the interval $\mu \pm 3\sigma = 12 + 3(2.19) = (5.43, 18.57)$ lies within the range 0 to 20. However, if we were to try to use the normal approximation with $n = 10$ and $p = .1$, the interval $\mu \pm 3\sigma$ is $1 \pm 3(.95)$, or $(-1.85, 3.85)$. As shown in Figure 4.25b, this interval is not contained within the range of x since $x = 0$ is the lower bound for a binomial random variable. Note in Figure 4.25b that the normal distribution will not "fit" in the range of x, and therefore it will not provide a good approximation to the binomial probabilities.

EXAMPLE 4.22

One problem with any product that is mass-produced (e.g., a graphing calculator) is quality control. The process must be monitored or audited to be sure the output of the process conforms to requirements. One monitoring method is *lot acceptance sampling*, in which items being produced are sampled at various stages of the production process and are carefully inspected. The lot of items from which the sample is drawn is then accepted or rejected, based on the number of defectives in the sample. Lots that are accepted may be sent forward for further processing or may be shipped to customers; lots that are rejected may be reworked or scrapped. For example, suppose a manufacturer of calculators chooses 200 stamped circuits from the day's production and determines x, the number of defective circuits in the sample. Suppose that up to a 6% rate of defectives is considered acceptable for the process.

FIGURE 4.25

Rule of Thumb for Normal Approximation to Binomial Probabilities

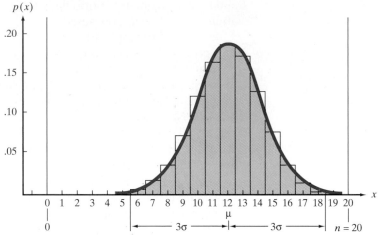

a. $n = 20$, $p = .6$: Normal approximation is good

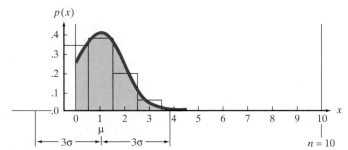

b. $n = 10$, $p = .1$: Normal approximation is poor

a. $\mu = 12$, $\sigma = 3.36$
b. .0129

a. Find the mean and standard deviation of x, assuming the defective rate is 6%.
b. Use the normal approximation to determine the probability that 20 or more defectives are observed in the sample of 200 circuits (i.e., find the approximate probability that $x \geq 20$).

Solution

a. The random variable x is binomial with $n = 200$ and the fraction defective $p = .06$. Thus,

$$\mu = np = 200(.06) = 12$$
$$\sigma = \sqrt{npq} = \sqrt{200(.06)(.94)} = \sqrt{11.28} = 3.36$$

We first note that

$$\mu \pm 3\sigma = 12 \pm 3(3.36) = 12 \pm 10.08 = (1.92, 22.08)$$

lies completely within the range from 0 to 200. Therefore, a normal probability distribution should provide an adequate approximation to this binomial distribution.

b. Using the rule of complements, $P(x \geq 20) = 1 - P(x \leq 19)$. To find the approximating area corresponding to $x \leq 19$, refer to Figure 4.26. Note that we want to include all the binomial probability histograms from 0 to 19, inclusive. Since the event is of the form $x \leq a$, the proper correction for continuity is $a + .5 = 19 + .5 = 19.5$. Thus, the z value of interest is

$$z = \frac{(a + .5) - \mu}{\sigma} = \frac{19.5 - 12}{3.36} = 2.23$$

Referring to Table III in Appendix A, we find that the area to the right of the mean 0 corresponding to $z = 2.23$ (see Figure 4.27) is .4871. So the area $A = P(z \leq 2.23)$ is

$$A = .5 + .4871 = .9871$$

Suggested Exercise 4.90

FIGURE 4.26

Normal Approximation to the Binomial Distribution with $n = 200$, $p = .06$

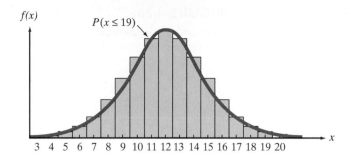

FIGURE 4.27

Standard Normal Distribution

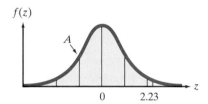

Thus, the normal approximation to the binomial probability we seek is

$$P(x \geq 20) = 1 - P(x \leq 19) \approx 1 - .9871 = .0129$$

In other words, the probability is extremely small that 20 or more defectives will be observed in a sample of 200 circuits—*if in fact the true fraction of defectives is* .06. If the manufacturer observes $x \geq 20$, the likely reason is that the process is producing more than the acceptable 6% defectives. The lot acceptance sampling procedure is another example of using the rare-event approach to make inferences. ■

The steps for approximating a binomial probability by a normal probability are given in the accompanying box.

Using a Normal Distribution to Approximate Binomial Probabilities

1. After you have determined n and p for the binomial distribution, calculate the interval

   $$\mu \pm 3\sigma = np \pm 3\sqrt{npq}$$

   If the interval lies in the range 0 to n, the normal distribution will provide a reasonable approximation to the probabilities of most binomial events.

2. Express the binomial probability to be approximated in the form $P(x \leq a)$ or $P(x \leq b) - P(x \leq a)$. For example,

   $$P(x < 3) = P(x \leq 2)$$
   $$P(x \geq 5) = 1 - P(x \leq 4)$$
   $$P(7 \leq x \leq 10) = P(x \leq 10) - P(x \leq 6)$$

3. For each value of interest a, the correction for continuity is $(a + .5)$, and the corresponding standard normal z value is

   $$z = \frac{(a + .5) - \mu}{\sigma} \quad \text{(see Figure 4.28)}$$

4. Sketch the approximating normal distribution and shade the area corresponding to the probability of the event of interest, as in Figure 4.28. Verify that the rectangles you have included in the shaded area correspond to the event probability you wish to approximate. Using Table III and the z value(s) you calculated in step 3, find the shaded area. This is the approximate probability of the binomial event.

FIGURE 4.28

Approximating Binomial Probabilities by Normal Probabilities

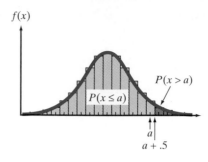

 EXERCISES 4.80–4.94

Learning the Mechanics

4.80 Assume that x is a binomial random variable with n and p as specified in parts **a–f**. For which cases would it be appropriate to use a normal distribution to approximate the binomial distribution?
 a. $n = 100, p = .01$ No
 b. $n = 20, p = .6$ Yes
 c. $n = 10, p = .4$ No
 d. $n = 1,000, p = .05$ Yes
 e. $n = 100, p = .8$ Yes
 f. $n = 35, p = .7$ Yes

4.81 Suppose x is a binomial random variable with $p = .4$ and $n = 25$.
 a. Would it be appropriate to approximate the probability distribution of x with a normal distribution? Explain.   Yes
 b. Assuming that a normal distribution provides an adequate approximation to the distribution of x, what are the mean and variance of the approximating normal distribution?   10; 6
 c. Use Table II of Appendix A to find the exact value of $P(x \geq 9)$.   .726
 d. Use the normal approximation to find $P(x \geq 9)$.

4.82 Assume that x is a binomial random variable with $n = 25$ and $p = .5$. Use Table II of Appendix A and the normal approximation to find the exact and approximate values, respectively, for the following probabilities:
 a. $P(x \leq 11)$ .345; .3446
 b. $P(x \geq 16)$ .115; .1151
 c. $P(8 \leq x \leq 16)$ .924; .9224

4.83 Assume that x is a binomial random variable with $n = 100$ and $p = .40$. Use a normal approximation to find the following:
 a. $P(x \leq 35)$ .1788
 b. $P(40 \leq x \leq 50)$ .5236
 c. $P(x \geq 38)$ .6950

4.84 Assume that x is a binomial random variable with $n = 1000$ and $p = .50$. Find each of the following probabilities:

 a. $P(x > 500)$ .4880
 b. $P(490 \leq x < 500)$ .2334
 c. $P(x > 550)$ 0

Applying the Concepts—Basic

4.85 In Exercise 4.33 (p. 188), you learned that only 5% of the nation's cigarette smokers ever enter into a treatment program to help them quit smoking (*USF Magazine*, Spring 2000). In a random sample of 200 smokers, let x be the number who enter into a treatment program.
 a. Find the mean of x. (This value should agree with your answer to Exercise 4.33**c**.)   10
 b. Find the standard deviation of x.   3.082
 c. Find the z-score for the value $x = 10.5$.   .16
 d. Find the approximate probability that the number of smokers in a sample of 200 who will enter into a treatment program is 11 or more.   .4364

4.86 In Exercise 4.34 (p. 188), you learned that 22% of all births in the U.S. occur by Caesarian section each year (*USA Today*, Sept. 19, 2000). In a random sample of 1,000 births this year, let x be the number that occur by Caesarian section.
 a. Find the mean of x. (This value should agree with your answer to Exercise 4.34**a**.)   220
 b. Find the standard deviation of x. (This value should agree with your answer to Exercise 4.34**b**.)   13.10
 c. Find the z-score for the value $x = 200.5$. -1.49
 d. Find the approximate probability that the number of Caesarian sections in a sample of 1,000 births is less than or equal to 200.   .0681

4.87 According to the *American Cancer Society*, melanoma, a form of skin cancer, kills 60% of Americans who suffer from the disease each year. Consider a sample of 10,000 melanoma patients.
 a. What are the expected value and variance of x, the number of the 10,000 melanoma patients who die of the affliction this year?   6,000; 2,400

b. Find the probability that *x* will exceed 6,100 patients per year.   .0202

c. Would you expect *x*, the number of patients dying of melanoma, to exceed 6,500 in any single year? Explain.   No, $P \approx 0$

4.88 According to *Time* (Oct. 11, 1999), 1% of all patients who undergo laser surgery to correct their vision have serious post-laser vision problems. In a sample of 100,000 patients, what is the approximate probability that fewer than 950 will experience serious post-laser vision problems?   .0537

Applying the Concepts—Intermediate

4.89 The computer chips in notebook and laptop computers are produced from semiconductor wafers. Certain semiconductor wafers are exposed to an environment that generates up to 100 possible defects per wafer. The number of defects per wafer, *x*, was found to follow a binomial distribution if the manufacturing process is stable and generates defects that are randomly distributed on the wafers (*IEEE Transactions on Semiconductor Manufacturing*, May 1995). Let *p* represent the probability that a defect occurs at any one of the 100 points of the wafer. For each of the following cases, determine whether the normal approximation can be used to characterize *x*.

a. $p = .01$ No
b. $p = .50$ Yes
c. $p = .90$ Yes

4.90 Refer to the FTC study of the pricing accuracy of supermarket electronic scanners, Exercise 4.37 (p. 188). Recall that the probability that a scanned item is priced incorrectly is $\frac{1}{30} = .033$.

a. Suppose 10,000 supermarket items are scanned. What is the approximate probability that you observe at least 100 items with incorrect prices?   1

b. Suppose 100 items are scanned and you are interested in the probability that fewer than five are incorrectly priced. Explain why the approximate method of part **a** may not yield an accurate estimate of the probability.

4.91 In Exercise 4.41 (p. 189) you learned that some researchers believe that one in every three women are victims of domestic abuse (*Annals of Internal Medicine*, Nov. 1995).

a. For a random sample of 150 women, what is the approximate probability that more than half are victims of domestic abuse? ≈ 0

b. For a random sample of 150 women, what is the approximate probability that fewer than 50 are victims of domestic abuse?   .4641

c. Would you expect to observe fewer than 30 domestically abused women in a sample of 150? Explain.

4.92 A manufacturer of CD-ROMs claims that 99.4% of its CDs are defect-free. A large software company that buys and uses large numbers of the CDs wants to verify this claim, so it selects 1,600 CDs to be tested. The tests reveal 12 CDs to be defective. Assuming that the manufacturer's claim is correct, what is the probability of finding 12 or more defective CDs in a sample of 1,600? Does your answer cast doubt on the manufacturer's claim? Explain.

Applying the Concepts—Advanced

4.93 According to *New Jersey Business* (Feb. 1996), Newark International Airport's new terminal handles an average of 3,000 international passengers an hour, but is capable of handling twice that number. Also, 80% of arriving international passengers pass through without their luggage being inspected and the remainder are detained for inspection. The inspection facility can handle 600 passengers an hour without unreasonable delays for the travelers.

a. When international passengers arrive at the rate of 1,500 per hour, what is the expected number of passengers who will be detained for luggage inspection?

b. In the future, it is expected that as many as 4,000 international passengers will arrive per hour. When that occurs, what is the expected number of passengers who will be detained for luggage inspection?   800

c. Refer to part **b**. Find the approximate probability that more than 600 international passengers will be detained for luggage inspection. (This is also the probability that travelers will experience unreasonable luggage inspection delays.) ≈ 1.0

4.94 The percentage of fat in the bodies of American men is an approximately normal random variable with mean equal to 15% and standard deviation equal to 2%.

a. If these values were used to describe the body fat of men in the United States Army and if 20% or more body fat is characterized as obese, what is the approximate probability that a random sample of 10,000 soldiers will contain fewer than 50 who would actually be characterized as obese?   .0559

b. If the army actually were to check the percentage of body fat for a random sample of 10,000 men and if only 30 contained 20% (or higher) body fat, would you conclude that the army was successful in reducing the percentage of obese men below the percentage in the general population? Explain your reasoning.

4.8 SAMPLING DISTRIBUTIONS

In previous sections we assumed that we knew the probability distribution of a random variable, and using this knowledge we were able to compute the mean, variance, and probabilities associated with the random variable. However, in most practical applications, the true mean and standard deviation are unknown quantities that would have to be estimated. Numerical quantities that describe probability distributions are called *parameters*. Thus, p, the probability of a success in a binomial experiment, and μ and σ, the mean and standard deviation of a normal distribution, are examples of parameters.

TEACHING TIP
Point out that in most situations the population parameters will be unknown values while the sample statistics will always be known.

DEFINITION 4.10

A **parameter** is a numerical descriptive measure of a population. Because it is based on the observations in the population, its value is almost always unknown.

We have also discussed the sample mean $\bar{x}$, sample variance s^2, sample standard deviation s, etc., which are numerical descriptive measures calculated from the sample. We will often use the information contained in these *sample statistics* to make inferences about the parameters of a population.

DEFINITION 4.11

A **sample statistic** is a numerical descriptive measure of a sample. It is calculated from the observations in the sample.

Note that the term *statistic* refers to a *sample* quantity and the term *parameter* refers to a *population* quantity.

In order to use sample statistics to make inferences about population parameters, we need to be able to evaluate their properties. Does one sample statistic contain more information than another about a population parameter? On what basis should we choose the "best" statistic for making inferences about a parameter? For example, if we want to estimate the population mean μ, we could use a number of sample statistics for our estimate. Two possibilities are the sample mean $\bar{x}$ and the sample median M. Which of these do you think will provide a better estimate of μ?

Before answering this question, consider the following example: Toss a fair die, and let x equal the number of dots showing on the up face. Suppose the die is tossed three times, producing the sample measurements 2, 2, 6. The sample mean is $\bar{x} = 3.33$ and the sample median is $M = 2$. Since the population mean of x is $\mu = 3.5$, you can see that for this sample of three measurements, the sample mean $\bar{x}$ provides an estimate that falls closer to μ than does the sample median (see Figure 4.29a). Now suppose we toss the die three more times and obtain the sample measurements 3, 4, 6. The mean and median of this sample are $\bar{x} = 4.33$ and $M = 4$, respectively. This time M is closer to μ (see Figure 4.29b).

FIGURE 4.29

Comparing the Sample Mean ($\bar{x}$) and Sample Median (M) as Estimators of the Population Mean (μ)

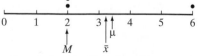

a. Sample 1: $\bar{x}$ is closer than M to μ

b. Sample 2: M is closer than $\bar{x}$ to μ

This simple example illustrates an important point: Neither the sample mean nor the sample median will *always* fall closer to the population mean. Consequently, we cannot compare these two sample statistics, or, in general, any two sample statistics, on the basis of their performance for a single sample. Instead, we need to recognize that sample statistics are themselves random variables, because different samples can lead to different values for the sample statistics. As random variables, sample statistics must be judged and compared on the basis of their probability distributions, i.e., the *collection* of values and associated probabilities of each statistic that would be obtained if the sampling experiment were repeated a *very large number of times*. We will illustrate this concept with another example.

Suppose it is known that in a certain part of Canada the daily high temperature recorded for all past months of January has a mean of $\mu = 10°F$ and a standard deviation of $\sigma = 5°F$. Consider an experiment consisting of randomly selecting 25 daily high temperatures from the records of past months of January and calculating the sample mean $\bar{x}$. If this experiment were repeated a very large number of times, the value of $\bar{x}$ would vary from sample to sample. For example, the first sample of 25 temperature measurements might have a mean $\bar{x} = 9.8$, the second sample a mean $\bar{x} = 11.4$, the third sample a mean $\bar{x} = 10.5$, etc. If the sampling experiment were repeated a very large number of times, the resulting histogram of sample means would be approximately the probability distribution of $\bar{x}$. If $\bar{x}$ is a good estimator of μ, we would expect the values of $\bar{x}$ to cluster around μ as shown in Figure 4.30. This probability distribution is called a *sampling distribution* because it is generated by repeating a sampling experiment a very large number of times.

TEACHING TIP
The repeated sampling necessary to generate a sampling distribution is one of the most difficult concepts for students to understand.

DEFINITION 4.12

The **sampling distribution** of a sample statistic calculated from a sample of n measurements is the probability distribution of the statistic.

FIGURE 4.30

Sampling Distribution for $\bar{x}$
Based on a Sample of
$n = 25$ Measurements

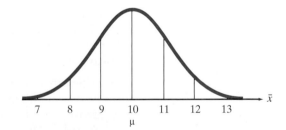

TEACHING TIP
Draw pictures of different sampling distributions as they relate to an unknown population parameter. Use the pictures to lay the groundwork for discussing the ideas of unbiased sampling distributions and minimum variance.

In actual practice, the sampling distribution of a statistic is obtained mathematically or (at least approximately) by simulating the sample on a computer using a procedure similar to that just described.

If $\bar{x}$ has been calculated from a sample of $n = 25$ measurements selected from a population with mean $\mu = 10$ and standard deviation $\sigma = 5$, the sampling distribution (Figure 4.30) provides information about the behavior of $\bar{x}$ in repeated sampling. For example, the probability that you will draw a sample of 25 measurements and obtain a value of $\bar{x}$ in the interval $9 \leq \bar{x} \leq 10$ will be the area under the sampling distribution over that interval.

Since the properties of a statistic are typified by its sampling distribution, it follows that to compare two sample statistics you compare their sampling distributions.

For example, if you have two statistics, A and B, for estimating the same parameter (for purposes of illustration, suppose the parameter is the population variance σ^2) and if their sampling distributions are as shown in Figure 4.31, you would choose statistic A in preference to statistic B. You would make this choice because the sampling distribution for statistic A centers over σ^2 and has less spread (variation) than the sampling distribution for statistic B. When you draw a single sample in a practical sampling situation, the probability is higher that statistic A will fall nearer σ^2.

FIGURE 4.31

Two Sampling Distributions for Estimating the Population Variance, σ^2

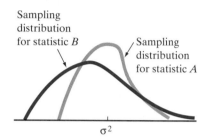

Remember that in practice we will not know the numerical value of the unknown parameter σ^2 so we will not know whether statistic A or statistic B is closer to σ^2 for a sample. We have to rely on our knowledge of the theoretical sampling distributions to choose the best sample statistic and then use it sample after sample. The procedure for finding the sampling distribution for a statistic is demonstrated next.

EXAMPLE 4.23

Consider a population consisting of the measurements 0, 3, and 12 and described by the probability distribution shown here. A random sample of $n = 3$ measurements is selected from the population.

x	0	3	12
$p(x)$	$\frac{1}{3}$	$\frac{1}{3}$	$\frac{1}{3}$

a. Find the sampling distribution of the sample mean $\bar{x}$.
b. Find the sampling distribution of the sample median M.

Solution

Every possible sample of $n = 3$ measurements is listed in Table 4.8 along with the sample mean and median. Also, because any one sample is as likely to be selected as any other (random sampling), the probability of observing any particular sample is $\frac{1}{27}$. The probability is also listed in Table 4.8.

a. From Table 4.8 you can see that $\bar{x}$ can assume the values 0, 1, 2, 3, 4, 5, 6, 8, 9, and 12. Because $\bar{x} = 0$ occurs in only one sample, $P(\bar{x} = 0) = \frac{1}{27}$. Similarly, $\bar{x} = 1$ occurs in three samples: $(0, 0, 3)$, $(0, 3, 0)$, and $(3, 0, 0)$. Therefore, $P(\bar{x} = 1) = \frac{3}{27} = \frac{1}{9}$. Calculating the probabilities of the remaining values of $\bar{x}$ and arranging them in a table, we obtain the probability distribution shown below. This is the sampling distribution for $\bar{x}$ because it specifies the probability associated with each possible value of $\bar{x}$.

$\bar{x}$	0	1	2	3	4	5	6	8	9	12
$p(\bar{x})$	$\frac{1}{27}$	$\frac{3}{27}$	$\frac{3}{27}$	$\frac{1}{27}$	$\frac{3}{27}$	$\frac{6}{27}$	$\frac{3}{27}$	$\frac{3}{27}$	$\frac{3}{27}$	$\frac{1}{27}$

TEACHING TIP
Use samples of $n = 2$ students in class and calculate the average age of the students in the samples. Discuss the variation in the sample means to illustrate what a sampling distribution represents.

TEACHING TIP
It is helpful to the student to point out that the repeated sampling nature of these problems will provide the necessary information that will be used when we take our one sample later in the book. Emphasize that we will need just one sample.

TABLE 4.8

Possible Samples	$\bar{x}$	M	Probability
0, 0, 0	0	0	$1/27$
0, 0, 3	1	0	$1/27$
0, 0, 12	4	0	$1/27$
0, 3, 0	1	0	$1/27$
0, 3, 3	2	3	$1/27$
0, 3, 12	5	3	$1/27$
0, 12, 0	4	0	$1/27$
0, 12, 3	5	3	$1/27$
0, 12, 12	8	12	$1/27$
3, 0, 0	1	0	$1/27$
3, 0, 3	2	3	$1/27$
3, 0, 12	5	3	$1/27$
3, 3, 0	2	3	$1/27$
3, 3, 3	3	3	$1/27$
3, 3, 12	6	3	$1/27$
3, 12, 0	5	3	$1/27$
3, 12, 3	6	3	$1/27$
3, 12, 12	9	12	$1/27$
12, 0, 0	4	0	$1/27$
12, 0, 3	5	3	$1/27$
12, 0, 12	8	12	$1/27$
12, 3, 0	5	3	$1/27$
12, 3, 3	6	3	$1/27$
12, 3, 12	9	12	$1/27$
12, 12, 0	8	12	$1/27$
12, 12, 3	9	12	$1/27$
12, 12, 12	12	12	$1/27$

b. In Table 4.8 you can see that the median M can assume one of the three values: 0, 3, or 12. The value $M = 0$ occurs in seven different samples. Therefore, $P(M = 0) = 7/27$. Similarly, $M = 3$ occurs in 13 samples and $M = 12$ occurs in seven samples. Therefore, the probability distribution (i.e., the sampling distribution) for the median M is as shown below.

M	0	3	12
p(M)	$7/27$	$13/27$	$7/27$

Example 4.23 demonstrates the procedure for finding the exact sampling distribution of a statistic when the number of different samples that could be selected from the population is relatively small. In the real world, populations often consist of a large number of different values, making samples difficult (or impossible) to enumerate. When this situation occurs, we may choose to obtain the approximate sampling distribution for a statistic by simulating the sampling over and over again and recording the proportion of times different values of the statistic occur. Example 4.24 illustrates this procedure.

EXAMPLE   4.24

Suggested Exercise 4.95

Suppose we perform the following experiment over and over again: Take a sample of 11 measurements from the distribution shown in Figure 4.32. This distribution is known as the **uniform distribution** over the interval $(0, 1)$. Calculate the two sample statistics

$$\bar{x} = \text{Sample mean} = \frac{\Sigma x}{11}$$

$M = \text{Median} = \text{Sixth sample measurement when the 11 measurements}$
$\text{are arranged in ascending order}$

Obtain approximations to the sampling distributions of $\bar{x}$ and M.

FIGURE 4.32

Uniform Distribution from 0 to 1

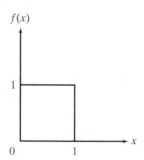

Solution

We use a computer to generate 1,000 samples, each with $n = 11$ observations. Then we compute $\bar{x}$ and M for each sample. Our goal is to obtain approximations to the sampling distributions of $\bar{x}$ and M to find out which sample statistic ($\bar{x}$ or M) contains more information about μ. (Note that, in this particular example, we *know* the population mean is $\mu = .5$.) The first 10 of the 1,000 samples generated are presented in Table 4.9. For instance, the first computer-generated sample from the uniform distribution (arranged in ascending order) contained the following measurements: .125, .138, .139, .217, .419, .506, .516, .757, .771, .786, and .919. The sample mean $\bar{x}$ and median M computed for this sample are

$$\bar{x} = \frac{.125 + .138 + \cdots + .919}{11} = .481$$

$M = \text{Sixth ordered measurement} = .506$

TABLE 4.9   First 10 Samples of $n = 11$ Measurements from a Uniform Distribution

Sample	Measurements										
1	.217	.786	.757	.125	.139	.919	.506	.771	.138	.516	.419
2	.303	.703	.812	.650	.848	.392	.988	.469	.632	.012	.065
3	.383	.547	.383	.584	.098	.676	.091	.535	.256	.163	.390
4	.218	.376	.248	.606	.610	.055	.095	.311	.086	.165	.665
5	.144	.069	.485	.739	.491	.054	.953	.179	.865	.429	.648
6	.426	.563	.186	.896	.628	.075	.283	.549	.295	.522	.674
7	.643	.828	.465	.672	.074	.300	.319	.254	.708	.384	.534
8	.616	.049	.324	.700	.803	.399	.557	.975	.569	.023	.072
9	.093	.835	.534	.212	.201	.041	.889	.728	.466	.142	.574
10	.957	.253	.983	.904	.696	.766	.880	.485	.035	.881	.732

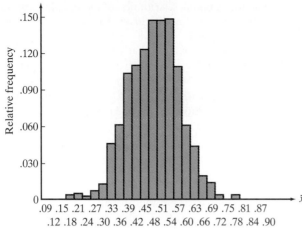

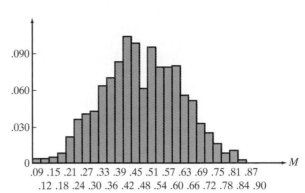

a. Sampling distribution for $\bar{x}$ (based on 1,000 samples of $n = 11$ measurements)

b. Sampling distribution for M (based on 1,000 samples of $n = 11$ measurements)

FIGURE 4.33

Relative Frequency Histograms for $\bar{x}$ and M, Example 4.24

The relative frequency histograms for $\bar{x}$ and M for the 1,000 samples of size $n = 11$ are shown in Figure 4.33.

You can see that the values of $\bar{x}$ tend to cluster around μ to a greater extent than do the values of M. Thus, on the basis of the observed sampling distributions, we conclude that $\bar{x}$ contains more information about μ than M does—at least for samples of $n = 11$ measurements from the uniform distribution. ∎

As noted earlier, many sampling distributions can be derived mathematically, but the theory necessary to do this is beyond the scope of this text. Consequently, when we need to know the properties of a statistic, we will present its sampling distribution and simply describe its properties. Several of the important properties we look for in sampling distributions are discussed in the next section.

 EXERCISES 4.95–4.101

Learning the Mechanics

4.95 The probability distribution shown here describes a population of measurements that can assume values of 0, 2, 4, and 6, each of which occurs with the same relative frequency:

x	0	2	4	6
$p(x)$	$1/4$	$1/4$	$1/4$	$1/4$

a. List all the different samples of $n = 2$ measurements that can be selected from this population.

b. Calculate the mean of each different sample listed in part **a**.

c. If a sample of $n = 2$ measurements is randomly selected from the population, what is the probability that a specific sample will be selected? $1/16$

d. Assume that a random sample of $n = 2$ measurements is selected from the population. List the different values of $\bar{x}$ found in part **b**, and find the probability of each. Then give the sampling distribution of the sample mean $\bar{x}$ in tabular form.

e. Construct a probability histogram for the sampling distribution of $\bar{x}$.

4.96 Simulate sampling from the population described in Exercise 4.95 by marking the values of x, one on each of four identical coins (or poker chips, etc.). Place the coins (marked 0, 2, 4, and 6) into a bag, randomly select one, and observe its value. Replace this coin, draw a second coin, and observe its value. Finally, calculate the mean $\bar{x}$ for this sample of $n = 2$ observations randomly selected from the population (Exercise 4.95, part **b**). Replace the coins, mix, and using the same procedure, select a sample of $n = 2$ observations from the population.

Record the numbers and calculate $\bar{x}$ for this sample. Repeat this sampling process until you acquire 100 values of $\bar{x}$. Construct a relative frequency distribution for these 100 sample means. Compare this distribution to the exact sampling distribution of $\bar{x}$ found in part **e** of Exercise 4.95. [*Note:* The distribution obtained in this exercise is an approximation to the exact sampling distribution. But, if you were to repeat the sampling procedure, drawing two coins not 100 times but 10,000 times, the relative frequency distribution for the 10,000 sample means would be almost identical to the sampling distribution of $\bar{x}$ found in Exercise 4.95, part **e**.]

4.97 Consider the population described by the probability distribution shown here.

x	1	2	3	4	5
$p(x)$	.2	.3	.2	.2	.1

The random variable x is observed twice. If these observations are independent, verify that the different samples of size 2 and their probabilities are as shown in the table below.

Sample	Probability	Sample	Probability
1, 1	.04	3, 4	.04
1, 2	.06	3, 5	.02
1, 3	.04	4, 1	.04
1, 4	.04	4, 2	.06
1, 5	.02	4, 3	.04
2, 1	.06	4, 4	.04
2, 2	.09	4, 5	.02
2, 3	.06	5, 1	.02
2, 4	.06	5, 2	.03
2, 5	.03	5, 3	.02
3, 1	.04	5, 4	.02
3, 2	.06	5, 5	.01
3, 3	.04		

a. Find the sampling distribution of the sample mean $\bar{x}$.
b. Construct a probability histogram for the sampling distribution of $\bar{x}$.
c. What is the probability that $\bar{x}$ is 4.5 or larger?
d. Would you expect to observe a value of $\bar{x}$ equal to 4.5 or larger? Explain.   No

4.98 Refer to Exercise 4.97 and find $E(x) = \mu$. Then use the sampling distribution of $\bar{x}$ found in Exercise 4.97 to find the expected value of $\bar{x}$. Note that $E(\bar{x}) = \mu$.

4.99 Refer to Exercise 4.97. Assume that a random sample of $n = 2$ measurements is randomly selected from the population.
a. List the different values that the sample median M may assume and find the probability of each. Then give the sampling distribution of the sample median.
b. Construct a probability histogram for the sampling distribution of the sample median and compare it with the probability histogram for the sample mean (Exercise 4.97, part **b**).

4.100 In Example 4.24 we use the computer to generate 1,000 samples, each containing $n = 11$ observations, from a uniform distribution over the interval from 0 to 1. For this exercise, use the computer to generate 500 samples, each containing $n = 15$ observations, from this population.
a. Calculate the sample mean for each sample. To approximate the sampling distribution of $\bar{x}$, construct a relative frequency histogram for the 500 values of $\bar{x}$.
b. Repeat part **a** for the sample median. Compare this approximate sampling distribution with the approximate sampling distribution of $\bar{x}$ found in part **a**.

4.101 Consider a population that contains values of x equal to 00, 01, 02, 03, . . . , 96, 97, 98, 99. Assume that these values of x occur with equal probability. Use the computer to generate 500 samples, each containing $n = 25$ measurements, from this population. Calculate the sample mean $\bar{x}$ and sample variance s^2 for each of the 500 samples.
a. To approximate the sampling distribution of $\bar{x}$, construct a relative frequency histogram for the 500 values of $\bar{x}$.
b. Repeat part **a** for the 500 values of s^2.

4.9   THE CENTRAL LIMIT THEOREM

Estimating the mean useful life of automobiles, the mean number of crimes per month in a large city, and the mean yield per acre of a new soybean hybrid are practical problems with something in common. In each case we are interested in making an inference about the mean μ of some population. As we mentioned in Chapter 2, the sample mean $\bar{x}$ is, in general, a good estimator of μ. We now develop pertinent information about the sampling distribution for this useful statistic.

E X A M P L E   4 . 2 5   Suppose a population has the uniform probability distribution given in Figure 4.34. The mean and standard deviation of this probability distribution are $\mu = .5$ and $\sigma = .29$. Now suppose a sample of 11 measurements is selected from this population. Describe the sampling distribution of the sample mean $\bar{x}$ based on the 1,000 sampling experiments discussed in Example 4.23.

FIGURE 4.34

Sampled Uniform Population

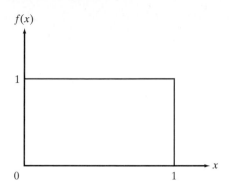

Solution

The sampling distribution is approximately with $\mu_{\bar{x}} = .5$ and $\sigma_{\bar{x}} = .1$

You will recall that in Example 4.23 we generated 1,000 samples of $n = 11$ measurements each. The relative frequency histogram for the 1,000 sample means is shown in Figure 4.35 with a normal probability distribution superimposed. You can see that this normal probability distribution approximates the computer-generated sampling distribution very well.

FIGURE 4.35

Relative Frequency Histogram for $\bar{x}$ in 1,000 Samples of $n = 11$ Measurements with Normal Distribution Superimposed

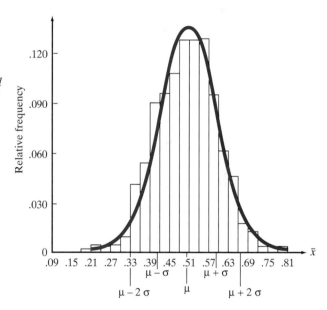

To fully describe a normal probability distribution, it is necessary to know its mean and standard deviation. Inspection of Figure 4.35 indicates that the mean of the distribution of $\bar{x}$, $\mu_{\bar{x}}$, appears to be very close to .5, the mean of the sampled uniform population. Furthermore, for a mound-shaped distribution such as that shown in Figure 4.35, almost all the measurements should fall within 3 standard deviations of the mean. Since the number of values of $\bar{x}$ is very large (1,000), the range of the observed $\bar{x}$'s divided by 6 (rather than 4) should give a reasonable approximation to the standard deviation of the sample means, $\sigma_{\bar{x}}$. The values of $\bar{x}$ range from about .2 to .8, so we calculate

$$\sigma_{\bar{x}} \approx \frac{\text{Range of } \bar{x}\text{'s}}{6} = \frac{.8 - .2}{6} = .1$$

To summarize our findings based on 1,000 samples, each consisting of 11 measurements from a uniform population, the sampling distribution of $\bar{x}$ appears to be approximately normal with a mean of about .5 and a standard deviation of <u>about .1</u>. ∎

TEACHING TIP
Emphasize that properties 1 and 2 are true regardless of the sample size selected. Students get confused once the Central Limit Theorem is introduced.

TEACHING TIP
Choose an example and use different values of n to illustrate the role that the sample size plays in determining the variation associated with the sampling distribution.

The sampling distribution of $\bar{x}$ has the properties given in the next box, assuming only that a random sample of n observations has been selected from *any* population.

Properties of the Sampling Distribution of $\bar{x}$

1. Mean of sampling distribution equals mean of sampled population. That is,
   $$\mu_{\bar{x}} = E(\bar{x}) = \mu.*$$
2. Standard deviation of sampling distribution equals

$$\frac{\text{Standard deviation of sampled population}}{\text{Square root of sample size}}$$

That is, $\sigma_{\bar{x}} = \sigma/\sqrt{n}.^{\dagger}$

The standard deviation $\sigma_{\bar{x}}$ is often referred to as the **standard error of the mean.**[‡]

You can see that our approximation to $\mu_{\bar{x}}$ in Example 4.25 was precise, since property 1 assures us that the mean is the same as that of the sampled population: .5. Property 2 tells us how to calculate the standard deviation of the sampling distribution of $\bar{x}$. Substituting $\sigma = .29$, the standard deviation of the sampled uniform distribution, and the sample size $n = 11$ into the formula for $\sigma_{\bar{x}}$, we find

$$\sigma_{\bar{x}} = \frac{\sigma}{\sqrt{n}} = \frac{.29}{\sqrt{11}} = .09$$

Thus, the approximation we obtained in Example 4.25, $\sigma_{\bar{x}} \approx .1$, is very close to the exact value, $\sigma_{\bar{x}} = .09$.

A third property, applicable when the sample size n is large, is contained in one of the most important theoretical results in statistics: the *Central Limit Theorem*.

Central Limit Theorem

Consider a random sample of n observations selected from a population (*any* population) with mean μ and standard deviation σ. Then, when n is sufficiently large, the sampling distribution of $\bar{x}$ will be approximately a normal distribution with mean $\mu_{\bar{x}} = \mu$ and standard deviation $\sigma_{\bar{x}} = \sigma/\sqrt{n}$. The larger the sample size, the better will be the normal approximation to the sampling distribution of $\bar{x}$.[§]

TEACHING TIP
Emphasize that the power of the Central Limit Theorem is that we no longer need to know the distribution of the population. The Central Limit Theorem applies to all types of population distributions.

Thus, for sufficiently large samples the sampling distribution of $\bar{x}$ is approximately normal. How large must the sample size n be so that the normal distribution provides a good approximation for the sampling distribution of $\bar{x}$? The answer depends on the shape of the distribution of the sampled population, as shown by Figure 4.36. Generally speaking, the greater the skewness of the sampled population

---

*When a sample statistic has a mean equal to the parameter estimated, the statistic is said to be an *unbiased estimate* of the parameter. From property 1, you can see that $\bar{x}$ is an unbiased estimate of μ.

[†]If the sample size n is large relative to the number N of elements in the population (e.g., 5% or more), $\sigma/\sqrt{n}$ must be multiplied by a finite population correction factor, $\sqrt{(N-n)/(N-1)}$. For most sampling situations, this correction factor will be close to 1 and can be ignored.

[‡]The variance of $\bar{x}$ is smallest among all unbiased estimators of μ. Thus, $\bar{x}$ is deemed the MVUE (i.e., minimum variance, unbiased estimator) for μ.

[§]Moreover, because of the Central Limit Theorem, the sum of a random sample of n observations, Σx, will possess a sampling distribution that is approximately normal for large samples. This distribution will have a mean equal to $n\mu$ and a variance equal to $n\sigma^2$. Proof of the Central Limit Theorem is beyond the scope of this book, but it can be found in many mathematical statistics texts.

FIGURE 4.36

*Sampling Distributions of $\bar{x}$
for Different Populations
and Different Sample Sizes*

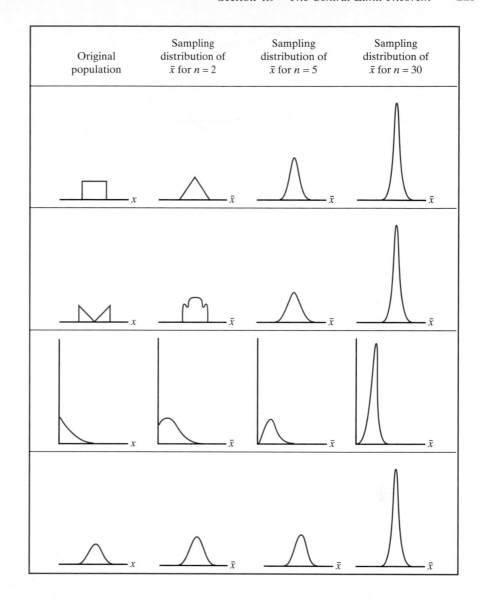

distribution, the larger the sample size must be before the normal distribution is an adequate approximation for the sampling distribution of $\bar{x}$. For most sampled populations, sample sizes of $n \geq 30$ will suffice for the normal approximation to be reasonable. We will use the normal approximation for the sampling distribution of $\bar{x}$ when the sample size is at least 30.

EXAMPLE 4.26

Suppose we have selected a random sample of $n = 25$ observations from a population with mean equal to 80 and standard deviation equal to 5. It is known that the population is not extremely skewed.

a. Sketch the relative frequency distributions for the population and for the sampling distribution of the sample mean, $\bar{x}$.

b. .0228   b. Find the probability that $\bar{x}$ will be larger than 82.

Solution

a. We do not know the exact shape of the population relative frequency distribution, but we do know that it should be centered about $\mu = 80$, its spread

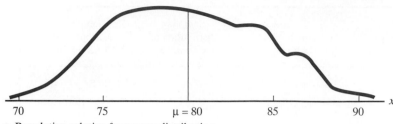

a. Population relative frequency distribution

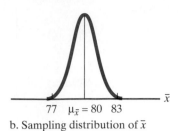

b. Sampling distribution of $\bar{x}$

FIGURE 4.37

A Population Relative
Frequency Distribution and
the Sampling Distribution for $\bar{x}$

should be measured by $\sigma = 5$, and it is not highly skewed. One possibility is shown in Figure 4.37a. From the Central Limit Theorem, we know that the sampling distribution of $\bar{x}$ will be approximately normal since the sampled population distribution is not extremely skewed. We also know that the sampling distribution will have mean and standard deviation

$$\mu_{\bar{x}} = \mu = 80 \quad \text{and} \quad \sigma_{\bar{x}} = \frac{\sigma}{\sqrt{n}} = \frac{5}{\sqrt{25}} = 1$$

The sampling distribution of $\bar{x}$ is shown in Figure 4.37b.

b. The probability that $\bar{x}$ will exceed 82 is equal to the highlighted area in Figure 4.38. To find this area, we need to find the z value corresponding to $\bar{x} = 82$. Recall that the standard normal random variable z is the difference between any normally distributed random variable and its mean, expressed in units of its standard deviation. Since $\bar{x}$ is a normally distributed random variable with mean $\mu_{\bar{x}} = \mu$ and standard deviation $\sigma_{\bar{x}} = \sigma/\sqrt{n}$, it follows that the standard normal z value corresponding to the sample mean, $\bar{x}$, is

$$z = \frac{(\text{Normal random variable}) - (\text{Mean})}{\text{Standard Deviation}} = \frac{\bar{x} - \mu_{\bar{x}}}{\sigma_{\bar{x}}}$$

Therefore, for $\bar{x} = 82$, we have

$$z = \frac{\bar{x} - \mu_{\bar{x}}}{\sigma_{\bar{x}}} = \frac{82 - 80}{1} = 2$$

The area A in Figure 4.38 corresponding to $z = 2$ is given in the table of areas under the normal curve (see Table III of Appendix A) as .4772. Therefore, the tail area corresponding to the probability that $\bar{x}$ exceeds 82 is

$$P(\bar{x} > 82) = P(z > 2) = .5 - .4772 = .0228$$

FIGURE 4.38

The Sampling Distribution
of $\bar{x}$

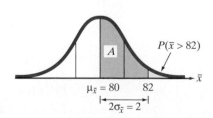

EXAMPLE 4.27

A manufacturer of automobile batteries claims that the distribution of the lengths of life of its best battery has a mean of 54 months and a standard deviation of 6 months. Suppose a consumer group decides to check the claim by purchasing a sample of 50 of these batteries and subjecting them to tests that determine battery life.

a. Assuming that the manufacturer's claim is true, describe the sampling distribution of the mean lifetime of a sample of 50 batteries.

b. .0094

b. Assuming that the manufacturer's claim is true, what is the probability the consumer group's sample has a mean life of 52 or fewer months?

Solution

a. Even though we have no information about the shape of the probability distribution of the lives of the batteries, we can use the Central Limit Theorem to deduce that the sampling distribution for a sample mean lifetime of 50 batteries is approximately normally distributed. Furthermore, the mean of this sampling distribution is the same as the mean of the sampled population, which is $\mu = 54$ months according to the manufacturer's claim. Finally, the standard deviation of the sampling distribution is given by

TEACHING TIP
Students often have trouble working with the denominator of the z-score. Emphasize that we must divide by the standard deviation of the sampling distribution, not the standard deviation of the population.

$$\sigma_{\bar{x}} = \frac{\sigma}{\sqrt{n}} = \frac{6}{\sqrt{50}} = .85 \text{ month}$$

Note that we used the claimed standard deviation of the sampled population, $\sigma = 6$ months. Thus, if we assume that the claim is true, the sampling distribution for the mean life of the 50 batteries sampled is as shown in Figure 4.39.

FIGURE 4.39

Sampling Distribution of $\bar{x}$ in Example 4.27 for $n = 50$

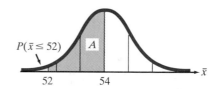

b. If the manufacturer's claim is true, the probability that the consumer group observes a mean battery life of 52 or fewer months for their sample of 50 batteries, $P(\bar{x} \leq 52)$, is equivalent to the highlighted area in Figure 4.39. Since the sampling distribution is approximately normal, we can find this area by computing the standard normal z value:

$$z = \frac{\bar{x} - \mu_{\bar{x}}}{\sigma_{\bar{x}}} = \frac{\bar{x} - \mu}{\sigma_{\bar{x}}} = \frac{52 - 54}{.85} = -2.35$$

TEACHING TIP
Draw a picture of the population distribution overlaid with the sampling distribution of the sample mean. Show that they are both centered at the same location, but the sampling distribution has much less spread than the population.

where $\mu_{\bar{x}}$, the mean of the sampling distribution of $\bar{x}$, is equal to μ, the mean of the lifetimes of the sampled population, and $\sigma_{\bar{x}}$ is the standard deviation of the sampling distribution of $\bar{x}$. Note that z is the familiar standardized distance (z-score) of Section 2.6 and, since $\bar{x}$ is approximately normally distributed, it will possess the standard normal distribution of Section 4.5.

The area A shown in Figure 4.39 between $\bar{x} = 52$ and $\bar{x} = 54$ (corresponding to $z = -2.35$) is found in Table III of Appendix A to be .4906. Therefore, the area to the left of $\bar{x} = 52$ is

$$P(\bar{x} \leq 52) = .5 - A = .5 - .4906 = .0094$$

Thus, the probability the consumer group will observe a sample mean of 52 or less is only .0094 if the manufacturer's claim is true. If the 50 tested batteries do exhibit a mean of 52 or fewer months, the consumer group will have strong evidence that the manufacturer's claim is untrue, because such an event is very unlikely to occur if the claim is true. (This is still another application of the *rare-event approach* to statistical inference.)

Suggested Exercise 4.109

We conclude this section with two final comments on the sampling distribution of $\bar{x}$. First, from the formula $\sigma_{\bar{x}} = \sigma/\sqrt{n}$, we see that the standard deviation of the sampling distribution of $\bar{x}$ gets smaller as the sample size n gets larger. For example, we computed $\sigma_{\bar{x}} = .85$ when $n = 50$ in Example 4.27. However, for $n = 100$ we obtain $\sigma_{\bar{x}} = \sigma/\sqrt{n} = 6/\sqrt{100} = .60$. This relationship will hold true for most of the sample statistics encountered in this text. That is: *The standard deviation of the sampling distribution decreases as the sample size increases.* Consequently, the larger the sample size, the more accurate the sample statistic (e.g., $\bar{x}$) is in estimating a population parameter (e.g., μ). We will use this result in Chapter 5 to help us determine the sample size needed to obtain a specified accuracy of estimation.

Our second comment concerns the Central Limit Theorem. In addition to providing a very useful approximation for the sampling distribution of a sample mean, the Central Limit Theorem offers an explanation for the fact that many relative frequency distributions of data possess mound-shaped distributions. Many of the measurements we take in various areas of research are really means or sums of a large number of small phenomena. For example, a year's growth of a pine seedling is the total of the many individual components that affect the plant's growth. Similarly, we can view the length of time a construction company takes to build a house as the total of the times taken to complete a multitude of distinct jobs, and we can regard the monthly demand for blood at a hospital as the total of the many individual patients' needs. Whether or not the observations entering into these sums satisfy the assumptions basic to the Central Limit Theorem is open to question. However, it is a fact that many distributions of data in nature are mound-shaped and possess the appearance of normal distributions.

 EXERCISES 4.102–4.117

Learning the Mechanics

4.102 Suppose a random sample of $n = 25$ measurements is selected from a population with mean μ and standard deviation σ. For each of the following values of μ and σ, give the values of $\mu_{\bar{x}}$ and $\sigma_{\bar{x}}$.
   a. $\mu = 10, \sigma = 3$ $10; .6$
   b. $\mu = 100, \sigma = 25$ $100; 5$
   c. $\mu = 20, \sigma = 40$ $20; 8$
   d. $\mu = 10, \sigma = 100$ $10; 20$

4.103 Consider the probability distribution shown here.

x	1	2	3	8
$p(x)$	.1	.4	.4	.1

   a. Find μ, σ^2, and σ. $2.9; 3.29; 1.814$
   b. Find the sampling distribution of $\bar{x}$ for random samples of $n = 2$ measurements from this distribution by listing all possible values of $\bar{x}$, and find the probability associated with each.
   c. Use the results of part **b** to calculate $\mu_{\bar{x}}$ and $\sigma_{\bar{x}}$. Confirm that $\mu_{\bar{x}} = \mu$ and that $\sigma_{\bar{x}} = \sigma/\sqrt{n} = \sigma/\sqrt{2}$.

4.104 Will the sampling distribution of $\bar{x}$ always be approximately normally distributed? Explain.   No

4.105 A random sample of $n = 64$ observations is drawn from a population with a mean equal to 20 and standard deviation equal to 16.
   a. Give the mean and standard deviation of the (repeated) sampling distribution of $\bar{x}$. 20.2
   b. Describe the shape of the sampling distribution of $\bar{x}$. Does your answer depend on the sample size?
   c. Calculate the standard normal z-score corresponding to a value of $\bar{x} = 15.5$. -2.25
   d. Calculate the standard normal z-score corresponding to $\bar{x} = 23$. 1.5

4.106 Refer to Exercise 4.105. Find the probability that
   a. $\bar{x}$ is less than 16 $.0228$
   b. $\bar{x}$ is greater than 23 $.0668$
   c. $\bar{x}$ is greater than 25 $.0062$
   d. $\bar{x}$ falls between 16 and 22 $.8185$
   e. $\bar{x}$ is less than 14 $.0013$

4.107 A random sample of $n = 100$ observations is selected from a population with $\mu = 30$ and $\sigma = 16$. Approximate the following probabilities:
   a. $P(\bar{x} \geq 28)$ $.8944$
   b. $P(22.1 \leq \bar{x} \leq 26.8)$ $.0228$
   c. $P(\bar{x} \leq 28.2)$ $.1292$
   d. $P(\bar{x} \geq 27.0)$ $.9699$

4.108 Consider a population that contains values of x equal to $0, 1, 2, \ldots, 97, 98, 99$. Assume that the values of x are equally likely. For each of the following values of n, use the computer to generate 500 random samples and calculate $\bar{x}$ for each sample. For each sample size, construct a relative frequency histogram of the 500 values of $\bar{x}$. What changes occur in the histograms as the value of n increases? What similarities exist? Use $n = 2, n = 5, n = 10, n = 30$, and $n = 50$.

Applying the Concepts—Basic

4.109 The ocean quahog is a type of clam found in the coastal waters of the mid-Atlantic states. A federal survey of offshore ocean quahog harvesting in New Jersey revealed an average catch per unit effort (CPUE) of 89.34 clams. The CPUE standard deviation was 7.74 (*Journal of Shellfish Research*, June 1995). Let $\bar{x}$ represent the mean CPUE for a sample of 35 attempts to catch ocean quahogs off the New Jersey shore.
   a. Compute $\mu_{\bar{x}}$ and $\sigma_{\bar{x}}$. Interpret their values.
   b. Sketch the sampling distribution of $\bar{x}$.
   c. Find $P(\bar{x} > 88)$.   .8461
   d. Find $P(\bar{x} < 87)$.   .0367

4.110 Refer to *The American Statistician* (May 2001) study of female students who suffer from bulimia, Exercise 2.27 (p. 39). Recall that each student completed a questionnaire from which a "fear of negative evaluation" (FNE) score was produced. (The higher the score, the greater the fear of negative evaluation.) Suppose the FNE scores of bulimic students have a distribution with mean $\mu = 18$ and standard deviation $\sigma = 5$. Now consider a random sample of 45 female students with bulimia.
   a. What is the probability that the sample mean FNE score is greater than 17.5?   .7486
   b. What is the probability that the sample mean FNE score is between 18 and 18.5?   .2486
   c. What is the probability that the sample mean FNE score is less than 18.5?   .7486

4.111 Refer to the *Chance* (Winter 2001) examination of Standardized Admission Test (SAT) scores of students who pay a private tutor to help them improve their results, Exercise 2.81 (p. 70). On the SAT-Mathematics test, these students had a mean score change of $+19$ points, with a standard deviation of 65 points. In a random sample of 100 students who pay a private tutor to help them improve their results, what is the likelihood that the sample mean score change is less than 10 points?   .0838

Applying the Concepts—Intermediate

4.112 The *College Student Journal* (Dec. 1992) investigated differences in traditional and nontraditional students, where nontraditional students are generally defined as those 25 years old or older. Based on the study results, we can assume that the population mean and standard deviation for the GPA of all nontraditional students is $\mu = 3.5$ and $\sigma = .5$. Suppose that a random sample of $n = 100$ nontraditional students is selected from the population of all nontraditional students, and the GPA of each student is determined. Then $\bar{x}$, the sample mean, will be approximately normally distributed (because of the Central Limit Theorem).
   a. Calculate $\mu_{\bar{x}}$ and $\sigma_{\bar{x}}$.   3.5; .05
   b. What is the approximate probability that the nontraditional student sample has a mean GPA between 3.40 and 3.60?   .9544
   c. What is the approximate probability that the sample of 100 nontraditional students has a mean GPA that exceeds 3.62?   .0082
   d. How would the sampling distribution of $\bar{x}$ change if the sample size n were doubled from 100 to 200? How do your answers to parts **b** and **c** change when the sample size is doubled?   .9954; 0

4.113 Interpersonal violence (e.g., rape) generally leads to psychological stress for the victim. *Clinical Psychology Review* (Vol. 15, 1995) reported on the results of all recently published studies of the relationship between interpersonal violence and psychological stress. The distribution of the time elapsed between the violent incident and the initial sign of stress has a mean of 5.1 years and a standard deviation of 6.1 years. Consider a random sample of $n = 150$ victims of interpersonal violence. Let $\bar{x}$ represent the mean time elapsed between the violent act and the first sign of stress for the sampled victims.
   a. Give the mean and standard deviation of the sampling distribution of $\bar{x}$.   5.1; .4981
   b. Will the sampling distribution of $\bar{x}$ be approximately normal? Explain.
   c. Find $P(\bar{x} > 5.5)$.   .2119
   d. Find $P(4 < \bar{x} < 5)$.   .4071

4.114 The Computer-Assisted Hypnosis Scale (CAHS) is designed to measure a person's susceptibility to hypnosis. In computer-assisted hypnosis, the computer serves as a facilitator of hypnosis by using digitized speech processing coupled with interactive involvement with the hypnotic subject. CAHS scores range from 0 (no susceptibility) to 12 (extremely high susceptibility). A study in *Psychological Assessment* (Mar. 1995) reported a mean CAHS score of 4.59 and a standard deviation of 2.95 for University of Tennessee undergraduates. Assume that $\mu = 4.29$ and $\sigma = 2.95$ for this population. Suppose a psychologist uses CAHS to test a random sample of 50 subjects.
   a. Would you expect to observe a sample mean CAHS score of $\bar{x} = 6$ or higher? Explain.   No, $p \approx 0$
   b. Suppose the psychologist actually observes $\bar{x} = 6.2$. Based on your answer to part **a**, make an inference about the population from which the sample was selected.

4.115 Last year a company began a program to compensate its employees for unused sick days, paying each employee a bonus of one-half the usual wage earned for each unused sick day. The question that naturally arises is: "Did this policy motivate employees to use fewer sick days?" *Before* last year, the number of sick days used by employees had a distribution with a mean of 7 days and a standard deviation of 2 days.

a. Assuming that these parameters did not change last year, find the approximate probability that the sample mean number of sick days used by 100 employees chosen at random was less than or equal to 6.4 last year.

b. How would you interpret the result if the sample mean for the 100 employees was 6.4?

Applying the Concepts—Advanced

4.116 The Test of Knowledge About Epilepsy (KAE), which is designed to measure attitudes toward persons with epilepsy, uses 20 multiple-choice items where all of the choices are incorrect. For each person, two scores (ranging from 0 to 20) are obtained, an attitude score (KAE-A) and a general knowledge score (KAE-GK). Based on a large-scale study of college students, the distribution of KAE-A scores has a mean of $\mu = 11.92$ and a standard deviation of $\sigma = 2.95$ while the distribution of KAE-GK scores has a mean of $\mu = 6.35$ and a standard deviation of $\sigma = 2.12$ (*Rehabilitative Psychology*, Spring 1995). Consider a random sample of 100 college students, and suppose you observe a sample mean KAE score of 6.5. Is this result more likely to be the mean of the attitude scores (KAE-A) or the general knowledge scores (KAE-GK)? Explain. KAE-GK

4.117 A soft drink bottler purchases glass bottles from a vendor. The bottles are required to have an internal pressure of at least 150 pounds per square inch (psi). A prospective bottle vendor claims that its production process yields bottles with a mean internal pressure of 157 psi and a standard deviation of 3 psi. The bottler strikes an agreement with the vendor that permits the bottler to sample from the vendor's production process to verify the vendor's claim. The bottler randomly selects 40 bottles from the last 10,000 produced, measures the internal pressure of each, and finds the mean pressure for the sample to be 1.3 psi below the process mean cited by the vendor.

a. Assuming the vendor's claim to be true, what is the probability of obtaining a sample mean this far or farther below the process mean? What does your answer suggest about the validity of the vendor's claim?

b. If the process standard deviation were 3 psi as claimed by the vendor, but the mean were 156 psi, would the observed sample result be more or less likely than in part **a**? What if the mean were 158 psi?

c. If the process mean were 157 psi as claimed, but the process standard deviation were 2 psi, would the sample result be more or less likely than in part **a**? What if instead the standard deviation were 6 psi?

QUICK REVIEW

Note: Starred (*) items are from the optional section in this chapter.

Key Terms

Bell curve   190
Bell-shaped distribution   190
Binomial experiment   177
Binomial random variable   177
Central Limit Theorem   224
Continuous probability
   distribution   190
Continuous random variable   166
Correction for continuity*   211
Cumulative binomial probabilities   183

Discrete probability
   distribution   168
Discrete random variable   166
Expected value   170
Frequency function   190
Normal distribution   190
Normal probability plot   204
Normal random variable   190
Parameter   216
Probability density function   190

Probability distribution   190
Random variable   164
Sample statistic   216
Sampling distribution   217
Standard error of the mean   224
Standard normal distribution   193
Standard normal random
   variable   193
Uniform distribution   220

Key Formulas

Discrete Random Variable	Probability Distribution	Mean (μ)	Variance (σ^2)
General, x	$p(x)$	$\sum_{\text{all } x} x p(x)$	$\sum_{\text{all } x} (x - \mu)^2 p(x)$ 170, 171
Binomial, x	$\binom{n}{x} p^x q^{n-x}$	np	npq 180, 182

Continuous Random Variable	Density Function	Mean	Standard Deviation
Normal, x	$f(x) = \dfrac{1}{\sigma\sqrt{2\pi}}e^{-(1/2)[(x-\mu)/\sigma]^2}$	μ	σ 191
Standard Normal, $z = \left(\dfrac{x-\mu}{\sigma}\right)$	$f(x) = \dfrac{1}{\sigma\sqrt{2\pi}}e^{-(1/2)z^2}$	$\mu = 0$	$\sigma = 1$ 193

	Mean	Standard Deviation	z-score
Sampling distribution of $\bar{x}$	$\mu_{\bar{x}} = \mu$	$\sigma_{\bar{x}} = \dfrac{\sigma}{\sqrt{n}}$	$z = \dfrac{\bar{x} - \mu_{\bar{x}}}{\sigma_{\bar{x}}} = \dfrac{\bar{x}-\mu}{\sigma/\sqrt{n}}$ 224

*Normal approximation to Binomial: $P(x \le a) = P\left[z \le \dfrac{(a + .5) - \mu}{\sigma}\right]$ 213

Language Lab

Symbol	Pronunciation	Description
$p(x)$		Probability distribution of the random variable x
S		The outcome of a binomial trial denoted a "success"
F		The outcome of a binomial trial denoted a "failure"
p		The probability of success (S) in a binomial trial
q		The probability of failure (F) in a binomial trial, where $q = 1 - p$
$f(x)$	f of x	Probability density function for a continuous random variable x
θ	theta	Population parameter (general)
$\mu_{\bar{x}}$	mu of x-bar	True mean of sampling distribution of $\bar{x}$
$\sigma_{\bar{x}}$	sigma of x-bar	True standard deviation of sampling distribution of $\bar{x}$
z_0	z-naught	Fixed value of standard normal random variable

STATISTICS IN ACTION
The Insomnia Pill

A research report published in the *Proceedings of the National Academy of Sciences* (Mar. 1994) brought encouraging news to insomniacs and long-distance travelers. Neuroscientists at the Massachusetts Institute of Technology (MIT) have been experimenting with melatonin—a hormone secreted by the pineal gland in the brain—as a sleep-inducing hormone. Since the hormone is naturally produced, it is considered nonaddictive. The researchers believe melatonin may also be effective in treating jet lag—the body's response to rapid travel across many time zones so that a daylight–darkness change disrupts sleep patterns.

In the MIT study, young male volunteers were given various doses of melatonin or a placebo (no dosage of melatonin). Then they were placed in a dark room at midday and told to close their eyes for 30 minutes. The variable of interest was the time (in minutes) elapsed before each volunteer fell asleep.

According to the lead investigator, Professor Richard Wurtman, "Our volunteers fall asleep in five or six minutes on melatonin, while those on placebo take about 15 minutes." Wurtman warns, however, that uncontrolled doses of melatonin could cause serious mood-altering side effects. (Melatonin is sold in some health food stores. However, sales are unregulated, and the purity and strength of the hormone are often uncertain.)

Focus

With the placebo (i.e., no hormone) the researchers found that the mean time to fall asleep was 15 minutes. Assume that with the placebo treatment $\mu = 15$ and $\sigma = 5$. Now, consider a random sample of 40 young males, each of whom is given a dose of the sleep-inducing hormone, melatonin. The times (in minutes) to fall asleep for these 40 males are listed in Table 4.10. Use the data to make an inference about the true value of μ for those taking the melatonin. (The data are stored in the file named INSOMNIA.) Does melatonin appear to be an effective drug against insomnia?

⊘ INSOMNIA

TABLE 4.10 Times (in Minutes) for 40 Male Volunteers to Fall Asleep*

6.4	6.0	3.2	4.4	6.2	1.7	5.1
5.9	1.6	4.4	16.2	4.8	8.3	7.5
4.8	3.3	4.0	6.2	6.3	5.0	6.3
5.1	6.4	15.6	3.4	3.1	6.1	6.0
5.0	1.8	6.1	4.5	4.5	1.5	4.7
7.6	8.2	4.9	6.1	3.9		

*These are simulated sleep times based on summary information provided in the MIT study.

SUPPLEMENTARY EXERCISES 4.118–4.144

Note: Starred (*) exercises refer to the optional section in this chapter.

Learning the Mechanics

4.118 For each of the following examples, decide whether x is a binomial random variable and explain your decision:

   a. A manufacturer of computer chips randomly selects 100 chips from each hour's production in order to estimate the proportion defectives. Let x represent the number of defectives in the 100 sampled chips.

   b. Of five applicants for a job, two will be selected. Although all applicants appear to be equally qualified, only three have the ability to fulfill the expectations of the company. Suppose that the two selections are made at random from the five applicants, and let x be the number of qualified applicants selected.

   c. A software developer establishes a support hotline for customers to call in with questions regarding use of the software. Let x represent the number of calls received on the support hotline during a specified workday.   Not binomial

   d. Florida is one of a minority of states with no state income tax. A poll of 1,000 registered voters is conducted to determine how many would favor a state income tax in light of the state's current fiscal condition. Let x be the number in the sample who would favor the tax.   Binomial

4.119 Suppose x is a binomial random variable. Find $p(x)$ for each of the following combinations of $x, n,$ and p.

   a. $x = 1, n = 3, p = .1$ .243
   b. $x = 4, n = 20, p = .3$ .131
   c. $x = 0, n = 2, p = .4$ .36
   d. $x = 4, n = 5, p = .5$ .157
   e. $n = 15, x = 12, p = .9$ .128
   f. $n = 10, x = 8, p = .6$ .121

4.120 Consider the discrete probability distribution shown here.

x	10	12	18	20
$p(x)$	.2	.3	.1	.4

   a. Calculate $\mu, \sigma^2,$ and σ.   15.4; 18.44; 4.294
   b. What is $P(x < 15)$?   .5
   c. Calculate $\mu \pm 2\sigma$.   (6.812, 23.988)
   d. What is the probability that x is in the interval $\mu \pm 2\sigma$?   1.00

4.121 Suppose x is a binomial random variable with $n = 20$ and $p = .7$.

   a. Find $P(x = 14)$.   .192
   b. Find $P(x \leq 12)$.   .228
   c. Find $P(x > 12)$.   .772
   d. Find $P(9 \leq x \leq 18)$.   .987
   e. Find $P(8 < x < 18)$.   .960
   f. Find $\mu, \sigma^2,$ and σ.   14; 4.2; 2.049
   g. What is the probability that x is in the interval $\mu \pm 2\sigma$?   .975

4.122 Find a z-score, call it z_0, such that

   a. $P(z \leq z_0) = .8708$ 1.13
   b. $P(z \geq z_0) = .0526$ 1.62
   c. $P(z \leq z_0) = .5$ 0
   d. $P(-z_0 \leq z \leq z_0) = .8164$ 1.33
   e. $P(z \geq z_0) = .8023$ −.85
   f. $P(z \geq z_0) = .0041$ 2.64

4.123 Find the following probabilities for the standard normal random variable z:

   a. $P(z \leq 2.1)$ .9821
   b. $P(z \geq 2.1)$ .0179
   c. $P(z \geq -1.65)$ .9505
   d. $P(-2.13 \leq z \leq -.41)$ .3243
   e. $P(-1.45 \leq z \leq 2.15)$ .9107
   f. $P(z \leq -1.43)$ .0764

4.124 The random variable x has a normal distribution with $\mu = 40$ and $\sigma^2 = 36$. Find a value of x, call it x_0, such that

   a. $P(x \geq x_0) = .5$ 40
   b. $P(x \leq x_0) = .9911$ 54.22
   c. $P(x \leq x_0) = .0028$ 23.38
   d. $P(x \geq x_0) = .0228$ 52
   e. $P(x \leq x_0) = .1003$ 32.32
   f. $P(x \geq x_0) = .7995$ 34.96

4.125 The random variable x has a normal distribution with $\mu = 70$ and $\sigma = 10$. Find the following probabilities:

   a. $P(x \leq 75)$ .6915
   b. $P(x \geq 90)$ .0228
   c. $P(60 \leq x \leq 75)$ .5328
   d. $P(x > 75)$ .3085
   e. $P(x = 75)$ 0
   f. $P(x \leq 95)$ .9938

*4.126 Assume that x is a binomial random variable with $n = 100$ and $p = .5$. Use the normal probability distribution to approximate the following probabilities:

   a. $P(x \leq 48)$ .3821
   b. $P(50 \leq x \leq 65)$ .5398
   c. $P(x \geq 70)$ 0
   d. $P(55 \leq x \leq 58)$ .1395
   e. $P(x = 62)$ .0045
   f. $P(x \leq 49$ or $x \geq 72)$ .4602

4.127 A random sample of $n = 68$ observations is selected from a population with $\mu = 19.6$ and $\sigma = 3.2$. Approximate each of the following probabilities.

   a. $P(\bar{x} \leq 19.6)$ .5
   b. $P(\bar{x} \leq 19)$ .0606
   c. $P(\bar{x} \geq 20.1)$ .0985
   d. $P(19.2 \leq \bar{x} \leq 20.6)$ .8436

4.128 A random sample of 40 observations is to be drawn from a large population of measurements. It is known that 30% of the measurements in the population are 1's, 20% are 2's, 20% are 3's, and 30% are 4's.

   a. Give the mean and standard deviation of the (repeated) sampling distribution of $\bar{x}$, the sample mean of the 40 observations.   2.5; .1904

b. Describe the shape of the sampling distribution of $\bar{x}$. Does your answer depend on the sample size?

4.129 A random sample of size n is to be drawn from a large population with mean 100 and standard deviation 10, and the sample mean $\bar{x}$ is to be calculated. To see the effect of different sample sizes on the standard deviation of the sampling distribution of $\bar{x}$, plot $\sigma/\sqrt{n}$ against n for $n = 1, 5, 10, 20, 30, 40,$ and 50.

Applying the Concepts—Basic

4.130 According to *Ladies Home Journal* (June 1988), 80% of married women would marry their current husbands again, if given the chance. Consider the number x in a random sample of ten married women who would marry their husbands again. Identify the discrete probability distribution that best models the distribution of x. Explain.   Binomial

4.131 In Exercise 2.11 (p. 27) you read about a study of book reviews in American history, geography, and area studies published in *Choice* magazine (*Library Acquisitions: Practice and Theory*, Vol. 19, 1995). The overall rating stated in each review was ascertained and recorded as follows: 1 = would not recommend, 2 = cautious or very little recommendation, 3 = little or no preference, 4 = favorable/recommended, 5 = outstanding/significant contribution. Based on a sample of 375 book reviews, the probability distribution of rating, x, is

Book Rating x	$p(x)$
1	.051
2	.099
3	.093
4	.635
5	.122

a. Is this a valid probability distribution?   Yes
b. What is the probability that a book reviewed in *Choice* has a rating of 1?   .051
c. What is the probability that a book reviewed in *Choice* has a rating of at least 4?   .757
d. What is the probability that a book reviewed in *Choice* has a rating of 2 or 3?   .192
e. Find $E(x)$.   3.678
f. Interpret the result, part **e**.

4.132 In baseball, a "no-hitter" is a regulation 9-inning game in which the pitcher yields no hits to the opposing batters. *Chance* (Summer 1994) reported on a study of no-hitters in Major League Baseball (MLB). The initial analysis focused on the total number of hits yielded per game per team for all 9-inning MLB games played between 1989 and 1993. The distribution of hits/9-innings is approximately normal with mean 8.72 and standard deviation 1.10.
a. What percentage of 9-inning MLB games result in fewer than 6 hits?   .0068
b. Demonstrate, statistically, why a no-hitter is considered an extremely rare occurrence in MLB.

4.133 All Florida high schools require their students to demonstrate competence in mathematics by scoring 70% or above on the FCAT mathematics achievement test. The FCAT math scores of those students taking the test for the first time are normally distributed with a mean of 77% and a standard deviation of 7.3%. What percentage of students who take the test for the first time will pass the test?

*4.134 The *Statistical Abstract of the United States* reports that 25% of the country's households are composed of one person. If 1,000 randomly selected homes are to participate in a Nielsen survey to determine television ratings, find the approximate probability that no more than 240 of these homes are one-person households.

4.135 This past year, an elementary school began using a new method to teach arithmetic to first graders. A standardized test, administered at the end of the year, was used to measure the effectiveness of the new method. The distribution of past scores on the standardized test produced a mean of 75 and a standard deviation of 10.
a. If the new method is no different from the old method, what is the approximate probability that the mean score $\bar{x}$ of a random sample of 36 students will be greater than 79?   .0082
b. What assumptions must be satisfied to make your answer valid?

Applying the Concepts—Intermediate

4.136 Refer to the *Professional Geographer* (Feb. 2000) study of urban and rural counties, Exercise 3.17 (p. 118). Forty-five percent of county commissioners in Nevada feel that "population concentration" is the single most important factor in identifying urban counties. Suppose 10 county commissioners are selected at random to serve on a Nevada review board that will consider redefining urban and rural areas. What is the probability that more than half of the commissioners will specify "population concentration" as the single most important factor in identifying urban counties?   .2616

4.137 The Food and Drug Administration (FDA) produces a quarterly report called *Total Diet Study*. The FDA's report covers more than 200 food items, each of which is analyzed for dangerous chemical compounds. A recent *Total Diet Study* reported that no pesticides at all were found on 65% of the domestically produced food samples (*Consumer's Research*, June 1995). Consider a random sample of 800 food items analyzed for the presence of pesticides.
a. Compute μ and σ for the random variable x, the number of food items found without any trace of pesticide.   520; 13.491
b. Based on a sample of 800 food items, is it likely you would observe less than half without any traces of pesticide? Explain.   No

4.138 Refer to the *Teaching Psychology* (May 1998) study of how external clues influence performance, Exercise 2.92 (p. 74). Recall that two different forms of a midterm psychology examination were given—one printed on blue paper and the other printed on red paper. Grading only the difficult questions, scores on the blue exam had a distribution with a mean of 53%

and a standard deviation of 15%, while scores on the red exam had a distribution with a mean of 39% and a standard deviation of 12%. Assuming that both distributions are approximately normal, on which exam form is a student more likely to score below 20% on the difficult questions, the blue or the red exam? (Compare your answer to 2.92c.)   Red exam

4.139 A chemical analysis of metamorphic rock in western Turkey was reported in *Geological Magazine* (May 1995). The trace amount (in parts per million) of the element rubidium was measured for 20 rock specimens. The data are listed here. Assess whether the sample data come from a normal population.

⬤ RUBIDIUM

164	286	355	308	277	330	323	370	241	402
301	200	202	341	327	285	277	247	213	424

Source: Bozkurt, E. *et al.* "Geochemistry and the tectonic significance of augen gneisses from the southern Menderes Massif (West Turkey)." *Geological Magazine,* Vol. 132, No. 3, May 1995, p. 291 (Table 1).

4.140 It is well known that children with developmental delays (i.e., mild mental retardation) are slower, cognitively, than normally developing children. Are their social skills also lacking? A study compared the social interactions of the two groups of children in a controlled playground environment (*American Journal on Mental Retardation*, Jan. 1992). One variable of interest was the number of intervals of "no play" by each child. Children with developmental delays had a mean of 2.73 intervals of "no play" and a standard deviation of 2.58 intervals. Based on this information, is it possible for the variable of interest to be normally distributed? Explain.   No

4.141 Over the past 20 years, the *Journal of Abnormal Psychology* has published numerous studies on the relationship between juvenile delinquency and poor verbal abilities. Assume that scores on a verbal IQ test have a population mean $\mu = 107$ and a population standard deviation $\sigma = 15$.
  a. What shape would you expect the sampling distribution of $\bar{x}$ for $n = 84$ juveniles to have? Does your answer depend on the shape of the distribution of verbal IQ scores for all juveniles?
  b. Assuming that the population mean and standard deviation for juveniles with no record of delinquency are the same as those for all juveniles, approximate the probability that the sample mean verbal IQ for $n = 84$ juveniles will be 110 or more. State any assumptions you make.   .0336
  c. In one cited study, the researchers found that a sample of $n = 84$ juveniles with no record of delinquency had a mean verbal IQ of $\bar{x} = 110$. Considering your answer to part **b**, do you think that the population mean and standard deviation for nondelinquent juveniles are the same as those for all juveniles? Explain.

4.142 Researchers at the Terry College of Business at the University of Georgia sampled 344 business students and asked them this question: "Over the course of your lifetime, what is the maximum number of years you expect to work for any one employer?" The sample resulted in $\bar{x} = 19.1$ years. Assume the sample of students was randomly selected from the 6,000 undergraduate students at the Terry College and that $\sigma = 6$ years.
  a. Describe the sampling distribution of $\bar{x}$.
  b. If the mean for the 6,000 undergraduate students is $\mu = 18.5$ years, find $P(\bar{x} > 19.1)$.   .0322
  c. If the mean for the 6,000 undergraduate students is $\mu = 19.5$ years, find $P(\bar{x} > 19.1)$.   .8925
  d. If $P(\bar{x} > 19.1) = .5$, what is μ?   19.1
  e. If $P(\bar{x} > 19.1) = .2$, is μ greater than or less than 19.1 years? Explain. $\mu < 19.1$

Applying the Concepts—Advanced

4.143 A team of soil scientists investigated the water retention properties of soil cores sampled from an uncropped field consisting of silt loam (*Soil Science*, Jan. 1995). At a pressure of .1 megapascal (MPa), the water content of the soil (measured in cubic meters of water per cubic meter of soil) was determined to be approximately normally distributed with $\mu = .27$ and $\sigma = .04$. In addition to water content readings at a pressure of .1 MPa, measurements were obtained at pressures 0, .005, .01, .03, and 1.5 MPa. Consider a soil core with a water content reading of .14. Is it likely that this reading was obtained at a pressure of .1 MPa? Explain.

*4.144 On January 28, 1986, the space shuttle *Challenger* exploded, killing all seven astronauts aboard. An investigation concluded that the explosion was caused by the failure of the O-ring seal in the joint between the two lower segments of the right solid rocket booster. In a report made one year prior to the catastrophe, the National Aeronautics and Space Administration (NASA) claimed that the probability of such a failure was about $^1/_{60,000}$, or about once in every 60,000 flights. But a risk-assessment study conducted for the Air Force at about the same time assessed the probability to be $^1/_{35}$, or about once in every 35 missions.
  a. The shuttle had flown 24 successful missions prior to the disaster. Assuming NASA's failure-rate estimate was accurate, compute the probability that no disasters would have occurred during 25 shuttle missions.   .9996
  b. Repeat part **a**, but use the Air Force's failure-rate estimate.   .4845
  c. What conditions must exist for the probabilities, parts **a** and **b**, to be valid?
  d. Given the events of January 28, 1986, which risk assessment—NASA's or the Air Force's—appears to be more appropriate? [*Hint:* Consider the complement of the events, parts **a** and **b**.]   Air Force

STUDENT PROJECTS

To understand the Central Limit Theorem and sampling distribution, consider the following experiment: Toss four identical coins, and record the number of heads observed. Then repeat this experiment four more times, so that you end up with a total of five observations for the random variable x, the number of heads when four coins are tossed.

Now derive and graph the probability distribution for x, assuming the coins are balanced. Note that the mean of this distribution is $\mu = 2$ and the standard deviation is $\sigma = 1$. This probability distribution represents the one from which you are drawing a random sample of five measurements.

Next, calculate the mean $\bar{x}$ of the five measurements—that is, calculate the mean number of heads you observed in five repetitions of the experiment. Although you have repeated the basic experiment five times, you have only one observed value of $\bar{x}$. To derive the probability distribution or sampling distribution of $\bar{x}$ empirically, you have to repeat the entire process (of tossing four coins five times) many times. Do it 100 times.

The approximate sampling distribution of $\bar{x}$ can be derived theoretically by making use of the Central Limit Theorem. We expect at least an approximate normal probability distribution with a mean $\mu = 2$ and a standard deviation

$$\sigma_{\bar{x}} = \frac{\sigma}{\sqrt{n}} = \frac{1}{\sqrt{5}} = .45$$

Count the number of your 100 $\bar{x}$'s that fall in each of the intervals in the accompanying figure. Use the normal probability distribution with $\mu = 2$ and $\sigma_{\bar{x}} = .45$ to calculate the expected number of the 100 $\bar{x}$'s in each of the intervals. How closely does the theory describe your experimental results?

Interval 1	Interval 2	Interval 3	Interval 4	Interval 5	Interval 6

1.10	1.55	2	2.45	2.90
$\mu - 2\sigma_{\bar{x}}$	$\mu - \sigma_{\bar{x}}$	μ	$\mu + \sigma_{\bar{x}}$	$\mu + 2\sigma_{\bar{x}}$

REFERENCES

Hogg, R. V., and Craig, A. T. *Introduction to Mathematical Statistics*, 5th ed. Upper Saddle River, N. J.: Prentice Hall, 1995.

Larsen, R. J., and Marx, M. L. *An Introduction to Mathematical Statistics and Its Applications*, 3rd ed. Upper Saddle River, N. J.: Prentice Hall, 2001.

Ramsey, P. P., and Ramsey, P. H. "Simple tests of normality in small samples." *Journal of Quality Technology*, Vol. 22, 1990, pp. 299–309.

Wackerly, D., Mendenhall, W., and Scheaffer, R. L. *Mathematical Statistics with Applications*, 6th ed. North Scituate, Mass.: Duxbury, 1999.

CHAPTER 5

Inferences Based on a Single Sample:

ESTIMATION WITH CONFIDENCE INTERVALS

Contents

Statistics in Action

Scallops, Sampling, and the Law

🖐 Where We've Been

We've learned that populations are characterized by numerical descriptive measures (called *parameters*) and that decisions about their values are based on sample statistics computed from sample data. Since statistics vary in a random manner from sample to sample, inferences based on them will be subject to uncertainty. This property is reflected in the sampling (probability) distribution of a statistic.

☞ Where We're Going

In this chapter, we'll put all the preceding material into practice; that is, we'll estimate population means and proportions based on a single sample selected from the population of interest. Most important, we use the sampling distribution of a sample statistic to assess the reliability of an estimate.

The estimation of the mean gas mileage for a new car model, the estimation of the expected life of a computer monitor, and the estimation of the mean yearly sales for companies in the steel industry are problems with a common element. In each case, we're interested in estimating the mean of a population of measurements. This important problem constitutes the primary topic of this chapter.

You'll see that different techniques are used for estimating a mean, depending on whether a sample contains a large or small number of measurements. Nevertheless, our objectives remain the same: We want to use the sample information to estimate the mean and to assess the reliability of the estimate.

First, we consider a method of estimating a population mean using a large sample (Section 5.1) and a small sample (Section 5.2). Then, we consider estimation of population proportions (Section 5.3). Finally, we see how to determine the sample sizes necessary for reliable estimates (Section 5.4).

TEACHING TIP
The key idea of this chapter revolves around the ability to generate an estimate with a corresponding measure of reliability.

5.1 LARGE-SAMPLE CONFIDENCE INTERVAL FOR A POPULATION MEAN

Suppose a large hospital wants to estimate the average length of time patients remain in the hospital. To accomplish this objective, the hospital administrators plan to randomly sample 100 of all previous patients' records and to use the sample mean, $\bar{x}$, of the lengths of stay to estimate μ, the mean of *all* patients' visits. The sample mean $\bar{x}$ represents a *point estimator* of the population mean μ. How can we assess the accuracy of this large-sample point estimator?

According to the Central Limit Theorem, the sampling distribution of the sample mean is approximately normal for large samples, as shown in Figure 5.1. Let us calculate the interval

$$\bar{x} \pm 2\sigma_{\bar{x}} = \bar{x} \pm \frac{2\sigma}{\sqrt{n}}$$

That is, we form an interval 4 standard deviations wide—from 2 standard deviations below the sample mean to 2 standard deviations above the mean. Prior to drawing the sample, what are the chances that this interval will enclose μ, the population mean?

FIGURE 5.1

Sampling Distribution of $\bar{x}$

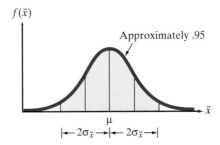

To answer this question, refer to Figure 5.1. If the 100 measurements yield a value of $\bar{x}$ that falls between the two lines on either side of μ—i.e., within 2 standard deviations of μ—then the interval $\bar{x} \pm 2\sigma_{\bar{x}}$ will contain μ; if $\bar{x}$ falls outside these boundaries, the interval $\bar{x} \pm 2\sigma_{\bar{x}}$ will not contain μ. Since the area under the normal curve (the sampling distribution of $\bar{x}$) between these boundaries is about .95 (more precisely, from Table III in Appendix A the area is .9544), we know that the interval $\bar{x} \pm 2\sigma_{\bar{x}}$ will contain μ with a probability approximately equal to .95.

For instance, consider the lengths of time spent in the hospital for 100 patients shown in Table 5.1. An SAS printout of summary statistics for the sample of 100 lengths of stay is shown in Figure 5.2.

TEACHING TIP
Explain that the population standard deviation is usually unknown and will need to be estimated with the sample standard deviation.

⊘ HOSPLOS

TABLE 5.1   Lengths of Stay (in Days) for 100 Patients

2	3	8	6	4	4	6	4	2	5
8	10	4	4	4	2	1	3	2	10
1	3	2	3	4	3	5	2	4	1
2	9	1	7	17	9	9	9	4	4
1	1	1	3	1	6	3	3	2	5
1	3	3	14	2	3	9	6	6	3
5	1	4	6	11	22	1	9	6	5
2	2	5	4	3	6	1	5	1	6
17	1	2	4	5	4	4	3	2	3
3	5	2	3	3	2	10	2	4	2

FIGURE 5.2

SAS Summary Statistics for the Lengths of 100 Hospital Stays

The MEANS Procedure

Analysis Variable : LOS

N	Mean	Std Dev	Minimum	Maximum
100	4.5300000	3.6775458	1.0000000	22.0000000

From the printout, we find $\bar{x} = 4.53$ days and $s = 3.68$ days. To achieve our objective, we must construct the interval

$$\bar{x} \pm 2\sigma_{\bar{x}} = 4.53 \pm 2\frac{\sigma}{\sqrt{100}}$$

TEACHING TIP
This repeated sampling interpretation is difficult for students to understand. Explain that the repeated intervals will move with the sample data, yet the parameter that they are estimating remains constant.

But now we face a problem. You can see that without knowing the standard deviation σ of the original population—that is, the standard deviation of the lengths of stay of *all* patients—we cannot calculate this interval. However, since we have a large sample ($n = 100$ measurements), we can approximate the interval by using the sample standard deviation s to approximate σ. Thus,

$$\bar{x} \pm 2\frac{\sigma}{\sqrt{100}} \approx \bar{x} \pm 2\frac{s}{\sqrt{100}} = 4.53 \pm 2\left(\frac{3.68}{10}\right) = 4.53 \pm .74$$

That is, we estimate the mean length of stay in the hospital for all patients to fall in the interval 3.79 to 5.27 days.

Can we be sure that μ, the true mean, is in the interval 3.79 to 5.27? We cannot be certain, but we can be reasonably confident that it is. This confidence is derived from the knowledge that if we were to draw repeated random samples of 100 measurements from this population and form the interval $\bar{x} \pm 2\sigma_{\bar{x}}$ each time, approximately 95% of the intervals would contain μ. We have no way of knowing (without looking at all the patients' records) whether our sample interval is one of the 95% that contain μ or one of the 5% that do not, but the odds certainly favor its containing μ. Consequently, the interval 3.79 to 5.27 provides a reliable estimate of the mean length of patient stay in the hospital.

The formula that tells us how to calculate an interval estimate based on sample data is called an *interval estimator*, or *confidence interval*. The probability, .95, that measures the confidence we can place in the interval estimate is called a *confidence coefficient*. The percentage, 95%, is called the *confidence level* for the interval estimate. It is not usually possible to assess precisely the reliability of point estimators because they are single points rather than intervals. So, because we prefer to use estimators for which a measure of reliability can be calculated, we will generally use interval estimators.

DEFINITION 5.1
An **interval estimator** (or **confidence interval**) is a formula that tells us how to use sample data to calculate an interval that estimates a population parameter.

DEFINITION 5.2
The **confidence coefficient** is the probability that an interval estimator encloses the population parameter—that is, the relative frequency with which the interval estimator encloses the population parameter when the estimator is used repeatedly a very large number of times. The **confidence level** is the confidence coefficient expressed as a percentage.

Now we have seen how an interval can be used to estimate a population mean. When we use an interval estimator, we can usually calculate the probability that the estimation *process* will result in an interval that contains the true value of the population mean. That is, the probability that the interval contains the parameter in repeated usage is usually known. Figure 5.3 shows what happens when 10 different samples are drawn from a population, and a confidence interval for μ is calculated from each. The location of μ is indicated by the vertical line in the figure. Ten confidence intervals, each based on one of 10 samples, are shown as horizontal line segments. Note that the confidence intervals move from sample to sample— sometimes containing μ and other times missing μ. *If our confidence level is 95%, in the long run, 95% of our sample confidence intervals will contain μ.*

FIGURE 5.3

Confidence Intervals for μ: 10 Samples

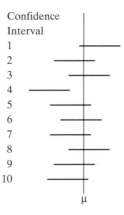

Suppose you wish to choose a confidence coefficient other than .95. Notice in Figure 5.1 that the confidence coefficient .95 is equal to the total area under the sampling distribution, less .05 of the area, which is divided equally between the two tails. Using this idea, we can construct a confidence interval with any desired confidence coefficient by increasing or decreasing the area (call it α) assigned to the tails of the sampling distribution (see Figure 5.4). For example, if we place area $\alpha/2$ in each tail and if $z_{\alpha/2}$ is the z value such that the area $\alpha/2$ will lie to its right, then the confidence interval with confidence coefficient $(1 - \alpha)$ is

$$\bar{x} \pm z_{\alpha/2}\sigma_{\bar{x}}$$

FIGURE 5.4

Locating $z_{\alpha/2}$ on the
Standard Normal Curve

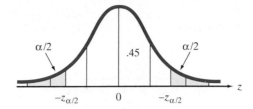

FIGURE 5.5

The z Value $(z_{.05})$
Corresponding to an Area
Equal to .05 in the Upper
Tail of the z-Distribution

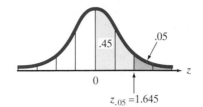

To illustrate, for a confidence coefficient of .90 we have $(1 - \alpha) = .90$, $\alpha = .10$, and $\alpha/2 = 0.5$; $z_{.05}$ is the z value that locates area .05 in the upper tail of the sampling distribution. Recall that Table III in Appendix A gives the areas between the mean and a specified z value. Since the total area to the right of the mean is .5, we find that $z_{.05}$ will be the z value corresponding to an area of $.5 - .05 = .45$ to the right of the mean (see Figure 5.5). This z value is $z_{.05} = 1.645$.

Confidence coefficients used in practice usually range from .90 to .99. The most commonly used confidence coefficients with corresponding values of α and $z_{\alpha/2}$ are shown in Table 5.2.

TABLE 5.2   Commonly Used Values of $z_{\alpha/2}$

Confidence Level 100(1 − α)	α	α/2	$z_{\alpha/2}$
90%	.10	.05	1.645
95%	.05	.025	1.96
99%	.01	.005	2.575

Large-Sample $100(1 - \alpha)$% Confidence Interval for μ

$$\bar{x} \pm z_{\alpha/2}\sigma_{\bar{x}} = \bar{x} \pm z_{\alpha/2}\frac{\sigma}{\sqrt{n}}$$

where $z_{\alpha/2}$ is the z value with an area $\alpha/2$ to its right (see Figure 5.4) and $\sigma_{\bar{x}} = \sigma/\sqrt{n}$. The parameter σ is the standard deviation of the sampled population and n is the sample size.

Note: When σ is unknown (as is almost always the case) and n is large (say, $n \geq 30$), the confidence interval is approximately equal to

$$\bar{x} \pm z_{\alpha/2}\left(\frac{s}{\sqrt{n}}\right)$$

where s is the sample standard deviation.

Assumptions: None, since the Central Limit Theorem guarantees that the sampling distribution of $\bar{x}$ is approximately normal.

FIGURE 5.6

*MINITAB Printout for
Example 5.1*

Variable	N	Mean	StDev	SE Mean	90.0% CI
NOSEATS	225	11.596	4.103	0.274	(11.144, 12.047)

■

E X A M P L E   5 . 1

⊘ AIRNOSHOWS
$11.6 \pm .45$

Unoccupied seats on flights cause airlines to lose revenue. Suppose a large airline wants to estimate its average number of unoccupied seats per flight over the past year. To accomplish this, the records of 225 flights are randomly selected, and the number of unoccupied seats is noted for each of the sampled flights. Descriptive statistics for the data are displayed in the MINITAB printout, Figure 5.6.

Estimate μ, the mean number of unoccupied seats per flight during the past year, using a 90% confidence interval.

Solution

The general form of the 90% confidence interval for a population mean is

$$\bar{x} \pm z_{\alpha/2}\sigma_{\bar{x}} = \bar{x} \pm z_{.05}\sigma_{\bar{x}} = \bar{x} \pm 1.645\left(\frac{\sigma}{\sqrt{n}}\right)$$

Suggested Exercise 5.6

From Figure 5.6, we find (after rounding) $\bar{x} = 11.6$. Since we do not know the value of σ (the standard deviation of the number of unoccupied seats per flight for all flights of the year), we use our best approximation—the sample standard deviation, $s = 4.1$, shown on the MINITAB printout. Then the 90% confidence interval is, approximately,

$$11.6 \pm 1.645\left(\frac{4.1}{\sqrt{225}}\right) = 11.6 \pm .45$$

or from 11.15 to 12.05. That is, at the 90% confidence level, we estimate the mean number of unoccupied seats per flight to be between 11.15 and 12.05 during the sampled year. This result is verified (except for rounding) on the right side of the MINITAB printout in Figure 5.6.

We stress that the confidence level for this example, 90%, refers to the procedure used. If we were to apply this procedure repeatedly to different samples, approximately 90% of the intervals would contain μ. Although we do not know for sure whether this particular interval (11.15, 12.05) is one of the 90% that contain μ or one of the 10% that do not, our knowledge of probability gives us "confidence" that the interval contains μ.

■

TEACHING TIP
Discuss the interpretation of the confidence interval in two parts: the practical side that gives the interval for μ and the theoretical part that involves repeated samples that gives the reliability to the interval.

The interpretation of confidence intervals for a population mean is summarized in the box on p. 243.

Sometimes, the estimation procedure yields a confidence interval that is too wide for our purposes. In this case, we will want to reduce the width of the interval to obtain a more precise estimate of μ. One way to accomplish this is to decrease the confidence coefficient, $1 - \alpha$. For example, consider the problem of estimating the mean length of stay, μ, for hospital patients. Recall that for a sample of 100 patients, $\bar{x} = 4.53$ days and $s = 3.68$ days. A 90% confidence interval for μ is

$$\bar{x} \pm 1.645(\sigma/\sqrt{n}) \approx 4.53 \pm (1.645)(3.68/\sqrt{100}) = 4.53 \pm .61$$

or (3.92, 5.14). You can see that this interval is narrower than the previously calculated 95% confidence interval, (3.79, 5.27). Unfortunately, we also have "less confidence" in the 90% confidence interval. An alternative method used to decrease the width of an interval without sacrificing "confidence" is to increase the sample size n. We demonstrate this method in Section 5.4.

TEACHING TIP
This notation is generally difficult for students. Use common levels of reliability like 95% and the *z* distribution to show where α and $(1 - \alpha)$ probabilities fall.

Suggested Exercise 5.18

Interpretation of a Confidence Interval for a Population Mean

When we form a $100(1 - \alpha)\%$ confidence interval for μ, we usually express our confidence in the interval with a statement such as, "We can be $100(1 - \alpha)\%$ confident that μ lies between the lower and upper bounds of the confidence interval," where for a particular application, we substitute the appropriate numerical values for the level of confidence and for the lower and upper bounds. *The statement reflects our confidence in the estimation process rather than in the particular interval that is calculated from the sample data.* We know that repeated application of the same procedure will result in different lower and upper bounds on the interval. Furthermore, we know that $100(1 - \alpha)\%$ of the resulting intervals will contain μ. There is (usually) no way to determine whether any particular interval is one of those that contain μ, or one that does not. However, unlike point estimators, confidence intervals have some measure of reliability, the confidence coefficient, associated with them. For that reason they are generally preferred to point estimators.

Using the TI-83 Graphing Calculator

Confidence Interval for a Population Mean
(known σ or $n \geq 30$)

Say we want to develop a **95%** confidence interval for the population mean from a sample size of **35** where we know the sample mean is **100** and the **population** deviation is **12**. For this problem we want **Z-Interval** because we are given sigma. On the calculator we will choose **"7:Zinterval"** from the **"TESTS"** menu.

Step 1   *Access the Statistical Tests Menu*
Press   **STAT** (this screen will appear).
Arrow right to **"TEST"**
Arrow down to **"7:ZInterval"**
Press   **ENTER**

```
EDIT CALC TESTS
1:Edit…
2:SortA(
3:SortD(
4:ClrList
5:SetUpEditor
```

Step 2   *Choose* **"Data"** *or* **"Stats"**

If you choose "Data" the calculator will read the data from LIST.

If you choose "Stats" then you will need to enter the settings yourself.

```
EDIT CALC TESTS
1:Z-Test…
2:T-Test…
3:2-SampZTest…
4:2-SampTTest…
5:1-PropZTest…
6:2-PropZTest…
7↓ZInterval…
```

Highlight **"Stats"**
Press **ENTER**

You must set:
σ: **12**
$\bar{x}$: **100**
n: **35**
C-Level: .95 (don't forget the decimal).

```
ZInterval
 Inpt:Data Stats
 σ:12
 x̄:100
 n:35
 C-Level:.95
 Calculate
```

Step 3   *View Display*
Next, highlight **"Calculate"**
and Press **ENTER**

As you can see our confidence interval is **(96.024, 103.98)**.

```
ZInterval
 (96.024,103.98)
 x̄:100
 n:35
```

EXERCISES 5.1–5.20

Learning the Mechanics

5.1 Find $z_{\alpha/2}$ for each of the following:
   a. $\alpha = .10$ 1.645
   b. $\alpha = .01$ 2.58
   c. $\alpha = .05$ 1.96
   d. $\alpha = .20$ 1.28

5.2 What is the confidence level of each of the following confidence intervals for μ?

   a. $\bar{x} \pm 1.96\left(\dfrac{\sigma}{\sqrt{n}}\right)$ 95%

   b. $\bar{x} \pm 1.645\left(\dfrac{\sigma}{\sqrt{n}}\right)$ 90%

   c. $\bar{x} \pm 2.575\left(\dfrac{\sigma}{\sqrt{n}}\right)$ 99%

   d. $\bar{x} \pm 1.282\left(\dfrac{\sigma}{\sqrt{n}}\right)$ 80%

   e. $\bar{x} \pm .99\left(\dfrac{\sigma}{\sqrt{n}}\right)$ 67.78%

5.3 A random sample of 64 observations from a population produced the following summary statistics:

$$\Sigma x_i = 700 \qquad \Sigma(x_i - \bar{x})^2 = 4{,}238$$

   a. Find a 95% confidence interval for μ.
   b. Interpret the confidence interval you found in part **a**.

5.4 A random sample of 90 observations produced a mean $\bar{x} = 25.9$ and a standard deviation $s = 2.7$.
   a. Find a 95% confidence interval for the population mean μ. $25.9 \pm .56$
   b. Find a 90% confidence interval for μ. $25.9 \pm .47$
   c. Find a 99% confidence interval for μ. $25.9 \pm .73$

5.5 A random sample of 100 observations from a normally distributed population possesses a mean equal to 83.2 and a standard deviation equal to 6.4.
   a. Find a 95% confidence interval for μ. 83.2 ± 1.25
   b. What do you mean when you say that a confidence coefficient is .95?
   c. Find a 99% confidence interval for μ. 83.2 ± 1.65
   d. What happens to the width of a confidence interval as the value of the confidence coefficient is increased while the sample size is held fixed?   Increases
   e. Would your confidence intervals of parts **a** and **c** be valid if the distribution of the original population was not normal? Explain.   Yes

5.6 A random sample of n measurements was selected from a population with unknown mean μ and standard deviation σ. Calculate a 95% confidence interval for μ for each of the following situations:
   a. $n = 75$, $\bar{x} = 28$, $s^2 = 12$ $28 \pm .784$
   b. $n = 200$, $\bar{x} = 102$, $s^2 = 22$ $102 \pm .65$
   c. $n = 100$, $\bar{x} = 15$, $s = .3$ $15 \pm .0588$

   d. $n = 100$, $\bar{x} = 4.05$, $s = .83$ $4.05 \pm .163$
   e. Is the assumption that the underlying population of measurements is normally distributed necessary to ensure the validity of the confidence intervals in parts **a–d**? Explain.

5.7 Explain the difference between an interval estimator and a point estimator for μ.

5.8 Explain what is meant by the statement, "We are 95% confident that an interval estimate contains μ."

5.9 Will a large-sample confidence interval be valid if the population from which the sample is taken is not normally distributed? Explain.

5.10 The mean and standard deviation of a random sample of n measurements are equal to 33.9 and 3.3, respectively.
   a. Find a 95% confidence interval for μ if $n = 100$.
   b. Find a 95% confidence interval for μ if $n = 400$.
   c. Find the widths of the confidence intervals found in parts **a** and **b**. What is the effect on the width of a confidence interval of quadrupling the sample size while holding the confidence coefficient fixed?

Applying the Concepts—Basic

5.11 The *Journal of the American Medical Association* (Apr. 21, 1993) reported on the results of a National Health Interview Survey designed to determine the prevalence of smoking among U.S. adults. Current smokers (more than 11,000 adults in the survey) were asked: "On the average, how many cigarettes do you now smoke a day?" The results yielded a mean of 20.0 cigarettes per day with an associated 95% confidence interval of $(19.7, 20.3)$.
   a. Interpret the 95% confidence interval.
   b. State any assumptions about the target population of current cigarette smokers that must be satisfied for inferences derived from the interval to be valid.
   c. A tobacco industry researcher claims that the mean number of cigarettes smoked per day by regular cigarette smokers is less than 15. Comment on this claim.

5.12 Refer to the *Applied Psycholinguistics* (June 1998) study of language skills of low-income children, Exercise 2.76 (p. 69). Each in a sample of 65 low-income children was administered the Communicative Development Inventory (CDI) exam. The sentence complexity scores had a mean of 7.62 and a standard deviation of 8.91.
   a. Construct a 90% confidence interval for the mean sentence complexity score of all low-income children.
   b. Interpret the interval, part **a**, in the words of the problem.
   c. Suppose we know that the true mean sentence complexity score of middle-income children is 15.55. Is there evidence that the true mean for low-income children differs from 15.55? Explain.   Yes

5.13 Refer to the *Chance* (Winter 2001) and National Education Longitudinal Survey (NELS) study of 265 students who paid a private tutor to help them improve their SAT scores, Exercise 2.81 (p. 70). The changes in both the SAT-Mathematics and SAT-Verbal scores for these students are reproduced in the table.

	SAT-Math	SAT-Verbal
Mean change in score	19	7
Standard deviation of score changes	65	49

a. Construct and interpret a 95% confidence interval for the population mean change in SAT-Mathematics score for students who pay a private tutor.
b. Repeat part **a** for the population mean change in SAT-Verbal score. 7 ± 5.90
c. Suppose the true population mean change in score on one of the SAT tests for all students who paid a private tutor is 15. Which of the two tests, SAT-Mathematics or SAT-Verbal, is most likely to have this mean change? Explain.   SAT-Mathematics

5.14 Donations to tax-exempt organizations such as the Red Cross, the Salvation Army, the YMCA, and the American Cancer Society not only go to the stated charitable purpose, but are used to cover fundraising expenses and overhead. For a sample of 30 charities, the table lists their charitable commitment, i.e., the percentage of their expenses that go toward the stated charitable purpose.
a. Give a point estimate for the mean charitable commitment of tax-exempt organizations.   74.97
b. Construct a 98% confidence interval for the mean charitable commitment. 74.97 ± 7.67
c. What assumption(s) must hold for the method of estimation used in part **b** to be appropriate?
d. Why is the confidence interval of part **b** a better estimator of the mean charitable commitment than the point estimator of part **a**?

Applying the Concepts—Intermediate

5.15 Animal behaviorists have discovered that the more domestic chickens peck at objects placed in their environment, the healthier the chickens seem to be. White string has been found to be a particularly attractive pecking stimulus. In one experiment, 72 chickens were exposed to a string stimulus. Instead of white string, blue-colored string was used. The number of pecks each chicken took at the blue string over a specified time interval was recorded. Summary statistics for the 72 chickens were: $\bar{x} = 1.13$ pecks, $s = 2.21$ pecks (*Applied Animal Behaviour Science*, Oct. 2000).
a. Estimate the population mean number of pecks made by chickens pecking at blue string using a 99% confidence interval. Interpret the result. $1.13 \pm .672$
b. Previous research has shown that $\mu = 7.5$ pecks if chickens are exposed to white string. Based on the results, part **a**, is there evidence that chickens are more apt to peck at white string than blue string? Explain.   Yes

5.16 Refer to *The Astronomical Journal* (July 1995) study of the velocity of light emitted from a galaxy in the universe, Exercise 2.78 (p. 69). A sample of 103 galaxies located in the galaxy cluster A2142 had a mean velocity of $\bar{x} = 27{,}117$ kilometers per second (km/s) and a standard deviation of $s = 1{,}280$ km/s. Suppose your goal is to make an inference about the population mean light velocity of galaxies in cluster A2142.
a. In part **b** of Exercise 2.78, you constructed an interval that captured approximately 95% of the galaxy velocities in the cluster. Explain why this interval is inappropriate for this inference.
b. Construct a 95% confidence interval for the mean light velocity emitted from all galaxies in cluster A2142. Interpret the result. $27{,}117 \pm 247.20$

5.17 The relationship between an employee's participation in the performance appraisal process and subsequent subordinate reactions toward the appraisal was investigated

⊘ CHARITY

Organization	Charitable Commitment	Organization	Charitable Commitment
American Cancer Society	62%	Multiple Sclerosis Association of America	56
American National Red Cross	91	Museum of Modern Art	79
Big Brothers Big Sisters of America	77	Nature Conservancy	77
Boy Scouts of America National Council	81	Paralyzed Veterans of America	50
Boys & Girls Clubs of America	81	Planned Parenthood Federation	81
CARE	91	Salvation Army	84
Covenant House	15	Shriners Hospital for Children	95
Disabled American Veterans	65	Smithsonian Institution	87
Ducks Unlimited	78	Special Olympics	72
Feed The Children	90	Trust for Public Land	88
Girl Scouts of the USA	83	United Jewish Appeal/Federation - NY	75
Goodwill Industries International	89	United States Olympic Committee	78
Habitat for Humanity International	81	United Way of New York City	85
Mayo Foundation	26	WGBH Educational Foundation	81
Mothers Against Drunk Drivers	71	YMCA of the USA	80

Source: "Look Before You Give," *Forbes*, Dec. 27, 1999, pp. 206–216.

MINITAB Output for Exercise 5.17

Variable	N	Mean	StDev	SE Mean	95.0% CI
CORR	34	0.4224	0.1998	0.0343	(0.3526, 0.4921)

in the *Journal of Applied Psychology* (Aug. 1998). In Chapter 9 we will discuss a quantitative measure of the relationship between two variables, called the coefficient of correlation r. The researchers obtained r for a sample of 34 studies that examined the relationship between appraisal participation and a subordinate's satisfaction with the appraisal. These correlations are listed in the table. (Values of r near $+1$ reflect a strong positive relationship between the variables.) A MINITAB printout showing a 95% confidence interval for the mean of the data is provided above. Locate the 95% confidence interval on the printout and interpret it in the words of the problem.

◉ CORR34

.50	.58	.71	.46	.63	.66	.31	.35	.51	.06	.35	.19
.40	.63	.43	.16	−.08	.51	.59	.43	.30	.69	.25	.20
.39	.20	.51	.68	.74	.65	.34	.45	.31	.27		

Source: Cawley, B. D., Keeping, L. M., and Levy, P. E. "Participation in the performance appraisal process and employee reactions: A meta-analytic review of field investigations." *Journal of Applied Psychology*, Vol. 83, No. 4, Aug. 1998, pp. 632–633 (Appendix).

5.18 Psychologists have found that twins, in their early years, tend to have lower IQs and pick up language more slowly than nontwins. The slower intellectual growth of most twins may be caused by benign parental neglect. Suppose it is desired to estimate the mean attention time given to twins per week by their parents. A sample of 50 sets of $2\frac{1}{2}$-year-old twin boys is taken, and at the end of 1 week the attention time given to each pair is recorded. The data (in hours) are listed in the accompanying table. An SAS printout gives summary statistics. Use this

information to find a 90% confidence interval for the mean attention time given to all twin boys by their parents. Interpret the confidence interval. 20.848 ± 3.121

Applying the Concepts—Advanced

5.19 Research reported in the *Journal of Psychology and Aging* (May 1992) studied the role that the age of workers has in determining their level of job satisfaction. The researcher hypothesized that both younger and older workers would have a higher job satisfaction rating than middle-age workers. Each of a sample of 1,686 adults was given a job satisfaction score based on answers to a series of questions. Higher job satisfaction scores indicate higher levels of job satisfaction. The data, arranged by age group, are summarized below.

	Age Group		
	Younger 18–24	**Middle-Age 25–44**	**Older 45–64**
$\bar{x}$	4.17	4.04	4.31
s	.75	.81	.82
n	241	768	677

a. Construct 95% confidence intervals for the mean job satisfaction scores of each age group. Carefully interpret each interval.

b. In the construction of three 95% confidence intervals, is it more or less likely that at least one of them will *not* contain the population mean it is intended to estimate than it is for a single confidence interval to miss the population mean? [*Hint:* Assume the three intervals are independent, and calculate the probability that at least one of them will not contain the population mean it estimates. Compare this

◉ ATTIMES

20.7	16.7	22.5	12.1	2.9	15.7	46.6	10.6	6.7	5.4
23.5	6.4	1.3	39.6	35.6	20.3	34.0	44.5	23.8	20.0
10.9	7.1	46.0	23.4	29.4	43.1	14.3	21.9	17.5	9.6
44.1	13.8	24.3	9.3	3.4	36.4	0.8	1.1	19.3	14.6
14.0	20.7	48.2	7.7	22.2	32.5	19.1	36.9	27.9	14.0

SAS Output for Exercise 5.18

Analysis Variable : ATTIME				
Mean	Std Dev	N	Minimum	Maximum
20.8480000	13.4138253	50	0.8000000	48.2000000

probability to the probability that a single interval fails to enclose the mean.]

c. Based on these intervals, does it appear that the researcher's hypothesis is supported? [*Caution:* We'll learn how to use sample information to compare population means in Chapter 7, and we'll return to this exercise at that time. Here, simply base your opinion on the individual confidence intervals you constructed in part **a**.]

5.20 According to scientists, the cockroach has had 300 million years to develop a resistance to destruction. In a study conducted by researchers for S. C. Johnson & Son, Inc. (manufacturers of Raid and Off), 5,000 roaches (the expected number in a roach-infested house) were released in the Raid test kitchen. One week later the kitchen was fumigated and 16,298 dead roaches were counted, a gain of 11,298 roaches for the 1-week period. Assume that none of the original roaches died during the 1-week period and that the standard deviation of x, the number of roaches produced per roach in a 1-week period, is 1.5. Use the number of roaches produced by the sample of 5,000 roaches to find a 95% confidence interval for the mean number of roaches produced per week for each roach in a typical roach-infested house.

5.2 SMALL-SAMPLE CONFIDENCE INTERVAL FOR A POPULATION MEAN

Federal legislation requires pharmaceutical companies to perform extensive tests on new drugs before they can be marketed. Initially, a new drug is tested on animals. If the drug is deemed safe after this first phase of testing, the pharmaceutical company is then permitted to begin human testing on a limited basis. During this second phase, inferences must be made about the safety of the drug based on information in very small samples.

Suppose a pharmaceutical company must estimate the average increase in blood pressure of patients who take a certain new drug. Assume that only six patients (randomly selected from the population of all patients) can be used in the initial phase of human testing. The use of a *small sample* in making an inference about μ presents two immediate problems when we attempt to use the standard normal z as a test statistic.

PROBLEM I

The shape of the sampling distribution of the sample mean $\bar{x}$ (and the z statistic) now depends on the shape of the population that is sampled. We can no longer assume that the sampling distribution of $\bar{x}$ is approximately normal, because the Central Limit Theorem ensures normality only for samples that are sufficiently large.

Solution to Problem 1

The sampling distribution of $\bar{x}$ (and z) is exactly normal even for relatively small samples if the sampled population is normal. It is approximately normal if the sampled population is approximately normal.

PROBLEM 2

TEACHING TIP
Stress to the students that the shape of the sampling distribution is unknown if the population is not normally distributed. It is useful here to introduce the topic of nonparametric statistics even if it will not be covered in the class.

The population standard deviation σ is almost always unknown. Although it is still true that $\sigma_{\bar{x}} = \sigma/\sqrt{n}$, the sample standard deviation s may provide a poor approximation for σ when the sample size is small.

Solution to Problem 2

Instead of using the standard normal statistic

$$z = \frac{\bar{x} - \mu}{\sigma_{\bar{x}}} = \frac{\bar{x} - \mu}{\sigma/\sqrt{n}}$$

which requires knowledge of or a good approximation to σ, we define and use the statistic

$$t = \frac{\bar{x} - \mu}{s/\sqrt{n}}$$

in which the sample standard deviation, s, replaces the population standard deviation, σ.

The distribution of the **t statistic** in repeated sampling was discovered by W. S. Gosset, a chemist in the Guinness brewery in Ireland, who published his discovery in 1908 under the pen name of Student. The main result of Gosset's work is that if we are sampling from a normal distribution, the t statistic has a sampling distribution very much like that of the z statistic: mound-shaped, symmetric, with mean 0. The primary difference between the sampling distributions of t and z is that the t statistic is more variable than the z, which follows intuitively when you realize that t contains two random quantities ($\bar{x}$ and s), whereas z contains only one ($\bar{x}$).

The actual amount of variability in the sampling distribution of t depends on the sample size n. A convenient way of expressing this dependence is to say that the t statistic has $(n - 1)$ **degrees of freedom (df)**. Recall that the quantity $(n - 1)$ is the divisor that appears in the formula for s^2. This number plays a key role in the sampling distribution of s^2 and appears in discussions of other statistics in later chapters. Particularly, the smaller the number of degrees of freedom associated with the t statistic, the more variable will be its sampling distribution.

In Figure 5.7 we show both the sampling distribution of z and the sampling distribution of a t statistic with 4 df. You can see that the increased variability of the t statistic means that the t value, t_α, that locates an area α in the upper tail of the t-distribution is larger than the corresponding value z_α. For any given value of α, the t value t_α increases as the number of degrees of freedom (df) decreases. Values of t that will be used in forming small-sample confidence intervals of μ are given in Table IV of Appendix A. A partial reproduction of this table is shown in Table 5.3.

TEACHING TIP
In order to use the t-table, it is always necessary to know the degrees of freedom of the distribution. When estimating a population mean, the degrees of freedom is always $(n - 1)$.

TEACHING TIP
Draw a graph with the z-distribution and several t-distributions on the same number line. Illustrate the role of the degrees of freedom by drawing t-distributions more similar to the z-distribution as the degrees of freedom get larger.

FIGURE 5.7

Standard Normal (z) Distribution and t-Distribution with 4 df

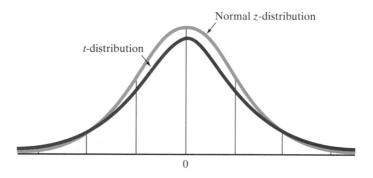

Note that t_α values are listed for various degrees of freedom where α refers to the tail area under the t-distribution to the right of t_α. For example, if we want the t value with an area of .025 to its right and 4 df, we look in the table under the column $t_{.025}$ for the entry in the row corresponding to 4 df. This entry is $t_{.025} = 2.776$, as shown in Figure 5.8. The corresponding standard normal z-score is $z_{.025} = 1.96$.

Note that the last row of Table IV, where df $= \infty$ (infinity), contains the standard normal z values. This follows from the fact that as the sample size n grows very large, s becomes closer to σ and thus t becomes closer in distribution to z. In fact, when df $= 29$, there is little difference between corresponding tabulated values of z and t. Thus, we choose the arbitrary cutoff of $n = 30$ (df $= 29$) to distinguish between the large-sample and small-sample inferential techniques.

TABLE 5.3   Reproduction of Part of Table IV in Appendix A

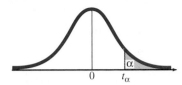

Degrees of Freedom	$t_{.100}$	$t_{.050}$	$t_{.025}$	$t_{.010}$	$t_{.005}$	$t_{.001}$	$t_{.0005}$
1	3.078	6.314	12.706	31.821	63.657	318.31	636.62
2	1.886	2.920	4.303	6.965	9.925	22.326	31.598
3	1.638	2.353	3.182	4.541	5.841	10.213	12.924
4	1.533	2.132	2.776	3.747	4.604	7.173	8.610
5	1.476	2.015	2.571	3.365	4.032	5.893	6.869
6	1.440	1.943	2.447	3.143	3.707	5.208	5.959
7	1.415	1.895	2.365	2.998	3.499	4.785	5.408
8	1.397	1.860	2.306	2.896	3.355	4.501	5.041
9	1.383	1.833	2.262	2.821	3.250	4.297	4.781
10	1.372	1.812	2.228	2.764	3.169	4.144	4.587
11	1.363	1.796	2.201	2.718	3.106	4.025	4.437
12	1.356	1.782	2.179	2.681	3.055	3.930	4.318
13	1.350	1.771	2.160	2.650	3.012	3.852	4.221
14	1.345	1.761	2.145	2.624	2.977	3.787	4.140
15	1.341	1.753	2.131	2.602	2.947	3.733	4.073
⋮	⋮	⋮	⋮	⋮	⋮	⋮	⋮
∞	1.282	1.645	1.960	2.326	2.576	3.090	3.291

FIGURE 5.8

The $t_{.025}$ Value in a t-Distribution with 4 df and the Corresponding $z_{.025}$ Value

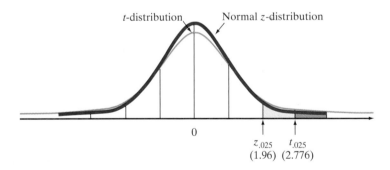

Returning to the example of testing a new drug, suppose that the six test patients have blood pressure increases of 1.7, 3.0, .8, 3.4, 2.7, and 2.1 points. How can we use this information to construct a 95% confidence interval for μ, the mean increase in blood pressure associated with the new drug for all patients in the population?

First, we know that we are dealing with a sample too small to assume that the sample mean $\bar{x}$ is approximately normally distributed by the Central Limit Theorem. That is, we do not get the normal distribution of $\bar{x}$ "automatically" from the Central Limit Theorem when the sample size is small. Instead, we must assume that the measured variable, in this case the increase in blood pressure, is normally distributed in order for the distribution of $\bar{x}$ to be normal.

Second, unless we are fortunate enough to know the population standard deviation σ, which in this case represents the standard deviation of *all* the patients' increases in blood pressure when they take the new drug, we cannot use the standard normal z statistic to form our confidence interval for μ. Instead, we must use the t-distribution, with $(n - 1)$ degrees of freedom.

In this case, $n - 1 = 5$ df, and the t value is found in Table 5.3 to be

$$t_{.025} = 2.571 \text{ with 5 df}$$

Recall that the large-sample confidence interval would have been of the form

$$\bar{x} \pm z_{\alpha/2}\sigma_{\bar{x}} = \bar{x} \pm z_{\alpha/2}\frac{\sigma}{\sqrt{n}} = \bar{x} \pm z_{.025}\frac{\sigma}{\sqrt{n}}$$

where 95% is the desired confidence level. To form the interval for a small sample from *a normal distribution, we simply substitute t for z and s for σ in the preceding formula:*

$$\bar{x} \pm t_{\alpha/2}\frac{s}{\sqrt{n}}$$

A STATISTIX printout showing descriptive statistics for the six blood pressure increases is displayed in Figure 5.9. After rounding, we have $\bar{x} = 2.283$ and $s = .950$.

Substituting these numerical values into the confidence interval formula, we get

$$2.283 \pm (2.571)\left(\frac{.950}{\sqrt{6}}\right) = 2.283 \pm .997$$

or 1.286 to 3.280 points. Note that this interval agrees (except for rounding) with the confidence interval highlighted on the STATISTIX printout in Figure 5.9.

TEACHING TIP
Point out that the interpretation of the small-sample confidence interval is exactly the same as the large-sample confidence interval. The only change is the method used to calculate the endpoints of the interval.

FIGURE 5.9

STATISTIX Analysis of Six Blood Pressure Increases

```
DESCRIPTIVE STATISTICS

                      BPINCREAS
N                             6
LO 95% CI                1.2868
MEAN                     2.2833
UP 95% CI                3.2798
SD                       0.9496
MINIMUM                  0.8000
MAXIMUM                  3.4000
```

We interpret the interval as follows: We can be 95% confident that the mean increase in blood pressure associated with taking this new drug is between 1.286 and 3.28 points. As with our large-sample interval estimates, our confidence is in the process, not in this particular interval. We know that if we were to repeatedly use this estimation procedure, 95% of the confidence intervals produced would contain the true mean μ, *assuming that the probability distribution of changes in blood pressure from which our sample was selected is normal.* The latter assumption is necessary for the small-sample interval to be valid.

What price did we pay for having to utilize a small sample to make the inference? First, we had to assume the underlying population is normally distributed, and if the assumption is invalid, our interval might also be invalid.* Second, we had to form the interval using a t value of 2.571 rather than a z value of 1.96, resulting in a wider interval to achieve the same 95% level of confidence. If the interval from 1.286 to 3.28 is too wide to be of much use, we know how to remedy the situation: increase the number of patients sampled in order to decrease the interval width (on average).

The procedure for forming a small-sample confidence interval is summarized in the next box.

---

*By *invalid*, we mean that the probability that the procedure will yield an interval that contains μ is not equal to $(1 - \alpha)$. Generally, if the underlying population is approximately normal, the confidence coefficient will approximate the probability that the interval contains μ.

TEACHING TIP
Discuss the benefits associated
with taking samples of size 30 or
more. Use this example assuming
$n = 100$ to compare the intervals
generated.

Small-Sample Confidence Interval* for μ

$$\bar{x} \pm t_{\alpha/2}\left(\frac{s}{\sqrt{n}}\right)$$

where $t_{\alpha/2}$ is based on $(n - 1)$ degrees of freedom

Assumption: A random sample is selected from a population with a relative frequency distribution that is approximately normal.

EXAMPLE 5.2

Some quality control experiments require *destructive sampling* (i.e., the test to determine whether the item is defective destroys the item) in order to measure some particular characteristic of the product. The cost of destructive sampling often dictates small samples. Suppose a manufacturer of printers for personal computers wishes to estimate the mean number of characters printed before the printhead fails. The printer manufacturer tests $n = 15$ printheads and records the number of characters printed until failure for each. These 15 measurements (in millions of characters) are listed in Table 5.4, followed by an SPSS summary statistics printout in Figure 5.10.

a. Form a 99% confidence interval for the mean number of characters printed before the printhead fails. Interpret the result.

b. What assumption is required for the interval, part **a**, to be valid? Is it reasonably satisfied?

a. $1.24 \pm .15$

⊘ PRINTHEAD

TABLE 5.4   Number of Characters (in Millions) for $n = 15$ Printhead Tests

1.13	1.55	1.43	.92	1.25	1.36	1.32	.85	1.07	1.48	1.20	1.33	1.18	1.22	1.29

FIGURE 5.10

SPSS Summary Statistics Printout for Data in Table 5.4

Descriptives			Statistic	Std. Error
NUMBER	Mean		1.2387	4.987E-02
	99% Confidence	Lower Bound	1.0902	
	Interval for Mean	Upper Bound	1.3871	
	5% Trimmed Mean		1.2430	
	Median		1.2500	
	Variance		3.731E-02	
	Std. Deviation		.1932	
	Minimum		.85	
	Maximum		1.55	
	Range		.70	
	Interquartile Range		.2300	
	Skewness		-.491	.580
	Kurtosis		.064	1.121

---

*The procedure given in the box assumes that the population standard deviation σ is unknown, which is almost always the case. If σ is known, we can form the small-sample confidence interval just as we would a large-sample confidence interval using a standard normal z value instead of t. However, we must still assume that the underlying population is approximately normal.

Solution

a. For this small sample ($n = 15$), we use the t statistic to form the confidence interval. We use a confidence coefficient of .99 and $n - 1 = 14$ degrees of freedom to find $t_{\alpha/2}$ in Table IV:

$$t_{\alpha/2} = t_{.005} = 2.977$$

[*Note:* The small sample forces us to extend the interval almost 3 standard deviations (of $\bar{x}$) on each side of the sample mean in order to form the 99% confidence interval.] From the SPSS printout, Figure 5.10, we find $\bar{x} = 1.24$ and $s = .19$. Substituting these (rounded) values into the confidence interval formula, we obtain:

$$\bar{x} \pm t_{.005}\left(\frac{s}{\sqrt{n}}\right) = 1.24 \pm 2.977\left(\frac{.19}{\sqrt{15}}\right)$$
$$= 1.24 \pm .15 \quad \text{or} \quad (1.09, 1.39)$$

This interval is highlighted on Figure 5.10. Thus, the manufacturer can be 99% confident that the printhead has a mean life of between 1.09 and 1.39 million characters. If the manufacturer were to advertise that the mean life of its printheads is (at least) 1 million characters, the interval would support such a claim. Our confidence is derived from the fact that 99% of the intervals formed in repeated applications of this procedure will contain μ.

b. Since n is small, we must assume that the number of characters printed before printhead failure is a random variable from a normal distribution. That is, we assume that the population from which the sample of 15 measurements is selected is distributed normally. One way to check this assumption is to graph the distribution of data in Table 5.4. If the sample data is approximately normal, then the population from which the sample is selected is very likely to be normal. A MINITAB stem-and-leaf plot for the sample data is displayed in Figure 5.11. The distribution is mound-shaped and nearly symmetric. Therefore, the assumption of normality appears to be reasonably satisfied. ■

Suggested Exercise 5.27

FIGURE 5.11

MINITAB Stem-and-Leaf Display of Data in Table 5.4

```
Stem-and-leaf of number    N = 15
Leaf Unit = 0.010

    1      8  5
    2      9  2
    3     10  7
    5     11  38
   (4)    12  0259
    6     13  236
    3     14  38
    1     15  5
```

We have emphasized throughout this section that an assumption that the population is approximately normally distributed is necessary for making small-sample inferences about μ when using the t statistic. Although many phenomena do have approximately normal distributions, it is also true that many random phenomena have distributions that are not normal or even mound-shaped. Empirical evidence acquired over the years has shown that the *t*-distribution is rather insensitive to moderate departures from normality. That is, use of the *t* statistic when sampling from slightly skewed mound-shaped populations generally produces credible results; however, for cases in which the distribution is distinctly nonnormal, we must either take a large sample or use a *nonparametric method*.

What Do You Do When the Population Relative Frequency Distribution Departs Greatly from Normality?

Answer: Use the nonparametric statistical method of optional Section 6.6.

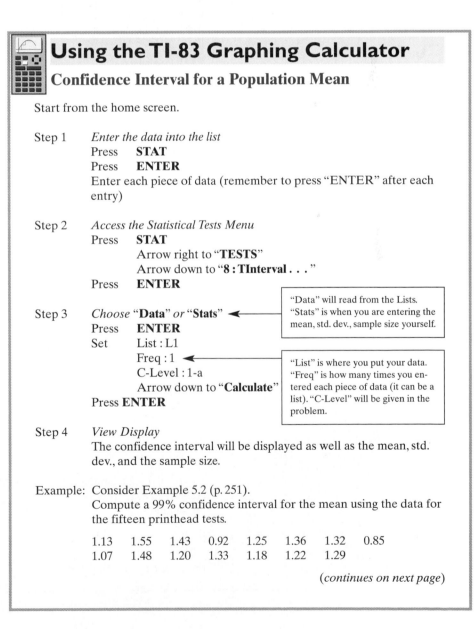

Using the TI-83 Graphing Calculator

Confidence Interval for a Population Mean

Start from the home screen.

Step 1   *Enter the data into the list*
Press   **STAT**
Press   **ENTER**
Enter each piece of data (remember to press "ENTER" after each entry)

Step 2   *Access the Statistical Tests Menu*
Press   **STAT**
Arrow right to "**TESTS**"
Arrow down to "**8 : TInterval . . .** "
Press   **ENTER**

Step 3   *Choose* "**Data**" *or* "**Stats**"   ◄——   "Data" will read from the Lists. "Stats" is when you are entering the mean, std. dev., sample size yourself.
Press   **ENTER**
Set   List : L1
Freq : 1   ◄——   "List" is where you put your data. "Freq" is how many times you entered each piece of data (it can be a list). "C-Level" will be given in the problem.
C-Level : 1-a
Arrow down to "**Calculate**"
Press **ENTER**

Step 4   *View Display*
The confidence interval will be displayed as well as the mean, std. dev., and the sample size.

Example: Consider Example 5.2 (p. 251).
Compute a 99% confidence interval for the mean using the data for the fifteen printhead tests.

1.13	1.55	1.43	0.92	1.25	1.36	1.32	0.85
1.07	1.48	1.20	1.33	1.18	1.22	1.29	

(*continues on next page*)

USING THE TI-83 GRAPHING CALCULATOR *(continued)*

Step 1   Enter the data into List1
Press   **STAT**
Press   **ENTER**
Enter each piece of data.

Step 2   Press   **STAT**
Arrow right to "**TESTS**"
Arrow down to "**8 : TInterval . . .**"
(you should be looking at this screen).

Press   **ENTER**

Step 3   Highlight "**Data**" then press **ENTER** (you will see this screen).

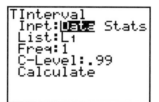

Match the settings on this screen then arrow down, highlight
"**Calculate**," and press **ENTER** (you will see this screen).

Step 4   As you can see from the screen our 99% confidence interval is
(1.0902, 1.3871). You will also notice it gives the mean, std. dev., and
the sample size.

Step 5   Return to the home screen.

Press   **CLEAR   ENTER**

EXERCISES 5.21–5.34

Learning the Mechanics

5.21 Explain the differences in the sampling distributions of $\bar{x}$ for large and small samples under the following assumptions.
   a. The variable of interest, x, is normally distributed.
   b. Nothing is known about the distribution of the variable x.

5.22 Suppose you have selected a random sample of $n = 7$ measurements from a normal distribution. Compare the standard normal z values with the corresponding t values if you were forming the following confidence intervals.
   a. 80% confidence interval   1.440
   b. 90% confidence interval   1.943
   c. 95% confidence interval   2.447
   d. 98% confidence interval   3.143
   e. 99% confidence interval   3.707
   f. Use the table values you obtained in parts **a–e** to sketch the z- and t-distributions. What are the similarities and differences?

5.23 Let t_0 be a specific value of t. Use Table IV in Appendix A to find t_0 values such that the following statements are true.
   a. $P(t \geq t_0) = .025$ where df $= 10$ 2.228
   b. $P(t \geq t_0) = .01$ where df $= 17$ 2.576
   c. $P(t \leq t_0) = .005$ where df $= 6$ -3.707
   d. $P(t \leq t_0) = .05$ where df $= 13$ -1.771

5.24 Let t_0 be a particular value of t. Use Table IV of Appendix A to find t_0 values such that the following statements are true.
   a. $P(-t_0 < t < t_0) = .95$ where df $= 16$ 2.120
   b. $P(t \leq -t_0 \text{ or } t \geq t_0) = .05$ where df $= 16$ 2.120
   c. $P(t \leq t_0) = .05$ where df $= 16$ -1.746
   d. $P(t \leq -t_0 \text{ or } t \geq t_0) = .10$ where df $= 12$ 1.782
   e. $P(t \leq -t_0 \text{ or } t \geq t_0) = .01$ where df $= 8$ 3.355

5.25 The following random sample was selected from a normal distribution: 4, 6, 3, 5, 9, 3.
   a. Construct a 90% confidence interval for the population mean μ. 5 ± 1.876
   b. Construct a 95% confidence interval for the population mean μ. 5 ± 2.394
   c. Construct a 99% confidence interval for the population mean μ. 5 ± 3.754
   d. Assume that the sample mean $\bar{x}$ and sample standard deviation s remain exactly the same as those you just calculated but that they are based on a sample of $n = 25$ observations rather than $n = 6$ observations. Repeat parts **a–c**. What is the effect of increasing the sample size on the width of the confidence intervals?

5.26 The following sample of 16 measurements was selected from a population that is approximately normally distributed:

91	80	99	110	95	106	78	121	106	100
97	82	100	83	115	104				

   a. Construct an 80% confidence interval for the population mean. 97.94 ± 4.24
   b. Construct a 95% confidence interval for the population mean and compare the width of this interval with that of part **a**. 97.94 ± 6.737
   c. Carefully interpret each of the confidence intervals and explain why the 80% confidence interval is narrower.

Applying the Concepts—Basic

5.27 Periodically, the Hillsborough County (Florida) Water Department tests the drinking water of homeowners for contaminants such as lead and copper. The lead and copper levels in water specimens collected in 1998 for a sample of 10 residents of the Crystal Lakes Manors subdivision are shown below.

⬤ LEADCOPP

Lead (µg/L)	Copper (mg/L)
1.32	.508
0	.279
13.1	.320
.919	.904
.657	.221
3.0	.283
1.32	.475
4.09	.130
4.45	.220
0	.743

Source: Hillsborough County Water Department Environmental Laboratory, Tampa, Florida.

   a. Construct a 99% confidence interval for the mean lead level in water specimens from Crystal Lake Manors. 2.8856 ± 4.034
   b. Construct a 99% confidence interval for the mean copper level in water specimens from Crystal Lake Manors. $.4083 \pm .2564$
   c. Interpret the intervals, parts **a** and **b**, in the words of the problem.
   d. Discuss the meaning of the phrase, "99% confident."

5.28 A group of Harvard University School of Public Health researchers studied the impact of cooking on the size of indoor air particles (*Environmental Science & Technology*, Sept. 1, 2000). The decay rate (measured as μm/hour) for fine particles produced from oven cooking or toasting was recorded on six randomly selected days. These six measurements are

⬤ DECAY

.95	.83	1.20	.89	1.45	1.12

Source: Abt, E., *et al.* "Relative contribution of outdoor and indoor particle sources to indoor concentrations." *Environmental Science & Technology,* Vol. 34, No. 17, Sept. 1, 2000 (Table 3).

a. Find and interpret a 95% confidence interval for the true average decay rate of fine particles produced from oven cooking or toasting. $1.073 \pm .243$

b. Explain what the phrase "95% confident" implies in the interpretation of part **a**.

c. What must be true about the distribution of the population of decay rates for the inference to be valid?

5.29 The *Australian Journal of Zoology* (Vol. 43, 1995) reported on a study of the diets and water requirements of spinifex pigeons. Sixteen pigeons were captured in the desert and the crop (i.e., stomach) contents of each examined. The accompanying table reports the weight (in grams) of dry seed in the crop of each pigeon. Use the accompanying SAS printout to find a 99% confidence interval for the average weight of dry seeds in the crops of spinifex pigeons inhabiting the Western Australian desert. Interpret the result.   (.61, 2.13)

⬤ PIGEONS

.457	3.751	.238	2.967	2.509	1.384	1.454	.818
.335	1.436	1.603	1.309	.201	.530	2.144	.834

Source: Excerpted from Williams, J.B., Bradshaw, D., and Schmidt, L. "Field metabolism and water requirements of spinifex pigeons *(Geophaps plumifera)* in Western Australia." *Australian Journal of Zoology,* Vol. 43, No. 1, 1995, p. 7 (Table 2).

SAS Output for Exercise 5.29

Sample Statistics for SEEDWT

N	Mean	Std. Dev.	Std. Error
16	1.37	1.03	0.26

99 % Confidence Interval for the Mean

Lower Limit: 0.61
Upper Limit: 2.13

Applying the Concepts—Intermediate

5.30 Patients with normal hearing in one ear and unaided sensorineural hearing loss in the other are characterized as suffering from unilateral hearing loss. In a study reported in the *American Journal of Audiology* (Mar. 1995), eight patients with unilateral hearing loss were fitted with a special wireless hearing aid in the "bad" ear. The absolute sound pressure level (SPL) was then measured near the eardrum of the ear when noise was produced at a frequency of 500 hertz. The SPLs of the eight patients, recorded in decibels, are listed below.

⬤ SOUND

73.0	80.1	82.8	76.8	73.5	74.3	76.0	68.1

Construct and interpret a 90% confidence interval for the true mean SPL of unilateral hearing loss patients when noise is produced at a frequency of 500 hertz.

5.31 It is customary practice in the United States to base roadway design on the 30th highest hourly volume in a year. Thus, all roadway facilities are expected to operate at acceptable levels of service for all but 29 hours of the year. The Florida Department of Transportation (DOT), however, has shifted from the 30th highest hour to the 100th highest hour as the basis for level-of-service determinators. Florida Atlantic University researcher Reid Ewing investigated whether this shift was warranted in the *Journal of STAR Research* (July 1994). The accompanying table gives the traffic counts at the 30th highest hour and the 100th highest hour of a recent year for 20 randomly selected DOT permanent count stations. MINITAB stem-and-leaf plots for the two variables as well as summary statistics and 95% confidence interval printouts are provided on p. 257.

⬤ TRAFFIC

Station	Type of Route	30th Highest Hour	100th Highest Hour
0117	small city	1,890	1,736
0087	recreational	2,217	2,069
0166	small city	1,444	1,345
0013	rural	2,105	2,049
0161	urban	4,905	4,815
0096	urban	2,022	1,958
0145	rural	594	548
0149	rural	252	229
0038	urban	2,162	2,048
0118	rural	1,938	1,748
0047	rural	879	811
0066	urban	1,913	1,772
0094	rural	3,494	3,403
0105	small city	1,424	1,309
0113	small city	4,571	4,425
0151	urban	3,494	3,359
0159	rural	2,222	2,137
0160	small city	1,076	989
0164	recreational	2,167	2,039
0165	recreational	3,350	3,123

Source: Ewing, R. "Roadway levels of service in an era of growth management." *Journal of STAR Research.* Vol. 3, July 1994, p. 103 (Table 2).

MINITAB Output for Exercise 5.31

```
Stem-and-leaf of HOUR30    N  = 20
Leaf Unit = 100

    1    0  2
    3    0  58
    6    1  044
    9    1  899
   (6)   2  011122
    5    2
    5    3  344
    2    3
    2    4
    2    4  59


Stem-and-leaf of HOUR100   N  = 20
Leaf Unit = 100

    1    0  2
    4    0  589
    6    1  33
   10    1  7779
   10    2  00001
    5    2
    5    3  134
    2    3
    2    4  4
    1    4  8


Variable           N     Mean     StDev   SE Mean        95.0% CI
HOUR30            20     2206      1224       274   (  1633,    2779)
HOUR100           20     2096      1203       269   (  1533,    2659)
```

a. Describe the population from which the sample data is selected.
b. Does the sample appear to be representative of the population? Explain.
c. Locate and interpret the 95% confidence interval for the mean traffic count at the 30th highest hour.
d. What assumption is necessary for the confidence interval to be valid? Does it appear to be satisfied? Explain.
e. Repeat parts **c** and **d** for the 100th highest hour.

5.32 In Exercise 2.25 (p. 39) you learned that postmortem interval (PMI) is the elapsed time between death and the performance of an autopsy on the cadaver. *Brain and Language* (June 1995) reported on the PMIs of 22 randomly selected human brain specimens obtained at autopsy. The data are reproduced in the table above.
a. Construct a 95% confidence interval for the true mean PMI of human brain specimens obtained at autopsy.

🔵 **BRAINPMI**

5.5	14.5	6.0	5.5	5.3	5.8	11.0	6.4
7.0	14.5	10.4	4.6	4.3	7.2	10.5	6.5
3.3	7.0	4.1	6.2	10.4	4.9		

Source: Hayes, T. L., and Lewis, D. A. "Anatomical specialization of the anterior motor speech area: Hemispheric differences in magnopyramidal neurons." *Brain and Language.* Vol. 49, No. 3, June 1995, p. 292 (Table 1).

b. Interpret the interval, part **a**.
c. What assumption is required for the interval, part **a**, to be valid? Is this assumption satisfied? Explain.
d. What is meant by the phrase "95% confidence"?

5.33 The "fear of negative evaluation" (FNE) scores for 11 bulimic female students and 14 normal female students, first presented in Exercise 2.27 (p. 39) are reproduced below. (Recall that the higher the score, the greater the fear of negative evaluation.)

🔵 **BULIMIA**

Bulimic students	21	13	10	20	25	19	16	21	24	13	14	
Normal students	13	6	16	13	8	19	23	18	11	19	7	10	15	20

Source: Randles, R. H. "On neutral responses (zeros) in the sign test and ties in the Wilcoxon-Mann-Whitney Test." *The American Statistician*, Vol. 55, No. 2, May 2001 (Figure 3).

a. Construct a 95% confidence interval for the mean FNE score of the population of bulimic female students. Interpret the result. 17.82 ± 3.31

b. Construct a 95% confidence interval for the mean FNE score of the population of normal female students. Interpret the result. 14.14 ± 3.05

c. What assumptions are required for the intervals of parts **a** and **b** to be statistically valid? Are these assumptions reasonably satisfied? Explain.

Applying the Concepts—Advanced

5.34 Scientists have discovered increased levels of the hormone adrenocorticotropin in people just before they awake from sleeping (*Nature*, Jan. 7, 1999). In the study, 15 subjects were monitored during their sleep after being told they would be woken at a particular time. One hour prior to the designated wake-up time, the adrenocorticotropin level (pg/ml) was measured in each, with the following results:

$$\bar{x} = 37.3 \qquad s = 13.9$$

a. Estimate the true mean adrenocorticotropin level of sleepers one hour prior to waking using a 95% confidence interval. 37.3 ± 7.70

b. Interpret the interval, part **a**, in the words of the problem.

c. The researchers also found that if the subjects were woken three hours earlier than they anticipated, the average adrenocorticotropin level was 25.5 pg/ml. Assume that $\mu = 25.5$ for all sleepers who are woken three hours earlier than expected. Use the interval, part **a**, to make an inference about the mean adrenocorticotropin level of sleepers under two conditions: one hour before anticipated wake-up time and three hours before anticipated wake-up time.

5.3   LARGE-SAMPLE CONFIDENCE INTERVAL FOR A POPULATION PROPORTION

The number of public opinion polls has grown at an astounding rate in recent years. Almost daily, the news media report the results of some poll. Pollsters regularly determine the percentage of people in favor of the president's welfare-reform program, the fraction of voters in favor of a certain candidate, the fraction of customers who favor a particular brand of wine, and the proportion of people who smoke cigarettes. In each case, we are interested in estimating the percentage (or proportion) of some group with a certain characteristic. In this section we consider methods for making inferences about population proportions when the sample is large.

EXAMPLE 5.3

$\hat{p} = .637$

Public opinion polls are conducted regularly to estimate the fraction of U.S. citizens who trust the president. Suppose 1,000 people are randomly chosen and 637 answer that they trust the president. How would you estimate the true fraction of *all* U.S. citizens who trust the president?

Solution

What we have really asked is how you would estimate the probability p of success in a binomial experiment, where p is the probability that a person chosen trusts the president. One logical method of estimating p for the population is to use the proportion of successes in the sample. That is, we can estimate p by calculating

$$\hat{p} = \frac{\text{Number of people sampled who trust the president}}{\text{Number of people sampled}}$$

where $\hat{p}$ is read "p hat." Thus, in this case,

$$\hat{p} = \frac{637}{1,000} = .637$$

To determine the reliability of the estimator $\hat{p}$, we need to know its sampling distribution. That is, if we were to draw samples of 1,000 people over and over again, each time calculating a new estimate $\hat{p}$, what would be the frequency distribution of all the $\hat{p}$ values? The answer lies in viewing $\hat{p}$ as the average, or mean, number of successes per trial over the n trials. If each success is assigned a value equal to 1 and a failure is assigned a value of 0, then the sum of all n sample observations is x, the total number of successes, and $\hat{p} = x/n$ is the average, or mean, number of successes per trial in the n trials. The Central Limit Theorem tells us that the relative frequency distribution of the sample mean for any population is approximately normal for sufficiently large samples.

The repeated sampling distribution of $\hat{p}$ has the characteristics listed in the next box and shown in Figure 5.12.

FIGURE 5.12

Sampling Distribution of $\hat{p}$

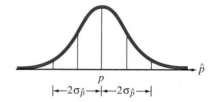

Sampling Distribution of $\hat{p}$

1. The mean of the sampling distribution of $\hat{p}$ is p; that is, $\hat{p}$ is an unbiased estimator of p.
2. The standard deviation of the sampling distribution of $\hat{p}$ is $\sqrt{pq/n}$; that is, $\sigma_p = \sqrt{pq/n}$, where $q = 1 - p$.
3. For large samples, the sampling distribution of $\hat{p}$ is approximately normal. A sample size is considered large if the interval $\hat{p} \pm 3\sigma_{\hat{p}}$ does not include 0 or 1. [*Note:* This requirement is almost equivalent to that given in Section 4.7 for approximating a binomial distribution with a normal one. The difference is that we assumed p to be known in Section 4.7; now we are trying to make inferences about an unknown p, so we use $\hat{p}$ to estimate p in checking the adequacy of the normal approximation.]

The fact that $\hat{p}$ is a "sample mean fraction of successes" allows us to form confidence intervals about p in a manner that is completely analogous to that used for large-sample estimation of μ.

Large-Sample Confidence Interval for p

$$\hat{p} \pm z_{\alpha/2}\sigma_{\hat{p}} = \hat{p} \pm z_{\alpha/2}\sqrt{\frac{pq}{n}} \approx \hat{p} \pm z_{\alpha/2}\sqrt{\frac{\hat{p}\hat{q}}{n}}$$

where $\hat{p} = \dfrac{x}{n}$ and $\hat{q} = 1 - \hat{p}$

Note: When n is large, $\hat{p}$ can approximate the value of p in the formula for $\sigma_{\hat{p}}$.

Thus, if 637 of 1,000 U.S. citizens say they trust the president, a 95% confidence interval for the proportion of *all* U.S. citizens who trust the president is

$$\hat{p} \pm z_{\alpha/2}\sigma_{\hat{p}} = .637 \pm 1.96\sqrt{\frac{pq}{1,000}}$$

where $q = 1 - p$. Just as we needed an approximation for σ in calculating a large-sample confidence interval for μ, we now need an approximation for p. As Table 5.5 shows, the approximation for p does not have to be especially accurate, because the value of $\sqrt{pq}$ needed for the confidence interval is relatively insensitive to changes in p. Therefore, we can use $\hat{p}$ to approximate p. Keeping in mind that $\hat{q} = 1 - \hat{p}$, we substitute these values into the formula for the confidence interval:

$$\hat{p} \pm 1.96\sqrt{pq/1,000} \approx \hat{p} \pm 1.96\sqrt{\hat{p}\hat{q}/1,000}$$
$$= .637 \pm 1.96\sqrt{(.637)(.363)/1,000} = .637 \pm .030$$
$$= (.607, .667)$$

Then we can be 95% confident that the interval from 60.7% to 66.7% contains the true percentage of *all* U.S. citizens who trust the president. That is, in repeated construction of confidence intervals, approximately 95% of all samples would produce confidence intervals that enclose p. Note that the guidelines for interpreting a confidence interval about μ also apply to interpreting a confidence interval for p because p is the "population mean fraction of successes" in a binomial experiment.

TABLE 5.5   Values of pq for Several Different Values of p

p	pq	$\sqrt{pq}$
.5	.25	.50
.6 or .4	.24	.49
.7 or .3	.21	.46
.8 or .2	.16	.40
.9 or .1	.09	.30

EXAMPLE 5.4

.531 ± .037

Many public polling agencies conduct surveys to determine the current consumer sentiment concerning the state of the economy. For example, the Bureau of Economic and Business Research (BEBR) at the University of Florida conducts quarterly surveys to gauge consumer sentiment in the Sunshine State. Suppose that BEBR randomly samples 484 consumers and finds that 257 are optimistic about the state of the economy. Use a 90% confidence interval to estimate the proportion of all consumers in Florida who are optimistic about the state of the economy. Based on the confidence interval, can BEBR infer that the majority of Florida consumers are optimistic about the economy?

Solution

The number, x, of the 484 sampled consumers who are optimistic about the Florida economy is a binomial random variable if we can assume that the sample was randomly selected from the population of Florida consumers and that the poll was conducted identically for each sampled consumer.

The point estimate of the proportion of Florida consumers who are optimistic about the economy is

$$\hat{p} = \frac{x}{n} = \frac{257}{484} = .531$$

We first check to be sure that the sample size is sufficiently large that the normal distribution provides a reasonable approximation for the sampling distribution of $\hat{p}$. We check the 3-standard-deviation interval around $\hat{p}$:

$$\hat{p} \pm 3\sigma_{\hat{p}} \approx \hat{p} \pm 3\sqrt{\frac{\hat{p}\hat{q}}{n}}$$

$$= .531 \pm 3\sqrt{\frac{(.531)(.469)}{484}} = .531 \pm .068 = (.463, .599)$$

Since this interval is wholly contained in the interval $(0, 1)$, we may conclude that the normal approximation is reasonable.

We now proceed to form the 90% confidence interval for p, the true proportion of Florida consumers who are optimistic about the state of the economy:

$$\hat{p} \pm z_{\alpha/2}\sigma_{\hat{p}} = \hat{p} \pm z_{\alpha/2}\sqrt{\frac{pq}{n}} \approx \hat{p} \pm z_{\alpha/2}\sqrt{\frac{\hat{p}\hat{q}}{n}}$$

$$= .531 \pm 1.645\sqrt{\frac{(.531)(.469)}{484}} = .531 \pm .037 = (.494, .568)$$

TEACHING TIP
Point out the similarities of these interpretations with those used when estimating a population mean.

Thus, we can be 90% confident that the proportion of all Florida consumers who are confident about the economy is between .494 and .568. As always, our confidence stems from the fact that 90% of all similarly formed intervals will contain the true proportion p and not from any knowledge about whether this particular interval does.

Can we conclude that the majority of Florida consumers are optimistic about the economy based on this interval? If we wished to use this interval to infer that a majority is optimistic, the interval would have to support the inference that p exceeds .5—that is, that more than 50% of the Florida consumers are optimistic about the economy. Note that the interval contains some values below .5 (as low as .494) as well as some above .5 (as high as .568). Therefore, we cannot conclude that the true value of p exceeds .5 based on this 90% confidence interval. ∎

Caution: Unless n is extremely large, the large-sample procedure presented in this section performs poorly when p is near 0 or 1. For example, suppose you want to estimate the proportion of people who die from a bee sting. This proportion is likely to be near 0 (say, $p \approx .001$). Confidence intervals for p based on a sample of size $n = 50$ will probably be misleading.

To overcome this potential problem, an *extremely* large sample size is required. Since the value of n required to satisfy "extremely large" is difficult to determine, statisticians (see Agresti & Coull, 1998) have proposed an alternative method, based on the Wilson (1927) point estimator of p. The procedure is outlined in the box. Researchers have shown that this confidence interval works well for any p even when the sample size n is very small.

Adjusted $(1 - \alpha)$ 100% Confidence Interval for a Population Proportion, p

$$p^* \pm z_{\alpha/2}\sqrt{\frac{p^*(1 - p^*)}{n + 4}}$$

where $p^* = \dfrac{x + 2}{n + 4}$ is the adjusted sample proportion of observations with the characteristic of interest, x is the number of successes in the sample, and n is the sample size.

E X A M P L E 5 . 5

.025 ± .021

According to *True Odds: How Risk Affects Your Everyday Life* (Walsh, 1997), the probability of being the victim of a violent crime is less than .01. Suppose that in a random sample of 200 Americans, 3 were victims of a violent crime. Estimate the true proportion of Americans who were victims of a violent crime using a 95% confidence interval.

Solution

Let p represent the true proportion of Americans who were victims of a violent crime. Since p is near 0, an "extremely large" sample is required to estimate its value using the usual large-sample method. Since we are unsure whether the sample size of 200 is large enough, we will apply Wilson's adjustment outlined in the box.

The number of "successes" (i.e., number of violent crime victims) in the sample is $x = 3$. Therefore, the adjusted sample proportion is

$$p^* = \frac{x + 2}{n + 4} = \frac{3 + 2}{200 + 4} = \frac{5}{204} = .025$$

Suggested Exercise 5.45

Note that this adjusted sample proportion is obtained by adding a total of four observations—two "successes" and two "failures"—to the sample data. Substituting $p^* = .025$ into the equation for a 95% confidence interval, we obtain

$$p^* \pm 1.96\sqrt{\frac{p^*(1 - p^*)}{n + 4}} = .025 \pm 1.96\sqrt{\frac{(.025)(.975)}{204}}$$
$$= .025 \pm .021$$

or (.004, .046). Consequently, we are 95% confident that the true proportion of Americans who are victims of a violent crime falls between .004 and .046. ∎

 ## Using the TI-83 Graphing Calculator

Confidence Intervals for Proportions

You can find confidence intervals for proportions using the **1-PropZInt** command in the **STAT TESTS** Menu.

Step 1   *Select the appropriate type of confidence interval*
Press **STAT** and highlight **TESTS**
Use the arrow keys to highlight **A:1-PropZInt**
Press **ENTER**

Step 2   *Determine your values for x, n, and confidence level*
Enter these values for ***x, n,*** and **C-Level**
where   ***x*** = number of successes
n = sample size
C-Level = level of confidence

Step 3   *Calculate the confidence interval*
Highlight **Calculate**
Press **ENTER**

USING THE TI-83 GRAPHING CALCULATOR (*continued***)**

Example   Suppose that 1100 U.S. citizens are randomly chosen and 532 answer that they favor a flat income tax rate. Use a 95% confidence interval to estimate the true proportion of citizens who favor a flat income tax rate.
In this example, $x = 532, n = 1100,$ and C-Level $= .95$
The TI-83 screens for this example are shown below.

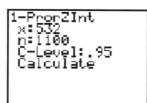   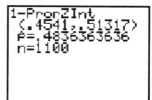

Thus, we can be 95% confident that the interval from 45.4% to 51.3% contains the true percentage of all U.S. citizens who favor a flat income tax rate.

 EXERCISES 5.35–5.50

Learning the Mechanics

5.35 Describe the sampling distribution of $\hat{p}$ based on large samples of size n. That is, give the mean, the standard deviation, and the (approximate) shape of the distribution of $\hat{p}$ when large samples of size n are (repeatedly) selected from the binomial distribution with probability of success p.

5.36 Explain the meaning of the phrase "$\hat{p}$ is an unbiased estimator of p."

5.37 A random sample of size $n = 196$ yielded $\hat{p} = .64$.
  a. Is the sample size large enough to use the methods of this section to construct a confidence interval for p? Explain.   Yes
  b. Construct a 95% confidence interval for p.
  c. Interpret the 95% confidence interval.
  d. Explain what is meant by the phrase "95% confidence interval."

5.38 A random sample of size $n = 144$ yielded $\hat{p} = .76$.
  a. Is the sample size large enough to use the methods of this section to construct a confidence interval for p? Explain.   Yes
  b. Construct a 90% confidence interval for p. $.76 \pm .059$
  c. What assumption is necessary to ensure the validity of this confidence interval?

5.39 For the binomial sample information summarized in each part, indicate whether the sample size is large enough to use the methods of this chapter to construct a confidence interval for p.
  a. $n = 500, \hat{p} = .05$ Yes

  b. $n = 100, \hat{p} = .05$ No
  c. $n = 10, \hat{p} = .5$ Yes
  d. $n = 10, \hat{p} = .3$ No

5.40 A random sample of 50 consumers taste-tested a new snack food. Their responses were coded (0: do not like; 1: like; 2: indifferent) and recorded as follows:

⊙ SNACK

1	0	0	1	2	0	1	1	0	0
0	1	0	2	0	2	2	0	0	1
1	0	0	0	0	1	0	2	0	0
0	1	0	0	1	0	0	1	0	1
0	2	0	0	1	1	0	0	0	1

  a. Use an 80% confidence interval to estimate the proportion of consumers who like the snack food.
  b. Provide a statistical interpretation for the confidence interval you constructed in part **a**.

5.41 A poll is taken in which 278 of 500 randomly sampled voters indicate their preference for a certain candidate. Use a 95% confidence interval to estimate the true fraction, p, of voters who prefer the candidate. $.556 \pm .044$

Applying the Concepts—Basic

5.42 According to a University of Michigan study, many adults have experienced lingering "fright" effects from a scary movie or TV show they saw as a teenager (*Tampa Tribune*, Mar. 10, 1999). In a survey of 150 college students, 39 said they still experience "residual anxiety" from a scary TV show or movie.

a. Give a point estimate, $\hat{p}$, for the true proportion of college students who experience "residual anxiety" from a scary TV show or movie.

b. Find a 95% confidence interval for p.

c. Interpret the interval, part **b**.

5.43 Refer to the *Chance* (Fall, 2000) study of 837 pottery pieces found at the ancient Greek settlement at Phylakopi, Exercise 2.8 (p. 26). Of these, 183 pieces were painted with either a curvilinear, geometric, or naturalistic decoration. Find a 90% confidence interval for the population proportion of all pottery artifacts at Phylakopi that are painted. Interpret the resulting interval.

5.44 The *American Journal of Orthopsychiatry* (July 1995) published an article on the prevalence of homelessness in the United States. A sample of 487 U.S. adults were asked to respond to the question: "Was there ever a time in your life when you did not have a place to live?" A 95% confidence interval for the true proportion who were/are homeless, based on the survey data, was determined to be (.045, .091). Fully interpret the result.

Applying the Concepts—Intermediate

5.45 Researchers at the University of Florida's Department of Exercise and Sport Sciences conducted a study of variety in exercise workouts (*Journal of Sport Behavior*, 2001). A sample of 114 men and women were randomly divided into three groups, 38 people per group. Group 1 members varied their exercise routine in workouts, group 2 members performed the same exercise at each workout, and group 3 members had no set schedule or regulations for their workouts.

a. By the end of the study, 14 people had dropped out of the first exercise group. Estimate the dropout rate for exercisers who vary their routine in workouts. Use a 90% confidence interval and interpret the result.

b. By the end of the study, 23 people had dropped out of the third exercise group. Estimate the dropout rate for exercisers who have no set schedule for their workouts. Use a 90% confidence interval and interpret the result.   .605 ± .130

5.46 The Gallup Organization surveyed 1,252 debit card-holders in the U.S. and found that 180 had used the debit card to purchase a product or service on the Internet (*Card Fax*, November 12, 1999).

a. Describe the population of interest to the Gallup Organization.

b. Is the sample size large enough to construct a valid confidence interval for the proportion of debit card-holders who have used their card in making purchases over the Internet?   Yes

c. Estimate the proportion referred to in part **b** using a 99% confidence interval. Interpret your result in the context of the problem.   .144 ± .026

d. If you had constructed a 90% confidence interval instead, would it be wider or narrower?   Narrower

5.47 By law, all new cars must be equipped with both driver-side and passenger-side safety air bags. There is concern, however, over whether air bags pose a danger for children sitting on the passenger side. In a National Highway Traffic Safety Administration (NHTSA) study of 55 people killed by the explosive force of air bags, 35 were children seated on the front-passenger side (*Wall Street Journal*, Jan. 22, 1997). This study led some car owners with children to disconnect the passenger-side air bag. Consider all fatal automobile accidents in which it is determined that air bags were the cause of death. Let p represent the true proportion of these accidents involving children seated on the front-passenger side.

a. Use the data from the NHTSA study to estimate p.

b. Construct a 99% confidence interval for p.

c. Interpret the interval, part **b**, in the words of the problem.

d. NHTSA investigators determined that 24 of 35 children killed by the air bags were not wearing seat belts or were improperly restrained. How does this information impact your assessment of the risk of an air bag fatality?

⬤ OILSPILL

5.48 Refer to the *Marine Technology* (Jan. 1995) study of the causes of fifty recent major oil spills from tankers and carriers, Exercise 2.15 (p. 29). The data are stored in the OILSPILL file.

a. Give a point estimate for the proportion of major oil spills that are caused by hull failure.

b. Form a 95% confidence interval for the estimate, part **a**. Interpret the result.   .24 ± .118

5.49 In Exercise 2.13 (p. 28) you learned about the signature whistles of bottlenose dolphins. Marine scientists categorize signature whistles by type—type a, type b, type c, etc. For one study of a sample of 185 whistles emitted from bottlenose dolphins in captivity, 97 were categorized as type a whistles (*Ethology*, July 1995).

a. Estimate the true proportion of bottlenose dolphin signature whistles that are type a whistles. Use a 99% confidence interval.   .524 ± .095

b. Interpret the interval, part **a**.

5.50 Refer to the Federal Trade Commission (FTC) 1998 "Price Check" study of electronic checkout scanners, Exercise 4.37 (p. 188). The FTC inspected 1,669 scanners at retail stores and supermarkets by scanning a sample of items at each store and determining if the scanned price was accurate. The FTC gives a store a "passing" grade if 98% or more of the items are priced accurately. Of the 1,669 stores in the study, 1,185 passed inspection.

a. Find a 90% confidence interval for the true proportion of retail stores and supermarkets with electronic scanners that pass the FTC price-check test. Interpret the result.   .710 ± .018

b. In 1996, the FTC found that 45% of the stores passed inspection. Use the interval, part **a**, to determine whether the proportion of stores that pass inspection in 1998 exceeds .45.

5.4   DETERMINING THE SAMPLE SIZE

Recall (Section 1.5) that one way to collect the relevant data for a study used to make inferences about the population is to implement a designed (planned) experiment. Perhaps the most important design decision faced by the analyst is to determine the size of the sample. We show in this section that the appropriate sample size for making an inference about a population mean or proportion depends on the desired reliability.

Estimating a Population Mean

Consider the example from Section 5.1 in which we estimated the mean length of stay for patients in a large hospital. A sample of 100 patients' records produced the 95% confidence interval $\bar{x} \pm 2\sigma_{\bar{x}} \approx 4.53 \pm .74$. Consequently, our estimate $\bar{x}$ was within .74 day of the true mean length of stay μ for all the hospital's patients at the 95% confidence level. That is, the 95% confidence interval for μ was $2(.74) = 1.48$ days wide when 100 accounts were sampled. This is illustrated in Figure 5.13a.

Now suppose we want to estimate μ to within .25 day with 95% confidence. That is, we want to narrow the width of the confidence interval from 1.48 days to .50 day, as shown in Figure 5.13b. How much will the sample size have to be increased to accomplish this? If we want the estimator $\bar{x}$ to be within .25 day of μ, we must have

$$2\sigma_{\bar{x}} = .25 \qquad \text{or, equivalently,} \qquad 2\left(\frac{\sigma}{\sqrt{n}}\right) = .25$$

The necessary sample size is obtained by solving this equation for n. To do this we need an approximation for σ. We have an approximation from the initial sample of 100 patients' records—namely, the sample standard deviation, $s = 3.68$. Thus,

$$2\left(\frac{\sigma}{\sqrt{n}}\right) \approx 2\left(\frac{s}{\sqrt{n}}\right) = 2\left(\frac{3.68}{\sqrt{n}}\right) = .25$$

$$\sqrt{n} = \frac{2(3.68)}{.25} = 29.44$$

$$n = (29.44)^2 = 866.71 \approx 867$$

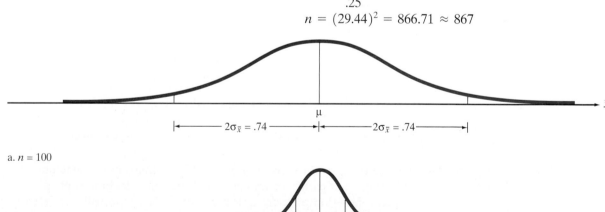

a. $n = 100$

b. $n = 867$

FIGURE 5.13

Relationship Between Sample Size and Width of Confidence Interval: Hospital Stay Example

Approximately 867 patients' records will have to be sampled to estimate the mean length of stay μ to within .25 day with (approximately) 95% confidence. The confidence interval resulting from a sample of this size will be approximately .50 day wide (see Figure 5.13b).

In general, we express the reliability associated with a confidence interval for the population mean μ by specifying the **bound, B,** within which we want to estimate μ with $100(1 - \alpha)$% confidence. The bound B then is equal to the half-width of the confidence interval, as shown in Figure 5.14.

The procedure for finding the sample size necessary to estimate μ to within a given bound B is given in the box.

FIGURE 5.14

Specifying the Bound B as the Half-Width of a Confidence Interval

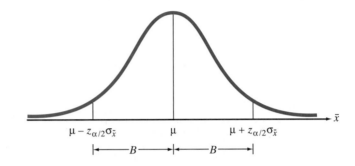

Sample Size Determination for $100(1 - \alpha)$% Confidence Intervals for μ

In order to estimate μ to within a bound B with $100(1 - \alpha)$% confidence, the required sample size is found as follows:

$$z_{\alpha/2}\left(\frac{\sigma}{\sqrt{n}}\right) = B$$

The solution can be written in terms of B as follows:

$$n = \frac{(z_{\alpha/2})^2\sigma^2}{B^2}$$

The value of σ is usually unknown. It can be estimated by the standard deviation s from a prior sample. Alternatively, we may approximate the range R of observations in the population, and (conservatively) estimate $\sigma \approx R/4$. In any case, you should round the value of n obtained *upward* to ensure that the sample size will be sufficient to achieve the specified reliability.

EXAMPLE 5.6

$n = 107$

Suppose the manufacturer of official NFL footballs uses a machine to inflate the new balls to a pressure of 13.5 pounds. When the machine is properly calibrated, the mean inflation pressure is 13.5 pounds, but uncontrollable factors cause the pressures of individual footballs to vary randomly from about 13.3 to 13.7 pounds. For quality control purposes, the manufacturer wishes to estimate the mean inflation pressure to within .025 pound of its true value with a 99% confidence interval. What sample size should be specified for the experiment?

Solution

We desire a 99% confidence interval that estimates μ to within $B = .025$ pound of its true value. For a 99% confidence interval, we have $z_{\alpha/2} = z_{.005} = 2.575$. To estimate

σ, we note that the range of observations is $R = 13.7 - 13.3 = .4$ and use $\sigma \approx R/4 = .1$. Now we use the formula derived in the box to find the sample size n:

$$n = \frac{(z_{\alpha/2})^2 \sigma^2}{B^2} \approx \frac{(2.575)^2(.1)^2}{(.025)^2} = 106.09$$

We round this up to $n = 107$. Realizing that σ was approximated by $R/4$, we might even advise that the sample size be specified as $n = 110$ to be more certain of attaining the objective of a 99% confidence interval with bound $B = .025$ pound or less. ∎

Sometimes the formula will lead to a solution that indicates a small sample size is sufficient to achieve the confidence interval goal. Unfortunately, the procedures and assumptions for small samples differ from those for large samples, as we discovered in Section 5.2. Therefore, if the formulas yield a small sample size ($n < 30$), one simple strategy is to select a sample size $n = 30$.

Estimating a Population Proportion

The method outlined above is easily applied to a population proportion p. For example, in Section 5.3 a pollster used a sample of 1,000 U.S. citizens to calculate a 95% confidence interval for the proportion who trust the president, obtaining the interval $.637 \pm .03$. Suppose the pollster wishes to estimate more precisely the proportion who trust the president, say, to within .015 with a 95% confidence interval.

The pollster wants a confidence interval with a bound B on the estimate of p of $B = .015$. The sample size required to generate such an interval is found by solving the following equation for n:

$$z_{\alpha/2}\sigma_{\hat{p}} = B \qquad \text{or} \qquad z_{\alpha/2}\sqrt{\frac{pq}{n}} = .015 \qquad \text{(see Figure 5.15)}$$

Since a 95% confidence interval is desired, the appropriate z value is $z_{\alpha/2} = z_{.025} = 1.96$. We must approximate the value of the product pq before we can solve the equation for n. As shown in Table 5.5, the closer the values of p and q to .5, the larger the product pq. Thus, to find a conservatively large sample size that will generate a confidence interval with the specified reliability, we generally choose an approximation of p close to .5. In the case of the proportion of U.S. citizens who trust the president, however, we have an initial sample estimate of $\hat{p} = .637$. A conservatively large estimate of pq can therefore be obtained by using, say, $p = .60$. We now substitute into the equation and solve for n:

$$1.96\sqrt{\frac{(.60)(.40)}{n}} = .015$$

$$n = \frac{(1.96)^2(.60)(.40)}{(.015)^2} = 4,097.7 \approx 4,098$$

FIGURE 5.15

Specifying the Bound B of a Confidence Interval for a Population Proportion p

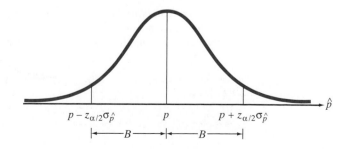

The pollster must sample about 4,098 U.S. citizens to estimate the percentage who trust the president with a confidence interval of width .03.

The procedure for finding the sample size necessary to estimate a population proportion p to within a given bound B is given in the box.

Sample Size Determination for $100(1 - \alpha)\%$ Confidence Interval for p

TEACHING TIP
Choose different values of p and q to show that the sample size is maximized (and considered conservative) when $p = q = .5$.

In order to estimate a binomial probability p to within a bound B with $100(1 - \alpha)\%$ confidence, the required sample size is found by solving the following equation for n:

$$z_{\alpha/2}\sqrt{\frac{pq}{n}} = B$$

The solution can be written in terms of B:

$$n = \frac{(z_{\alpha/2})^2(pq)}{B^2}$$

Since the value of the product pq is unknown, it can be estimated by using the sample fraction of successes, $\hat{p}$, from a prior sample. Remember (Table 5.5) that the value of pq is at its maximum when p equals .5, so that you can obtain conservatively large values of n by approximating p by .5 or values close to .5. In any case, you should round the value of n obtained *upward* to ensure that the sample size will be sufficient to achieve the specified reliability.

EXAMPLE 5.7

$n = 2,436$

Suggested Exercise 5.58

Suppose a large telephone manufacturer that entered the post-regulation market quickly has an initial problem with excessive customer complaints and consequent returns of the phones for repair or replacement. The manufacturer wants to estimate the magnitude of the problem in order to design a quality control program. How many telephones should be sampled and checked in order to estimate the fraction defective, p, to within .01 with 90% confidence?

Solution

In order to estimate p to within a bound of .01, we set the half-width of the confidence interval equal to $B = .01$, as shown in Figure 5.16.

The equation for the sample size n requires an estimate of the product pq. We could most conservatively estimate $pq = .25$ (i.e., use $p = .5$), but this may be overly conservative when estimating a fraction defective. A value of .1, corresponding to 10% defective, will probably be conservatively large for this application. The solution is therefore

FIGURE 5.16

Specified Reliability for Estimate of Fraction Defective in Example 5.7

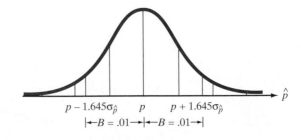

$$n = \frac{(z_{\alpha/2})^2(pq)}{B^2} = \frac{(1.645)^2(.1)(.9)}{(.01)^2} = 2{,}435.4 \approx 2{,}436$$

Thus, the manufacturer should sample 2,436 telephones in order to estimate the fraction defective, p, to within .01 with 90% confidence. Remember that this answer depends on our approximation for pq, where we used .09. If the fraction defective is closer to .05 than .10, we can use a sample of 1,286 telephones (check this) to estimate p to within .01 with 90% confidence. ∎

The cost of sampling will also play an important role in the final determination of the sample size to be selected to estimate either μ or p. Although more complex formulas can be derived to balance the reliability and cost considerations, we will solve for the necessary sample size and note that the sampling budget may be a limiting factor. Consult the references for a more complete treatment of this problem.

STATISTICS IN ACTION
Scallops, Sampling, and the Law

Arnold Bennett, a Sloan School of Management professor at the Massachusetts Institute of Technology (MIT), describes a recent legal case in which he served as a statistical "expert" in *Interfaces* (Mar.–Apr. 1995). The case involved a ship that fishes for scallops off the coast of New England. In order to protect baby scallops from being harvested, the U.S. Fisheries and Wildlife Service requires that "the average meat per scallop weigh at least $\frac{1}{36}$ of a pound." The ship was accused of violating this weight standard. Bennett lays out the scenario:

The vessel arrived at a Massachusetts port with 11,000 bags of scallops, from which the harbormaster randomly selected 18 bags for weighing. From each such bag, his agents took a large scoopful of scallops; then, to estimate the bag's average meat per scallop, they divided the total weight of meat in the scoopful by the number of scallops it contained. Based on the 18 [numbers] thus generated, the harbormaster estimated that each of the ship's scallops possessed an average $\frac{1}{39}$ of a pound of meat (that is, they were about seven percent lighter than the minimum requirement). Viewing this outcome as conclusive evidence that the weight standard had been violated, federal authorities at once confiscated *95 percent* of the catch (which they then sold at auction). The fishing voyage was thus transformed into a financial catastrophe for its participants.

Bennett provided the actual scallop weight measurements for each of the 18 sampled bags in the article. The data (saved in the SCALLOPS file) are listed in Table 5.6.

For ease of exposition, Bennett expressed each number as a multiple of $\frac{1}{36}$ of a pound, the minimum permissible average weight per scallop. Consequently, numbers below one indicate individual bags that do not meet the standard.

The ship's owner filed a lawsuit against the federal government, declaring that his vessel had fully complied with the weight standard. A Boston law firm was hired to represent the owner in legal proceedings and Bennett was retained by the firm to provide statistical litigation support and, if necessary, expert witness testimony.

Focus

a. Recall that the harbormaster sampled only 18 of the ship's 11,000 bags of scallops. One of the questions the lawyers asked Bennett was: "Can a reliable estimate of the mean weight of all the scallops be obtained from a sample of size 18?" Give your opinion on this issue.
b. As stated in the article, the government's decision rule is to confiscate a scallop catch if the sample mean weight of the scallops is less than $\frac{1}{36}$ of a pound. Do you see any flaws in this rule?
c. Develop your own procedure for determining whether a ship is in violation of the minimum weight restriction. Apply your rule to the data in Table 5.6. Draw a conclusion about the ship in question.

TABLE 5.6 Scallop Weight Measurements for 18 Bags Sampled

SCALLOPS

| .93 | .88 | .85 | .91 | .91 | .84 | .90 | .98 | .88 |
| .89 | .98 | .87 | .91 | .92 | .99 | 1.14 | 1.06 | .93 |

Source: Bennett, A. "Misapplications review: Jail terms." *Interfaces*, Vol. 25, No. 2, Mar.–Apr. 1995, p. 20.

TEACHING TIP
Suggestions for class discussion of the above case can be found in the *Instructor's Notes*

EXERCISES 5.51–5.67

Learning the Mechanics

5.51 If you wish to estimate a population mean to within a bound $B = .2$ using a 95% confidence interval and you know from prior sampling that σ^2 is approximately equal to 5.4, how many observations would you have to include in your sample? 519

5.52 If nothing is known about p, .5 can be substituted for p in the sample-size formula for a population proportion. But when this is done, the resulting sample size may be larger than needed. Under what circumstances will using $p = .5$ in the sample-size formula yield a sample size larger than needed to construct a confidence interval for p with a specified bound and a specified confidence level?

5.53 Suppose you wish to estimate a population mean correct to within a bound $B = .15$ with probability equal to .90. You do not know σ^2, but you know that the observations will range in value between 31 and 39.

    a. Find the approximate sample size that will produce the desired accuracy of the estimate. You wish to be conservative to ensure that the sample size will be ample to achieve the desired accuracy of the estimate. [*Hint:* Using your knowledge of data variation from Section 2.5, assume that the range of the observations will equal 4σ.] 482

    b. Calculate the approximate sample size making the less conservative assumption that the range of the observations is equal to 6σ. 214

5.54 In each case, find the approximate sample size required to construct a 95% confidence interval for p that has bound $B = .06$.

    a. Assume p is near .3 225

    b. Assume you have no prior knowledge about p, but you wish to be certain that your sample is large enough to achieve the specified accuracy for the estimate. 267

5.55 The following is a 90% confidence interval for p: (.26, .54). How large was the sample used to construct this interval?

5.56 It costs you $10 to draw a sample of size $n = 1$ and measure the attribute of interest. You have a budget of $1,200.

    a. Do you have sufficient funds to estimate the population mean for the attribute of interest with a 95% confidence interval 4 units in width? Assume $\sigma = 12$. no

    b. If a 90% confidence level were used, would your answer to part **a** change? Explain. yes

5.57 Suppose you wish to estimate the mean of a normal population using a 95% confidence interval, and you know from prior information that $\sigma^2 \approx 1$.

    a. To see the effect of the sample size on the width of the confidence interval, calculate the width of the confidence interval for $n = 16, 25, 49, 100,$ and 400.

    b. Plot the width as a function of sample size n on graph paper. Connect the points by a smooth curve and note how the width decreases as n increases.

Applying the Concepts—Basic

5.58 A gigantic warehouse located in Tampa, Florida, stores approximately 60 million empty aluminum beer and soda cans. Recently, a fire occurred at the warehouse. The smoke from the fire contaminated many of the cans with blackspot, rendering them unusable. A University of South Florida statistician was hired by the insurance company to estimate p, the true proportion of cans in the warehouse that were contaminated by the fire. How many aluminum cans should be randomly sampled to estimate the true proportion to within .02 with 90% confidence? 1,692

5.59 The EPA wants to test a randomly selected sample of n water specimens and estimate μ, the mean daily rate of pollution produced by a mining operation. If the EPA wants a 95% confidence interval estimate with a bound on the error of 1 milligram per liter (mg/L), how many water specimens are required in the sample? Assume prior knowledge indicates that pollution readings in water samples taken during a day are approximately normally distributed with a standard deviation equal to 5 (mg/L). 97

Applying the Concepts—Intermediate

5.60 Refer to Exercise 2.49 (p. 53) and the *Teaching of Psychology* (May 1998) study in which students assisted in the training of zoo animals. A sample of 15 psychology students rated "The Training Game" as a "great" method of understanding the animal's perspective during training on a 7-1 point scale (where 1 = strongly disagree and 7 = strongly agree). The mean response was 5.87 with a standard deviation of 1.51.

    a. Construct a 95% confidence interval for the true mean response of the students. $5.87 \pm .836$

    b. Suppose you want to reduce the width of the 95% confidence interval to half the size obtained in part **a**. How many students are required in the sample to obtain the desired confidence interval width? 51

5.61 Is the bottled water you drink safe? According to an article in *U.S. News & World Report* (April 12, 1999), the Natural Resources Defense Council warns that the bottled water you are drinking may contain more bacteria and other potentially carcinogenic chemicals than allowed by state and federal regulations. Of the more than 1,000 bottles studied, nearly one-third exceeded government levels. Suppose that the Natural Resources Defense Council wants an updated estimate of the population proportion of bottled water that violates at least one government standard. Determine the sample size (number of bottles) needed to estimate this proportion to within ±0.01 with 99% confidence. 14,785

5.62 The chemical benzalkonium chloride (BAC) is an antibacterial agent that is added to some asthma medications to prevent contamination. Researchers at the University of Florida College of Pharmacy have discovered that adding BAC to asthma drugs can cause airway constriction in patients. In a sample of 18 asthmatic patients, each of whom received a heavy dose of BAC, 10 experienced a significant drop in breathing capacity (*Journal of Allergy and Clinical Immunology*, Jan. 2001). Based on this information, a 95% confidence interval for the true percentage of asthmatic patients who experience breathing difficulties after taking BAC is (.326, .785).

  a. Why might the confidence interval lead to an erroneous inference?

  b. How many asthma patients must be included in the study to estimate the true percentage who experience a significant drop in breathing capacity to within 4% using a 95% confidence interval? 593

5.63 Refer to Exercise 5.28 (p. 255). Suppose that we want to estimate the average decay rate of fine particles produced from oven cooking or toasting to within .04 with 95% confidence. How large a sample should be selected? Use the standard deviation obtained in Exercise 5.28. 129

5.64 According to a Food and Drug Administration (FDA) study, a cup of coffee contains an average of 115 milligrams (mg) of caffeine, with the amount per cup ranging from 60 to 180 mg. Suppose you want to repeat the FDA experiment to obtain an estimate of the mean caffeine content in a cup of coffee correct to within 5 mg with 95% confidence. How many cups of coffee would have to be included in your sample?

5.65 The United States Golf Association (USGA) tests all new brands of golf balls to ensure that they meet USGA specifications. One test conducted is intended to measure the average distance traveled when the ball is hit by a machine called "Iron Byron." Suppose the USGA wishes to estimate the mean distance for a new brand to within 1 yard with 90% confidence. Assume that past tests have indicated that the standard deviation of the distances Iron

Byron hits golf balls is approximately 10 yards. How many golf balls should be hit by Iron Byron to achieve the desired accuracy in estimating the mean? 271

Applying the Concepts—Advanced

5.66 Destructive sampling, in which the test to determine whether an item is defective destroys the item, is generally expensive, and the high costs involved often prohibit large sample sizes. For example, suppose the National Highway Traffic Safety Administration (NHTSA) wishes to determine the proportion of new tires that will fail when subjected to hard braking at a speed of 60 miles per hour. NHTSA can obtain the tires for $25 (wholesale price) each. Suppose the budget for the experiment is $10,000, and NHTSA wishes to estimate the percentage that will fail to within .02 with 95% confidence. Assuming that the entire $10,000 can be spent on tires (ignoring other costs) and that the true fraction that will fail is approximately .05, can NHTSA attain its goal while staying within the budget? Explain. $11,425 is needed

5.67 It costs more to produce defective items—since they must be scrapped or reworked—than it does to produce nondefective items. This simple fact suggests that manufacturers should ensure the quality of their products by perfecting their production processes instead of depending on inspection of finished products (Deming, 1986). In order to better understand a particular metal stamping process, a manufacturer wishes to estimate the mean length of items produced by the process during the past 24 hours.

  a. How many parts should be sampled in order to estimate the population mean to within .1 millimeter (mm) with 90% confidence? Previous studies of this machine have indicated that the standard deviation of lengths produced by the stamping operation is about 2 mm.

  b. Time permits the use of a sample size no larger than 100. If a 90% confidence interval for μ is constructed using $n = 100$, will it be wider or narrower than would have been obtained using the sample size determined in part **a**? Explain.

  c. If management requires that μ be estimated to within .1 mm and that a sample size of no more than 100 be used, what is (approximately) the maximum confidence level that could be attained for a confidence interval that meets management's specifications?

QUICK REVIEW

Key Terms

Bound on the error of estimation 266
Confidence coefficient 240
Confidence interval 240
Confidence level 240

Degrees of freedom 248
Interval estimator 240
t statistic 248
Wilson adjustment for estimating p 261

Key Formulas

$\hat{\theta} \pm (z_{\alpha/2})\sigma_{\hat{\theta}}$ Large-sample confidence interval for population parameter θ where $\hat{\theta}$ and $\sigma_{\hat{\theta}}$ are obtained from the table below.

Parameter θ	Estimator $\hat{\theta}$	Standard Error $\sigma_{\hat{\theta}}$	
Mean, μ	$\bar{x}$	$\sigma/\sqrt{n}$	241
Proportion, p	$\hat{p}$	$\sqrt{\dfrac{pq}{n}}$	259

$\bar{x} \pm t_{\alpha/2}\left(\dfrac{s}{\sqrt{n}}\right)$ Small-sample confidence interval for population mean μ 251

$p^* = \dfrac{x + 2}{n + 4}$ Adjusted estimator of p 261

$n = \dfrac{(z_{\alpha/2})^2 \sigma^2}{B^2}$ Determining the sample size n for estimating μ 266

$n = \dfrac{(z_{\alpha/2})^2 (pq)}{B^2}$ Determining the sample size n for estimating p 268

Language Lab

Symbol	Pronunciation	Description
θ	theta	General population parameter
μ	mu	Population mean
p		Population proportion
B		Bound on error of estimation
α	alpha	$(1 - \alpha)$ represents the confidence coefficient
$z_{\alpha/2}$	z of alpha over 2	z value used in a $100(1 - \alpha)\%$ large-sample confidence interval
$t_{\alpha/2}$	t of alpha over 2	t value used in a $100(1 - \alpha)\%$ small-sample confidence interval
$\bar{x}$	x-bar	Sample mean; point estimate of μ
$\hat{p}$	p-hat	Sample proportion; point estimate of p
σ	sigma	Population standard deviation
s		Sample standard deviation; point estimate of σ
$\sigma_{\bar{x}}$	sigma of $\bar{x}$	Standard deviation of sampling distribution of $\bar{x}$
$\sigma_{\hat{p}}$	sigma of $\hat{p}$	Standard deviation of sampling distribution of $\hat{p}$
p^*		Wilson's adjusted estimate of p

 SUPPLEMENTARY EXERCISES 5.68–5.90

Note: List the assumptions necessary for the valid implementation of the statistical procedures you use in solving all these exercises.

Learning the Mechanics

5.68 Let t_0 represent a particular value of t from Table IV of Appendix A. Find the table values such that the following statements are true.
  a. $P(t \le t_0) = .05$ where df = 17 -1.740
  b. $P(t \ge t_0) = .005$ where df = 14 2.977
  c. $P(t \le -t_0 \text{ or } t \ge t_0) = .10$ where df = 6 1.943
  d. $P(t \le -t_0 \text{ or } t \ge t_0) = .01$ where df = 22 2.819

5.69 In each of the following instances, determine whether you would use a z or t statistic (or neither) to form a 95% confidence interval; then look up the appropriate z or t value.
  a. Random sample of size $n = 21$ from a normal distribution with unknown mean μ and standard deviation σ.
  b. Random sample of size $n = 175$ from a normal distribution with unknown mean μ and standard deviation σ 1.96

c. Random sample of size $n = 12$ from a normal distribution with unknown mean μ and standard deviation $\sigma = 5$ 1.96

d. Random sample of size $n = 65$ from a distribution about which nothing is known   1.96

e. Random sample of size $n = 8$ from a distribution about which nothing is known   Neither

5.70 A random sample of 225 measurements is selected from a population, and the sample mean and standard deviation are $\bar{x} = 32.5$ and $s = 30.0$, respectively.

a. Use a 99% confidence interval to estimate the mean of the population, μ. 32.5 ± 5.16

b. How large a sample would be needed to estimate μ to within .5 with 99% confidence?   23,964

c. What is meant by the phrase "99% confidence" as it is used in this exercise?

Applying the Concepts—Basic

5.71 The Centers for Disease Control and Prevention (CDCP) in Atlanta, Georgia, conducts an annual survey of the general health of the U.S. population as part of its Behavioral Risk Factor Surveillance System (*New York Times*, Mar. 29, 1995). Using random-digit dialing, the CDCP telephones U.S. citizens over 18 years of age and asks them the following four questions:

(1) Is your health generally excellent, very good, good, fair, or poor?   p

(2) How many days during the previous 30 days was your physical health not good because of injury or illness? μ

(3) How many days during the previous 30 days was your mental health not good because of stress, depression, or emotional problems? μ

(4) How many days during the previous 30 days did your physical or mental health prevent you from performing your usual activities? μ

Identify the parameter of interest for each question.

5.72 Refer to Exercise 5.71. According to the CDCP, 89,582 of 102,263 adults interviewed stated their health was good, very good, or excellent.

a. Use a 99% confidence interval to estimate the true proportion of U.S. adults who believe their health to be good to excellent. Interpret the interval.

b. Why might the estimate, part **a**, be overly optimistic (i.e., biased high)?

5.73 Refer to the *Journal of the American Medical Association* (Apr. 21, 1993) report on the prevalence of cigarette smoking among U.S. adults, Exercise 5.11 (p. 244). Of the 43,732 survey respondents, 11,239 indicated that they were current smokers and 10,539 indicated they were former smokers.

a. Construct and interpret a 90% confidence interval for the percentage of U.S. adults who currently smoke cigarettes. $.257 \pm .003$

b. Construct and interpret a 90% confidence interval for the percentage of U.S. adults who are former cigarette smokers. $.241 \pm .003$

5.74 Refer to the *Environmental Science & Technology* (Sept. 1, 2000) study on the ammonia levels near the exit ramp of a San Francisco highway tunnel, Exercise 2.45 (p. 52). The ammonia concentration (parts per million) data for eight randomly selected days during afternoon drive-time are reproduced in the table. Find a 99% confidence interval for the population mean daily ammonia level in air in the tunnel. Interpret your result. $1.471 \pm .079$

⊚ AMMONIA

1.53	1.50	1.37	1.51	1.55	1.42	1.41	1.48

5.75 A health researcher wishes to estimate the mean number of cavities per child for children under the age of 12 who live in a specified environment. The number of cavities per child for a random sample of 35 children under the age of 12 has a mean of 2 and a standard deviation of 1.5. Construct a 90% confidence interval for the mean number of cavities per child under the age of 12 who lives in the sampled environment. $2 \pm .47$

Applying the Concepts—Intermediate

5.76 Pulse rate is an important measure of the fitness of a person's cardiovascular system. The mean pulse rate for all U.S. adult males is approximately 72 heart beats per minute. A random sample of 21 U.S. adult males who jog at least 15 miles per week had a mean pulse rate of 52.6 beats per minute and a standard deviation of 3.22 beats per minute.

a. Find a 95% confidence interval for the mean pulse rate of all U.S. adult males who jog at least 15 miles per week. 52.6 ± 1.47

b. Interpret the interval found in part **a**. Does it appear that jogging at least 15 miles per week reduces the mean pulse rate for adult males?

c. What assumptions are required for the validity of the confidence interval and the inference, part **b**?

5.77 In a study reported in *The Wall Street Journal* (April 4, 1999), the Tupperware Corporation surveyed 1,007 U.S. workers. Of the people surveyed, 665 indicated that they take their lunch to work with them. Of these 665 taking their lunch, 200 reported that they take it in brown bags.

a. Find a 95% confidence interval estimate of the population proportion of U.S. workers who take their lunch to work with them. Interpret the interval.

b. Consider the population of U.S. workers who take their lunch to work with them. Find a 95% confidence interval estimate of the population proportion who take brown-bag lunches. Interpret the interval.

5.78 Refer to the *Psychology and Aging* (Dec. 2000) study of the roles that elderly people feel are the most important to them, Exercise 2.6 (p. 26). Recall that in a national sample of 1,102 adults, 65 years or older, 424 identified "spouse" as their most salient role. Use a 95% confidence interval to estimate the true percentage of adults, 65 years or older, who feel being a spouse is their most salient role. Give a practical interpretation of this interval.   .385 ± .029

5.79 Tropical swarm-founding wasps rely on female workers to raise their offspring. One possible explanation for this strange behavior is inbreeding, which increases relatedness among the wasps, presumably making it easier for the workers to pick out their closest relatives as propagators of their own genetic material. To test this theory, 197 swarm-founding wasps were captured in Venezuela, frozen at $-70°$ C, and then subjected to a series of genetic tests (*Science*, Nov. 1988). The data were used to generate an inbreeding coefficient, x, for each wasp specimen, with the following results: $\bar{x} = .044$ and $s = .884$.

   a. Construct a 99% confidence interval for the mean inbreeding coefficient of this species of wasp.

   b. A coefficient of 0 implies that the wasp has no tendency to inbreed. Use the confidence interval, part **a**, to make an inference about the tendency for this species of wasp to inbreed.

5.80 A company is interested in estimating μ, the mean number of days of sick leave taken by all its employees. The firm's statistician selects at random 100 personnel files and notes the number of sick days taken by each employee. The following sample statistics are computed: $\bar{x} = 12.2$ days, $s = 10$ days.

   a. Estimate μ using a 90% confidence interval.

   b. How many personnel files would the statistician have to select in order to estimate μ to within 2 days with a 99% confidence interval?   167

5.81 *IEEE Transactions* (June 1990) presented a hybrid algorithm for solving a polynomial zero-one mathematical programming problem. The algorithm incorporates a mixture of pseudo-Boolean concepts and time-proven implicit enumeration procedures. Fifty-two random problems were solved using the hybrid algorithm; the times to solution (CPU time in seconds) are listed in the table. Use the accompanying SAS printout to estimate, with 95% confidence, the mean solution time for the hybrid algorithm. Interpret the result.

🖉 MATCPU

.045	1.055	.136	1.894	.379	.136	.336	.258	1.070
.506	.088	.242	1.639	.912	.412	.361	8.788	.579
1.267	.567	.182	.036	.394	.209	.445	.179	.118
.333	.554	.258	.182	.070	3.985	.670	3.888	.136
.091	.600	.291	.327	.130	.145	4.170	.227	.064
.194	.209	.258	3.046	.045	.049	.079		

Source: Snyder, W. S., and Chrissis, J. W. "A hybrid algorithm for solving zero-one mathematical programming problems." *IEEE Transactions*, Vol. 22, No. 2, June 1990, p. 166 (Table 1). The Institute of Industrial Engineers, 25 Technology Park/Atlanta, Norcross, GA 30092. (770) 449-0461.

5.82 A meteorologist wishes to estimate the mean amount of snowfall per year in Spokane, Washington. A random sample of the recorded snowfalls for 20 years produces a sample mean equal to 54 inches and a standard deviation of 9.59 inches.

SAS Output for Exercise 5.81

Sample Statistics for soltime

N	Mean	Std. Dev.	Std. Error
52	0.81	1.50	0.21

Hypothesis Test

Null hypothesis: Mean of soltime = 0
Alternative: Mean of soltime ^= 0

t Statistic	Df	Prob > t
3.892	51	0.0003

95 % Confidence Interval for the Mean

Lower Limit: 0.39
Upper Limit: 1.23

   a. Estimate the true mean amount of snowfall in Spokane using a 99% confidence interval.

   b. If you were purchasing snow-removal equipment for a city, what numerical descriptive measure of the distribution of depth of snowfall would be of most interest to you? Would it be the mean?

5.83 A company purchases large quantities of naphtha in 50-gallon drums. Because the purchases are ongoing, small shortages in the drums can represent a sizable loss to the company. The weights of the drums vary slightly from drum to drum, so the weight of the naphtha is determined by removing it from the drums and measuring it. Suppose the company samples the contents of 20 drums, measures the naphtha in each, and calculates $\bar{x} = 49.70$ gallons and $s = .32$ gallon. Find a 95% confidence interval for the mean number of gallons of naphtha per drum. What assumptions are necessary to assure the validity of the confidence interval?   49.70 ± .1498

5.84 Before approval is given for the use of a new insecticide, the United States Department of Agriculture (USDA) requires that several tests be performed to see how the substance will affect wildlife. In particular, the USDA would like to know the proportion of starlings that will die after being exposed to the insecticide. A random sample of 80 starlings were caught and fed a diet of regular food that had been treated with the substance. After 10 days, 10 starlings had died. Use a 99% confidence interval to estimate the true proportion of starlings that will be killed by the substance.   .125 ± .095

5.85 In Exercise 2.139 (p. 105) you considered data on suicides in correctional facilities, published in the *American Journal of Psychiatry* (July 1995). The researchers wanted to know what factors increase the risk of suicide by those incarcerated in urban jails. The data on 37 suicides that occurred from 1967 to 1992 in Wayne County Jail, Detroit, Michigan, are reproduced in the table. Answer each of the questions posed below, using a 95% confidence interval for the target parameter implied in the question. Fully interpret each confidence interval.

a. What proportion of suicides at the jail are committed by inmates charged with murder/manslaughter?

b. What proportion of suicides at the jail are committed at night? $.703 \pm .147$

c. What is the average length of time an inmate is in jail before committing suicide? 41.405 ± 21.493

d. What percentage of suicides are committed by white inmates? $.378 \pm 1.56$

5.86 Refer to the *American Psychologist* (July 1995) study of applications for university positions in experimental psychology, Exercise 2.47 (p. 53). One of the objectives of the analysis was to identify problems and errors (e.g., not submitting at least three letters of recommendation) in the application packages. In a sample of 148 applications for the positions, 30 applicants failed to provide *any* letters of recommendation. Estimate, with 90% confidence, the true fraction of applicants for a university position in experimental psychology that fail to provide letters of recommendation. Interpret the result.

5.87 Recently, a case of salmonella (bacterial) poisoning was traced to a particular brand of ice cream bar, and the

SUICIDE

Victim	Days in Jail Before Suicide	Marital Status	Race	Murder/Manslaughter Charge	Time of Suicide
1	3	Married	W	Yes	11 pm–7 am
2	4	Single	W	Yes	11 pm–7 am
3	5	Single	NW	Yes	3 pm–11 pm
4	7	Widowed	NW	Yes	11 pm–7 am
5	10	Single	NW	Yes	3 pm–11 am
6	15	Married	NW	Yes	11 pm–7 am
7	15	Divorced	NW	Yes	11 pm–7 am
8	19	Married	NW	Yes	7 am–3 pm
9	22	Single	NW	Yes	11 pm–7 am
10	29	Married	W	Yes	11 pm–7 am
11	31	Single	NW	Yes	11 pm–7 am
12	85	Single	NW	Yes	3 pm–11 pm
13	126	Married	NW	Yes	11 pm–7 am
14	221	Divorced	W	Yes	11 pm–7 am
15	14	Single	NW	No	11 pm–7 am
16	22	Single	W	No	3 pm–11 pm
17	42	Single	W	No	7 am–3 pm
18	122	Single	NW	No	11 pm–7 am
19	309	Married	NW	No	11 pm–7 am
20	69	Married	NW	No	11 pm–7 am
21	143	Single	NW	No	11 pm–7 am
22	6	Married	NW	No	11 pm–7 am
23	1	Single	NW	No	7 am–3 pm
24	1	Single	NW	No	11 pm–7 am
25	1	Single	W	No	11 pm–7 am
26	2	Single	W	No	7 am–3 pm
27	3	Single	NW	No	11 pm–7 am
28	4	Married	W	No	3 pm–11 pm
29	4	Single	W	No	3 pm–11 pm
30	4	Single	NW	No	11 pm–7 am
31	6	Single	W	No	7 am–3 pm
32	6	Married	NW	No	11 pm–7 am
33	10	Single	W	No	11 pm–7 am
34	18	Single	NW	No	11 pm–7 am
35	26	Married	W	No	11 pm–7 am
36	41	Single	NW	No	11 pm–7 am
37	86	Married	W	No	11 pm–7 am

Source: DuRand, C. J., *et al.* "A quarter century of suicide in a major urban jail: Implications for community psychiatry." *American Journal of Psychiatry,* Vol. 152, No. 7, July 1995, p. 1078 (Table 1).

manufacturer removed the bars from the market. Despite this response, many consumers refused to purchase *any* brand of ice cream bars for some period of time after the event (McClave, personal consulting). One manufacturer conducted a survey of consumers 6 months after the poisoning. A sample of 244 ice cream bar consumers were contacted, and 23 of them indicated that they would not purchase ice cream bars because of the potential for food poisoning.

a. What is the point estimate of the true fraction of the entire market who refuse to purchase bars 6 months after the poisoning?   .094

b. Is the sample size large enough to use the normal approximation for the sampling distribution of the estimator of the binomial probability? Justify your response.   Yes

c. Construct a 95% confidence interval for the true proportion of the market who still refuse to purchase ice cream bars 6 months after the event.   .094 ± .037

d. Interpret both the point estimate and confidence interval in terms of this application.

5.88 Refer to Exercise 5.87. Suppose it is now 1 year after the poisoning was traced to ice cream bars. The manufacturer wishes to estimate the proportion who still will not purchase bars to within .02 using a 95% confidence interval. How many consumers should be sampled?

5.89 A common hazardous compound found in contaminated soil is benzo(a)pyrene [B(a)p]. An experiment was conducted to determine the effectiveness of a method designed to remove B(a)p from soil (*Journal of Hazardous Materials*, June 1995). Three soil specimens contaminated with a known amount of B(a)p were treated with a toxin that inhibits microbial growth. After 95 days of incubation, the percentage of B(a)p removed from each soil specimen was measured. The experiment produced the following summary statistics: $\bar{x} = 49.3$ and $s = 1.5$.

a. Use a 99% confidence interval to estimate the mean percentage of B(a)p removed from a soil specimen in which the toxin was used.   49.3 ± 8.60

b. Interpret the interval in terms of this application.

c. What assumption is necessary to ensure the validity of this confidence interval?

d. How many soil specimens must be sampled to estimate the mean percentage removed to within .5 using a 99% confidence interval?   35

5.90 The accounting firm Price Waterhouse periodically monitors the U.S. Postal Service's performance. One parameter of interest is the percentage of mail delivered on time. In a sample of 332,000 mailed items, Price Waterhouse determined that 282,200 items were delivered on time (*Tampa Tribune*, Mar. 26, 1995). Use this information to estimate with 99% confidence the true percentage of items delivered on time by the U.S. Postal Service. Interpret the result.   .85 ± .002

STUDENT PROJECTS

Choose a population pertinent to your major area of interest—a population that has an unknown mean or, if the population is binomial, that has an unknown probability of success. For example, a marketing major may be interested in the proportion of consumers who prefer a certain product. A sociology major may be interested in estimating the proportion of people in a certain socioeconomic group or the mean income of people living in a particular part of a city. A political science major may wish to estimate the proportion of an electorate in favor of a certain candidate, a certain amendment, or a certain presidential policy. A pre-med student might want to find the average length of time patients stay in the hospital or the average number of people treated daily in the emergency room. We could continue with examples, but the point should be clear—choose something of interest to you.

Define the parameter you want to estimate and conduct a *pilot study* to obtain an initial estimate of the parameter of interest and, more importantly, an estimate of the variability associated with the estimator. A pilot study is a small experiment (perhaps 20 to 30 observations) used to gain some information about the population of interest. The purpose of the study is to help plan more elaborate future experiments. Using the results of your pilot study, determine the sample size necessary to estimate the parameter to within a reasonable bound (of your choice) with a 95% confidence interval.

REFERENCES

Agresti, A., and Coull, B. A. "Approximate is better than 'exact' for interval estimation of binomial proportions." *The American Statistician*, Vol. 52, No. 2, May 1998, pp. 119–126.

Deming, W. E. *Out of the Crisis*. Cambridge, Mass.: M.I.T. Center for Advanced Study of Engineering, 1986.

Mendenhall, W., Beaver, R. J., and Beaver, B. *Introduction to Probability and Statistics*, 10th ed. North Scituate, Mass.: Duxbury, 1999.

Wilson, E. B. "Probable inference, the law of succession, and statistical inference." *Journal of the American Statistical Association*, Vol. 22, 1927, pp. 209–212.

CHAPTER 6

Inferences Based on a Single Sample:

TESTS OF HYPOTHESIS

Contents

Statistics in Action

March Madness: Handicapping the NCAA Basketball Tourney

✋ *Where We've Been*

We saw how to use sample information to estimate population parameters in Chapter 5. The sampling distribution of a statistic is used to assess the reliability of an estimate, which we express in terms of a confidence interval.

☞ *Where We're Going*

We'll see how to utilize sample information to test what the value of a population parameter may be. This type of inference is called a *test of hypothesis*. We'll also see how to conduct a test of hypothesis about a population mean μ and a population proportion p. And, just as with estimation, we'll stress the measurement of the reliability of the inference. An inference without a measure of reliability is little more than a guess.

Suppose you wanted to determine whether the mean level of a driver's blood alcohol exceeds the legal limit after two drinks, or whether the majority of registered voters approve of the president's performance. In both cases you are interested in making an inference about how the value of a parameter relates to a specific numerical value. Is it less than, equal to, or greater than the specified number? This type of inference, called a **test of hypothesis**, is the subject of this chapter.

We introduce the elements of a test of hypothesis in Section 6.1. We then show how to conduct a large-sample test of hypothesis about a population mean in Sections 6.2 and 6.3. In Section 6.4 we utilize small samples to conduct tests about means, and in optional Section 6.6 we consider an alternative nonparametric test. Large-sample tests about binomial probabilities are the subject of Section 6.5.

6.1   THE ELEMENTS OF A TEST OF HYPOTHESIS

Suppose building specifications in a certain city require that the average breaking strength of residential sewer pipe be more than 2,400 pounds per foot of length (i.e., per linear foot). Each manufacturer who wants to sell pipe in this city must demonstrate that its product meets the specification. Note that we are again interested in making an inference about the mean μ of a population. However, in this example we are less interested in estimating the value of μ than we are in testing a *hypothesis* about its value. That is, *we want to decide whether the mean breaking strength of the pipe exceeds 2,400 pounds per linear foot.*

TEACHING TIP
Use word problem examples to
teach the student how to develop
the null and alternative
hypotheses.

The method used to reach a decision is based on the rare-event concept explained in earlier chapters. We define two hypotheses: (1) The **null hypothesis** is that which represents the status quo to the party performing the sampling experiment—the hypothesis that will be accepted unless the data provide convincing evidence that it is false. (2) The **alternative**, or **research**, **hypothesis** is that which will be accepted only if the data provide convincing evidence of its truth. From the point of view of the city conducting the tests, the null hypothesis is that the manufacturer's pipe does *not* meet specifications unless the tests provide convincing evidence otherwise. The null and alternative hypotheses are therefore

> *Null hypothesis* (H_0): $\mu \leq 2{,}400$
> (i.e., the manufacturer's pipe does not meet specifications)
>
> *Alternative (research) hypothesis* (H_a): $\mu > 2{,}400$
> (i.e., the manufacturer's pipe meets specifications)

How can the city decide when enough evidence exists to conclude that the manufacturer's pipe meets specifications? Since the hypotheses concern the value of the population mean μ, it is reasonable to use the sample mean $\bar{x}$ to make the inference, just as we did when forming confidence intervals for μ in Sections 5.1 and 5.2. The city will conclude that the pipe meets specifications only when the sample mean $\bar{x}$ convincingly indicates that the population mean exceeds 2,400 pounds per linear foot.

"Convincing" evidence in favor of the alternative hypothesis will exist when the value of $\bar{x}$ exceeds 2,400 by an amount that cannot be readily attributed to sampling variability. To decide, we compute a **test statistic**, which is the z value that measures the distance between the value of $\bar{x}$ and the value of μ specified in the null hypothesis. When the null hypothesis contains more than one value of μ, as in this case (H_0: $\mu \leq 2{,}400$), we use the value of μ closest to the values specified

TEACHING TIP
Discuss the role of the test statistic in determining which hypothesis is correct. Show how different sample data will lead to different test statistics.

in the alternative hypothesis. The idea is that if the hypothesis that μ *equals* 2,400 can be rejected in favor of $\mu > 2,400$, then μ *less than or equal to* 2,400 can certainly be rejected. Thus, the test statistic is

$$z = \frac{\bar{x} - 2,400}{\sigma_{\bar{x}}} = \frac{\bar{x} - 2,400}{\sigma/\sqrt{n}}$$

Note that a value of $z = 1$ means that $\bar{x}$ is 1 standard deviation above $\mu = 2,400$; a value of $z = 1.5$ means that $\bar{x}$ is 1.5 standard deviations above $\mu = 2,400$, etc. How large must z be before the city can be convinced that the null hypothesis can be rejected in favor of the alternative and conclude that the pipe meets specifications?

If you examine Figure 6.1, you will note that the chance of observing $\bar{x}$ more than 1.645 standard deviations above 2,400 is only .05—*if in fact the true mean μ is 2,400*. Thus, if the sample mean is more than 1.645 standard deviations above 2,400, either H_0 is true and a relatively rare event has occurred (.05 probability) or H_a is true and the population mean exceeds 2,400. Since we would most likely reject the notion that a rare event has occurred, we would reject the null hypothesis ($\mu \leq 2,400$) and conclude that the alternative hypothesis ($\mu > 2,400$) is true. What is the probability that this procedure will lead us to an incorrect decision?

FIGURE 6.1

The Sampling Distribution of $\bar{x}$, Assuming $\mu = 2,400$

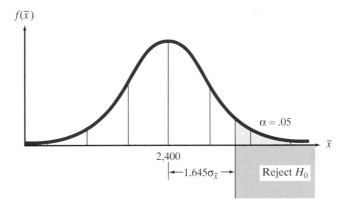

Such an incorrect decision—deciding that the null hypothesis is false when in fact it is true—is called a **Type I decision error**. As indicated in Figure 6.1, the risk of making a Type I error is denoted by the symbol α. That is,

$$\alpha = P(\text{Type I error})$$
$$= P(\text{Rejecting the null hypothesis when in fact the null hypothesis is true})$$

In our example

$$\alpha = P(z > 1.645 \text{ when in fact } \mu = 2,400) = .05$$

TEACHING TIP
Students always have difficulty understanding α and how it affects the rejection region of the test of hypothesis. Be aware of these difficulties and be prepared with plenty of examples.

We now summarize the elements of the test:

$$H_0: \mu \leq 2,400$$
$$H_a: \mu > 2,400$$
$$\text{Test statistic: } z = \frac{\bar{x} - 2,400}{\sigma_{\bar{x}}}$$
$$\text{Rejection region: } z > 1.645, \text{ which corresponds to } \alpha = .05$$

Note that the **rejection region** refers to the values of the test statistic for which we will *reject the null hypothesis*.

To illustrate the use of the test, suppose we test 50 sections of sewer pipe and find the mean and standard deviation for these 50 measurements to be:

$$\bar{x} = 2,460 \text{ pounds per linear foot} \qquad s = 200 \text{ pounds per linear foot}$$

As in the case of estimation, we can use s to approximate σ when s is calculated from a large set of sample measurements.

The test statistic is

$$z = \frac{\bar{x} - 2,400}{\sigma_{\bar{x}}} = \frac{\bar{x} - 2,400}{\sigma/\sqrt{n}} \approx \frac{\bar{x} - 2,400}{s/\sqrt{n}}$$

Substituting $\bar{x} = 2,460$, $n = 50$, and $s = 200$, we have

$$z \approx \frac{2,460 - 2,400}{200/\sqrt{50}} = \frac{60}{28.28} = 2.12$$

Therefore, the sample mean lies $2.12\sigma_{\bar{x}}$ above the hypothesized value of $\mu = 2,400$ as shown in Figure 6.2. Since this value of z exceeds 1.645, it falls in the rejection region. That is, we reject the null hypothesis that $\mu = 2,400$ and conclude that $\mu > 2,400$. Thus, it appears that the company's pipe has a mean strength that exceeds 2,400 pounds per linear foot.

FIGURE 6.2

Location of the Test Statistic for a Test of the Hypothesis
$H_0: \mu = 2,400$

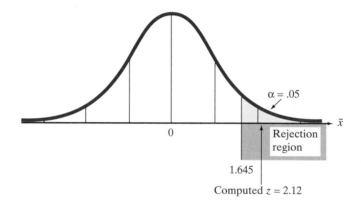

How much faith can be placed in this conclusion? What is the probability that our statistical test could lead us to reject the null hypothesis (and conclude that the company's pipe meets the city's specifications) when in fact the null hypothesis is true? The answer is $\alpha = .05$. That is, we selected the level of risk, α, of making a Type I error when we constructed the test. Thus, the chance is only 1 in 20 that our test would lead us to conclude the manufacturer's pipe satisfies the city's specifications when in fact the pipe does *not* meet specifications.

Now, suppose the sample mean breaking strength for the 50 sections of sewer pipe turned out to be $\bar{x} = 2,430$ pounds per linear foot. Assuming that the sample standard deviation is still $s = 200$, the test statistic is

$$z = \frac{2,430 - 2,400}{2/\sqrt{50}} = \frac{30}{28.28} = 1.06$$

Therefore, the sample mean $\bar{x} = 2,430$ is only 1.06 standard deviations above the null hypothesized value of $\mu = 2,400$. As shown in Figure 6.3, this value does not fall into the rejection region ($z > 1.645$). Therefore, we know that we cannot reject H_0 using $\alpha = .05$. Even though the sample mean exceeds the city's specification of 2,400 by 30 pounds per linear foot, it does not exceed the specification by enough to provide *convincing* evidence that the *population mean* exceeds 2,400.

FIGURE 6.3

Location of the Test Statistic when $\bar{x} = 2,430$

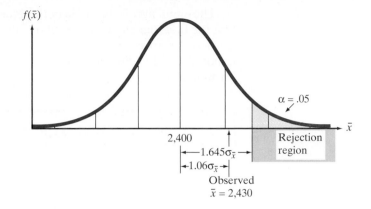

Should we accept the null hypothesis H_0: $\mu \leq 2,400$ and conclude that the manufacturer's pipe does not meet specifications? To do so would be to risk a **Type II error**—that of concluding that the null hypothesis is true (the pipe does not meet specifications) when in fact it is false (the pipe does meet specifications). We denote the probability of committing a Type II error by β. Unfortunately, β is often difficult to determine precisely. Rather than make a decision (accept H_0) for which the probability of error (β) is unknown, we avoid the potential Type II error by avoiding the conclusion that the null hypothesis is true. Instead, we will simply state that *the sample evidence is insufficient to reject H_0 at $\alpha = .05$.* Since the null hypothesis is the "status-quo" hypothesis, the effect of not rejecting H_0 is to maintain the status quo. In our pipe-testing example, the effect of having insufficient evidence to reject the null hypothesis that the pipe does not meet specifications is probably to prohibit the utilization of the manufacturer's pipe unless and until there is sufficient evidence that the pipe does meet specifications. That is, until the data indicate convincingly that the null hypothesis is false, we usually maintain the status quo implied by its truth.

TABLE 6.1   Conclusions and Consequences for a Test of Hypothesis

Conclusion	True State of Nature	
	H_0 **True**	H_a **True**
Accept H_0 (Assume H_0 True)	Correct decision	Type II error (probability β)
Reject H_0 (Assume H_a True)	Type I error (probability α)	Correct decision

Table 6.1 summarizes the four possible outcomes of a test of hypothesis. The "true state of nature" columns in Table 6.1 refer to the fact that either the null hypothesis H_0 is true or the alternative hypothesis H_a is true. Note that the true state of nature is unknown to the researcher conducting the test. The "conclusion" rows in Table 6.1 refer to the action of the researcher, assuming that he or she will conclude either that H_0 is true or that H_a is true, based on the results of the sampling experiment. Note that a Type I error can be made *only* when the alternative hypothesis is accepted (equivalently, when the null hypothesis is rejected), and a Type II error can be made *only* when the null hypothesis is accepted. Our policy will be to make a decision only when we know the probability of making the error that corresponds to that decision. Since α is usually specified by the analyst (typically .05 in research), we will generally be able to reject H_0 (accept H_a) when the sample evidence supports that decision. However, since β is usually not specified, *we will generally avoid the decision to accept H_0, preferring instead to state that the sample evidence is insufficient to reject H_0 when the test statistic is not in the rejection region.*

The elements of a test of hypothesis are summarized in the following box. Note that the first four elements are all specified *before* the sampling experiment is performed. In no case will the results of the sample be used to determine the hypotheses—the data are collected to test the predetermined hypotheses, not to formulate them.

Elements of a Test of Hypothesis

1. *Null hypothesis* (H_0): A theory about the values of one or more population parameters. The theory generally represents the status quo, which we accept until it is proven false.

2. *Alternative (research) hypothesis* (H_a): A theory that contradicts the null hypothesis. The theory generally represents that which we will accept only when sufficient evidence exists to establish its truth.

3. *Test statistic:* A sample statistic used to decide whether to reject the null hypothesis.

4. *Rejection region:* The numerical values of the test statistic for which the null hypothesis will be rejected. The rejection region is chosen so that the probability is α that it will contain the test statistic when the null hypothesis is true, thereby leading to a Type I error. The value of α is usually chosen to be small (e.g., .01, .05, or .10), and is referred to as the **level of significance** of the test.

5. *Assumptions:* Clear statement(s) of any assumptions made about the population(s) being sampled.

6. *Experiment and calculation of test statistic:* Performance of the sampling experiment and determination of the numerical value of the test statistic.

7. *Conclusion:*

   a. If the numerical value of the test statistic falls in the rejection region, we reject the null hypothesis and conclude that the alternative hypothesis is true. We know that the hypothesis-testing process will lead to this conclusion incorrectly (Type I error) only $100\alpha\%$ of the time when H_0 is true.

   b. If the test statistic does not fall in the rejection region, we do not reject H_0. Thus, we reserve judgment about which hypothesis is true. We do not conclude that the null hypothesis is true because we do not (in general) know the probability β that our test procedure will lead to an incorrect acceptance of H_0 (Type II error).

 EXERCISES 6.1–6.16

Learning the Mechanics

6.1 Which hypothesis, the null or the alternative, is the status-quo hypothesis? Which is the research hypothesis?

6.2 Which element of a test of hypothesis is used to decide whether to reject the null hypothesis in favor of the alternative hypothesis?   Test statistic

6.3 What is the level of significance of a test of hypothesis?

6.4 What is the difference between Type I and Type II errors in hypothesis testing? How do α and β relate to Type I and Type II errors?

6.5 List the four possible results of the combinations of decisions and true states of nature for a test of hypothesis.

6.6 We (generally) reject the null hypothesis when the test statistic falls in the rejection region, but we do not accept the null hypothesis when the test statistic does not fall in the rejection region. Why?

6.7 If you test a hypothesis and reject the null hypothesis in favor of the alternative hypothesis, does your test prove that the alternative hypothesis is correct? Explain.   No

Applying the Concepts—Basic

6.8 American Express Consulting reported in *USA Today* (June 15, 2001) that 80% of U.S. companies have formal, written travel and entertainment policies for their employees. Give the null hypothesis for testing the claim made by American Express Consulting.

6.9 According to the *Journal of Advanced Nursing* (Jan. 2001), 45% of senior women (i.e., women over the age of 65) use herbal therapies to prevent or treat health problems. Also, of senior women who use herbal therapies, they use an average of 2.5 herbal products in a year.
 a. Give the null hypothesis for testing the first claim by the journal. $H_0: p = .45;$ $H_a: p \neq .45$
 b. Give the null hypothesis for testing the second claim by the journal. $H_0: \mu = 2.5;$ $H_a: \mu \neq 2.5$

6.10 *Science* (Jan. 1, 1999) reported that the mean listening time of 7-month-old infants exposed to a three-syllable sentence (e.g., "ga ti ti") is 9 seconds. Set up the null and alternative hypotheses for testing the claim.

6.11 Infants exposed to cocaine in their mother's womb are thought to be at a high risk for major birth defects. However, according to a University of Florida study, more than 75% of all cocaine-exposed babies suffer no major problems (*Explore*, Fall 1998). Set up the null and alternative hypotheses if you want to support the findings. $H_0: p = .75; H_a: p > .75$

6.12 The national student loan default rate has dropped steadily over the last decade. *USF Magazine* (Spring, 1999) reported the default rate (i.e., the proportion of college students who default on their loans) at .10 in fiscal year 1996. Set up the null and alternative hypotheses if you want to determine if the student loan default rate in 2000 is less than .10. $H_0: p = .10; H_a: p < .10$

Applying the Concepts—Intermediate

6.13 According to a University of Florida wildlife ecology and conservation researcher, the average level of mercury uptake in wading birds in the Everglades has declined over the past several years (*UF News*, Dec. 15, 2000). In 1994, the average level was 15 parts per million.
 a. Give the null and alternative hypotheses for testing whether the average level in 2000 was less than 15 ppm.
 b. Describe a Type I error for this test.
 c. Describe a Type II error for this test.

6.14 The Computer-Assisted Hypnosis Scale (CAHS) is designed to measure a person's susceptibility to hypnosis. CAHS scores range from 0 (no susceptibility to hypnosis) to 12 (extremely high susceptibility to hypnosis). *Psychological Assessment* (Mar. 1995) reported that University of Tennessee undergraduates had a mean CAHS score of $\mu = 4.6$. Suppose you want to test whether undergraduates at your college or university are more susceptible to hypnosis than University of Tennessee undergraduates.
 a. Set up H_0 and H_a for the test.
 b. Describe a Type I error for this test.
 c. Describe a Type II error for this test.

Applying the Concepts—Advanced

6.15 A group of physicians subjected the *polygraph* (or *lie detector*) to the same careful testing given to medical diagnostic tests. They found that if 1,000 people were subjected to the polygraph and 500 told the truth and 500 lied, the polygraph would indicate that approximately 185 of the truth-tellers were liars and that approximately 120 of the liars were truth-tellers (*Discover*, 1986).
 a. In the application of a polygraph test, an individual is presumed to be a truth-teller (H_0) until "proven" a liar (H_a). In this context, what is a Type I error? A Type II error?
 b. According to the study, what is the probability (approximately) that a polygraph test will result in a Type I error? A Type II error?

6.16 Sometimes, the outcome of a jury trial defies the "commonsense" expectations of the general public (e.g., the O. J. Simpson verdict in the "Trial of the Century"). Such a verdict is more acceptable if we understand that the jury trial of an accused murderer is analogous to the statistical hypothesis-testing process. The null hypothesis in a jury trial is that the accused is innocent. (The status-quo hypothesis in the U.S. system of justice is innocence, which is assumed to be true until proven *beyond a reasonable doubt*.) The alternative hypothesis is guilt, which is accepted only when sufficient evidence exists to establish its truth. If the vote of the jury is unanimous in favor of guilt, the null hypothesis of innocence is rejected and the court concludes that the accused murderer is guilty. Any vote other than a unanimous one for guilt results in a "not guilty" verdict. The court never accepts the null hypothesis; that is, the court never declares the accused "innocent." A "not guilty" verdict (as in the O. J. Simpson case) implies that the court could not find the defendant guilty *beyond a reasonable doubt*.
 a. Define Type I and Type II errors in a murder trial.
 b. Which of the two errors is the more serious? Explain.
 c. The court does not, in general, know the values of α and β; but ideally, both should be small. One of these probabilities is assumed to be smaller than the other in a jury trial. Which one, and why? α
 d. The court system relies on the belief that the value of α is made very small by requiring a unanimous vote before guilt is concluded. Explain why this is so.
 e. For a jury prejudiced against a guilty verdict as the trial begins, will the value of α increase or decrease? Explain.   Decrease
 f. For a jury prejudiced against a guilty verdict as the trial begins, will the value of β increase or decrease? Explain.   Increase

6.2 LARGE-SAMPLE TEST OF HYPOTHESIS ABOUT A POPULATION MEAN

In Section 6.1 we learned that the null and alternative hypotheses form the basis for a test of hypothesis inference. The null and alternative hypotheses may take one of several forms. In the sewer pipe example we tested the null hypothesis that the population mean strength of the pipe is less than or equal to 2,400 pounds per linear foot against the alternative hypothesis that the mean strength exceeds 2,400. That is, we tested

$$H_0: \mu \le 2{,}400$$
$$H_a: \mu > 2{,}400$$

This is a **one-tailed** (or **one-sided**) **statistical test** because the alternative hypothesis specifies that the population parameter (the population mean μ, in this example) is strictly greater than a specified value (2,400, in this example). If the null hypothesis had been $H_0: \mu \ge 2{,}400$ and the alternative hypothesis had been $H_a: \mu < 2{,}400$, the test would still be one-sided, because the parameter is still specified to be on "one side" of the null hypothesis value. Some statistical investigations seek to show that the population parameter is *either larger or smaller* than some specified value. Such an alternative hypothesis is called a **two-tailed** (or **two-sided**) **hypothesis**.

While alternative hypotheses are always specified as strict inequalities, such as $\mu < 2{,}400$, $\mu > 2{,}400$, or $\mu \ne 2{,}400$, null hypotheses are usually specified as equalities, such as $\mu = 2{,}400$. *Even when the null hypothesis is an inequality, such as $\mu \le 2{,}400$, we specify $H_0: \mu = 2{,}400$, reasoning that if sufficient evidence exists to show that $H_a: \mu > 2{,}400$ is true when tested against $H_0: \mu = 2{,}400$, then surely sufficient evidence exists to reject $\mu < 2{,}400$ as well.* Therefore, the null hypothesis is specified as the value of μ closest to a one-sided alternative hypothesis and as the only value *not* specified in a two-tailed alternative hypothesis. The steps for selecting the null and alternative hypotheses are summarized in the box on p. 285.

The rejection region for a two-tailed test differs from that for a one-tailed test. When we are trying to detect departure from the null hypothesis in *either* direction, we must establish a rejection region in both tails of the sampling distribution of the test statistic. Figures 6.4a and 6.4b show the one-tailed rejection regions for lower- and upper-tailed tests, respectively. The two-tailed rejection region is illustrated in Figure 6.4c. Note that a rejection region is established in each tail of the sampling distribution for a two-tailed test.

The rejection regions corresponding to typical values selected for α are shown in Table 6.2 for one-and two-tailed tests. Note that the smaller α you select, the more evidence (the larger z) you will need before you can reject H_0.

FIGURE 6.4

Rejection Regions Corresponding to One- and Two-Tailed Tests

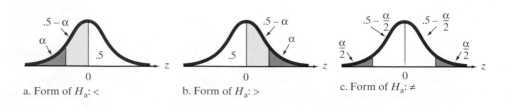

a. Form of H_a: $<$ b. Form of H_a: $>$ c. Form of H_a: $\ne$

Steps for Selecting the Null and Alternative Hypotheses

TEACHING TIP
We use the equal sign in the null hypothesis for both one- and two-tailed tests. Writing the null hypothesis as $\leq$ (or $\geq$) for the one-tailed tests allows the student to understand that one of the two hypotheses must be correct. This is helpful in understanding the Type I and Type II errors.

1. Select the *alternative hypothesis* as that which the sampling experiment is intended to establish. The alternative hypothesis will assume one of three forms:

   a. One-tailed, upper-tailed   *Example: H_a: $\mu > 2{,}400$*

   b. One-tailed, lower-tailed   *Example: H_a: $\mu < 2{,}400$*

   c. Two-tailed   *Example: H_a: $\mu \neq 2{,}400$*

2. Select the *null hypothesis* as the status quo, that which will be presumed true unless the sampling experiment conclusively establishes the alternative hypothesis. The null hypothesis will be specified as that parameter value closest to the alternative in one-tailed tests, and as the complementary (or only unspecified) value in two-tailed tests.

   Example: $H_0: \mu = 2{,}400$

TEACHING TIP
Use the normal table to show the student where and how these critical values are found.

TABLE 6.2   Rejection Regions for Common Values of α

	Alternative Hypotheses		
	Lower-Tailed	**Upper-Tailed**	**Two-Tailed**
$\alpha = .10$	$z < -1.28$	$z > 1.28$	$z < -1.645$ or $z > 1.645$
$\alpha = .05$	$z < -1.645$	$z > 1.645$	$z < -1.96$ or $z > 1.96$
$\alpha = .01$	$z < -2.33$	$z > 2.33$	$z < -2.575$ or $z > 2.575$

EXAMPLE 6.1

The effect of drugs and alcohol on the nervous system has been the subject of considerable research. Suppose a research neurologist is testing the effect of a drug on response time by injecting 100 rats with a unit dose of the drug, subjecting each to a neurological stimulus, and recording its response time. The neurologist knows that the mean response time for rats not injected with the drug (the "control" mean) is 1.2 seconds. She wishes to test whether the mean response time for drug-injected rats differs from 1.2 seconds. Set up the test of hypothesis for this experiment, using $\alpha = .01$.

Solution

Since the neurologist wishes to detect whether the mean response time, μ, for drug-injected rats differs from the control mean of 1.2 seconds in *either* direction—that is, $\mu < 1.2$ or $\mu > 1.2$—we conduct a two-tailed statistical test. Following the procedure for selecting the null and alternative hypotheses, we specify as the alternative hypothesis that the mean differs from 1.2 seconds, since determining whether the drug-injected mean differs from the control mean is the purpose of the experiment. The null hypothesis is the presumption that drug-injected rats have the same mean response time as control rats unless the research indicates otherwise. Thus,

$$H_0: \mu = 1.2$$
$$H_a: \mu \neq 1.2 \qquad (\text{i.e., } \mu < 1.2 \text{ or } \mu > 1.2)$$

The test statistic measures the number of standard deviations between the observed value of $\bar{x}$ and the null hypothesized value $\mu = 1.2$:

$$\textit{Test statistic:} \quad z = \frac{\bar{x} - 1.2}{\sigma_{\bar{x}}}$$

The rejection region must be designated to detect a departure from $\mu = 1.2$ in *either* direction, so we will reject H_0 for values of z that are either too small (negative) or too large (positive). To determine the precise values of z that comprise the rejection region, we first select α, the probability that the test will lead to incorrect rejection of the null hypothesis. Then we divide α equally between the lower and upper tail of the distribution of z, as shown in Figure 6.5. In this example, $\alpha = .01$, so $\alpha/2 = .005$ is placed in each tail. The areas in the tails correspond to $z = -2.575$ and $z = 2.575$, respectively (from Table 6.2):

Rejection region: $z < -2.575$ or $z > 2.575$ (see Figure 6.5)

Suggested Exercise 6.26

Assumptions: Since the sample size of the experiment is large enough ($n > 30$), the Central Limit Theorem will apply, and no assumptions need be made about the population of response time measurements. The sampling distribution of the sample mean response of 100 rats will be approximately normal regardless of the distribution of the individual rats' response times.

■

FIGURE 6.5

Two-Tailed Rejection Region: $\alpha = .01$

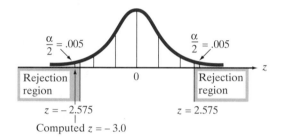

Note that the test in Example 6.1 is set up *before* the sampling experiment is conducted. The data are not used to develop the test. Evidently, the neurologist wants to conclude that the mean response time for the drug-injected rats differs from the control mean only when the evidence is very convincing, because the value of α has been set quite low at .01. If the experiment results in the rejection of H_0, she can be 99% confident that the mean response time of the drug-injected rats differs from the control mean.

Once the test is set up, she is ready to perform the sampling experiment and conduct the test. The test is performed in Example 6.2.

■

E X A M P L E   6 . 2

Refer to the neurological response-time test set up in Example 6.1. Suppose the sampling experiment is conducted with the following results:

$n = 100$ response times for drug-injected rats
$\bar{x} = 1.05$ seconds $s = .5$ second

$z = -3.0$, Reject H_0

Use the results of the sampling experiment to conduct the test of hypothesis.

Solution

Since the test is completely specified in Example 6.1, we simply substitute the sample statistics into the test statistic:

$$z = \frac{\bar{x} - 1.2}{\sigma_{\bar{x}}} = \frac{\bar{x} - 1.2}{\sigma/\sqrt{n}} = \frac{1.05 - 1.2}{\sigma/\sqrt{100}}$$

$$\approx \frac{1.05 - 1.2}{s/10} = \frac{-.15}{.5/10} = -3.0$$

The implication is that the sample mean, 1.05, is (approximately) 3 standard deviations below the null hypothesized value of 1.2 in the sampling distribution of $\bar{x}$. You can see in Figure 6.5 that this value of z is in the lower-tail rejection region, which consists of all values of $z < -2.575$. This sampling experiment provides sufficient evidence to reject H_0 and conclude, at the $\alpha = .01$ level of significance, that the mean response time for drug-injected rats differs from the control mean of 1.2 seconds. It appears that the rats receiving an injection of this drug have a mean response time that is less than 1.2 seconds. ▪

The setup of a large-sample test of hypothesis about a population mean is summarized in the following box. Both the one- and two-tailed tests are shown.

Large-Sample Test of Hypothesis About μ

ONE-TAILED TEST	TWO-TAILED TEST
$H_0: \mu = \mu_0$	$H_0: \mu = \mu_0$
$H_a: \mu < \mu_0$ (or $H_a: \mu > \mu_0$)	$H_a: \mu \neq \mu_0$
Test statistic: $z = \dfrac{\bar{x} - \mu_0}{\sigma_{\bar{x}}}$	Test statistics: $z = \dfrac{\bar{x} - \mu_0}{\sigma_{\bar{x}}}$
Rejection region: $z < -z_\alpha$ (or $z > z_\alpha$ when $H_a: \mu > \mu_0$)	Rejection region: $z < -z_{\alpha/2}$ or $z > z_{\alpha/2}$
where z_α is chosen so that $\quad P(z > z_\alpha) = \alpha$	where $z_{\alpha/2}$ is chosen so that $\quad P(z > z_{\alpha/2}) = \alpha/2$

Assumptions: No assumptions need to be made about the probability distribution of the population because the Central Limit Theorem assures us that, for large samples, the test statistic will be approximately normally distributed regardless of the shape of the underlying probability distribution of the population.
Note: μ_0 is the symbol for the numerical value assigned to μ under the null hypothesis.

Once the test has been set up, the sampling experiment is performed and the test statistic is calculated. The next box contains possible conclusions for a test of hypothesis, depending on the result of the sampling experiment.

Two final points about the test of hypothesis in Example 6.2 apply to all statistical tests:

1. Since z is less than -2.575, it is tempting to state our conclusion at a significance level lower than $\alpha = .01$. We resist the temptation because the level of α is determined *before* the sampling experiment is performed. If we decide that we are willing to tolerate a 1% Type I error rate, the result of the sampling experiment should have no effect on that decision. *In general, the same data should not be used both to set up and to conduct the test.*

2. When we state our conclusion at the .01 level of significance, we are referring to the failure rate of the *procedure*, not the result of this particular test. We know that the test procedure will lead to the rejection of the null hypothesis only 1% of the time when in fact $\mu = 1.2$. *Therefore, when the test statistic falls in the rejection region, we infer that the alternative $\mu \neq 1.2$ is true and express our confidence in the procedure by quoting the α level of significance, or the $100(1 - \alpha)\%$ confidence level.*

Possible Conclusions for a Test of Hypothesis

TEACHING TIP
Emphasize the danger of
accepting H_0.

1. If the calculated test statistic falls in the rejection region, reject H_0 and conclude that the alternative hypothesis H_a is true. State that you are rejecting H_0 at the α level of significance. Remember that the confidence is in the testing *process*, not the particular result of a single test.

2. If the test statistic does not fall in the rejection region, conclude that the sampling experiment does not provide sufficient evidence to reject H_0 at the α level of significance. (Generally, we will not "accept" the null hypothesis unless the probability β of a Type II error has been calculated. Consult the chaper references for some advanced methods of calculating β.)

 EXERCISES 6.17–6.29

Learning the Mechanics

6.17 For each of the following rejection regions, sketch the sampling distribution for z and indicate the location of the rejection region.
   a. $z > 1.96$
   b. $z > 1.645$
   c. $z > 2.575$
   d. $z < -1.28$
   e. $z < -1.645$ or $z > 1.645$
   f. $z < -2.575$ or $z > 2.575$
   g. For each of the rejection regions specified in parts **a–f**, what is the probability that a Type I error will be made?

6.18 Suppose you are interested in conducting the statistical test of $H_0: \mu = 200$ against $H_a: \mu > 200$, and you have decided to use the following decision rule: Reject H_0 if the sample mean of a random sample of 100 items is more than 215. Assume that the standard deviation of the population is 80.
   a. Express the decision rule in terms of z.
   b. Find α, the probability of making a Type I error, by using this decision rule.   .0301

6.19 A random sample of 100 observations from a population with standard deviation 60 yielded a sample mean of 110.
   a. Test the null hypothesis that $\mu = 100$ against the alternative hypothesis that $\mu > 100$ using $\alpha = .05$. Interpret the results of the test. $z = 1.67$
   b. Test the null hypothesis that $\mu = 100$ against the alternative hypothesis that $\mu \neq 100$ using $\alpha = .05$. Interpret the results of the test. $z = 1.67$
   c. Compare the results of the two tests you conducted. Explain why the results differ.

6.20 A random sample of 64 observations produced the following summary statistics: $\bar{x} = .323$ and $s^2 = .034$.
   a. Test the null hypothesis that $\mu = .36$ against the alternative hypothesis that $\mu < .36$ using $\alpha = .10$.
   b. Test the null hypothesis that $\mu = .36$ against the alternative hypothesis that $\mu \neq .36$ using $\alpha = .10$. Interpret the result. $z = -1.61$

Applying the Concepts—Basic

6.21 In Exercise 4.55 (p. 201) you learned that the mean alkalinity level of water specimens collected from the Han River in Seoul, Korea, is 50 milligrams per liter (*Environmental Science & Engineering*, Sept. 1, 2000). Consider a random sample of 100 water specimens collected from a tributary of the Han River. Suppose the mean and standard deviation of the alkalinity levels for the sample are $\bar{x} = 67.8$ mpl and $s = 14.4$ mpl. Is there sufficient evidence (at $\alpha = .01$) to indicate that the population mean alkalinity level of water in the tributary exceeds 50 mpl? $z = 12.36$

6.22 A group of St. Louis University researchers administered the Vividness of Visual Imagery Questionnaire (VVIQ) to a sample of 51 college students on varsity sports teams (American Statistical Association Joint Meetings, Aug. 1996). The mean and standard deviation of the VVIQ scores were 65.60 and 19.38, respectively. Scores range from 16 to 80, with a lower score representing greater vividness. In the general population, the mean VVIQ score is 67. Conduct a test to determine whether the mean VVIQ score of college athletes is less than 67. Test using $\alpha = .01$. $z = -0.52$

6.23 A study reported in the *Journal of Occupational and Organizational Psychology* (Dec. 1992) investigated the relationship of employment status to mental health. Each of a sample of 49 unemployed men was given a mental health examination using the General Health Questionnaire (GHQ). The GHQ is a widely recognized measure of present mental health, with lower values indicating better mental health. The mean and standard deviation of the GHQ scores were $\bar{x} = 10.94$ and $s = 5.10$, respectively.
   a. Specify the appropriate null and alternative hypotheses if we wish to test the research hypothesis that the mean GHQ score for all unemployed men exceeds 10. Is the test one-tailed or two-tailed? Why?
   b. If we specify $\alpha = .05$, what is the appropriate rejection region for this test? $z > 1.645$

c. Conduct the test, and state your conclusion clearly in the language of this exercise. $z = 1.29$

Applying the Concepts—Intermediate

6.24 During the National Football League (NFL) season, Las Vegas oddsmakers establish a point spread on each game for betting purposes. For example, the Baltimore Ravens were established as 3-point favorites over the New York Giants in the 2001 Super Bowl. The final scores of NFL games were compared against the final point spreads established by the oddsmakers in *Chance* (Fall 1998). The difference between the game outcome and point spread (called a point-spread error) was calculated for 240 NFL games. The mean and standard deviation of the point-spread errors are $\bar{x} = -1.6$ and $s = 13.3$. Use this information to test the hypothesis that the true mean point-spread error for all NFL games is 0. Conduct the test at $\alpha = .01$ and interpret the result.

6.25 *Psychological Assessment* (Mar. 1995) published the results of a study of World War II aviators who were captured by German forces after they were shot down. Having located a total of 239 World War II aviator POW survivors, the researchers asked each veteran to participate in the study; 33 responded to the letter of invitation. Each of the 33 POW survivors were administered the Minnesota Multiphasic Personality Inventory, one component of which measures level of post-traumatic stress disorder (PTSD). [*Note:* The higher the score, the higher the level of PTSD.] The aviators produced a mean PTSD score of $\bar{x} = 9.00$ and a standard deviation of $s = 9.32$.

a. Set up the null and alternative hypotheses for determining whether the true mean PTSD score of all World War II aviator POWs is less than 16. [*Note:* The value, 16, represents the mean PTSD score established for Vietnam POWs.]

b. Conduct the test, part **a**, using $\alpha = .10$. What are the practical implications of the test? $z = -4.31$

c. Discuss the representativeness of the sample used in the study and its ramifications.

6.26 Current technology uses X-rays and lasers for inspection of solder-joint defects on printed circuit boards (PCBs) (*Quality Congress Transactions*, 1986). A particular manufacturer of laser-based inspection equipment claims that its product can inspect on average at least 10 solder joints per second when the joints are spaced .1 inch apart. The equipment was tested by a potential buyer on 48 different PCBs. In each case, the

equipment was operated for exactly 1 second. The number of solder joints inspected on each run follows:

🔵 PCB

10	9	10	10	11	9	12	8	8	9	6	10
7	10	11	9	9	13	9	10	11	10	12	8
9	9	9	7	12	6	9	10	10	8	7	9
11	12	10	0	10	11	12	9	7	9	9	10

a. The potential buyer wants to know whether the sample data refute the manufacturer's claim. Specify the null and alternative hypotheses that the buyer should test. $H_0: \mu = 10; \ H_a: \mu < 10$

b. In the context of this exercise, what is a Type I error? A Type II error?

c. Conduct the hypothesis test you described in part **a**, and interpret the test's results in the context of this exercise. Use $\alpha = .05$ and the SPSS descriptive statistics printout at the bottom of the page. $z = -2.33$

6.27 Humerus bones from the same species of animal tend to have approximately the same length-to-width ratios. When fossils of humerus bones are discovered, archeologists can often determine the species of animal by examining the length-to-width ratios of the bones. It is known that species A exhibits a mean ratio of 8.5. Suppose 41 fossils of humerus bones were unearthed at an archeological site in East Africa, where species A is believed to have lived. (Assume that the unearthed bones were all from the same unknown species.) The length-to-width ratios of the bones were calculated and listed, as shown in the table. Summary statistics for the data set are shown in the STATISTIX printout on p. 290.

a. Test whether the population mean ratio of all bones of this particular species differs from 8.5. Use $\alpha = .01$.

b. What are the practical implications of the test, part **a**?

🔵 BONES

10.73	8.89	9.07	9.20	10.33	9.98	9.84	9.59
8.48	8.71	9.57	9.29	9.94	8.07	8.37	6.85
8.52	8.87	6.23	9.41	6.66	9.35	8.86	9.93
8.91	11.77	10.48	10.39	9.39	9.17	9.89	8.17
8.93	8.80	10.02	8.38	11.67	8.30	9.17	12.00
9.38							

SPSS Output for Exercise 6.26

Descriptive Statistics

	N	Minimum	Maximum	Mean	Std. Deviation
NUMBER	48	.00	13.00	9.2917	2.1033
Valid N (listwise)	48				

	DESCRIPTIVE STATISTICS
	LWRATIO
N	41
MEAN	9.2576
SD	1.2036
MINIMUM	6.2300
MEDIAN	9.2000
MAXIMUM	12.000

6.28 *Environmental Science & Technology* (Oct. 1993) reported on a study of contaminated soil in The Netherlands. Seventy-two 400-gram soil specimens were sampled, dried, and analyzed for the contaminant cyanide. The cyanide concentration [in milligrams per kilogram (mg/kg) of soil] of each soil specimen was determined using an infrared microscopic method. The sample resulted in a mean cyanide level of $\bar{x} = 84$ mg/kg and a standard deviation of $s = 80$ mg/kg.

a. Test the hypothesis that the true mean cyanide level in soil in The Netherlands exceeds 100 mg/kg. Use $\alpha = .10$. $z = 1.70$

b. Would you reach the same conclusion as in part **a** using $\alpha = .05$? Using $\alpha = .01$? Why can the conclusion of a test change when the value of α is changed?

Applying the Concepts—Advanced

6.29 The *Community Mental Health Journal* (Aug. 2000) presented the results of a survey of over 6,000 clients of the Department of Mental Health and Addiction Services (DMHAS) in Connecticut. One of the many variables measured for each mental health patient was frequency of social interaction (on a 5-point scale, where 1 = very infrequently, 3 = occasionally, and 5 = very frequently). The 6,681 clients who were evaluated had a mean social interaction score of 2.95 with a standard deviation of 1.10.

a. Conduct a hypothesis test (at $\alpha = .01$) to determine if the true mean social interaction score of all Connecticut mental health patients differs from 3.

b. Examine the results of the study from a practical view, then discuss why "statistical significance" does not always imply "practical significance."

c. Because the variable of interest is measured on a 5-point scale, it is unlikely that the population of ratings will be normally distributed. Consequently, some analysts may perceive the test, part **a**, to be invalid and search for alternative methods of analysis. Defend or refute this position.

6.3   OBSERVED SIGNIFICANCE LEVELS: p-VALUES

According to the statistical test procedure described in Section 6.2, the rejection region and, correspondingly, the value of α are selected prior to conducting the test, and the conclusions are stated in terms of rejecting or not rejecting the null hypothesis. A second method of presenting the results of a statistical test is one that reports the extent to which the test statistic disagrees with the null hypothesis and leaves to the reader the task of deciding whether to reject the null hypothesis. This measure of disagreement is called the *observed significance level* (or *p-value*) for the test.

> **DEFINITION 6.1**
> The **observed significance level**, or **p-value**, for a specific statistical test is the probability (assuming H_0 is true) of observing a value of the test statistic that is at least as contradictory to the null hypothesis, and supportive of the alternative hypothesis, as the actual one computed from the sample data.

For example, the value of the test statistic computed for the sample of $n = 50$ sections of sewer pipe was $z = 2.12$. Since the test is one-tailed—i.e., the alternative (research) hypothesis of interest is $H_a: \mu > 2,400$—values of the test statistic even more contradictory to H_0 than the one observed would be values larger than $z = 2.12$. Therefore, the observed significance level (p-value) for this test is

$$p\text{-value} = P(z \geq 2.12)$$

FIGURE 6.6

Finding the p-Value for an Upper-Tailed Test when z = 2.12

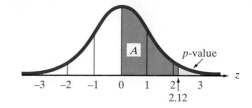

or, equivalently, the area under the standard normal curve to the right of $z = 2.12$ (see Figure 6.6).

The area A in Figure 6.6 is given in Table III in Appendix A as .4830. Therefore, the upper-tail area corresponding to $z = 2.12$ is

$$p\text{-value} = .5 - .4830 = .0170$$

TEACHING TIP
Use examples in which you change the alternative hypothesis from a one-tailed to a two-tailed test. Illustrate what happens to the *p*-value when the change is made.

Consequently, we say that these test results are "very significant"; that is, they disagree rather strongly with the null hypothesis, $H_0: \mu = 2{,}400$, and favor $H_a: \mu > 2{,}400$. The probability of observing a z value as large as 2.12 is only .0170, if in fact the true value of μ is 2,400.

If you are inclined to select $\alpha = .05$ for this test, then you would reject the null hypothesis because the *p*-value for the test, .0170, is less than .05. In contrast, if you choose $\alpha = .01$, you would not reject the null hypothesis because the *p*-value for the test is larger than .01. Thus, the use of the observed significance level is identical to the test procedure described in the preceding sections except that the choice of α is left to you.

The steps for calculating the *p*-value corresponding to a test statistic for a population mean are given in the box.

TEACHING TIP
One big benefit of the *p*-value approach to making conclusions is that a conclusion can easily be made for any choice of α. There is no need to re-compute a rejection region for each choice of α.

Steps for Calculating the *p*-Value for a Test of Hypothesis

1. Determine the value of the test statistic z corresponding to the result of the sampling experiment.

2. a. If the test is one-tailed, the *p*-value is equal to the tail area beyond z in the same direction as the alternative hypothesis. Thus, if the alternative hypothesis is of the form $>$, the *p*-value is the area to the right of, or above, the observed z value. Conversely, if the alternative is of the form $<$, the *p*-value is the area to the left of, or below, the observed z value. (See Figure 6.7.)

   b. If the test is two-tailed, the *p*-value is equal to twice the tail area beyond the observed z value in the direction of the sign of z. That is, if z is positive, the *p*-value is twice the area to the right of, or above, the observed z value. Conversely, if z is negative, the *p*-value is twice the area to the left of, or below, the observed z value. (See Figure 6.8.)

FIGURE 6.7

Finding the p-Value for a One-Tailed Test

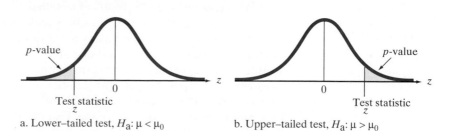

a. Lower–tailed test, $H_a: \mu < \mu_0$ b. Upper–tailed test, $H_a: \mu > \mu_0$

FIGURE 6.8

Finding the p-Value for a Two-
Tailed Test: p-Value = 2(p/2)

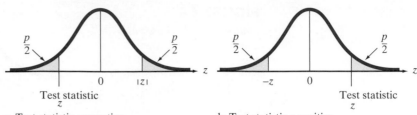

a. Test statistic z negative   b. Test statistic z positive

EXAMPLE 6.3

$p = .0026$

Find the observed significance level for the test of the mean response time for drug-injected rats in Examples 6.1 and 6.2.

Solution

Example 6.1 presented a two-tailed test of the hypothesis

$$H_0: \mu = 1.2 \text{ seconds}$$

against the alternative hypothesis

$$H_a: \mu \neq 1.2 \text{ seconds}$$

The observed value of the test statistic in Example 6.2 was $z = -3.0$, and any value of z less than -3.0 or greater than $+3.0$ (because this is a two-tailed test) would be even more contradictory to H_0. Therefore, the observed significance level for the test is

$$p\text{-value} = P(z < -3.0 \text{ or } z > +3.0)$$

Thus, we calculate the area below the observed z value, $z = -3.0$, and double it. Consulting Table III in Appendix A, we find that $P(z < -3.0) = .5 - .4987 = .0013$. Therefore, the p-value for this two-tailed test is

$$2P(z < -3.0) = 2(.0013) = .0026$$

We can interpret this p-value as a strong indication that the mean reaction time of drug-injected rats differs from the control mean ($\mu \neq 1.2$), since we would observe a test statistic this extreme or more extreme only 26 in 10,000 times if the drug-injected mean were equal to the control mean ($\mu = 1.2$). The extent to which the mean differs from 1.2 could be better determined by calculating a confidence interval for μ.

TEACHING TIP
This interpretation works for one-tailed and two-tailed tests. There's no need to adjust α as the adjustments have already been made in the calculation of the p-value.

When publishing the results of a statistical test of hypothesis in journals, case studies, reports, etc., many researchers make use of p-values. Instead of selecting α beforehand and then conducting a test, as outlined in this chapter, the researcher computes (usually with the aid of a statistical software package) and reports the value of the appropriate test statistic and its associated p-value. It is left to the reader of the report to judge the significance of the result—i.e., the reader must determine whether to reject the null hypothesis in favor of the alternative hypothesis, based on the reported p-value. Usually, the null hypothesis is rejected if the observed significance level is *less than* the fixed significance level, α, chosen by the reader. The inherent advantage of reporting test results in this manner is twofold: (1) Readers are permitted to select the maximum value of α that they would be willing to tolerate if they actually carried out a standard test of hypothesis in the manner outlined in this chapter, and (2) a measure of the degree of significance of the result (i.e., the *p*-value) is provided.

Reporting Test Results as *p*-Values: How to Decide Whether to Reject H_0

Suggested Exercise 6.39

1. Choose the maximum value of α that you are willing to tolerate.
2. If the observed significance level (*p*-value) of the test is less than the chosen value of α, reject the null hypothesis. Otherwise, do not reject the null hypothesis.

■

E X A M P L E   6 . 4

The lengths of stay (in days) for 100 randomly selected hospital patients, first presented in Table 5.1, are reproduced in Table 6.3. Suppose we want to test the hypothesis that the true mean length of stay (LOS) at the hospital is less than 5 days, that is,

$$H_0: \mu = 5$$
$$H_a: \mu < 5$$

Fail to reject H_0

Assuming $\sigma = 3.68$, use the data in the table to conduct the test at $\alpha = .05$.

🌐 HOSPLOS

TABLE 6.3   Lengths of Stay for 100 Hospital Patients

2	3	8	6	4	4	6	4	2	5
8	10	4	4	4	2	1	3	2	10
1	3	2	3	4	3	5	2	4	1
2	9	1	7	17	9	9	9	4	4
1	1	1	3	1	6	3	3	2	5
1	3	3	14	2	3	9	6	6	3
5	1	4	6	11	22	1	9	6	5
2	2	5	4	3	6	1	5	1	6
17	1	2	4	5	4	4	3	2	3
3	5	2	3	3	2	10	2	4	2

Solution

Instead of conducting the test by hand, we will use a statistical software package. The data were entered into a computer and MINITAB was used to conduct the analysis. The MINITAB printout for the lower-tailed test is displayed in Figure 6.9. Both the

FIGURE 6.9

MINITAB Printout for the Lower-Tailed Test in Example 6.4

```
Test of mu = 5 vs mu < 5
The assumed sigma = 3.68

Variable          N       Mean      StDev     SE Mean
LOS             100      4.530      3.678       0.368

Variable       95.0% Upper Bound           Z          P
LOS                        5.135       -1.28      0.101
```

test statistic, $z = -1.28$, and *p*-value of the test, $p = .101$, are highlighted on the MINITAB printout. Since the *p*-value exceeds our selected α value, $\alpha = .05$, we cannot reject the null hypothesis. Hence, there is insufficient evidence (at $\alpha = .05$) to conclude that the true mean LOS at the hospital is less than 5 days. ■

Note: MINITAB, SAS, and STATISTIX all provide an option for selecting one-tailed or two-tailed tests and report the appropriate *p*-value. Some statistical software packages such as SPSS will conduct only two-tailed tests of hypothesis. For these packages, you obtain the *p*-value for a one-tailed test as follows:

TEACHING TIP
It is helpful for the student to understand the *p*-value associated with each of the three possible hypotheses for a given test statistic. This understanding will allow the student to take the *p*-value from a computer printout and adjust it accordingly.

$$p = \frac{\text{Reported } p\text{-value}}{2} \quad \text{if} \begin{cases} H_a \text{ is of form } > \text{ and } z \text{ is positive} \\ H_a \text{ is of form } < \text{ and } z \text{ is negative} \end{cases}$$

$$p = 1 - \left(\frac{\text{Reported } p\text{-value}}{2}\right) \quad \text{if} \begin{cases} H_a \text{ is of form } > \text{ and } z \text{ is negative} \\ H_a \text{ is of form } < \text{ and } z \text{ is positive} \end{cases}$$

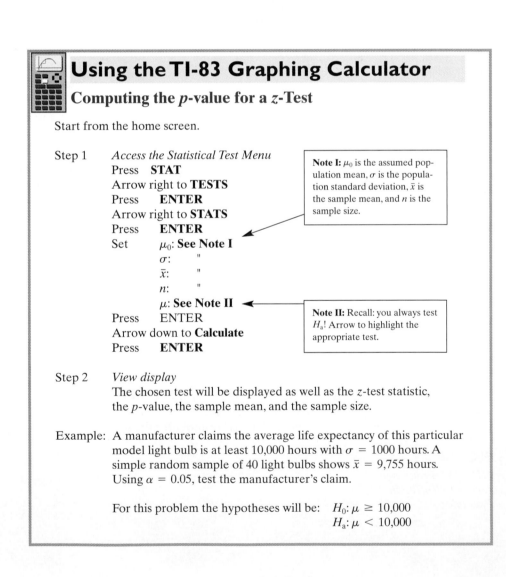

Using the TI-83 Graphing Calculator

Computing the *p*-value for a *z*-Test

Start from the home screen.

Step 1   *Access the Statistical Test Menu*
Press   **STAT**
Arrow right to **TESTS**
Press   **ENTER**
Arrow right to **STATS**
Press   **ENTER**
Set μ_0: **See Note I**
      σ: "
      $\bar{x}$:   "
      n: "
      μ: **See Note II**
Press   ENTER
Arrow down to **Calculate**
Press   **ENTER**

Note I: μ_0 is the assumed population mean, σ is the population standard deviation, $\bar{x}$ is the sample mean, and n is the sample size.

Note II: Recall: you always test H_a! Arrow to highlight the appropriate test.

Step 2   *View display*
The chosen test will be displayed as well as the *z*-test statistic, the *p*-value, the sample mean, and the sample size.

Example: A manufacturer claims the average life expectancy of this particular model light bulb is at least 10,000 hours with $\sigma = 1000$ hours. A simple random sample of 40 light bulbs shows $\bar{x} = 9,755$ hours. Using $\alpha = 0.05$, test the manufacturer's claim.

For this problem the hypotheses will be: $H_0: \mu \geq 10,000$
                                             $H_a: \mu < 10,000$

USING THE TI-83 GRAPHING CALCULATOR (*continued*)

Step 1   *Access the Statistical Test Menu*
Enter your setting for this problem (see the screen below left).

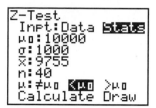
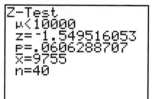

Arrow down to Calculate
Press **ENTER** (above right screen will be displayed).

Step 2   *View display*
As you can see, the *p*-value is 0.061. Recall we reject H_0 only if $p < \alpha$. Therefore *do not* reject H_0.

Step 3   Clear the screen for the next problem. Return to the home screen. Press **CLEAR**

 EXERCISES 6.30–6.44

Learning the Mechanics

6.30 If a hypothesis test were conducted using $\alpha = .05$, for which of the following *p*-values would the null hypothesis be rejected?
a. .06
b. .10
c. .01
d. .001
e. .251
f. .042

6.31 For each α and observed significance level (*p*-value) pair, indicate whether the null hypothesis would be rejected.
a. $\alpha = .05$, *p*-value $= .10$ Fail to reject H_0
b. $\alpha = .10$, *p*-value $= .05$ Reject H_0
c. $\alpha = .01$, *p*-value $= .001$ Reject H_0
d. $\alpha = .025$, *p*-value $= .05$ Fail to reject H_0
e. $\alpha = .10$, *p*-value $= .45$ Fail to reject H_0

6.32 An analyst tested the null hypothesis $\mu \geq 20$ against the alternative hypothesis $\mu < 20$. The analyst reported a *p*-value of .06. What is the smallest value of α for which the null hypothesis would be rejected?

6.33 In a test of $H_0: \mu = 100$ against $H_a: \mu > 100$, the sample data yielded the test statistic $z = 2.17$. Find the *p*-value for the test. $p = .0150$

6.34 In a test of $H_0: \mu = 100$ against $H_a: \mu \neq 100$, the sample data yielded the test statistic $z = 2.17$. Find the *p*-value for the test. $p = .0300$

6.35 In a test of the hypothesis $H_0: \mu = 50$ versus $H_a: \mu > 50$, a sample of $n = 100$ observations possessed mean $\bar{x} = 49.4$ and standard deviation $s = 4.1$. Find and interpret the *p*-value for this test. $p = .9279$

6.36 In a test of the hypothesis $H_0: \mu = 10$ versus $H_a: \mu \neq 10$, a sample of $n = 50$ observations possessed mean $\bar{x} = 10.7$ and standard deviation $s = 3.1$. Find and interpret the *p*-value for this test. $p = .1096$

6.37 Consider a test of $H_0: \mu = 75$ performed using the computer. SPSS reports a two-tailed *p*-value of .1032. Make the appropriate conclusion for each of the following situations:
a. $H_a: \mu < 75$, $z = -1.63$, $\alpha = .05$ Fail to reject H_0
b. $H_a: \mu < 75$, $z = 1.63$, $\alpha = .10$ Fail to reject H_0
c. $H_a: \mu > 75$, $z = 1.63$, $\alpha = .10$ Reject H_0
d. $H_a: \mu \neq 75$, $z = -1.63$, $\alpha = .01$ Fail to reject H_0

Applying the Concepts—Basic

6.38 Refer to the *Psychological Assessment* study of World War II aviator POWs, Exercise 6.25 (p. 289). You tested whether the true mean post-traumatic stress disorder score of World War II aviator POWs is less than 16. Recall that $\bar{x} = 9.00$ and $s = 9.32$ for a sample of $n = 33$ POWs.
  a. Compute the *p*-value of the test. $p \approx 0$
  b. In general, do large *p*-values or small *p*-values support the alternative hypothesis? Explain.

6.39 According to *USA Today* (Dec. 30, 1999), the average age of viewers of MSNBC cable television news programming is 50 years. A random sample of 50 U.S. households that receive cable television programming yielded the following additional information about the ages of MSNBC news viewers: $\bar{x} = 51.3$ years and $s = 7.1$ years. Consider a test to determine whether the average age of all MSNBC's viewers is greater than 50 years.
  a. Find the observed significance level (*p*-value) of the test and interpret its value in the context of the problem. $p = .0985$
  b. Refer to part **a**. Would the *p*-value have been larger or smaller if $\bar{x}$ had been larger? Explain.   Smaller

6.40 Refer to Exercise 6.23, (p. 288), in which a random sample of 49 unemployed men were administered the General Health Questionnaire (GHQ). The sample mean and standard deviation were 10.94 and 5.10, respectively. Denoting the population mean GHQ for unemployed workers by μ, we wish to test the null hypothesis $H_0: \mu = 10$ versus the one-tailed alternative $H_a: \mu > 10$. When the data are analyzed in MINITAB, the results are as shown in the printout below. What conclusion would you reach about the test based on the MINITAB analysis?

6.41 In Exercise 6.27 (p. 289), you tested $H_0: \mu = 8.5$ versus $H_a: \mu \neq 8.5$, where μ is the population mean length-to-width ratio of humerus bones of a particular species of animal. A STATISTIX printout for the hypothesis test is provided (below, right). Locate the *p*-value on the printout and interpret its value. $p = .0002$

Applying the Concepts—Intermediate

6.42 Refer to the *Applied Animal Behaviour Science* (Oct. 2000) study of domestic chickens exposed to a pecking stimulus, Exercise 5.15 (p. 245). Recall that the average number of pecks a chicken takes at a white string over a specified time interval is known to be $\mu = 7.5$ pecks.

In an experiment where 72 chickens were exposed to blue string, the average number of pecks was $\bar{x} = 1.13$ pecks, with a standard deviation of $s = 2.21$ pecks.
  a. On average, are chickens more apt to peck at white string than at blue string? Conduct the appropriate test of hypothesis using $\alpha = .05$. $z = -24.46$
  b. Compare your answer to part **a** with your answer to Exercise 5.15b.
  c. Find the *p*-value for the test and interpret its value.

6.43 Research published in *Nature* (Aug. 27, 1998) revealed that people are more attracted to "feminized" faces, regardless of gender. In one experiment, 50 human subjects viewed both a Japanese female and Caucasian male face on a computer. Using special computer graphics, each subject could morph the faces (by making them more feminine or more masculine) until they attained the "most attractive" face. The level of feminization x (measured as a percentage) was measured.
  a. For the Japanese female face, $\bar{x} = 10.2\%$ and $s = 31.3\%$. The researchers used this sample information to test the null hypothesis of a mean level of feminization equal to 0%. Verify that the test statistic is equal to 2.3.
  b. Refer to part **a**. The researchers reported the *p*-value of the test as $p \approx .02$. Verify and interpret this result.
  c. For the Caucasian male face, $\bar{x} = 15.0\%$ and $s = 25.1\%$. The researchers reported the test statistic (for the test of the null hypothesis stated in part **a**) as 4.23 with an associated *p*-value of approximately 0. Verify and interpret these results. $p \approx 0$

6.44 In Exercise 6.26 (p. 289), you tested $H_0: \mu \geq 10$ versus $H_a: \mu < 10$, where μ is the average number of solder joints inspected per second when the joints are spaced .1 inch apart. An SPSS printout of the hypothesis test is shown on p. 297.
  a. Locate the two-tailed *p*-value of the test shown on the printout. $p = .024$
  b. Adjust the *p*-value for the one-tailed test (if necessary) and interpret its value.

MINITAB Output for Exercise 6.40

```
Test of mu = 10 vs mu > 10

Variable          N      Mean    StDev    SE Mean
GHQ              49     10.94     5.10       0.73

Variable     95.0% Lower Bound        T        P
GHQ                      8.776     1.29    .1016
```

STATISTIX Output for Exercise 6.41

```
ONE-SAMPLE T TEST FOR LWRATIO

NULL HYPOTHESIS: MU =  8.5
ALTERNATIVE HYP: MU <> 8.5

MEAN            9.2576
STD ERROR       0.1880
MEAN - H0       0.7576
LO 95% CI       0.3777
UP 95% CI       1.1375
T                 4.03
DF                  40
P               0.0002

CASES INCLUDED 41    MISSING CASES 0
```

SPSS Output for Exercise 6.44

					95% Confidence Interval of the Difference	
	t	df	Sig. (2-tailed)	Mean Difference	Lower	Upper
NUMBER	-2.333	47	.024	-.7083	-1.3191	-9.76E-02

One-Sample Test

Test Value = 10

6.4   SMALL-SAMPLE TEST OF HYPOTHESIS ABOUT A POPULATION MEAN

Most water-treatment facilities monitor the quality of their drinking water on an hourly basis. One variable monitored is pH, which measures the degree of alkalinity or acidity in the water. A pH below 7.0 is acidic, one above 7.0 is alkaline, and a pH of 7.0 is neutral. One water-treatment plant has a target pH of 8.5 (most try to maintain a slightly alkaline level). The mean and standard deviation of 1 hour's test results, based on 17 water samples at this plant, are

$$\bar{x} = 8.42 \qquad s = .16$$

Does this sample provide sufficient evidence that the mean pH level in the water differs from 8.5?

This inference can be placed in a test of hypothesis framework. We establish the target pH as the null hypothesized value and then utilize a two-tailed alternative that the true mean pH differs from the target:

$$H_0: \mu = 8.5$$
$$H_a: \mu \neq 8.5$$

Recall from Section 5.3 that when we are faced with making inferences about a population mean using the information in a small sample, two problems emerge:

1. The normality of the sampling distribution for $\bar{x}$ does not follow from the Central Limit Theorem when the sample size is small. We must assume that the distribution of measurements from which the sample was selected is approximately normally distributed in order to ensure the approximate normality of the sampling distribution of $\bar{x}$.

2. If the population standard deviation σ is unknown, as is usually the case, then we cannot assume that s will provide a good approximation for σ when the sample size is small. Instead, we must use the t-distribution rather than the standard normal z-distribution to make inferences about the population mean μ.

Therefore, as the test statistic of a small-sample test of a population mean, we use the t statistic:

$$\textit{Test statistic:}\ t = \frac{\bar{x} - \mu_0}{s/\sqrt{n}} = \frac{\bar{x} - 8.5}{s/\sqrt{n}}$$

where μ_0 is the null hypothesized value of the population mean, μ. In our example, $\mu_0 = 8.5$.

To find the rejection region, we must specify the value of α, the probability that the test will lead to rejection of the null hypothesis when it is true, and then consult the t table (Table IV of Appendix A). Using $\alpha = .05$, the two-tailed rejection region is

$$Rejection\ region: t_{\alpha/2} = t_{.025} = 2.120 \text{ with } n - 1 = 16 \text{ degrees of freedom}$$

$$\text{Reject } H_0 \text{ if } t < -2.120 \text{ or } t > 2.120$$

The rejection region is shown in Figure 6.10.

We are now prepared to calculate the test statistic and reach a conclusion:

$$t = \frac{\bar{x} - \mu_0}{s/\sqrt{n}} = \frac{8.42 - 8.50}{.16/\sqrt{17}} = \frac{-.08}{.039} = -2.05$$

Since the calculated value of t does not fall in the rejection region (Figure 6.10), we cannot reject H_0 at the $\alpha = .05$ level of significance. Thus, the water-treatment plant should not conclude that the mean pH differs from the 8.5 target based on the sample evidence.

FIGURE 6.10

Two-Tailed Rejection Region for Small-Sample t-Test

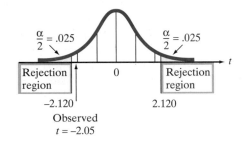

It is interesting to note that the calculated t value, -2.05, is *less than* the .05 level z value, -1.96. The implication is that if we had *incorrectly* used a z statistic for this test, we would have rejected the null hypothesis at the .05 level, concluding that the mean pH level differs from 8.5. The important point is that the statistical procedure to be used must always be closely scrutinized and all the assumptions understood. Many statistical lies are the result of misapplications of otherwise valid procedures.

The technique for conducting a small-sample test of hypothesis about a population mean is summarized in the following box.

TEACHING TIP
Point out the similarities between this box and the one from the large-sample section. Emphasize the assumption that is now necessary.

Small-Sample Test of Hypothesis About μ

ONE-TAILED TEST

$H_0: \mu = \mu_0$

$H_a: \mu < \mu_0$ (or $H_a: \mu > \mu_0$)

Test statistic: $t = \dfrac{\bar{x} - \mu_0}{s/\sqrt{n}}$

Rejection region: $t < -t_{\alpha}$
   (or $t > t_{\alpha}$ when $H_a: \mu > \mu_0$)

TWO-TAILED TEST

$H_0: \mu = \mu_0$

$H_a: \mu \neq \mu_0$

Test statistic: $t = \dfrac{\bar{x} - \mu_0}{s/\sqrt{n}}$

Rejection region: $t < -t_{\alpha/2}$ or $t > t_{\alpha/2}$

where t_{α} and $t_{\alpha/2}$ are based on $(n - 1)$ degrees of freedom
Assumption: A random sample is selected from a population with a relative frequency distribution that is approximately normal.

EXAMPLE 6.5

A major car manufacturer wants to test a new engine to determine whether it meets new air-pollution standards. The mean emission μ of all engines of this type must be less than 20 parts per million of carbon. Ten engines are manufactured for testing purposes, and the emission level of each is determined. The data (in parts per million) are listed below:

⊘ EMISSIONS

15.6	16.2	22.5	20.5	16.4	19.4	16.6	17.9	12.7	13.9

Do the data supply sufficient evidence to allow the manufacturer to conclude that this type of engine meets the pollution standard? Assume that the manufacturer is willing to risk a Type I error with probability $\alpha = .01$.

$t = -3.002$, reject H_0

Solution

The manufacturer wants to support the research hypothesis that the mean emission level μ for all engines of this type is less than 20 parts per million. The elements of this small-sample one-tailed test are

$$H_0: \mu = 20$$
$$H_a: \mu < 20$$
$$\text{Test statistic: } t = \frac{\bar{x} - 20}{s/\sqrt{n}}$$

Assumption: The relative frequency distribution of the population of emission levels for all engines of this type is approximately normal.

Rejection region: For $\alpha = .01$ and df $= n - 1 = 9$, the one-tailed rejection region (see Figure 6.11) is $t < -t_{.01} = -2.821$.

Suggested Exercise 6.55

To calculate the test statistic, we entered the data into a computer and analyzed it using SAS. The SAS printout is shown in Figure 6.12. From the printout, we obtain $\bar{x} = 17.17$ and $s = 2.98$. Substituting these values into the test statistic formula, we get (rounding)

$$t = \frac{\bar{x} - 20}{s/\sqrt{n}} = \frac{17.17 - 20}{2.98/\sqrt{10}} = -3.00$$

Since the calculated t falls in the rejection region (see Figure 6.11), the manufacturer concludes that $\mu < 20$ parts per million and the new engine type meets the pollution standard. Are you satisfied with the reliability associated with this inference? The probability is only $\alpha = .01$ that the test would support the research hypothesis if in fact it were false.

FIGURE 6.11

A t-Distribution with 9 df and the Rejection Region for Example 6.5

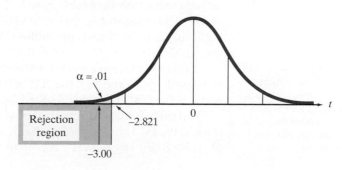

FIGURE 6.12

SAS Analysis of 10 Emission Levels

One Sample t-test for a Mean

Sample Statistics for EMIT

N	Mean	Std. Dev.	Std. Error
10	17.17	2.98	0.94

Hypothesis Test

Null hypothesis: Mean of EMIT => 20
Alternative: Mean of EMIT <   20

t Statistic	Df	Prob > t
-3.002	9	0.0075

EXAMPLE 6.6

.005 < p < .01

Find the observed significance level for the test described in Example 6.5. Interpret the result.

Solution

The test of Example 6.5 was a lower-tailed test: $H_0: \mu = 20$ versus $H_a: \mu < 20$. Since the value of t computed from the sample data was $t = -3.00$, the observed significance level (or p-value) for the test is equal to the probability that t would assume a value less than or equal to -3.00 if in fact H_0 were true. This is equal to the area in the lower tail of the t-distribution (highlighted in Figure 6.13).

FIGURE 6.13

The Observed Significance Level for the Test of Example 6.5

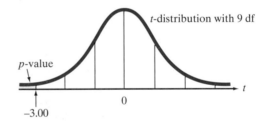

t-distribution with 9 df

p-value

-3.00 0 t

One way to find this area—i.e., the p-value for the test—is to consult the t-table (Table IV in Appendix A). Unlike the table of areas under the normal curve, Table IV gives only the t values corresponding to the areas .100, .050, .025, .010, .005, .001, and .0005. Therefore, we can only approximate the p-value for the test. Since the observed t value was based on 9 degrees of freedom, we use the df = 9 row in Table IV and move across the row until we reach the t values that are closest to the observed $t = -3.00$. [*Note:* We ignore the minus sign.] The t values corresponding to p-values of .010 and .005 are 2.821 and 3.250, respectively. Since the observed t value falls between $t_{.010}$ and $t_{.005}$, the p-value for the test lies between .005 and .010. In other words, $.005 < p\text{-value} < .01$. Thus, we would reject the null hypothesis, $H_0: \mu = 20$ parts per million, for any value of α larger than .01 (the upper bound of the p-value).

A second, more accurate, way to obtain the p-value is to use a statistical software package to conduct the test of hypothesis. Both the test statistic (-3.002) and p-value (.0075) are highlighted on the SAS printout, Figure. 6.12.

You can see that the actual p-value of the test falls within the bounds obtained from Table IV. Thus, the two methods agree; we will reject $H_0: \mu = 20$ in favor of $H_a: \mu < 20$ for any α level larger than .01.

Small-sample inferences typically require more assumptions and provide less information about the population parameter than do large-sample inferences. Nevertheless, the *t*-test is a method of testing a hypothesis about a population mean of a normal distribution when only a small number of observations is available.

> ## What Can Be Done if the Population Relative Frequency Distribution Departs Greatly from Normal?
>
> *Answer:* Use the nonparametric statistical method described in optional Section 6.6.

 EXERCISES 6.45–6.60

Learning the Mechanics

6.45 Under what circumstances should you use the *t*-distribution in testing a hypothesis about a population mean?

6.46 In what ways are the distributions of the *z* statistic and *t* statistic alike? How do they differ?

6.47 For each of the following rejection regions, sketch the sampling distribution of *t*, and indicate the location of the rejection region on your sketch:
a. $t > 1.440$ where df $= 6$
b. $t < -1.782$ where df $= 12$
c. $t < -2.060$ or $t > 2.060$ where df $= 25$

6.48 For each of the rejection regions defined in Exercise 6.47, what is the probability that a Type I error will be made? .10, .05, .05

6.49 A random sample of *n* observations is selected from a normal population to test the null hypothesis that $\mu = 10$. Specify the rejection region for each of the following combinations of H_a, α, and *n*:
a. $H_a: \mu \neq 10; \alpha = .05; n = 14$ $|t| > 2.160$
b. $H_a: \mu > 10; \alpha = .01; n = 24$ $t > 2.500$
c. $H_a: \mu > 10; \alpha = .10; n = 9$ $t > 1.397$
d. $H_a: \mu < 10; \alpha = .01; n = 12$ $t < -2.718$
e. $H_a: \mu \neq 10; \alpha = .10; n = 20$ $|t| > 1.729$
f. $H_a: \mu < 10; \alpha = .05; n = 4$ $t < 2.353$

6.50 The following sample of six measurements was randomly selected from a normally distributed population: 1, 3, −1, 5, 1, 2.

a. Test the null hypothesis that the mean of the population is 3 against the alternative hypothesis, $\mu < 3$. Use $\alpha = .05$. $t = -1.40$
b. Test the null hypothesis that the mean of the population is 3 against the alternative hypothesis, $\mu \neq 3$. Use $\alpha = .05$. $t = 1.40$
c. Find the observed significance level for each test.

6.51 A sample of five measurements, randomly selected from a normally distributed population, resulted in the following summary statistics: $\bar{x} = 4.8$, $s = 1.3$.
a. Test the null hypothesis that the mean of the population is 6 against the alternative hypothesis, $\mu < 6$. Use $\alpha = .05$. $t = -2.064$
b. Test the null hypothesis that the mean of the population is 6 against the alternative hypothesis, $\mu \neq 6$. Use $\alpha = .05$. $t = -2.064$
c. Find the observed significance level for each test.

6.52 MINITAB is used to conduct a *t*-test for the null hypothesis $H_0: \mu = 1,000$ versus the alternative hypothesis $H_a: \mu > 1,000$ based on a sample of 17 observations. The software's output is shown below.
a. What assumptions are necessary for the validity of this procedure? $p = .0382$
b. Interpret the results of the test.
c. Suppose the alternative hypothesis had been the two-tailed $H_a: \mu \neq 1,000$. If the *t* statistic were unchanged, what would the *p*-value be for this test? Interpret the *p*-value for the two-tailed test.

MINITAB Output for Exercise 6.52

Test of mu = 1000 vs mu > 1000

Variable	N	Mean	StDev	SE Mean
X	17	1020	43.54	10.56

Variable	95.0% Lower Bound	T	P
X	981.62	1.89	.0382

Applying the Concepts—Basic

6.53 Refer to the *Psychological Reports* (Aug. 1997) study of college students who completed the Dental Anxiety Scale, Exercise 4.54 (p. 201). Recall that scores range from 0 (no anxiety) to 20 (extreme anxiety). Summary statistics for the scores of the 27 students who completed the questionnaire follow: $\bar{x} = 10.7$, $s = 3.6$. Conduct a test of hypothesis to determine if the mean Dental Anxiety Scale score for the population of college students differs from $\mu = 11$. Use $\alpha = .05$. $t = -.43$

6.54 The Cleveland Casting Plant produces iron automotive castings for Ford Motor Company. When the process is stable, the target pouring temperature of the molten iron is 2,550 degrees (*Quality Engineering*, Vol. 7, 1995). The pouring temperatures (in degrees Fahrenheit) for a random sample of 10 crankshafts produced at the plant are listed in the table. Conduct a test to determine whether the true mean pouring temperature differs from the target setting. Test using $\alpha = .01$. Use the accompanying SAS printout to carry out the test. $t = 1.210$

🖰 IRONTEMP

2,543	2,541	2,544	2,620	2,560	2,559	2,562
2,553	2,552	2,553				

Source: Price, B., & Barth, B. "A structural model relating process inputs and final product characteristics." *Quality Engineering*, Vol. 7, No. 4, 1995, p. 696 (Table 2).

SAS Output for Exercise 6.54

One Sample t-test for a Mean

Sample Statistics for IRONTEMP

N	Mean	Std. Dev.	Std. Error
10	2558.70	22.75	7.19

Hypothesis Test

Null hypothesis: Mean of IRONTEMP = 2550
Alternative: Mean of IRONTEMP ^= 2550

t Statistic	Df	Prob > t
1.210	9	0.2573

6.55 In Exercise 5.29 (p. 256), you analyzed data from a study of the diets of spinifex pigeons. The data, extracted from the *Australian Journal of Zoology*, are reproduced in the table. Each measurement in the data set represents the weight (in grams) of the crop content of a spinifex pigeon. MINITAB was used to conduct a test of the null hypothesis $H_0: \mu = 1$, where μ represents the mean weight in the crops of all spinifex pigeons. The resulting printout is shown at the bottom of the page.

🖰 PIGEONS

.457	3.751	.238	2.967	2.509	1.384	1.454	.818
.335	1.436	1.603	1.309	.201	.530	2.144	.834

a. Identify the alternative hypothesis specified on the printout. $H_a: \mu \neq 1$

b. Locate the value of the test statistic and give its value.

c. Find and interpret the *p*-value of the test shown on the printout.

Applying the Concepts—Intermediate

6.56 A study was conducted to evaluate the effectiveness of a new mosquito repellent designed by the U.S. Army to be applied as camouflage face paint (*Journal of the Mosquito Control Association*, June 1995). The repellent was applied to the forearms of five volunteers and then they were exposed to fifteen active mosquitos for a ten-hour period. The percentage of the forearm surface area protected from bites (called percent repellency) was calculated for each of the five volunteers. For one color of paint (loam), the following summary statistics were obtained:

$$\bar{x} = 83\% \quad s = 15\%$$

a. The new repellent is considered effective if it provides a percent repellency of at least 95. Conduct a test to determine whether the mean repellency percentage of the new mosquito repellent is less than 95. Test using $\alpha = .10$. $t = -1.79$

b. What assumptions are required for the hypothesis test in part **a** to be valid?

6.57 The Mississippi Department of Transportation collected data on the number of cracks (called *crack intensity*) in an undivided two-lane highway using van-mounted state-of-the-art video technology (*Journal of Infrastructure Systems*, Mar. 1995). The mean number of cracks found in a sample of eight 50-meter sections of the highway was

MINITAB Output for Exercise 6.55

Test of mu = 1 vs mu not = 1

Variable	N	Mean	StDev	SE Mean
WEIGHT	16	1.373	1.034	0.258

Variable	95.0% CI	T	P
WEIGHT	(0.822, 1.924)	1.44	0.169

$\bar{x} = .210$, with a variance of $s^2 = .011$. Suppose the American Association of State Highway and Transportation Officials (AASHTO) recommends a maximum mean crack intensity of .100 for safety purposes. Test the hypothesis that the true mean crack intensity of the Mississippi highway exceeds the AASHTO recommended maximum. Use $\alpha = .01$. $t = 2.97$

6.58 One of the most feared predators in the ocean is the great white shark. It is known that the white shark grows to a mean length of 21 feet; however, one marine biologist believes that great white sharks off the Bermuda coast grow much longer owing to unusual feeding habits. To test this claim, some full-grown great white sharks were captured off the Bermuda coast, measured, and then set free. However, because the capture of sharks is difficult, costly, and very dangerous, only three specimens were sampled. Their lengths were 24, 20, and 22 feet.

a. Do the data provide sufficient evidence to support the marine biologist's claim? Use $\alpha = .10$. $t = .87$

b. Give the approximate observed significance level for the test in part **a**, and interpret its value. $p > .10$

c. What assumptions must be made in order to carry out the test?   Normal population

d. Do you think these assumptions are likely to be satisfied in this sampling situation?

6.59 The SCL-90-R is a 90-item symptom inventory checklist designed to reflect the psychological status of an individual. Each symptom (e.g., obsessive-compulsive behavior) is scored on a scale of 0 (none) to 4 (extreme). The total of these scores yields an individual's Positive Symptom Total (PST). "Normal" individuals are known to have a mean PST of about 40. The *Journal of Head Trauma Rehabilitation* (Apr. 1995) reported that a sample of 23 patients diagnosed with mild to moderate traumatic brain injury had a mean PST score of $\bar{x} = 48.43$ and a standard deviation of $s = 20.76$. Is there sufficient evidence to claim that the true mean PST score of all patients with mild to moderate traumatic brain injury exceeds the "normal" value of 40? Test using $\alpha = .05$.

Applying the Concepts—Advanced

6.60 "Hot Tamales" are chewy, cinnamon-flavored candies. A bulk vending machine is known to dispense, on average, 15 Hot Tamales per bag. *Chance* (Fall 2000) published an article on a classroom project in which students were required to purchase bags of Hot Tamales from the machine and count the number of candies per bag. One student group claimed they purchased five bags that had the following candy counts: 25, 23, 21, 21, and 20. There was some question as to whether the students had fabricated the data. Use a hypothesis test to gain insight into whether or not the data collected by the students are fabricated. Use a level of significance that gives the benefit of the doubt to the students. $t = 7.83$

6.5   LARGE-SAMPLE TEST OF HYPOTHESIS ABOUT A POPULATION PROPORTION

Inferences about population proportions (or percentages) are often made in the context of the probability, p, of "success" for a binomial distribution. We saw how to use large samples from binomial distributions to form confidence intervals for p in Section 5.3. We now consider tests of hypotheses about p.

Consider, for example, a method currently used by doctors to screen women for possible breast cancer. The method fails to detect cancer in 20% of the women who actually have the disease. Suppose a new method has been developed that researchers hope will detect cancer more accurately. This new method was used to screen a random sample of 140 women known to have breast cancer. Of these, the new method failed to detect cancer in 12 women. Does this sample provide evidence that the failure rate of the new method differs from the one currently in use?

We first view this as a binomial experiment with 140 screened women as the trials and failure to detect breast cancer as "Success" (in binomial terminology). Let p represent the probability that the new method fails to detect the breast cancer. If the new method is no better than the current one, then the failure rate is $p = .2$. On the other hand, if the new method is either better or worse than the current method, then the failure rate is either smaller or larger than 20%; that is, $p \neq .2$.

We can now place the problem in the context of a test of hypothesis:

$$H_0: p = .2$$
$$H_a: p \neq .2$$

Recall that the sample proportion, $\hat{p}$, is really just the sample mean of the outcomes of the individual binomial trials and, as such, is approximately normally distributed (for large samples) according to the Central Limit Theorem. Thus, for large samples we can use the standard normal z as the test statistic:

$$\textit{Test statistic:} \quad z = \frac{\text{Sample proportion } - \text{ Null hypothesized proportion}}{\text{Standard deviation of sample proportion}}$$

$$= \frac{\hat{p} - p_0}{\sigma_{\hat{p}}}$$

where we use the symbol p_0 to represent the null hypothesized value of p.

Rejection region: We use the standard normal distribution to find the appropriate rejection region for the specified value of α. Using $\alpha = .05$, the two-tailed rejection region is

$$z < -z_{\alpha/2} = -z_{.025} = -1.96 \quad \text{or} \quad z > z_{\alpha/2} = z_{.025} = 1.96$$

See Figure 6.14.

FIGURE 6.14

Rejection Region for Breast Cancer Example

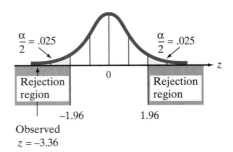

We are now prepared to calculate the value of the test statistic. Before doing so, we want to be sure that the sample size is large enough to ensure that the normal approximation for the sampling distribution of $\hat{p}$ is reasonable. To check this, we calculate a 3-standard-deviation interval around the null hypothesized value, p_0, which is assumed to be the true value of p until our test procedure proves otherwise. Recall that $\sigma_{\hat{p}} = \sqrt{pq/n}$ and that we need an estimate of the product pq in order to calculate a numerical value of the test statistic z. Since the null hypothesized value is generally the accepted-until-proven-otherwise value, we use the value of $p_0 q_0$ (where $q_0 = 1 - p_0$) to estimate pq in the calculation of z. Thus,

$$\sigma_{\hat{p}} = \sqrt{\frac{pq}{n}} \approx \sqrt{\frac{p_0 q_0}{n}} = \sqrt{\frac{(.2)(.8)}{140}} = .034$$

and the 3-standard-deviation interval around p_0 is

$$p_0 \pm 3\sigma_{\hat{p}} \approx .2 \pm 3(.034) = (.166, .234)$$

As long as this interval does not contain 0 or 1 (i.e., is completely contained in the interval 0 to 1), as is the case here, the normal distribution will provide a reasonable approximation for the sampling distribution of $\hat{p}$.

Returning to the hypothesis test at hand, the proportion of the screenings that failed to detect breast cancer is

$$\hat{p} = \frac{12}{140} = .086$$

Finally, we calculate the number of standard deviations (the z value) between the sampled and hypothesized value of the binomial proportion:

$$z = \frac{\hat{p} - p_0}{\sigma_{\hat{p}}} = \frac{\hat{p} - p_0}{\sqrt{p_0 q_0 / n}} = \frac{.086 - .2}{.034} = \frac{-.114}{.034} = -3.36$$

The implication is that the observed sample proportion is (approximately) 3.36 standard deviations below the null hypothesized proportion .2 (Figure 6.14). Therefore, we reject the null hypothesis, concluding at the .05 level of significance that the true failure rate of the new method for detecting breast cancer differs from .20. Since $\hat{p} = .086$, it appears that the new method is better (i.e., has a smaller failure rate) than the method currently in use. (To estimate the magnitude of the failure rate for the new method, a confidence interval can be constructed.)

Suggested Exercise 6.71

The test of hypothesis about a population proportion p is summarized in the next box. Note that the procedure is entirely analogous to that used for conducting large-sample tests about a population mean.

Large-Sample Test of Hypothesis About p

TEACHING TIP
Point out the similarities between this box and the one from the large-sample section for testing population means. Emphasize the sample size assumption that is necessary now.

ONE-TAILED TEST	TWO-TAILED TEST
H_0: $p = p_0$	H_0: $p = p_0$
H_a: $p < p_0$ (or H_a: $p > p_0$)	H_a: $p \neq p_0$

Test statistic: $z = \dfrac{\hat{p} - p_0}{\sigma_{\hat{p}}}$ *Test statistic:* $z = \dfrac{\hat{p} - p_0}{\sigma_{\hat{p}}}$

where $p_0 =$ hypothesized value of p, $\sigma_{\hat{p}} = \sqrt{p_0 q_0 / n}$ and $q_0 = 1 - p_0$

Rejection region: $z < -z_\alpha$ *Rejection region:* $z < -z_{\alpha/2}$ or $z > z_{\alpha/2}$
(or $z > z_\alpha$ when H_a: $p > p_0$)

Assumption: The experiment is binomial, and the sample size is large enough that the interval $p_0 \pm 3\sigma_{\hat{p}}$ does not include 0 or 1.

■ EXAMPLE 6.7

z = −1.35, Fail to reject H_0

The reputations (and hence sales) of many businesses can be severely damaged by shipments of manufactured items that contain a large percentage of defectives. For example, a manufacturer of alkaline batteries may want to be reasonably certain that fewer than 5% of its batteries are defective. Suppose 300 batteries are randomly selected from a very large shipment; each is tested and 10 defective batteries are found. Does this provide sufficient evidence for the manufacturer to conclude that the fraction defective in the entire shipment is less than .05? Use $\alpha = .01$.

Solution

Before conducting the test of hypothesis, we check to determine whether the sample size is large enough to use the normal approximation for the sampling distribution of $\hat{p}$. The criterion is tested by the interval

$$p_0 \pm 3\sigma_{\hat{p}} = p_0 \pm 3\sqrt{\frac{p_0 q_0}{n}} = .05 \pm 3\sqrt{\frac{(.05)(.95)}{300}}$$

$$= .05 \pm .04 \quad \text{or} \quad (.01, .09)$$

FIGURE 6.15

Rejection Region for
Example 6.7

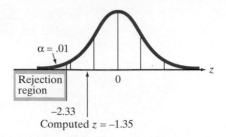

$\alpha = .01$

Rejection region

0

z

-2.33

Computed $z = -1.35$

Since the interval lies within the interval $(0, 1)$, the normal approximation will be adequate.

The objective of the sampling is to determine whether there is sufficient evidence to indicate that the fraction defective, p, is less than .05. Consequently, we will test the null hypothesis that $p = .05$ against the alternative hypothesis that $p < .05$. The elements of the test are

H_0: $p = .05$
H_a: $p < .05$

Test statistic: $z = \dfrac{\hat{p} - p_0}{\sigma_{\hat{p}}}$

Rejection region: $z < -z_{.01} = -2.33$ (see Figure 6.15)

We now calculate the test statistic:

$$z = \frac{\hat{p} - .05}{\sigma_{\hat{p}}} = \frac{(10/300) - .05}{\sqrt{p_0 q_0/n}} = \frac{.033 - .05}{\sqrt{p_0 q_0/300}}$$

Notice that we use p_0 to calculate $\sigma_{\hat{p}}$ because, in contrast to calculating $\sigma_{\hat{p}}$ for a confidence interval, the test statistic is computed on the assumption that the null hypothesis is true—that is, $p = p_0$. Therefore, substituting the values for $\hat{p}$ and p_0 into the z statistic, we obtain

$$z \approx \frac{-.017}{\sqrt{(.05)(.95)/300}} = \frac{-.017}{.0126} = -1.35$$

As shown in Figure 6.15, the calculated z value does not fall in the rejection region. Therefore, there is insufficient evidence at the .01 level of significance to indicate that the shipment contains fewer than 5% defective batteries.   ■

EXAMPLE 6.8

$p = .0885$

In Example 6.7 we found that we did not have sufficient evidence, at the $\alpha = .01$ level of significance, to indicate that the fraction defective p of alkaline batteries was less than $p = .05$. How strong was the weight of evidence favoring the alternative hypothesis (H_a: $p < .05$)? Find the observed significance level (p-value) for the test.

Solution

The computed value of the test statistic z was $z = -1.35$. Therefore, for this lower-tailed test, the observed significance level is

$$p\text{-value} = P(z \le -1.35)$$

This lower-tail area is shown in Figure 6.16. The area between $z = 0$ and $z = 1.35$ is given in Table III in Appendix A as .4115. Therefore, the observed significance level is $.5 - .4115 = .0885$. Note that this probability is quite small. Although we

FIGURE 6.16

The Observed Significance Level for Example 6.8

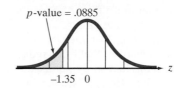

p-value = .0885

−1.35 0

z

did not reject H_0: $p = .05$ at $\alpha = .01$, the probability of observing a z value as small as or smaller than -1.35 is only .0885 if in fact H_0 is true. Therefore, we would reject H_0 if we choose $\alpha = .10$ (since the observed significance level is less than .10), and we would not reject H_0 (the conclusion of Example 6.7) if we choose $\alpha = .05$ or $\alpha = .01$

∎

Small-sample test procedures are also available for p. These are omitted from our discussion because most surveys use samples that are large enough to employ the large-sample tests presented in this section.

 EXERCISES 6.61–6.75

Learning the Mechanics

6.61 For the binomial sample sizes and null hypothesized values of p in each part, determine whether the sample size is large enough to use the normal approximation methodology presented in this section to conduct a test of the null hypothesis H_0: $p = p_0$.

a. $n = 500$, $p_0 = .05$ Yes
b. $n = 100$, $p_0 = .99$ No
c. $n = 50$, $p_0 = .2$ Yes
d. $n = 20$, $p_0 = .2$ No
e. $n = 10$, $p_0 = .4$ No

6.62 Suppose a random sample of 100 observations from a binomial population gives a value of $\hat{p} = .69$ and you wish to test the null hypothesis that the population parameter p is equal to .75 against the alternative hypothesis that p is less than .75.

a. Noting that $\hat{p} = .69$, what does your intuition tell you? Does the value of $\hat{p}$ appear to contradict the null hypothesis?
b. Use the large-sample z-test to test H_0: $p = .75$ against the alternative hypothesis H_a: $p < .75$. Use $\alpha = .05$. How do the test results compare with your intuitive decision from part **a**? $z = -1.39$
c. Find and interpret the observed significance level of the test you conducted in part **b**. $p = .0823$

6.63 Suppose the sample in Exercise 6.62 has produced $\hat{p} = .84$ and we wish to test H_0: $p = .9$ against the alternative H_a: $p < .9$.

a. Calculate the value of the z statistic for this test.
b. Note that the numerator of the z statistic ($\hat{p} - p_0 = .84 - .90 = -.06$) is the same as for Exercise 6.62. Considering this, why is the absolute value of z for this exercise larger than that calculated in Exercise 6.62?
c. Complete the test using $\alpha = .05$ and interpret the result.   Reject H_0

d. Find the observed significance level for the test and interpret its value. $p = .0228$

6.64 A random sample of 100 observations is selected from a binomial population with unknown probability of success p. The computed value of $\hat{p}$ is equal to .74.

a. Test H_0: $p = .65$ against H_a: $p > .65$. Use $\alpha = .01$.
b. Test H_0: $p = .65$ against H_a: $p > .65$. Use $\alpha = .10$.
c. Test H_0: $p = .90$ against H_a: $p \neq .90$. Use $\alpha = .05$.
d. Form a 95% confidence interval for p. $.74 \pm .09$
e. Form a 99% confidence interval for p. $.74 \pm .11$

6.65 Refer to Exercise 5.40 (p. 263), in which 50 consumers taste-tested a new snack food.

a. Test H_0: $p = .5$ against H_a: $p > .5$, where p is the proportion of customers who do not like the snack food. Use $\alpha = .10$. $z = 1.13$
b. Report the observed significance level of your test.

6.66 A statistics student will use statistical software to test the null hypothesis H_0: $p = .5$ against the one-tailed alternative, H_a: $p > .5$. A sample of 500 observations are input into the computer, which returns the following result: $z = .44$, one-tailed p-value $= .3300$.

a. The student concludes, based on the p-value, that there is a 33% chance that the alternative hypothesis is true. Do you agree? If not, correct the interpretation.   No
b. How would the p-value change if the alternative hypothesis were two-tailed, H_a: $p \neq .5$? Interpret this p-value. p-value $= .6600$

Applying the Concepts—Basic

6.67 Examining data collected on 835 males from the National Youth Survey (a longitudinal survey of a random sample of U.S. households), researchers at Carnegie Mellon University found that 401 of the male youths were raised in a single-parent family (*Sociological Methods & Research*, Feb. 2001). Does this information allow you to conclude that more than 45% of male youths are raised in a single-parent family? Test at $\alpha = .05$.

6.68 Pond's Age-Defying Complex, a cream with alpha-hydroxy acid, advertises that it can reduce wrinkles and improve the skin. In a study published in *Archives of Dermatology* (June 1996), 33 women over age 40 used a cream with alpha-hydroxy acid for twenty-two weeks. At the end of the study period, 23 of the women exhibited skin improvement (as judged by a dermatologist).

    a. Is this evidence that the cream will improve the skin of more than 60% of women over age 40? Test using $\alpha = .05$. $z = 1.14$

    b. Find and interpret the *p*-value of the test.

6.69 According to *Forbes* (Dec. 13, 1999), Creative Good, a New York consulting firm, claims that 40% of shoppers fail in their attempts to purchase merchandise online because Web sites are too complex. Another consulting firm asked a random sample of 60 online shoppers to each test a different randomly selected e-commerce Web site. Only 15 reported sufficient frustration with their sites to deter making a purchase. Do these data provide sufficient evidence to reject the claim made by Creative Good? Conduct the appropriate hypothesis test using $\alpha = .01$. $z = -2.37$

Applying the Concepts—Intermediate

6.70 Any sentence that contains an animate noun as a direct object and another noun as a second object (e.g., "Sue offered Ann a cookie.") is termed a double object dative (DOD). The connection between certain verbs and a DOD was investigated in *Applied Psycholinguistics* (June 1998). The subjects were 35 native English speakers who were enrolled in an introductory English composition course at a Hawaiian community college. After viewing pictures of a family, each subject was asked to write a sentence about the family using a specified verb. Of the 35 sentences using the verb "buy," 10 had a DOD structure. Conduct a test to determine if the true fraction of sentences with the verb "buy" that are DODs is less than $\frac{1}{3}$. Use $\alpha = .05$. $z = -.59$

6.71 Refer to the *Nature* (Aug. 27, 1998) study of facial characteristics that are deemed attractive, Exercise 6.43 (p. 296). In another experiment, 67 human subjects viewed side-by-side an image of a Caucasian male face and the same image 50% masculinized. Each subject was asked to select the facial image that they deemed more attractive. Fifty-eight of the 67 subjects felt that masculinization of face shape decreased attractiveness of the male face. The researchers used this sample information to test whether the subjects showed preference for either the unaltered or morphed male face.

    a. Set up the null and alternative hypotheses for this test. $H_0: p = .5$; $H_a: p \neq 5$

    b. Compute the test statistic. $z = 5.99$

    c. The researchers reported *p*-value ≈ 0 for the test. Do you agree?   Yes

    d. Make the appropriate conclusion in the words of the problem. Use $\alpha = .01$.   Reject H_0

6.72 *Meteoritics* (Mar. 1995) reported the results of a study of lunar soil evolution. Data were obtained from the Apollo 16 mission to the moon, during which a 62-cm core was extracted from the soil near the landing site. Monominer-alic grains of lunar soil were separated out and examined for coating with dust and glass fragments. Each grain was then classified as coated or uncoated. Of interest is the "coat index," that is, the proportion of grains that are coated. According to soil evolution theory, the coat index will exceed .5 at the top of the core, equal .5 in the middle of the core, and fall below .5 at the bottom of the core. Use the summary data in the accompanying table to test each part of the 3-part theory. Use $\alpha = .05$ for each test.

	Location (depth)		
	Top (4.25 cm)	**Middle (28.1 cm)**	**Bottom (54.5 cm)**
Number of grains sampled	84	73	81
Number coated	64	35	29

Source: Basu, A., and McKay, D. S. "Lunar soil evolution processes and Apollo 16 core 60013/60014." *Meteoritics*, Vol. 30, No. 2, Mar. 1995, p. 166 (Table 2).

6.73 The *placebo effect* describes the phenomenon of improvement in the condition of a patient taking a placebo—a pill that looks and tastes real but contains no medically active chemicals. Physicians at a clinic in La Jolla, California, gave what they thought were drugs to 7,000 asthma, ulcer, and herpes patients. Although the doctors later learned that the drugs were really placebos, 70% of the patients reported an improved condition (*Forbes*, May 22, 1995). Use this information to test (at $\alpha = .05$) the placebo effect at the clinic. Assume that if the placebo is ineffective, the probability of a patient's condition improving is .5. $z = 33.47$

Applying the Concepts—Advanced

6.74 In Exercise 4.36 (p. 188) you read about a nationwide survey that claimed that 60% of parents with young children condone spanking their child as a regular form of punishment (*Tampa Tribune*, Oct. 5, 2000). In a random sample of 100 parents with young children, how many parents would need to say they condone spanking as a form of punishment in order to refute the claim?

6.75 To get their names on the ballot of a local election, political candidates often must obtain petitions bearing the signatures of a minimum number of registered voters. In Pinellas County, Florida, a certain political candidate obtained petitions with 18,200 signatures (*St. Petersburg Times*, Apr. 7, 1992). To verify that the names on the petitions were signed by actual registered voters, election officials randomly sampled 100 of the names and checked each for authenticity. Only two were invalid signatures.

    a. Is 98 out of 100 verified signatures sufficient to believe that more than 17,000 of the total 18,200 signatures are valid? $z = 1.85$

    b. Repeat part **a** if only 16,000 valid signatures are required. $z = 3.09$

6.6 A NONPARAMETRIC TEST ABOUT A POPULATION MEDIAN (OPTIONAL)

In Sections 6.2–6.4 we utilized the z and t statistics for testing hypotheses about a population mean. The z statistic is appropriate for large random samples selected from "general" populations—that is, with few limitations on the probability distribution of the underlying population. The t statistic was developed for small-sample tests in which the sample is selected at random from a *normal* distribution. The question is: How can we conduct a test of hypothesis when we have a small sample from a *nonnormal* distribution?

The answer is: Use a *distribution-free* procedure that requires fewer or less stringent assumptions about the underlying population—called a *nonparametric* method.

> **DEFINITION 6.2**
> **Distribution-free tests** are statistical tests that do not rely on any underlying assumptions about the probability distribution of the sampled population.
>
> **DEFINITION 6.3**
> The branch of inferential statistics devoted to distribution-free tests is called **nonparametrics**.

The **sign test** is a relatively simple nonparametric procedure for testing hypotheses about the central tendency of a nonnormal probability distribution. Note that we used the phrase *central tendency* rather than *population mean*. This is because the sign test, like many nonparametric procedures, provides inferences about the population *median* rather than the population mean μ. Denoting the population median by the Greek letter η, we know (Chapter 2) that η is the 50th percentile of the distribution (Figure 6.17) and as such is less affected by the skewness of the distribution and the presence of outliers (extreme observations). Since the nonparametric test must be suitable for all distributions, not just the normal, it is reasonable for nonparametric tests to focus on the more robust (less sensitive to extreme values) measure of central tendency, the median.

FIGURE 6.17

Location of the Population Median, η

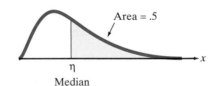

For example, increasing numbers of both private and public agencies are requiring their employees to submit to tests for substance abuse. One laboratory that conducts such testing has developed a system with a normalized measurement scale, in which values less than 1.00 indicate "normal" ranges and values equal to or greater than 1.00 are indicative of potential substance abuse. The lab reports a normal result as long as the median level for an individual is less than 1.00. Eight independent measurements of each individual's sample are made. One individual's results are shown in Table 6.4.

⊘ SUBABUSE

TABLE 6.4   Substance Abuse Test Results

.78	.51	3.79	.23	.77	.98	.96	.89

If the objective is to determine whether the *population* median (that is, the true median level if an infinitely large number of measurements were made on the same individual sample) is less than 1.00, we establish that as our alternative hypothesis and test

$$H_0: \eta = 1.00$$
$$H_a: \eta < 1.00$$

The one-tailed sign test is conducted by counting the number of sample measurements that "favor" the alternative hypothesis—in this case, the number that are less than 1.00. If the null hypothesis is true, we expect approximately half of the measurements to fall on each side of the hypothesized median and if the alternative is true, we expect significantly more than half to favor the alternative—that is, to be less than 1.00. Thus,

TEACHING TIP
Point out that the median is useful to test because the population contains 50% of the observations both above and below its value.

Test statistic: S = Number of measurements less than 1.00, the null hypothesized median

If we wish to conduct the test at the $\alpha = .05$ level of significance, the rejection region can be expressed in terms of the observed significance level, or p-value, of the test:

Rejection region: p-value $\leq .05$

In this example, $S = 7$ of the 8 measurements are less than 1.00. To determine the observed significance level associated with this outcome, we note that the number of measurements less than 1.00 is a binomial random variable (check the binomial characteristics presented in Chapter 4), and *if* H_0 *is true*, the binomial probability p that a measurement lies below (or above) the median 1.00 is equal to .5 (Figure 6.17). What is the probability that a result is *as contrary to or more contrary to* H_0 than the one observed? That is, what is the probability that 7 *or more* of 8 binomial measurements will result in Success (be less than 1.00) if the probability of Success is .5? Binomial Table II in Appendix A (using $n = 8$ and $p = .5$) indicates that

Suggested Exercise 6.83

$$P(x \geq 7) = 1 - P(x \leq 6) = 1 - .965 = .035$$

Thus, the probability that at least 7 of 8 measurements would be less than 1.00 *if the true median were 1.00* is only .035. The p-value of the test is therefore .035.

This p-value can also be obtained by using a statistical software package. The MINITAB printout of the analysis is shown in Figure 6.18, with the p-value highlighted on the printout. Since $p = .035$ is less than $\alpha = .05$, we conclude that this sample provides sufficient evidence to reject the null hypothesis. The implication of this rejection is that the laboratory can conclude at the $\alpha = .05$ level of significance that the true median level for the tested individual is less than 1.00. However, we note that one of the measurements greatly exceeds the others, with a value of 3.79, and deserves special attention. Note that this large measurement is an outlier that would make the use of a t-test and its

FIGURE 6.18
MINITAB Printout of Sign Test

```
Sign test of median = 1.000 versus  <  1.000

                N  Below  Equal  Above        P    Median
READING         8      7      0      1   0.0352    0.8350
```

concomitant assumption of normality dubious. The only assumption necessary to ensure the validity of the sign test is that the probability distribution of measurements is continuous.

The use of the sign test for testing hypotheses about population medians is summarized in the next box.

<table>
<tr><td colspan="2">

Sign Test for a Population Median η

</td></tr>
<tr>
<td>

ONE-TAILED TEST

$H_0: \eta = \eta_0$

$H_a: \eta > \eta_0$ [or $H_a: \eta < \eta_0$]

Test statistic:

S = Number of sample measurements greater than η_0 [or S = number of measurements less than η_0]

Observed significance level:

p-value = $P(x \geq S)$

</td>
<td>

TWO-TAILED TEST

$H_0: \eta = \eta_0$

$H_a: \eta \neq \eta_0$

Test statistic:

S = Larger of S_1 and S_2, where S_1 is the number of measurements less than η_0 and S_2 is the number of measurements greater than η_0

Observed significance level:

p-value = $2P(x \geq S)$

</td>
</tr>
<tr>
<td colspan="2">

where x has a binomial distribution with parameters n and $p = .5$. (Use Table II, Appendix A.)

Rejection region: Reject H_0 if p-value $\leq \alpha$

Assumption: The sample is selected randomly from a continuous probability distribution. [*Note:* No assumptions need to be made about the shape of the probability distribution.]

</td>
</tr>
</table>

TEACHING TIP
Point out that no assumptions concerning the population are necessary. This procedure is ideal when the parametric test cannot be used to test μ.

Recall that the normal probability distribution provides a good approximation for the binomial distribution when the sample size is large. For tests about the median of a distribution, the null hypothesis implies that $p = .5$, and the normal distribution provides a good approximation if $n \geq 10$. (Samples with $n \geq 10$ satisfy the condition that $np \pm 2\sqrt{npq}$ is contained in the interval 0 to n.) Thus, we can use the standard normal z-distribution to conduct the sign test for large samples. The large-sample sign test is summarized in the next box.

EXAMPLE 6.9

$z = 1.565$; fail to reject H_0

A manufacturer of compact disk (CD) players has established that the median time to failure for its players is 5,250 hours of utilization. A sample of 20 CDs from a competitor is obtained, and they are continuously tested until each fails. The 20 failure times range from 5 hours (a "defective" player) to 6,575 hours, and 14 of the 20 exceed 5,250 hours. Is there evidence that the median failure time of the competitor differs from 5,250 hours? Use $\alpha = .10$.

Solution

The null and alternative hypotheses of interest are

$$H_0: \eta = 5,250 \text{ hours}$$
$$H_a: \eta \neq 5,250 \text{ hours}$$

> ## Large-Sample Sign Test for a Population Median η
>
> **ONE-TAILED TEST**   **TWO-TAILED TEST**
>
> $H_0: \eta = \eta_0$ $H_0: \eta = \eta_0$
>
> $H_a: \eta > \eta_0$ $H_a: \eta \neq \eta_0$
> [or $H_a: \eta < \eta_0$]
>
> $$\text{Test statistic:} \quad z = \frac{(S - .5) - .5n}{.5\sqrt{n}}$$
>
> [*Note:* S is calculated as shown in the previous box. We subtract .5 from S as the "correction for continuity." The null hypothesized mean value is $np = .5n$, and the standard deviation is
>
> $$\sqrt{npq} = \sqrt{n(.5)(.5)} = .5\sqrt{n}$$
>
> See Chapter 4 for details on the normal approximation to the binomial distribution.]
>
> *Rejection region:* $z > z_\alpha$ *Rejection region:* $z > z_{\alpha/2}$
>
> where tabulated z values can be found in Table III, Appendix A.

Since $n \geq 10$, we use the standard normal z statistic:

$$\text{Test statistic:} \quad z = \frac{(S - .5) - .5n}{.5\sqrt{n}}$$

where S is the maximum of S_1, the number of measurements greater than 5,250, and S_2, the number of measurements less than 5,250.

$$\text{Rejection region:} \, z > 1.645, \text{ where } z_{\alpha/2} = z_{.05} = 1.645$$

Assumptions: The distribution of the failure times is continuous (time is a continuous variable), but nothing is assumed about the shape of its probability distribution.

Since the number of measurements exceeding 5,250 is $S_2 = 14$, then the number of measurements less than 5,250 is $S_1 = 6$. Consequently, $S = 14$, the greater of S_1 and S_2. The calculated z statistic is therefore

$$z = \frac{(S - .5) - .5n}{.5\sqrt{n}} = \frac{13.5 - 10}{.5\sqrt{20}} = \frac{3.5}{2.236} = 1.565$$

The value of z is not in the rejection region, so we cannot reject the null hypothesis at the $\alpha = .10$ level of significance. Thus, the CD manufacturer should not conclude, on the basis of this sample, that its competitor's CDs have a median failure time that differs from 5,250 hours.   ∎

    The one-sample nonparametric sign test for a median provides an alternative to the *t*-test for small samples from nonnormal distributions. However, if the distribution is approximately normal, the *t*-test provides a more powerful test about the central tendency of the distribution.

EXERCISES 6.76–6.85

Learning the Mechanics

6.76 What is the probability that a randomly selected observation exceeds the
  a. Mean of a normal distribution? .5
  b. Median of a normal distribution? .5
  c. Mean of a nonnormal distribution? Unknown
  d. Median of a nonnormal distribution? .5

6.77 Use Table II of Appendix A to calculate the following binomial probabilities:
  a. $P(x \geq 6)$ when $n = 7$ and $p = .5$ .063
  b. $P(x \geq 5)$ when $n = 9$ and $p = .5$ .500
  c. $P(x \geq 8)$ when $n = 8$ and $p = .5$ .004
  d. $P(x \geq 10)$ when $n = 15$ and $p = .5$. Also use the normal approximation to calculate this probability, then compare the approximation with the exact value.
  e. $P(x \geq 15)$ when $n = 25$ and $p = .5$. Also use the normal approximation to calculate this probability, then compare the approximation with the exact value.

6.78 Consider the following sample of 10 measurements:

🌐 LM6_78

| 8.4 | 16.9 | 15.8 | 12.5 | 10.3 | 4.9 | 12.9 | 9.8 | 23.7 | 7.3 |

Use these data to conduct each of the following sign tests using the binomial tables (Table II, Appendix A) and $\alpha = .05$.
  a. $H_0: \eta = 9$ versus $H_a: \eta > 9$ $p = .172$
  b. $H_0: \eta = 9$ versus $H_a: \eta \neq 9$ $p = .344$
  c. $H_0: \eta = 20$ versus $H_a: \eta < 20$ $p = .011$
  d. $H_0: \eta = 20$ versus $H_a: \eta \neq 20$ $p = .022$
  e. Repeat each of the preceding tests using the normal approximation to the binomial probabilities. Compare the results.
  f. What assumptions are necessary to ensure the validity of each of the preceding tests?

6.79 Suppose you wish to conduct a test of the research hypothesis that the median of a population is greater than 80. You randomly sample 25 measurements from the population and determine that 16 of them exceed 80. Set up and conduct the appropriate test of hypothesis at the .10 level of significance. Be sure to specify all necessary assumptions. $p = .115$

Applying the Concepts—Basic

6.80 In *The American Statistician* (May 2001), the nonparametric sign test was used to analyze data on the quality of white shrimp. One measure of shrimp quality is cohesiveness. Since freshly caught shrimp are usually stored on ice, there is concern that cohesiveness will deteriorate after storage. For a sample of 20 newly caught white shrimp, cohesiveness was measured both before storage and after storage on ice for two weeks. The difference in the cohesiveness measurements (before minus after) was obtained for each shrimp. If storage has no effect on cohesiveness, the population median of the differences will be 0. If cohesiveness deteriorates after storage, the population median of the differences will be positive.
  a. Set up the null and alternative hypotheses to test whether cohesiveness will deteriorate after storage.
  b. In the sample of 20 shrimp, there were 13 positive differences. Use this value to find the *p*-value of the test.
  c. Make the appropriate conclusion (in the words of the problem) if $\alpha = .05$.

6.81 Refer to the *Environmental Science & Technology* (Sept. 1, 2000) study of ammonia levels near the exit ramp of a San Francisco highway tunnel, Exercise 2.45 (p. 52). The daily ammonia concentrations (parts per million) on eight randomly selected days during afternoon drive-time are reproduced in the table. Suppose you want to determine if the median daily ammonia concentration for all afternoon drive-time days exceeds 1.5 ppm.

🌐 AMMONIA

| 1.53 | 1.50 | 1.37 | 1.51 | 1.55 | 1.42 | 1.41 | 1.48 |

  a. Set up the null and alternative hypotheses for the test. $H_0: \eta = 15; H_a: \eta > 1.5$
  b. Find the value of the test statistic. $S = 3$
  c. Find the *p*-value of the test. $p = .855$
  d. Give the appropriate conclusion (in the words of the problem) if $\alpha = .05$.

6.82 The table on p. 314 lists the lymphocyte (LYMPHO) count results from hematology tests administered to a sample of 50 West Indian or African workers. Test (at $\alpha = .05$) the hypothesis that the median lymphocyte count of all West Indian/African workers exceeds 20. Use the MINITAB printout below to make your conclusion.

```
MINITAB Output for Exercise 6.82
```

Sign test of median = 20.00 versus > 20.00

	N	Below	Equal	Above	P	Median
LYMPHO	50	19	6	25	0.2257	20.50

LYMPHO

Case Number	LYMPHO	Case Number	LYMPHO	Case Number	LYMPHO
1	14	18	28	35	11
2	15	19	17	36	25
3	19	20	14	37	30
4	23	21	8	38	32
5	17	22	25	39	17
6	20	23	37	40	22
7	21	24	20	41	20
8	16	25	15	42	20
9	27	26	9	43	20
10	34	27	16	44	26
11	26	28	18	45	40
12	28	29	17	46	22
13	24	30	23	47	61
14	26	31	43	48	12
15	23	32	17	49	20
16	9	33	23	50	35
17	18	34	31		

Source: Royston, J. P. "Some techniques for assessing multivariate normality based on the Shapiro-Wilk W." *Applied Statistics*, Vol. 32, No. 2, pp. 121–133.

Applying the Concepts—Intermediate

6.83 The biting rate of a particular species of fly was investigated in a study reported in the *Journal of the American Mosquito Control Association* (Mar. 1995). Biting rate was defined as the number of flies biting a volunteer during 15 minutes of exposure. This species of fly is known to have a median biting rate of 5 bites per 15 minutes on Stanbury Island, Utah. However, it is theorized that the median biting rate is higher in bright, sunny weather. To test this theory, 122 volunteers were exposed to the flies during a sunny day on Stanbury Island. Of these volunteers, 95 experienced biting rates greater than 5.

   a. Set up the null and alternative hypotheses for the test.

   b. Calculate the approximate *p*-value of the test. [*Hint:* Use the normal approximation for a binomial probability.] $p \approx 0$

   c. Make the appropriate conclusion at $\alpha = .01$.

6.84 In a study of the excessive transitory migration of guppy populations, 40 adult female guppies were placed in the left compartment of an experimental aquarium tank which was divided in half by a glass plate. After the plate was removed, the numbers of fish passing through the slit from the left compartment to the right one, and vice versa, were monitored every minute for 30 minutes (*Zoological Science*, Vol. 6, 1989). If an equilibrium is reached, the researchers would expect the median number of fish remaining in the left compartment to be 20. The data for the 30 observations (that is, numbers of fish in the left compartment at the end of each one-minute interval) are shown in the table below. Use the large-sample sign test to determine whether the median is less than 20. Test using $\alpha = .05$.

6.85 According to the National Restaurant Association, hamburgers are the number-one selling fast-food item in the United States. An economist studying the fast-food buying habits of Americans paid graduate students to stand outside two suburban McDonald's restaurants near Boston and ask departing customers whether they spent more than $2.25 on hamburger products for their lunch. Twenty answered yes; 50 said no; and 10 refused to answer the question (*Newark Star-Ledger*, Mar. 17, 1997).

   a. Is there sufficient evidence to conclude that the median amount spent for hamburgers at lunch at McDonald's is less than $2.25? $z = 2.12$

   b. Does your conclusion apply to all Americans who eat lunch at McDonald's? Justify your answer.   No

   c. What assumptions must hold to ensure the validity of your test in part **a**?

GUPPY

16	11	12	15	14	16	18	15	13	15
14	14	16	13	17	17	14	22	18	19
17	17	20	23	18	19	21	17	21	17

Source: Terami, H., and Watanabe, M. "Excessive transitory migration of guppy populations. III. Analysis of perception of swimming space and a mirror effect." *Zoological Science*, Vol. 6, 1989, p. 977 (Figure 2).

Key Terms

Note: Starred (*) terms are from the optional section in this chapter.

Alternative (research)
   hypothesis 278
Conclusion 288
Distribution-free tests* 309
Level of significance 282
Lower-tailed test 285
Nonparametrics* 309

Null hypothesis 278
Observed significance level
   (*p*-value) 290
One-tailed statistical test 284
Rejection region 279
Sign test* 309
Test of hypothesis 278

Test statistic 278
Two-tailed hypothesis 284
Two-tailed test 285
Type I error 279
Type II error 281
Upper-tailed test 285

STATISTICS IN ACTION
March Madness: Handicapping the NCAA Basketball Tourney

For three weeks each March, the National Collegiate Athletic Association (NCAA) holds its annual men's basketball championship tournament. The 64 best college basketball teams in the nation play a single-elimination tournament—a total of 63 games—to determine the NCAA champion. Due to its extreme popularity (all 63 games are televised by CBS), the media refers to the tournament as "March Madness."

The NCAA groups the 64 teams into four regions (East, South, Midwest, and West) of 16 teams each. The teams in each region are ranked (seeded) from 1 to 16 based on their performance during the regular season. The NCAA considers such factors as overall record, strength of schedule, average margin of victory, wins on the road, and conference affiliation to determine a team's seeding. In the first round, the number one seed plays the number 16 seed, the second seed plays the fifteenth seed, the third seed plays the fourteenth seed, etc. (See Figure 6.19.) Winners continue playing until the champion is determined.

Tournament followers, from hardcore gamblers to the casual fan who enters the office betting pool, have a strong interest in handicapping the games. Obviously, predicting

FIGURE 6.19

*The Design of a 16-Team
NCAA Regional Tournament*

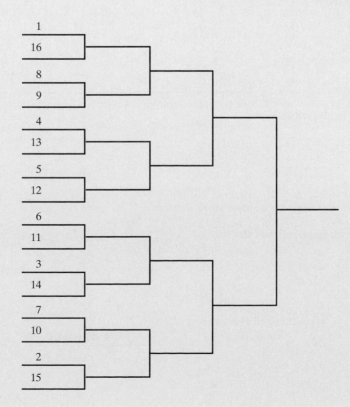

STATISTICS IN ACTION (*continued*)

the eventual champion is of prime interest. However, knowing who will win each game and the margin of victory may be just as important. To provide insight into this phenomenon, statisticians Hal Stern and Barbara Mock analyzed data from 13 NCAA tournaments and published their results in *Chance* (Winter 1998). The results of first-round games are summarized in Table 6.5.

Focus

a. A common perception among fans, media, and gamblers is that the higher seeded team has a better than 50-50 chance of winning a first-round game. Is there evidence to support this perception? Conduct the appropriate test for each matchup. What trends do you observe?

b. Is there evidence to support the claim that a 1-, 2-, 3-, or 4-seeded team will win by an average of more than 10 points in first-round games? Conduct the appropriate test for each matchup.

c. Is there evidence to support the claim that a 5-, 6-, 7-, or 8-seeded team will win by an average of less than five-points in first-round games? Conduct the appropriate test for each matchup.

d. (Optional) For each matchup, test the null hypothesis that the standard deviation of the victory margin is 11 points.

e. The researchers also calculated the difference between the game outcome (victory margin, in points) and point spread established by Las Vegas oddsmakers for a sample of 360 recent NCAA tournament games. The mean difference is .7 and the standard deviation of the difference is 11.3. If the true mean difference is 0, then the point spread can be considered a good predictor of the game outcome. Use this sample information to test the hypothesis that the point spread, on average, is a good predictor of the victory margin in NCAA tournament games.

TABLE 6.5   Summary of First-Round NCAA Tournament Games, 1985–1997

Matchup (Seeds)	Number of Games	Number Won by Favorite (Higher Seed)	Margin of Victory (Points)	
			Mean	Standard Deviation
1 vs 16	52	52	22.9	12.4
2 vs 15	52	49	17.2	11.4
3 vs 14	52	41	10.6	12.0
4 vs 13	52	42	10.0	12.5
5 vs 12	52	37	5.3	10.4
6 vs 11	52	36	4.3	10.7
7 vs 10	52	35	3.2	10.5
8 vs 9	52	22	−2.1	11.0

Source: Stern, H. S., & Mock, B. "College basketball upsets: Will a 16-seed ever beat a 1-seed?" *Chance*, Vol. 11, No. 1, Winter 1998, p. 29 (Table 3).

TEACHING TIP

Suggestions for class discussion of the above case can be found in the *Instructor's Notes*.

Key Formulas

Note: Starred (*) formulas are from the optional section in this chapter.

For testing $H_0: \theta = \theta_0$, the **large-sample test statistic** is

$$z = \frac{\hat{\theta} - \theta_0}{\sigma_{\hat{\theta}}}$$

where $\hat{\theta}$, θ_0, and $\sigma_{\hat{\theta}}$ are obtained from the table below:

Parameter, θ	Hypothesized Parameter Value, θ_0	Estimator, $\hat{\theta}$	Standard Error of Estimator, $\sigma_{\hat{\theta}}$	
μ	μ_0	$\bar{x}$	$\dfrac{\sigma}{\sqrt{n}}$	287
p	p_0	$\hat{p}$	$\sqrt{\dfrac{p_0 q_0}{n}}$	305

For testing $H_0: \mu = \mu_0$, the **small-sample test statistic** is

$$t = \frac{\bar{x} - \mu_0}{s/\sqrt{n}} \quad 298$$

*For testing $H_0: \eta = \eta_0$, the nonparametric test statistic is

$$S = \text{the number of sample measurement greater than (or less than) } \eta \text{ (small samples)}$$

$$z = \frac{(S - .5) - .5n}{.5\sqrt{n}} \text{ (large samples)} \quad 312$$

Language Lab

Symbol	Pronunciation	Description
H_0	H-oh	Null hypothesis
H_a	H-a	Alternative hypothesis
α	alpha	Probability of Type I error
β	beta	Probability of Type II error
S		Test statistic for sign test
η	eta	Population median

 SUPPLEMENTARY EXERCISES 6.86–6.107

Note: List the assumptions necessary for the valid implementation of the statistical procedures you use in solving all these exercises. Starred (*) exercises refer to the optional section in this chapter.

Learning the Mechanics

6.86 Specify the differences between a large-sample and small-sample test of hypothesis about a population mean μ. Focus on the assumptions and test statistics.

6.87 Which of the elements of a test of hypothesis can and should be specified *prior* to analyzing the data that are to be utilized to conduct the test? H_0, H_a, α

6.88 A random sample of 20 observations selected from a normal population produced $\bar{x} = 72.6$ and $s^2 = 19.4$.
   a. Form a 90% confidence interval for the population mean. 72.6 ± 1.70
   b. Test $H_0: \mu = 80$ against $H_a: \mu < 80$. Use $\alpha = .05$.
   c. Test $H_0: \mu = 80$ against $H_a: \mu \neq 80$. Use $\alpha = .01$.
   d. Form a 99% confidence interval for μ. 72.6 ± 2.82
   e. How large a sample would be required to estimate μ to within 1 unit with 95% confidence? $n = 75$

6.89 *Complete the following statement:* The smaller the p-value associated with a test of hypothesis, the stronger the support for the _____ hypothesis. Explain your answer.

6.90 *Complete the following statement:* The larger the p-value associated with a test of hypothesis, the stronger the support for the _____ hypothesis. Explain your answer.   Null

6.91 A random sample of $n = 200$ observations from a binomial population yields $\hat{p} = .29$.
   a. Test $H_0: p = .35$ against $H_a: p < .35$. Use $\alpha = .05$.
   b. Test $H_0: p = .35$ against $H_a: p \neq .35$. Use $\alpha = .05$.
   c. Form a 95% confidence interval for p. $.29 \pm .063$
   d. Form a 99% confidence interval for p. $.29 \pm .083$
   e. How large a sample would be required to estimate p to within .05 with 99% confidence? $n = 549$

6.92 A random sample of 175 measurements possessed a mean $\bar{x} = 8.2$ and a standard deviation $s = .79$.
   a. Form a 95% confidence interval for μ. $8.2 \pm .12$
   b. Test $H_0: \mu = 8.3$ against $H_a: \mu \neq 8.3$. Use $\alpha = .05$.
   c. Test $H_0: \mu = 8.4$ against $H_a: \mu \neq 8.4$. Use $\alpha = .05$.

6.93 A t-test is conducted for the null hypothesis $H_0: \mu = 10$ versus the alternative $H_a: \mu > 10$ for a random sample of $n = 17$ observations. The data are analyzed using MINITAB, with the results shown below.

MINITAB Output for Exercise 6.93

```
Test of mu = 10 vs mu > 10

Variable        N        Mean      StDev     SE Mean
X              17       12.50       8.78        2.13


Variable          95.0% CI              T          P
X            (  8.00,    17.00)       1.17      .1288
```

a. Interpret the *p*-value.   *p* = .1288
b. What assumptions are necessary for the validity of this test?   Normal population
c. Calculate and interpret the *p*-value assuming the alternative hypothesis was instead $H_a: \mu \neq 10$.

Applying the Concepts—Basic

6.94 An article in the *Annals of the Association of American Geographers* (June 1992) revealed that only 133 of 337 randomly selected residences in Los Angeles County were protected by earthquake insurance.
  a. What are the appropriate null and alternative hypotheses to test the research hypothesis that less than 40% of the residents of Los Angeles County were protected by earthquake insurance?
  b. Do the data provide sufficient evidence to support the research hypothesis? Use $\alpha = .10$. $z = -.19$
  c. Calculate and interpret the *p*-value for the test.

6.95 When a new drug is formulated, the pharmaceutical company must subject it to lengthy and involved testing before receiving the necessary permission from the Food and Drug Administration (FDA) to market the drug. The FDA requires the pharmaceutical company to provide substantial evidence that the new drug is safe for potential consumers.
  a. If the new drug testing were to be placed in a test of hypothesis framework, would the null hypothesis be that the drug is safe or unsafe? The alternative hypothesis?
  b. Given the choice of null and alternative hypotheses in part **a**, describe Type I and Type II errors in terms of this application. Define α and β in terms of this application.
  c. If the FDA wants to be very confident that the drug is safe before permitting it to be marketed, is it more important that α or β be small? Explain. α

*6.96 The American Cancer Society funds medical research focused on finding treatments for various forms of cancer. One new treatment is being tested to determine whether the average remission time for a particularly aggressive form of cancer is extended by the treatment. Assume that current treatments have been shown to provide a median remission time of 4.5 years. Seven patients with this cancer are given the new treatment, and their remission times (in years) are as follows:

🖳 REMISSION

5.3	7.3	3.6	5.2	6.1	4.8	8.4

Use the sign test to determine if the median remission time is increased by the new treatment. Test at $\alpha = .05$.

6.97 In a British study, 12 healthy college students, deprived of one night's sleep, received an array of tests intended to measure thinking time, fluency, flexibility, and origi-

nality of thought. The overall test scores of the sleep-deprived students were compared to the average score expected from students who received their accustomed sleep (*Sleep*, Jan. 1989). Suppose the overall scores of the 12 sleep-deprived students had a mean of $\bar{x} = 63$ and a standard deviation of 17. (Lower scores are associated with a decreased ability to think creatively.)
  a. Test the hypothesis that the true mean score of sleep-deprived subjects is less than 80, the mean score of subjects who received sleep prior to taking the test. Use $\alpha = .05$. $t = -3.46$
  b. What assumption is required for the hypothesis test of part **a** to be valid?

6.98 "Take the Pepsi Challenge" was a marketing campaign used by the Pepsi-Cola Company. Coca-Cola drinkers participated in a blind taste test in which they tasted unmarked cups of Pepsi and Coke and were asked to select their favorite. Pepsi claimed that "in recent blind taste tests, more than half the Diet Coke drinkers surveyed said they preferred the taste of Diet Pepsi" (*Consumer's Research*, May 1993). Suppose 100 Diet Coke drinkers took the Pepsi Challenge and 56 preferred the taste of Diet Pepsi. Test the hypothesis that more than half of all Diet Coke drinkers will select Diet Pepsi in a blind taste test. Use $\alpha = .05$. $z = 1.20$

Applying the Concepts—Intermediate

6.99 In order to be effective, a certain mechanical component used in a spacecraft must have a mean length of life greater than 1,100 hours. Owing to the prohibitive cost of this component, only three were tested under simulated space conditions. The lifetimes (hours) of the components were recorded and the computed statistics were $\bar{x} = 1,173.6$ and, $s = 36.3$. These data were analyzed using MINITAB, as shown below.

MINITAB Output for Exercise 6.99

Test of mu = 1100 vs mu > 1100

Variable	N	Mean	StDev	SE Mean
X	3	1173.6	36.3	20.96

Variable	95.0% CI	T	P
X	(1083.5, 1263.7)	3.51	.0362

  a. Verify that the software has correctly calculated the *t* statistic, and use Table IV of Appendix A to determine whether the *p*-value is in the appropriate range.
  b. Interpret the *p*-value.   *p* = .0362
  c. What assumptions are necessary for the validity of this test?
  d. Which type of error, I or II, is of greater concern for this test? Explain.   Type I
  e. Would you recommend that this component be passed as meeting specifications?

6.100 Medical tests have been developed to detect many serious diseases. A medical test is designed to minimize the probability that it will produce a "false positive" or a "false negative." A false positive is a positive test result for an individual who does not have the disease, whereas a false negative is a negative test result for an individual who does have the disease.

   a. If we treat a medical test for a disease as a statistical test of hypothesis, what are the null and alternative hypotheses for the medical test?

   b. What are the Type I and Type II errors for the test? Relate each to false positives and false negatives.

   c. Which of these errors has graver consequences? Considering this error, is it more important to minimize α or β? Explain.

6.101 Refer to the *Science* (Nov. 1988) study of inbreeding in tropical swarm-founding wasps, Exercise 5.79 (p. 274). A sample of 197 wasps, captured, frozen, and subjected to a series of genetic tests, yielded a sample mean inbreeding coefficient of $\bar{x} = .044$ with a standard deviation of $s = .884$. Recall that if the wasp has no tendency to inbreed, the true mean inbreeding coefficient μ for the species will equal 0.

   a. Test the hypothesis that the true mean inbreeding coefficient μ for this species of wasp exceeds 0. Use $\alpha = .05$. $z = 0.70$

   b. Compare the inference, part **a**, to the inference obtained in Exercise 5.79 using a confidence interval. Do the inferences agree? Explain.

6.102 A consumer protection group is concerned that a ketchup manufacturer is filling its 20-ounce family-size containers with less than 20 ounces of ketchup. The group purchases 10 family-size bottles of this ketchup, weighs the contents of each, and finds that the mean weight is equal to 19.86 ounces, and the standard deviation is equal to .22 ounce.

   a. Do the data provide sufficient evidence for the consumer group to conclude that the mean fill per family-size bottle is less than 20 ounces? Test using $\alpha = .05$. $t = -2.01$

   b. If the test in part **a** were conducted on a periodic basis by the company's quality control department, is the consumer group more concerned about making a Type I error or a Type II error? (The probability of making this type of error is called the *consumer's risk*.)

   c. The ketchup company is also interested in the mean amount of ketchup per bottle. It does not wish to overfill them. For the test conducted in part **a**, which type of error is more serious from the company's point of view—a Type I error or a Type II error? (The probability of making this type of error is called the *producer's risk*.)   Type I

6.103 The EPA sets a limit of 5 parts per million (ppm) on PCB (a dangerous substance) in water. A major manufacturing firm producing PCB for electrical insulation discharges small amounts from the plant. The company management, attempting to control the PCB in its discharge, has given instructions to halt production if the mean amount of PCB in the effluent exceeds 3 ppm. A random sample of 50 water specimens produced the following statistics: $\bar{x} = 3.1$ ppm and $s = .5$ ppm.

   a. Do these statistics provide sufficient evidence to halt the production process? Use $\alpha = .01$. $z = 1.41$

   b. If you were the plant manager, would you want to use a large or a small value for α for the test in part **a**?

6.104 Research reported in the *Journal of Psychology* (Mar. 1991) studied the personality characteristics of obese individuals. One variable, the locus of control (LOC), measures the individual's degree of belief that he or she has control over situations. High scores on the LOC scale indicate less perceived control. For one sample of 19 obese adolescents, the mean LOC score was 10.89 with a standard deviation of 2.48. Suppose we wish to test whether the mean LOC score for all obese adolescents exceeds 10, the average for "normal" individuals.

   a. Specify the null and alternative hypotheses for this test. $H_0: \mu = 10; H_a: \mu > 10$

   b. The data are analyzed by computer, and the following STATISTIX output is obtained. Check the calculation of the t statistic, and determine whether the p-value is in the correct range according to Table IV of Appendix A. $.05 < p < .10$

STATISTIX Output for Exercise 6.104

```
ONE-SAMPLE T TEST FOR LOC


NULL HYPOTHESIS: MU =    10
ALTERNATIVE HYP: MU >    10


MEAN           10.89
STD ERROR      0.569
MEAN - H0      0.89
T              1.564
DF                18
P             0.0676


CASES INCLUDED 19     MISSING CASES 0
```

6.105 The National Science Foundation, in a survey of 2,237 engineering graduate students who earned their Ph.D. degrees, found that 607 were U.S. citizens; the majority (1,630) of the Ph.D. degrees were awarded to foreign nationals (*Science*, Sept. 24, 1993). Conduct a test to determine whether the true percentage of engineering Ph.D. degrees awarded to foreign nationals exceeds 50%. Use $\alpha = .01$. $z = 21.66$

*6.106 Many water treatment facilities supplement the natural fluoride concentration with hydrofluosilicic acid in order to reach a target concentration of fluoride in drinking water. Certain levels are thought to enhance dental health, but very high concentrations can be dangerous. Suppose that one such treatment plant targets .75 milligrams per liter (mg/L) for their water. The plant tests 25 samples each day to determine whether the median level differs from the target.

a. Set up the null and alternative hypotheses.

b. Set up the test statistic and rejection region using $\alpha = .10$.

c. Explain the implication of a Type I error in the context of this application. A Type II error.

d. Suppose that one day's samples result in 18 values that exceed .75 mg/L. Conduct the test and state the appropriate conclusion in the context of this application. $p = .044$

e. When it was suggested to the plant's supervisor that a *t*-test should be used to conduct the daily test, she replied that the probability distribution of the fluoride concentrations was "heavily skewed to the right." Show graphically what she meant by this, and explain why this is a reason to prefer the sign test to the *t*-test.

Applying the Concepts—Advanced

6.107 If a manufacturer (the vendee) buys all items of a particular type from a particular vendor, the manufacturer is practicing *sole sourcing* (Schonberger and Knod, *Operations Managment*, 1994). As part of a sole-sourcing arrangement, a vendor agrees to periodically supply its vendee with sample data from its production process. The vendee uses the data to investigate whether the mean length of rods produced by the vendor's production process is truly 5.0 millimeters (mm) or more, as claimed by the vendor and desired by the vendee.

a. If the production process has a standard deviation of .01 mm, the vendor supplies $n = 100$ items to the vendee, and the vendee uses $\alpha = .05$ in testing $H_0: \mu = 5.0$ mm against $H_a: \mu < 5.0$ mm, what is the probability that the vendee's test will fail to reject the null hypothesis when in fact $\mu = 4.9975$ mm? What is the name given to this type of error? .1949

b. Refer to part **a**. What is the probability that the vendee's test will reject the null hypothesis when in fact $\mu = 5.0$? What is the name given to this type of error? .05

c. What is the probability that the test detects a departure of .0025 mm below the specified mean rod length of 5.0 mm? (This probability is called the **power** of the test.) .8051

STUDENT PROJECTS

The "efficient market" theory postulates that the best predictor of a stock's price at some point in the future is the current price of the stock (with some adjustments for inflation and transaction costs, which we shall assume to be negligible for the purpose of this exercise). To test this theory, select a random sample of 25 stocks on the New York Stock Exchange and record the closing prices on the last days of two recent consecutive months. Calculate the increase or decrease in the stock price over the 1-month period.

a. Define μ as the mean change in price of all stocks over a 1-month period. Set up the appropriate null and alternative hypotheses in terms of μ.

b. Use the sample of the 25 stock price differences to conduct the test of hypothesis established in part a. Use $\alpha = .05$.

c. Tabulate the number of rejections and nonrejections of the null hypothesis in your class. If the null hypothesis were true, how many rejections of the null hypothesis would you expect among those in your class? How does this expectation compare with the actual number of nonrejections? What does the result of this exercise indicate about the efficient market theory?

REFERENCES

Conover, W. J. *Practical Nonparametric Statistics*, 2nd ed. New York: Wiley, 1980.

Daniel, W. W. *Applied Nonparametric Statistics*, 2nd ed. Boston: PWS-Kent, 1990.

Gibbons, J. D. *Nonparametric Statistical Inference*, 2nd ed. New York: McGraw-Hill, 1985.

Hollander, M., and Wolfe, D. A. *Nonparametric Statistical Methods*. New York: Wiley, 1973.

Marascuilo, L. A., and McSweeney, M. *Nonparametric and Distribution-Free Methods for the Social Sciences*. Monterey, Ca.: Brooks/Cole, 1977.

Snedecor, G. W., and Cochran, W. G. *Statistical Methods*, 7th ed. Ames: Iowa State University Press, 1980.

Wackerly, D., Mendenhall, W., and Scheaffer, R. *Mathematical Statistics with Applications*, 6th ed. North Scituate, Mass.: Duxbury, 2002.

Comparing Population Means

Contents

Statistics in Action

On the Trail of the Cockroach

🐾 *Where We've Been*

We explored two methods for making statistical inferences, confidence intervals and tests of hypotheses based on single samples, in Chapters 5 and 6. In particular, we gave confidence intervals and tests of hypotheses concerning a population mean μ and a population proportion p, and we learned how to select the sample size necessary to obtain a specified amount of information concerning a parameter.

☞ *Where We're Going*

Now that we've learned to make inferences about a single population, we'll learn how to compare two (or more) populations. For example, we may wish to compare the mean gas mileages for two models of automobiles, or the mean reaction times of men and women to a visual stimulus. In this chapter we'll see how to decide whether differences exist and how to estimate the differences between population means. We will learn how to compare population proportions in Chapter 8.

Many experiments involve a comparison of population means. For instance, a sociologist may want to estimate the difference in mean life expectancy between inner-city and suburban residents. A consumer group may want to test whether two major brands of food freezers differ in the mean amount of electricity they use. In this chapter we consider techniques for using two (or more) samples to compare the means of the populations from which they were selected.

7.1   COMPARING TWO POPULATION MEANS: INDEPENDENT SAMPLING

TEACHING TIP
Discuss the two methods of data collection for comparing means. Use examples that illustrate both the independent and matched pairs sampling designs.

Many of the same procedures that are used to estimate and test hypotheses about a single parameter can be modified to make inferences about two parameters. Both the z and t statistics may be adapted to make inferences about the difference between two population means.

In this section we develop both large-sample and small-sample methodologies for comparing two population means. In the large-sample case we use the z statistic, while in the small-sample case we use the t statistic.

Large Samples

EXAMPLE 7.1

A dietitian has developed a diet that is low in fats, carbohydrates, and cholesterol. Although the diet was initially intended to be used by people with heart disease, the dietitian wishes to examine the effect this diet has on the weights of obese people. Two random samples of 100 obese people each are selected, and one group of 100 is placed on the low-fat diet. The other 100 are placed on a diet that contains approximately the same quantity of food but is not as low in fats, carbohydrates, and cholesterol. For each person, the amount of weight lost (or gained) in a 3-week period is recorded. The data, saved in the DIETSTUDY file, are listed in Table 7.1. Form a 95% confidence interval for the difference between the population mean weight losses for the two diets. Interpret the result.

Solution

Recall that the general form of a large-sample confidence interval for a single mean μ is $\bar{x} \pm z_{\alpha/2}\sigma_{\bar{x}}$. That is, we add and subtract $z_{\alpha/2}$ standard deviations of the sample estimate, $\bar{x}$, to the value of the estimate. We employ a similar procedure to form the confidence interval for the difference between two population means.

TEACHING TIP
Throughout this chapter, point out similarities with one-sample inferences. Try to show the student that our general ideas developed in Chapters 5 and 6 extend to the two sample inferences in Chapter 7.

Let μ_1 represent the mean of the conceptual population of weight losses for all obese people who could be placed on the low-fat diet. Let μ_2 be similarly defined for the other diet. We wish to form a confidence interval for $(\mu_1 - \mu_2)$. An intuitively appealing estimator for $(\mu_1 - \mu_2)$ is the difference between the sample means, $(\bar{x}_1 - \bar{x}_2)$. Thus, we will form the confidence interval of interest by

$$(\bar{x}_1 - \bar{x}_2) \pm z_{\alpha/2}\sigma_{(\bar{x}_1 - \bar{x}_2)}$$

Assuming the two samples are independent, the standard deviation of the difference between the sample means is

$$\sigma_{(\bar{x}_1 - \bar{x}_2)} = \sqrt{\frac{\sigma_1^2}{n_1} + \frac{\sigma_2^2}{n_2}} \approx \sqrt{\frac{s_1^2}{n_1} + \frac{s_2^2}{n_2}}$$

⊘ DIETSTUDY
TABLE 7.1 Diet Study Data, Example 7.1

Weight Losses for Lowfat Diet

8	10	10	12	9	3	11	7	9	2
21	8	9	2	2	20	14	11	15	6
13	8	10	12	1	7	10	13	14	4
8	12	8	10	11	19	0	9	10	4
11	7	14	12	11	12	4	12	9	2
4	3	3	5	9	9	4	3	5	12
3	12	7	13	11	11	13	12	18	9
6	14	14	18	10	11	7	9	7	2
16	16	11	11	3	15	9	5	2	6
5	11	14	11	6	9	4	17	20	10

Weight Losses for Regular Diet

6	6	5	5	2	6	10	3	9	11
14	4	10	13	3	8	8	13	9	3
4	12	6	11	12	9	8	5	8	7
6	2	6	8	5	7	16	18	6	8
13	1	9	8	12	10	6	1	0	13
11	2	8	16	14	4	6	5	12	9
11	6	3	9	9	14	2	10	4	13
8	1	1	4	9	4	1	1	5	6
14	0	7	12	9	5	9	12	7	9
8	9	8	10	5	8	0	3	4	8

FIGURE 7.1

SPSS Summary Statistics for Diet Study

Group Statistics					
	DIET	N	Mean	Std. Deviation	Std. Error Mean
WTLOSS	LOWFAT	100	9.31	4.67	.47
	REGULAR	100	7.40	4.04	.40

Summary statistics for the diet data are displayed in the SPSS printout, Figure 7.1. Note that $\bar{x}_1 = 9.31$, $\bar{x}_2 = 7.40$, $s_1 = 4.67$, and $s_2 = 4.04$. Using these values and noting that $\alpha = .05$ and $z_{.025} = 1.96$, we find that the 95% confidence interval is, approximately,

TEACHING TIP
Remind the students what the phrase "95% confident" means. Draw a picture of the repeated intervals to refresh their memories.

$$(9.31 - 7.40) \pm 1.96\sqrt{\frac{(4.67)^2}{100} + \frac{(4.04)^2}{100}} = 1.91 \pm (1.96)(.62) = 1.91 \pm 1.22$$

or (.69, 3.13). Using this estimation procedure over and over again for different samples, we know that approximately 95% of the confidence intervals formed in this manner will enclose the difference in population means ($\mu_1 - \mu_2$). Therefore, we are highly confident that the mean weight loss for the low-fat diet is between .69 and 3.13 pounds more than the mean weight loss for the other diet. With this information, the dietitian better understands the potential of the low-fat diet as a weight-reducing diet.

■

FIGURE 7.2
Sampling Distribution of
$(\bar{x}_1 - \bar{x}_2)$

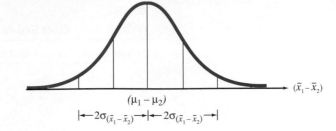

$(\mu_1 - \mu_2)$

$|\!\leftarrow\!2\sigma_{(\bar{x}_1-\bar{x}_2)}\!\rightarrow\!|\!\leftarrow\!2\sigma_{(\bar{x}_1-\bar{x}_2)}\!\rightarrow\!|$

$(\bar{x}_1 - \bar{x}_2)$

The justification for the procedure used in Example 7.1 to estimate $(\mu_1 - \mu_2)$ relies on the properties of the sampling distribution of $(\bar{x}_1 - \bar{x}_2)$. The performance of the estimator in repeated sampling is pictured in Figure 7.2, and its properties are summarized in the next box.

Properties of the Sampling Distribution of $(\bar{x}_1 - \bar{x}_2)$

1. The mean of the sampling distribution of $(\bar{x}_1 - \bar{x}_2)$ is $(\mu_1 - \mu_2)$.
2. If the two samples are independent, the standard deviation of the sampling distribution is

$$\sigma_{(\bar{x}_1-\bar{x}_2)} = \sqrt{\frac{\sigma_1^2}{n_1} + \frac{\sigma_2^2}{n_2}}$$

   where σ_1^2 and σ_2^2 are the variances of the two populations being sampled and n_1 and n_2 are the respective sample sizes. We also refer to $\sigma_{(\bar{x}_1-\bar{x}_2)}$ as the **standard error** of the statistic $(\bar{x}_1 - \bar{x}_2)$.
3. The sampling distribution of $(\bar{x}_1 - \bar{x}_2)$ is approximately normal for *large samples* by the Central Limit Theorem.

TEACHING TIP
Explain that the sample variances can be used to estimate the population variances when the sample sizes are both large.

In Example 7.1, we noted the similarity in the procedures for forming a large-sample confidence interval for one population mean and a large-sample confidence interval for the difference between two population means. When we are testing hypotheses, the procedures are again very similar. The general large-sample procedures for forming confidence intervals and testing hypotheses about $(\mu_1 - \mu_2)$ are summarized in the next two boxes.

Large Sample Confidence Interval for $(\mu_1 - \mu_2)$

$$(\bar{x}_1 - \bar{x}_2) \pm z_{\alpha/2}\sigma_{(\bar{x}_1-\bar{x}_2)} = (\bar{x}_1 - \bar{x}_2) \pm z_{\alpha/2}\sqrt{\frac{\sigma_1^2}{n_1} + \frac{\sigma_2^2}{n_2}}$$

Assumptions: The two samples are randomly selected in an independent manner from the two populations. The sample sizes, n_1 and n_2, are large enough so that $\bar{x}_1$ and $\bar{x}_2$ both have approximately normal sampling distributions and so that s_1^2 and s_2^2 provide good approximations to σ_1^2 and σ_2^2. This will be true if $n_1 \geq 30$ and $n_2 \geq 30$.

TEACHING TIP
When writing the confidence interval formula, replace the symbols with the following words: point estimate $\pm$ critical value $\times$ standard error. All of the confidence intervals for means and proportions follow this general form.

Large-Sample Test of Hypothesis for $(\mu_1 - \mu_2)$

ONE-TAILED TEST	TWO-TAILED TEST
$H_0: (\mu_1 - \mu_2) = D_0$	$H_0: (\mu_1 - \mu_2) = D_0$
$H_a: (\mu_1 - \mu_2) < D_0$	$H_a: (\mu_1 - \mu_2) \neq D_0$
$\quad$ [or $H_a: (\mu_1 - \mu_2) > D_0$]	

where D_0 = Hypothesized difference between the means (this difference is often hypothesized to be equal to 0)

Test statistic:

$$z = \frac{(\bar{x}_1 - \bar{x}_2) - D_0}{\sigma_{(\bar{x}_1 - \bar{x}_2)}} \quad \text{where} \quad \sigma_{(\bar{x}_1 - \bar{x}_2)} = \sqrt{\frac{\sigma_1^2}{n_1} + \frac{\sigma_2^2}{n_2}}$$

Rejection region: $z < -z_\alpha$	*Rejection region:* $\lvert z \rvert > z_{\alpha/2}$
$\quad$ [or $z > z_\alpha$ when	
$\quad$ $H_a: (\mu_1 - \mu_2) > D_0$]	

Assumptions: Same as for the large-sample confidence interval.

EXAMPLE 7.2

$z = 3.09$, Reject H_0

Refer to the study of obese people on a low-fat diet and a regular diet, Example 7.1. Another way to compare the mean weight losses for the two different diets is to conduct a test of hypothesis. Use the information on the SPSS printout, Figure 7.1, to conduct the test. Use $\alpha = .05$.

Solution

Again, we let μ_1 and μ_2 represent the population mean weight losses of obese people on the low-fat diet and regular diet, respectively. If one diet is more effective in reducing the weights of obese people, then either $\mu_1 < \mu_2$ or $\mu_2 < \mu_1$; that is, $\mu_1 \neq \mu_2$.

Thus, the elements of the test are as follows:

$H_0: (\mu_1 - \mu_2) = 0 \qquad$ (i.e., $\mu_1 = \mu_2$; note that $D_0 = 0$ for this hypothesis test)

$H_a: (\mu_1 - \mu_2) \neq 0 \qquad$ (i.e., $\mu_1 \neq \mu_2$)

Test statistic: $z = \dfrac{(\bar{x}_1 - \bar{x}_2) - D_0}{\sigma_{(\bar{x}_1 - \bar{x}_2)}} = \dfrac{\bar{x}_1 - \bar{x}_2 - 0}{\sigma_{(\bar{x}_1 - \bar{x}_2)}}$

Rejection region: $z < -z_{\alpha/2} = -1.96$ or $z > z_{\alpha/2} = 1.96 \qquad$ (see Figure 7.3)

Substituting the summary statistics given in Figure 7.1 into the test statistic, we obtain

Suggested Exercise 7.13

$$z = \frac{(\bar{x}_1 - \bar{x}_2) - 0}{\sigma_{(\bar{x}_1 - \bar{x}_2)}} = \frac{9.31 - 7.40}{\sqrt{\dfrac{\sigma_1^2}{n_1} + \dfrac{\sigma_2^2}{n_2}}}$$

$$\approx \frac{1.91}{\sqrt{\dfrac{s_1^2}{n_1} + \dfrac{s_2^2}{n_2}}} = \frac{1.91}{\sqrt{\dfrac{(4.67)^2}{100} + \dfrac{(4.04)^2}{100}}} = \frac{1.91}{.617} = 3.09$$

As you can see in Figure 7.3, the calculated z value clearly falls in the rejection region. Therefore, the samples provide sufficient evidence, at $\alpha = .05$, for the dietitian to conclude that the mean weight losses for the two diets differ.

FIGURE 7.3

Rejection Region for
Example 7.2

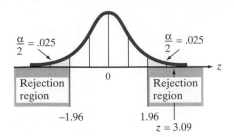

$\frac{\alpha}{2} = .025$ $\frac{\alpha}{2} = .025$

Rejection region   Rejection region

−1.96   1.96
z = 3.09

This conclusion agrees with the inference drawn from the 95% confidence interval in Example 7.1. However, the confidence interval provides more information on the mean weight losses. From the hypothesis test, we know only that the two means differ; that is, $\mu_1 \neq \mu_2$. From the confidence interval in Example 7.1, we found that the mean weight loss μ_1 of the low-fat diet was between .69 and 3.13 pounds more than the mean weight loss μ_2 of the regular diet. In other words, the test tells us that the means differ, but the confidence interval tells us how large the difference is. Both inferences are made with the same degree of reliability—namely, with 95% confidence (or, at $\alpha = .05$.) ∎

EXAMPLE 7.3

$p = .002$

Find the observed significance level for the test in Example 7.2. Interpret the result.

Solution

The alternative hypothesis in Example 7.2, $H_a\colon \mu_1 - \mu_2 \neq 0$, required a two-tailed test using

$$z = \frac{\bar{x}_1 - \bar{x}_2}{\sigma_{(\bar{x}_1 - \bar{x}_2)}}$$

as a test statistic. Since the z value calculated from the sample data was 3.09, the observed significance level (p-value) for the two-tailed test is the probability of observing a value of z at least as contradictory to the null hypothesis as $z = 3.09$; that is,

$$p\text{-value} = 2 \cdot P(z \geq 3.09)$$

This probability is computed assuming H_0 is true and is equal to the highlighted area shown in Figure 7.4.

The tabulated area corresponding to $z = 3.09$ in Table III of Appendix A is .4990. Therefore,

$$P(z \geq 3.09) = .5 - .4990 = .0010$$

and the observed significance level for the test is

$$p\text{-value} = 2(.001) = .002$$

FIGURE 7.4

The Observed Significance
Level for Example 7.2

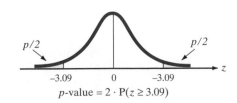

$p/2$ $p/2$

−3.09   0   −3.09
p-value = 2 · P(z ≥ 3.09)

FIGURE 7.5

MINITAB Printout for the Hypothesis Test of Example 7.2

```
Two-sample T for LOWFAT vs REGULAR

            N     Mean    StDev   SE Mean
LOWFAT    100     9.31     4.67     0.47
REGULAR   100     7.40     4.04     0.40


Difference = mu LOWFAT - mu REGULAR
Estimate for difference:  1.910
95% CI for difference: (0.693, 3.127)
T-Test of difference = 0 (vs not =): T-Value = 3.10   P-Value = 0.002
DF = 193
```

Since our selected α value, .05, exceeds this *p*-value, we have sufficient evidence to reject $H_0: \mu_1 - \mu_2 = 0$.

The *p*-value of the test is more easily obtained from a statistical software package. A MINITAB printout for the hypothesis test is displayed in Figure 7.5. The *p*-value, highlighted on the printout, is .002. This value agrees with our calculated *p*-value.

$\blacksquare$

Small Samples

When comparing two population means with small samples (say, $n_1 < 30$ and $n_2 < 30$), the methodology of the previous three examples is invalid. The reason? When the sample sizes are small, estimates of σ_1^2 and σ_2^2 are unreliable and the Central Limit Theorem (which guarantees that the z statistic is normal) can no longer be applied. But as in the case of a single mean (Section 6.4), we use the familiar Student's *t*-distribution described in Chapter 5.

To use the t-distribution, both sampled populations must be approximately normally distributed with equal population variances, and the random samples must be selected independently of each other. The normality and equal variances assumptions imply relative frequency distributions for the populations that would appear as shown in Figure 7.6.

FIGURE 7.6

Assumptions for the Two-Sample t: (1) Normal Populations, (2) Equal Variances

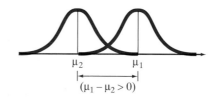

Since we assume the two populations have equal variances ($\sigma_1^2 = \sigma_2^2 = \sigma^2$), it is reasonable to use the information contained in both samples to construct a **pooled sample estimator σ^2** for use in confidence intervals and test statistics. Thus, if s_1^2 and s_2^2 are the two sample variances (both estimating the variance σ^2 common to both populations), the pooled estimator of σ^2, denoted as s_p^2, is

$$s_p^2 = \frac{(n_1 - 1)s_1^2 + (n_2 - 1)s_2^2}{(n_1 - 1) + (n_2 - 1)} = \frac{(n_1 - 1)s_1^2 + (n_2 - 1)s_2^2}{n_1 + n_2 - 2}$$

or

$$s_p^2 = \frac{\overbrace{\sum (x_1 - \bar{x}_1)^2}^{\text{From sample 1}} + \overbrace{\sum (x_2 - \bar{x}_2)^2}^{\text{From sample 2}}}{n_1 + n_2 - 2}$$

where x_1 represents a measurement from sample 1 and x_2 represents a measurement from sample 2. Recall that the term *degrees of freedom* was defined in Section 5.2 as 1 less than the sample size. Thus, in this case, we have $(n_1 - 1)$ degrees of freedom for sample 1 and $(n_2 - 1)$ degrees of freedom for sample 2. Since we are pooling the information on σ^2 obtained from both samples, the degrees of freedom associated with the pooled variance s_p^2 is equal to the sum of the degrees of freedom for the two samples, namely, the denominator of s_p^2; that is, $(n_1 - 1) + (n_2 - 1) = n_1 + n_2 - 2$.

Note that the second formula given for s_p^2 shows that the pooled variance is simply a *weighted average* of the two sample variances s_1^2 and s_2^2. The weight given each variance is proportional to its degrees of freedom. If the two variances have the same number of degrees of freedom (i.e., if the sample sizes are equal), then the pooled variance is a simple average of the two sample variances. The result is an average or "pooled" variance that is a better estimate of σ^2 than either s_1^2 or s_2^2 alone.

Both the confidence interval and the test of hypothesis procedures for comparing two population means with small samples are summarized in the accompanying box.

TEACHING TIP
An easy way to remember the degrees of freedom for the *t*-distribution is to look in the denominator of the estimate of the variance. In this case, df $= n_1 + n_2 - 2$.

TEACHING TIP
Emphasize that these analyses will work *only* when the stated assumptions are valid.

Small-Sample Confidence Interval for $(\mu_1 - \mu_2)$: Independent Samples

$$(\bar{x}_1 - \bar{x}_2) \pm t_{\alpha/2} \sqrt{s_p^2 \left(\frac{1}{n_1} + \frac{1}{n_2} \right)}$$

where $s_p^2 = \dfrac{(n_1 - 1)s_1^2 + (n_2 - 1)s_2^2}{n_1 + n_2 - 2}$

and $t_{\alpha/2}$ is based on $(n_1 + n_2 - 2)$ degrees of freedom.

Assumptions: 1. Both sampled populations have relative frequency distributions that are approximately normal.

2. The population variances are equal.

3. The samples are randomly and independently selected from the populations.

Small-Sample Test of Hypothesis for $(\mu_1 - \mu_2)$: Independent Samples

ONE-TAILED TEST	**TWO-TAILED TEST**
$H_0: (\mu_1 - \mu_2) = D_0$	$H_0: (\mu_1 - \mu_2) = D_0$
$H_a: (\mu_1 - \mu_2) < D_0$	$H_a: (\mu_1 - \mu_2) \neq D_0$
[or $H_a: (\mu_1 - \mu_2) > D_0$]	

Test statistic:
$$t = \frac{(\bar{x}_1 - \bar{x}_2) - D_0}{\sqrt{s_p^2 \left(\dfrac{1}{n_1} + \dfrac{1}{n_2} \right)}}$$

Rejection region: $t < -t_{\alpha}$ *Rejection region:* $|t| > t_{\alpha/2}$
[or $t > t_{\alpha}$ when
$H_a: (\mu_1 - \mu_2) > D_0$]

where t_{α} and $t_{\alpha/2}$ are based on $(n_1 + n_2 - 2)$ degrees of freedom.
Assumptions: Same as for the small-sample confidence interval for $(\mu_1 - \mu_2)$ in the previous box.

EXAMPLE 7.4

Suppose you wish to compare a new method of teaching reading to "slow learners" to the current standard method. You decide to base this comparison on the results of a reading test given at the end of a learning period of 6 months. Of a random sample of 22 slow learners, 10 are taught by the new method and 12 are taught by the standard method. All 22 children are taught by qualified instructors under similar conditions for a 6-month period. The results of the reading test at the end of this period are given in Table 7.2.

⊘ READING

TABLE 7.2   Reading Test Scores for Slow Learners

New Method				Standard Method			
80	80	79	81	79	62	70	68
76	66	71	76	73	76	86	73
70	85			72	68	75	66

a. 4.07 ± 5.47

a. Use the data in the table to estimate the true mean difference between the test scores for the new method and the standard method. Use a 95% confidence interval.
b. Interpret the interval, part **a**.
c. What assumptions must be made in order that the estimate be valid? Are they reasonably satisfied?

Solution

a. For this experiment, let μ_1 and μ_2 represent the mean reading test scores of slow learners taught with the new and standard methods, respectively. Then, the objective is to obtain a 95% confidence interval for $(\mu_1 - \mu_2)$.

The first step in constructing the confidence interval is to obtain summary statistics (e.g., $\bar{x}$ and s) on reading test scores for each method. The data of Table 7.2 was entered into a computer, and SAS used to obtain these descriptive statistics. The SAS printout appears in Figure 7.7. Note that $\bar{x}_1 = 76.4$, $s_1 = 5.8348$, $\bar{x}_2 = 72.333$, and $s_2 = 6.3437$

FIGURE 7.7

SAS Output for Example 7.4

Two Sample t-test for the Means of SCORE within METHOD

Sample Statistics

Group	N	Mean	Std. Dev.	Std. Error
NEW	10	76.4	5.8348	1.8451
STD	12	72.33333	6.3437	1.8313

Hypothesis Test

Null hypothesis: Mean 1 - Mean 2 = 0
Alternative: Mean 1 - Mean 2 ^= 0

If Variances Are	t statistic	Df	Pr > t
Equal	1.552	20	0.1364
Not Equal	1.564	19.77	0.1336

95% Confidence Interval for the Difference between Two Means

Lower Limit	Upper Limit
-1.40	9.53

Next, we calculate the pooled estimate of variance:

$$s_p^2 = \frac{(n_1 - 1)s_1^2 + (n_2 - 1)s_2^2}{n_1 + n_2 - 2}$$

$$= \frac{(10 - 1)(5.8348)^2 + (12 - 1)(6.3437)^2}{10 + 12 - 2} = 37.45$$

where s_p^2 is based on $(n_1 + n_2 - 2) = (10 + 12 - 2) = 20$ degrees of freedom. Also, we find $t_{\alpha/2} = t_{.025} = 2.086$ (based on 20 degrees of freedom) from Table IV of Appendix A.

Finally, the 95% confidence interval for $(\mu_1 - \mu_2)$, the difference between mean test scores for the two methods, is

Suggested Exercise 7.19

$$(\bar{x}_1 - \bar{x}_2) \pm t_{\alpha/2}\sqrt{s_p^2\left(\frac{1}{n_1} + \frac{1}{n_2}\right)} = (76.4 - 72.33) \pm t_{.025}\sqrt{37.45\left(\frac{1}{10} + \frac{1}{12}\right)}$$

$$= 4.07 \pm (2.086)(2.62)$$
$$= 4.07 \pm 5.47$$

TEACHING TIP
When interpreting the confidence interval for $\mu_1 - \mu_2$, it is usually important to identify if the value 0 is contained in the confidence interval. Discuss why the 0 value is significant in terms of interpretations.

or $(-1.4, 9.54)$. This interval agrees (except for rounding) with the one shown at the bottom of the SAS printout, Figure 7.7.

b. The interval can be interpreted as follows. With a confidence coefficient equal to .95, we estimate the difference in mean test scores between using the new method of teaching and using the standard method to fall in the interval -1.4 to 9.54. In other words, we estimate (with 95% confidence) the mean test score for the new method to be anywhere from 1.4 points less than to 9.54 points more than the mean test score for the standard method. Although the sample means seem to suggest that the new method is associated with a higher mean test score, there is insufficient evidence to indicate that $(\mu_1 - \mu_2)$ differs from 0 because the interval includes 0 as a possible value for $(\mu_1 - \mu_2)$. To show a difference in mean test scores (if it exists), you could increase the sample size and thereby narrow the width of the confidence interval for $(\mu_1 - \mu_2)$. An alternative is to design the experiment differently. This possibility is discussed in the next section.

c. To properly use the small-sample confidence interval, the following assumptions must be satisfied:

(1) The samples are randomly and independently selected from the populations of slow learners taught by the new method and the standard method.

(2) The test scores are normally distributed for both teaching methods.

(3) The variance of the test scores are the same for the two populations, that is, $\sigma_1^2 = \sigma_2^2$.

The first assumption is satisfied, based on the information provided about the sampling procedure in the problem description. To check the plausibility of the remaining two assumptions, we resort to graphical methods. Figure 7.8 is a portion of the SPSS printout that gives normal probability plots for the test scores of the two samples of slow learners. The near straight-line trends on both plots indicate that the score distributions are approximately mound-shaped and symmetric. Consequently, each sample data set appears to come from a population that is approximately normal.

One way to check assumption #3 is to examine box plots for the sample data.* Figure 7.9 is an SPSS printout that shows side-by-side vertical box plots for the test scores in the two samples. Recall, from Section 2.9, that the box plot represents the "spread" of a data set. The two box plots appear to have about

---

*Another, more formal, method is to test the null hypothesis $H_0: \sigma_1^2 = \sigma_2^2$. Consult the chapter references to learn more about this test.

FIGURE 7.8

SPSS Normal Probability Plots for Example 7.4

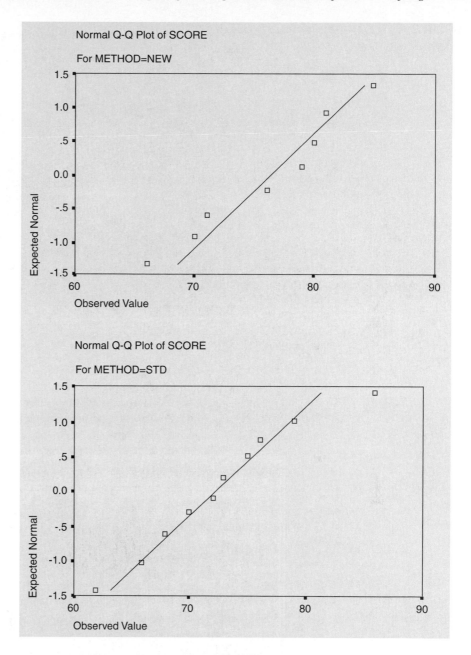

the same spread; thus, the samples appear to come from populations with approximately the same variance.

All three assumptions, then, appear to be reasonably satisfied for this application of the small-sample confidence interval.

The two-sample t statistic is a powerful tool for comparing population means when the assumptions are satisfied. It has also been shown to retain its usefulness when the sampled populations are only approximately normally distributed. And when the sample sizes are equal, the assumption of equal population variances can be relaxed. That is, if $n_1 = n_2$, then σ_1^2 and σ_2^2 can be quite different and the test statistic will still possess, approximately, a Student's t-distribution. In the case where $\sigma_1^2 \neq \sigma_2^2$ and $n_1 \neq n_2$, an approximate small-sample confidence interval or test can be obtained by modifying the degrees of freedom associated with the t-distribution.

FIGURE 7.9

SPSS Box Plots for Example 7.4

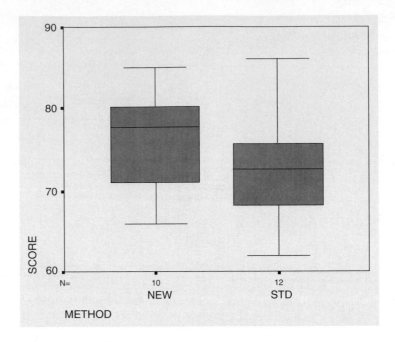

The next box gives the approximate small-sample procedures to use when the assumption of equal variances is violated. The test for the "unequal sample sizes" case is based on Satterthwaite's (1946) approximation.

Approximate Small-Sample Procedures when $\sigma_1^2 \neq \sigma_2^2$

EQUAL SAMPLE SIZES ($n_1 = n_2 = n$)

Confidence interval: $\qquad\qquad\qquad (\bar{x}_1 - \bar{x}_2) \pm t_{\alpha/2}\sqrt{(s_1^2 + s_2^2)/n}$

Test statistic for H_0: $(\mu_1 - \mu_2) = 0$: $\qquad t = (\bar{x}_1 - \bar{x}_2)/\sqrt{(s_1^2 + s_2^2)/n}$
where t is based on $\nu = n_1 + n_2 - 2 = 2(n - 1)$ degrees of freedom.

UNEQUAL SAMPLE SIZES ($n_1 \neq n_2$)

Confidence interval: $\qquad\qquad\qquad (\bar{x}_1 - \bar{x}_2) \pm t_{\alpha/2}\sqrt{(s_1^2/n_1) + (s_2^2/n_2)}$

Test statistic for H_0: $(\mu_1 - \mu_2) = 0$: $\qquad t = (\bar{x}_1 - \bar{x}_2)/\sqrt{(s_1^2/n_1) + (s_2^2/n_2)}$
where t is based on degrees of freedom equal to

$$\nu = \frac{(s_1^2/n_1 + s_2^2/n_2)^2}{\dfrac{(s_1^2/n_1)^2}{n_1 - 1} + \dfrac{(s_2^2/n_2)^2}{n_2 - 1}}$$

Note: The value of ν will generally not be an integer. Round ν down to the nearest integer to use the t table.

When the assumptions are not clearly satisfied, you can select larger samples from the populations or you can use other available statistical tests (nonparametric statistical tests).

What Should You Do If the Assumptions Are Not Satisfied?

Answer: If you are concerned that the assumptions are not satisfied, use the Wilcoxon rank sum test for independent samples to test for a shift in population distributions. See optional Section 7.4.

 EXERCISES 7.1–7.24

Learning the Mechanics

7.1 In order to compare the means of two populations, independent random samples of 400 observations are selected from each population, with the following results:

Sample 1	Sample 2
$\bar{x}_1 = 5{,}275$	$\bar{x}_2 = 5{,}240$
$s_1 = 150$	$s_2 = 200$

a. Use a 95% confidence interval to estimate the difference between the population means ($\mu_1 - \mu_2$). Interpret the confidence interval. 35 ± 24.5

b. Test the null hypothesis $H_0: (\mu_1 - \mu_2) = 0$ versus the alternative hypothesis $H_a: (\mu_1 - \mu_2) \neq 0$. Give the significance level of the test, and interpret the result.

c. Suppose the test in part **b** was conducted with the alternative hypothesis $H_a: (\mu_1 - \mu_2) > 0$. How would your answer to part **b** change? $p = .0026$

d. Test the null hypothesis $H_0: (\mu_1 - \mu_2) = 25$ versus the alternative $H_a: (\mu_1 - \mu_2) \neq 25$. Give the significance level, and interpret the result. Compare your answer to the test conducted in part **b**. $p = .4238$

e. What assumptions are necessary to ensure the validity of the inferential procedures applied in parts **a–d**?

7.2 Independent random samples of 100 observations each are chosen from two normal populations with the following means and standard deviations:

Population 1	Population 2
$\mu_1 = 14$	$\mu_2 = 10$
$\sigma_1 = 4$	$\sigma_2 = 3$

Let $\bar{x}_1$ and $\bar{x}_2$ denote the two sample means.

a. Give the mean and standard deviation of the sampling distribution of $\bar{x}_1$. $14; .4$

b. Give the mean and standard deviation of the sampling distribution of $\bar{x}_2$. $10; .3$

c. Suppose you were to calculate the difference ($\bar{x}_1 - \bar{x}_2$) between the sample means. Find the mean and standard deviation of the sampling distribution of ($\bar{x}_1 - \bar{x}_2$). $4; .5$

d. Will the statistic ($\bar{x}_1 - \bar{x}_2$) be normally distributed? Explain.   Yes

7.3 Two populations are described in each of the following cases. In which cases would it be appropriate to apply the small-sample t-test to investigate the difference between the population means?

a. Population 1: Normal distribution with variance σ_1^2 Population 2: Skewed to the right with variance $\sigma_2^2 = \sigma_1^2$ No

b. Population 1: Normal distribution with variance σ_1^2 Population 2: Normal distribution with variance $\sigma_2^2 \neq \sigma_1^2$ No

c. Population 1: Skewed to the left with variance σ_1^2 Population 2: Skewed to the left with variance $\sigma_2^2 = \sigma_1^2$

d. Population 1: Normal distribution with variance σ_1^2 Population 2: Normal distribution with variance $\sigma_2^2 = \sigma_1^2$ Yes

7.4 Assume that $\sigma_1^2 = \sigma_2^2 = \sigma^2$. Calculate the pooled estimator of σ^2 for each of the following cases:

a. $s_1^2 = 200, s_2^2 = 180, n_1 = n_2 = 25$ $s_p^2 = 190$

b. $s_1^2 = 25, s_2^2 = 40, n_1 = 20, n_2 = 10$

c. $s_1^2 = .20, s_2^2 = .30, n_1 = 8, n_2 = 12$ $s_p^2 = .2611$

d. $s_1^2 = 2{,}500, s_2^2 = 1{,}800, n_1 = 16, n_2 = 17$

e. Note that the pooled estimate is a weighted average of the sample variances. To which of the variances does the pooled estimate fall nearer in each of the above cases?

7.5 Independent random samples from normal populations produced the results shown below:

🔵 LM7_5

Sample 1	Sample 2
1.2	4.2
3.1	2.7
1.7	3.6
2.8	3.9
3.0	

a. Calculate the pooled estimate of σ^2. $s_p^2 = .5989$

b. Do the data provide sufficient evidence to indicate that $\mu_2 > \mu_1$? Test using $\alpha = .10$. $t = -2.39$

c. Find a 90% confidence interval for ($\mu_1 - \mu_2$).

d. Which of the two inferential procedures, the test of hypothesis in part **b** or the confidence interval in part **c**, provides more information about ($\mu_1 - \mu_2$)?

MINITAB Output for Exercise 7.7

```
Two-sample T for density

            N       Mean      StDev     SE Mean
X1         233        473         84       15.26
X2         312        485         93       17.66


Difference = mu (X1) - mu (X2)
Estimate for difference: -8.0
95% CI for difference: (-22.93, 6.93)
T-Test of difference = 0 (vs not =): T-Value = -1.58   P-Value = 0.115
DF = 543
```

7.6 Two independent random samples have been selected, 100 observations from population 1 and 100 from population 2. Sample means $\bar{x}_1 = 70$ and $\bar{x}_2 = 50$ were obtained. From previous experience with these populations, it is known that the variances are $\sigma_1^2 = 100$ and $\sigma_2^2 = 64$.

a. Find $\sigma_{(\bar{x}_1 - \bar{x}_2)}$.   1.2806

b. Sketch the approximate sampling distribution $(\bar{x}_1 - \bar{x}_2)$ assuming $(\mu_1 - \mu_2) = 5$.

c. Locate the observed value of $(\bar{x}_1 - \bar{x}_2)$ on the graph you drew in part **b**. Does it appear that this value contradicts the null hypothesis $H_0: (\mu_1 - \mu_2) = 5$?   Yes

d. Use the z-table to determine the rejection region for the test of $H_0: (\mu_1 - \mu_2) = 5$ against $H_a: (\mu_1 - \mu_2) \neq 5$. Use $\alpha = .05$.

e. Conduct the hypothesis test of part **d** and interpret your result. $z = 11.71$

f. Construct a 95% confidence interval for $(\mu_1 - \mu_2)$. Interpret the interval. 20 ± 2.51

g. Which inference provides more information about the value of $(\mu_1 - \mu_2)$—the test of hypothesis in part **e** or the confidence interval in part **f**?

7.7 Independent random samples are selected from two populations and used to test the hypothesis $H_0: (\mu_1 - \mu_2) = 0$ against the alternative $H_a: (\mu_1 - \mu_2) \neq 0$. A total of 233 observations from population 1 and 312 from population 2 are analyzed by MINITAB, with the result shown above.

a. Interpret the results of the computer analysis.

b. If the alternative hypothesis had been $H_a: (\mu_1 - \mu_2) < 0$, how would the p-value change? Interpret the p-value for this one-tailed test. $p = 0.575$

7.8 Independent random samples selected from two normal populations produced the sample means and standard deviations shown below:

Sample 1	Sample 2
$n_1 = 17$	$n_2 = 12$
$\bar{x}_1 = 5.4$	$\bar{x}_2 = 7.9$
$s_1 = 3.4$	$s_2 = 4.8$

a. The test $H_0: (\mu_1 - \mu_2) = 0$ against $H_a: (\mu_1 - \mu_2) \neq 0$ was conducted using SAS, with the results shown below. Check and interpret the results.

SAS Output for Exercise 7.8

Two Sample t-test for the Means of X within SAMPLE

Sample Statistics

Group	N	Mean	Std. Dev.	Std. Error
SAM1	17	5.4	3.4	4.123
SAM2	12	7.9	4.8	3.464

Hypothesis Test

Null hypothesis: Mean 1 - Mean 2 = 0
Alternative: Mean 1 - Mean 2 ^= 0

If Variances Are	t statistic	Df	Pr > t
Equal	-1.646	27	0.1114
Not Equal	-1.550	18.57	0.1386

95% Confidence Interval for the Difference between Two Means

Lower Limit	Upper Limit
-5.48	0.48

b. Check and interpret the 95% confidence interval for $(\mu_1 - \mu_2)$ shown on the printout.

Applying the Concepts—Basic

7.9 Are children who repeat a grade in elementary school shorter on average, than their peers? To answer this question, researchers compared the heights of Australian school children who repeated a grade to those who did not (*The Archives of Disease in Childhood*, Apr. 2000). All height measurements were standardized using z-scores. A summary of the results, by gender, is shown in the table.

	Never Repeated	**Repeated a Grade**
Boys	$n = 1,349$	$n = 86$
	$\bar{x} = .30$	$\bar{x} = -.04$
	$s = .97$	$s = 1.17$
Girls	$n = 1,366$	$n = 43$
	$\bar{x} = .22$	$\bar{x} = .26$
	$s = 1.04$	$s = .94$

Source: Wake, M., Coghlan, D., and Hesketh, K. "Does height influence progression through primary school grades?" *The Archives of Disease in Childhood*, Vol. 82, Apr. 2000 (Table 3).

a. Conduct a test of hypothesis to determine whether the average height of Australian boys who repeated a grade is less than the average height of boys who never repeated. Use $\alpha = .05$. $z = -2.64$

b. Repeat part **a** for Australian girls. $z = .27$

c. Summarize the results of the hypothesis tests in the words of the problem.

7.10 A group of University of Florida psychologists investigated the effects of age and gender on the short-term memory of adults (*Cognitive Aging Conference,* Apr. 1996). Each in a sample of 152 adults was asked to place 20 common household items (e.g., eyeglasses, keys, hat, hammer) into the rooms of a computer-image house. After performing some unrelated activities, each subject was asked to recall the locations of the objects they had placed. The number of correct responses (out of 20) was recorded.

a. The researchers theorized that women will have a higher mean recall score than men. Set up the null and alternative hypotheses to test this theory.

b. Refer to part **a.** The 43 men in the study had a mean recall score of 13.5, while the 109 women had a mean recall score of 14.4. The observed significance level for comparing these two means was found to be $p = .0001$. Interpret this value.

c. The researchers also hypothesized that younger adults would have a higher mean recall score than older adults. Set up H_0 and H_a to test this theory.

d. The observed significance level for the test of part **c** was reported as $p = .0001$. Interpret this result.

7.11 The "fear of negative evaluation" (FNE) scores for 11 female students known to suffer from the eating disorder bulimia and 14 female students with normal eating habits, first presented in Exercise 2.27 (p. 39), are reproduced below. (Recall that the higher the score, the greater the fear of negative evaluation.)

a. Find a 95% confidence interval for the difference between the population means of the FNE scores for bulimic and normal female students. Interpret the result.

b. What assumptions are required for the interval of part **a** to be statistically valid? Are these assumptions reasonably satisfied? Explain.

7.12 Do the early questions in an exam affect your performance on later questions? Research reported in the *Journal of Psychology* (Jan. 1991) investigated this relationship by conducting the following experiment: 140 college students were randomly assigned to two groups. Group A took an exam that contained 15 difficult questions followed by five moderate questions. Group B took an exam that contained 15 easy questions followed by the same five moderate questions. One of the goals of the study was to compare the average score on the moderate questions for the two groups.

a. The researchers hypothesized that the students who had the easy questions would score better than the students who had the difficult questions. Set up the appropriate null and alternative hypotheses to test their research hypothesis, defining any symbols you use. $H_0: \mu_1 - \mu_2 = 0; H_a: \mu_1 - \mu_2 < 0$

b. Suppose the p-value for this test was reported as .3248. What conclusion would you reach based on this p-value?

c. What assumptions are necessary to ensure the validity of this conclusion?

7.13 On average, do males outperform females in mathematics? To answer this question, psychologists at the University of Minnesota compared the scores of male and female eighth-grade students who took a basic skills mathematics achievement test (*American Educational Research Journal*, Fall 1998). One form of the test consisted of 68 multiple-choice questions. A summary of the test scores is displayed in the table.

	Males	**Females**
Sample size	1,764	1,739
Mean	48.9	48.4
Standard deviation	12.96	11.85

Source: Bielinski, J., and Davison, M. L. "Gender differences by item difficulty interactions in multiple-choice mathematics items." *American Educational Research Journal,* Vol. 35, No. 3, Fall 1998, p. 464 (Table 1).

🌐 BULIMIA

Bulimic students	21	13	10	20	25	19	16	21	24	13	14		
Normal students	13	6	16	13	8	19	23	18	11	19	7	10	15	20

Source: Randles, R. H. "On neutral responses (zeros) in the sign test and ties in the Wilcoxon-Mann-Whitney test." *The American Statistician*, Vol. 55, No. 2, May 2001 (Figure 3).

a. Is there evidence of a difference between the true mean mathematics test scores of male and female eighth-graders? $z = 1.19$

b. Use a 90% confidence interval to estimate the true difference in mean test scores between males and females. Does the confidence interval support the result of the test you conducted in part **a**? $.5 \pm .690$

c. What assumptions about the distributions of the populations of test scores are necessary to ensure the validity of the inferences you made in parts **a** and **b**?

d. What is the observed significance level of the test you conducted in part **a**? $p = .2340$

Applying the Concepts—Intermediate

⊘ EVOS

7.14 Refer to the *Journal of Agricultural, Biological, and Environmental Statistics* (Sept. 2000) impact study of a tanker oil spill on the seabird population in Alaska, Exercise 2.135 (p. 102). Recall that for each of 96 shoreline locations (called transects), the number of seabirds found, the length (in kilometers) of the transect, and whether or not the transect was in an oiled area were recorded. (The data are saved in the EVOS file.) Observed seabird density is defined as the observed count divided by the length of the transect. A comparison of the mean densities of oiled and unoiled transects is displayed in the MINITAB printout below. Use this information to make an inference about the difference in the population mean seabird densities of oiled and unoiled transects.

7.15 Teachers Involve Parents in Schoolwork (TIPS) is an interactive homework process designed to improve the quality of homework assignments for elementary, middle, and high school students. TIPS homework assignments require students to conduct interactions with family partners (parents, guardians, etc.) while completing the homework. Frances Van Voorhis (Johns Hopkins University) conducted a study to investigate the effects of TIPS in science, mathematics, and language arts homework assignments (April, 2001). Each in a sample of 128 middle school students was assigned to complete TIPS homework assignments, while 98 students in a second sample were assigned traditional, non-interactive, homework assignments (called ATIPS). At the end of the study, all students reported on the level of family involvement in their homework on a

4-point scale (0 = never, 1 = rarely, 2 = sometimes, 3 = frequently, 4 = always). Three scores were recorded for each student: one for science homework, one for math homework, and one for language arts homework. The data for the study are saved in the HWSTUDY file. (The first 5 and last 5 observations in the data set are listed in the following table.)

⊘ HWSTUDY (First and last 5 observations)

Homework Condition	Science	Math	Language
ATIPS	1	0	0
ATIPS	0	1	1
ATIPS	0	1	0
ATIPS	1	2	0
ATIPS	1	1	2
TIPS	2	3	2
TIPS	1	4	2
TIPS	2	4	2
TIPS	4	0	3
TIPS	2	0	1

Source: Van Voorhis, F. L., "Teachers' use of interactive homework and its effects on family involvement and science achievement of middle grade students." Paper presented at the annual meeting of the American Educational Research Association, Seattle, April 2001.

a. Conduct an analysis to compare the mean level of family involvement in science homework assignments of TIPS and ATIPS students. Use $\alpha = .05$. Make a practical conclusion. $t = -7.24$

b. Repeat part **a** for mathematics homework assignments. $t = -.050$

c. Repeat part **a** for language arts homework assignments. $t = -1.25$

d. What assumptions are necessary for the inferences of parts **a**–**c** to be valid? Are they reasonably satisfied?

7.16 Two methods of castrating male piglets were investigated in *Applied Animal Behaviour Science* (Nov. 1, 2000). Method 1 involved an incision of the spermatic cords while Method 2 involved pulling and severing the cords. Forty-nine male piglets were randomly allocated to one of the two methods. During castration, the researchers measured the number of high frequency vocal responses (squeals) per second over a 5-second period. The data are summarized in the table. Conduct a test of hypoth-

MINITAB Output for Problem 7.14

```
Two-sample T for DENSITY

OIL        N       Mean      StDev     SE Mean
no         36      3.27      6.70          1.1
yes        60      3.50      5.97         0.77


Difference = mu (no) - mu (yes)
Estimate for difference: -0.22
95% CI for difference: (-2.93, 2.49)
T-Test of difference = 0 (vs not =): T-Value = -0.16   P-Value = 0.871
DF = 67
```

esis to determine whether the population mean number of high frequency vocal responses differs for piglets castrated by the two methods. Use $\alpha = .05$. $t = 1.56$

	Method 1	Method 2
Sample size	24	25
Mean number of squeals	.74	.70
Standard deviation	.09	.09

Source: Taylor, A. A., and Weary, D. M. "Vocal responses of piglets to castration: Identifying procedural sources of pain." *Applied Animal Behaviour Science*, Vol. 70, No. 1, November 1, 2000.

7.17 Each week, *Billboard* magazine ranks the top 200 selling albums (or CDs) in the commercial music industry. Prior to 1991, *Billboard*'s charting methodology was based on reported album sales at a sample of record stores. In 1991, *Billboard* began using the SoundScan system to chart album sales. With SoundScan, record sales were pulled directly from barcode-reading sales registers across the country. The impact of SoundScan on *Billboard* magazine's ratings was investigated in *Organizational Science* (May–June, 2000). Based on summary statistics provided in the article, the table lists the (simulated) number of new artists entering the weekly *Billboard* chart for 12 weeks prior to the introduction of SoundScan and for 12 weeks after the introduction of SoundScan.

⊙ BILLBOARD

Pre-SoundScan	1	1	0	2	3	4	3	4	5	5	2	3
Post-SoundScan	0	2	3	1	3	1	0	1	4	2	2	1

a. Find a 90% confidence interval for the difference between the mean number of new artists entering the *Billboard* chart per week prior to SoundScan and the mean number of new artists per week after SoundScan was introduced. Interpret the result.

b. *Organizational Science* reported the *p*-value for testing the difference between the two means as $p < .1$. Find the actual *p*-value of a two-tailed test and interpret it. Does the result agree with the reported value? $.05 < p < .10$

7.18 To help students organize global information about people, places, and environments, geographers encourage them to develop "mental maps" of the world. A series of lessons was designed to aid students in the development of mental maps (*Journal of Geography,* May/June 1997). In one experiment, a class of 24 seventh-grade geography students was given mental map lessons while a second class of 20 students received traditional instruction. All students were asked to sketch a map of the world, and each portion of the map was evaluated for accuracy using a 5-point scale (1 = low accuracy, 5 = high accuracy).

a. The mean accuracy scores of the two groups of seventh-graders were compared using a test of hypothesis. State H_0 and H_a for a test to determine whether the mental map lessons improve a student's ability to sketch a world map.

b. What assumptions (if any) are required for the test to be statistically valid?

c. The observed significance level of the test for comparing the mean accuracy scores for continents drawn is .0507. Interpret this result.

d. The observed significance level of the test for comparing the mean accuracy scores for labeling oceans is .7371. Interpret the result.

e. The observed significance level of the test for comparing the mean accuracy scores for the entire map is .0024. Interpret the result.

7.19 Bear gallbladder is used in Chinese medicine to treat inflamation. Due to the difficulty of obtaining bear gallbladder, Chinese medical researchers are searching for a more readily available source of animal bile. A study in the *Journal of Ethnopharmacology* (June 1995) examined pig gallbladder as an effective substitute for bear gallbladder. Twenty male mice were divided randomly into two groups: 10 were given a dosage of bear bile and 10 were given a dosage of pig bile. All the mice then received an injection of croton oil in the left ear lobe to induce inflammation. Four hours later, both the left and right ear lobes were weighed, with the difference (in milligrams) representing the degree of swelling. Summary statistics on the degree of swelling are provided in the following table.

Bear Bile	Pig Bile
$n_1 = 10$	$n_2 = 10$
$\bar{x}_1 = 9.19$	$\bar{x}_2 = 9.71$
$s_1 = 4.17$	$s_2 = 3.33$

a. Compare the mean degree of swelling for mice treated with bear and pig bile with a 95% confidence interval. Interpret the result. $-.52 \pm 3.546$

b. What assumptions are necessary for the inference, part **a**, to be valid?

c. Another group of 10 mice (the control group) was given normal saline instead of bear or pig bile. The degree of swelling for this control group produced the following summary statistics: $\bar{x} = 18.30, s = 3.38$. Compare the mean degree of swelling for the control group to the mice treated with pig bile using a 95% confidence interval. Interpret the results.

7.20 Hermaphrodites are animals that possess the reproductive organs of both sexes. *Genetical Research* (June 1995) published a study of the mating systems of hermaphroditic snail species. The mating habits of the snails were classified into two groups: (1) self-fertilizing (selfing) snails that mate with snails of the same sex and (2) cross-fertilizing (outcrossing) snails that mate with snails of the opposite sex. One variable of interest in the study was the effective population size of the snail species. The means and standard deviations of the effective population size for independent random samples of 17 outcrossing snail species and 5 selfing snail species are given in the table on p. 338. Compare the mean effective population sizes of the two types of snail species with a 90% confidence interval. Interpret the result. $761 \pm 1,688.19$

	Effective Population Size		
Snail Mating System	**Sample Size**	**Mean**	**Standard Deviation**
Outcrossing	17	4,894	1,932
Selfing	5	4,133	1,890

Source: Jarne, P. "Mating system, bottlenecks, and genetic polymorphism in hermaphroditic animals." *Genetical Research,* Vol. 65, No. 3, June 1995, p. 197 (Table 4).

7.21 Refer to the *Journal of Communication Disorders* (Mar. 1995) study of specifically language-impaired (SLI) children, Exercise 2.51 (p. 53). The data on deviation intelligence quotient (DIQ) for 10 SLI children and 10 younger, normally developing (YND) children are reproduced in the table. Use the methodology of this section to compare the mean DIQ of the two groups of children. (Use $\alpha = .10$.) What do you conclude?

⊘ SLI

SLI Children			**YND Children**		
86	87	84	110	90	105
94	86	107	92	92	96
89	98	95	86	100	92
110			90		

7.22 Do cocaine abusers have radically different personalities than nonabusing college students? This was one of the questions researched in *Psychological Assessment* (June 1995). A personality questionnaire (ZKPQ) was administered to a sample of 450 cocaine abusers and a sample of 589 college students. The ZKPQ yields scores (measured on a 20-point scale) on each of five dimensions: impulsive–sensation seeking, sociability, neuroticism–anxiety, aggression–hostility, and activity. The results are summarized in the table. Compare the mean ZKPQ scores of the two groups on each dimension using a statistical test of hypothesis. Interpret the results at $\alpha = .01$.

	Cocaine Abusers (n = 450)		**College Students (n = 589)**	
ZKPQ Dimension	**Mean**	**Std. Dev.**	**Mean**	**Std. Dev.**
Impulsive–sensation seeking	9.4	4.4	9.5	4.4
Sociability	10.4	4.3	12.5	4.0
Neuroticism–anxiety	8.6	5.1	9.1	4.6
Aggression–hostility	8.6	3.9	7.3	4.1
Activity	11.1	3.4	8.0	4.1

Source: Ball, S. A. "The validity of an alternative five-factor measure of personality in cocaine abusers." *Psychological Assessment,* Vol. 7, No. 2, June 1995, p. 150 (Table 1).

Applying the Concepts—Advanced

7.23 An investigation of ethnic differences in reports of pain perception was presented at the annual meeting of the American Psychosomatic Society (March 2001). A sample of 55 blacks and 159 whites participated in the study. Subjects rated (on a 13-point scale) the intensity and unpleasantness of pain felt when a bag of ice was placed on their foreheads for two minutes. (Higher ratings correspond to higher pain intensity.) A summary of the results is provided in the accompanying table.

	Blacks	**Whites**
Sample size	55	159
Mean pain intensity	8.2	6.9

a. Why is it dangerous to draw a statistical inference from the summarized data? Explain.

b. Give values of the missing sample standard deviations that would lead you to conclude (at $\alpha = .05$) that blacks, on average, have a higher pain intensity rating than whites.

c. Give values of the missing sample standard deviations that would lead you to an inconclusive decision (at $\alpha = .05$) regarding whether blacks or whites have a higher mean intensity rating.

7.24 The *Florida Scientist* (Summer/Autumn 1991) reported on a study of the feeding habits of sea urchins. A sample of 20 urchins were captured from Biscayne Bay (Miami), placed in marine aquaria, then starved for 48 hours. Each sea urchin was then fed a 5-cm blade of turtle grass. Ten of the urchins received only green blades, while the other half received only decayed blades. (Assume that the two samples of sea urchins— 10 urchins per sample—were randomly and independently selected.) The ingestion time, measured from the time the blade first made contact with the urchin's mouth to the time the urchin had finished ingesting the blade, was recorded. A summary of the results is provided in the next table.

	Green Blades	**Decayed Blades**
Number of sea urchins	10	10
Mean ingestion time (hours)	3.35	2.36
Standard deviation (hours)	.79	.47

Source: Montague, J. R., *et al.* "Laboratory measurement of ingestion rate for the sea urchin *Lytechinus variegatus.*" *Florida Scientist,* Vol. 54, Nos. 3/4, Summer/Autumn, 1991 (Table 1).

According to the researchers, "the difference in rates at which the urchins ingested the blades suggest that green, unblemished turtle grass may not be a particularly palatable food compared with decayed turtle grass. If so, urchins in the field may find it more profitable to selectively graze on decayed portions of the leaves." Do the results support this conclusion? Yes, .99 ± .504

7.2 COMPARING TWO POPULATION MEANS: PAIRED DIFFERENCE EXPERIMENTS

TEACHING TIP
Review the discussion of the two types of sampling designs for comparing the means of two populations. Discuss the main advantage of the paired difference experiment.

In Example 7.4, we compared two methods of teaching reading to slow learners by means of a 95% confidence interval. Suppose it is possible to measure the slow learners' "reading IQs" *before* they are subjected to a teaching method. Eight pairs of slow learners with similar reading IQs are found, and one member of each pair is randomly assigned to the standard teaching method while the other is assigned to the new method. The data are given in Table 7.3. Do the data support the hypothesis that the population mean reading test score for slow learners taught by the new method is greater than the mean reading test score for those taught by the standard method?

◉ PAIREDSCORES

TABLE 7.3   Reading Test Scores for Eight Pairs of Slow Learners

Pair	New Method (1)	Standard Method (2)
1	77	72
2	74	68
3	82	76
4	73	68
5	87	84
6	69	68
7	66	61
8	80	76

We want to test

$$H_0: (\mu_1 - \mu_2) = 0$$
$$H_a: (\mu_1 - \mu_2) > 0$$

One way to conduct this test is to use the t statistic for two independent samples (Section 7.1). The analysis is shown on the STATISTIX printout, Figure 7.10. The test statistic, $t = 1.26$, is highlighted on the printout as well as the p-value of the test, $p = .1149$. At $\alpha = .10$, the p-value exceeds α. Thus, from *this* analysis we might conclude that we do not have sufficient evidence to infer a difference in the mean test scores for the two methods.

If you carefully examine the data in Table 7.3, however, you will find this result difficult to accept. The test score of the new method is larger than the corresponding test score for the standard method *for every one of the eight pairs of slow learners*.

FIGURE 7.10

STATISTIX Analysis of Data in Table 7.3

```
TWO-SAMPLE T TESTS FOR NEW VS STANDARD

                          SAMPLE
VARIABLE        MEAN       SIZE        S.D.         S.E.
----------    -------    -------     -------      -------
NEW           76.000        8        6.9282       2.4495
STANDARD      71.625        8        7.0089       2.4780
DIFFERENCE     4.3750

NULL HYPOTHESIS:   DIFFERENCE = 0
ALTERNATIVE HYP:   DIFFERENCE > 0

ASSUMPTION               T        DF        P       95% CI FOR DIFFERENCE
-------------------    -----    ------    ------    ---------------------
EQUAL VARIANCES         1.26      14      0.1149    (-3.0982,  11.848)
UNEQUAL VARIANCES       1.26      14.0    0.1149    (-3.0983,  11.848)


                         F      NUM DF    DEN DF      P
TESTS FOR EQUALITY     ------    ------    ------    ------
      OF VARIANCES      1.02        7         7      0.4882

CASES INCLUDED 16      MISSING CASES 0
```

This, in itself, seems to provide strong evidence to indicate that μ_1 exceeds μ_2. Why, then, did the *t*-test fail to detect this difference? The answer is: *The independent samples t-test is not a valid procedure to use with this set of data.*

The *t*-test is inappropriate because the assumption of independent samples is invalid. We have randomly chosen *pairs of test scores*, and thus, once we have chosen the sample for the new method, we have *not* independently chosen the sample for the standard method. The dependence between observations within pairs can be seen by examining the pairs of test scores, which tend to rise and fall together as we go from pair to pair. This pattern provides strong visual evidence of a violation of the assumption of independence required for the two-sample *t*-test of Section 7.1. Note, also, that

$$s_p^2 = \frac{(n_1 - 1)s_1^2 + (n_2 - 1)s_2^2}{n_1 + n_2 - 2} = \frac{(8 - 1)(6.93)^2 + (8 - 1)(7.01)^2}{8 + 8 - 2} = 48.55$$

Thus, there is a *large variation within samples* (reflected by the large value of s_p^2) in comparison to the relatively *small difference between the sample means*. Because s_p^2 is so large, the *t*-test of Section 7.1 is unable to detect a difference between μ_1 and μ_2.

TABLE 7.4   Differences in Reading Test Scores

Pair	New Method	Standard Method	Difference (New Method–Standard Method)
1	77	72	5
2	74	68	6
3	82	76	6
4	73	68	5
5	87	84	3
6	69	68	1
7	66	61	5
8	80	76	4

We now consider a valid method of analyzing the data of Table 7.3. In Table 7.4 we add the column of differences between the test scores of the pairs of slow learners. We can regard these differences in test scores as a random sample of differences for all pairs (matched on reading IQ) of slow learners, past and present. Then we can use this sample to make inferences about the mean of the population of differences, μ_D, which is equal to the difference $(\mu_1 - \mu_2)$. That is, the mean of the population (and sample) of differences equals the difference between the population (and sample) means. Thus, our test becomes

$$H_0: \mu_D = 0 \quad (\mu_1 - \mu_2 = 0)$$
$$H_a: \mu_D > 0 \quad (\mu_1 - \mu_2 > 0)$$

The test statistic is a one-sample *t* (Section 7.1), since we are now analyzing a single sample of differences for small *n*:

$$\textit{Test statistic:} \quad t = \frac{\bar{x}_D - 0}{s_D/\sqrt{n_D}}$$

where $\bar{x}_D$ = Sample mean difference
$\quad\quad\quad s_D$ = Sample standard deviation of differences
$\quad\quad\quad n_D$ = Number of differences = Number of pairs

Assumptions: The population of differences in test scores is approximately normally distributed. The sample differences are randomly selected from the population differences. [*Note:* We do not need to make the assumption that $\sigma_1^2 = \sigma_2^2$.]
Rejection region: At significance level $\alpha = .05$, we will reject H_0 if $t > t_{.05}$, where $t_{.05}$ is based on $(n_D - 1)$ degrees of freedom.

FIGURE 7.11

Rejection Region for
Example 7.4

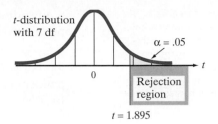

$t = 1.895$

Referring to Table IV in Appendix A, we find the t value corresponding to $\alpha = .05$ and $n_D - 1 = 8 - 1 = 7$ df to be $t_{.05} = 1.895$. Then we will reject the null hypothesis if $t > 1.895$ (see Figure 7.11). Note that the number of degrees of freedom decreases from $n_1 + n_2 - 2 = 14$ to 7 when we use the paired difference experiment rather than the two independent random samples design.

Summary statistics for the $n = 8$ differences are shown in the MINITAB printout, Figure 7.12. Note that $\bar{x}_D = 4.375$ and $s_D = 1.685$. Substituting these values into the formula for the test statistic, we have

$$t = \frac{\bar{x}_D - 0}{s_D / \sqrt{n_D}} = \frac{4.375}{1.685 / \sqrt{8}} = 7.34$$

FIGURE 7.12

MINITAB Paired Difference
Analysis of Data in Table 7.4

```
Paired T for NEW - STANDARD


               N      Mean      StDev     SE Mean
NEW            8     76.00       6.93        2.45
STANDARD       8     71.63       7.01        2.48
Difference     8     4.375       1.685       0.596


95% lower bound for mean difference: 3.246
T-Test of mean difference = 0 (vs > 0): T-Value = 7.34   P-Value = 0.000
```

Because this value of t falls in the rejection region, we conclude (at $\alpha = .05$) that the population mean test score for slow learners taught by the new method exceeds the population mean score for those taught by the standard method. We can reach the same conclusion by noting that the p-value of the test, highlighted in Figure 7.12, is much smaller than $\alpha = .05$.

This kind of experiment, in which observations are paired and the differences are analyzed, is called a **paired difference experiment**. In many cases, a paired difference experiment can provide more information about the difference between population means than an independent samples experiment. The idea is to compare population means by comparing the differences between pairs of experimental units (objects, people, etc.) that were very similar prior to the experiment. The differencing removes sources of variation that tend to inflate σ^2. For example, when two children are taught to read by two different methods, the observed difference in achievement may be due to a difference in the effectiveness of the two teaching methods *or* it may be due to differences in the initial reading levels and IQs of the two children (random error). To reduce the effect of differences in the children on the observed differences in reading achievement, the two methods of reading are imposed on two children who are more likely to possess similar intellectual capacity, namely children with nearly equal IQs. The effect of this pairing is to remove the larger source of variation that would be present if children with different abilities were randomly assigned to the two samples. Making comparisons within groups of similar experimental units is called **blocking**, and the paired difference experiment is a simple example of a **randomized block experiment**. In our example, pairs of children with matching IQ scores represent the blocks.

Some other examples for which the paired difference experiment might be appropriate are the following:

1. Suppose you want to estimate the difference $(\mu_1 - \mu_2)$ in mean price per gallon between two major brands of premium gasoline. If you choose two independent random samples of stations for each brand, the variability in price due to geographic location may be large. To eliminate this source of variability you could choose pairs of stations of similar size, one station for each brand, in close geographic proximity and use the sample of differences between the prices of the brands to make an inference about $(\mu_1 - \mu_2)$.

2. Suppose a college placement center wants to estimate the difference $(\mu_1 - \mu_2)$ in mean starting salaries for men and women graduates who seek jobs through the center. If it independently samples men and women, the starting salaries may vary because of their different college majors and differences in grade point averages. To eliminate these sources of variability, the placement center could match male and female job-seekers according to their majors and grade point averages. Then the differences between the starting salaries of each pair in the sample could be used to make an inference about $(\mu_1 - \mu_2)$.

3. Suppose you wish to estimate the difference $(\mu_1 - \mu_2)$ in mean absorption rate into the bloodstream for two drugs that relieve pain. If you independently sample people, the absorption rates might vary because of age, weight, sex, blood pressure, etc. In fact, there are many possible sources of nuisance variability, and pairing individuals who are similar in all the possible sources would be quite difficult. However, it may be possible to obtain two measurements *on the same person*. First, we administer one of the two drugs and record the time until absorption. After a sufficient amount of time, the other drug is administered and a second measurement on absorption time is obtained. The differences between the measurements for each person in the sample could then be used to estimate $(\mu_1 - \mu_2)$. This procedure would be advisable only if the amount of time allotted between drugs is sufficient to guarantee little or no carryover effect. Otherwise, it would be better to use different people matched as closely as possible on the factors thought to be most important.

The hypothesis-testing procedures and the method of forming confidence intervals for the difference between two means using a paired difference experiment are summarized in the next two boxes for both large and small *n*.

Paired Difference Confidence Interval for $\mu_D = \mu_1 - \mu_2$

LARGE SAMPLE

$$\bar{x}_D \pm z_{\alpha/2} \frac{\sigma_D}{\sqrt{n_D}} \approx \bar{x}_D \pm z_{\alpha/2} \frac{s_D}{\sqrt{n_D}}$$

Assumption: The sample differences are randomly selected from the population of differences.

SMALL SAMPLE

$$\bar{x}_D \pm t_{\alpha/2} \frac{s_D}{\sqrt{n_D}}$$

where $t_{\alpha/2}$ is based on $(n_D - 1)$ degrees of freedom
Assumptions:

1. The relative frequency distribution of the population of differences is normal.
2. The sample differences are randomly selected from the population of differences.

Paired Difference Test of Hypothesis for $\mu_D = \mu_1 - \mu_2$

ONE-TAILED TEST

$H_0: \mu_D = D_0$

$H_a: \mu_D < D_0$

[or $H_a: \mu_D > D_0$]

TWO-TAILED TEST

$H_0: \mu_D = D_0$

$H_a: \mu_D \neq D_0$

LARGE SAMPLE

Test statistic: $z = \dfrac{\bar{x}_D - D_0}{\sigma_D/\sqrt{n_D}} \approx \dfrac{\bar{x}_D - D_0}{s_D/\sqrt{n_D}}$

Rejection region: $z < -z_\alpha$ *Rejection region:* $|z| > z_{\alpha/2}$
[or $z > z_\alpha$ when $H_a: \mu_D > D_0$]

Assumption: The differences are randomly selected from the population of differences.

SMALL SAMPLE

Test statistic: $t = \dfrac{\bar{x}_D - D_0}{s_D/\sqrt{n_D}}$

Rejection region: $t < -t_\alpha$ *Rejection region:* $|t| > t_{\alpha/2}$
[or $t > t_\alpha$ when $H_a: \mu_D > D_0$]
where t_α and $t_{\alpha/2}$ are based on $(n_D - 1)$ degrees of freedom

Assumptions:

1. The relative frequency distribution of the population of differences is normal.
2. The differences are randomly selected from the population of differences.

EXAMPLE 7.5

400 ± 311

An experiment is conducted to compare the starting salaries of male and female college graduates who find jobs. Pairs are formed by choosing a male and a female with the same major and similar grade point averages (GPA). Suppose a random sample of 10 pairs is formed in this manner and the starting annual salary of each person is recorded. The results are shown in Table 7.5. Compare the mean starting salary, μ_1, for males to the mean starting salary, μ_2, for females using a 95% confidence interval. Interpret the results.

PAIREDGRADS
TABLE 7.5 Data on Annual Salaries for Matched Pairs of College Graduates

Pair	Male	Female	Difference (Male – Female)	Pair	Male	Female	Difference (Male – Female)
1	$29,300	$28,800	$ 500	6	$37,800	$38,000	$ −200
2	41,500	41,600	−100	7	69,500	69,200	300
3	40,400	39,800	600	8	41,200	40,100	1,100
4	38,500	38,500	0	9	38,400	38,200	200
5	43,500	42,600	900	10	59,200	58,500	700

FIGURE 7.13a

SAS Descriptive Statistics for Differences in Table 7.5

The MEANS Procedure

Analysis Variable : DIFF

N	Mean	Std Dev	Minimum	Maximum
10	400.0000000	434.6134937	-200.0000000	1100.00

FIGURE 7.13b

SAS Paired Difference Analysis of Annual Salaries

Two Sample Paired t-test for the Means of MALE and FEMALE

Sample Statistics

Group	N	Mean	Std. Dev.	Std. Error
MALE	10	43930	11665	3688.8
FEMALE	10	43530	11617	3673.6

Hypothesis Test

Null hypothesis: Mean of (MALE - FEMALE) = 0
Alternative: Mean of (MALE - FEMALE) ^= 0

t Statistic	Df	Prob > t
2.910	9	0.0173

95% Confidence Interval for the Difference between Two Paired Means

Lower Limit	Upper Limit
89.10	710.90

Solution

Since the data on annual salary are collected in pairs of males and females matched on GPA and major, a paired difference experiment is performed. To conduct the analysis, we first compute the differences between the salaries, as shown in Table 7.5. Summary statistics for these $n = 10$ differences are displayed in the SAS printout, Figure 7.13a.

The 95% confidence interval for $\mu_D = (\mu_1 - \mu_2)$ for this small sample is

$$\bar{x}_D \pm t_{\alpha/2}\frac{s_D}{\sqrt{n_D}}$$

where $t_{\alpha/2} = t_{.025} = 2.262$ (obtained from Table IV, Appendix A) is based on $n - 1 = 9$ degrees of freedom. Substituting the values of $\bar{x}_D$ and s_D shown on the printout, we obtain

Suggested Exercise 7.26

$$\bar{x}_D \pm 2.262\frac{s_D}{\sqrt{n_D}} = 400 \pm 2.262\left(\frac{434.613}{\sqrt{10}}\right)$$
$$= 400 \pm 310.88 \approx 400 \pm 311 = (\$89, \$711)$$

[*Note:* This interval is also shown highlighted on the SAS printout, Figure 7.13b.] Our interpretation is that the true mean difference between the starting salaries of males and females falls between $89 and $711, with 95% confidence. Since the interval falls above 0, we infer that $\mu_1 - \mu_2 > 0$; that is, the mean salary for males exceeds the mean salary for females. ∎

FIGURE 7.14

SAS Analysis of Data in Table 7.5, Assuming Independent Samples

Two Sample t-test for the Means of MALE and FEMALE

Sample Statistics

Group	N	Mean	Std. Dev.	Std. Error
MALE	10	43930	11665	3688.8
FEMALE	10	43530	11617	3673.6

Hypothesis Test

Null hypothesis: Mean 1 - Mean 2 = 0
Alternative: Mean 1 - Mean 2 ^= 0

If Variances Are	t statistic	Df	Pr > t
Equal	0.077	18	0.9396
Not Equal	0.077	18.00	0.9396

95% Confidence Interval for the Difference between Two Means

Lower Limit	Upper Limit
-10537.5	11337.50

To measure the amount of information about $(\mu_1 - \mu_2)$ gained by using a paired difference experiment in Example 7.5 rather than an independent samples experiment, we can compare the relative widths of the confidence intervals obtained by the two methods. A 95% confidence interval for $(\mu_1 - \mu_2)$ using the paired difference experiment is, from Example 7.5, ($89, $711). If we analyzed the same data as though this were an independent samples experiment,* we would first obtain the descriptive statistics shown in the SAS printout, Figure 7.14. Then we substitute the sample means and standard deviations shown on the printout into the formula for a 95% confidence interval for $(\mu_1 - \mu_2)$ using independent samples:

$$(\bar{x}_1 - \bar{x}_2) \pm t_{.025} \sqrt{s_p^2 \left(\frac{1}{n_1} + \frac{1}{n_2} \right)}$$

where

$$s_p^2 = \frac{(n_1 - 1)s_1^2 + (n_2 - 1)s_2^2}{n_1 + n_2 - 2}$$

SAS performed these calculations and obtained the interval ($-10,537.50, $11,337.50). This interval is highlighted on Figure 7.14.

Notice that the independent samples interval includes 0. Consequently, if we were to use this interval to make an inference about $(\mu_1 - \mu_2)$, we would incorrectly conclude that the mean starting salaries of males and females do not differ! You can see that the confidence interval for the independent sampling experiment is about five times wider than for the corresponding paired difference confidence interval. Blocking out the variability due to differences in majors and grade point

TEACHING TIP
Point out that the paired difference experiment is a very useful tool in statistics and should be attempted whenever it is appropriate. The disadvantage discussed here is very small relative to the large benefits the experiment can provide.

---

*This is done only to provide a measure of the increase in the amount of information obtained by a paired design in comparison to an unpaired design. Actually, if an experiment is designed using pairing, an unpaired analysis would be invalid because the assumption of independent samples would not be satisfied.

averages significantly increases the information about the difference in male and female mean starting salaries by providing a much more accurate (smaller confidence interval for the same confidence coefficient) estimate of $(\mu_1 - \mu_2)$.

You may wonder whether conducting a paired difference experiment is always superior to an independent samples experiment. The answer is: Most of the time, but not always. We sacrifice half the degrees of freedom in the t statistic when a paired difference design is used instead of an independent samples design. This is a loss of information, and unless this loss is more than compensated for by the reduction in variability obtained by blocking (pairing), the paired difference experiment will result in a net loss of information about $(\mu_1 - \mu_2)$. Thus, we should be convinced that the pairing will significantly reduce variability before performing the paired difference experiment. Most of the time this will happen.

One final note: The pairing of the observations is determined *before* the experiment is performed (that is, by the *design* of the experiment). A paired difference experiment is *never* obtained by pairing the sample observations after the measurements have been acquired.

What Do You Do When the Assumption of a Normal Distribution for the Population of Differences Is Not Satisfied?

TEACHING TIP
In general, nonparametric analyses are useful whenever the assumptions are not satisfied.

Answer: Use the Wilcoxon signed rank test for the paired difference design (optional Section 7.5).

 EXERCISES 7.25–7.40

Learning the Mechanics

7.25 A paired difference experiment yielded n_D pairs of observations. In each case, what is the rejection region for testing $H_0: \mu_D = 2$ against $H_a: \mu_D > 2$?
  a. $n_D = 10, \alpha = .05$ $t > 1.833$
  b. $n_D = 20, \alpha = .10$ $t > 1.328$
  c. $n_D = 5, \alpha = .025$ $t > 2.776$
  d. $n_D = 9, \alpha = .01$ $t > 2.896$

7.26 A paired difference experiment produced the following data:

$$n_D = 16 \quad \bar{x}_1 = 143 \quad \bar{x}_2 = 150 \quad \bar{x}_D = -7 \quad s_D^2 = 64$$

  a. Determine the values of t for which the null hypothesis, $\mu_1 - \mu_2 = 0$, would be rejected in favor of the alternative hypothesis, $\mu_1 - \mu_2 < 0$. Use $\alpha = .10$.
  b. Conduct the paired difference test described in part **a**. Draw the appropriate conclusions. $t = -3.5$
  c. What assumptions are necessary so that the paired difference test will be valid?
  d. Find a 90% confidence interval for the mean difference μ_D. -7 ± 3.506
  e. Which of the two inferential procedures, the confidence interval of part **d** or the test of hypothesis of

part **b**, provides more information about the difference between the population means?

7.27 The data for a random sample of six paired observations are shown in the table.

● LM7_27

Pair	Sample from Population 1	Sample from Population 2
1	7	4
2	3	1
3	9	7
4	6	2
5	4	4
6	8	7

  a. Calculate the difference between each pair of observations by subtracting observation 2 from observation 1. Use the differences to calculate $\bar{x}_D$ and s_D^2.
  b. If μ_1 and μ_2 are the means of populations 1 and 2, respectively, express μ_D in terms of μ_1 and μ_2.
  c. Form a 95% confidence interval for μ_D. 2 ± 1.484
  d. Test the null hypothesis $H_0: \mu_D = 0$ against the alternative hypothesis $H_a: \mu_D \neq 0$. Use $\alpha = .05$.

7.28 The data for a random sample of 10 paired observations are shown in the next table.

LM7_28

Pair	Population 1	Population 2
1	19	24
2	25	27
3	31	36
4	52	53
5	49	55
6	34	34
7	59	66
8	47	51
9	17	20
10	51	55

a. If you wish to test whether these data are sufficient to indicate that the mean for population 2 is larger than that for population 1, what are the appropriate null and alternative hypotheses? Define any symbols you use. $H_0: \mu_D = 0; H_a: \mu_D < 0$

STATISTIX Output for Exercise 7.28

```
PAIRED T TEST FOR X1 - X2

NULL HYPOTHESIS: DIFFERENCE =   0
ALTERNATIVE HYP: DIFFERENCE <> 0

MEAN          -3.7000
STD ERROR      0.7000
LO 95% CI     -5.2835
UP 95% CI     -2.1165
T             -5.29
DF             9
P              0.0005

CASES INCLUDED 10      MISSING CASES 0
```

b. The data are analyzed using STATISTIX, with the output shown above. Interpret the test results.
c. The STATISTIX output also includes a confidence interval. Interpret this interval. $(-5.28, -2.12)$
d. What assumptions are necessary to ensure the validity of this analysis?

7.29 A paired difference experiment yielded the data shown in the table.

LM7_29

Pair	x	y	Pair	x	y
1	55	44	5	75	62
2	68	55	6	52	38
3	40	25	7	49	31
4	55	56			

a. Test $H_0: \mu_D = 10$ against $H_a: \mu_D \neq 10$, where $\mu_D = (\mu_1 - \mu_2)$. Use $\alpha = .05$. $t = .81$
b. Report the *p*-value for the test you conducted in part **a**. Interpret the *p*-value. $p > .20$

Applying the Concepts—Basic

7.30 Does winning an Academy of Motion Picture Arts and Sciences award lead to long-term mortality for movie actors? In an article in the *Annals of Internal Medicine* (May 15, 2001), researchers identified 762 Academy Award winners and matched each one with another actor of the same sex who was in the same winning film and was born in the same era. The life expectancy (age) of each pair of actors was compared.
a. Explain why the data should be analyzed as a paired difference experiment.
b. Set up the null hypothesis for a test to compare the mean life expectancies of Academy Award winners and non-winners.
c. The sample mean life expectancies of Academy Award winners and non-winners were reported as 79.7 years and 75.8 years, respectively. The *p*-value for comparing the two population means was reported as $p = .003$. Interpret this value in the context of the problem.

7.31 Hypersexual behavior caused by traumatic brain injury (TBI) is often treated with the drug Depo-Provera. In one clinical study, eight young male TBI patients who exhibited hypersexual behavior were treated weekly with 400 milligrams of Depo-Provera for six months (*Journal of Head Trauma Rehabilitation,* June 1995). The testosterone levels (in nanograms per deciliter) of the patients both prior to treatment and at the end of the 6-month treatment period are given in the table below.

DEPOPROV

Patient	Pretreatment	After 6 months on Depo-Provera
1	849	96
2	903	41
3	890	31
4	1,092	124
5	362	46
6	900	53
7	1,006	113
8	672	174

Source: Emory, L. E., Cole, C. M., and Meyer, W. J. "Use of Depo-Provera to control sexual aggression in persons with traumatic brain injury." *Journal of Head Trauma Rehabilitation,* Vol. 10, No. 3, June 1995, p. 52 (Table 2).

a. Construct a 99% confidence interval for the true mean difference between the pretreatment and after-treatment testosterone levels of young male TBI patients. 749.5 ± 278.068

b. Use the interval, part **a**, to make an inference about the effectiveness of Depo-Provera in reducing testosterone levels of young male TBI patients.

7.32 Researchers at the University of South Alabama compared the attitudes of male college students toward their fathers with their attitudes toward their mothers (*Journal of Genetic Psychology*, March 1998). Each of a sample of 13 males was asked to complete the following statement about each of their parents: My relationship with my father (mother) can best be described as: (1) Awful, (2) Poor, (3) Average, (4) Good, or (5) Great. The following data were obtained:

⊘FMATTITUDES

Student	Attitude toward Father	Attitude toward Mother
1	2	3
2	5	5
3	4	3
4	4	5
5	3	4
6	5	4
7	4	5
8	2	4
9	4	5
10	5	4
11	4	5
12	5	4
13	3	3

Source: Adapted from Vitulli, W. F., and Richardson, D. K. "College student's attitudes toward relationships with parents: A five-year comparative analysis." *Journal of Genetic Psychology*, Vol. 159, No. 1, (March 1998), pp. 45–52.

a. Specify the appropriate hypotheses for testing whether male students' attitudes toward their fathers differ from their attitudes toward their mothers, on average. $H_0: \mu_D = 0$; $H_a: \mu_D \neq 0$

b. Conduct the test of part **a** at $\alpha = .05$. Interpret the results in the context of the problem. $t = -1.08$

Applying the Concepts—Intermediate

7.33 According to *Webster's New World Dictionary*, a tongue-twister is "a phrase that is hard to speak rapidly." Do tongue-twisters have an effect on the length of time it takes to read silently? To answer this question, 42 undergraduate psychology students participated in a reading experiment (*Memory & Cognition*, Sept. 1997). Two lists, each comprised of 600 words, were constructed. One list contained a series of tongue-twisters and the other list (called the *control*) did not contain any tongue-twisters. Each student read both lists and the length of time (in minutes) required to complete the lists was recorded. The researchers used a test of hypothesis to compare the mean reading response times for the tongue-twister and control lists.

List Type	Response Time (minutes)	
	Mean	**Standard Deviation**
Tongue-twister	6.59	1.94
Control	6.34	1.92
Difference	.25	.78

Source: Robinson, D. H., and Katayama, A. D. "At-lexical, articulatory interference in silent reading: The 'upstream' tongue-twister effect." *Memory & Cognition*, Vol. 25, No. 5, Sept. 1997, p. 663.

a. Set up the null hypothesis for the test.

b. Use the information in the accompanying table to find the test statistic and *p*-value of the test.

c. Give the appropriate conclusion. Use $\alpha = .05$.

7.34 When searching for an item (e.g., a roadside traffic sign, a lost earring, or a tumor in a mammogram), common sense dictates that you will not re-examine items previously rejected. However, researchers at Harvard Medical School found that a visual search has no memory (*Nature*, Aug. 6, 1998). In their experiment, nine subjects searched for the letter "T" mixed among several letters "L." Each subject conducted the search under two conditions: random and static. In the random condition, the location of the letters were changed every 111 milliseconds; in the static condition, the location of the letters remained unchanged. In each trial, the reaction time (i.e., the amount of time it took the subject to locate the target letter) was recorded in milliseconds.

a. One goal of the research is to compare the mean reaction times of subjects in the two experimental conditions. Explain why the data should be analyzed as a paired-difference experiment.

b. If a visual search has no memory, then the main reaction times in the two conditions will not differ. Specify H_0 and H_a for testing the "no memory" theory.

c. The test statistic was calculated as $t = 1.52$ with *p*-value $= .15$. Make the appropriate conclusion.

7.35 A study published in *Clinical Kinesiology* (Spring 1995) was designed to examine the metabolic and cardiopulmonary responses during exercise of persons diagnosed with multiple sclerosis (MS). Leg cycling and arm cranking exercises were performed by 10 MS patients and 10 healthy (non-MS) control subjects. Each member of the control group was selected based on gender, age, height, and weight to match (as closely as possible) with one member of the MS group. Consequently, the researchers compared the MS and non-MS groups by matched-pairs *t*-tests on such outcome variables as oxygen uptake, carbon dioxide output, and peak aerobic power. The data on the matching variables used in the experiment are shown in the shown in the table on p. 349.

a. Visually examine the data table. Does it appear that the researchers have successfully matched the MS and non-MS subjects? Yes

◉ MSSTUDY

	MS Subjects				Non-MS Subjects			
Matched Pair	**Gender**	**Age (years)**	**Height (cm)**	**Weight (kg)**	**Gender**	**Age (years)**	**Height (cm)**	**Weight (kg)**
1	M	48	171.0	80.8	M	45	173.0	76.3
2	F	34	158.5	75.0	F	34	158.0	75.6
3	F	34	167.6	55.5	F	34	164.5	57.7
4	M	38	167.0	71.3	M	34	161.3	70.0
5	M	45	182.5	90.9	M	39	179.0	96.0
6	F	42	166.0	72.4	F	42	167.0	77.8
7	M	32	172.0	70.5	M	34	165.8	74.7
8	F	35	166.5	55.3	F	43	165.1	71.4
9	F	33	166.5	57.9	F	31	170.1	60.4
10	F	46	175.0	79.9	F	43	175.0	77.9

Source: Ponichtera-Mulcare, J. A., *et al.* "Maximal aerobic exercise of individuals with multiple sclerosis using three modes of ergometry." *Clinical Kinesiology,* Vol. 49, No. 1, Spring 1995, p. 7 (Table 1).

b. Conduct a test to determine whether the mean ages of the MS and non-MS groups are significantly different. Use $\alpha = .05$. $t = .65$

c. SPSS printouts for comparing the mean heights and weights of the two groups as shown below. Interpret the results. $p = .21; p = .149$

d. Based on the results, parts **b** and **c**, comment on the validity of the *t*-tests performed by the researchers on the outcome variables.

7.36 Twins, because they share an identical genotype, make ideal subjects for investigating the degree to which various environmental conditions impact personality. The classical method of studying this phenomenon and the subject of an interesting book by Susan Farber (*Identical Twins Reared Apart,* New York: Basic Books, 1981) is the study of identical twins separated early in life and reared apart. Much of Farber's discussion focuses on a comparison of IQ scores. The data for this analysis appear in the table on p. 350. One member (A) of each of the $n = 32$ pairs of twins was reared by a natural parent; the other member (B) was reared by a relative or some other person. Is there a significant difference between the average IQ scores of identical twins, where one member of the pair is reared by the natural parents and the other member of the pair is not? Use $\alpha = .05$ to make your conclusion. $t = -1.848$

SPSS Output for Exercise 7.35

Paired Samples Statistics

		Mean	N	Std. Deviation	Std. Error Mean
Pair 1	MSHT	169.2600	10	6.3969	2.0229
	NONMSHT	167.8800	10	6.4175	2.0294
Pair 2	MSWT	70.9500	10	11.7751	3.7236
	NONMSWT	73.7800	10	10.5141	3.3248

Paired Samples Test

		Paired Differences					t	df	Sig. (2-tailed)
		Mean	Std. Deviation	Std. Error Mean	95% Confidence Interval of the Difference Lower	Upper			
Pair 1	MSHT - NONMSHT	1.3800	3.2303	1.0215	-.9308	3.6908	1.351	9	.210
Pair 2	MSWT - NONMSWT	-2.8300	5.6698	1.7930	-6.8859	1.2259	-1.578	9	.149

● TWINSIQ

Pair ID	Twin A	Twin B	Pair ID	Twin A	Twin B
112	113	109	228	100	88
114	94	100	232	100	104
126	99	86	236	93	84
132	77	80	306	99	95
136	81	95	308	109	98
148	91	106	312	95	100
170	111	117	314	75	86
172	104	107	324	104	103
174	85	85	328	73	78
180	66	84	330	88	99
184	111	125	338	92	111
186	51	66	342	108	110
202	109	108	344	88	83
216	122	121	350	90	82
218	97	98	352	79	76
220	82	94	416	97	98

Source: Adapted from *Identical Twins Reared Apart*, by Susan L. Farber 1981 by Basic Books, Inc.

7.37 A study reported in the *Journal of Psychology* (Mar. 1991) measures the change in female students' self-concepts as they move from high school to college. A sample of 133 Boston College female students were selected for the study. Each was asked to evaluate several aspects of her life at two points in time: at the end of her senior year of high school, and during her sophomore year of college. Each student was asked to evaluate where she believed she stood on a scale that ranged from top 10% of class (1) to lowest 10% of class (5). The results for three of the traits evaluated are reported in the table below.

Trait	n	Senior Year of High School $\bar{x}$	Sophomore Year of College $\bar{x}$
Leadership	133	2.09	2.33
Popularity	133	2.48	2.69
Intellectual self-confidence	133	2.29	2.55

   a. What null and alternative hypotheses would you test to determine whether the mean self-concept of females decreases between the senior year of high school and the sophomore year of college as measured by each of these three traits?

   b. Are these traits more appropriately analyzed using an independent samples test or a paired difference test? Explain.   Paired difference test

   c. Noting the size of the sample, what assumptions are necessary to ensure the validity of the tests?

   d. The article reports that the leadership test results in a *p*-value greater than .05, while the tests for popularity and intellectual self-confidence result in *p*-values less than .05. Interpret these results.

7.38 Merck Research Labs conducted an experiment to evaluate the effect of a new drug using the single-T swim maze. Nineteen impregnated dam rats were allocated a dosage of 12.5 milligrams of the drug. One male and one female rat pup were randomly selected from each resulting litter to perform in the swim maze. Each pup was placed in the water at one end of the maze and allowed to swim until it escaped at the opposite end. If the pup failed to escape after a certain period of time, it was placed at the beginning of the maze and given another chance. The experiment was repeated until each pup accomplished three successful escapes. The accompanying table reports the number of swims required by each pup to perform three successful escapes. Is there sufficient evidence of a difference between the mean number of swims required by male and female pups? Use the accompanying MINITAB printout to conduct the test (at $\alpha = .10$). Comment on the assumptions required for the test to be valid.

● RATPUPS

Litter	Male	Female	Litter	Male	Female
1	8	5	11	6	5
2	8	4	12	6	3
3	6	7	13	12	5
4	6	3	14	3	8
5	6	5	15	3	4
6	6	3	16	8	12
7	3	8	17	3	6
8	5	10	18	6	4
9	4	4	19	9	5
10	4	4			

Source: Thomas E. Bradstreet, Merck Research Labs, BL 3–2, West Point, PA 19486.

MINITAB Output for Exercise 7.38

```
Paired T for male - female

                   N      Mean     StDev     SE Mean
male               19     5.895    2.378     0.546
female             19     5.526    2.458     0.564
Difference         19     0.368    3.515     0.806

95% CI for mean difference: (-1.326, 2.063)
T-Test of mean difference = 0 (vs not = 0): T-Value = 0.46   P-Value = 0.653
```

Applying the Concepts—Advanced

7.39 A *homophone* is a word whose pronunciation is the same as that of another word having a different meaning and spelling (e.g., *nun* and *none, doe* and *dough,* etc.). *Brain and Language* (Apr. 1995) reported on a study of homophone spelling in patients with Alzheimer's disease. Twenty Alzheimer's patients were asked to spell 24 homophone pairs given in random order, then the number of homophone confusions (e.g., spelling *doe* given the context, *bake bread dough*) was recorded for each patient. One year later, the same test was given to the same patients. The data for the study are provided in the table. The researchers posed the following question: "Do Alzheimer's patients show a significant increase in mean homophone confusion errors over time?" Perform an analysis of the data to answer the researchers' question. Use the relevant information in the SAS printout. What assumptions are necessary for the procedure used to be valid? Are they satisfied? $t = -2.306$

SAS output for Exercise 7.39

Two Sample Paired t-test for the Means of TIME1 and TIME2

Sample Statistics

Group	N	Mean	Std. Dev.	Std. Error
TIME1	20	4.15	3.4985	0.7823
TIME2	20	5.8	4.2128	0.942

Hypothesis Test

Null hypothesis: Mean of (TIME1 - TIME 2) => 0
Alternative: Mean of (TIME1 - TIME 2) < 0

t Statistic	Df	Prob > t
-2.306	19	0.0163

95% Confidence Interval for the Difference between Two Paired Means

Lower Limit	Upper Limit
-3.15	-0.15

⊘ HOMOPHON

Patient	Time 1	Time 2
1	5	5
2	1	3
3	0	0
4	1	1
5	0	1
6	2	1
7	5	6
8	1	2
9	0	9
10	5	8
11	7	10
12	0	3
13	3	9
14	5	8
15	7	12
16	10	16
17	5	5
18	6	3
19	9	6
20	11	8

Source: Neils, J., Roeltgen, D. P., and Constantinidou, F. "Decline in homophone spelling associated with loss of semantic influence on spelling in Alzheimer's disease." *Brain and Language,* Vol. 49, No. 1, Apr. 1995, p. 36 (Table 3).

7.40 Inositol is a complex cyclic alcohol found to be effective against clinical depression. Medical researchers believe inositol may also be used to treat panic disorder. To test this theory, a double-blind, placebo-controlled study of 21 patients diagnosed with panic disorder was conducted (*American Journal of Psychiatry,* July 1995). Patients completed daily panic diaries recording the occurrence of their panic attacks. The data for a week in which patients received a glucose placebo and for a week when they were treated with inositol are provided in the table. [*Note:* Neither the patients nor the treating physicians knew which week the placebo was given.] Analyze the data and interpret the results. Comment on the validity of the assumptions. $t = 1.77$

⊘ INOSITOL

Patient	Placebo	Inositol
1	0	0
2	2	1
3	0	3
4	1	2
5	0	0
6	10	5
7	2	0
8	6	4
9	1	1
10	1	0
11	1	3
12	0	2
13	3	1
14	3	1
15	3	4
16	4	2
17	6	4
18	15	21
19	28	8
20	30	0
21	13	0

Source: Benjamin, J., *et al.* "Double-blind, placebo-controlled, crossover trial of inositol treatment for panic disorder." *American Journal of Psychiatry,* Vol. 152, No. 7, July 1995, p. 1085 (Table 1).

7.3   DETERMINING THE SAMPLE SIZE

You can find the appropriate sample size to estimate the difference between a pair of means with a specified degree of reliability by using the method described in Section 5.4. That is, to estimate the difference between a pair of means correct to within B units with probability $(1 - \alpha)$, let $z_{\alpha/2}$ standard deviations of the sampling distribution of the estimator equal B. Then solve for the sample size. To do this, you have to solve the problem for a specific ratio between n_1 and n_2. Most often, you will want to have equal sample sizes, that is, $n_1 = n_2 = n$. We will illustrate the procedure with two examples.

EXAMPLE 7.6

$n_1 = n_2 = 769$

New fertilizer compounds are often advertised with the promise of increased crop yields. Suppose we want to compare the mean yield μ_1 of wheat when a new fertilizer is used to the mean yield μ_2 with a fertilizer in common use. The estimate of the difference in mean yield per acre is to be correct to within .25 bushel with a confidence coefficient of .95. If the sample sizes are to be equal, find $n_1 = n_2 = n$, the number of 1-acre plots of wheat assigned to each fertilizer.

Solution

To solve the problem, you need to know something about the variation in the bushels of yield per acre. Suppose from past records you know the yields of wheat possess a range of approximately 10 bushels per acre. You could then approximate $\sigma_1 = \sigma_2 = \sigma$ by letting the range equal 4σ. Thus,

$$4\sigma \approx 10 \text{ bushels}$$
$$\sigma \approx 2.5 \text{ bushels}$$

The next step is to solve the equation

TEACHING TIP
When determining the sample sizes for two samples, we set the condition that both sample sizes will be equal. The n we find is the sample size necessary for *both* samples.

$$z_{\alpha/2}\sigma_{(\bar{x}_1 - \bar{x}_2)} = B \quad \text{or} \quad z_{\alpha/2}\sqrt{\frac{\sigma_1^2}{n_1} + \frac{\sigma_2^2}{n_2}} = B$$

for n, where $n = n_1 = n_2$. Since we want the estimate to lie within $B = .25$ of $(\mu_1 - \mu_2)$ with confidence coefficient equal to .95, we have $z_{\alpha/2} = z_{.025} = 1.96$. Then, letting $\sigma_1 = \sigma_2 = 2.5$ and solving for n, we have

$$1.96\sqrt{\frac{(2.5)^2}{n} + \frac{(2.5)^2}{n}} = .25$$

$$1.96\sqrt{\frac{2(2.5)^2}{n}} = .25$$

$$n = 768.32 \approx 769 \text{ (rounding up)}$$

Consequently, you will have to sample 769 acres of wheat for each fertilizer to estimate the difference in mean yield per acre to within .25 bushel. Since this would necessitate extensive and costly experimentation, you might decide to allow a larger bound (say, $B = .50$ or $B = 1$) in order to reduce the sample size, or you might decrease the confidence coefficient. The point is that we can obtain an idea of the experimental effort necessary to achieve a specified precision in our final estimate by determining the approximate sample size *before* the experiment is begun.

EXAMPLE 7.7

A laboratory manager wishes to compare the difference in the mean reading of two instruments, A and B, designed to measure the potency (in parts per million) of an antibiotic. To conduct the experiment, the manager plans to select n_D specimens of the antibiotic from a vat and to measure each specimen with both instruments.

The difference $(\mu_A - \mu_B)$ will be estimated based on the n_D paired differences $(x_A - x_B)$ obtained in the experiment. If preliminary measurements suggest that the differences will range between plus or minus 10 parts per million, how many differences will be needed to estimate $(\mu_A - \mu_B)$ correct to within 1 part per million with confidence equal to .99?

Solution

The estimator for $(\mu_A - \mu_B)$, based on a paired difference experiment, is $\bar{x}_D = (\bar{x}_A - \bar{x}_B)$ and

$$\sigma_{\bar{x}_D} = \frac{\sigma_D}{\sqrt{n_D}}$$

Thus, the number n_D of pairs of measurements needed to estimate $(\mu_A - \mu_B)$ to within 1 part per million can be obtained by solving for n_D in the equation

$$z_{\alpha/2}\left(\frac{\sigma_D}{\sqrt{n_D}}\right) = B$$

where $z_{.005} = 2.58$ and $B = 1$. To solve this equation for n_D, we need to have an approximate value for σ_D.

We are given the information that the differences are expected to range from -10 to 10 parts per million. Letting the range equal $4\sigma_D$, we find

$$\text{Range} = 20 \approx 4\sigma_D$$

$$\sigma_D \approx 5$$

Substituting $\sigma_D = 5$, $B = 1$, and $z_{.005} = 2.58$ into the equation and solving for n_D, we obtain

$$2.58\left(\frac{5}{\sqrt{n_D}}\right) = 1$$

$$n_D = [(2.58)(5)]^2$$

$$= 166.41$$

Therefore, it will require $n_D = 167$ pairs of measurements to estimate $(\mu_A - \mu_B)$ correct to within 1 part per million using the paried difference experiment. ∎

The box summarizes the procedures for determining the sample sizes necessary for estimating $(\mu_1 - \mu_2)$ for the case $n_1 = n_2 = 2$ and for estimating μ_D.

TEACHING TIP
When determining sample sizes, always use conservative (large) estimates for the population variances. In addition, always round the sample sizes found to the next higher integer.

Determination of Sample Size for Comparing Two Means

Independent Random Samples

To estimate $(\mu_1 - \mu_2)$ to within a given bound B with probability $(1 - \alpha)$, use the following formula to solve for equal sample sizes that will achieve the desired reliability:

$$n_1 = n_2 = \frac{(z_{\alpha/2})^2(\sigma_1^2 + \sigma_2^2)}{B^2}$$

You will need to substitute estimates for the values of σ_1^2 and σ_2^2 before solving for the sample size. These estimates might be sample variances s_1^2 and s_2^2 from prior sampling (e.g., a pilot study), or from an educated (and conservatively large) guess based on the range—that is, $s \approx R/4$.

(continues on next page)

Determination of Sample Size for Comparing Two Means (*continued*)

Paired Difference Experiment

To estimate μ_D to within a given bound B with probability $(1 - \alpha)$, use the following formula to solve for n:

$$n = \frac{(z_{\alpha/2})^2 \sigma_D^2}{B^2}$$

You will need to substitute an estimate of σ_D^2 before solving for the sample size. This estimate might be the sample variance s_D^2 from prior sampling (e.g., a pilot study), or from an educated (and conservatively large) guess based on the range—that is, $s \approx R/4$.

 EXERCISES 7.41–7.49

Learning the Mechanics

7.41 Suppose you want to estimate the difference between two population means correct to within 2.2 with probability .95. If prior information suggests that the population variances are approximately equal to $\sigma_1^2 = \sigma_2^2 = 15$ and you want to select independent random samples of equal size from the populations, how large should the sample sizes, n_1 and n_2, be? $n_1 = n_2 = 24$

7.42 Find the appropriate values of n_1 and n_2 (assume $n_1 = n_2$) needed to estimate $(\mu_1 - \mu_2)$ with:
  a. A bound on the error of estimation equal to 3.2 with 95% confidence. From prior experience it is known that $\sigma_1 \approx 15$ and $\sigma_2 \approx 17$. $n_1 = n_2 = 193$
  b. A bound on the error of estimation equal to 8 with 99% confidence. The range of each population is 60.
  c. A 90% confidence interval of width 1.0. Assume that $\sigma_1^2 \approx 5.8$ and $\sigma_2^2 \approx 7.5$. $n_1 = n_2 = 144$

7.43 Enough money has been budgeted to collect $n = 100$ paired observations from populations 1 and 2 in order to estimate $(\mu_1 - \mu_2)$. Prior information indicates that $\sigma_D = 12$. Have sufficient funds been allocated to construct a 90% confidence interval for $(\mu_1 - \mu_2)$ of width 5 or less? Justify your answer. Yes

Applying the Concepts—Basic

7.44 A high school counselor wants to estimate the difference in mean income per day between high school graduates who have a college education and those who have not gone on to college. Suppose it is decided to compare the daily incomes of 30-year-olds, and the range of daily incomes for both groups is approximately $200 per day. How many people from each group should be sampled in order to estimate the true difference between mean daily incomes correct to within $10 per day with probability .9? Assume that $n_1 = n_2$.

7.45 Refer to *The American Statistician* (May 2001) study comparing the "fear of negative evaluation" (FNE) scores for bulimic and normal female students, Exercise 7.11 (p. 335). Suppose you want to estimate $(\mu_B - \mu_N)$, the difference between the population means of the FNE scores for bulimic and normal female students, using a 95% confidence interval with a bound on the error of estimation of two points. Find the sample sizes required to obtain such an estimate. Assume equal sample sizes of $\sigma_B^2 = \sigma_N^2 = 25$. $n_1 = n_2 = 49$

7.46 Refer to the *Journal of Genetic Psychology* (Mar. 1998) study comparing male students' attitudes toward their fathers and mothers, Exercise 7.32 (p. 348). Suppose you want to estimate $\mu_D = (\mu_F - \mu_M)$, the difference between the population mean attitudes toward fathers and mothers, using a 90% confidence interval with a bound on the error of estimation of .3. Find the number of male students required to obtain such an estimate. Assume the variance of the paired differences is $\sigma_D^2 = 1$.

Applying the Concepts—Intermediate

7.47 Is housework hazardous to your health? A study in the *Public Health Reports* (July–Aug. 1992) compares the life expectancies of 25-year-old white women in the labor force to those who are housewives. How large a sample would have to be taken from each group in order to be 95% confident that the estimate of difference in mean life expectancies for the two groups is within 1 year of the true difference? Assume that equal sample sizes will be selected from the two groups, and that the standard deviation for both groups is approximately 15 years.

7.48 Refer to the *Journal of Ethnopharmacology*, Exercise 7.19 (page 337). Recall that mice with inflamed ear lobes were randomly divided into two groups; one group was treated with bear bile while the other was treated with normal saline. Assuming equal-sized groups, how many mice should be included in each group in order to estimate the difference in the mean degree of swelling to within 2 milligrams with 95% confidence? Use the fact that the degree of swelling standard deviations for the bear bile and normal saline groups are approximately 4 milligrams and 3 milligrams, respectively.

7.49 In seeking a free agent NFL running back, a general manager is looking for a player with high mean yards gained per carry and a small standard deviation. Suppose the GM wishes to compare the mean yards gained per carry for two free agents based on independent random samples of their yards gained per carry. Data from last year's pro football season indicate that $\sigma_1 = \sigma_2 \approx 5$ yards. If the GM wants to estimate the difference in means correct to within 1 yard with probability equal to .9, how many runs would have to be observed for each player? (Assume equal sample sizes.)

7.4 A NONPARAMETRIC TEST FOR COMPARING TWO POPULATIONS: INDEPENDENT SAMPLING (OPTIONAL)

TEACHING TIP
Present the rank sum test as an alternative to the independent comparison of means procedure presented in Section 7.1.

Suppose two independent random samples are to be used to compare two populations and the *t*-test of Section 7.1 is inappropriate for making the comparison. We may be unwilling to make assumptions about the form of the underlying population probability distributions or we may be unable to obtain exact values of the sample measurements. If the data can be ranked in order of magnitude for either of these situations, the nonparametric **Wilcoxon rank sum test** (developed by Frank Wilcoxon) can be used to test the hypothesis that the probability distributions associated with the two populations are equivalent.

Suppose, for example, an experimental psychologist wants to compare reaction times for adult males under the influence of drug A to those under the influence of drug B. Experience has shown that the populations of reaction time measurements often possess probability distributions that are skewed to the right, as shown in Figure 7.15. Consequently, a *t*-test should not be used to compare the mean reaction times for the two drugs because the normality assumption that is required for the *t*-test may not be valid.

FIGURE 7.15

Typical Probability Distribution of Reaction Times

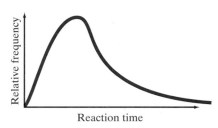

Suppose the psychologist randomly assigns seven subjects to each of two groups, one group to receive drug A and the other to receive drug B. The reaction time for each subject is measured at the completion of the experiment. These data (with the exception of the measurement for one subject in group A who was eliminated from the experiment for personal reasons) are shown in Table 7.6.

The population of reaction times for either of the drugs, say drug A, is that which could conceptually be obtained by giving drug A to all adult males. To compare the probability distributions for populations A and B, *we first rank the sample observations as though they were all drawn from the same population.* That is, we pool the measurements from both samples and then rank all the measurements from the smallest (a rank of 1) to the largest (a rank of 13). The results of this ranking process are also shown in Table 7.6.

Suggested Exercise 7.59

⊘ REACTION
TABLE 7.6 **Reaction Times of Subjects Under the Influence of Drug A or B**

Drug A		Drug B	
Reaction Time (seconds)	**Rank**	**Reaction Time (seconds)**	**Rank**
1.96	4	2.11	6
2.24	7	2.43	9
1.71	2	2.07	5
2.41	8	2.71	11
1.62	1	2.50	10
1.93	3	2.84	12
		2.88	13

If the two populations were identical, we would expect the ranks to be *randomly mixed* between the two samples. If, on the other hand, one population tends to have longer reaction times than the other, we would expect the larger ranks to be mostly in one sample and the smaller ranks mostly in the other. Thus, the test statistic for the Wilcoxon test is based on the totals of the ranks for each of the two samples—that is, on the **rank sums**. When the sample sizes are equal, for example, the greater the difference in the rank sums, the greater will be the weight of evidence to indicate a difference between the probability distributions of the populations. In the reaction times example, we denote the rank sum for drug A by T_1 and that for drug B by T_2. Then

$$T_1 = 4 + 7 + 2 + 8 + 1 + 3 = 25$$
$$T_2 = 6 + 9 + 5 + 11 + 10 + 12 + 13 = 66$$

The sum of T_1 and T_2 will always equal $n(n + 1)/2$, where $n = n_1 + n_2$. So, for this example, $n_1 = 6, n_2 = 7$, and

$$T_1 + T_2 = \frac{13(13 + 1)}{2} = 91$$

TEACHING TIP
Point out that the smaller size sample is the one that is used to sum the ranks when finding the test statistic.

Since $T_1 + T_2$ is fixed, a small value for T_1 implies a large value for T_2 (and vice versa) and a large difference between T_1 and T_2. Therefore, the smaller the value of one of the rank sums, the greater the evidence to indicate that the samples were selected from different populations.

The test statistic for this test is the rank sum for the smaller sample; or, in the case where $n_1 = n_2$, either rank sum can be used. Values that locate the rejection region for this rank sum are given in Table V of Appendix A. A partial reproduction of this table is shown in Table 7.7. The columns of the table represent n_1, the first sample size, and the rows represent n_2, the second sample size. *The T_L and T_U entries in the table are the boundaries of the lower and upper regions, respectively, for the rank sum associated with the sample that has fewer measurements.* If the sample sizes n_1 and n_2 are the same, either rank sum may be used as the test statistic. To illustrate, suppose $n_1 = 8$ and $n_2 = 10$. For a two-tailed test with $\alpha = .05$, we consult part a of the table and find that the null hypothesis will be rejected if the rank sum of sample 1 (the sample with fewer measurements), T, is less than or equal to $T_L = 54$ *or* greater than or equal to $T_U = 98$. The Wilcoxon rank sum test is summarized in the next box.

TEACHING TIP
Use several examples to help the student become comfortable with using the Wilcoxon Rank Sum table.

TABLE 7.7   Reproduction of Part of Table V in Appendix A: Critical Values for the Wilcoxon Rank Sum Test

$\alpha = .025$ one-tailed; $\alpha = .05$ two-tailed

n_2 \ n_1	3		4		5		6		7		8		9		10	
	T_L	T_U	T_L	T_U	T_L	T_U	T_L	T_U	T_L	T_U	T_L	T_U	T_L	T_U	T_L	T_U
3	5	16	6	18	6	21	7	23	7	26	8	28	8	31	9	33
4	6	18	11	25	12	28	12	32	13	35	14	38	15	41	16	44
5	6	21	12	28	18	37	19	41	20	45	21	49	22	53	24	56
6	7	23	12	32	19	41	26	52	28	56	29	61	31	65	32	70
7	7	26	13	35	20	45	28	56	37	68	39	73	41	78	43	83
8	8	28	14	38	21	49	29	61	39	73	49	87	51	93	54	98
9	8	31	15	41	22	53	31	65	41	78	51	93	63	108	66	114
10	9	33	16	44	24	56	32	70	43	83	54	98	66	114	79	131

Wilcoxon Rank Sum Test: Independent Samples*

Let D_1 and D_2 represent the probability distributions for populations 1 and 2, respectively.

ONE-TAILED TEST

H_0: D_1 and D_2 are identical

H_a: D_1 is shifted to the right of D_2

[or H_a: D_1 is shifted to the left of D_2]

Test statistic:

T_1, if $n_1 < n_2$; T_2, if $n_2 < n_1$ (Either rank sum can be used if $n_1 = n_2$.)

Rejection region:

T_1: $T_1 \geq T_U$ [or $T_1 \leq T_L$]

T_2: $T_2 \leq T_L$ [or $T_2 \geq T_U$]

TWO-TAILED TEST

H_0: D_1 and D_2 are identical

H_a: D_1 is shifted either to the left or to the right of D_2

Test statistic:

T_1, if $n_1 < n_2$; T_2, if $n_2 < n_1$ (Either rank sum can be used if $n_1 = n_2$.) We will denote this rank sum as T.

Rejection region:

$T \leq T_L$ or $T \geq T_U$

where T_L and T_U are obtained from Table V of Appendix A.

Assumptions: 1. The two samples are random and independent.

2. The two probability distributions from which the samples are drawn are continuous.

Ties: Assign tied measurements the average of the ranks they would receive if they were unequal but occurred in successive order. For example, if the third-ranked and fourth-ranked measurements are tied, assign each a rank of $(3 + 4)/2 = 3.5$.

Note that the assumptions necessary for the validity of the Wilcoxon rank sum test do not specify the shape or type of probability distribution. However, the distributions are assumed to be continuous so that the probability of tied measurements is 0 (see Chapter 4), and each measurement can be assigned a unique rank. In practice, however, rounding of continuous measurements will sometimes produce ties. As long as the number of ties is small relative to the sample sizes, the Wilcoxon test procedure will still have approximate significance level α. The test is not recommended to compare discrete distributions for which many ties are expected.

■ **EXAMPLE 7.8**

Do the data given in Table 7.6 provide sufficient evidence to indicate a shift in the probability distributions for drugs A and B, that is, that the probability distribution corresponding to drug A lies either to the right or left of the probability distribution corresponding to drug B? Test at the .05 level of significance.

$T_1 = 25$, $p = .007$; Reject H_0

Solution

H_0: The two populations of reaction times corresponding to drug A and drug B have the same probability distribution

---

*Another statistic used for comparing two populations based on independent random samples is the *Mann–Whitney U statistic*. The *U* statistic is a simple function of the rank sums. It can be shown that the Wilcoxon rank sum test and the Mann–Whitney *U*-test are equivalent.

H_a: The probability distribution for drug A is shifted to the right or left of the probability distribution corresponding to drug B.*

Test statistic: Since drug A has fewer subjects than drug B, the test statistic is T_1, the rank sum of drug A's reaction times.

Rejection region: Since the test is two-sided, we consult part a of Table V for the rejection region corresponding to $\alpha = .05$. We will reject H_0 for $T_1 \leq T_L$ or $T_1 \geq T_U$. Thus, we will reject H_0 if $T_1 \leq 28$ or $T_1 \geq 56$.

Since T_1, the rank sum of drug A's reaction times in Table 7.6, is 25, it is in the rejection region (see Figure 7.16).[†] Therefore, there is sufficient evidence to reject H_0. This same conclusion can be reached using a statistical software package. The SAS printout of the analysis is shown in Figure 7.17. Both the test statistic ($T_1 = 25$) and one-tailed p-value ($p = .007$) are highlighted on the printout. The one-tailed p-value is less than $\alpha = .05$, leading us to reject H_0.

Our conclusion is that the probability distributions for drugs A and B are not identical. In fact, it appears that drug B tends to be associated with reaction times that are larger than those associated with drug A (because T_1 falls in the lower tail of the rejection region). ∎

FIGURE 7.16

Alternative Hypothesis and Rejection Region for Example 7.8

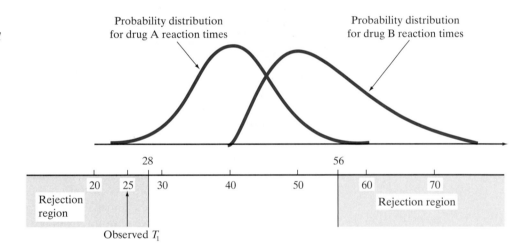

Table V in Appendix A gives values of T_L and T_U for values of n_1 and n_2 less than or equal to 10. When both sample sizes, n_1 and n_2, are 10 or larger, the sampling distribution of T_1 can be approximated by a normal distribution with mean and variance

$$E(T_1) = \frac{n_1(n_1 + n_2 + 1)}{2} \quad \text{and} \quad \sigma_{T_1}^2 = \frac{n_1 n_2(n_1 + n_2 + 1)}{12}$$

Therefore, for $n_1 \geq 10$ and $n_2 \geq 10$ we can conduct the Wilcoxon rank sum test using the familiar z-test of Section 7.1. The test is summarized in the next box.

---

*The alternative hypotheses in this chapter will be stated in terms of a difference in the *location* of the distributions. However, since the shapes of the distributions may also differ under H_a, some of the figures (e.g., Figure 7.16) depicting the alternative hypothesis will show probability distributions with different shapes.

[†]Figure 7.16 depicts only one side of the two-sided alternative hypothesis. The other would show the distribution for drug A shifted to the right of the distribution for drug B.

FIGURE 7.17

SAS Printout for Example 7.8

The NPAR1WAY Procedure

Wilcoxon Scores (Rank Sums) for Variable REACTIME
Classified by Variable DRUG

DRUG	N	Sum of Scores	Expected Under HO	Std Dev Under HO	Mean Score
A	6	25.0	42.0	7.0	4.166667
B	7	66.0	49.0	7.0	9.428571

Wilcoxon Two-Sample Test

Statistic	25.0000

Normal Approximation

Z	-2.4286
One-Sided Pr < Z	0.0076
Two-Sided Pr > |Z|	0.0152

t Approximation

One-Sided Pr < Z	0.0159
Two-Sided Pr > |Z|	0.0318

Exact Test

One-Sided Pr <= S	0.0070
Two-Sided Pr >= |S - Mean|	0.0140

Kruskal-Wallis Test

Chi-Square	5.8980
DF	1
Pr > Chi-Square	0.0152

The Wilcoxon Rank Sum Test for Large Samples ($n_1 \geq 10$ and $n_2 \geq 10$)

Let D_1 and D_2 represent the probability distributions for populations 1 and 2, respectively.

ONE-TAILED TEST

H_0: D_1 and D_2 are identical

H_a: D_1 is shifted to the right of D_2
(or H_a: D_1 is shifted to the left of D_2)

TWO-TAILED TEST

H_0: D_1 and D_2 are identical

H_a: D_1 is shifted to the right or to the left of D_2

Test statistic: $z = \dfrac{T_1 - \dfrac{n_1(n_1 + n_2 + 1)}{2}}{\sqrt{\dfrac{n_1 n_2(n_1 + n_2 + 1)}{12}}}$

Rejection region:

$z > z_\alpha$ (or $z < -z_\alpha$)

Rejection region:

$|z| > z_{\alpha/2}$

EXERCISES 7.50–7.61

Learning the Mechanics

7.50 Specify the test statistic and the rejection region for the Wilcoxon rank sum test for independent samples in each of the following situations:

a. H_0: Two probability distributions, 1 and 2, are identical
H_a: Probability distribution for population 1 is shifted to the right or left of the probability distribution for population 2
$n_1 = 7, n_2 = 8, \alpha = .10$ $T_1 \leq 41$ or $T_1 \geq 71$

b. H_0: Two probability distributions, 1 and 2, are identical
H_a: Probability distribution for population 1 is shifted to the right of the probability distribution for population 2
$n_1 = 6, n_2 = 6, \alpha = .05$ $T_1 \geq 50$

c. H_0: Two probability distributions, 1 and 2, are identical
H_a: Probability distribution for population 1 is shifted to the left of the probability distribution for population 2
$n_1 = 7, n_2 = 10, \alpha = .025$ $T_1 \leq 43$

d. H_0: Two probability distributions, 1 and 2, are identical
H_a: Probability distribution for population 1 is shifted to the right or left of the probability distribution for population 2
$n_1 = 20, n_2 = 20, \alpha = .05$ $|z| > 1.96$

7.51 Suppose you want to compare two treatments, A and B. In particular, you wish to determine whether the distribution for population B is shifted to the right of the distribution for population A. You plan to use the Wilcoxon rank sum test.

a. Specify the null and alternative hypotheses you would test.

b. Suppose you obtained the following independent random samples of observations on experimental units subjected to the two treatments:

🖉 LM7_51

| Sample A | 37, | 40, | 33, | 29, | 42, | 33, | 35, | 28, | 34 |
| Sample B | 65, | 35, | 47, | 52 |

Conduct a test of the hypotheses described in part **a**. Test using $\alpha = .05$. $T_2 = 42.5$

7.52 Random samples of sizes $n_1 = 16$ and $n_2 = 12$ were drawn from populations 1 and 2, respectively. The measurements obtained are listed in the table.

🖉 LM7_52

Sample 1				Sample 2		
9.0	15.6	25.6	31.1	10.1	11.1	13.5
21.1	26.9	24.6	20.0	12.0	18.2	10.3
24.8	16.5	26.0	25.1	9.2	7.0	14.2
17.2	30.1	18.7	26.1	15.8	13.6	13.2

a. Conduct a hypothesis test to determine whether the probability distribution for population 2 is shifted to the left of the probability distribution for population 1. Use $\alpha = .05$. $z = -3.76$

b. What is the approximate p-value of the test of part **a**?

7.53 Independent random samples are selected from two populations. The data are shown in the table.

🖉 LM7_53

Sample 1		Sample 2		
15	16	5	9	5
10	13	12	8	10
12	8	9	4	

a. Use the Wilcoxon rank sum test to determine whether the data provide sufficient evidence to indicate a shift in the locations of the probability distributions of the sampled populations. Test using $\alpha = .05$. $T_1 = 62.5$

b. Do the data provide sufficient evidence to indicate that the probability distribution for population 1 is shifted to the right of the probability distribution for population 2? Use the Wilcoxon rank sum test with $\alpha = .05$. $T_1 = 62.5$

7.54 Suppose you wish to compare two treatments, A and B, based on independent random samples of 15 observations selected from each of the two populations. If $T_1 = 173$, do the data provide sufficient evidence to indicate that distribution A is shifted to the left of distribution B? Test using $\alpha = .05$. $z = -2.47$

Applying the Concepts—Basic

7.55 Refer to exercise 7.11 (p. 335). The "fear of negative evaluation" (FNE) scores for 11 female students known to suffer from the eating disorder bulimia and 14 female students with normal eating habits are reproduced on p. 361. (Recall that the higher the score, the greater the fear of negative evaluation.) Suppose you want to determine if the distribution of the FNE scores for bulimic female students is shifted above the corresponding distribution for female students with normal eating habits.

a. Specify H_0 and H_a for the test.

b. Rank all 25 FNE scores in the data set from smallest to largest.

c. Sum the ranks of the 11 FNE scores for bulimic students. $T_1 = 174.5$

d. Sum the ranks of the 14 FNE scores for students with normal eating habits. $T_2 = 150.5$

e. Give the rejection region for a nonparametric test of the data if $\alpha = .10$. $z > 1.28$

f. Conduct the test and give the conclusion in the words of the problem. $z = 1.72$

BULIMIA

Bulimic students	21	13	10	20	25	19	16	21	24	13	14			
Normal students	13	6	16	13	8	19	23	18	11	19	7	10	15	20

Source: Randles, R. H. "On neutral responses (zeros) in the sign test and ties in the Wilcoxon–Mann–Whitney test." *The American Statistician*, Vol. 55, No. 2, May 2001 (Figure 3).

7.56 Refer to the *Organizational Science* (May–June 2000) study of new artists on *Billboard* magazine's chart of the top selling music CDs, Exercise 7.17 (p. 337). Recall that in 1991, *Billboard* began using the SoundScan system to chart CD and album sales. The number of new artists entering the weekly *Billboard* chart for 12 weeks prior to the introduction of SoundScan and for 12 weeks after the introduction of SoundScan are reproduced in the table.

BILLBOARD

Pre-SoundScan	1	1	0	2	3	4	3	4	5	5	2	3
Post-SoundScan	0	2	3	1	3	1	0	1	4	2	2	1

a. In Exercise 7.17 you analyzed the data using the Student's *t*-distribution. Why might a nonparametric procedure be a better test to apply?

b. Use nonparametrics to test for a shift in the distributions of number of new artists entering the *Billboard* chart prior to and after SoundScan. Test at $\alpha = .10$. $z = 1.67$

7.57 In a comparison of visual acuity of deaf and hearing children, eye movement rates are taken on 10 deaf and 10 hearing children. The data are shown in the table. A clinical psychologist believes that deaf children have greater visual acuity than hearing children. (The larger a child's eye movement rate, the more visual acuity the child possesses.)

EYEMOVE

Deaf Children		Hearing Children	
2.75	1.95	1.15	1.23
3.14	2.17	1.65	2.03
3.23	2.45	1.43	1.64
2.30	1.83	1.83	1.96
2.64	2.23	1.75	1.37

a. Use the Wilcoxon rank sum procedure to test the psychologist's claim at $\alpha = .05$. $T_1 = 150.5$

b. Conduct the test by using the large-sample approximation for the Wilcoxon rank sum test. Compare the results with those found in part **a**.

Applying the Concepts—Intermediate

7.58 The data in the table, extracted from *Technometrics* (Feb. 1986), represent daily accumulated stream flow and precipitation (in inches) for two U.S. Geological

Survey stations in Colorado. Conduct a test to determine whether the distributions of daily accumulated stream flow and precipitation for the two stations differ in location. Use $\alpha = .10$. Why is a nonparametric test appropriate for these data? $T_1 = 71$

COLORAIN

Station 1			Station 2		
127.96	108.91	100.85	114.79	85.54	280.55
210.07	178.21	85.89	109.11	117.64	145.11
203.24	285.37		330.33	302.74	95.36

Source: Gastwirth, J. L., and Mahmoud, H. "An efficient robust nonparametric test for scale change for data from a gamma distribution." *Technometrics*, Vol. 28, No. 1, Feb. 1986, p. 83 (Table 2).

7.59 University of Queensland researchers sampled private sector and public sector organizations in Australia to study the planning undertaken by their information systems departments (*Management Science*, July 1996). As part of the process they asked each sample organization how much it had spent on information systems and technology in the previous fiscal year as a percentage of the organization's total revenues. The results are reported in the table.

INFOSYS

Private Sector	Public Sector
2.58%	5.40%
5.05	2.55
.05	9.00
2.10	10.55
4.30	1.02
2.25	5.11
2.50	24.42
1.94	1.67
2.33	3.33

Adapted from Hann, J., and Weber, R. "Information systems planning: A model and empirical tests." *Management Science*, Vol. 42, No. 2 (July, 1996), pp. 1043–1064.

a. Do the two sampled populations have identical probability distributions or is the distribution for public sector organizations in Australia located to the right of Australia's private sector firms? Test using $\alpha = .05$. $T_2 = 105$

b. Is the *p*-value for the test less than or greater than .05? Justify your answer. $p \approx .05$

c. What assumption must be met to ensure the validity of the test you conducted in part **a**?

7.60 The term *brood-parasitic intruder* is used to describe birds that search for and lay eggs in the nests built by another bird species. For example, the brown-headed cowbird is known to be a brood parasite of the smaller-sized willow flycatcher. Ornithologists theorize that those flycatchers that recognize but do not vocally react to cowbird calls are more apt to defend their nests and less likely to be found and parasitized. In a study published in *The Condor* (May 1995), each of 13 active flycatcher nests was categorized as parasitized (if at least one cowbird egg was present) or nonparasitized. Cowbird songs were taped and played back while the flycatcher pairs were sitting in the nest prior to incubation. The vocalization rate (number of calls per minute) of each flycatcher pair was recorded. The data for the two groups of flycatchers are given in the table. Do the data suggest (at $\alpha = .05$) that the vocalization rates of parasitized flycatchers are higher than those of nonparasitized flycatchers? Use the accompanying SAS printout to form your conclusion. $p = .0064$

⊘ COWBIRD

Parasitized	Not Parasitized
2.00	1.00
1.25	1.00
8.50	0
1.10	3.25
1.25	1.00
3.75	.25
5.50	

Source: Uyehara, J. C., and Narins, P. M. "Nest defense by Willow Flycatchers to brood-parasitic intruders." *The Condor*, Vol. 97, No. 2, May 1995, p. 364 (Figure 1).

7.61 Refer to the impact study of the interactive Teachers Involve Parents in Schoolwork (TIPS) program, Exercise 7.15 (p. 336). Recall that 128 middle school students were assigned to complete TIPS homework assignments, while 98 students were assigned traditional, noninteractive, homework assignments (ATIPS). At the

SAS Output for Exercise 7.60

The NPAR1WAY Procedure

Wilcoxon Scores (Rank Sums) for Variable VOCRATE
Classified by Variable PARGROUP

PARGROUP	N	Sum of Scores	Expected Under HO	Std Dev Under HO	Mean Score
PAR	7	66.0	49.0	6.951757	9.428571
NOT	6	25.0	42.0	6.951757	4.166667

Average scores were used for ties.

Wilcoxon Two-Sample Test

Statistic (S)	25.0000

Normal Approximation
Z	-2.4454
One-Sided Pr < Z	0.0072
Two-Sided Pr > |Z|	0.0145

t Approximation
One-Sided Pr < Z	0.0154
Two-Sided Pr > |Z|	0.0309

Exact Test
One-Sided Pr <= S	0.0064
Two-Sided Pr >= |S - Mean|	0.0128

Kruskal-Wallis Test

Chi-Square	5.9801
DF	1
Pr > Chi-Square	0.0145

end of the study, all students reported on the level of family involvement in their homework on a 5-point scale (0 = Never, 1 = Rarely, 2 = Sometimes, 3 = Frequently, 4 = Always). The data for the science, math, and language arts homework are saved in the HWSTUDY file. (The first five and last five observations in the data set are reproduced in the table.)

a. Why might a nonparametric test be the most appropriate test to apply in order to compare the levels of family involvement in homework assignments of TIPS and ATIPS students?

b. Conduct a nonparametric analysis to compare the science homework assignments of TIPS and ATIPS students. Use $\alpha = .05$. $p < .0001$

c. Repeat part **a** for mathematics homework assignments. $p = .6930$

d. Repeat part **a** for language arts homework assignments. $p = .1777$

⊘HWSTUDY

Homework Condition	Science	Math	Language
ATIPS	1	0	0
ATIPS	0	1	1
ATIPS	0	1	0
ATIPS	1	2	0
ATIPS	1	1	2
⋮	⋮	⋮	⋮
TIPS	2	3	2
TIPS	1	4	2
TIPS	2	4	2
TIPS	4	0	3
TIPS	2	0	1

Source: Van Voorhis, F. L. "Teachers' use of interactive homework and its effects on family involvement and science achievement of middle grade students." Paper presented at the annual meeting of the American Educational Research Association, Seattle, April 2001.

7.5 A NONPARAMETRIC TEST FOR COMPARING TWO POPULATIONS: PAIRED DIFFERENCE EXPERIMENTS (OPTIONAL)

Nonparametric techniques may also be employed to compare two probability distributions when a paired difference design is used. For example, consumer preferences for two competing products are often compared by having each of a sample of consumers rate both products. Thus, the ratings have been paired on each consumer. Here is an example of this type of experiment.

For some paper products, softness is an important consideration in determining consumer acceptance. One method of determining softness is to have judges give a sample of the products a softness rating. Suppose each of 10 judges is given a sample of two products that a company wants to compare. Each judge rates the softness of each product on a scale from 1 to 10, with higher ratings implying a softer product. The results of the experiment are shown in Table 7.8.

⊘SOFTPAPER
TABLE 7.8   Softness Ratings of Paper

Judge	Product A	B	Difference (A − B)	Absolute Value of Difference	Rank of Absolute Value
1	6	4	2	2	5
2	8	5	3	3	7.5
3	4	5	−1	1	2
4	9	8	1	1	2
5	4	1	3	3	7.5
6	7	9	−2	2	5
7	6	2	4	4	9
8	5	3	2	2	5
9	6	7	−1	1	2
10	8	2	6	6	10

T_+ = Sum of positive ranks = 46
T_- = Sum of negative ranks = 9

Since this is a paired difference experiment, we analyze the differences between the measurements (see Section 7.2). However, a nonparametric approach developed by Wilcoxon requires that we calculate the ranks of the absolute values of the differences between the measurements, that is, the ranks of the differences after removing any minus signs. *Note that tied absolute differences are assigned the average of the ranks they would receive if they were unequal but successive measurements.* After the absolute differences are ranked, the sum of the ranks of the positive differences of the original measurements, T_+, and the sum of the ranks of the negative differences of the original measurements, T_-, are computed.

We are now prepared to test the nonparametric hypotheses:

H_0: The probability distributions of the ratings for products A and B are identical.

H_a: The probability distributions of the ratings differ (in location) for the two products. (Note that this is a two-sided alternative and that it implies a two-tailed test.)

Test statistic: $T = $ Smaller of the positive and negative rank sums T_+ and T_-

The smaller the value of T, the greater the evidence to indicate that the two probability distributions differ in location. The rejection region for T can be determined by consulting Table VI in Appendix A (part of the table is shown in Table 7.9). This table gives a value T_0 for both one-tailed and two-tailed tests for each value of n, the number of matched pairs. For a two-tailed test with $\alpha = .05$, we will reject H_0 if $T \leq T_0$. You can see in Table 7.9 that the value of T_0 that locates the boundary of the rejection region for the judges' ratings for $\alpha = .05$ and $n = 10$ pairs of observations is 8. Thus, the rejection region for the test (see Figure 7.18) is

Rejection region: $T \leq 8$ for $\alpha = .05$

TABLE 7.9 **Reproduction of Part of Table VI of Appendix A: Critical Values for the Wilcoxon Paired Difference Signed Rank Test**

One-Tailed	Two-Tailed	$n = 5$	$n = 6$	$n = 7$	$n = 8$	$n = 9$	$n = 10$
$\alpha = .05$	$\alpha = .10$	1	2	4	6	8	11
$\alpha = .025$	$\alpha = .05$		1	2	4	6	8
$\alpha = .01$	$\alpha = .02$			0	2	3	5
$\alpha = .005$	$\alpha = .01$				0	2	3
		$n = 11$	$n = 12$	$n = 13$	$n = 14$	$n = 15$	$n = 16$
$\alpha = .05$	$\alpha = .10$	14	17	21	26	30	36
$\alpha = .025$	$\alpha = .05$	11	14	17	21	25	30
$\alpha = .01$	$\alpha = .02$	7	10	13	16	20	24
$\alpha = .005$	$\alpha = .01$	5	7	10	13	16	19
		$n = 17$	$n = 18$	$n = 19$	$n = 20$	$n = 21$	$n = 22$
$\alpha = .05$	$\alpha = .10$	41	47	54	60	68	75
$\alpha = .025$	$\alpha = .05$	35	40	46	52	59	66
$\alpha = .01$	$\alpha = .02$	28	33	38	43	49	56
$\alpha = .005$	$\alpha = .01$	23	28	32	37	43	49
		$n = 23$	$n = 24$	$n = 25$	$n = 26$	$n = 27$	$n = 28$
$\alpha = .05$	$\alpha = .10$	83	92	101	110	120	130
$\alpha = .025$	$\alpha = .05$	73	81	90	98	107	117
$\alpha = .01$	$\alpha = .02$	62	69	77	85	93	102
$\alpha = .005$	$\alpha = .01$	55	61	68	76	84	92

FIGURE 7.18

Rejection Region for Paired Difference Experiment

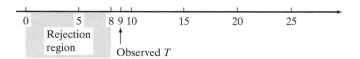

Since the smaller rank sum for the paper data, $T_- = 9$, does not fall within the rejection region, the experiment has not provided sufficient evidence to indicate that the two paper products differ with respect to their softness ratings at the $\alpha = .05$ level.

Note that if a significance level of $\alpha = .10$ had been used, the rejection region would have been $T \leq 11$ and we would have rejected H_0. In other words, the samples do provide evidence that the probability distributions of the softness ratings differ at the $\alpha = .10$ significance level.

The **Wilcoxon signed rank test** is summarized in the following box. Note that the difference measurements are assumed to have a continuous probability distribution so that the absolute differences will have unique ranks. Although tied (absolute) differences can be assigned ranks by averaging, the number of ties should be small relative to the number of observations to ensure the validity of the test.

Suggested Exercise 7.68

Wilcoxon Signed Rank Test for a Paired Difference Experiment

Let D_1 and D_2 represent the probability distributions for populations 1 and 2, respectively.

ONE-TAILED TEST

H_0: D_1 and D_2 are identical

H_a: D_1 is shifted to the right of D_2
  [or H_a: D_1 is shifted to the left of D_2]

TWO-TAILED TEST

H_0: D_1 and D_2 are identical

H_a: D_1 is shifted either to the left
  or to the right of D_2

Calculate the difference within each of the n matched pairs of observations. Then rank the absolute value of the n differences from the smallest (rank 1) to the highest (rank n) and calculate the rank sum T_- of the negative differences and the rank sum T_+ of the positive differences. [*Note:* Differences equal to 0 are eliminated, and the number n of differences is reduced accordingly.]

Test statistic:

T_-, the rank sum of the negative differences [or T_+, the rank sum of the positive differences]

Test statistic:

T, the smaller of T_+ or T_-

Rejection region:

$T_- \leq T_0$ [or $T_+ \leq T_0$]

Rejection region:

$T \leq T_0$

where T_0 is given in Table VI in Appendix A.

Assumptions: 1. The sample of differences is randomly selected from the population of differences.

2. The probability distribution from which the sample of paired differences is drawn is continuous.

Ties: Assign tied absolute differences the average of the ranks they would receive if they were unequal but occurred in successive order. For example, if the third-ranked and fourth-ranked differences are tied, assign both a rank of $(3 + 4)/2 = 3.5$.

EXAMPLE 7.9

$T_+ = 15.5; p = .2168$, fail to reject H_0

Suppose the police commissioner in a small community must choose between two plans for patrolling the town's streets. Plan A, the less expensive plan, uses voluntary citizen groups to patrol certain high-risk neighborhoods. In contrast, plan B would utilize police patrols. As an aid in reaching a decision, both plans are examined by 10 trained criminologists, each of whom is asked to rate the plans on a scale from 1 to 10. (High ratings imply a more effective crime prevention plan.) The city will adopt plan B (and hire extra police) only if the data provide sufficient evidence that criminologists tend to rate plan B more effective than plan A.

The results of the survey are shown in Table 7.10. Do the data provide evidence at the $\alpha = .05$ level that the distribution of ratings for plan B lies above that for plan A?

CRIMEPLAN

TABLE 7.10   Effectiveness Ratings by 10 Qualified Crime Prevention Experts

| Crime Prevention Expert | Plan | | Difference | Rank of Absolute Difference |
	A	B	(A − B)	
1	7	9	−2	4.5
2	4	5	−1	2
3	8	8	0	(Eliminated)
4	9	8	1	2
5	3	6	−3	6
6	6	10	−4	7.5
7	8	9	−1	2
8	10	8	2	4.5
9	9	4	5	9
10	5	9	−4	7.5

Positive rank sum $= T_+ = 15.5$

Solution

The null and alternative hypotheses are

H_0: The two probability distributions of effectiveness ratings are identical

H_a: The effectiveness ratings of the more expensive plan (B) tend to exceed those of plan A

Observe that the alternative hypothesis is one-sided (that is, we only wish to detect a shift in the distribution of the B ratings to the right of the distribution of A ratings) and therefore it implies a one-tailed test of the null hypothesis (see Figure 7.19).

FIGURE 7.19

The Alternative Hypothesis for Example 7.9

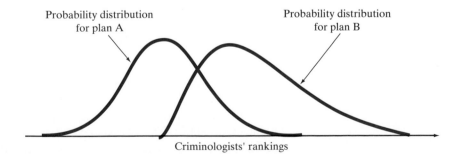

Probability distribution for plan A

Probability distribution for plan B

Criminologists' rankings

If the alternative hypothesis is true, the B ratings will tend to be larger than the paired A ratings, more negative differences in pairs will occur, T_- will be large, and T_+ will be small. Because Table VI is constructed to give lower-tail values of T_0, we will use T_+ as the test statistic and reject H_0 for $T_+ \leq T_0$.

The differences in ratings for the pairs (A − B) are shown in Table 7.10. Note that one of the differences equals 0. Consequently, we eliminate this pair from the ranking and reduce the number of pairs to $n = 9$. Looking in Table VI, we have $T_0 = 8$ for a one-tailed test with $\alpha = .05$ and $n = 9$. Therefore, the test statistic and rejection region for the test are

$$\text{Test statistic:} \quad T_+, \text{ the positive rank sum}$$

$$\text{Rejection region:} \quad T_+ \leq 8$$

Summing the ranks of the positive differences from Table 7.10, we find $T_+ = 15.5$. Since this value exceeds the critical value, $T_0 = 8$, we conclude that this sample provides insufficient evidence at the $\alpha = .05$ level to support the alternative hypothesis. The commissioner *cannot* conclude that the plan utilizing police patrols tends to be rated higher than the plan using citizen volunteers. That is, on the basis of this study, extra police will not be hired.

A STATISTIX printout of the analysis, shown in Figure 7.20, confirms this conclusion. Both the test statistic and p-value are highlighted on the printout. Since the p-value, .2168, exceeds $\alpha = .05$, we fail to reject H_0.

FIGURE 7.20

STATISTIX Printout for Example 7.9

```
WILCOXON SIGNED RANK TEST FOR A - B


SUM OF NEGATIVE RANKS                                        -29.500
SUM OF POSITIVE RANKS                                         15.500


EXACT PROBABILITY OF A RESULT AS OR MORE
EXTREME THAN THE OBSERVED RANKS (1 TAILED P-VALUE)   0.2168


NORMAL APPROXIMATION WITH CONTINUITY CORRECTION        0.770
TWO TAILED P-VALUE FOR NORMAL APPROXIMATION            0.4413


TOTAL NUMBER OF VALUES THAT WERE TIED        7
NUMBER OF ZERO DIFFERENCES DROPPED           1
MAX. DIFF. ALLOWED BETWEEN TIES       0.00001


CASES INCLUDED 9    MISSING 1
```

As is the case for the rank sum test for independent samples, the sampling distribution of the signed rank statistic can be approximated by a normal distribution when the number n of paired observations is large (say $n \geq 25$). The large-sample z-test is summarized in the next box.

Wilcoxon Signed Ranks Test for Large Samples ($n \geq 25$)

Let D_1 and D_2 represent the probability distributions for populations 1 and 2, respectively.   *(continues on next page)*

Wilcoxon Signed Ranks Test for Large Samples ($n \geq 25$) *(continued)*

ONE-TAILED TEST

H_0: D_1 and D_2 are identical

H_a: D_1 is shifted to the right of D_2
[or H_a: D_1 is shifted to the left of D_2]

TWO-TAILED TEST

H_0: D_1 and D_2 are identical

H_a: D_1 is shifted either to the left
or to the right of D_2

Test statistic: $$z = \frac{T_+ - [n(n+1)/4]}{\sqrt{[n(n+1)(2n+1)]/24}}$$

Rejection region:

$z > z_\alpha$ [or $z < -z_\alpha$]

Rejection region:

$|z| > z_{\alpha/2}$

Assumptions: The sample size n is greater than or equal to 25. Differences equal to 0 are eliminated and the number *n* of differences is reduced accordingly. Tied absolute differences receive ranks equal to the average of the ranks they would have received had they not been tied.

 EXERCISES 7.62–7.72

Learning the Mechanics

7.62 Specify the test statistic and the rejection region for the Wilcoxon signed rank test for the paired difference design in each of the following situations:

a. H_0: Two probability distributions, A and B, are identical

H_a: Probability distribution for population A is shifted to the right or left of the probability distribution for population B

$n = 20, \alpha = .10$

b. H_0: Two probability distributions, A and B, are identical

H_a: Probability distribution for population A is shifted to the right of the probability distribution for population B

$n = 39, \alpha = .05$

c. H_0: Two probability distributions, A and B, are identical

H_a: Probability distribution for population A is shifted to the left of the probability distribution for population B

$n = 7, \alpha = .005$

7.63 Suppose you want to test a hypothesis that two treatments, A and B, are equivalent against the alternative hypothesis that the responses for A tend to be larger than those for B. You plan to use a paired difference experiment and to analyze the resulting data using the Wilcoxon signed rank test.

⊘ LM7_63

Pair	A	B	Pair	A	B
1	54	45	6	77	75
2	60	45	7	74	63
3	98	87	8	29	30
4	43	31	9	63	59
5	82	71	10	80	82

a. Specify the null and alternative hypotheses you would test.

b. Suppose the paired difference experiment yielded the data in the table. Conduct the test of part **a**. Test using $\alpha = .025$. $T_- = 3.5$

7.64 Suppose you wish to test a hypothesis that two treatments, A and B, are equivalent against the alternative that the responses for A tend to be larger than those for B.

a. If the number of pairs equals 25, give the rejection region for the large-sample Wilcoxon signed rank test for $\alpha = .05$. $z > 1.645$

b. Suppose that $T_+ = 273$. State your test conclusions.

c. Find the *p*-value for the test and interpret it.

7.65 A paired difference experiment with $n = 30$ pairs yielded $T_+ = 354$.

   a. Specify the null and alternative hypotheses that should be used in conducting a hypothesis test to determine whether the probability distribution for population A is located to the right of that for population B.

   b. Conduct the test of part **a** using $\alpha = .05$.

   c. What is the approximate *p*-value of the test of part **b**?

   d. What assumptions are necessary to ensure the validity of the test you performed in part **b**?

7.66 A random sample of nine pairs of measurements is shown in the table.

Pair	Sample Data from Population 1	Sample Data from Population 2
1	8	7
2	10	1
3	6	4
4	10	10
5	7	4
6	8	3
7	4	6
8	9	2
9	8	4

   a. Use the Wilcoxon signed rank test to determine whether the data provide sufficient evidence to indicate that the probability distribution for population 1 is shifted to the right of the probability distribution for population 2. Test using $\alpha = .05$.

   b. Use the Wilcoxon signed rank test to determine whether the data provide sufficient evidence to indicate that the probability distribution for population 1 is shifted either to the right or to the left of the probability distribution for population 2. Test using $\alpha = .05$. $T_- = 2.5$

Applying the Concepts—Basic

7.67 Refer to the *Journal of Genetic Psychology* (March 1998) study of attitudes of male college students toward their parents, Exercise 7.32 (p. 348). Recall that each of 13 students was asked to complete the following statement: My relationship with my father (mother) can best be described as: (1) Awful, (2) Poor, (3) Average, (4) Good, or (5) Great. The study data are reproduced in the next table. The researchers want to compare male students' attitudes toward their fathers with their attitudes toward their mothers.

   a. Why is a nonparametric test applicable for analyzing this data set?

   b. Specify H_0 and H_a for the test.

   c. Compute and rank the differences between the father and mother attitudes.

   d. Sum the ranks of the positive and negative differences. $T_- = 44; T_+ = 22$

   e. Find the rejection region for the test at $\alpha = .05$.

⊘ **FMATTITUDES**

Student	Attitude toward Father	Attitude toward Mother
1	2	3
2	5	5
3	4	3
4	4	5
5	3	4
6	5	4
7	4	5
8	2	4
9	4	5
10	5	4
11	4	5
12	5	4
13	3	3

Source: Adapted from Vitulli, W. F., and Richardson, D. K. "College student's attitudes toward relationships with parents: A five-year comparative analysis." *Journal of Genetic Psychology*, Vol. 159, No. 1, (March 1998), pp. 45–52.

   f. Give the conclusion of the test in the context of the problem. Compare your answer to the test results of Exercise 7.32.

7.68 A study published in the *Journal of Business Communications* (Fall 1985) found that video teleconferencing may be a more effective method of dealing with complex group problem-solving tasks than face-to-face meetings. Ten groups of four people each were randomly assigned both to a specific communication setting (face-to-face or video teleconferencing) and to one of two specific complex problems. Upon completion of the problem-solving task, the same groups were placed in the alternative communication setting and asked to complete the second problem-solving task. The percentage of each problem task correctly completed was recorded for each group, with the results given in the accompanying table.

⊘ **FACEVT**

Group	Face-to-Face	Video Teleconferencing
1	65%	75%
2	82	80
3	54	60
4	69	65
5	40	55
6	85	90
7	98	98
8	35	40
9	85	89
10	70	80

   a. What type of experimental design was used in this study?   Paired difference problem

   b. Specify the null and alternative hypotheses that should be used in determining whether the data provide sufficient evidence to conclude that the problem-solving performance of video teleconferencing groups is superior to that of groups that interact face-to-face.

SPSS Output for Exercise 7.69

Ranks

		N	Mean Rank	Sum of Ranks
SECBORN - FRSTBORN	Negative Ranks	7[a]	5.93	41.50
	Positive Ranks	4[b]	6.13	24.50
	Ties	1[c]		
	Total	12		

a. SECBORN < FRSTBORN
b. SECBORN > FRSTBORN
a. FRSTBORN = SECBORN

Test Statistics[b]

	SECBORN-FRSTBORN
Z	-.756[a]
Asymp. Sig. (2-tailed)	.449

a. Based on positive ranks.
b. Wilcoxon Signed Ranks Test

c. Conduct the hypothesis test of part **b**. Use $\alpha = .05$. Interpret the results of your test in the context of the problem. $T_+ = 3.5$

d. What is the *p*-value of the test in part **c**?

7.69 Twelve sets of identical twins are given psychological tests to determine whether the firstborn of the twins tends to be more aggressive than the secondborn. The test scores are shown in the table, where the higher score indicates greater aggressiveness. Do the data provide sufficient evidence (at $\alpha = .05$) to indicate that the firstborn of a pair of twins is more aggressive than the other? Use the accompanying SPSS printout above to make your conclusion. $p = .2245$

In a survey of atlas users (*Journal of Geography,* May/June 1995), a large sample of high school teachers in British Columbia ranked 12 thematic atlas topics for usefulness. The consensus rankings of the teachers (based on the percentage of teachers who responded they "would definitely use" the topic) are given in the accompanying table. These teacher rankings were compared to the rankings a group of university geography alumni made three years earlier. Compare the distributions of theme rankings for the two groups with an appropriate nonparametric test. Use $\alpha = .05$. Interpret the results practically. $T_+ = 27.5$

AGGTWINS

Set	Firstborn	Secondborn
1	86	88
2	71	77
3	77	76
4	68	64
5	91	96
6	72	72
7	77	65
8	91	90
9	70	65
10	71	80
11	88	81
12	87	72

Applying the Concepts—Intermediate

7.70 The regional atlas is an important educational resource that is updated on a periodic basis. One of the most critical aspects of a new atlas design is its thematic content.

ATLAS

	Rankings	
Theme	High School Teachers	Geography Alumni
Tourism	10	2
Physical	2	1
Transportation	7	3
People	1	6
History	2	5
Climate	6	4
Forestry	5	8
Agriculture	7	10
Fishing	9	7
Energy	2	8
Mining	10	11
Manufacturing	12	12

Source: Keller, C. P., *et al.* "Planning the next generation of regional atlases: Input from educators." *Journal of Geography,* Vol. 94, No. 3, May/June 1995, p. 413 (Table 1).

7.71 Hypoglycemia is a condition in which blood sugar is below normal limits. To compare the effectiveness of two compounds, X and Y, for treating hypoglycemia, each compound is applied to half the diaphragms of each of seven white mice. Blood glucose uptake in milligrams per gram of tissue is measured for each half, producing the results listed in the table below. Do the data provide sufficient evidence to indicate that one of the compounds tends to produce higher blood sugar uptake readings than the other? Test using $\alpha = .10$.

⊘ HYPOGLY

Mouse	X	Y	Mouse	X	Y
1	4.7	5.1	5	7.0	6.1
2	3.3	4.6	6	4.7	4.1
3	8.5	8.7	7	5.2	5.1
4	3.9	3.6			

7.72 Eleven prisoners of war during the war in Croatia were evaluated for neurological impairment after their release from a Serbian detention camp (*Collegium Antropologicum*, June 1997). All 11 released POWs received blows to the head and neck and/or loss of consciousness during imprisonment. Neurological impairment was assessed by measuring the amplitude of the visual evoked potential (VEP) in both eyes at two points in time: 157 days and 379 days after release from prison. (The higher the VEP value, the greater the neurological impairment.) The data for the 11 POWs are shown in the table. Determine whether the VEP measurements of POWs 379 days after their release tend to be greater than the VEP measurements of POWs 157 days after their release. Test using $\alpha = .05$.

⊘ POWVEP

POW	157 Days After Release	379 Days After Release
1	2.46	3.73
2	4.11	5.46
3	3.93	7.04
4	4.51	4.73
5	4.96	4.71
6	4.42	6.19
7	1.02	1.42
8	4.30	8.70
9	7.56	7.37
10	7.07	8.46
11	8.00	7.16

Source: Vrca, A., *et al.* "The use of visual evoked potentials to follow-up prisoners of war after release from detention camps." *Collegium Antropologicum*, Vol. 21, No. 1, June 1997, p. 232. (Data simulated from information provided in Table 3.)

7.6 COMPARING THREE OR MORE POPULATION MEANS: ANALYSIS OF VARIANCE (OPTIONAL)

Suppose we are interested in comparing the means of three or more populations. For example, we may want to compare the mean SAT scores of seniors at three different high schools. Or, we could compare the mean income per household of residents in four census districts. Since the methods of Sections 7.1–7.5 apply to two populations only, we require an alternative technique. In this optional section, we discuss a method for comparing two or more populations based on independent random samples, called an **analysis of variance (ANOVA)**.

In the jargon of ANOVA, "treatments" represent the groups or populations of interest. Thus, the primary objective of an analysis of variance is to compare the treatment (or population) means. If we denote the true means of the p treatments as $\mu_1, \mu_2, \ldots, \mu_p$, then we will test the null hypothesis that the treatment means are all equal against the alternative that at least two of the treatment means differ:

$$H_0: \mu_1 = \mu_2 = \cdots = \mu_p$$

H_a: At least two of the p treatment means differ

To conduct a statistical test of these hypotheses, we will use the means of the independent random samples selected from the treatment populations using the completely randomized design. That is, we compare the p sample means $\bar{x}_1, \bar{x}_2, \ldots, \bar{x}_p$.

For example, suppose you select independent random samples of five female and five male high school seniors and obtain sample mean SAT scores of 550 and 590, respectively. Can we conclude that males score 40 points higher, on average, than females? To answer this question, we must consider the amount of sampling variability among the experimental units (students). If the scores are as depicted in the dot plot shown in Figure 7.21, then the difference between the means is small relative to the sampling variability of the scores within the treatments, Female and Male. We would be inclined not to reject the null hypothesis of equal population means in this case.

FIGURE 7.21

Dot Plot of SAT Scores: Difference Between Means Dominated by Sampling Variability

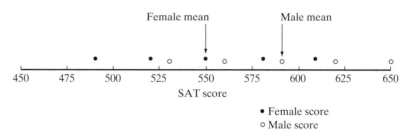

In contrast, if the data are as depicted in the dot plot of Figure 7.22, then the sampling variability is small relative to the difference between the two means. We would be inclined to favor the alternative hypothesis that the population means differ in this case.

You can see that the key is to compare the difference between the treatment means to the amount of sampling variability. To conduct a formal statistical test of the hypothesis requires numerical measures of the difference between the treatment means and the sampling variability within each treatment. The variation between the treatment means is measured by the **Sum of Squares for Treatments (SST)**, which is calculated by squaring the distance between each treatment mean and the overall mean of *all* sample measurements, multiplying each squared distance by the number of sample measurements for the treatment, and adding the results over all treatments:

$$\text{SST} = \sum_{i=1}^{p} n_i(\bar{x}_i - \bar{x})^2 = 5(550 - 570)^2 + 5(590 - 570)^2 = 4,000$$

where we use $\bar{x}$ to represent the overall mean response of all sample measurements, that is, the mean of the combined samples. The symbol n_i is used to denote the sample size for the *i*th treatment. You can see that the value of SST is 4,000 for the two samples of five female and five male SAT scores depicted in Figures 7.21 and 7.22.

Next, we must measure the sampling variability within the treatments. We call this the **Sum of Squares for Error** (SSE) because it measures the variability around the treatment means that is attributed to sampling error. Suppose the 10 mea-

FIGURE 7.22

Dot Plot of SAT Scores: Difference Between Means Large Relative to Sampling Variability

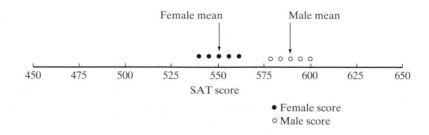

surements in the first dot plot (Figure 7.21) are 490, 520, 550, 580, and 610 for females and 530, 560, 590, 620, and 650 for males. Then the value of SSE is computed by summing the squared distance between each response measurement and the corresponding treatment mean, and then adding the squared differences over all measurements in the entire sample:

$$SSE = \sum_{j=1}^{n_1} (x_{1j} - \bar{x}_1)^2 + \sum_{j=1}^{n_2} (x_{2j} - \bar{x}_2)^2 + \ldots + \sum_{j=1}^{n_p} (x_{pj} - \bar{x}_p)^2$$

where the symbol x_{1j} is the jth measurement in sample 1, x_{2j} is the jth measurement in sample 2, and so on. This rather complex-looking formula can be simplified by recalling the formula for the sample variance, s^2, given in Chapter 2:

$$s^2 = \sum_{i=1}^{n} \frac{(x_i - \bar{x})^2}{n - 1}$$

Note that each sum in SSE is simply the numerator of s^2 for that particular treatment. Consequently, we can rewrite SSE as

$$SSE = (n_1 - 1)s_1^2 + (n_2 - 1)s_2^2 + \cdots + (n_p - 1)s_p^2$$

where $s_1^2, s_2^2, \cdots, s_p^2$ are the sample variances for the p treatments. For our samples of SAT scores, we find $s_1^2 = 2{,}250$ (for females) and $s_2^2 = 2{,}250$ (for males); then we have

$$SSE = (5 - 1)(2{,}250) + (5 - 1)(2{,}250) = 18{,}000$$

To make the two measurements of variability comparable, we divide each by the degrees of freedom to convert the sums of squares to mean squares. First, the **Mean Square for Treatments** (MST), which measures the variability *among* the treatment means, is equal to

$$MST = \frac{SST}{p - 1} = \frac{4{,}000}{2 - 1} = 4{,}000$$

where the number of degrees of freedom for the p treatments is $(p - 1)$. Next, the **Mean Square for Error** (MSE), which measures the sampling variability *within* the treatments, is

$$MSE = \frac{SSE}{n - p} = \frac{18{,}000}{10 - 2} = 2{,}250$$

Finally, we calculate the ratio of MST to MSE—an **F statistic**:

$$F = \frac{MST}{MSE} = \frac{4{,}000}{2{,}250} = 1.78$$

TEACHING TIP
Rejecting the null hypothesis only indicates to us that differences exist between the population means. It does not specify where these differences occur.

Values of the F statistic near 1 indicate that the two sources of variation, between treatment means and within treatments, are approximately equal. In this case, the difference between the treatment means may well be attributable to sampling error, which provides little support for the alternative hypothesis that the population treatment means differ. Values of F well in excess of 1 indicate that the variation among treatment means well exceeds that within means and therefore support the alternative hypothesis that the population treatment means differ.

When does F exceed 1 by enough to reject the null hypothesis that the means are equal? This depends on the sampling distribution of F. When the null hypothesis is true, the sampling distribution of F is the **F-distribution** with $\nu_1 = (p - 1)$ numerator degrees of freedom and $\nu_2 = (n - p)$ denominator degrees of freedom, respectively. An F-distribution with 7 and 9 df is shown in Figure 7.23. As you can see, the distribution is skewed to the right, since MST/MSE cannot be less than 0 but can increase without bound.

FIGURE 7.23

An F-Distribution with 7 numerator and 9 Denominator Degrees of Freedom

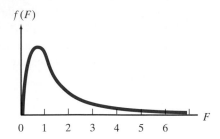

We need to be able to find F values corresponding to the tail areas of this distribution in order to establish the rejection region for our test of hypothesis. The upper-tail F values for $\alpha = .10, .05, .025,$ and $.01$ can be found in Tables VII, VIII, IX, and X of Appendix A. Table VIII is partially reproduced in Table 7.11. It gives F values that correspond to $\alpha = .05$ upper-tail areas for different degrees of freedom ν_1 (columns) and ν_2 (rows). Thus, if the numerator degrees of freedom is $\nu_1 = 7$ and the denominator degrees of freedom is $\nu_2 = 9$, we look in the seventh column and ninth row to find (shaded) $F_{.05} = 3.29$.

Returning to the SAT score example, the F statistic has $\nu_1 = (2 - 1) = 1$ numerator degree of freedom and $\nu_2 = (10 - 2) = 8$ denominator degrees of freedom. Thus, for $\alpha = .05$ we find (Table VIII of Appendix A)

$$F_{.05} = 5.32$$

TABLE 7.11   **Reproduction of Part of Table VIII in Appendix A: Percentage Points of the *F*-distribution, $\alpha = .05$**

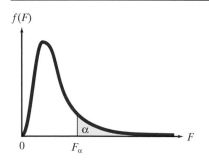

ν_2 \ ν_1	1	2	3	4	5	6	7	8	9
1	161.4	199.5	215.7	224.6	230.2	234.0	236.8	238.9	240.5
2	18.51	19.00	19.16	19.25	19.30	19.33	19.35	19.37	19.38
3	10.13	9.55	9.28	9.12	9.01	8.94	8.89	8.85	8.81
4	7.71	6.94	6.59	6.39	6.26	6.16	6.09	6.04	6.00
5	6.61	5.79	5.41	5.19	5.05	4.95	4.88	4.82	4.77
6	5.99	5.14	4.76	4.53	4.39	4.28	4.21	4.15	4.10
7	5.59	4.74	4.35	4.12	3.97	3.87	3.79	3.73	3.68
8	5.32	4.46	4.07	3.84	3.69	3.58	3.50	3.44	3.39
9	5.12	4.26	3.86	3.63	3.48	3.37	3.29	3.23	3.18
10	4.96	4.10	3.71	3.48	3.33	3.22	3.14	3.07	3.02
11	4.84	3.98	3.59	3.36	3.20	3.09	3.01	2.95	2.90
12	4.75	3.89	3.49	3.26	3.11	3.00	2.91	2.85	2.80
13	4.67	3.81	3.41	3.18	3.03	2.92	2.83	2.77	2.71
14	4.60	3.74	3.34	3.11	2.96	2.85	2.76	2.70	2.65

NUMERATOR DEGREES OF FREEDOM

Denominator Degrees of Freedom

The implication is that MST would have to be 5.32 times greater than MSE before we could conclude at the .05 level of significance that the two population treatment means differ. Since the data yielded $F = 1.78$, our initial impressions for the dot plot in Figure 7.21 are confirmed—there is insufficient information to conclude that the mean SAT scores differ for the populations of female and male high school seniors. The rejection region and the calculated F value are shown in Figure 7.24.

FIGURE 7.24

Rejection Region and Calculated F Values for SAT Score Samples

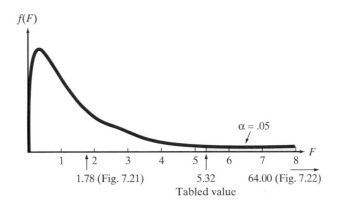

In contrast, consider the dot plot in Figure 7.22. Since the means are the same as in the first example, 550 and 590, respectively, the variation between the means is the same, MST = 4,000. But the variation within the two treatments appears to be considerably smaller. The observed SAT scores are 540, 545, 550, 555, and 560 for females and 580, 585, 590, 595, and 600 for males. These values yield $s_1^2 = 62.5$ and $s_2^2 = 62.5$. Thus, the variation within the treatments is measured by

$$\text{SSE} = (5 - 1)(62.5) + (5 - 1)(62.5)$$
$$= 500$$

$$\text{MSE} = \frac{\text{SSE}}{n - p} = \frac{500}{8} = 62.5$$

Then the F-ratio is

$$F = \frac{\text{MST}}{\text{MSE}} = \frac{4,000}{62.5} = 64.0$$

Again, our visual analysis of the dot plot is confirmed statistically: $F = 64.0$ well exceeds the tabled F value, 5.32, corresponding to the .05 level of significance. We would therefore reject the null hypothesis at that level and conclude that the SAT mean score of males differs from that of females.

Recall that we performed a hypothesis test for the difference between two means in Section 7.1 using a two-sample t statistic for two independent samples. When two independent samples are being compared, the t- and F-tests are equivalent. To see this, recall the formula

$$t = \frac{\bar{x}_1 - \bar{x}_2}{\sqrt{s_p^2 \left(\dfrac{1}{n_1} + \dfrac{1}{n_2}\right)}} = \frac{590 - 550}{\sqrt{(62.5)\left(\dfrac{1}{5} + \dfrac{1}{5}\right)}} = \frac{40}{5} = 8$$

where we used the fact that $s_p^2 = \text{MSE}$, which you can verify by comparing the formulas. Note that the calculated F for these samples ($F = 64$) equals the square of the calculated t for the same samples ($t = 8$). Likewise, the tabled F value (5.32) equals the square of the tabled t value at the two-sided .05 level of significance ($t_{.025} = 2.306$ with 8 df). Since both the rejection region and the calculated values are related in the same way, the tests are equivalent. Moreover, the assumptions that must be met to ensure the validity of the t- and F-tests are the same:

1. The probability distributions of the populations of responses associated with each treatment must all be normal.
2. The probability distributions of the populations of responses associated with each treatment must have equal variances.
3. The samples of experimental units selected for the treatments must be random and independent.

In fact, the only real difference between the tests is that the F-test can be used to compare *more than two* treatment means, whereas the t-test is applicable to two samples only. The ***F-test*** is summarized in the accompanying box.

**Anova Test to Compare p Treatment Means:
Independent Sampling**

$H_0: \mu_1 = \mu_2 = \cdots = \mu_p$
$H_a:$ At least two treatment means differ

Test statistic: $\quad F = \dfrac{\text{MST}}{\text{MSE}}$

Assumptions: 1. Samples are selected randomly and independently from the respective populations.
           2. All p population probability distributions are normal.
           3. The p population variances are equal.

Rejection region: $F > F_\alpha$, where F_α is based on $\nu_1 = (p - 1)$ numerator degrees of freedom (associated with MST) and $\nu_2 = (n - p)$ denominator degrees of freedom (associated with MSE).

Computational formulas for MST and MSE are given in Appendix C. We will rely on statistical software to compute the F statistic, concentrating on the interpretation of the results rather than their calculations.

EXAMPLE 7.10

Suppose the USGA wants to compare the mean distances associated with four different brands of golf balls when struck with a driver. A independent sampling design is employed, with Iron Byron, the USGA's robotic golfer, using a driver to hit a random sample of 10 balls of each brand in a random sequence. The distance is recorded for each hit, and the results are shown in Table 7.12, organized by brand.

a. Set up the test to compare the mean distances for the four brands. Use $\alpha = .10$.
b. Use SAS to obtain the test statistic and p-value. Interpret the results.

a. $H_0: \mu_1 = \mu_2 = \mu_3 = \mu_4$
b. $F = 43.99; p = .0001;$
   Reject H_0

⊘ GOLFCRD

TABLE 7.12   Results of Completely Randomized Design: Iron Byron Driver

	Brand A	Brand B	Brand C	Brand D
	251.2	263.2	269.7	251.6
	245.1	262.9	263.2	248.6
	248.0	265.0	277.5	249.4
	251.1	254.5	267.4	242.0
	260.5	264.3	270.5	246.5
	250.0	257.0	265.5	251.3
	253.9	262.8	270.7	261.8
	244.6	264.4	272.9	249.0
	254.6	260.6	275.6	247.1
	248.8	255.9	266.5	245.9
Sample means	250.8	261.1	270.0	249.3

Solution

a. To compare the mean distances of the four brands, we first specify the hypotheses to be tested. Denoting the population mean of the ith brand by μ_i, we test

$H_0: \mu_1 = \mu_2 = \mu_3 = \mu_4$
$H_a:$ The mean distances differ for at least two of the brands.

The test statistic compares the variation among the four treatment (Brand) means to the sampling variability within each of the treatments.

$$\textit{Test statistic:} \quad F = \frac{\text{MST}}{\text{MSE}}$$

$$\textit{Rejection region:} \quad F > F_\alpha = F_{.10} \text{ with } \nu_1 = (p - 1) = 3 \text{ df}$$
$$\text{and } \nu_2 = (n - p) = 36 \text{ df}$$

From Table VII of Appendix A, we find $F_{.10} \approx 2.25$ for 3 and 36 df. Thus, we will reject H_0 if $F > 2.25$. (See Figure 7.26.)

The assumptions necessary to ensure the validity of the test are as follows:
1. The samples of 10 golf balls for each brand are selected randomly and independently.
2. The probability distributions of the distances for each brand are normal.
3. The variances of the distance probability distributions for each brand are equal.

b. The SAS printout for the data in Table 7.12 resulting from this completely randomized design is given in Figure 7.25. The Total Sum of Squares is designated the **Corrected Total**, and it is partitioned into the **Model** and **Error Sums of Squares**. The bottom part of the printout further partitions the **Model** component into the factors that comprise the model. In this single-factor experiment, the Model and Brand sums of squares are the same. The **Sum of Squares** column is headed **ANOVA SS**.

The values of the mean squares, MST and MSE (highlighted on the printout), are 931.46 and 21.18, respectively. The F-ratio, 43.99, also highlighted on the printout, exceeds the tabled value of 2.25. We therefore reject the null hypothesis at the .10 level of significance, concluding that at least two of the brands differ with respect to mean distance traveled when struck by the driver.

TEACHING TIP
Have the students' state exactly how the *p*-value is interpreted. Stress that more work is needed if any specific information is desired concerning the relationship of the four population means.

FIGURE 7.25

SAS Analysis of Variance Printout for Golf Ball Distance Data: Completely Randomized Design

The ANOVA Procedure

Class Level Information

Class	Levels	Values
BRAND	4	A B C D

Number of observations   40

Dependent Variable: DISTANCE

Source	DF	Sum of Squares	Mean Square	F Value	Pr > F
Model	3	2794.388750	931.462917	43.99	<.0001
Error	36	762.301000	21.175028		
Corrected Total	39	3556.689750			

R-Square	Coeff Var	Root MSE	DISTANCE Mean
0.785671	1.785118	4.601633	257.7775

Source	DF	Anova SS	Mean Square	F Value	Pr > F
BRAND	3	2794.388750	931.462917	43.99	<.0001

The observed significance level of the *F*-test is also highlighted on the printout: *p*-value < .0001. This is the area to the right of the calculated *F* value and it implies that we would reject the null hypothesis that the means are equal at any α level of .0001 or greater. The result of the test is displayed in Figure 7.26.

FIGURE 7.26

F-Test for Completely Randomized Design: Golf Ball Experiment

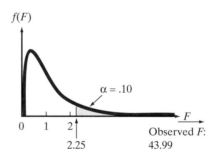

The results of an analysis of variance (ANOVA) can be summarized in a simple tabular format similar to that obtained from the SAS program in Example 7.10. The general form of the table is shown in Table 7.13, where the symbols df, SS, and MS stand for degrees of freedom, Sum of Squares, and Mean Square, respectively. Note that the two sources of variation, Treatments and Error, add to the Total Sum of Squares, SS(Total). The ANOVA summary table for Example 7.10 is given in Table 7.14.

Suppose the *F*-test results in a rejection of the null hypothesis that the treatment means are equal. Is the analysis complete? Usually, the conclusion that at least two of the treatment means differ leads to other questions. Which of the means differ, and by how much? For example, the *F*-test in Example 7.10 leads to the conclusion that at least two of the brands of golf balls have different mean distances traveled when struck with a driver. Now the question is, which of the brands differ? How are the brands ranked with respect to mean distance?

TABLE 7.13 General ANOVA Summary Table for a Completely Randomized Design

Source	df	SS	MS	F
Treatments	$p - 1$	SST	$MST = \dfrac{SST}{p-1}$	$\dfrac{MST}{MSE}$
Error	$n - p$	SSE	$MSE = \dfrac{SSE}{n-p}$	
Total	$n - 1$	SS(Total)		

TABLE 7.14 ANOVA Summary Table for Example 7.10

Source	df	SS	MS	F	p-Value
Brands	3	2,794.39	931.46	43.99	.0001
Error	36	762.30	21.18		
Total	39	3,556.69			

One way to obtain this information is to construct a confidence interval for the difference between the means of any pair of treatments using the method of Section 7.1. For example, if a 95% confidence interval for $\mu_A - \mu_C$ in Example 7.10 is found to be $(-24, -13)$, we are confident that the mean distance for Brand C exceeds the mean for Brand A (since all differences in the interval are negative). Constructing these confidence intervals for all possible brand pairs allows you to rank the brand means. A method for conducting these *multiple comparisons*—one that controls for Type I errors—is beyond the scope of this introductory text. Consult the references to learn about this methodology.

Using the TI-83 Graphing Calculator
"One-Way ANOVA" on the TI-83

Start from the home screen.

Step 1   Enter each data set into its own list. (i.e., sample 1 into List1, sample 2 into List2, sample 3 into List3, etc.)

Step 2   *Access the "Statistical Test" Menu*
Press **STAT**
Arrow right to **TEST**
Arrow up to **F: ANOVA(**
Press **ENTER**

Step 3   Enter each list, separated by commas, for which you want to perform the analysis. (e.g., L1, L2, L3, L4)
Press **ENTER**

Step 4   *View Display*
The calculator will display the *F*-test statistic, as well as the *p*-value, the factor degrees of freedom, sum of squares, mean square, and by arrowing down the error degrees of freedom, sum of squares, mean square, and the pooled standard deviation.

(continues on next page)

USING THE TI-83 GRAPHING CALCULATOR *(continued)*

Example   Below are four different samples. At the $\alpha = 0.05$ level of
significance, test whether the four population means are equal. The
null hypothesis will be $H_0: \mu_1 = \mu_2 = \mu_3 = \mu_4$. The alternative
hypothesis is H_a: At least one mean is different.

SAMPLE 1	SAMPLE 2	SAMPLE 3	SAMPLE 4
60	59	55	58
61	52	55	58
56	51	52	55

Step 1   Enter the data into the lists.

Step 2   Access the Statistical Test Menu.

Step 3   Enter each list for which you wish to perform the analysis (see
screen below left).

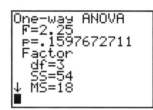

Step 4   Press   **ENTER** and view screen (see screen above right.) As you
can see from the screen the *p*-value is 0.1598 which is *not less than*
0.05 therefore we should *not reject* $H_0: \mu_1 = \mu_2 = \mu_3 = \mu_4$. The
differences are not significant.

Step 5   Clear the screen for the next problem. Return to the home screen
Press **CLEAR.**

E X A M P L E   7 . 1 1

Refer to the completely randomized design ANOVA conducted in Example 7.10.
Are the assumptions required for the test approximately satisfied?

Solution

The assumptions for the test are repeated below.

1. The samples of golf balls for each brand are selected randomly and
   independently.
2. The probability distributions of the distances for each brand are normal.
3. The variances of the distance probability distributions for each brand are equal.

Since the sample consisted of 10 randomly selected balls of each brand and
the robotic golfer Iron Byron was used to drive all the balls, the first assumption
of independent random samples is satisfied. To check the next two assumptions, we
will employ two graphical methods presented in Chapter 2: stem-and-leaf displays
and dot plots. A MINITAB stem-and-leaf display for the sample distances of each
brand of golf ball is shown in Figure 7.27, and a MINITAB dot plot in Figure 7.28.

The normality assumption can be checked by examining the stem-and-leaf
displays in Figure 7.27. With only 10 sample measurements for each brand, however, the displays are not very informative. More data would need to be collected for each brand before we could assess whether the distances come from

FIGURE 7.27

MINITAB Stem-and-Leaf Displays for Distance Data: Completely Randomized Design

```
Stem-and-leaf of BrandA     N = 10
Leaf Unit = 1.0

   2   24  45
   2   24
   4   24  88
  (3)  25  011
   3   25  3
   2   25  4
   1   25
   1   25
   1   26  0


Stem-and-leaf of BrandB     N = 10
Leaf Unit = 1.0

   2   25  45
   3   25  7
   3   25
   4   26  0
  (3)  26  223
   3   26  445


Stem-and-leaf of BrandC     N = 10
Leaf Unit = 1.0

   1   26  3
   2   26  5
   4   26  67
   5   26  9
   5   27  00
   3   27  2
   2   27  5
   1   27  7


Stem-and-leaf of BrandD     N = 10
Leaf Unit = 1.0

   1   24  2
   2   24  5
   4   24  67
  (3)  24  899
   3   25  11
   1   25
   1   25
   1   25
   1   25
   1   26  1
```

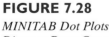
Suggested Exercise 7.83

FIGURE 7.28

MINITAB Dot Plots for Distance Data: Completely Randomized Design

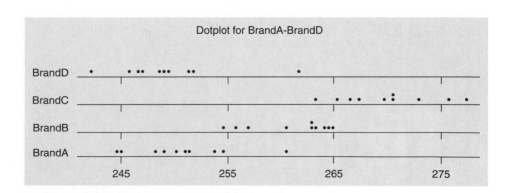
Dotplot for BrandA-BrandD

normal distributions. Fortunately, analysis of variance has been shown to be a very **robust method** when the assumption of normality is not satisfied exactly: That is, *moderate departures from normality do not have much effect on the significance level of the ANOVA F-test or on confidence coefficients*. Rather than spend the time, energy, or money to collect additional data for this experiment in order to verify the normality assumption, we will rely on the robustness of the ANOVA methodology.

Dot plots are a convenient way to obtain a rough check on the assumption of equal variances. With the exception of a possible outlier for Brand D, the dot plots in Figure 7.28 show that the spread of the distance measurements is about the same for each brand. Since the sample variances appear to be the same, the assumption of equal population variances for the brands is probably satisfied. Although robust with respect to the normality assumption, ANOVA is *not robust* with respect to the equal variances assumption. Departures from the assumption of equal population variances can affect the associated measures of reliability (e.g., *p*-values and confidence levels). Fortunately, the effect is slight when the sample sizes are equal, as in this experiment. ∎

Although graphs can be used to check the ANOVA assumptions, as in Example 7.11, no measures of reliability can be attached to these graphs. When you have a plot that is unclear as to whether or not an assumption is satisfied, you can use formal statistical tests that are beyond the scope of this text. Consult the references at the end of the chapter for information on these tests. When the validity of the ANOVA assumptions is in doubt, nonparametric statistical methods are useful.

EXERCISES 7.73–7.87

Learning the Mechanics

7.73 Use Tables VII, VIII, IX, and X of Appendix A to find each of the following F values:
  a. $F_{.05}, \nu_1 = 3, \nu_2 = 4$ 6.59
  b. $F_{.01}, \nu_1 = 3, \nu_2 = 4$ 16.69
  c. $F_{.10}, \nu_1 = 20, \nu_2 = 40$ 1.61
  d. $F_{.025}, \nu_1 = 12, \nu_2 = 9$ 3.87

7.74 Consider dot plots a. and b. shown below. In which dot plot is the difference between the sample means small relative to the variability within the sample observations? Justify your answer.

7.75 Refer to Exercise 7.74. Assume that the two samples represent independent, random samples corresponding to two treatments in a completely randomized design.
  a. Calculate the treatment means, i.e., the means of samples 1 and 2, for both dot plots.
  b. Use the means to calculate the Sum of Squares for Treatments (SST) for each dot plot.
  c. Calculate the sample variance for each sample and use these values to obtain the Sum of Squares for Error (SSE) for each dot plot.
  d. Calculate the Total Sum of Squares [SS(Total)] for the two dot plots by adding the Sums of Squares for

Plots for Exercise 7.74

a.
| | 5 | 6 | 7 | 8 | 9 | 10 | 11 | 12 | 13 | 14 | 15 | 16 | 17 | 18 |

b.
| | 5 | 6 | 7 | 8 | 9 | 10 | 11 | 12 | 13 | 14 | 15 | 16 | 17 | 18 |

  • Sample 1
  □ Sample 2

Treatment and Error. What percentage of SS(Total) is accounted for by the treatments—that is, what percentage of the Total Sum of Squares is the Sum of Squares for Treatment—in each case?

e. Convert the Sum of Squares for Treatment and Error to mean squares by dividing each by the appropriate number of degrees of freedom. Calculate the *F*-ratio of the Mean Square for Treatment (MST) to the Mean Square for Error (MSE) for each dot plot. 37.5; 5.21

f. Use the *F*-ratios to test the null hypothesis that the two samples are drawn from populations with equal means. Use $\alpha = .05$.

g. What assumptions must be made about the probability distributions corresponding to the responses for each treatment in order to ensure the validity of the *F*-tests conducted in part **f**?

7.76 Refer to Exercises 7.74 and 7.75. Conduct a two-sample *t*-test (Section 7.1) of the null hypothesis that the two treatment means are equal for each dot plot. Use $\alpha = .05$ and two-tailed tests. In the course of the test, compare each of the following with the *F*-tests in Exercise 7.75:

a. The pooled variances and the MSEs

b. The *t*- and the *F*-test statistics

c. The tabled values of *t* and *F* that determine the rejection regions

d. The conclusions of the *t*- and *F*-tests

e. The assumptions that must be made in order to ensure the validity of the *t*- and *F*-tests

7.77 Below is a MINITAB printout for an experiment utilizing independent random sampling.

MINITAB Output for Exercise 7.77

```
Analysis of Variance for RESPONSE

Source    DF      SS        MS        F       P
FACTOR     3    57258     19086    14.80   0.000
Error     34    43836      1289
Total     37   101094
```

a. How many treatments are involved in the experiment? What is the total sample size?

b. Conduct a test of the null hypothesis that the treatment means are equal. Use $\alpha = .01$. $F = 14.80$

c. What additional information is needed in order to be able to compare specific pairs of treatment means? Sample means

7.78 A partially completed ANOVA table for an independent sampling design is shown here:

Source	df	SS	MS	F
Treatments	6	18.4		
Error				
Total	41	45.2		

a. Complete the ANOVA table.

b. How many treatments are involved in the experiment? 7

c. Do the data provide sufficient evidence to indicate a difference among the population means? Test using $\alpha = .10$. $F = 4.01$

d. Find the approximate observed significance level for the test in part **c**, and interpret it. $p < .01$

Applying the Concepts—Basic

7.79 Robotics researchers investigated whether robots could be trained to behave like ants in an ant colony (*Nature*, Aug. 2000). Robots were trained and randomly assigned to "colonies" (i.e., groups) consisting of 3, 6, 9, or 12 robots. The robots were assigned the task of foraging for "food" and to recruit another robot when they identified a resource-rich area. One goal of the experiment was to compare the mean energy expended (per robot) of the four different colony sizes.

a. What type of experimental design was employed?

b. Identify the treatments and the dependent variable.

c. Set up the null and alternative hypotheses of the test. $H_0: \mu_1 = \mu_2 = \mu_3 = \mu_4$

d. The following ANOVA results were reported: $F = 7.70$, numerator df = 3, denominator df = 56, *p*-value $< .001$. Conduct the test at a significance level of $\alpha = .05$ and interpret the result. Reject H_0

7.80 An article in the *American Journal of Political Science* (Jan. 1998) examined the attitudes of three groups of professionals that influence U.S. policy. Random samples of 100 scientists, 100 journalists, and 100 government officials were asked about the safety of nuclear power plants. Responses were made on a seven-point scale, where 1 = very unsafe and 7 = very safe. The mean safety scores for the groups are: scientists, 4.1; journalists, 3.7; government officials, 4.2.

a. Identify the response variable for this study.

b. How many treatments are included in this study? Describe them. 3

c. Specify the null and alternative hypotheses that should be used to investigate whether there are differences in the attitudes of scientists, journalists, and government officials regarding the safety of nuclear power plants. $H_0: \mu_1 = \mu_2 = \mu_3$

d. The MSE for the sample data is 2.355. At least how large must MST be in order to reject the null hypothesis of the test of part **a** using $\alpha = .05$?

e. If the MST = 11.280, what is the approximate *p*-value of the test of part **a**? $p < .01$

7.81 Refer to *The Archives of Disease in Childhood* (Apr. 2000) study of whether height influences a child's progression through elementary school, Exercise 7.9 (p. 335). Within each grade, Australian school children were divided into equal thirds (tertiles) based on age (youngest third, middle third, and oldest third). The researchers compared the average heights of the three groups using an analysis of variance. (All height measurements were standardized using *z*-scores.) A summary of the results for all grades combined, by gender, is shown in the table on p. 384.

	Sample Size	Youngest Tertile Mean Height	Middle Tertile Mean Height	Oldest Tertile Mean Height	F Value	p-Value
Boys	1439	0.33	0.33	0.16	4.57	0.01
Girls	1409	0.27	0.18	0.21	0.85	0.43

Source: Wake, M., Coghlan, D., and Hesketh, K. "Does height influence progression through primary school grades?" *The Archives of Disease in Childhood*, Vol. 82, Apr. 2000 (Table 2).

a. What is the null hypothesis for the ANOVA of the boys' data? $H_0: \mu_1 = \mu_2 = \mu_3$

b. Interpret the results of the test, part **a**. Use $\alpha = .05$.

c. Repeat parts **a** and **b** for the girls' data.

d. Summarize the results of the hypothesis tests in the words of the problem.

7.82 The *Journal of Hazardous Materials* (July 1995) published the results of a study of the chemical properties of three different types of hazardous organic solvents used to clean metal parts: aromatics, chloroalkanes, and esters. One variable studied was sorption rate, measured as mole percentage. Independent samples of solvents from each type were tested and their sorption rates were recorded, as shown in the table. A MINITAB analysis of variance of the data is provided.

⊘ SORPRATE

Aromatics		Chloroalkanes		Esters		
1.06	.95	1.58	1.12	.29	.43	.06
.79	.65	1.45	.91	.06	.51	.09
.82	1.15	.57	.83	.44	.10	.17
.89	1.12	1.16	.43	.61	.34	.60
1.05				.55	.53	.17

Source: Reprinted from *Journal of Hazardous Materials*, Vol. 42, No. 2, J. D. Ortego *et al.*, "A review of polymeric geosynthetics used in hazardous waste facilities." p. 142 (Table 9), July 1995, Elsevier Science-NL, Sara Burgerhartstraat 25, 1055 KV Amsterdam, The Netherlands.

MINITAB Output for Exercise 7.82

```
Analysis of Variance for SORPRATE

Source     DF        SS        MS       F       P
SOLVENT     2    3.3054    1.6527   24.51   0.000
Error      29    1.9553    0.0674
Total      31    5.2608
```

a. Construct an ANOVA table from the MINITAB printout.

b. Is there evidence of differences among the mean sorption rates of the three organic solvent types? Test using $\alpha = .10$. $F = 24.51$

Applying the Concepts—Intermediate

7.83 Studies conducted at the University of Melbourne (Australia) indicate that there may be a difference between the pain thresholds of blondes and brunettes. Men and women of various ages were divided into four categories according to hair color: light blond, dark blond, light brunette, and dark brunette. The purpose of the experiment was to determine whether hair color is related to the amount of pain evoked by common types of mishaps and assorted types of trauma. Each person in the experiment was given a pain threshold score based on his or her performance in a pain sensitivity test (the higher the score, the higher the person's pain tolerance). SAS was used to conduct the analysis of variance of the data listed in the table on p. 385. The SAS printout is provided below.

SAS Output for Exercise 7.83

The ANOVA Procedure

Class Level Information

Class	Levels	Values
COLOR	4	DkBlond DkBrunet LtBlond LtBrunet

Number of observations 19

Dependent Variable: PAIN

Source	DF	Sum of Squares	Mean Square	F Value	Pr > F
Model	3	1360.726316	453.575439	6.79	0.0041
Error	15	1001.800000	66.786667		
Corrected Total	18	2362.526316			

R-Square	Coeff Var	Root MSE	PAIN Mean
0.575962	17.08184	8.172311	47.84211

Source	DF	Anova SS	Mean Square	F Value	Pr > F
COLOR	3	1360.726316	453.575439	6.79	0.0041

⊘ HAIRPAIN

Light Blond	Dark Blond	Light Brunette	Dark Brunette
62	63	42	32
60	57	50	39
71	52	41	51
55	41	37	30
48	43		35

a. Based on the given information, what type of experimental design appears to have been employed?

b. Using the SAS printout, conduct a test to determine whether the mean pain thresholds differ among people possessing the four types of hair color. Use $\alpha = .05$. $F = 6.79$

c. What is the observed significance level for the test in part **b**? Interpret it. $p = .0041$

d. What assumptions must be met in order to ensure the validity of the inferences you made in part **b**?

7.84 Psychologists at Lancaster University (United Kingdom) evaluated three methods of name retrieval in a controlled setting (*Journal of Experimental Psychology-Applied*, June 2000.) A sample of 139 students was randomly divided into three groups, and each group of students used a different method to learn the names of the other students in the group. Group 1 used the "simple name game," where the first student states his/her full name, the second student announces his/her name and the name of the first student, the third student says his/her name and the names of the first two students, etc. Group 2 used the "elaborate name game," a modification of the simple name game where the students not only state their names but also their favorite activity (e.g., sports).

⊘ NAMEGAME
Simple Name Game

24	43	38	65	35	15	44	44	18	27	0	38	50	31
7	46	33	31	0	29	0	0	52	0	29	42	39	26
51	0	42	20	37	51	0	30	43	30	99	39	35	19
24	34	3	60	0	29	40	40						

Elaborate Name Game

39	71	9	86	26	45	0	38	5	53	29	0	62	0
1	35	10	6	33	48	9	26	83	33	12	5	0	0
25	36	39	1	37	2	13	26	7	35	3	8	55	50

Pairwise Introductions

5	21	22	3	32	29	32	0	4	41	0	27	5	9
66	54	1	15	0	26	1	30	2	13	0	2	17	14
5	29	0	45	35	7	11	4	9	23	4	0	8	2
18	0	5	21	14									

Source: Morris, P.E., and Fritz, C.O. "The name game: Using retrieval practice to improve the learning of names." *Journal of Experimental Psychology-Applied*, Vol. 6, No. 2, June 2000 (data simulated from Figure 1).

Group 3 used "pairwise introductions," where students are divided into pairs and each student must introduce the other member of the pair. One year later, all subjects were sent pictures of the students in their group and asked to state the full name of each. The researchers measured the percentage of names recalled for each student respondent. The data (simulated based on summary statistics provided in the research article) are shown in the table. Conduct an analysis of variance to determine whether the mean percentages of names recalled differ for the three name retrieval methods. Use $\alpha = .05$. $F = 7.687$

7.85 What do people infer from facial expressions of emotion? This was the research question of interest in an article published in the *Journal of Nonverbal Behavior* (Fall 1996). A sample of 36 introductory psychology students was randomly divided into six groups. Each group was assigned to view one of six slides showing a person making a facial expression.* The six expressions were (1) angry, (2) disgusted, (3) fearful, (4) happy, (5) sad, and (6) neutral faces. After viewing the slides, the students rated the degree of dominance they inferred from the facial expression (on a scale ranging from -15 to $+15$). The data (simulated from summary information provided in the article) are listed in the table. Conduct an analysis of variance to determine whether the mean dominance ratings differ among the six facial expressions. Use $\alpha = .10$. $F = 3.96$

⊘ FACES

Angry	Disgusted	Fearful	Happy	Sad	Neutral
2.10	.40	.82	1.71	.74	1.69
.64	.73	-2.93	$-.04$	-1.26	$-.60$
.47	$-.07$	$-.74$	1.04	-2.27	$-.55$
.37	$-.25$	.79	1.44	$-.39$	.27
1.62	.89	$-.77$	1.37	-2.65	$-.57$
$-.08$	1.93	-1.60	.59	$-.44$	-2.16

⊘ OILSPILL

7.86 Refer to the *Marine Technology* (Jan. 1995) study of major ocean oil spills by tanker vessels, Exercise 2.15 (p. 29). The spillage amounts (thousands of metric tons) and cause of accident for 48 tankers are saved in the OILSPILL file. (*Note:* Delete the two tankers with oil spills of unknown causes.) Conduct an analysis of variance (at $\alpha = .01$) to compare the mean spillage amounts for the four accident types: (1) collision, (2) grounding, (3) fire/explosion, and (4) hull failure. Interpret your results. $F = .53$

Applying the Concepts—Advanced

7.87 Do you experience episodes of excessive eating accompanied by being overweight? If so, you may suffer from binge eating disorder. Cognitive-behavioral therapy (CBT), in which patients are taught how to make

* In the actual experiment, each group viewed all six facial expression slides and the design employed was a Latin Square (beyond the scope of this text).

changes in specific behavior patterns (e.g., exercise, eat only low-fat foods), can be effective in treating the disorder. A group of Stanford University researchers investigated the effectiveness of interpersonal therapy (IPT) as a second level of treatment for binge eaters (*Journal of Consulting and Clinical Psychology*, June 1995). The researchers employed a design that randomly assigned a sample of 41 overweight individuals diagnosed with binge eating disorder to either a treatment group (30 subjects) or a control group (11 subjects). Subjects in the treatment group received 12 weeks of cognitive-behavior therapy, then were subdivided into two groups. Those who responded successfully to CBT (17 subjects) were assigned to a weight loss therapy (WLT) program for the next 12 weeks. Those CBT subjects who did not respond to treatment (13 subjects) received 12 weeks of IPT. The subjects in the control group received no therapy of any type. Thus, the study ultimately consisted of three groups of overweight binge eaters: the CBT-WLT group, the CBT-IPT group, and the control group. One outcome (response) variable measured for each subject was the number of binge eating episodes per week, x. Summary statistics for each of the three groups at the end of the 24-week period are shown in the table. The data were analyzed as an independent sampling design with three treatments (CBT-WLT, CBT-IPT, and Control). Although the ANOVA tables were not provided in the article, sufficient information is provided in the table to reconstruct them.

a. Compute SST for the ANOVA using the formula on p. 372:

$$SST = \sum_{i=1}^{3} n_i(\bar{x}_i - \bar{x})^2$$

where $\bar{x}$ is the overall mean number of binges per week of all 41 subjects. [*Hint:* $\bar{x} = (\sum_{i=1}^{3} n_i\bar{x}_i)/41$].

b. Recall that SSE for the ANOVA can be written as

$$SSE = (n_1 - 1)s_1^2 + (n_2 - 1)s_2^2 + (n_3 - 1)s_3^2$$

where s_1^2, s_2^2, and s_3^2 are the sample variances associated with the three treatments. Compute SSE for the ANOVA.   SSE = 77.24

c. Use the results, parts **a** and **b,** to construct the ANOVA table.

d. Is there sufficient evidence (at $\alpha = .01$) of differences among the mean number of binges experienced per week by the subjects in the three groups?

e. Comment on the validity of the ANOVA assumptions. How might this impact the results of the study?

f. Comment on the use of an independent sampling design for this study. Have subjects been randomly and independently assigned to each group? How might this impact the results of the study?

	CBT-WLT	CBT-IPT	Control
Sample size	17	13	11
Mean number of binges per week	0.2	1.9	2.9
Standard deviation	0.4	1.7	2.0

Source: Agras, W. S., *et al.* "Does interpersonal therapy help patients with binge eating disorder who fail to respond to cognitive-behavioral therapy?" *Journal of Consulting and Clinical Psychology*, Vol. 63, No. 3, June 1995, p. 358 (Table 1).

STATISTICS IN ACTION
On the Trail of the Cockroach

Entomologists have long established that insects such as ants, bees, caterpillars, and termites use chemical or "odor" trails for navigation. These trails are used as highways between sources of food and the insect nest. Until recently, however, "bug" researchers believed that navigational behavior of cockroaches scavenging for food was random and not linked to a chemical trail.

One of the first researchers to challenge the "random-walk" theory for cockroaches was professor and entomologist Dini Miller of Virginia Tech University. According to Miller, "the idea that roaches forage randomly means that they would have to come out of their hiding places every night and bump into food and water by accident. But roaches never seem to go hungry." Since cockroaches had never before been evaluated for trail following behavior, Miller designed an experiment to test a cockroach's ability to follow a trail to their fecal material (*Explore*, Research at the University of Florida, Fall 1998).

First, Dr. Miller developed a methanol extract from roach feces—called a pheromone. She theorized that "pheromones are communication devices between cockroaches. If you have an infestation and have a lot of fecal material around, it advertises, 'Hey, this is a good cockroach place.'" Then, she created a chemical trail with the pheromone on a strip of white chromatography paper and placed the paper at the bottom of a plastic, V-shaped container, 122 square centimeters in size. German cockroaches were released into the container at the beginning of the trail, one at a time, and a video surveillance camera was used to monitor the roach's movements.

In addition to the trail containing the fecal extract (the treatment), a trail using methanol only was created. This second trail served as a "control" to compare back against the treated trail. Since Dr. Miller also wanted to determine if trail following ability differed among cockroaches of different age, sex, and reproductive stage, four roach groups were utilized in the experiment: adult males, adult females, gravid

STATISTICS IN ACTION (continued)

(pregnant) females, and nymphs (immatures). Twenty roaches of each type were randomly assigned to the treatment trail and 10 of each type were randomly assigned to the control trail. Thus, a total of 120 roaches were used in the experiment. A layout of the design is illustrated in Figure 7.29.

The movement pattern of each cockroach tested was translated into *xy* coordinates every one-tenth of a second by the Dynamic Animal Movement Analyzer (DAMA) program. Miller measured the perpendicular distance of each *xy* coordinate from the trail and then averaged these distances, or deviations, for each cockroach. The average trail deviations (measured in pixels*) for 120 cockroaches in the study are listed in Table 7.15 and saved in the ROACH file.

_____
*1 pixel ≈ 2 centimeters

Focus

Conduct the appropriate analysis of the data. Use the results to answer the following research questions (not necessarily in the order presented):

(1) Is there evidence that cockroaches within a group exhibit the ability to follow fecal extract trails? In other words, for a particular roach type, is the average trail deviation for the extract smaller than the average trail deviation for the control?

(2) Does fecal extract trail following ability differ among cockroaches of different age, sex, and reproductive status?

Write up the results of your analysis in a report and present it to your class.

FIGURE 7.29   **Layout of Experimental Design for Cockroach Study**

		Roach Type			
		Adult Male	**Adult Female**	**Gravid Female**	**Nymph**
Trail	**Extract**	$n = 20$	$n = 20$	$n = 20$	$n = 20$
	Control	$n = 10$	$n = 10$	$n = 10$	$n = 10$

⊘ROACH

TABLE 7.15   **Average Trail Deviations for 120 Cockroaches**

Adult Males		Adult Females		Gravid Females		Nymphs	
Extract	**Control**	**Extract**	**Control**	**Extract**	**Control**	**Extract**	**Control**
3.1	42.0	7.2	70.2	78.7	54.6	7.7	132.9
6.2	22.7	17.3	49.0	70.3	54.3	27.7	19.7
34.0	93.1	9.1	40.5	79.9	63.5	18.4	32.1
2.1	17.5	13.2	13.3	51.0	52.6	47.6	66.4
2.4	78.1	101.2	31.8	13.6	95.1	22.4	126.0
4.4	74.1	4.6	116.0	20.4	117.9	8.3	131.1
2.4	50.3	18.1	164.0	51.2	53.3	13.5	50.7
7.6	8.9	73.0	30.2	27.5	84.0	45.6	93.8
5.5	11.3	4.8	44.3	63.1	103.5	8.4	59.4
6.9	82.0	20.5	72.6	4.8	53.0	3.3	25.6
25.4		51.6		23.4		51.2	
2.2		5.8		48.2		10.4	
2.5		27.8		13.3		32.0	
4.9		2.8		57.4		6.9	
18.5		4.4		65.4		32.6	
4.6		3.2		10.5		23.8	
7.7		3.6		59.9		5.1	
3.2		1.7		38.4		3.8	
2.4		29.8		27.0		3.1	
1.5		21.7		76.6		2.8	

Source: Dini Miller, Department of Entomology, Virginia Polytechnic Institute and State University.

_____

TEACHING TIP

Suggestions for class discussion of the above case can be found in the *Instructor's Notes*.

> ██████ QUICK REVIEW

Key Terms

Analysis of variance (ANOVA)* 371
Blocking 341
F-distribution* 373
F-statistic* 376
Mean square for error (MSE)* 373
Mean square for treatment (MST)* 373
Paired difference experiment 341
Pooled sample estimator 327

Rank sum* 356
Randomized block experiment 341
Robust method 382
Standard error 324
Sum of squares for treatment (SST)* 372
Sum of squares for error (SSE)* 372
Wilcoxon rank sum test* 355
Wilcoxon signed rank test* 365

Note: Starred (*) terms are from the optional sections in this chapter.

Key Formulas

Note: Starred (*) formulas are from the optional sections in this chapter.

$(1 - \alpha)100\%$ confidence interval for θ: (see table)

   Large samples: $\hat{\theta} \pm z_{\alpha/2}\sigma_{\hat{\theta}}$

   Small samples: $\hat{\theta} \pm t_{\alpha/2}\sigma_{\hat{\theta}}$

For testing $H_0: \theta = D_0$: (see table)

   Large samples: $z = \dfrac{\hat{\theta} - D_0}{\sigma_{\hat{\theta}}}$

   Small samples: $t = \dfrac{\hat{\theta} - D_0}{\sigma_{\hat{\theta}}}$

Parameter, θ	Estimator, $\hat{\theta}$	Standard Error of Estimator, $\sigma_{\hat{\theta}}$	Estimated Standard Error	
$\mu_1 - \mu_2$ (independent samples)	$\bar{x}_1 - \bar{x}_2$	$\sqrt{\dfrac{\sigma_1^2}{n_1} + \dfrac{\sigma_2^2}{n_2}}$	Large n: $\sqrt{\dfrac{s_1^2}{n_1} + \dfrac{s_2^2}{n_2}}$	324, 325
			Small n: $\sqrt{s_p^2\left(\dfrac{1}{n_1} + \dfrac{1}{n_2}\right)}$	328
μ_D (paired sample)	$\bar{x}_D$	$\dfrac{\sigma_D}{\sqrt{n_D}}$	$\dfrac{s_D}{\sqrt{n_D}}$	342, 343

Pooled sample variance:	$s_p^2 = \dfrac{(n_1 - 1)s_1^2 + (n_2 - 1)s_2^2}{n_1 + n_2 - 2}$	328
Determining the sample size for estimating $\mu_1 - \mu_2$:	$n_1 = n_2 = \dfrac{(z_{\alpha/2})^2(\sigma_1^2 + \sigma_2^2)}{B^2}$	353
	$\mu_D: n = \dfrac{(z_{\alpha12})^2\sigma_D^2}{B^2}$	354
*Wilcoxon rank sum (large sample)	$z = \dfrac{T_1 - \dfrac{n_1(n_1 + n_2 + 1)}{2}}{\sqrt{\dfrac{n_1 n_2(n_1 + n_2 + 1)}{12}}}$	359
*Wilcoxon signed ranks (large sample):	$z = \dfrac{T_1 - \dfrac{n(n + 1)}{4}}{\sqrt{\dfrac{n(n + 1)(2n + 1)}{24}}}$	368
*ANOVA F-test:	$F = \dfrac{MST}{MSE}$	376

Language Lab

Symbol	Pronunciation	Description
$\mu_1 - \mu_2$	mu-1 minus mu-2	Difference between population means
$\bar{x}_1 - \bar{x}_2$	x-bar-1 minus x-bar-2	Difference between sample means
$\sigma_{(\bar{x}_1 - \bar{x}_2)}$	sigma of x-bar-1 minus x-bar-2	Standard deviation of the sampling distribution of $(\bar{x}_1 - \bar{x}_2)$
s_p^2	s-p squared	Pooled sample variance
D_0	D naught	Hypothesized value of difference
μ_D	mu D	Difference between population means, paired data
$\bar{x}_D$	x-bar D	Mean of sample differences
s_D	s-D	Standard deviation of sample differences
n_D	n-D	Number of differences in sample
$*T_1$	t-1	Sum of ranks of observations in sample 1
$*T_2$	t-2	Sum of ranks of observations in sample 2
$*T_L$	t-l	Critical lower Wilcoxon rank sum value
$*T_U$	t-u	Critical upper Wilcoxon rank sum value
$*T_+$	t-plus	Sum of ranks of positive differences of paired observations
$*T_-$	t-minus	Sum of ranks of negative differences of paired observations
$*T_0$	t-naught	Critical value of Wilcoxon signed rank test
$*F_\alpha$	F-alpha	Critical value of F associated with tail area α
$*\nu_1$	nu-1	Numerator degrees of freedom for F statistic
$*\nu_2$	nu-2	Denominator degrees of freedom for F statistic
*ANOVA		Analysis of variance
*SST		Sum of squares for treatments
*SSE		Sum of squares for error
*MST		Mean squares for treatments
*MSE		Mean square for error

 SUPPLEMENTARY EXERCISES 7.88–7.112

Note: Starred (*) exercises refer to the optional sections in this chapter.

Learning the Mechanics

7.88 Independent random samples were selected from two normally distributed populations with means μ_1 and μ_2, respectively. The sample sizes, means, and variances are shown in the following table.

Sample 1	Sample 2
$n_1 = 12$	$n_2 = 14$
$\bar{x}_1 = 17.8$	$\bar{x}_2 = 15.3$
$s_1^2 = 74.2$	$s_2^2 = 60.5$

a. Test $H_0: (\mu_1 - \mu_2) = 0$ against $H_a: (\mu_1 - \mu_2) > 0$. Use $\alpha = .05$. $t = .78$

b. Form a 99% confidence interval for $(\mu_1 - \mu_2)$.

c. How large must n_1 and n_2 be if you wish to estimate $(\mu_1 - \mu_2)$ to within 2 units with 99% confidence? Assume that $n_1 = n_2$. $n_1 = n_2 = 225$

7.89 Two independent random samples are taken from two populations. The results of these samples are summarized in the following table.

Sample 1	Sample 2
$n_1 = 135$	$n_2 = 148$
$\bar{x}_1 = 12.2$	$\bar{x}_2 = 8.3$
$s_1^2 = 2.1$	$s_2^2 = 3.0$

a. Form a 90% confidence interval for $(\mu_1 - \mu_2)$.

b. Test $H_0: (\mu_1 - \mu_2) = 0$ against $H_a: (\mu_1 - \mu_2) \neq 0$. Use $\alpha = .01$. $z = 20.60$

c. What sample size would be required if you wish to estimate $(\mu_1 - \mu_2)$ to within .2 with 90% confidence? Assume that $n_1 = n_2$. $n_1 = n_2 = 346$

*7.90 An independent sampling design is utilized to compare four treatment means. The data are shown in the table (p. 390).

⊘ LM7_90

Treatment 1	Treatment 2	Treatment 3	Treatment 4
8	6	9	12
10	9	10	13
9	8	8	10
10	8	11	11
11	7	12	11

a. Given that SST = 36.95 and SS(Total) = 62.55, complete an ANOVA table for this experiment.

b. Is there evidence that the treatment means differ? Use $\alpha = .10$. $F = 7.70$

7.91 A random sample of five pairs of observations were selected, one observation from a population with mean μ_1, the other from a population with mean μ_2. The data are shown in the accompanying table.

⊘ LM7_91

Pair	Value from Population 1	Value from Population 2
1	28	22
2	31	27
3	24	20
4	30	27
5	22	20

a. Test the null hypothesis $H_0: \mu_D = 0$ against $H_a: \mu_D \neq 0$, where $\mu_D = \mu_1 - \mu_2$. Use $\alpha = .05$.

b. Form a 95% confidence interval for μ_D. 3.8 ± 1.84

c. When are the procedures you used in parts **a** and **b** valid?

7.92 List the assumptions necessary for each of the following inferential techniques:

a. Large-sample inferences about the difference $(\mu_1 - \mu_2)$ between population means using a two-sample z statistic

b. Small-sample inferences about $(\mu_1 - \mu_2)$ using an independent samples design and a two-sample t statistic

c. Small-sample inferences about $(\mu_1 - \mu_2)$ using a paired difference design and a single-sample t statistic to analyze the differences

***d.** Inferences about three population means, μ_1, μ_2, and μ_3

7.93 Suppose you use the SPSS statistical program package to test the null hypothesis $H_0: (\mu_1 - \mu_2) = 0$ against $H_a: (\mu_1 - \mu_2) \neq 0$ with independent samples of size 12 and 10, respectively. Assuming you are using $\alpha = .05$, what conclusion would you reach in each of the following instances of observed significance level reported by the program? (Recall that SPSS reports the two-tailed p-value.)

a. .0429   Reject H_0

b. .1984   Do not reject H_0

c. .0001   Reject H_0

d. .0344   Reject H_0

e. .0545   Do not reject H_0

f. .9633   Do not reject H_0

g. What assumptions are necessary to ensure the validity of this test?

***7.94** Two independent random samples produced the measurements listed in the table. Do the data provide sufficient evidence to conclude that there is a difference between the locations of the probability distributions for the sampled populations? Test using $\alpha = .05$.

⊘ LM7_94

Sample 1		Sample 2	
1.2	1.0	1.5	1.9
1.9	1.8	1.3	2.7
.7	1.1	2.9	3.5
2.5			

***7.95** A random sample of nine pairs of observations are recorded on two variables, x and y. The data are shown in the table. Do the data provide sufficient evidence to indicate that the probability distribution for x is shifted to the right of that for y? Test using $\alpha = .05$. $T_- = 1.5$

⊘ LM7_95

Pair	x	y	Pair	x	y
1	19	12	6	29	10
2	27	19	7	16	16
3	15	7	8	22	10
4	35	25	9	16	18
5	13	11			

Applying the Concepts—Basic

⊘ OILSPILL

7.96 Refer to the *Marine Technology* (Jan. 1995) study of major oil spills from tankers and carriers, Exercise 2.15 (p. 29). The data for the 50 spills are saved in the OILSPILL file.

a. Construct a 90% confidence interval for the difference between the mean spillage amount of accidents caused by collision and the mean spillage amount of accidents caused by fire/explosion. Interpret the result.

b. Conduct a test of hypothesis to compare the mean spillage amount of accidents caused by grounding to the corresponding mean of accidents caused by hull failure. Use $\alpha = .05$. $t = -.279$

c. Refer to parts **a** and **b**. State any assumptions required for the inferences derived from the analyses to be valid. Are these assumptions reasonably satisfied?

7.97 Nontraditional university students are generally defined as those at least 25 years old. A study reported in the *College Student Journal* (Dec. 1992) compared traditional and nontraditional students on a number of factors, including grade point average (GPA). The table summarizes the information from the sample.

	GPA	
	Traditional Students	**Nontraditional Students**
n	94	73
$\bar{x}$	2.90	3.50
s	.50	.50

a. What are the appropriate null and alternative hypotheses if we want to test whether the mean GPAs of traditional and nontraditional students differ?

b. Conduct the test using $\alpha = .01$, and interpret the result. $z = -7.69$

c. What assumptions are necessary to ensure the validity of the test?

7.98 Dr. Philip Lieberman, a neuroscientist at Brown University, recently conducted a field experiment to gauge the effect of high altitude on a person's ability to think critically (*New York Times*, Aug. 23, 1995). The subjects of the experiment were five male members of an American expedition climbing Mount Everest. At the base camp, Lieberman read sentences to the climbers while they looked at simple pictures in a book. The length of time (in seconds) it took for each climber to match the picture with a sentence was recorded. Using a radio, Lieberman repeated the task when the climbers reached a camp 5 miles above sea level. At this altitude, he noted that the climbers took 50% longer to complete the task.

a. What is the variable measured for this experiment?

b. What are the experimental units? Climbers

c. Discuss how the data should be analyzed.

*7.99 The *Journal of Testing and Evaluation* (July 1992) published an investigation of the mean compression strength of corrugated fiberboard shipping containers. Comparisons were made for boxes of five different sizes: A, B, C, D, and E. Twenty identical boxes of each size were tested and the peak compression strength (pounds) was recorded for each box. The figure shows the sample means for the five box types as well as the variation around each sample mean.

a. Refer to box types B and D. Based on the graph, does it appear that the mean compressive strengths of these two box types are significantly different? Explain.

b. Based on the graph, does it appear that the mean compressive strengths of all five box types are significantly different? Explain.

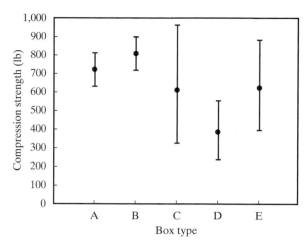

Source: Singh, S. P., *et al.* "Compression of single-wall corrugated shipping containers using fixed and floating test platens." *Journal of Testing and Evaluation*, Vol. 20, No. 4, July 1992, p. 319 (Figure 3). Copyright American Society for Testing and Materials.

7.100 Some power plants are located near rivers or oceans so that the available water can be used for cooling the condensers. Suppose that, as part of an environmental impact study, a power company wants to estimate the difference in mean water temperature between the discharge of its plant and the offshore waters. How many sample measurements must be taken at each site in order to estimate the true difference between means to within .2°C with 95% confidence? Assume that the range in readings will be about 4°C at each site and the same number of readings will be taken at each site.

7.101 A plant that purifies its liquid waste discharges the water into a local river. An EPA inspector has collected six water specimens of the discharge of the plant and six water specimens in the river upstream from the plant. The bacteria counts for each of the 12 specimens are reported in the accompanying table.

⬤ BACTERIA

Plant Discharge			Upstream		
30	36	33	29	30	26
28	29	34	27	31	32

a. Why might the bacteria counts shown tend to be approximately normally distributed?

b. What are the appropriate null and alternative hypotheses to test whether the mean bacteria count for the plant discharge exceeds that for the upstream location? Be sure to define any symbols you use.

c. The data are analyzed using SPSS. Carefully interpret the SPSS output shown on p. 392. $p = .074$

d. What assumptions are necessary to ensure the validity of this test?

*e. Reanalyze the data using the appropriate nonparametric test.

`SPSS Output for Exercise 7.101`

	LOCATION	N	Mean	Std. Deviation	Std. Error Mean
BACTERIA	PLANT	6	31.67	3.14	1.28
	UPSTREAM	6	29.17	2.32	.95

Group Statistics

Independent Samples Test

		Levene's Test for Equality of Variances		t-test for Equality of Means					95% Confidence Interval of the Difference	
		F	Sig.	t	df	Sig. (2-tailed)	Mean Difference	Std. Error Difference	Lower	Upper
BACTERIA	Equal variances assumed	1.563	.240	1.569	10	.148	2.50	1.59	-1.05	6.05
	Equal variances not assumed			1.569	9.197	.150	2.50	1.59	-1.09	6.09

Applying the Concepts—Intermediate

7.102 A pupillometer is a device used to observe changes in an individual's pupil dilations as he or she is exposed to different visual stimuli. The Design and Market Research Laboratories of the Container Corporation of America used a pupillometer to evaluate consumer reaction to different silverware patterns for one of its clients. Suppose 15 consumers were chosen at random, and each was shown two different silverware patterns. The pupillometer readings (in millimeters) for each consumer are shown below.

⊙ PUPILL

Consumer	Pattern 1	Pattern 2
1	1.00	.80
2	.97	.66
3	1.45	1.22
4	1.21	1.00
5	.77	.81
6	1.32	1.11
7	1.81	1.30
8	.91	.32
9	.98	.91
10	1.46	1.10
11	1.85	1.60
12	.33	.21
13	1.77	1.50
14	.85	.65
15	.15	.05

a. Which type of experiment does this represent—independent samples or paired difference? Explain.

b. Use a 90% confidence interval to estimate the difference in mean pupil dilation per consumer for silverware patterns 1 and 2. Interpret the confidence interval, assuming that the pupillometer indeed measures consumer interest. $.24 \pm .073$

c. Test the hypothesis that the mean dilation differs for the two patterns. Use $\alpha = .10$. Does the test conclusion support your interpretation of the confidence interval in part **b**? $t = 5.76$

d. What assumptions are necessary to ensure the validity of the inferences in parts **b** and **c**?

7.103 A traditional approach to teaching basic nursing skills was compared with an innovative approach in the *Journal of Nursing Education* (Jan. 1992). The innovative approach utilizes Vee heuristics and concept maps to link theoretical concepts with practical skills. Forty-two students enrolled in an upper-division nursing course participated in the study. Half (21) were randomly assigned to labs that utilized the innovative approach. After completing the course, all students were given short-answer questions about scientific principles underlying each of 10 nursing skills. The objective of the research is to compare the mean scores of the two groups of students.

a. What is the appropriate test to use to compare the two groups?

b. Are any assumptions required for the test?

c. One question dealt with the use of clean/sterile gloves. The mean scores for this question were 3.28 (traditional) and 3.40 (innovative). Is there sufficient information to perform the test? No

d. Refer to part **c**. The *p*-value for the test was reported as $p = .79$. Interpret this result.

e. Another question concerned the choice of a stethoscope. The mean scores of the two groups were 2.55 (traditional) and 3.60 (innovative) with an associated *p*-value of .02. Interpret these results.

7.104 Can a person control certain body functions if that person is trained in a program of *biofeedback* exercises? An experiment is conducted to show that blood pressure levels can be consciously reduced in people trained in this program. The blood pressure measurements (in millimeters of mercury) listed in the table represent readings before and after the biofeedback training of six subjects.

⊘ BIOFEED

Subject	Before	After
1	136.9	130.2
2	201.4	180.7
3	166.8	149.6
4	150.0	153.2
5	173.2	162.6
6	169.3	160.1

a. If we want to test whether the mean blood pressure decreases after the training, what are the appropriate null and alternative hypotheses? Define any symbols you use. $H_0: \mu_D = 0; H_a: \mu_D > 0$

b. Conduct the test of part **a** at $\alpha = .05$. Interpret these results. $t = 2.98$

c. Find a 95% confidence interval for the parameter of interest. Interpret this interval.

***d.** Reanalyze the data using an appropriate nonparametric test.

***7.105** The Minnesota Multiphasic Personality Inventory (MMPI) is a questionnaire used to gauge personality type. *Psychological Assessment* (Mar. 1995) published a study that investigated the effectiveness of the MMPI scales in detecting deliberately distorted responses. A independent sampling design with four treatments was employed. The treatments consisted of independent random samples of females in the following four groups: nonforensic psychiatric patients ($n_1 = 65$), forensic psychiatric patients ($n_2 = 28$), college students who were requested to respond honestly ($n_3 = 140$), and college students who were instructed to provide "fake bad" responses ($n_4 = 45$). All 278 participants were given the MMPI and the scores on several scales designed to assess response distortion were recorded for each. Each scale was treated as a response variable, and an analysis of variance was conducted. The ANOVA *F*-values are reported in the following table.

Response (Scale) Variable	ANOVA *F*-Value
Infrequency	155.8
Obvious	49.7
Subtle	10.3
Obvious-subtle	45.4
Dissimulation	39.1

For each response variable, determine whether the mean scores of the four groups completing the MMPI differ significantly. Use $\alpha = .5$ for each test.

7.106 Excess postexercise oxygen consumption (EPOC) describes energy expended during the body's recovery period immediately following aerobic exercise. *The Journal of Sports Medicine and Physical Fitness* (Dec. 1994) published a study designed to investigate the effect of fitness level on the magnitude and duration of EPOC. Ten healthy young adult males volunteered for the study. Five of these were endurance-trained and comprised the fit group; the other five were not engaged in any systematic training and comprised the sedentary group. Each volunteer engaged in a weight-supported exercise on a cycle ergometer until 300 kilocalories were expended. The magnitude (in kilocalories) and duration (in minutes) of the EPOC of each exerciser were measured. The study results are summarized in the table below.

Variable		Fit ($n = 5$)	Sedentary ($n = 5$)	*p*-value
Magnitude (kcal)	Mean	12.2	12.2	.998
	Std. dev.	3.1	4.3	
Duration (min)	Mean	16.6	20.4	.344
	Std. dev.	3.1	7.8	

Source: Sedlock, D. A. "Fitness levels and postexercise energy expenditure." *The Journal of Sports Medicine and Physical Fitness*, Vol. 34, No. 4, Dec. 1994, p. 339 (Table III).

a. Conduct a test of hypothesis to determine whether the true mean magnitude of EPOC differs for fit and sedentary young adult males. Use $\alpha = .10$.

b. The *p*-value for the test, part **a**, is given in the table. Interpret this value. $p = .998$

c. Conduct a test of hypothesis to determine whether the true mean duration of EPOC differs for fit and sedentary young adult males. Use $\alpha = .10$.

d. The *p*-value for the test, part **c**, is given in the table. Interpret this value.

7.107 Students enrolled in music classes at the University of Texas (Austin) participated in a study to compare the observations and teacher evaluations of music education majors and non-music majors (*Journal of Research in Music Education*, Winter 1991). Independent random samples of 100 music majors and 100 nonmajors rated the overall performance of their teacher using a 6-point scale, where 1 = lowest rating and 6 = highest rating. Use the information in the table to compare the mean teacher ratings of the two groups of music students with a 95% confidence interval. Interpret the result.

	Music Majors	Non-Music Majors
Sample size	100	100
Mean "overall" rating	4.26	4.59
Standard deviation	.81	.78

Source: Duke, R. A., and Blackman, M. D. "The relationship between observers' recorded teacher behavior and evaluation of music instruction." *Journal of Research in Music Education*, Vol. 39, No. 4, Winter 1991 (Table 2).

7.108 For each of the following studies, give the parameter of interest and state any assumptions that are necessary for the inferences to be invalid.

a. To investigate a possible link between jet lag and memory impairment, a University of Bristol (England) neurologist recruited 20 female flight attendants who worked flights across several time zones. Half of the attendants had only a short recovery time between flights and half had a long recovery time between flights. The average size of the right temporal lobe of the brain for the short-recovery group was significantly smaller than the average size of the right temporal lobe of the brain for the long-recovery group (*Tampa Tribune*, May 23, 2001).

b. A University of Florida animal sciences professor has discovered that feeding chickens corn oil causes them to produce larger eggs (*UF News*, April 11, 2001). The weight of eggs produced by each of a sample of chickens on a regular feed diet was recorded. Then, the same chickens were fed a diet supplemented by corn oil, and the weight of eggs produced by each was recorded. The mean weight of the eggs produced with corn oil was 3 grams heavier than the mean weight produced with the regular diet.

*__7.109__ A major razor blade manufacturer advertises that its twin-blade disposable razor will "get you more shaves" than any single-blade disposable razor on the market. A rival blade company that has been very successful in selling single-blade razors wishes to test this claim. Marketing managers in this company randomly sampled eight single-blade shavers and eight twin-blade shavers, and counted the number of shaves that each got before a change of blades was indicated. The results are shown in the table.

RAZORS

Twin Blades		Single Blade	
8	15	10	13
17	10	6	14
9	6	3	5
11	12	7	7

a. Do the data support the twin-blade manufacturer's claim? Use a nonparametric test at $\alpha = .05$. $T_1 = 82$

b. Do you think this experiment was designed in the best possible way? If not, what design might have been better? No, paired difference design

c. What assumptions are necessary for the test performed in part **a** to be valid? Do the assumptions seem reasonable for this application?

7.110 Marine biochemists at the University of Tokyo studied the properties of crustacean striated muscles (*The Journal of Experimental Zoology*, Sept. 1993). The main purpose of the experiment was to compare the biochemical properties of fast and slow muscles of crayfish. Using crayfish obtained from a local supplier, the researchers excised twelve fast-muscle fiber bundles and tested each fiber bundle for uptake of calcium. Twelve slow-muscle fiber bundles were excised from a second sample of crayfish, and calcium uptake was measured. The results of the experiment are summarized here. (All calcium measurements are in moles per milligram.) Analyze the data using a 95% confidence interval. Make an inference about the difference between the calcium uptake means of fast and slow muscles. $.20 \pm .066$

Fast Muscle	Slow Muscle
$n_1 = 12$	$n_2 = 12$
$\bar{x}_1 = .57$	$\bar{x}_2 = .37$
$s_1 = .104$	$s_2 = .035$

Source: Ushio, H., and Watabe, S. "Ultrastructural and biochemical analysis of the sarcoplasmic reticulum from crayfish fast and slow striated muscles." *The Journal of Experimental Zoology*, Vol. 267, Sept. 1993, p. 16 (Table 1). Copyright © 1993 John Wiley & Sons, Inc. Reprinted with permission.

Applying the Concepts—Advanced

7.111 Researchers at Rochester Institute of Technology investigated the use of isolation timeout as a behavioral management technique (*Exceptional Children*, Feb. 1995). Subjects for the study were 155 emotionally disturbed students enrolled in a special education facility. The students were randomly assigned to one of two types of classrooms—Option II classrooms (one teacher, one paraprofessional, and a maximum of 12 students) and Option III classrooms (one teacher, one paraprofessional, and a maximum of 6 students). Over the academic year the number of behavioral incidents resulting in an isolation timeout was recorded for each student. Summary statistics for the two groups of students are shown in the following table.

	Option II	Option III
Number of students	100	55
Mean number of timeout incidents	78.67	102.87
Standard deviation	59.08	69.33

Source: Costenbader, V., and Reading-Brown, M. "Isolation timeout used with students with emotional disturbance." *Exceptional Children*, Vol. 61, No.4, Feb. 1995, p. 359 (Table 3).

Do you agree with the following statement: "On average, students in Option III classrooms had significantly more timeout incidents than students in Option II classrooms?"

*__7.112__ A research psychologist wishes to investigate the difference in maze test scores for a strain of laboratory mice trained under different laboratory conditions. The experiment is conducted using 18 randomly selected mice of this strain, with six receiving no training at all (control group), six trained under condition 1, and six trained under condition 2. Then each of the mice is given a test score between 0 and 100, depending on its performance in a test maze. The experiment produced the results shown in the following table. Analyze the data and interpret the results.

MICE

Control	Condition 1	Condition 2
58	73	53
32	70	74
59	68	72
64	71	62
55	60	58
49	62	61

STUDENT PROJECTS

We have now discussed two methods of collecting data to compare two population means. In many experimental situations a decision must be made either to collect two independent samples or to conduct a paired difference experiment. The importance of this decision cannot be overemphasized, since the amount of information obtained and the cost of the experiment are both directly related to the method of experimentation that is chosen.

Choose two populations (pertinent to your major area) which have unknown means and for which you could both collect two independent samples and collect paired observations. Before conducting the experiment, state which method of sampling you think will provide more information (and why). Compare the two methods, first performing the independent sampling procedure by collecting 10 observations from each population (a total of 20 measurements), then performing the paired difference experiment by collecting 10 pairs of observations.

Construct two 95% confidence intervals, one for each experiment you conduct. Which method provides the narrower confidence interval and hence more information on this performance of the experiment? Does this result agree with your preliminary expectations?

REFERENCES

Agresti, A., and Agresti, B. F. *Statistical Methods for the Social Sciences*, 2nd ed. San Francisco: Dellen, 1986.

Conover, W. J. *Practical Nonparametric Statistics*, 2nd ed. New York: Wiley, 1980.

Daniel, W. W. *Applied Nonparametric Statistics*, 2nd ed. Boston: PWS-Kent, 1990.

Dunn, O. J. "Multiple comparisons using rank sums." *Technometrics*, Vol. 6, 1964.

Gibbons, J. D. *Nonparametric Statistical Inference*, 2nd ed. New York: McGraw-Hill, 1985.

Hollander, M., and Wolfe, D. A. *Nonparametric Statistical Methods*. New York: Wiley, 1973.

Kruskal, W. H., and Wallis, W. A. "Use of ranks in one-criterion variance analysis." *Journal of the American Statistical Association*, Vol. 47, 1952.

Lehmann, E. L. *Nonparametrics: Statistical Methods Based on Ranks*. San Francisco: Holden-Day, 1975.

Marascuilo, L. A., and McSweeney, M. *Nonparametric and Distribution-Free Methods for the Social Sciences*. Monterey, Ca.: Brooks/Cole, 1977.

Mendenhall, W. *Introduction to Linear Models and the Design and Analysis of Experiments*. Belmont, Calif.: Wadsworth, 1968.

Mendenhall, W., Beaver, R. J., and Beaver, B. M. *Introduction to Probability and Statistics*, 10th ed. North Scituate, Mass.: Duxbury, 1999.

Miller, R. G. *Simultaneous Statistical Inference*. New York: Springer-Verlag, 1981.

Satterthwaite, F. W. "An approximate distribution of estimates of variance components." *Biometrics Bulletin*, Vol. 2, 1946, pp. 110–114.

Scheffé, H. *The Analysis of Variance*. New York: Wiley, 1959.

Snedecor, G. W., and Cochran, W. *Statistical Methods*, 7th ed. Ames: Iowa State University Press, 1980.

Steel, R. G. D., and Torrie, J. H. *Principles and Procedures of Statistics*, 2nd ed. New York: McGraw-Hill, 1980.

Tukey, J. W. "Comparing individual means in the analysis of variance." *Biometrics*, Vol. 5, 1949, pp. 99–114.

Wilcoxon, F., and Wilcox, R. A. "Some rapid approximate statistical procedures." The American Cyanamid Co., 1964.

Winer, B. J. *Statistical Principles in Experimental Design*, 2d ed. New York: McGraw-Hill, 1971.

Comparing Population Proportions

Contents

Statistics in Action

The Level of Agreement Among Movie Reviewers: Thumbs Up or Thumbs Down?

🖐 Where We've Been

Chapter 7 presented both parametric and nonparametric methods for comparing two or more population means.

☛ Where We're Going

In this chapter, we will consider a problem of comparable importance—comparing two or more population proportions. The need to compare population proportions arises because many business and social experiments involve questioning people and classifying their responses. We will learn how to determine whether the proportion of consumers favoring product A differs from the proportion favoring product B, and we will learn how to estimate the difference with a confidence interval.

Many experiments are conducted in the biological, physical, and social sciences to compare two or more proportions. Those conducted in business and the social sciences to sample the opinions of people are called **sample surveys**. For example, a state government might wish to estimate the difference between the proportions of people in two regions of the state who would qualify for a new welfare program. Or, after an innovative process change, an engineer might wish to determine whether the proportion of defective items produced by a manufacturing process was less than the proportion of defectives produced before the change. In Section 8.1 we show you how to test hypotheses about the difference between two population proportions based on independent random sampling. We will also show how to find a confidence interval for the difference. Then, in Section 8.3 we will compare more than two population proportions, and in Section 8.4 we will present a related problem.

8.1   COMPARING TWO POPULATION PROPORTIONS: INDEPENDENT SAMPLING

Suppose a presidential candidate wants to compare the preference of registered voters in the northeastern United States (NE) to those in the southeastern United States (SE). Such a comparison would help determine where to concentrate campaign efforts. The candidate hires a professional pollster to randomly choose 1,000 registered voters in the northeast and 1,000 in the southeast and interview each to learn her or his voting preference. The objective is to use this sample information to make an inference about the difference $(p_1 - p_2)$ between the proportion p_1 of *all* registered voters in the northeast and the proportion p_2 of *all* registered voters in the southeast who plan to vote for the presidential candidate.

The two samples represent independent binomial experiments. (See Section 4.3 for the characteristics of binomial experiments.) The binomial random variables are the numbers x_1 and x_2 of the 1,000 sampled voters in each area who indicate they will vote for the candidate. The results are summarized below.

NE	SE
$n_1 = 1,000$	$n_2 = 1,000$
$x_1 = 546$	$x_2 = 475$

We can now calculate the sample proportions $\hat{p}_1$ and $\hat{p}_2$ of the voters in favor of the candidate in the northeast and southeast, respectively:

$$\hat{p} = \frac{x_1}{n_1} = \frac{546}{1,000} = .546 \quad \hat{p}_2 = \frac{x_2}{n_2} = \frac{475}{1,000} = .475$$

The difference between the sample proportions $(\hat{p}_1 - \hat{p}_2)$ makes an intuitively appealing point estimator of the difference between the population $(p_1 - p_2)$. For our example, the estimate is

$$(\hat{p}_1 - \hat{p}_2) = .546 - .475 = .071$$

To judge the reliability of the estimator $(\hat{p}_1 - \hat{p}_2)$, we must observe its performance in repeated sampling from the two populations. That is, we need to know the sampling distribution of $(\hat{p}_1 - \hat{p}_2)$. The properties of the sampling distribution are given in the following box. Remember that $\hat{p}_1$ and $\hat{p}_2$ can be viewed as means of the number of successes per trial in the respective samples, so the Central Limit Theorem applies when the sample sizes are large.

Properties of the Sampling Distribution of $(\hat{p}_1 - \hat{p}_2)$

1. The mean of the sampling distribution of $(\hat{p}_1 - \hat{p}_2)$ is $(p_1 - p_2)$ that is,

$$E(\hat{p}_1 - \hat{p}_2) = p_1 - p_2$$

   Thus, $(\hat{p}_1 - \hat{p}_2)$ is an unbiased estimator of $(p_1 - p_2)$.

2. The standard deviation of the sampling distribution of $(\hat{p}_1 - \hat{p}_2)$ is

$$\sigma_{(\hat{p}_1-\hat{p}_2)} = \sqrt{\frac{p_1 q_1}{n_1} + \frac{p_2 q_2}{n_2}}$$

3. If the sample sizes n_1 and n_2 are large (see Section 5.3 for a guideline), the sampling distribution of $(\hat{p}_1 - \hat{p}_2)$ is approximately normal.

Since the distribution of $(\hat{p}_1 - \hat{p}_2)$ in repeated sampling is approximately normal, we can use the z statistic to derive confidence intervals for $(p_1 - p_2)$ or to test a hypothesis about $(p_1 - p_2)$.

For the voter example, a 95% confidence interval for the difference $(p_1 - p_2)$ is

$$(\hat{p}_1 - \hat{p}_2) \pm 1.96 \sigma_{(\hat{p}_1-\hat{p}_2)} \quad \text{or} \quad (\hat{p}_1 - \hat{p}_2) \pm 1.96\sqrt{\frac{p_1 q_1}{n_1} + \frac{p_2 q_2}{n_2}}$$

The quantities $p_1 q_1$ and $p_2 q_2$ must be estimated in order to complete the calculation of the standard deviation, $\sigma_{(\hat{p}_1-\hat{p}_2)}$, and hence the calculation of the confidence interval. In Section 5.3 we showed that the value of pq is relatively insensitive to the value chosen to approximate p. Therefore, $\hat{p}_1 \hat{q}_1$ and $\hat{p}_2 \hat{q}_2$ will provide satisfactory estimates to approximate $p_1 q_1$ and $p_2 q_2$, respectively. Then

$$\sqrt{\frac{p_1 q_1}{n_1} + \frac{p_2 q_2}{n_2}} \approx \sqrt{\frac{\hat{p}_1 \hat{q}_1}{n_1} + \frac{\hat{p}_2 \hat{q}_2}{n_2}}$$

and we will approximate the 95% confidence interval by

$$(\hat{p}_1 - \hat{p}_2) \pm 1.96\sqrt{\frac{\hat{p}_1 \hat{q}_1}{n_1} + \frac{\hat{p}_2 \hat{q}_2}{n_2}}$$

Substituting the sample quantities yields

$$(.546 - .475) \pm 1.96\sqrt{\frac{(.546)(.454)}{1,000} + \frac{(.475)(.525)}{1,000}}$$

or $.071 \pm .044$. Thus, we are 95% confident that the interval from .027 to .115 contains $(p_1 - p_2)$.

We infer that there are between 2.7% and 11.5% more registered voters in the northeast than in the southeast who plan to vote for the presidential candidate. It seems that the candidate should direct a greater campaign effort in the southeast compared to the northeast.

The general form of a confidence interval for the difference $(p_1 - p_2)$ between population proportions is given in the following box.

TEACHING TIP
Point out that the *t*-distribution is not used with proportions because the underlying assumption of a normal distribution will never be true.

Large-Sample $100(1 - \alpha)\%$ Confidence Interval for $(p_1 - p_2)$

$$(\hat{p}_1 - \hat{p}_2) \pm z_{\alpha/2}\sigma_{(\hat{p}_1-\hat{p}_2)} = (\hat{p}_1 - \hat{p}_2) \pm z_{\alpha/2}\sqrt{\frac{p_1 q_1}{n_1} + \frac{p_2 q_2}{n_2}}$$

$$\approx (\hat{p}_1 - \hat{p}_2) \pm z_{\alpha/2}\sqrt{\frac{\hat{p}_1\hat{q}_1}{n_1} + \frac{\hat{p}_2\hat{q}_2}{n_2}}$$

Assumption: The two samples are independent random samples. Both samples should be large enough that the normal distribution provides an adequate approximation to the sampling distribution of $\hat{p}_1$ and $\hat{p}_2$ (see Section 6.5).

The *z* statistic,

$$z = \frac{(\hat{p}_1 - \hat{p}_2) - (p_1 - p_2)}{\sigma_{(\hat{p}_1-\hat{p}_2)}}$$

is used to test the null hypothesis that $(p_1 - p_2)$ equals some specified difference, say D_0. For the special case where $D_0 = 0$, that is, where we want to test the null hypothesis $H_0: (p_1 - p_2) = 0$ (or, equivalently, $H_0: p_1 = p_2$), the best estimate of $p_1 = p_2 = p$ is obtained by dividing the total number of successes $(x_1 + x_2)$ for the two samples by the total number of observations $(n_1 + n_2)$; that is,

$$\hat{p} = \frac{x_1 + x_2}{n_1 + n_2} \quad \text{or} \quad \hat{p} = \frac{n_1\hat{p}_1 + n_2\hat{p}_2}{n_1 + n_2}$$

The second equation shows that $\hat{p}$ is a weighted average of $\hat{p}_1$ and $\hat{p}_2$, with the larger sample receiving more weight. If the sample sizes are equal, then $\hat{p}$ is a simple average of the two sample proportions of successes.

We now substitute the weighted average $\hat{p}$ for both p_1 and p_2 in the formula for the standard deviation of $(\hat{p}_1 - \hat{p}_2)$:

$$\sigma_{(\hat{p}_1-\hat{p}_2)} = \sqrt{\frac{p_1 q_1}{n_1} + \frac{p_2 q_2}{n_2}} \approx \sqrt{\frac{\hat{p}\hat{q}}{n_1} + \frac{\hat{p}\hat{q}}{n_2}} = \sqrt{\hat{p}\hat{q}\left(\frac{1}{n_1} + \frac{1}{n_2}\right)}$$

The test is summarized in the next box.

TEACHING TIP
No matched pairs experiment with proportions exists for this type of analysis.

Large-Sample Test of Hypothesis About $(p_1 - p_2)$

ONE-TAILED TEST

$H_0: (p_1 - p_2) = 0$*

$H_a: (p_1 - p_2) < 0$
[or $H_a: (p_1 - p_2) > 0$]

TWO-TAILED TEST

$H_0: (p_1 - p_2) = 0$

$H_a: (p_1 - p_2) \neq 0$

Test statistic:
$$z = \frac{(\hat{p}_1 - \hat{p}_2)}{\sigma_{(\hat{p}_1-\hat{p}_2)}}$$

Rejection region: $z < -z_\alpha$
[or $z > z_\alpha$ when $H_a: (p_1 - p_2) > 0$]

Rejection region: $|z| > z_{\alpha/2}$

Note: $\sigma_{(\hat{p}_1-\hat{p}_2)} = \sqrt{\frac{p_1 q_1}{n_1} + \frac{p_2 q_2}{n_2}} \approx \sqrt{\hat{p}\hat{q}\left(\frac{1}{n_1} + \frac{1}{n_2}\right)}$ where $\hat{p} = \frac{x_1 + x_2}{n_1 + n_2}$

Assumption: Same as for large-sample confidence interval for $(p_1 - p_2)$. (See previous box.)

---

*The test can be adapted to test for a difference $D_0 \neq 0$. Because most applications call for a comparison of p_1 and p_2, implying $D_0 = 0$, we will confine our attention to this case.

EXAMPLE 8.1

In the past decade intensive antismoking campaigns have been sponsored by both federal and private agencies. Suppose the American Cancer Society randomly sampled 1,500 adults in 1992 and then sampled 1,750 adults in 2002 to determine whether there was evidence that the percentage of smokers had decreased. The results of the two sample surveys are shown in the table, where x_1 and x_2 represent the numbers of smokers in the 1992 and 2002 samples, respectively. Do these data indicate that the fraction of smokers decreased over this 10-year period? Use $\alpha = .05$.

$z = 2.37$; Reject H_0

1992	2002
$n_1 = 1{,}500$	$n_2 = 1{,}750$
$x_1 = 555$	$x_2 = 578$

Solution

If we define p_1 and p_2 as the true proportions of adult smokers in 1992 and 2002, the elements of our test are

$$H_0 : (p_1 - p_2) = 0$$
$$H_a : (p_1 - p_2) > 0$$

(The test is one-tailed since we are interested only in determining whether the proportion of smokers *decreased*.)

$$\text{Test statistic: } z = \frac{(\hat{p}_1 - \hat{p}_2) - 0}{\sigma_{(\hat{p}_1 - \hat{p}_2)}}$$

Rejection region using $\alpha = .05$:
$$z > z_\alpha = z_{.05} = 1.645 \qquad \text{(see Figure 8.1)}$$

FIGURE 8.1

Rejection Region for Example 8.1

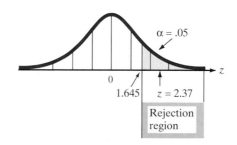

Suggested Exercise 8.12

We now calculate the sample proportions of smokers

$$\hat{p}_1 = \frac{555}{1{,}500} = .37 \qquad \hat{p}_2 = \frac{578}{1{,}750} = .33$$

Then

$$z = \frac{(\hat{p}_1 - \hat{p}_2) - 0}{\sigma_{(\hat{p}_1 - \hat{p}_2)}} \approx \frac{(\hat{p}_1 - \hat{p}_2)}{\sqrt{\hat{p}\hat{q}\left(\dfrac{1}{n_1} + \dfrac{1}{n_2}\right)}}$$

where

$$\hat{p} = \frac{x_1 + x_2}{n_1 + n_2} = \frac{555 + 578}{1,500 + 1,750} = .349$$

Note that $\hat{p}$ is a weighted average of $\hat{p}_1$ and $\hat{p}_2$, with more weight given to the larger (2002) sample.

Thus, the computed value of the test statistic is

$$z = \frac{.37 - .33}{\sqrt{(.349)(.651)\left(\frac{1}{1,500} + \frac{1}{1,750}\right)}} = \frac{.040}{.0168} = 2.37$$

There is sufficient evidence at the $\alpha = .05$ level to conclude that the proportion of adults who smoke has decreased over the 1992–2002 period. We could place a confidence interval on $(p_1 - p_2)$ if we were interested in estimating the extent of the decrease. ∎

EXAMPLE 8.2

$p = .009$

Use a statistical software package to conduct the test in Example 8.1.

Solution

We entered the sample sizes (n_1 and n_2) and numbers of successes (x_1 and x_2) into MINITAB and obtained the printout shown in Figure 8.2. The test statistic for this one-tailed test, $z = 2.37$, is shaded on the printout, as well as the p-value of the test. Note that p-value = .009 is smaller than $\alpha = .05$. Consequently, we have strong evidence to reject H_0 and conclude that p_1 exceeds p_2. ∎

FIGURE 8.2

MINITAB Output for Test of Two Proportions

```
Test and CI for Two Proportions

Sample      X      N   Sample p
1         555   1500   0.370000
2         578   1750   0.330286


Estimate for p(1) - p(2):  0.0397143
95% lower bound for p(1) - p(2): 0.0121024
Test for p(1) - p(2) = 0 (vs > 0): Z = 2.37   P-Value = 0.009
```

 EXERCISES 8.1–8.18

Learning the Mechanics

8.1 Explain why the Central Limit Theorem is important in finding an approximate distribution for $(\hat{p}_1 - \hat{p}_2)$.

8.2 In each case, determine whether the sample sizes are large enough to conclude that the sampling distribution of $(\hat{p}_1 - \hat{p}_2)$ is approximately normal.
  a. $n_1 = 10, n_2 = 12, \hat{p}_1 = .50, \hat{p}_2 = .50$ Yes
  b. $n_1 = 10, n_2 = 12, \hat{p}_1 = .10, \hat{p}_2 = .08$ No
  c. $n_1 = n_2 = 30, \hat{p}_1 = .20, \hat{p}_2 = .30$ No
  d. $n_1 = 100, n_2 = 200, \hat{p}_1 = .05, \hat{p}_2 = .09$ No
  e. $n_1 = 100, n_2 = 200, \hat{p}_1 = .95, \hat{p}_2 = .91$ No

8.3 For each of the following values of α, find the values of z for which $H_0: (p_1 - p_2) = 0$ would be rejected in favor of $H_a: (p_1 - p_2) < 0$.
  a. $\alpha = .01$ $z < -2.33$
  b. $\alpha = .025$ $z < -1.96$
  c. $\alpha = .05$ $z < -1.645$
  d. $\alpha = .10$ $z < -1.28$

8.4 Independent random samples, each containing 800 observations, were selected from two binomial populations. The samples from populations 1 and 2 produced 320 and 400 successes, respectively.

a. Test $H_0: (p_1 - p_2) = 0$ against $H_a: (p_1 - p_2) \neq 0$. Use $\alpha = .05$. $z = -4.02$

b. Test $H_0: (p_1 - p_2) = 0$ against $H_a: (p_1 - p_2) \neq 0$. Use $\alpha = .01$. $z = -4.02$

c. Test $H_0: (p_1 - p_2) = 0$ against $H_a: (p_1 - p_2) < 0$. Use $\alpha = .01$. $z = -4.02$

d. Form a 90% confidence interval for $(p_1 - p_2)$.

8.5 Construct a 95% confidence interval for $(p_1 - p_2)$ in each of the following situations:

a. $n_1 = 400$, $\hat{p}_1 = .65$; $n_2 = 400$, $\hat{p}_2 = .58$
b. $n_1 = 180$, $\hat{p}_1 = .31$; $n_2 = 250$, $\hat{p}_2 = .25$
c. $n_1 = 100$, $\hat{p}_1 = .46$; $n_2 = 120$, $\hat{p}_2 = .61$

8.6 Sketch the sampling distribution of $(\hat{p}_1 - \hat{p}_2)$ based on independent random samples of $n_1 = 100$ and $n_2 = 200$ observations from two binomial populations with success probabilities $p_1 = .1$ and $p_2 = .5$, respectively.

8.7 Random samples of size $n_1 = 50$ and $n_2 = 60$ were drawn from populations 1 and 2, respectively. The samples yielded $\hat{p}_1 = .4$ and $\hat{p}_2 = .2$. Test $H_0: (p_1 - p_2) = .1$ against $H_a: (p_1 - p_2) > .1$ using $\alpha = .05$. $z = 1.16$

Applying the Concepts—Basic

8.8 Refer to the *Journal of Sport Behavior* (2001) study of variety in exercise workouts, Exercise 5.45 (p. 264). One group of 38 people varied their exercise routine in workouts while a second group of 38 exercisers had no set schedule or regulations for their workouts. By the end of the study, 14 people had dropped out of the first exercise group and 23 had dropped out of the second group.

a. Find the dropout rates (i.e., the percentage of exercisers who had dropped out of the exercise group) for each of the two groups of exercisers. .368; .605

b. Find a 90% confidence interval for the difference between the dropout rates of the two groups of exercisers. $-.237 \pm .183$

c. Give a practical interpretation of the confidence interval, part **c**.

8.9 The *Journal of the American Medical Association* (April 18, 2001) published a study of the effectiveness of using extracts of the herbal medicine, St. John's wort, in treating major depression. In an 8-week randomized, controlled trial, 200 patients diagnosed with major depression were divided into two groups: one group ($n_1 = 98$) received St. John's wort extract while the other group ($n_2 = 102$) received a placebo (no drug). At the end of the study period, 14 of the St. John's wort patients were in remission compared to 5 of the placebo patients.

a. Compute the proportion of the St. John's wort patients who were in remission. .143

b. Compute the proportion of the placebo patients who were in remission. .049

c. If St. John's wort is effective in treating major depression, then the proportion of St. John's wort patients in remission will exceed the proportion of placebo patients in remission. At $\alpha = .01$, is St. John's wort effective in treating major depression?

d. Repeat part **c**, but use $\alpha = .10$. $z = 2.27$

e. Explain why the choice of α is critical for this study.

8.10 The *Journal of Fish Biology* (Aug. 1990) reported on a study to compare the incidence of parasites (tapeworms) in species of Mediterranean and Atlantic fish. In the Mediterranean Sea, 588 brill were captured and dissected, and 211 were found to be infected by the parasite. In the Atlantic Ocean, 123 brill were captured and dissected, and 26 were found to be infected. Compare the proportions of infected brill at the two capture sites using a 90% confidence interval. Interpret the interval.

Applying the Concepts—Intermediate

8.11 A University of South Florida biologist conducted an experiment to determine whether increased levels of carbon dioxide kill leaf-eating moths (*USF Magazine*, Winter 1999). Moth larvae were placed in open containers filled with oak leaves. Half the containers had normal carbon dioxide levels while the other half had double the normal level of carbon dioxide. Ten percent of the larvae in the containers with high carbon dioxide levels died, compared to 5 percent in the containers with normal levels. Assume that 80 moth larvae were placed, at random, in each of the two types of containers. Do the experimental results demonstrate that an increased level of carbon dioxide is effective in killing a higher percentage of leaf-eating moth larvae? Test using $\alpha = .01$. $z = 1.201$

8.12 When female undergraduates switch from science, mathematics, and engineering (SME) majors into disciplines that are not science-based, are their reasons different from those of their male counterparts? This question was investigated in *Science Education* (July 1995). A sample of 335 junior/senior undergraduates—172 females and 163 males—at two large research universities were identified as "switchers," that is, they left a declared SME major for a non-SME major. Each student listed one or more factors that contributed to the switching decision.

a. Of the 172 females in the sample, 74 listed lack or loss of interest in SME (i.e., "turned off" by science) as a major factor, compared to 72 of the 163 males. Conduct a test (at $\alpha = .10$) to determine whether the proportion of female switchers who give "lack of interest in SME" as a major reason for switching differs from the corresponding proportion of males.

b. Thirty-three of the 172 females in the sample indicated that they were discouraged or lost confidence because of low grades in SME during their early years, compared to 44 of 163 males. Construct a 90% confidence interval for the difference between the proportions of female and male switchers who lost confidence due to low grades in SME. Interpret the result.

8.13 Trying to think of a word you know, but can't instantly retrieve, is called the "tip of the tongue" phenomenon. *Psychology and Aging* (Sept. 2001) published a study of this phenomenon in senior citizens. The researchers compared 40 people between 60 and 72 years of age with 40

MINITAB Output for Exercise 8.13

```
Test and CI for Two Proportions

Sample      X      N    Sample p
1          31     40    0.775000
2          22     40    0.550000


Estimate for p(1) - p(2):  0.225
95% CI for p(1) - p(2): (0.0237157, 0.426284)
Test for p(1) - p(2) = 0 (vs not = 0): Z = 2.13   P-Value = 0.033
```

between 73 and 83 years of age. When primed with the initial syllable of a missing word (e.g., seeing the word "include" to help recall the word "incisor"), the younger seniors had a higher recall rate. Suppose 31 of the 40 seniors in the younger group could recall the word when primed with the initial syllable, while only 22 of the 40 seniors could recall the word. The recall rates of the two groups were compared, with the results shown in the accompanying MINITAB printout. Interpret these results. Does one group of elderly people have a significantly higher recall rate than the other?

8.14 Geneticists at Duke University Medical Center have identified the E2F1 transcription factor as an important component of cell proliferation control (*Nature,* Sept. 23, 1993). The researchers induced DNA synthesis in two batches of serum-starved cells. Each cell in one batch was microinjected with the E2F1 gene, while the cells in the second batch (the controls) were not exposed to E2F1. After 30 hours, the number of cells in each batch that exhibited altered growth was determined. The results of the experiment are summarized in the table.

	Control	**E2F1-Treated Cells**
Total number of cells	158	92
Number of growth-altered cells	15	41

Source: Johnson, D. G., *et al.* "Expression of transcription factor E2F1 induces quiescent cells to enter S phase." *Nature,* Vol. 365, No. 6444, Sept. 23, 1993, p. 351 (Table 1).

  a. Compare the percentages of cells exhibiting altered growth in the two batches with a 90% confidence interval. $-3.51 \pm .093$
  b. Use the interval, part **a**, to make an inference about the ability of the E2F1 transcription factor to induce cell growth.

8.15 Scientists have linked a catastrophic decline in the number of frogs inhabiting the world to ultraviolet radiation (*Tampa Tribune,* Mar. 1, 1994). The Pacific tree frog, however, is not believed to be in decline, possibly because it manufactures an enzyme that protects its eggs from ultraviolet radiation. Researchers at Oregon State University compared the hatching rates of two groups of Pacific tree frog eggs. One group of eggs was shielded with ultraviolet-blocking sun shades, while the second group was not. The number of eggs successfully hatched

in each group is provided in the following table. Compare the hatching rates of the two groups of Pacific tree frog eggs with a test of hypothesis. Use $\alpha = .01$.

	Sun-Shaded Eggs	**Unshaded Eggs**
Total number	70	80
Number hatched	34	31

8.16 Can stress management help heart patients reduce their risk of heart attacks or heart surgery? In a study published in the *Archives of Internal Medicine* (Oct. 27, 1997), 107 heart patients were randomly divided into three groups. One group participated in a stress management program, the second group underwent an exercise program, and the third group received usual heart care from their doctors. After three years, the number in each group suffering "cardiac events" (e.g., heart attack, bypass surgery, or angioplasty) was recorded. The results are shown in the next table.

	Management	**Exercise**	**Usual Care**
Number of heart patients	33	34	40
Number who suffer a "cardiac event"	3	7	12

  a. Use a 99% confidence interval to compare the proportion of stress management patients who suffer a cardiac event to the proportion of exercise patients who suffer a cardiac event. What inference can you make? $-.115 \pm .220$
  b. Repeat part **a,** but compare the stress management group to the usual care group. $-.209 \pm .227$
  c. Repeat part **a,** but compare the exercise group to the usual care group. $-.094 \pm .258$

Applying the Concepts—Advanced

8.17 Do you have an insatiable craving for chocolate or some other food? Since many people apparently do, psychologists are designing scientific studies to examine the phenomenon. According to the *New York Times* (Feb. 22, 1995), one of the largest studies of food cravings involved a survey of 1,000 McMaster University (Canada) students. The survey revealed that 97% of the women in the study acknowledged specific food cravings while only 67% of the men did. Assume that 600 of the respondents were women and 400 were men.

a. Is there sufficient evidence to claim that the true proportion of women who acknowledge having food cravings exceeds the corresponding proportion of men? Test using $\alpha = .01$. $z = 13.02$

b. Why is it dangerous to conclude from the study that women have a higher incidence of food cravings than men? $z = 3.55$

8.18 Can taking an antidepressant drug help cigarette smokers kick their habit? The *New England Journal of Medicine* (Oct. 23, 1997) published a study in which 615 smokers (all who wanted to give up smoking) were randomly assigned to receive either Zyban (an antidepressant) or a placebo (a dummy pill) for six weeks. Of the 309 patients who received Zyban, 71 were not smoking one year later. Of the 306 patients who received a placebo, 37 were not smoking one year later. Conduct a test of hypothesis (at $\alpha = .05$) to answer the research question posed above.

8.2   DETERMINING THE SAMPLE SIZE

The sample sizes n_1 and n_2 required to compare two population proportions can be found in a manner similar to the method described in Section 7.3 for comparing two population means. We will assume equal sized, i.e., $n_1 = n_2 = n$, and then choose n so that $(\hat{p}_1 - \hat{p}_2)$ will differ from $(p_1 - p_2)$ by no more than a bound B with a specified probability. We will illustrate the procedure with an example.

E X A M P L E   8 . 3

A production supervisor suspects that a difference exists between the proportions p_1 and p_2 of defective items produced by two different machines. Experience has shown that the proportion defective for each of the two machines is in the neighborhood of .03. If the supervisor wants to estimate the difference in the proportions to within .005 using a 95% confidence interval, how many items must be randomly sampled from the production of each machine? (Assume that the supervisor wants $n_1 = n_2 = n$.)

Solution

In this sampling problem, $B = .005$, and for the specified level of reliability, $z_{\alpha/2} = z_{.025} = 1.96$. Then, letting $p_1 = p_2 = .03$ and $n_1 = n_2 = n$, we find the required sample size per machine by solving the following equation for n:

$$z_{\alpha/2}\sigma_{(\hat{p}_1 - \hat{p}_2)} = B$$

or

$$z_{\alpha/2}\sqrt{\frac{p_1 q_1}{n_1} + \frac{p_2 q_2}{n_2}} = B$$

$$1.96\sqrt{\frac{(.03)(.97)}{n} + \frac{(.03)(.97)}{n}} = .005$$

$$1.96\sqrt{\frac{2(.03)(.97)}{n}} = .005$$

$$n = 8{,}943.2$$

You can see that this may be a tedious sampling procedure. If the supervisor insists on estimating $(p_1 - p_2)$ correct to within .005 with 95% confidence, approximately 9,000 items will have to be inspected for each machine. ∎

You can see from the calculations in Example 8.3 that $\sigma_{(\hat{p}_1 - \hat{p}_2)}$ (and hence the solution, $n_1 = n_2 = n$) depends on the actual (but unknown) values of p_1 and p_2. In fact, the required sample size $n_1 = n_2 = n$ is largest when $p_1 = p_2 = .5$. Therefore, if you have no prior information on the approximate values of p_1 and p_2, use $p_1 = p_2 = .5$ in the formula for $\sigma_{(\hat{p}_1 - \hat{p}_2)}$. If p_1 and p_2 are in fact close to .5, then the values of n_1 and n_2 that you have calculated will be correct. If p_1 and p_2 differ substantially from .5, then your solutions for n_1 and n_2 will be larger than needed. Consequently, using $p_1 = p_2 = .5$ when solving for n_1 and n_2 is a conservative procedure because the sample sizes n_1 and n_2 will be at least as large as (and probably larger than) needed.

The procedure for determining sample sizes necessary for estimating $(p_1 - p_2)$ for the case $n_1 = n_2$ is given in the following box.

Determination of Sample Size for Estimating $p_1 - p_2$

To estimate $(p_1 - p_2)$ to within a given bound B with probability $(1 - \alpha)$, use the following formula to solve for equal sample sizes that will achieve the desired reliability:

$$n_1 = n_2 = \frac{(z_{\alpha/2})^2(p_1q_1 + p_2q_2)}{B^2}$$

You will need to substitute estimates for the values of p_1 and p_2 before solving for the sample size. These estimates might be based on prior samples, obtained from educated guesses or, most conservatively, specified as $p_1 = p_2 = .5$.

 EXERCISES 8.19–8.24

Learning the Mechanics

8.19 Assuming that $n_1 = n_2$, find the sample sizes needed to estimate $(p_1 - p_2)$ for each of the following situations:
  a. Bound = .01 with 99% confidence. Assume that $p_1 \approx .4$ and $p_2 \approx .7$. $n_1 = n_2 = 29,954$
  b. A 90% confidence interval of width .05. Assume there is no prior information available to obtain approximate values of p_1 and p_2. $n_1 = n_2 = 2,165$
  c. Bound = .03 with 90% confidence. Assume that $p_1 \approx .2$ and $p_2 \approx .3$. $n_1 = n_2 = 1,113$

Applying the Concepts—Basic

8.20 A pollster wants to estimate the difference between the proportions of men and women who favor a particular national candidate using a 90% confidence interval of width .04. Suppose the pollster has no prior information about the proportions. If equal numbers of men and women are to be polled, how large should the sample sizes be?

8.21 Refer to the *Journal of Sport Behavior* (2001) study comparing the dropout rates of two groups of exercisers, Exercise 8.8 (p. 403). Suppose you want to reduce the bound on the error of estimating $(p_1 - p_2)$ with a 90% confidence interval to .1. Determine the number of exercisers to be sampled from each group in order to obtain such an estimate. Assume equal sample sizes, $p_1 \approx .4$, and $p_2 \approx .6$. $n_1 = n_2 = 130$

Applying the Concepts—Intermediate

8.22 All cable television companies carry at least one home shopping channel. Who uses these home shopping ser-vices? Are the shoppers primarily men or women? Suppose you want to estimate the difference in the proportions of men and women who say they have used or expect to use televised home shopping using an 80% confidence interval of width .06 or less.
  a. Approximately how many people should be included in your samples? $n_1 = n_2 = 911$
  b. Suppose you want to obtain individual estimates for the two proportions of interest. Will the sample size found in part **a** be large enough to provide estimates of each proportion correct to within .02 with 90% confidence? Justify your answer.

8.23 A manufacturer of large-screen televisions wants to compare with a competitor the proportions of its best sets that need repair within 1 year. If it is desired to estimate the difference between proportions to within .05 with 90% confidence, and if the manufacturer plans to sample twice as many buyers (n_1) of its sets as buyers (n_2) of the competitor's sets, how many buyers of each brand must be sampled? Assume that the proportion of sets that need repair will be about .2 for both brands.

8.24 Refer to Exercise 8.16 (p. 404), where you compared groups of patients who suffer "cardiac events." Consider the stress management and exercise groups of patients. How large would the samples have to be to estimate the difference in the proportions who suffer a cardiac event to within .1 with 95% confidence? Assume the sample sizes for the two groups will be equal, and use the estimated proportions from Exercise 8.16.

8.3   COMPARING POPULATION PROPORTIONS: MULTINOMIAL EXPERIMENT

Recall from Section 1.4 (p. 9) that observations on a qualitative variable can only be categorized. For example, consider the highest level of education attained by a professional hockey player. Level of education is a qualitative variable with several categories, including some high school, high school diploma, some college, college undergraduate degree, and graduate degree. If we were to record education level for all professional hockey players, the result of the categorization would be a count of the numbers of players falling in the respective categories.

When the qualitative variable of interest results in one of two responses (e.g., yes or no, success or failure, favor or do not favor), the data—called *counts*—can be analyzed using the binomial probability distribution discussed in Section 4.3. However, qualitative variables, such as level of education, that allow for more than two categories for a response are much more common, and these must be analyzed using a different method. Qualitative data with more than two levels often result from a **multinomial experiment**. The characteristics for a multinomial experiment with *k* outcomes are described in the box. You can see that the binomial experiment of Chapter 4 is a multinomial experiment with $k = 2$.

Properties of the Multinomial Experiment

1. The experiment consists of *n* identical trials.

2. There are *k* possible outcomes to each trial. These outcomes are sometimes called **classes, categories**, or **cells**.

3. The probabilities of the *k* outcomes, denoted by $p_1, p_2, \ldots, p_k$, remain the same from trial to trial, where $p_1 + p_2 + \cdots + p_k = 1$.

4. The trials are independent.

5. The random variables of interest are the **cell counts**, $n_1, n_2, \ldots, n_k$, of the number of observations that fall in each of the *k* categories.

E X A M P L E   8 . 4

Yes

Consider the problem of determining the highest level of education attained by each of a sample of $n = 40$ National Hockey League (NHL) players. Suppose we categorize level of education into one of five categories—some high school, high school diploma, some college, college undergraduate degree, and graduate degree—and count the number of the 40 players that fall into each category. Is this a multinomial experiment to a reasonable degree of approximation?

Solution

Checking the five properties of a multinomial experiment shown in the box, we have

1. The experiment consists of $n = 40$ identical trials, where each trial is to determine the education level of an NHL player.
2. There are $k = 5$ possible outcomes to each trial corresponding to the five education-level responses.

3. The probabilities of the $k = 5$ outcomes, p_1, p_2, p_3, p_4, and p_5, remain the same from trial to trial (to a reasonable degree of approximation), where p_i represents the true probability that an NHL player attains level-of-education category i.
4. The trials are independent, that is, the education level attained by one NHL player does not affect the level attained by any other player.
5. We are interested in the count of the number of hockey players who fall into each of the five education-level categories. These five cell counts are denoted n_1, n_2, n_3, n_4, and n_5.

Thus, the properties of a multinomial experiment are satisfied.   ∎

In this section, we consider a multinomial experiment with k outcomes that corresponds to the categories of a *single* qualitative variable. The results of such an experiment are summarized in a **one-way table**. The term "one-way" is used because only one variable is classified. Typically, we want to make inferences about the true percentages that occur in the k categories based on the sample information in the one-way table.

To illustrate, suppose three political candidates are running for the same elective position. Prior to the election, we conduct a survey to determine the voting preferences of a random sample of 150 eligible voters. The qualitative variable of interest is *preferred candidate*, which has three possible outcomes: candidate 1, candidate 2, and candidate 3. Suppose the number of voters preferring each candidate is tabulated and the resulting count data appear as in Table 8.1.

TABLE 8.1   Results of Voter-Preference Survey

Candidate		
1	**2**	**3**
61	53	36

Note that our voter-preference survey satisfies the properties of a multinomial experiment for the qualitative variable, preferred candidate. The experiment consists of randomly sampling $n = 150$ voters from a large population of voters containing an unknown proportion p_1 who favor candidate 1, a proportion p_2 who favor candidate 2, and a proportion p_3 who favor candidate 3. Each voter sampled represents a single trial that can result in one of three outcomes: The voter will favor candidate 1, 2, or 3 with probabilities p_1, p_2, and p_3, respectively. (Assume that all voters will have a preference.) The voting preference of any single voter in the sample does not affect the preference of another; consequently, the trials are independent. And, finally, you can see that the recorded data are the numbers of voters in each of the three voter-preference categories. Thus, the voter-preference survey satisfies the five properties of a multinomial experiment.

In the voter-preference survey, and in most practical applications of the multinomial experiment, the k outcome probabilities $p_1, p_2, \ldots, p_k$ are unknown and we want to use the survey data to make inferences about their values. The unknown probabilities in the voter-preference survey are

$$p_1 = \text{Proportion of all voters who favor candidate 1}$$

$$p_2 = \text{Proportion of all voters who favor candidate 2}$$

$$p_3 = \text{Proportion of all voters who favor candidate 3}$$

TEACHING TIP

The null hypothesis to be tested allows for the unique specification of the population proportions (e.g., $H_0: p_1 = .2, p_2 = .3, p_3 = .5$). In many cases, however, the appropriate test specifies that all proportions are equal.

To decide whether the voters have a preference for any of the candidates, we will want to test the null hypothesis that the candidates are equally preferred (that is, $p_1 = p_2 = p_3 = \frac{1}{3}$) against the alternative hypothesis that one candidate is preferred (that is, at least one of the probabilities p_1, p_2, and p_3 exceeds $\frac{1}{3}$). Thus, we want to test

$H_0: p_1 = p_2 = p_3 = \frac{1}{3}$ (no preference)

H_a: At least one of the proportions exceeds $\frac{1}{3}$ (a preference exists)

If the null hypothesis is true and $p_1 = p_2 = p_3 = \frac{1}{3}$, the expected value (mean value) of the number of voters who prefer candidate 1 is given by

$$E(n_1) = np_1 = (n)\tfrac{1}{3} = (150)\tfrac{1}{3} = 50$$

Similarly, $E_{(n_2)} = E_{(n_3)} = 50$ if the null hypothesis is true and no preference exists.

The following test statistic—the **chi-square test**—measures the degree of disagreement between the data and the null hypothesis:

$$\chi^2 = \frac{[n_1 - E(n_1)]^2}{E(n_1)} + \frac{[n_2 - E(n_2)]^2}{E(n_2)} + \frac{[n_3 - E(n_3)]^2}{E(n_3)}$$

$$= \frac{(n_1 - 50)^2}{50} + \frac{(n_2 - 50)^2}{50} + \frac{(n_3 - 50)^2}{50}$$

Note that the farther the observed numbers n_1, n_2, and n_3 are from their expected value (50), the larger χ^2 will become. That is, large values of χ^2 imply that the null hypothesis is false.

We have to know the distribution of χ^2 in repeated sampling before we can decide whether the data indicate that a preference exists. When H_0 is true, χ^2 can be shown to have (approximately) a **chi-square (χ^2) distribution**. The shape of the chi-square distribution depends on the degrees of freedom (df) associated with Table 8.1. For this one-way classification, the χ^2 distribution has $(k - 1)$ degrees of freedom.*

Critical values of χ^2 are provided in Table XI of Appendix A, a portion of which is shown in Table 8.2. To illustrate, the rejection region for the voter-preference survey for $\alpha = .05$ and $k - 1 = 3 - 1 = 2$ df is

Rejection region: $\chi^2 > \chi^2_{.05}$

This value of $\chi^2_{.05}$ (found in Table XI) is 5.99147. (See Figure 8.3.) The computed value of the test statistic is

$$\chi^2 = \frac{(n_1 - 50)^2}{50} + \frac{(n_2 - 50)^2}{50} + \frac{(n_3 - 50)^2}{50}$$

$$= \frac{(61 - 50)^2}{50} + \frac{(53 - 50)^2}{50} + \frac{(36 - 50)^2}{50} = 6.52$$

---

*The derivation of the degrees of freedom for χ^2 involves the number of linear restrictions imposed on the count data. In the present case, the only constraint is that $\sum n_i = n$, where n (the sample size) is fixed in advance. Therefore, df $= k - 1$. For other cases, we will give the degrees of freedom for each usage of χ^2 and refer the interested reader to the references for more detail.

TABLE 8.2 **Reproduction of Part of Table XI in Appendix A**

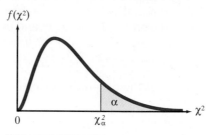

Degrees of Freedom	$\chi^2_{.100}$	$\chi^2_{.050}$	$\chi^2_{.025}$	$\chi^2_{.010}$	$\chi^2_{.005}$
1	2.70554	3.84146	5.02389	6.63490	7.87944
2	4.60517	5.99147	7.37776	9.21034	10.5966
3	6.25139	7.81473	9.34840	11.3449	12.8381
4	7.77944	9.48773	11.1433	13.2767	14.8602
5	9.23635	11.0705	12.8325	15.0863	16.7496
6	10.6446	12.5916	14.4494	16.8119	18.5476
7	12.0170	14.0671	16.0128	18.4753	20.2777

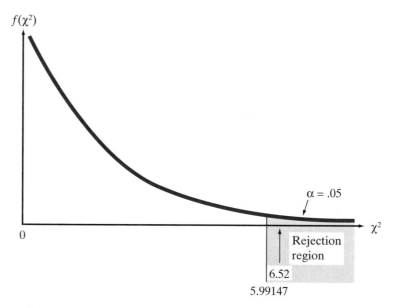

FIGURE 8.3

Rejection Region for Voter-Preference Survey

Since the computed $\chi^2 = 6.52$ exceeds the critical value of 5.99147, we conclude at the $\alpha = .05$ level of significance that there does exist a voter preference for one or more of the candidates.

Now that we have evidence to indicate that the proportions p_1, p_2, and p_3 are unequal, we can make inferences concerning their individual values using the methods of Section 5.3. [*Note:* We cannot use the methods of Section 8.1 to compare two proportions because the cell counts are dependent random variables.] The general form for a test of a hypothesis concerning multinomial probabilities is shown in the next box.

TEACHING TIP
Use plenty of in-class examples to illustrate the one-way test. Lay the groundwork for the test of independence in the two-way analysis in the next section.

A Test of a Hypothesis About Multinomial Probabilities: One-way Table

$H_0: p_1 = p_{1,0}, p_2 = p_{2,0}, \ldots, p_k = p_{k,0}$

where $p_{1,0}, p_{2,0}, \ldots, p_{k,0}$ represent the hypothesized values of the multinomial probabilities

H_a: At least one of the multinomial probabilities does not equal its hypothesized value

Test statistic: $\chi^2 = \sum \dfrac{[n_i - E(n_i)]^2}{E(n_i)}$

where $E(n_i) = np_{i,0}$ is the **expected cell count**, that is, the expected number of outcomes of type i assuming that H_0 is true. The total sample size is n.

Rejection region: $\chi^2 > \chi_\alpha^2$, where χ_α^2 has $(k - 1)$ df

Assumptions: 1. A multinomial experiment has been conducted. This is generally satisfied by taking a random sample from the population of interest.

2. The sample size n will be large enough so that for every cell, the expected cell count $E(n_i)$ will be equal to 5 or more.*

EXAMPLE 8.5

$\chi^2 = 13.249$; Reject H_0

Suppose an educational television station has broadcast a series of programs on the physiological and psychological effects of smoking marijuana. Now that the series is finished, the station wants to see whether the citizens within the viewing area have changed their minds about how possession of marijuana should be considered legally. Before the series was shown, it was determined that 7% of the citizens favored legalization, 18% favored decriminalization, 65% favored the existing law (an offender could be fined or imprisoned), and 10% had no opinion.

A summary of the opinions (after the series was shown) of a random sample of 500 people in the viewing area is given in Table 8.3. Test at the $\alpha = .01$ level to see whether these data indicate that the distribution of opinions differs significantly from the proportions that existed before the educational series was aired.

TABLE 8.3   Distribution of Opinions About Marijuana Possession

Legalization	Decriminalization	Existing Laws	No Opinion
39	99	336	26

Solution

Define the proportions after the airing to be

p_1 = Proportion of citizens favoring legalization
p_2 = Proportion of citizens favoring decriminalization
p_3 = Proportion of citizens favoring existing laws
p_4 = Proportion of citizens with no opinion

Then the null hypothesis representing no change in the distribution of percentages is

$H_0: p_1 = .07, p_2 = .18, p_3 = .65, p_4 = .10$

*The assumption that all expected cell counts are at least 5 is necessary in order to ensure that the χ^2 approximation is appropriate. Exact methods for conducting the test of a hypothesis exist and may be used for small expected cell counts, but these methods are beyond the scope of this text.

and the alternative is

H_a: At least one of the proportions differs from its null hypothesized value

Test statistic: $\chi^2 = \sum \dfrac{[n_i - E(n_i)]^2}{E(n_i)}$

where

$E(n_1) = np_{1,0} = 500(.07) = 35$
$E(n_2) = np_{2,0} = 500(.18) = 90$
$E(n_3) = np_{3,0} = 500(.65) = 325$
$E(n_4) = np_{4,0} = 500(.10) = 50$

Since all these values are larger than 5, the χ^2 approximation is appropriate. Also, if the citizens in the sample were randomly selected, the properties of the multinomial probability distribution are satisfied.

Rejection region: For $\alpha = .01$ and df $= k - 1 = 3$, reject H_0 if $\chi^2 > \chi^2_{.01}$, where (from Table XI in Appendix A) $\chi^2_{.01} = 11.3449$.

We now calculate the test statistic:

$$\chi^2 = \frac{(39 - 35)^2}{35} + \frac{(99 - 90)^2}{90} + \frac{(336 - 325)^2}{325} + \frac{(26 - 50)^2}{50} = 13.249$$

Since this value exceeds the table value of χ^2 (11.3449), the data provide sufficient evidence ($\alpha = .01$) that the opinions on legalization of marijuana have changed since the series was aired.

The χ^2-test can also be conducted using an available statistical software package. Figure 8.4 is an SPSS printout of the analysis of the data in Table 8.3. The test statistic and *p*-value of the test are highlighted on the printout. Since $\alpha = .01$ exceeds $p = .004$, there is sufficient evidence to reject H_0.   ∎

FIGURE 8.4

SPSS Analysis of Data in Table 8.3

OPINION

	Observed N	Expected N	Residual
LEGAL	39	35.0	4.0
DECRIM	99	90.0	9.0
EXISTLAW	336	325.0	11.0
NONE	26	50.0	-24.0
Total	500		

Test Statistics

	OPINION
Chi-Square[a]	13.249
df	3
Asymp. Sig.	.004

[a] 0 cells (.0%) have expected frequencies less than 5. The minimum expected cell fequency is 35.0.

Suggested Exercise 8.32

If we focus on one particular outcome of a multinomial experiment, we can use the methods developed in Section 5.3 for a binomial proportion to establish a confidence interval for any one of the multinomial probabilities.* For example, if we want a 95% confidence interval for the proportion of citizens in the viewing area who have no opinion about the issue, we calculate

$$\hat{p}_4 \pm 1.96\sigma_{\hat{p}_4}$$

---

*Note that focusing on one outcome has the effect of lumping the other $(k - 1)$ outcomes into a single group. Thus, we obtain, in effect, two outcomes—or a binomial experiment.

where

$$\hat{p}_4 = \frac{n_4}{n} = \frac{26}{500} = .052 \quad \text{and} \quad \sigma_{\hat{p}_4} \approx \sqrt{\frac{\hat{p}_4(1 - \hat{p}_4)}{n}}$$

Thus, we get

$$.052 \pm 1.96\sqrt{\frac{(.052)(.948)}{500}} = .052 \pm .019$$

or (.033, .071). Thus, we estimate that between 3.3% and 7.1% of the citizens now have no opinion on the issue of marijuana legalization. The series of programs may have helped citizens who formerly had no opinion on the issue to form an opinion, since it appears that the proportion of "no opinions" is now less than 10%.

 EXERCISES 8.25–8.36

Learning the Mechanics

8.25 Use Table XI of Appendix A to find each of the following χ^2 values:
  a. $\chi^2_{.05}$ for df = 10   18.3070
  b. $\chi^2_{.990}$ for df = 50   29.7067
  c. $\chi^2_{.10}$ for df = 16   23.5418
  d. $\chi^2_{.005}$ for df = 50   79.4900

8.26 a. What are the characteristics of a multinomial experiment? Compare the characteristics to those of a binomial experiment.
  b. What conditions must n satisfy to make the χ^2-test valid?

8.27 Find the rejection region for a one-dimensional χ^2-test of a null hypothesis concerning $p_1, p_2, \ldots, p_k$ if
  a. $k = 3; \alpha = .05$ $\chi^2 > 5.99147$
  b. $k = 5; \alpha = .10$ $\chi^2 > 7.77944$
  c. $k = 4; \alpha = .01$ $\chi^2 > 11.3449$

8.28 A multinomial experiment with $k = 3$ cells and $n = 320$ produced the data shown in the following table. Do these data provide sufficient evidence to contradict the null hypothesis that $p_1 = .25, p_2 = .25,$ and $p_3 = .50$? Test using $\alpha = .05$. $\chi^2 = 8.075$

	Cell		
	1	**2**	**3**
n_i	78	60	182

8.29 A multinomial experiment with $k = 4$ cells and $n = 205$ produced the data shown in the table.

	Cell			
	1	**2**	**3**	**4**
n_i	43	56	59	47

  a. Do these data provide sufficient evidence to conclude that the multinomial probabilities differ? Test using $\alpha = .05$.
  b. What are the Type I and Type II errors associated with the test of part **a**?
  c. Construct a 95% confidence interval for the multinomial probability associated with cell 3.

Applying the Concepts—Basic

8.30 There has been a recent trend for professional sports franchises in Major League Baseball (MLB), the National Football League (NFL), the National Basketball Association (NBA), and the National Hockey League (NHL) to build new stadiums and ballparks in urban, downtown venues. An article in *Professional Geographer* (Feb. 2000) investigated whether there has been a significant suburban-to-urban shift in the location of major sport facilities. In 1985, 40% of all major sport facilities were located downtown, 30% in central city, and 30% in suburban areas. In contrast, of the 113 major sports franchises that existed in 1997, 58 were built downtown, 26 in central city, and 29 in a suburban area.
  a. Describe the qualitative variable of interest in the study. Give the levels (categories) associated with the variable.
  b. Give the null hypothesis for a test to determine whether the proportions of major sports facilities in downtown, central city, and suburban areas in 1997 are the same as in 1985.
  c. If the null hypothesis, part **b**, is true, how many of the 113 sports facilities in 1997 would you expect to be located in downtown, central city, and suburban areas, respectively?   45.2; 33.9; 33.9
  d. Find the value of the chi-square statistic for testing the null hypothesis, part **b**. $\chi^2 = 6.17$

e. Find the (approximate) *p*-value of the test and give the appropriate conclusion in the words of the problem. Assume $\alpha = .05$.

8.31 Refer to the *Chance* (Fall 2000) study of ancient Greek pottery, Exercise 2.8 (p. 26). Recall that 837 pottery pieces were uncovered at the excavation site. The table describing the types of pottery found is reproduced here.

Pot Category	Number Found
Burnished	133
Monochrome	460
Painted	183
Other	61
Total	**837**

Source: Berg, I., and Bliedon, S. "The pots of Phyiakopi: Applying statistical techniques to archaeology." *Chance*, Vol. 13, No. 4, Fall 2000.

a. Describe the qualitative variable of interest in the study. Give the levels (categories) associated with the variable.

b. Assume the four types of pottery occur with equal probability at the excavation site. What are the values of p_1, p_2, p_3, and p_4, the probabilities associated with the four pottery types? .25; .25; .25; .25

c. Give the null and alternative hypotheses for testing whether one type of pottery is more likely to occur at the site than any of the other types.

d. Use the STATISTIX printout provided below to find the test statistic for testing the hypotheses, part **c**.

e. Find and interpret the *p*-value of the test. State the conclusion in the words of the problem if you use $\alpha = .10$. $p = .0000$

8.32 *Inc. Technology* (Mar. 18, 1997) reported the results of an Equifax/Harris Consumer Privacy Survey in which 328 Internet users indicated their level of agreement with the following statement: "The government needs to be able to scan Internet messages and user commu-nications to prevent fraud and other crimes." The number of users in each response category is summarized below.

Agree Strongly	Agree Somewhat	Disagree Somewhat	Disagree Strongly
59	108	82	79

a. Specify the null and alternative hypotheses you would use to determine if the opinions of Internet users are evenly divided among the four categories.

b. Conduct the test of part **a** using $\alpha = .05$.

c. In the context of this exercise, what is a Type I error? A Type II error?

d. What assumptions must hold in order to ensure the validity of the test you conducted in part **b**?

Applying the Concepts—Intermediate

8.33 Each year, approximately 1.3 million people in the United States suffer adverse drug effects (ADEs), that is, unintended injuries caused by prescribed medication. A study in the *Journal of the American Medical Association* (July 5, 1995) identified the cause of 247 ADEs that occurred at two Boston hospitals. The researchers found that dosing errors (that is, wrong dosage prescribed and/or dispensed) were the most common. The table summarizes the proximate cause of 95 ADEs that resulted from a dosing error. Conduct a test (at $\alpha = .10$) to determine whether the true percentages of ADEs in the five "cause" categories are different. Use the SPSS printout on p. 415 to arrive at your decision. $p = .003$

Wrong Dosage Cause	Number of ADEs
(1) Lack of knowledge of drug	29
(2) Rule violation	17
(3) Faulty dose checking	13
(4) Slips	9
(5) Other	27

STATISTIX Output for Exercise 8.31

```
MULTINOMIAL TEST

HYPOTHESIZED PROPORTIONS VARIABLE: HYPPROP
OBSERVED FREQUENCIES VARIABLE:        FOUND


           HYPOTHESIZED   OBSERVED    EXPECTED     CHI-SQUARE
CATEGORY   PROPORTION     FREQUENCY   FREQUENCY    CONTRIBUTION
    1        0.25000        133        209.25        27.79
    2        0.25000        460        209.25       300.48
    3        0.25000        183        209.25         3.29
    4        0.25000         61        209.25       105.03


OVERALL CHI-SQUARE      436.59
P-VALUE                 0.0000
DEGREES OF FREEDOM           3
```

SPSS Output for Exercise 8.33

CAUSE

	Observed N	Expected N	Residual
LACKKNOW	29	19.0	10.0
RULEVIOLATE	17	19.0	-2.0
FAULTYDOSE	13	19.0	-6.0
SLIPS	9	19.0	-10.0
OTHER	27	19.0	8.0
Total	95		

Test Statistics

	CAUSE
Chi-Square[a]	16.000
df	4
Asymp. Sig.	.003

a. 0 cells (.0%) have expected frequencies less than 5. The minimum expected cell fequency is 19.0.

8.34 In education, the term *instructional technology* refers to products such as computers, spreadsheets, CD-ROMs, videos, and presentation software. How frequently do professors use instructional technology in the classroom? To answer this question, researchers at Western Michigan University surveyed 306 of their fellow faculty (*Educational Technology*, Mar.–Apr. 1995). Responses to the frequency-of-technology use in teaching were recorded as "weekly to every class," "once a semester to monthly," or "never." The faculty responses (number in each response category) for the three technologies are summarized in the table.

Technology	Weekly	Once a Semester/ Monthly	Never
Computer spreadsheets	58	67	181
Word processing	168	61	77
Statistical software	37	82	187

a. Determine whether the percentages in the three frequency-of-use response categories differ for computer spreadsheets. Use $\alpha = .01$. $\chi^2 = 92.176$

b. Repeat part **a** for word processing. $\chi^2 = 65.314$

c. Repeat part **a** for statistical software.

d. Construct a 99% confidence interval for the true percentage of faculty who never use computer spreadsheets in the classroom. Interpret the interval.

8.35 Interferons are proteins produced naturally by the human body that help fight infections and regulate the immune system. A drug developed from interferons, called Avonex, is now available for treating patients with multiple sclerosis (MS). In a clinical study, 85 MS patients received weekly injections of Avonex over a 2-year period. The number of exacerbations (i.e., flare-ups of symptoms) was recorded for each patient and is summarized in the next table. For MS patients who take a placebo (no drug) over a similar 2-year period, it is known from previous studies that 26% will experience no exacerbations, 30% one exacerbation, 11% two exacerbations, 14% three exacerbations, and 19% four or more exacerbations.

Number of Exacerbations	Number of Patients
0	32
1	26
2	15
3	6
4 or more	6

Source: Biogen, Inc., 1997.

a. Conduct a test to determine whether the exacerbation distribution of MS patients who take Avonex differs from the percentages reported for placebo patients. Test using $\alpha = .05$. $\chi^2 = 17.16$

b. Find a 95% confidence interval for the true percentage of Avonex MS patients who are exacerbation-free during a 2-year period. $.376 \pm .103$

c. Refer to part **b**. Is there evidence that Avonex patients are more likely to have no exacerbations than placebo patients? Explain.

8.36 *Nature* (Sept. 1993) reported on a study of animal and plant species "hotspots" in Great Britain. A hotspot is defined as a 10-km^2 area that is species-rich, that is, is heavily populated by the species of interest. Analogously, a coldspot is a 10-km^2 area that is species-poor. The table gives the number of butterfly hotspots and the number of butterfly coldspots in a sample of 2,588 10-km^2 areas. In theory, 5% of the areas should be butterfly hotspots and 5% should be butterfly coldspots, while the remaining areas (90%) are neutral. Test the theory using $\alpha = .01$. $\chi^2 = 2.764$

Butterfly hotspots	123
Butterfly coldspots	147
Neutral areas	2,318
Total	**2,588**

Source: Prendergast, J. R., *et al.* "Rare species, the coincidence of diversity hotspots and conservation strategies." *Nature*, Vol. 365, No. 6444, Sept. 23, 1993, p. 335 (Table 1).

8.4 CONTINGENCY TABLE ANALYSIS

In Section 8.1, we introduced the multinomial probability distribution and considered data classified according to a single criterion. We now consider multinomial experiments in which the data are classified according to two criteria, that is, *classification with respect to two qualitative factors.*

For example, consider a study in the *Journal of Marketing* (Fall 1992) on the impact of using celebrities in television advertisements. The researchers investigated the relationship between gender of a viewer and the viewer's brand awareness. Three hundred TV viewers were asked to identify products advertised by male celebrity spokespersons. The data are summarized in the **two-way table** shown in Table 8.4. This table is called a **contingency table**; it presents multinomial count data classified on two scales, or **dimensions**, **of classification**—namely, gender of viewer and brand awareness.

TABLE 8.4   Contingency Table for Marketing Example

		Gender		
		Male	**Female**	**Totals**
Brand Awareness	**Could Identify Product**	95	41	136
	Could Not Identify Product	55	109	164
	Totals	150	150	300

The symbols representing the cell counts for the multinomial experiment in Table 8.4 are shown in Table 8.5a; and the corresponding cell, row, and column probabilities are shown in Table 8.5b. Thus, n_{11} represents the number of viewers who are male and could identify the brand and p_{11} represents the corresponding cell probability. Note the symbols for the row and column totals and also the symbols for the probability totals. The latter are called **marginal probabilities** for each row and column. The marginal probability p_{r1} is the probability that a TV viewer identifies the product; the marginal probability p_{c1} is the probability that the TV viewer is male. Thus,

$$p_{r1} = p_{11} + p_{12} \text{ and } p_{c1} = p_{11} + p_{21}$$

Thus, we can see that this really is a multinomial experiment with a total of 300 trials, $(2)(2) = 4$ cells or possible outcomes, and probabilities for each cell as shown in Table 8.5b. If the 300 TV viewers are randomly chosen, the trials are considered independent and the probabilities are viewed as remaining constant from trial to trial.

Suppose we want to know whether the two classifications, gender and brand awareness, are dependent. That is, if we know the gender of the TV viewer, does

TABLE 8.5a   Observed Counts for Contingency Table 8.4

		Gender		
		Male	**Female**	**Totals**
Brand Awareness	**Could Identify Product**	n_{11}	n_{12}	r_1
	Could Not Identify Product	n_{21}	n_{22}	r_2
	Totals	c_1	c_2	n

TABLE 8.5b   Probabilities for Contingency Table 8.4

		Gender		
		Male	**Female**	**Totals**
Brand Awareness	**Could Identify Product**	p_{11}	p_{12}	p_{r1}
	Could Not Identify Product	p_{21}	p_{22}	p_{r2}
	Totals	p_{c1}	p_{c2}	1

that information give us a clue about the viewer's brand awareness? In a probabilistic sense we know (Chapter 3) that independence of events A and B implies $P(AB) = P(A)P(B)$. Similarly, in the contingency table analysis, if the **two classifications are independent**, the probability that an item is classified in any particular cell of the table is the product of the corresponding marginal probabilities. Thus, under the hypothesis of independence, in Table 8.5b, we must have

$$p_{11} = p_{r1}p_{c1} \quad p_{12} = p_{r1}p_{c2}$$
$$p_{21} = p_{r2}p_{c1} \quad p_{22} = p_{r2}p_{c2}$$

To test the hypothesis of independence, we use the same reasoning employed in the one-dimensional tests of Section 8.3. First, we calculate the *expected*, or *mean, count in each cell* assuming that the null hypothesis of independence is true. We do this by noting that the expected count in a cell of the table is just the total number of multinomial trials, n, times the cell probability. Recall that n_{ij} represents the **observed count** in the cell located in the ith row and jth column. Then the expected cell count for the upper-left-hand cell (first row, first column) is

$$E(n_{11}) = np_{11}$$

or, when the null hypothesis (the classifications are independent) is true,

$$E(n_{11}) = np_{r1}p_{c1}$$

Since these true probabilities are not known, we estimate p_{r1} and p_{c1} by the same proportions $\hat{p}_{r1} = r_1/n$ and $\hat{p}_{c1} = c_1/n$. Thus, the estimate of the expected value $E(n_{11})$ is

$$\hat{E}(n_{11}) = n\left(\frac{r_1}{n}\right)\left(\frac{c_1}{n}\right) = \frac{r_1 c_1}{n}$$

Similarly, for each i, j,

$$\hat{E}(n_{ij}) = \frac{(\text{Row total})(\text{Column total})}{\text{Total sample size}}$$

Thus,

$$\hat{E}(n_{12}) = \frac{r_1 c_2}{n}$$

$$\hat{E}(n_{21}) = \frac{r_2 c_1}{n}$$

$$\hat{E}(n_{22}) = \frac{r_2 c_2}{n}$$

Using the data in Table 8.4, we find

$$\hat{E}(n_{11}) = \frac{r_1 c_1}{n} = \frac{(136)(150)}{300} = 68$$

$$\hat{E}(n_{12}) = \frac{r_1 c_2}{n} = \frac{(136)(150)}{300} = 68$$

$$\hat{E}(n_{21}) = \frac{r_2 c_1}{n} = \frac{(164)(150)}{300} = 82$$

$$\hat{E}(n_{22}) = \frac{r_2 c_2}{n} = \frac{(164)(150)}{300} = 82$$

The observed data and the estimated expected values (in parentheses) are shown in Table 8.6.

TABLE 8.6 Observed and Estimated Expected (in Parentheses) Counts

		Gender		
		Male	**Female**	**Totals**
Brand Awareness	**Could Identify Product**	95 (68)	41 (68)	136
	Could Not Identify Product	55 (82)	109 (82)	164
	Totals	150	150	300

We now use the χ^2 statistic to compare the observed and expected (estimated) counts in each cell of the contingency table:

$$\chi^2 = \frac{[n_{11} - \hat{E}(n_{11})]^2}{\hat{E}(n_{11})} + \frac{[n_{12} - \hat{E}(n_{12})]^2}{\hat{E}(n_{12})} + \frac{[n_{21} - \hat{E}(n_{21})]^2}{\hat{E}(n_{21})} + \frac{[n_{22} - \hat{E}(n_{22})]^2}{\hat{E}(n_{22})}$$

$$= \sum \frac{[n_{ij} - \hat{E}(n_{ij})]^2}{\hat{E}(n_{ij})}$$

Note: The use of $\sum$ in the context of a contingency table analysis refers to a sum over all cells in the table.

Substituting the data of Table 8.6 into this expression, we get

$$\chi^2 = \frac{(95 - 68)^2}{68} + \frac{(41 - 68)^2}{68} + \frac{(55 - 82)^2}{82} + \frac{(109 - 82)^2}{82} = 39.22$$

TEACHING TIP
Point out the similarities between the test for independence and the one-way test in the last section. Use a computer example to illustrate how the *p*-value can be used in both types of problems.

Large values of χ^2 imply that the observed counts do not closely agree and hence that the hypothesis of independence is false. To determine how large χ^2 must be before it is too large to be attributed to chance, we make use of the fact that the sampling distribution of χ^2 is approximately a χ^2 probability distribution when the classifications are independent.

When testing the null hypothesis of independence in a two-way contingency table, the appropriate degrees of freedom will be $(r - 1)(c - 1)$, where r is the number of rows and c is the number of columns in the table.

For the brand awareness example, the degrees of freedom for χ^2 is $(r - 1)(c - 1) = (2 - 1)(2 - 1) = 1$. Then, for $\alpha = .05$, we reject the hypothesis of independence when

$$\chi^2 > \chi^2_{.05} = 3.84146$$

Since the computed $\chi^2 = 39.22$ exceeds the value 3.84146, we conclude that viewer gender and brand awareness are dependent events.

The pattern of **dependence** can be seen more clearly by expressing the data as percentages. We first select one of the two classifications to be used as the base variable. In the above example, suppose we select gender of the TV viewer as the classificatory variable to be the base. Next, we represent the responses for each level of the second categorical variable (brand awareness in our example) as a percentage of the subtotal for the base variable. For example, from Table 8.6 we convert the response for males who identify the brand (95) to a percentage of the total number of male viewers (150). That is,

$$(^{95}/_{150})100\% = 63.3\%$$

TABLE 8.7   **Percentage of TV Viewers Who Identify Brand, by Gender**

		Gender		
		Male	**Female**	**Totals**
Brand Awareness	**Could Identify Product**	63.3	27.3	45.3
	Could Not Identify Product	36.7	72.7	54.7
	Totals	100	100	100

The conversions of all Table 8.6 entries are similarly computed, and the values are shown in Table 8.7. The value shown at the right of each row is the row's total expressed as a percentage of the total number of responses in the entire table. Thus, the percentage of TV viewers who identify the product is: $\left(\frac{136}{300}\right)100\% = 45.3\%$ (rounded to the nearest percent).

If the gender and brand awareness variables are independent, then the percentages in the cells of the table are expected to be approximately equal to the corresponding row percentages. Thus, we would expect the percentage who identify the brand for each gender to be approximately 45% if the two variables are independent. The extent to which each gender's percentage departs from this value determines the dependence of the two classifications, with greater variability of the row percentages meaning a greater degree of dependence. A plot of the percentages helps summarize the observed pattern. In the SPSS bar graph in Figure 8.5, we show the gender of the viewer (the base variable) on the horizontal axis, and the percentage of TV viewers who identify the brand on the vertical axis. The "expected" percentage under the assumption of independence is shown as a dotted horizontal line.

Figure 8.5 clearly indicates the reason that the test resulted in the conclusion that the two classifications in the contingency table are dependent. The percentage of male TV viewers who identify the brand promoted by a male celebrity is more than twice as high as the percentage of female TV viewers who identify the

FIGURE 8.5

SPSS Bar Graph Showing Percentage of Viewers Who Identified the TV Product

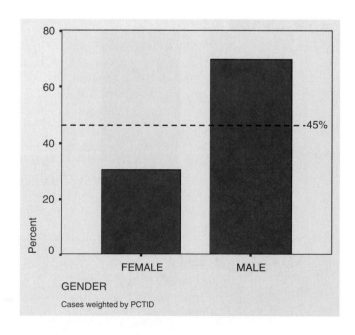

TABLE 8.8   General $r \times c$ Contingency Table

		Column				Row Totals
		1	**2**	$\cdots$	**c**	
Row	**1**	n_{11}	n_{12}	$\cdots$	n_{1c}	r_1
	2	n_{21}	n_{22}	$\cdots$	n_{2c}	r_2
	$\vdots$	$\vdots$	$\vdots$		$\vdots$	$\vdots$
	r	n_{r1}	n_{r2}	$\cdots$	n_{rc}	r_r
Column Totals		c_1	c_2	$\cdots$	c_c	n

brand. Statistical measures of the degree of dependence and procedures for making comparisons of pairs of levels for classifications are available. They are beyond the scope of this text, but can be found in the references. We will, however, utilize descriptive summaries such as Figure 8.5 to examine the degree of dependence exhibited by the sample data.

The general form of a two-way contingency table containing r rows and c columns (called an $r \times c$ contingency table) is shown in Table 8.8. Note that the observed count in the *(ij)* cell is denoted by n_{ij}, the ith row total is r_i, the jth column total is c_j, and the total sample size is n. Using this notation, we give the general form of the contingency table test for independent classifications in the next box.

General Form of a Two-way (Contingency) Table Analysis: A Test for Independence

H_0: The two classifications are independent
H_a: The two classifications are dependent

Test statistic: $\chi^2 = \sum \dfrac{[n_{ij} - \hat{E}(n_{ij})]^2}{\hat{E}(n_{ij})}$

where $\hat{E}(n_{ij}) = \dfrac{r_i c_j}{n}$

Rejection region: $\chi^2 > \chi_\alpha^2$, where χ_α^2 has $(r - 1)(c - 1)$ df

Assumptions: 1. The n observed counts are a random sample from the population of interest. We may then consider this to be a multinomial experiment with $r \times c$ possible outcomes.

2. The sample size, n, will be large enough so that, for every cell, the expected count, $E(n_{ij})$, will be equal to 5 or more.

EXAMPLE 8.6

a. $\chi^2 = 7.135$, fail to reject H_0

A social scientist wants to determine whether the marital status (divorced or not divorced) of U.S. men is independent of their religious affiliation (or lack thereof). A sample of 500 U.S. men is surveyed and the results are tabulated as shown in Table 8.9.

a. Test to see whether there is sufficient evidence to indicate that the marital status of men who have been or are currently married is dependent on religious affiliation. Test $\alpha = .01$.

b. Graph the data and describe the patterns revealed. Is the result of the test supported by the graph?

TABLE 8.9   Survey Results (Observed Counts), Example 8.6

		Religious Affiliation					
		A	**B**	**C**	**D**	**None**	**Totals**
Marital Status	**Divorced**	39	19	12	28	18	116
	Never divorced	172	61	44	70	37	384
	Totals	211	80	56	98	55	500

Solution

a. The first step is to calculate estimated expected cell frequencies under the assumption that the classifications are independent. Rather than compute these values by hand, we resort to a computer. The SAS printout of the analysis of Table 8.9 is displayed in Figure 8.6. Each cell in Figure 8.6 contains the observed (top) and expected (bottom) frequency in that cell. Note that $\hat{E}(n_{11})$, the estimated expected count for the Divorced, A cell, is 48.952. Similarly, the estimated expected count for the Divorced, B cell, is $\hat{E}(n_{12}) = 18.56$. Since all the estimated expected cell frequencies are greater than 5, the χ^2 approximation for the test statistic is appropriate. Assuming the men chosen were randomly selected from all married or previously married American men, the characteristics of the multinomial probability distribution are satisfied.

Suggested Exercise 8.49

FIGURE 8.6

SAS Contingency Table Printout

The FREQ Procedure

Table of MARITAL by RELIGION

MARITAL / Frequency Expected Col Pct	A	B	C	D	NONE	Total
DIVORCED	39 / 48.952 / 18.48	19 / 18.56 / 23.75	12 / 12.992 / 21.43	28 / 22.736 / 28.57	18 / 12.76 / 32.73	116
NEVER	172 / 162.05 / 81.52	61 / 60.44 / 76.25	44 / 43.008 / 78.57	70 / 75.264 / 71.43	37 / 42.24 / 67.27	384
Total	211	80	56	98	55	500

Statistics for Table of MARITAL by RELIGION

Statistic	DF	Value	Prob
Chi-Square	4	7.1355	0.1289
Likelihood Ratio Chi-Square	4	6.9854	0.1367
Mantel-Haenszel Chi-Square	1	6.4943	0.0108
Phi Coefficient		0.1195	
Contingency Coefficient		0.1186	
Cramer's V		0.1195	

Sample Size = 500

FIGURE 8.7

SAS Bar Graph Showing
Percentage of Divorced Males
by Religion

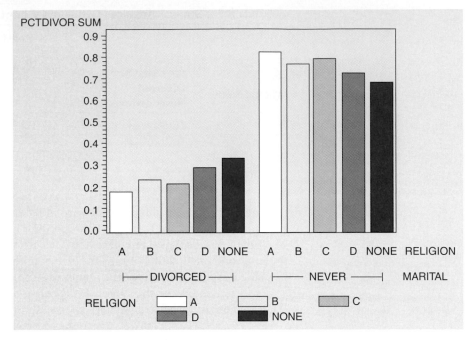

The null and alternative hypotheses we want to test are

H_0: The marital status of U.S. men and their religious affiliation are independent
H_a: The marital status of U.S. men and their religious affiliation are dependent

The test statistic, $\chi^2 = 7.135$, is highlighted at the bottom of the printout, as is the observed significance level (p-value) of the test. Since $\alpha = .01$ is less than $p = .129$, we fail to reject H_0; that is, we cannot conclude that the marital status of U.S. men depends on their religious affiliation. (Note that we could not reject H_0 even with $\alpha = .10$.)

b. The marital status frequencies are expressed as percentages of the number of men in each religious affiliation category and highlighted in the SAS printout, Figure 8.6. The expected percentage of divorced men under the assumption of independence is $\left(\frac{116}{500}\right)100\% = 23\%$. A SAS graph of the percentages is shown in Figure 8.7. Note that the percentages of divorced men (see the bars in the "DIVORCED" block of the SAS graph) deviate only slightly from that expected under the assumption of independence, supporting the result of the test in part a. That is, neither the descriptive bar graph nor the statistical test provides evidence that the male divorce rate depends on (varies with) religious affiliation. ∎

 Using the TI-83 Graphing Calculator

Finding p-Values for Contingency Tables on the TI-83

Start from the home screen

Step 1   Access the Matrix Menu and enter your observed values as Matrix [A], and your expected values as Matrix [B].

Press   MATRIX
Arrow right to **EDIT**

USING THE TI-83 GRAPHING CALCULATOR *(continued)*

> **Press   ENTER**
> Enter the dimensions of your observed matrix
> Enter all values
> **Press   MATRIX**
> Arrow right to **EDIT**
> Arrow down to **2:[B]**
> **Press   ENTER**
> Enter the dimensions of the expected values matrix
> (*it will be the same size as your observed matrix above*)
> Enter all values

Step 2   Access the Statistical Tests Menu and perform the χ^2 test.

> **Press   STAT**
> Arrow right to **TESTS**
> Arrow up to **C: χ^2 − Tests . . .**
> **Press   ENTER**
> Enter the names for the observed and expected matrices
> (*Press Matrix arrow to desired Matrix, Press ENTER*)
> Arrow down to **Calculate**
> **Press   ENTER**

Step 3   *View Display*
Reject H_0 if the *p*-value $< \alpha$.

Example   Consider Example 8.6 from the Text.

Our observed matrix will be $[A] = \begin{vmatrix} 39 & 19 & 12 & 28 & 18 \\ 172 & 61 & 44 & 70 & 37 \end{vmatrix}$

Our expected matrix will be $[B] =$

$$\begin{vmatrix} 48.952 & 18.56 & 12.992 & 22.736 & 12.76 \\ 162.05 & 60.44 & 43.008 & 75.264 & 42.24 \end{vmatrix}$$

The hypotheses for this problem are as follows:

> H_0: The matrix entries represent independent events.
> H_a: The matrix entries represent events that *are not* independent.

Step 1   Access the Matrix Menu and enter all values

> **Press   MATRIX**
> Arrow right to **EDIT**
> **Press   ENTER**
> Enter the dimensions of the observed matrix (see screens
> below).

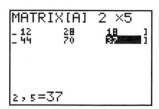

(continues on next page)

USING THE T1-83 GRAPHING CALCULATOR *(continued)*

Enter all values (see screens above).

Press **MATRIX**
Arrow right to **EDIT**
Press **ENTER**
Enter the dimensions of the expected matrix (same as above).
Enter all expected values* (see screen below).

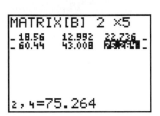

Each expected value entry is the product of (that row total)* (that column total) /grand total.

Step 2   Perform the χ^2 − Test.

Press **STAT**
Arrow right to **TESTS**
Arrow up to **C:** χ^2 − **Test . . .** (see screen)

Press **ENTER(see screen)**
Set Observed: **[A]**
Expected: **[B]**
Arrow down to **Calculate**
Press **ENTER**(see screen)

Step 3   *View Display*
As you can see from the screen, the *p*-value is 0.129. Since the *p*-value is *not less* than α we *do not* reject H_0. There is no statistical evidence at the $\alpha = 0.01$ level to conclude that the male divorce rate depends on religious affiliation.

Step 4   Clear screen for the next problem. Return to the home screen.
Press   **CLEAR**

EXERCISES 8.37–8.51

Learning the Mechanics

8.37 Find the rejection region for a test of independence of two classifications where the contingency table contains r rows and c columns and
  a. $r = 5$, $c = 5$, $\alpha = .05$ $\chi^2 > 26.2962$
  b. $r = 3$, $c = 6$, $\alpha = .10$ $\chi^2 > 15.9871$
  c. $r = 2$, $c = 3$, $\alpha = .01$ $\chi^2 > 9.21034$

8.38 Consider the accompanying 2×3 (i.e., $r = 2$ and $c = 3$) contingency table.

		Column		
		1	**2**	**3**
Row	**1**	9	34	53
	2	16	30	25

  a. Specify the null and alternative hypotheses that should be used in testing the independence of the row and column classifications.
  b. Specify the test statistic and the rejection region that should be used in conducting the hypothesis test of part **a**. Use $\alpha = .01$. $\chi^2 > 9.21034$
  c. Assuming the row classification and the column classification are independent, find estimates for the expected cell counts.
  d. Conduct the hypothesis test of part **a**. Interpret your result. $\chi^2 = 8.71$

8.39 Refer to Exercise 8.38.
  a. Convert the frequency responses to percentages by calculating the percentage of each column total falling in each row. Also convert the row totals to percentages of the total number of responses. Display the percentages in a table.
  b. Create a bar graph with row 1 percentage on the vertical axis and column number on the horizontal axis. Show the row 1 total percentage as a horizontal line on the graph.
  c. What pattern do you expect to see if the rows and columns are independent? Does the plot support the result of the test of independence in Exercise 8.38?

8.40 Test the null hypothesis of independence of the two classifications, A and B, of the 3×3 contingency table shown here. Test using $\alpha = .05$. $\chi^2 = 12.33$

		B		
		B₁	**B₂**	**B₃**
	A₁	40	72	42
A	**A₂**	63	53	70
	A₃	31	38	30

8.41 Refer to Exercise 8.40. Convert the responses to percentages by calculating the percentage of each B class total falling into each A classification. Also, calculate the percentage of the total number of responses that constitute each of the A classification totals.
  a. Create a bar graph with row A_1 percentage on the vertical axis and B classification on the horizontal axis. Does the graph support the result of the test of hypothesis in Exercise 8.40? Explain.
  b. Repeat part **a** for the row A_2 percentages.
  c. Repeat part **a** for the row A_3 percentages.

Applying the Concepts—Basic

8.42 The *Journal of the American Medical Association* (Apr. 18, 2001) published the results of a study of alcohol consumption in patients suffering from acute myocardial infarction (AMI). The patients were classified according to average number of alcoholic drinks per week and whether or not they had congestive heart failure. A summary of the results for 1,913 AMI patients is shown in the table.

 AMI

		Alcohol Consumption		
		Abstainers	**Less than 7 drinks/week**	**7 or more drinks/week**
Congestive Heart Failure	**Yes**	146	106	29
	No	750	590	292
	Totals	896	696	321

Source: Mukamal, K. J., *et al.* "Prior alcohol consumption and mortality following acute myocardial infarction." *Journal of the American Medical Association,* Vol. 285, No. 15, Apr. 18, 2001 (Table 1).

  a. Find the sample proportion of abstainers with congestive heart failure.   .163
  b. Find the sample proportion of moderate drinkers (patients who have less than 7 drinks per week) with congestive heart failure.   .152
  c. Find the sample proportion of heavy drinkers (patients who have 7 or more drinks per week) with congestive heart failure.   .090
  d. Compare the sample proportions, parts **a–c**. Does it appear that the proportion of AMI patients with congestive heart failure depends on alcohol consumption?
  e. Give the null hypothesis for testing whether the proportion of AMI patients with congestive heart failure depends on alcohol consumption.
  f. Use the MINITAB printout on p. 426 to conduct the test, part **e**. Test at $\alpha = .05$. $p = .006$

8.43 One of the most common musculoskeletal disorders is lumbar disk disease (LDD). Medical researchers

MINITAB Output for Exercise 8.42

```
Expected counts are printed below observed counts

        ABSTAIN    LESS7   7ORMORE   Total
   1       146       106        29     281
        131.61    102.24     47.15

   2       750       590       292    1632
        764.39    593.76    273.85

 Total     896       696       321    1913

Chi-Sq =  1.573 +  0.139 +  6.988 +
          0.271 +  0.024 +  1.203 = 10.197
DF = 2, P-Value = 0.006
```

reported finding a common genetic risk factor for LDD (*Journal of the American Medical Association*, Apr. 11, 2001). The study included 171 Finnish patients diagnosed with LDD (the patient group) and 321 Finnish individuals without LDD (the control group). Of the 171 LDD patients, 21 were discovered to have the genetic trait. Of the 321 people in the control group, 15 had the genetic trait.

a. Consider the two categorical variables, group and presence/absence of genetic trait. Form a 2 × 2 contingency table for these variables.

b. Conduct a test to determine whether the genetic trait occurs at a higher rate in LDD patients than in the controls without LDD. Use $\alpha = .01$. $\chi^2 = 9.52$

c. Construct a bar graph that will visually support your conclusion in part **b**.

8.44 The *American Journal of Public Health* (July 1995) reported on a population-based study of trauma in Hispanic children. One of the objectives of the study was to compare the use of protective devices in motor vehicles used to transport Hispanic and non-Hispanic white children. On the basis of data collected from the San Diego County Regionalized Trauma System, 792 children treated for injuries sustained in vehicular accidents were classified according to ethnic status (Hispanic or non-Hispanic white) and seatbelt usage (worn or not worn) during the accident. The data are summarized in the table.

⊘ TRAUMA

	Hispanic	Non-Hispanic White	Totals
Seatbelts worn	31	148	179
Seatbelts not worn	283	330	613
Totals	314	478	792

Source: Matteneci, R.M., *et al.* "Trauma among Hispanic children: A population-based study in a regionalized system of trauma care." *American Journal of Public Health*, Vol. 85, No. 7, July 1995, p. 1007 (Table 2).

a. Calculate the sample proportion of injured Hispanic children who were not wearing seatbelts during the accident. .901

b. Calculate the sample proportion of injured non-Hispanic white children who were not wearing seatbelts during the accident. .690

c. Compare the two sample proportions, parts **a** and **b**. Do you think the true population proportions differ?

d. Conduct a test to determine whether seatbelt usage in motor vehicle accidents depends on ethnic status in the San Diego County Regionalized Trauma System. Use $\alpha = .01$. $\chi^2 = 48.2$

e. Construct a 99% confidence interval for the difference between the proportions, parts **a** and **b**. Interpret the interval. .211 ± .070

Applying the Concepts—Intermediate

8.45 University of Maryland professor Ted R. Gurr examined the political strategies used by ethnic groups worldwide in their fight for minority rights (*Political Science & Politics*, June 2000). Each in a sample of 275 ethnic groups was classified according to world region and highest level of political action reported. The data are summarized in the accompanying contingency table. Conduct a test at $\alpha = .10$ to determine whether political strategy of ethnic groups depends on world region. Support your answer with a graph.

⊘ ETHNIC

		No Political Action	**Mobilization, Mass Action**	**Terrorism, Rebellion, Civil War**
			Political Strategy	
	Latin American	24	31	7
	Post-Communist	32	23	4
World Region	South, Southeast, East Asia	11	22	26
	Africa/Middle East	39	36	20

Source: Gurr, T. R. "Nonviolence in ethnopolitics: Strategies for the attainment of group rights and autonomy." *Political Science & Politics*, Vol. 33, No. 2, June 2000 (Table 1).

8.46 An article in *Sociological Methods & Research* (May 2001) analyzed the data presented in the table. A sample of 262 Kansas pig farmers were classified according to their education level (college or not) and size of their pig farm (number of pigs). Conduct a test to determine whether a pig farmer's education level has an impact on the size of the pig farm. Use $\alpha = .05$ and support your answer with a graph. $\chi^2 = 2.142$

⊘PIGFARM

		Education Level		
		No College	**College**	**Totals**
	<1,000 pigs	42	53	95
Farm Size	**1,000–2,000 pigs**	27	42	69
	2,001–5,000 pigs	22	20	42
	>5,000 pigs	27	29	56
	Totals	118	144	262

Source: Agresti, A., and Liu, I. "Strategies for modeling a categorical variable allowing multiple category choices." *Sociological Methods & Research,* Vol. 29, No. 4, May 2001 (Table I).

8.47 A field study was conducted to identify the natural predators of the gypsy moth (*Environmental Entomology*, June 1995). For one part of the study, 24 black-capped chickadees (common wintering birds) were captured in mist nets and individually caged. Each bird was offered a mass of gypsy moth eggs attached to a piece of bark. Half the birds were offered no other food (no choice) and half were offered a variety of other naturally occurring foods such as spruce and pine seeds (choice). The numbers of birds that did and did not feed on the gypsy moth egg mass are given in the table. Analyze the data in the table to determine if a relationship exists between food choice and whether or not the chickadees fed on gypsy moth eggs. Use $\alpha = .10$.

⊘MOTH

	Fed on Egg Mass	
	Yes	**No**
Choice of foods	2	10
No choice	8	4

8.48 In order to evaluate their situational awareness, fighter aircraft pilots participate in battle simulations. At a random point in the trial, the simulator is frozen and data on situation awareness are immediately collected. The simulation is then continued until, ultimately, performance (e.g., number of kills) is measured. A study reported in *Human Factors* (Mar. 1995) investigated whether temporarily stopping the simulation results in any change in pilot performance. Trials were designed so that some simulations were stopped to collect situation awareness data while others were not stopped. Each trial was then classified according to the number of kills made by the pilot. The data for 180 trials are summarized in the contingency table, and an SPSS printout of the analysis is provided on p. 428. Fully interpret the results. $p = .704$

⊘SIMKILLS

	Number of Kills					
	0	**1**	**2**	**3**	**4**	**Totals**
Stops	32	33	19	5	2	91
No Stops	24	36	18	8	3	89
Totals	56	69	37	13	5	180

8.49 A standardized procedure for determining a person's susceptibility to hypnosis is the Stanford Hypnotic Susceptibility Scale, Form C (SHSS:C). Recently, a new method called the Computer-Assisted Hypnosis Scale (CAHS), which uses a computer as a facilitator of hypnosis, has been developed. Each scale classifies a person's hypnotic susceptibility as low, medium, high, and very high. Researchers at the University of Tennessee compared the two scales by administering both tests to each of 130 undergraduate volunteers (*Psychological Assessment*, Mar. 1995). The hypnotic classifications are summarized in the table below. A chi-square analysis of the data in the contingency table is shown in the SAS printout on p. 429. Note that the expected cell counts are provided in the computer printout.

a. Check to see if the assumption of expected cell counts of 5 or more is satisfied. Should you proceed with the analysis? Explain.   No

⊘HYPNOSIS

		CAHS Level				
		Low	**Medium**	**High**	**Very High**	**Totals**
	Low	32	14	2	0	48
SHSS: C Level	**Medium**	11	14	6	0	31
	High	6	14	19	3	42
	Very High	0	2	4	3	9
	Totals	49	44	31	6	130

Source: Grant, C. D., and Nash, M. R. "The Computer-Assisted Hypnosis Scale: Standardization and norming of a computer-administered measure of hypnotic ability." *Psychological Assessment*, Vol. 7, No. 1, Mar. 1995, p. 53 (Table 4).

SPSS Output for Exercise 8.48

STOPS *KILLS Crosstabulation

			KILLS					Total
			0	1	2	3	4	
STOPS	NO	Count	24	36	18	8	3	89
		Expected Count	27.7	34.1	18.3	6.4	2.5	89.0
	YES	Count	32	33	19	5	2	91
		Expected Count	28.3	34.9	18.7	6.6	2.5	91.0
Total		Count	56	69	37	13	5	180
		Expected Count	56.0	69.0	37.0	13.0	5.0	180.0

Chi-Square Tests

	Value	df	Asymp. Sig. (2-sided)
Pearson Chi-Square	2.171[a]	4	.704
Likelihood Ratio	2.182	4	.702
Linear-by-Linear Association	1.401	1	.237
N of Valid Cases	180		

a. 2 cells (20.0%) have expected count less than 5. The minimum expected count is 2.47.

b. One way to satisfy the assumption, part **a**, is to combine the data for two or more categories in the contingency table (e.g., high and very high). Form a new contingency table by combining the data for the high and very high categories in both the rows and columns.

c. Calculate the expected cell counts in the new contingency table, part **c**. Is the assumption now satisfied?

d. Perform the chi-square test on the new contingency table. Use $\alpha = .05$. Interpret the results.

8.50 The National Gang Crime Research Center (NGCRC) has developed a 6-level gang classification system for both adults and juveniles. The six categories are shown in the table below. The classification system was developed as a potential predictor of a gang member's propensity for violence when in prison, jail, or a correctional facility. To test this theory, the NGCRC collected data on approximately 10,000 confined offenders and assigned each a score using the gang classification system (*Journal of Gang Research*, Winter 1997). One of several other variables measured by the NGCRC was whether or not the offender has ever carried a homemade weapon (e.g., knife) while in custody.

The data on gang score and homemade weapon are summarized in the table. Conduct a test to determine if carrying a homemade weapon in custody depends on gang classification score. (Use $\alpha = .01$.) Support your conclusion with a graph.

Applying the Concepts—Advanced

8.51 New, effective AIDS vaccines are now being developed using the process of "sieving," that is, sifting out infections with some strains of HIV. Harvard School of Public Health statistician Peter Gilbert demonstrated how to

◉ HIVVAC1

		MN Strain		
		Positive	Negative	Totals
Patient Group	Unvaccinated	22	9	31
	Vaccinated	2	5	7
	Totals	24	14	38

Source: Gilbert, P. "Developing an AIDS vaccine by sieving." *Chance*, Vol. 13, No. 4, Fall 2000.

◉ GANGS

Gang Classification Score	Homemade Weapon Carried	
	Yes	No
0 (Never joined a gang, no close friends in a gang)	255	2,551
1 (Never joined a gang, 1–4 close friends in a gang)	110	560
2 (Never joined a gang, 5 or more friends in a gang)	151	636
3 (Inactive gang member)	271	959
4 (Active gang member, no position of rank)	175	513
5 (Active gang member, holds position of rank)	476	831

Source: Knox, G. W., *et al.* "A gang classification system for corrections." *Journal of Gang Research*, Vol. 4, No. 2, Winter 1997, p. 54 (Table 4).

SAS Output for Exercise 8.49

The FREQ Procedure

Table of SHSSC by CAHS

SHSSC   CAHS

Frequency Expected Col Pct	High	Low	Medium	VeryHi	Total
High	19 10.015	6 15.831	14 14.215	3 1.9385	42
Low	2 11.446	32 18.092	14 16.246	0 2.2154	48
Medium	6 7.3923	11 11.685	14 10.492	0 1.4308	31
VeryHi	4 2.1462	0 3.3923	2 3.0462	3 0.4154	9
Total	31	49	44	6	130

Statistics for Table of SHSSC by CAHS

Statistic	DF	Value	Prob
Chi-Square	9	60.1026	<.0001
Likelihood Ratio Chi-Square	9	59.6382	<.0001
Mantel-Haenszel Chi-Square	1	2.3505	0.1252
Phi Coefficient		0.6799	
Contingency Coefficient		0.5623	
Cramer's V		0.3926	

WARNING: 44% of the cells have expected counts less than 5. Chi-Square may not be a valid test.

Sample Size = 130

test the efficacy of an HIV vaccine in *Chance* (Fall 2000). As an example, Gilbert reported the results of VaxGen's preliminary HIV vaccine trial using the 2×2 table on p. 428. The vaccine was designed to eliminate a particular strain of the virus, called the "MN strain." The trial consisted of 7 AIDS patients vaccinated with the new drug and 31 AIDS patients who were treated with a placebo (no vaccination). The table shows the number of patients who tested positive and negative for the MN strain in the trial follow-up period.

a. Conduct a test to determine whether the vaccine is effective in treating the MN strain of HIV. Use $\alpha = .05$.

b. Are the assumptions for the test, part **a**, satisfied? What are the consequences if the assumptions are violated?

c. In the case of a 2×2 contingency table, R.A. Fisher (1935) developed a procedure for computing the

exact *p*-value for the test (called *Fishers exact test*). The method utilizes a probability of the form:

$$\frac{\binom{7}{2}\binom{31}{22}}{\binom{38}{24}}$$

This represents the probability that 2 out of 7 vaccinated AIDS patients test positive and 22 out of 31 unvaccinated patients test positive, that is, the probability of the table result given that the null hypothesis of independence is true. Compute this probability (called the *probability of the contingency table*).

d. Refer to part **c**. Two contingency tables (with the same marginal totals as the original table) that are more contradictory to the null hypothesis of independence than the observed table are shown at right. First, explain why these tables provide more evidence to reject H_0 than the original table. Then, compute the probability of each table using the hypergeometric formula.   .0057; .0003

e. The *p*-value of Fisher's exact test is the probability of observing a result at least as contradictory to the null hypothesis as the observed contingency table, given the same marginal totals. Sum the probabilities of parts **c** and **d** to obtain the *p*-value of Fisher's exact test. (To verify your calculations, check the *p*-value at the bottom of the accompanying SAS printout labeled **Left-sided Pr < = F**.) Interpret this value in the context of the vaccine trial.

⊘ HIVVAC2

		MN Strain		
		Positive	Negative	Totals
Patient Group	Unvaccinated	23	8	31
	Vaccinated	1	6	7
	Totals	24	14	38

⊘ HIVVAC3

		MN Strain		
		Positive	Negative	Totals
Patient Group	Unvaccinated	24	7	31
	Vaccinated	0	7	7
	Totals	24	14	38

SAS Output for Exercise 8.51

The FREQ Procedure

Table of group by MNSTRAIN

GROUP   MNSTRAIN

Frequency Expected	NEG	POS	Total
UNVACC	9 11.421	22 19.579	31
VACC	5 2.5789	2 4.4211	7
Total	14	24	38

Statistics for Table of group by MNSTRAIN

Statistic	DF	Value	Prob
Chi-Square	1	4.4112	0.0357
Likelihood Ratio Chi-Square	1	4.2893	0.0384
Continuity Adj. Chi-Square	1	2.7773	0.0956
Mantel-Haenszel Chi-Square	1	4.2952	0.0382
Phi Coefficient		-0.3407	
Contingency Coefficient		0.3225	
Cramer's V		-0.3407	

WARNING: 50% of the cells have expected counts less than 5. Chi-Square may not be a valid test.

Fisher's Exact Test

Cell (1, 1) Frequency (F)	9
Left-sided Pr <= F	0.0498
Right-sided Pr >= F	0.9940
Table Probability (P)	0.0438
Two-sided Pr <= P	0.0772

Sample Size = 38

STATISTICS IN ACTION
The Level of Agreement Among Movie Reviewers: Thumbs Up or Thumbs Down?

For over 20 years, movie critics Gene Siskel of the *Chicago Tribune* and Roger Ebert of the *Chicago Sun-Times* evaluated the latest movie releases on national television, first on PBS with *Sneak Previews*, then in syndication with *Siskel and Ebert at the Movies*. During their entertaining 30-minute shows, Siskel (now deceased) and Ebert would face each other across a theater aisle and discuss the merits or faults of a movie. After engaging in a lively debate about the film, the reviewers would give the movie either a "thumbs up" or a "thumbs down." To film advertisers and producers, "two thumbs up" from Siskel and Ebert would almost assure a money-making movie.

The popularity of Siskel and Ebert's television shows have encouraged other well-known critics to imitate the face-to-face format. For example, Michael Medved and Jeffrey Lyons star on PBS's *Sneak Previews*, and Gene Shalit (NBC's *Today Show*) and Rex Reed (*New York Observer*) face off in a syndicated TV show. (Currently, Roger Ebert teams with Richard Roeper, Siskel's replacement.) The confrontational nature of these programs leave the viewer with the impression that the critics disagree about most movies. But how strongly do these popular movie reviewers really disagree?

To answer this question, University of Florida statisticians Alan Agresti and Larry Winner collected data on the reviews of movie critics (*Chance*, Spring 1997). They built a database from the weekly reviews that were published in *Variety* magazine during 1996. Each critic's review was categorized as Pro ("thumbs up"), Con ("thumbs down") or Mixed (neutral opinion). Consequently, for the movies evaluated by Siskel and Ebert, each film has a Siskel rating (Pro, Con, or Mixed) and an Ebert rating (Pro, Con, or Mixed).

⊘ SISEBERT contains the data for 160 movie reviews by Siskel and Ebert

⊘ MEDLYONS contains the data for 123 movie reviews by Medved and Lyons

Focus

a. Conduct an analysis to determine whether the movie reviews of Gene Siskel and Roger Ebert are independent. Support your conclusion with a graph.

b. Based on your analysis, part a, do Siskel and Ebert tend to agree or disagree? Explain.

c. Repeat parts a and b for the movie reviews of Michael Medved and Jeffrey Lyons.

QUICK REVIEW

Key Terms

Categories   407
Cell   407
Cell counts   407
Chi-square distribution   409
Chi-square test   409
Classes   407
Contingency table   416

Dependence   418
Dimensions of classification   416
Expected cell count   411
Independence of two classifications   417
Marginal probabilities   416
Multinomial experiment   407

Observed cell count   417
One-way table   408
Sample survey   398
Sampling distribution of $\hat{p}_1 - \hat{p}_2$ 399
Two-way table   416

Key Formulas

Confidence interval for $p_1 - p_2$:

$$(\hat{p}_1 - \hat{p}_2) \pm z_{\alpha/2} \sqrt{\frac{\hat{p}_1\hat{q}_1}{n_1} + \frac{\hat{p}_2\hat{q}_2}{n_2}} \quad 400$$

Test statistic for $H_0: p_1 - p_2 = 0$:

$$z = \frac{(\hat{p}_1 - \hat{p}_2) - 0}{\sqrt{\hat{p}\hat{q}\left(\frac{1}{n_2} + \frac{1}{n_2}\right)}} \quad 400$$

where $\hat{p} = \dfrac{x_1 + x_2}{n_1 + n_2}$

$\hat{q} = 1 - \hat{p}$

Sample size for estimating $(p_1 - p_2)$:

$$n_1 = n_2 = \frac{(z_{\alpha/2})^2(p_1q_1 + p_2q_2)}{B^2} \quad 406$$

Test statistic for one-way table:

$$\chi^2 = \sum \frac{[n_i - E(n_i)]^2}{E(n_i)} \quad 411$$

where n_i = count for cell i

$E(n_i) = np_{i,0}$

$p_{i,0}$ = hypothesized value of p_i in H_0

Test statistic for two-way table: $\chi^2 = \sum \dfrac{[n_{ij} - \hat{E}(n_{ij})]^2}{\hat{E}(n_{ij})}$ 400

where n_{ij} = count for cell in row i, column j

$\hat{E}(n_{ij}) = r_i c_j / n$

r_i = total for row i

c_j = total for column j

n = total sample size

Language Lab

Symbol	Pronunciation	Description
$p_1 - p_2$	*p*-1 minus *p*-2	Difference between population proportions
$\hat{p}_1 - \hat{p}_2$	*p*-1 hat minus *p*-2 hat	Difference between sample proportions
$\sigma_{(\hat{p}_1 - \hat{p}_2)}$	sigma of *p*-1 hat minus *p*-2 hat	Standard deviation of the sampling distribution of $(\hat{p}_1 - \hat{p}_2)$
D_0	*D* naught	Hypothesized value of $\hat{p}_1 - \hat{p}_2$
$p_{i,0}$	*p-i*-zero	Value of multinomial probability p_i hypothesized in H_0
χ^2	Chi-square	Test statistic used in analysis of count data
n_i	*n-i*	Number of observed outcomes in cell i of one-way table
$E(n_i)$	Expected value of n_i	Expected number of outcomes in cell i of one-way table when H_0 is true
p_{ij}	*p-i-j*	Probability of an outcome in row i and column j of a two-way contingency table
n_{ij}	*n-i-j*	Number of observed outcomes in row i and column j of a two-way contingency table
$\hat{E}(n_{ij})$	Estimated expected value of n_{ij}	Estimated expected number of outcomes in row i and column j of a two-way contingency table
r_i	*r-i*	Total number of outcomes in row i of a contingency table
c_j	*c-j*	Total number of outcomes in column j of a contingency table

 SUPPLEMENTARY EXERCISES 8.52–8.69

Learning the Mechanics

8.52 Independent random samples were selected from two binomial populations. The size and number of observed successes for each sample are shown in the table.

Sample 1	Sample 2
$n_1 = 200$	$n_2 = 200$
$x_1 = 110$	$x_2 = 130$

  a. Test $H_0: (p_1 - p_2) = 0$ against $H_a: (p_1 - p_2) < 0$. Use $\alpha = .10$. $z = -2.04$

  b. Form a 95% confidence interval for $(p_1 - p_2)$.

  c. What sample sizes would be required if we wish to use a 95% confidence interval of width .01 to estimate $(p_1 - p_2)$? $n_1 = n_2 = 72{,}991$

8.53 A random sample of 150 observations was classified into the categories shown in the table.

	Category				
	1	**2**	**3**	**4**	**5**
n_i	28	35	33	25	29

  a. Do the data provide sufficient evidence that the categories are not equally likely? Use $\alpha = .10$.

  b. Form a 90% confidence interval for p_2, the probability that an observation will fall in category 2.

8.54 A random sample of 250 observations was classified according to the row and column categories shown in the table on p. 433.

		Column		
		1	**2**	**3**
Row	**1**	20	20	10
	2	10	20	70
	3	20	50	30

a. Do the data provide sufficient evidence to conclude that the rows and columns are dependent? Test using $\alpha = .05$. $\chi^2 = 54.14$

b. Would the analysis change if the row totals were fixed before the data were collected? No

c. Do the assumptions required for the analysis to be valid differ according to whether the row (or column) totals are fixed? Explain. Yes

d. Convert the table entries to percentages by using each column total as a base and calculating each row response as a percentage of the corresponding column total. In addition, calculate the row totals and convert them to percentages of all 250 observations.

e. Create a bar graph with row 1 percentages on the vertical axis and the column number on the horizontal axis. Draw a horizontal line corresponding to the row 1 total percentage. Does the graph support the result of the test conducted in part **a**?

f. Repeat part **e** for the row 2 percentages.

g. Repeat part **e** for the row 3 percentages.

Applying the Concepts—Basic

8.55 According to research reported in the *Journal of the National Cancer Institute* (Apr. 1991), eating foods high in fiber may help protect against breast cancer. The researchers randomly divided 120 laboratory rats into four groups of 30 each. All rats were injected with a drug that causes breast cancer, then each rat was fed a diet of fat and fiber for 15 weeks. However, the levels of fat and fiber varied from group to group. At the end of the feeding period, the number of rats with cancer tumors was determined for each group. The data are summarized in the table at the bottom of the page.

a. Does the sampling appear to satisfy the assumptions for a multinomial experiment? Explain.

b. Calculate the expected cell counts for the contingency table.

c. Calculate the χ^2 statistic. $\chi^2 = 12.9$

d. Is there evidence to indicate that diet and presence/absence of cancer are independent? Test using $\alpha = .05$.

e. Compare the percentage of rats on a high-fat/no-fiber diet with cancer to the percentage of rats on a high-fat/fiber diet with cancer using a 95% confidence interval. Interpret the result. $.233 \pm .200$

8.56 Does marriage hinder the career progression of women to a greater degree than that of men? According to *The Journal of Applied Psychology* (Vol. 77, 1992), in a random sample of 795 male managers and 223 female managers from 20 Fortune 500 corporations, 86% of the male managers and 45% of the female managers were married.

a. Use a 95% confidence interval to estimate the difference between the proportions of men and women managers who are married.

b. Interpret the interval. State your conclusion in terms of the populations from which the samples were drawn.

8.57 The *Journal of Intellectual Disability Research* (Feb. 1995) published a longitudinal study of hearing impairment in a group of elderly patients with intellectual disability. The hearing function of each patient was screened each year over a 10-year period. At the study's conclusion, the hearing loss of each patient was categorized as severe, moderate, mild, or none. The classifications of the 28 surviving patients are summarized in the table below.

⊘ HEARIMP

Hearing Loss	Number of Patients
None	7
Mild	7
Moderate	9
Severe	5
Total	28

a. Conduct a test to determine whether the true proportions of intellectually disabled elderly patients in each of the hearing-loss categories differ. Use $\alpha = .05$. $\chi^2 = 1.143$

b. Use a 90% confidence interval to estimate the proportion of disabled elderly patients with severe hearing loss. $.1786 \pm .1191$

⊘ TUMORS

		Diet				
		High-Fat/No-Fiber	**High-Fat/Fiber**	**Low-Fat/No-Fiber**	**Low-Fat/Fiber**	**Totals**
Cancer Tumors	**Yes**	27	20	19	14	80
	No	3	10	11	16	40
	Totals	30	30	30	30	120

Source: Tampa Tribune, Apr. 3, 1991.

8.58 According to one geneticist's theory, a crossing of red and white snapdragons should produce offspring that are 25% red, 50% pink, and 25% white. An experiment conducted to test the theory produced 30 red, 78 pink, and 36 white offspring in 144 crossings. Do the data provide sufficient evidence to contradict the geneticist's theory? $\chi^2 = 1.5$

8.59 A sociologist was interested in knowing whether sons have a tendency to choose the same occupation as their fathers. To investigate this question, 500 males were polled and questioned concerning their occupations and those of their fathers. A summary of the numbers of father–son pairs falling in each occupational category is shown in the table below. Do the data provide sufficient evidence at $\alpha = .05$ to indicate a dependence between a son's choice of occupation and his father's occupation? Use the accompanying SAS printout to conduct the analysis.

Data for Exercise 8.59

⊘ FATHSON

		Son			
		Professional or Business	**Skilled**	**Unskilled**	**Farmer**
Father	**Professional or Business**	55	38	7	0
	Skilled	79	71	25	0
	Unskilled	22	75	38	10
	Farmer	15	23	10	32

SAS Output for Exercise 8.59

The FREQ Procedure

Table of FATHER by SON

FATHER   SON

Frequency Expected	FARMER	PROF/BUS	SKILLED	UNSKILL	Total
FARMER	32	15	23	10	80
	6.72	27.36	33.12	12.8	
PROF/BUS	0	55	38	7	100
	8.4	34.2	41.4	16	
SKILLED	0	79	71	25	175
	14.7	59.85	72.45	28	
UNSKILL	10	22	75	38	145
	12.18	49.59	60.03	23.2	
Total	42	171	207	80	500

Statistics for Table of FATHER by SON

Statistic	DF	Value	Prob
Chi-Square	9	180.8739	<.0001
Likelihood Ratio Chi-Square	9	160.8323	<.0001
Mantel-Haenszel Chi-Square	1	52.0404	<.0001
Phi Coefficient		0.6015	
Contingency Coefficient		0.5154	
Cramer's V		0.3473	

Sample Size = 500

QUAKE

	Counties			
	Contra Costa	**Santa Clara**	**Los Angeles**	**San Bernardino**
Sample Size	521	556	337	372
Number with Earthquake Insurance	117	222	133	109

Applying the Concepts—Intermediate

8.60 An article in the *Annals of the Association of American Geographers* (June 1992) investigated many factors that are considered when purchasing earthquake insurance. One factor investigated was the proximity to a major earthquake fault. Surveys were mailed to residents in four California counties. The data collected are shown in the table above.

    a. State the hypothesis necessary to test the theory that the proportion of earthquake-insured residents differs for the four sampled counties.

    b. Do the data provide evidence of a difference in the proportion of earthquake-insured residents for the four sampled counties? Test using $\alpha = .05$.

    c. Los Angeles County is the closest of the four to a major earthquake fault. Calculate 95% confidence intervals for the difference in the proportions of earthquake-insured residents in Los Angeles County and each of the other counties.

    d. Do these results support the contention that closer proximities to major earthquake faults result in higher proportions of earthquake-insured residents?

    e. How large would the samples from Los Angeles and San Bernardino counties have to be in order to estimate the difference between earthquake-insured proportions to within .03 with 95% confidence?

8.61 Inositol—a complex alcohol nutrient found in breast milk—has been found to reduce the risk of lung and eye damage in premature infants. The study, published in the *New England Journal of Medicine* (May 7, 1992), involved 220 premature infants. The infants were randomly divided into two groups; half (110) of the infants received an intravenous feeding of inositol, while the other half received a standard diet. The researchers found that 14 of the inositol-fed infants suffered retinopathy of prematurity, an eye injury that results from the high oxygen levels needed to compensate for poorly developed lungs. In contrast, 29 of the infants on the standard diet developed this disease. Analyze the results with a test of hypothesis. Use $\alpha = .01$.

8.62 Operation Crossroads was a 1946 military exercise in which atomic bombs were detonated over empty target ships in the Pacific Ocean. The Navy assigned sailors to wash down the test ships immediately after the atomic blasts. The National Academy of Science reported "that the overall death rate among Operation Crossroads sailors was 4.6% higher than among a comparable group

of sailors. ... However, this increase was not statistically significant." (*Tampa Tribune*, Oct. 30, 1996.)

    a. Describe the parameter of interest in the National Academy of Science study.

    b. Interpret the statement: "This increase was not statistically significant."

8.63 An experimenter wishes to determine whether there is a relationship between hair color and eye color. One hundred people are randomly sampled and the eyes and hair of each person are judged to be light or dark. A summary of the number of people in each of the four categories is shown in the table. Do the data provide sufficient evidence at $\alpha = .10$ to indicate a relationship between eye and hair color? Use the MINITAB printout provided here to perform the analysis.

HAIREYE

	Light Hair	**Dark Hair**	**Totals**
Light Eyes	31	21	52
Dark Eyes	14	34	48
Totals	45	55	100

MINITAB Output for Exercise 8.63

```
Expected counts are printed below observed counts

        LGHTHAIR DARKHAIR      Total
   1         31       21         52
          23.40    28.60

   2         14       34         48
          21.60    26.40

Total        45       55        100

Chi-Sq =  2.468 +  2.020 +
          2.674 +  2.188 = 9.350
DF = 1, P-Value = 0.002
```

8.64 Marketing researchers often use the purchases of a sample of households, called a *scanner panel*, to understand the purchasing patterns and preferences of consumers. When shopping, these households present a magnetic identification card that permits their purchase data to be identified and aggregated. Researchers recently studied the extent to which panel households' purchase behavior is representative of the population of households shopping at the same stores (*Journal of Marketing Research*, Nov. 1996). The table below reports the peanut butter purchase data collected by A.C. Nielson Company for the panel of 2,500 households in Sioux Falls, South Dakota, over a 102-week period. The market share percentages in the right-hand column are derived from all peanut butter purchases at the same 15 stores at which the panel shopped during the same 102-week period.

⊘SCANNER

Brand	Size	Number of Purchases by Household Panel	Market Shares
Jif	18 oz.	3,165	20.10%
Jif	28	1,892	10.10
Jif	40	726	5.42
Peter Pan	10	4,079	16.01
Skippy	18	6,206	28.65
Skippy	28	1,627	12.38
Skippy	40	1,420	7.32
Total		19,115	

Source: Gupta, S., Chintagunta, P., Kaul, A., and Wittink, D. "Do household scanner data provide representative inferences from brand choices: A comparison with store data." *Journal of Marketing Research*, Vol. 33, Nov. 1996, p. 393 (Table 6).

a. Do the data provide sufficient evidence to conclude that the purchases of the household panel are representative of the population of households? Test using $\alpha = .05$. $\chi^2 = 880.5$

b. What assumptions must hold to ensure the validity of the testing procedure you used in part **a**?

c. Find the approximate *p*-value for the test of part **a** and interpret it in the context of the problem.

8.65 Doctors have theorized that renegade antibodies in blood may be a major underlying cause of stroke. To investigate the causal link between antibodies and stroke, medical researchers examined two groups of patients for renegade antibodies (*Tampa Tribune*, Feb. 2, 1992). One group of 255 patients had suffered a first stroke while the other group of 257 patients were hospitalized for other reasons. Twenty-five of the stroke patients had antibodies in their blood compared to 11 of the other patients. Analyze the data for the medical researchers. What conclusions can you draw from the analysis?

8.66 The *American Journal on Mental Retardation* (Jan. 1992) published a study of the social interactions of two groups of children. Independent random samples of 15 children with and 15 children without developmental delays (i.e., mild mental retardation) were the subjects of the experiment. After observing the children during "freeplay," the number of children who exhibited disruptive behavior (e.g., ignoring or rejecting other children, taking toys from another child) was recorded for each group. The data are summarized in the two-way table at the bottom of the page. Analyze the data and interpret the results.

8.67 *Human Factors* (Dec. 1988) published a study of color brightness as a body orientation clue. Ninety college students, reclining on their backs in the dark, were disoriented when positioned on a rotating platform under a slowly rotating disk that blocked their field of vision. The subjects were asked to say "Stop" when they felt as if they were right-side up. The position of the brightness pattern on the disk in relation to each student's body orientation was then recorded. Subjects selected only three disk brightness patterns as subjective vertical clues: (1) brighter side up, (2) darker side up, and (3) brighter and darker sides aligned on either side of the subjects' heads. The frequency counts for the experiment are given in the accompanying table. Conduct a test to compare the proportions of subjects that fall in the three disk-orientation categories. Assume you want to determine whether the three proportions differ. Use $\alpha = .05$. $\chi^2 = 39.267$

Disk Orientation		
Brighter Side Up	**Darker Side Up**	**Bright and Dark Sides Aligned**
58	15	17

Applying the Concepts—Advanced

8.68 The *Journal of Adlerian Theory, Research and Practice* (Mar. 1986) investigated the role lifestyles play in forming a marriage. Adlerian psychologists group people into one of the following five lifestyle types:

1. *Pleasing priority:* a strong desire to make others happy and win their approval
2. *Achieving priority:* industrial, responsible, orderly, with an active approach to life (particularly in the workplace)

⊘DISRUPT

	Disruptive Behavior	Nondisruptive Behavior	Totals
With Developmental Delays	12	3	15
Without Developmental Delays	5	10	15
Total	17	13	30

Source: Kopp, C. B., Baker, B., and Brown, K. W. "Social skills and their correlates: Preschoolers with developmental delays." *American Journal on Mental Retardation*, Vol. 96, No. 4, Jan. 1992.

● LIFESTYLE

		Wife's Lifestyle					
		Pleasing	**Outdoing**	**Avoiding**	**Control**	**Achieving**	**Totals**
Husband's Lifestyle	**Pleasing**	9	6	6	3	9	33
	Outdoing	7	11	12	11	5	46
	Avoiding	8	8	6	11	5	38
	Control	8	10	7	15	11	51
	Achieving	4	6	7	10	7	34
	Totals	36	41	38	50	37	202

Source: Evans, T. D., and Bozarth, J. "Pairing of personality profiles in marriage." *Journal of Adlerian Theory, Research and Practice*, Vol. 42, No. 1, March 1986, pp. 59–64.

3. *Outdoing priority:* a strong desire to be on top, in a position of superiority over others; likes to be considered the best at whatever he/she does

4. *Suppressing (control) priority:* maintains control over his/her emotions; discloses very little of himself/herself

5. *Avoiding priority:* is easily hurt; avoids any potentially painful (physical or emotional) situation

Data on both wife's and husband's lifestyles were obtained for 202 married couples living in family housing at the University of Georgia. The results are summarized in the table above.

a. The key question of the study is: "Do the personality profiles contribute to marriage pairings?" In other words, is there a relationship between the personality priorities of the husband and wife pairs? Analyze the data for the researchers. $\chi^2 = 13.715$

b. If no relationship between the personality priorities of the husband and wife pairs is found, part **a**, then analyze the data for husbands and wives separately. Are there differences in the proportions of husbands who fall in the five personality priorities? Are there differences in the proportions of wives who fall in the five personality profiles?

8.69 A statistical analysis is to be done on a set of data consisting of 1,000 monthly salaries. The analysis requires the assumption that the sample was drawn from a normal distribution. A preliminary test, called the χ^2 *goodness-of-fit test*, can be used to help determine whether it is reasonable to assume that the sample is from a normal distribution. Suppose the mean and standard deviation of the 1,000 salaries are hypothesized to be $1,200 and $200, respectively. Using the standard normal table, we can approximate the probability of a salary being in the intervals listed in the next table. The third column represents the expected number of the 1,000 salaries to be found in each interval if the sample was drawn from a normal distribution with $\mu = \$1,200$ and $\sigma = \$200$. Suppose the last column contains the actual observed frequencies in the sample. Large differences between the observed and expected frequencies cast doubt on the normality assumption.

Interval	Probability	Expected Frequency	Observed Frequency
Less than $800	.023	23	26
$800 < $1,000	.136	136	146
$1,000 < $1,200	.341	341	361
$1,200 < $1,400	.341	341	311
$1,400 < $1,600	.136	136	143
$1,600 or above	.023	23	13

a. Compute the χ^2 statistic based on the observed and expected frequencies. $\chi^2 = 9.647$

b. Find the tabulated χ^2 value when $\alpha = .05$ and there are 5 degrees of freedom. (There are $k - 1 = 5$ df associated with this χ^2 statistic.)

c. Based on the χ^2 statistic and the tabulated χ^2 value, is there evidence that the salary distribution is non-normal? No

d. Find the approximate observed significance level for the test in part **c**. $.05 < p < .10$

STUDENT PROJECTS

Many researchers rely on surveys to estimate the proportions of experimental units in populations that possess certain specified characteristics. A political scientist may want to estimate the proportion of an electorate in favor of a certain legislative bill.

A social scientist may be interested in the proportions of people in a geographical region who fall in certain socioeconomic classifications. A psychologist might want to compare the proportions of patients who have different psychological disorders.

Choose a specific topic, similar to those described above, that interests you. Clearly define the population of interest, identify data categories of specific interest, and identify the proportions associated with them. Now *guesstimate* the proportions of the population that you think fall in each of the categories. For example, you might guess that all the proportions are equal, or that the first proportion is twice as large as the second but equal to the third, etc.

You are now ready to collect the data by obtaining a random sample from your population of interest. Select a sample size so that all expected cell counts are at least 5 (preferably larger), then collect the data.

Use the count data you have obtained to test the null hypothesis that the true proportions in the population are equal to your presampling guesstimates. Would failure to reject this null hypothesis imply that your guesstimates are correct?

REFERENCES

Agresti, A. *Categorical Data Analysis.* New York: Wiley, 1990.

Cochran, W. G. "The χ^2 test of goodness of fit." *Annals of Mathematical Statistics*, 1952, 23.

Conover, W. J. *Practical Nonparametric Statistics*, 2nd ed. New York: Wiley, 1980.

Fisher, R. A. "The logic of inductive inference (with discussion)." *Journal of the Royal Statistical Society*, Vol. 98, 1935, pp. 39–82.

Hollander, M., and Wolfe, D. A. *Nonparametric Statistical Methods.* New York: Wiley, 1973.

Savage, I. R. "Bibliography of nonparametric statistics and related topics." *Journal of the American Statistical Association*, 1953, 48.

Simple Linear Regression

Contents

Statistics in Action

Can "Dowsers" Really Detect Water?

👆 Where We've Been

We've learned how to estimate and test hypotheses about population parameters based on a random sample of observations from the population. We've also seen how to extend these methods to allow for a comparison of parameters from two or more populations.

☞ Where We're Going

Suppose we want to predict the rainfall at a given location on a given day. We could select a single random sample of n daily rainfalls, use the methods of Chapter 5 to estimate the mean daily rainfall μ, and then use this quantity to predict the day's rainfall. A better method uses information that is available to any forecaster, e.g., barometric pressure and cloud cover. If we measure barometric pressure and cloud cover at the same time as daily rainfall, we establish the relationship between these variables—one that lets us use these variables for prediction. This chapter covers the simplest situation—relating two variables.

In Chapters 5–7 we described methods for making inferences about population means. The mean of a population has been treated as a *constant*, and we have shown how to use sample data to estimate or to test hypotheses about this constant mean. In many applications, the mean of a population is not viewed as a constant, but rather as a variable. For example, the mean sale price of residences in a large city last year can be treated as a constant and might be equal to $150,000. But we might also treat the mean sale price as a variable that depends on the square feet of living space in the residence. For example, the relationship might be

$$\text{Mean sale price} = \$30{,}000 + \$60(\text{Square feet})$$

This formula implies that the mean sale price of 1,000-square-foot homes is $90,000, the mean sale price of 2,000-square-foot homes is $150,000, and the mean sale price of 3,000-square-foot homes is $210,000.

What do we gain by treating the mean as a variable rather than a constant? In many practical applications we will be dealing with highly variable data, data for which the standard deviation is so large that a constant mean is almost "lost" in a sea of variability. For example, if the mean residential sale price is $150,000 but the standard deviation is $75,000, then the actual sale prices will vary considerably, and the mean price is not a very meaningful or useful characterization of the price distribution. On the other hand, if the mean sale price is treated as a variable that depends on the square feet of living space, the standard deviation of sale prices for any given size of home might be only $10,000. In this case, the mean price will provide a much better characterization of sale prices when it is treated as a variable rather than a constant.

TEACHING TIP
Discuss other examples of when one variable can be predicted by the value of another variable.

In this chapter we discuss situations in which the mean of the population is treated as a variable, dependent on the value of another variable. The dependence of residential sale price on the square feet of living space is one illustration. Other examples include the dependence of mean reaction time on the amount of a drug in the bloodstream, the dependence of mean starting salary of a college graduate on the student's GPA, and the dependence of mean number of years to which a criminal is sentenced on the number of previous convictions.

In this chapter we discuss the simplest of all models relating a population mean to another variable, the *straight-line model*. We show how to use the sample data to estimate the straight-line relationship between the mean value of one variable, *y*, as it relates to a second variable, *x*. The methodology of estimating and using a straight-line relationship is referred to as *simple linear regression analysis*.

9.1   PROBABILISTIC MODELS

An important consideration when taking a drug is how it may affect one's perception or general awareness. Suppose you want to model the length of time it takes to respond to a stimulus (a measure of awareness) as a function of the percentage of a certain drug in the bloodstream. The first question to be answered is this: "Do you think an exact relationship exists between these two variables?" That is, do you think it is possible to state the exact length of time it takes an individual (subject) to respond if the amount of the drug in the bloodstream is known? We think you will agree with us that this is *not* possible for several reasons. The reaction time depends on many variables other than the percentage of the drug in the bloodstream—for example, the time of day, the amount of sleep the subject had the night before, the subject's visual acuity, the subject's general reaction time without the drug, and the subject's age would all probably affect reaction time. Even if many variables are included in a model, it is still unlikely that we would be able to predict *exactly* the subject's reaction time. There will almost certainly be some variation in response times due strictly to *random phenomena* that cannot be modeled or explained.

If we were to construct a model that hypothesized an exact relationship between variables, it would be called a **deterministic model**. For example, if we believe that y, the reaction time (in seconds), will be exactly one and one-half times x, the amount of drug in the blood, we write

$$y = 1.5x$$

Suggested Exercise 9.7

This represents a **deterministic relationship** between the variables y and x. It implies that y can always be determined exactly when the value of x is known. *There is no allowance for error in this prediction.*

If, on the other hand, we believe there will be unexplained variation in reaction times—perhaps caused by important but unincluded variables or by random phenomena—we discard the deterministic model and use a model that accounts for this **random error**. This **probabilistic model** includes both a deterministic component and a random error component. For example, if we hypothesize that the response time y is related to the percentage of drug x by

$$y = 1.5x + \text{Random error}$$

we are hypothesizing a **probabilistic relationship** between y and x. Note that the deterministic component of this probabilistic model is $1.5x$.

Figure 9.1a shows the possible responses for five different values of x, the percentage of drug in the blood, when the model is deterministic. All the responses must fall exactly on the line because a deterministic model leaves no room for error.

Figure 9.1b shows a possible set of responses for the same values of x when we are using a probabilistic model. Note that the deterministic part of the model (the straight line itself) is the same. Now, however, the inclusion of a random error component allows the response times to vary from this line. Since we know that the response time does vary randomly for a given value of x, the probabilistic model provides a more realistic model for y than does the deterministic model.

FIGURE 9.1

Possible Reaction Times, y, for Five Different Drug Percentages, x

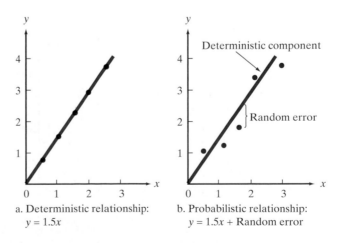

a. Deterministic relationship: $y = 1.5x$

b. Probabilistic relationship: $y = 1.5x + \text{Random error}$

TEACHING TIP
Emphasize that the regression
model allows individual data
values to fall around the
regression line but the mean
values fall on the regression line.

General Form of Probabilistic Models

y = Deterministic component + Random error

where y is the variable of interest. We always assume that the mean value of the random error equals 0. This is equivalent to assuming that the mean value of y, $E(y)$, equals the deterministic component of the model; that is,

$E(y)$ = Deterministic component

In this chapter we present the simplest of probabilistic models—the **straight-line model**—which derives its name from the fact that the deterministic portion of the model graphs as a straight line. Fitting this model to a set of data is an example of **regression analysis**, or **regression modeling**. The elements of the straight-line model are summarized in the next box.

TEACHING TIP
Point out that the slope
interpretation of β_1 is only valid
when using a simple linear
regression model. As the models
become more complicated, this
interpretation will change.

A First-Order (Straight-Line) Probabilistic Model

$y = \beta_0 + \beta_1 x + \varepsilon$

where y = **Dependent** *or* **response variable** (variable to be modeled)

x = **Independent** *or* **predictor variable** (variable used as a predictor of y)*

$E(y) = \beta_0 + \beta_1 x$ = Deterministic component

ε (epsilon) = Random error component

β_0 (beta zero) = **y-intercept of the line**, that is, the point at which the line intersects or cuts through the y-axis (see Figure 9.2)

β_1 (beta one) = **Slope of the line**, that is, the amount of increase (or decrease) in the deterministic component of y for every 1-unit increase in x. [As you can see in Figure 9.2, $E(y)$ increases by the amount β_1 as x increases from 2 to 3.]

FIGURE 9.2

The Straight-Line Model

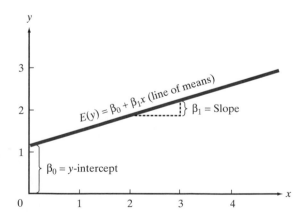

---

*The word *independent* should not be interpreted in a probabilistic sense, as defined in Chapter 3. The phrase *independent variable* is used in regression analysis to refer to a predictor variable for the response y.

In the probabilistic model, the deterministic component is referred to as the **line of means**, because the mean of y, $E(y)$, is equal to the straight-line component of the model. That is,

$$E(y) = \beta_0 + \beta_1 x$$

Note that the Greek symbols β_0 and β_1, respectively, represent the y-intercept and slope of the model. They are population parameters that will be known only if we have access to the entire population of (x, y) measurements. Together with a specific value of the independent variable x, they determine the mean value of y, which is just a specific point on the line of means (Figure 9.2).

The values of β_0 and β_1 will be unknown in almost all practical applications of regression analysis. The process of developing a model, estimating the unknown parameters, and using the model can be viewed as the five-step procedure shown in the next box.

<table>
<tr><td>Step 1</td><td>Hypothesize the deterministic component of the model that relates the mean, $E(y)$, to the independent variable x (Section 9.1).</td></tr>
<tr><td>Step 2</td><td>Use the sample data to estimate unknown parameters in the model (Section 9.2).</td></tr>
<tr><td>Step 3</td><td>Specify the probability distribution of the random error term and estimate the standard deviation of this distribution (Sections 9.3 and 9.4).</td></tr>
<tr><td>Step 4</td><td>Statistically evaluate the usefulness of the model (Sections 9.5, 9.6, and 9.7).</td></tr>
<tr><td>Step 5</td><td>When satisfied that the model is useful, use it for prediction, estimation, and other purposes (Section 9.8).</td></tr>
</table>

TEACHING TIP
Point out that these are basically the same steps necessary to complete a multiple regression analysis as well.

 EXERCISES 9.1–9.9

Learning the Mechanics

9.1 In each case, graph the line that passes through the given points.
 a. $(1, 1)$ and $(5, 5)$
 b. $(0, 3)$ and $(3, 0)$
 c. $(-1, 1)$ and $(4, 2)$
 d. $(-6, -3)$ and $(2, 6)$

9.2 Give the slope and y-intercept for each of the lines graphed in Exercise 9.1.

9.3 The equation for a straight line (deterministic) is

$$y = \beta_0 + \beta_1 x$$

If the line passes through the point $(-2, 4)$, then $x = -2, y = 4$ must satisfy the equation; that is,

$$4 = \beta_0 + \beta_1(-2)$$

Similarly, if the line passes through the point $(4, 6)$, then $x = 4, y = 6$ must satisfy the equation; that is,

$$6 = \beta_0 + \beta_1(4)$$

Use these two equations to solve for β_0 and β_1; then find the equation of the line that passes through the points $(-2, 4)$ and $(4, 6)$. $\beta_1 = \frac{1}{3}; \beta_0 = \frac{14}{3}$

9.4 Refer to Exercise 9.3. Find the equations of the lines that pass through the points listed in Exercise 9.1.

9.5 Plot the following lines:
 a. $y = 4 + x$
 b. $y = 5 - 2x$
 c. $y = -4 + 3x$
 d. $y = -2x$
 e. $y = x$
 f. $y = .50 + 1.5x$

9.6 Give the slope and y-intercept for each of the lines defined in Exercise 9.5.

9.7 Why do we generally prefer a probabilistic model to a deterministic model? Give examples for which the two types of models might be appropriate.

9.8 What is the line of means?

9.9 If a straight-line probabilistic relationship relates the mean $E(y)$ to an independent variable x, does it imply that every value of the variable y will always fall exactly on the line of means? Why or why not?

9.2   FITTING THE MODEL: THE LEAST SQUARES APPROACH

After the straight-line model has been hypothesized to relate the mean $E(y)$ to the independent variable x, the next step is to collect data and to estimate the (unknown) population parameters, the y-intercept β_0 and the slope β_1.

To begin with a simple example, suppose an experiment involving five subjects is conducted to determine the relationship between the percentage of a certain drug in the bloodstream and the length of time it takes to react to a stimulus. The results are shown in Table 9.1. (The number of measurements and the measurements themselves are unrealistically simple in order to avoid arithmetic confusion in this introductory example.) This set of data will be used to demonstrate the five-step procedure of regression modeling given in Section 9.1. In this section we hypothesize the deterministic component of the model and estimate its unknown parameters (steps 1 and 2). The model assumptions and the random error component (step 3) are the subjects of Sections 9.3 and 9.4, whereas Sections 9.5–9.7 assess the utility of the model (step 4). Finally, we use the model for prediction and estimation (step 5) in Section 9.8.

⊘STIMULUS

TABLE 9.I   Reaction Time Versus Drug Percentage

Subject	Amount of Drug x (%)	Reaction Time y (seconds)
1	1	1
2	2	1
3	3	2
4	4	2
5	5	4

Step 1   *Hypothesize the deterministic component of the probabilistic model.* As stated before, we will consider only straight-line models in this chapter. Thus the complete model to relate mean response time $E(y)$ to drug percentage x is given by

$$E(y) = \beta_0 + \beta_1 x$$

Step 2   *Use sample data to estimate unknown parameters in the model.* This step is the subject of this section—namely, how can we best use the information in the sample of five observations in Table 9.1 to estimate the unknown y-intercept β_0 and slope β_1?

To determine whether a linear relationship between y and x is plausible, it is helpful to plot the sample data in a **scattergram**. Recall (Section 2.9) that a scattergram locates each of the five data points on a graph, as shown in Figure 9.3. Note that the scattergram suggests a general tendency for y to increase as x increases. If you place a ruler on the scattergram, you will see that a line may be

FIGURE 9.3

Scattergram for Data in Table 9.1

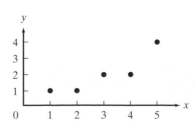

FIGURE 9.4

*Visual Straight Line Fitted
to the Data in Figure 9.3*

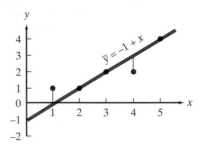

drawn through three of the five points, as shown in Figure 9.4. To obtain the equation of this visually fitted line, note that the line intersects the y-axis at $y = -1$, so the y-intercept is -1. Also, y increases exactly 1 unit for every 1-unit increase in x, indicating that the slope is $+1$. Therefore, the equation is

$$\tilde{y} = -1 + 1(x) = -1 + x$$

where $\tilde{y}$ is used to denote the predicted y from the visual model.

One way to decide quantitatively how well a straight line fits a set of data is to note the extent to which the data points deviate from the line. For example, to evaluate the model in Figure 9.4, we calculate the magnitude of the *deviations*, i.e., the differences between the observed and the predicted values of y. These deviations, or **errors of prediction**, are the vertical distances between observed and predicted values (see Figure 9.4). The observed and predicted values of y, their differences, and their squared differences are shown in Table 9.2. Note that the *sum of errors* equals 0 and the *sum of squares of the errors* (SSE), which gives greater emphasis to large deviations of the points from the line, is equal to 2.

**TABLE 9.2   Comparing Observed and Predicted Values
for the Visual Model**

x	y	$\tilde{y} = -1 + x$	$(y - \tilde{y})$	$(y - \tilde{y})^2$
1	1	0	$(1 - 0) = \quad 1$	1
2	1	1	$(1 - 1) = \quad 0$	0
3	2	2	$(2 - 2) = \quad 0$	0
4	2	3	$(2 - 3) = -1$	1
5	4	4	$(4 - 4) = \quad 0$	0
			Sum of errors = $\quad$ 0	Sum of squared errors (SSE) = 2

You can see by shifting the ruler around the graph that it is possible to find many lines for which the sum of errors is equal to 0, but it can be shown that there is one (and only one) line for which the SSE is a *minimum*. This line is called the **least squares line**, the **regression line**, or the **least squares prediction equation**. The methodology used to obtain this line is called the **method of least squares**.

To find the least squares prediction equation for a set of data, assume that we have a sample of n data points consisting of pairs of values of x and y, say (x_1, y_1), $(x_2, y_2), \ldots, (x_n, y_n)$. For example, the $n = 5$ data points shown in Table 9.2 are $(1, 1), (2, 1), (3, 2), (4, 2),$ and $(5, 4)$. The fitted line, which we will calculate based on the five data points, is written as

$$\hat{y} = \hat{\beta}_0 + \hat{\beta}_1 x$$

The "hats" indicate that the symbols below them are estimates: $\hat{y}$ (y-hat) is an estimator of the mean value of y, $E(y)$, and a predictor of some future value of y; and $\hat{\beta}_0$ and $\hat{\beta}_1$ are estimators of β_0 and β_1, respectively.

For a given data point, say the point (x_i, y_i), the observed value of y is y_i and the predicted value of y would be obtained by substituting x_i into the prediction equation:

$$\hat{y}_i = \hat{\beta}_0 + \hat{\beta}_1 x_i$$

And the deviation of the ith value of y from its predicted value is

$$(y_i - \hat{y}_i) = [y_i - (\hat{\beta}_0 + \hat{\beta}_1 x_i)]$$

Then the sum of squares of the deviations of the y-values about their predicted values for all the n data points is

$$\text{SSE} = \sum [y_i - (\hat{\beta}_0 + \hat{\beta}_1 x_i)]^2$$

The quantities $\hat{\beta}_0$ and $\hat{\beta}_1$ that make the SSE a minimum are called the **least squares estimates** of the population parameters β_0 and β_1, and the prediction equation $\hat{y} = \hat{\beta}_0 + \hat{\beta}_1 x$ is called the *least squares line*.

DEFINITION 9.1

The **least squares line** is one that has the following two properties:

1. the sum of the errors (SE) equals 0
2. the sum of squared errors (SSE) is smaller than that for any other straight-line model

---

The values of $\hat{\beta}_0$ and $\hat{\beta}_1$ that minimize the SSE are (proof omitted) given by the formulas in the next box.*

Formulas for the Least Squares Estimates

Slope: $\hat{\beta}_1 = \dfrac{\text{SS}_{xy}}{\text{SS}_{xx}}$

y-intercept: $\hat{\beta}_0 = \bar{y} - \hat{\beta}_1 \bar{x}$

where $\text{SS}_{xy} = \sum (x_i - \bar{x})(y_i - \bar{y}) = \sum x_i y_i - \dfrac{\left(\sum x_i\right)\left(\sum y_i\right)}{n}$

$\text{SS}_{xx} = \sum (x_i - \bar{x})^2 = \sum x_i^2 - \dfrac{\left(\sum x_i\right)^2}{n}$

$n = $ Sample size

Preliminary computations for finding the least squares line for the drug reaction example are presented in Table 9.3. We can now calculate

---

*Students who are familiar with calculus should note that the values of β_0 and β_1 that minimize SSE $= \Sigma(y_i - \hat{y}_i)^2$ are obtained by setting the two partial derivatives $\partial \text{SSE}/\partial \beta_0$ and $\partial \text{SSE}/\partial \beta_1$ equal to 0. The solutions to these two equations yield the formulas shown in the box. Furthermore, we denote the *sample* solutions to the equations by $\hat{\beta}_0$ and $\hat{\beta}_1$, where the "hat" denotes that these are sample estimates of the true population intercept β_0 and slope β_1.

TABLE 9.3 **Preliminary Computations for the Drug Reaction Example**

x_i	y_i	x_i^2	x_iy_i
1	1	1	1
2	1	4	2
3	2	9	6
4	2	16	8
5	4	25	20
Totals $\sum x_i = 15$	$\sum y_i = 10$	$\sum x_i^2 = 55$	$\sum x_iy_i = 37$

$$SS_{xy} = \sum x_iy_i - \frac{\left(\sum x_i\right)\left(\sum y_i\right)}{5} = 37 - \frac{(15)(10)}{5} = 37 - 30 = 7$$

$$SS_{xx} = \sum x_i^2 - \frac{\left(\sum x_i\right)^2}{5} = 55 - \frac{(15)^2}{5} = 55 - 45 = 10$$

Suggested Exercise 9.13

Then the slope of the least squares line is

$$\hat{\beta}_1 = \frac{SS_{xy}}{SS_{xx}} = \frac{7}{10} = .7$$

and the *y*-intercept is

$$\hat{\beta}_0 = \bar{y} - \hat{\beta}_1\bar{x} = \frac{\sum y_i}{5} - \hat{\beta}_1\frac{\sum x_i}{5}$$

$$= \frac{10}{5} - (.7)\left(\frac{15}{5}\right) = 2 - (.7)(3) = 2 - 2.1 = -.1$$

The least squares line is thus

$$\hat{y} = \hat{\beta}_0 + \hat{\beta}_1x = -.1 + .7x$$

The graph of this line is shown in Figure 9.5.

The predicted value of *y* for a given value of *x* can be obtained by substituting into the formula for the least squares line. Thus, when $x = 2$ we predict *y* to be

$$\hat{y} = -.1 + .7x = -.1 + .7(2) = 1.3$$

We show how to find a prediction interval for *y* in Section 9.8.

The observed and predicted values of *y*, the deviations of the *y* values about their predicted values, and the squares of these deviations are shown in Table 9.4. Note that the sum of squares of the deviations, SSE, is 1.10, and (as we would expect) this is less than the SSE $= 2.0$ obtained in Table 9.2 for the visually fitted line.

FIGURE 9.5

The Line $\hat{y} = -.1 + .7x$
Fitted to the Data

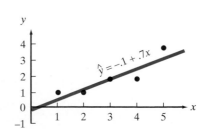

TABLE 9.4 **Comparing Observed and Predicted Values for the Least Squares Prediction Equation**

x	y	$\hat{y} = -.1 + .7x$	$(y - \hat{y})$	$(y - \hat{y})^2$
1	1	.6	$(1 - .6) = .4$	.16
2	1	1.3	$(1 - 1.3) = -.3$	.09
3	2	2.0	$(2 - 2.0) = 0$	.00
4	2	2.7	$(2 - 2.7) = -.7$	.49
5	4	3.4	$(4 - 3.4) = .6$	.36
			Sum of errors = 0	SSE = 1.10

The calculations required to obtain $\hat{\beta}_0$, $\hat{\beta}_1$, and SSE in simple linear regression, although straightforward, can become rather tedious. Even with the use of a pocket calculator, the process is laborious and susceptible to error, especially when the sample size is large. Fortunately, the use of statistical computer software can significantly reduce the labor involved in regression calculations. The SAS, SPSS, MINITAB, and STATISTIX outputs for the simple linear regression of the data in Table 9.1 are displayed in Figure 9.6a–d. The values of $\hat{\beta}_0$ and $\hat{\beta}_1$ are highlighted on the printouts. These values, $\hat{\beta}_0 = -.1$ and $\hat{\beta}_1 = .7$, agree exactly with our hand-calculated values. The value of SSE = 1.10 is also highlighted on the printouts.

Whether you use a hand calculator or a computer, it is important that you be able to interpret the intercept and slope in terms of the data being utilized to fit the model. In the drug reaction example, the estimated y-intercept, $\hat{\beta}_0 = -.1$, appears to imply that the estimated mean reaction time is equal to $-.1$ second when the amount of drug, x, is equal to 0%. Since negative reaction times are not possible, this seems to make the model nonsensical. However, *the model parameters should be interpreted only within the sampled range of the independent variable*—in this case, for amounts of drug in the bloodstream between 1% and 5%. Thus, the y-intercept—which is, by definition, at $x = 0$ (0% drug)—is not within the range of the sampled values of x and is not subject to meaningful interpretation.

TEACHING TIP
The interpretation of β_0 only makes sense if the value of $x = 0$ is possible and the data collected includes values of x over this range.

FIGURE 9.6a

SAS Printout for Time–Drug Regression

The REG Procedure
Dependent Variable: TIME_Y

Analysis of Variance

Source	DF	Sum of Squares	Mean Square	F Value	Pr > F
Model	1	4.90000	4.90000	13.36	0.0354
Error	3	1.10000	0.36667		
Corrected Total	4	6.00000			

Root MSE		0.60553	R-Square	0.8167	
Dependent Mean		2.00000	Adj R-Sq	0.7556	
Coeff Var		30.27650			

Parameter Estimates

Variable	DF	Parameter Estimate	Standard Error	t Value	Pr > |t|
Intercept	1	-0.10000	0.63509	-0.16	0.8849
DRUG_X	1	0.70000	0.19149	3.66	0.0354

FIGURE 9.6b

SPSS Printout for Time–Drug Regression

Model Summary

Model	R	R Square	Adjusted R Square	Std. Error of the Estimate
1	.904[a]	.817	.756	.61

a. Predictors: (Constant), DRUG_X

ANOVA[b]

Model		Sum of Squares	df	Mean Square	F	Sig.
1	Regression	4.900	1	4.900	13.364	.035[a]
	Residual	1.100	3	.367		
	Total	6.000	4			

a. Predictors: (Constant), DRUG_X
b. Dependent Variable: Time_Y

Coefficients[a]

Model		Unstandardized Coefficients B	Std. Error	Standardized Coefficients Beta	t	Sig.
1	(Constant)	-1.00E-01	.635		-.157	.885
	DRUG_X	.700	.191	.904	3.656	.035

a. Dependent Variable: Time_Y

FIGURE 9.6c

MINITAB Printout for Time–Drug Regression

```
The regression equation is
TIME_Y = - 0.100 + 0.700 DRUG_X


Predictor        Coef      SE Coef          T         P
Constant      -0.1000       0.6351      -0.16     0.885
DRUG_X         0.7000       0.1915       3.66     0.035


S = 0.6055       R-Sq = 81.7%     R-Sq(adj) = 75.6%


Analysis of Variance


Source            DF          SS          MS         F         P
Regression         1      4.9000      4.9000     13.36     0.035
Residual Error     3      1.1000      0.3667
Total              4      6.0000
```

FIGURE 9.6d

STATISTIX Printout for Time–Drug Regression

```
UNWEIGHTED LEAST SQUARES LINEAR REGRESSION OF TIME_Y

PREDICTOR
VARIABLES    COEFFICIENT    STD ERROR     STUDENT'S T        P
---------    -----------    ---------    -------------    ------

CONSTANT      -0.10000       0.63509         -0.16        0.8849
DRUG_X         0.70000       0.19149          3.66        0.0354


R-SQUARED              0.8167      RESID. MEAN SQUARE  (MSE)    0.36667
ADJUSTED R-SQUARED     0.7556      STANDARD DEVIATION           0.60553


SOURCE          DF      SS           MS           F        P
----------      ---    ----------   ----------   -----    ------

REGRESSION       1     4.90000      4.90000      13.36    0.0354
RESIDUAL         3     1.10000      0.36667
TOTAL            4     6.00000


CASES INCLUDED 5    MISSING CASES 0
```

Using the TI-83 Graphing Calculator

Straight-Line (Linear) Regression on the TI-83

A.   Finding the Least Squares Regression Equation

Step 1   *Enter the data*
Press **STAT   1**   for   **STAT Edit**
Enter your *x*-data in **L1** and your *y*-data in **L2**

Step 2   *Find the equation*
Press **STAT** and highlight **CALC**
Press **4** for **LinReg(ax + b)**
Press **ENTER**

The screen will show the values for *a* and *b* in the equation
$y = ax + b$.

Example   The figures below show a table of data entered on the TI-83 and the regression equation obtained using the steps given above.

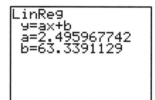

B.   Finding *r* and r^2

(*Note:* We discuss these statistics in Sections 9.6 and 9.7.)

If *r* and r^2 do not already appear on the LinReg screen from part A,

Step 1   *Turn the diagnostics feature on*
Press **2nd 0** for **CATALOG**
Press the x^{-1} key for **D**
Press the down arrow key until **DiagnosticsOn** is highlighted
Press **ENTER** twice

Step 2   *Find the regression equation as shown in part A above*
The values for *r* and r^2 will appear on the screen as well.

Example: The figures below show a table of data entered on the TI-83 and the regression equation, *r*, and r^2 obtained using the steps given above.

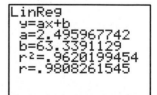

USING THE TI-83 GRAPHING CALCULATOR *(continued)*

C.   Graphing the Least Squares Line With the Scatterplot

Step 1   *Enter the data as shown in part A above*

Step 2   *Set up the data plot*
Press   **2nd**   **Y=**   for **STATPLOT**
Press   **1**   for **PLOT1**

Use the arrow and **ENTER** keys to set up the screen as shown below.

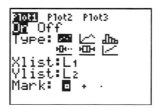

Step 3   *Find the regression equation as shown in part A above*

Step 4   *Enter the regression equation*
Press **Y=**
Press **VARS 5** for **Statistics** . . .
Highlight **EQ** and press **ENTER**

You should see the regression equation in the **Y=** window.

Step 5   *View the scatterplot and regression line*
Press ZOOM 9 for ZoomStat

You should see the data graphed along with the regression line.

Example   The figures below show a table of data entered on the TI-83, and the graph of the scatterplot and least squares line obtained using the steps given above.

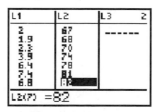 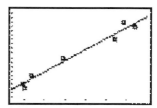

The slope of the least squares line, $\hat{\beta} = .7$, implies that for every unit increase of x, the mean value of y is estimated to increase by .7 unit. In terms of this example, for every 1% increase in the amount of drug in the bloodstream, the mean reaction time is estimated to increase by .7 second *over the sampled range of drug amounts from 1% to 5%*. Thus, the model does not imply that increasing the drug amount from 5% to 10% will result in an increase in mean reaction time of 3.5 seconds, because the range of x in the sample does not extend to 10% ($x = 10$). In fact, 10% might be such a high concentration that the drug would kill the subject! Be careful to interpret the estimated parameters only within the sampled range of x.

Even when the interpretations of the estimated parameters are meaningful, we need to remember that they are only estimates based on the sample. As such, their values will typically change in repeated sampling. How much confidence do we have that the estimated slope, $\hat{\beta}_1$, accurately approximates the true slope, β_1? This requires statistical inference, in the form of confidence intervals and tests of hypotheses, which we address in Section 9.5.

To summarize, we defined the best-fitting straight line to be the one that minimizes the sum of squared errors around the line, and we called it the least squares line. We should interpret the least squares line only within the sampled range of the independent variable. In subsequent sections we show how to make statistical inferences about the model.

 EXERCISES 9.10–9.24

Learning the Mechanics

9.10 The following table is similar to Table 9.3. It is used for making the preliminary computations for finding the least squares line for the given pairs of x and y values.
  a. Complete the table.
  b. Find SS_{xy}. -26.286
  c. Find SS_{xx}. 33.714
  d. Find $\hat{\beta}_1$. $-.7797$
  e. Find $\bar{x}$ and $\bar{y}$. $3.429, 4.429$
  f. Find $\hat{\beta}_0$. 7.102
  g. Find the least squares line. $\hat{y} = 7.102 - .7797x$

x_i	y_i	x_i^2	x_iy_i
7	2	—	—
4	4	—	—
6	2	—	—
2	5	—	—
1	7	—	—
1	6	—	—
3	5	—	—
Totals $\sum x_i =$	$\sum y_i =$	$\sum x_i^2 =$	$\sum x_iy_i =$

9.11 Refer to Exercise 9.10. After the least squares line has been obtained, the table below (which is similar to Table 9.4) can be used for (1) comparing the observed and the predicted values of y, and (2) computing SSE.

x	y	$\hat{y}$	$(y - \hat{y})$	$(y - \hat{y})^2$
7	2	—	—	—
4	4	—	—	—
6	2	—	—	—
2	5	—	—	—
1	7	—	—	—
1	6	—	—	—
3	5	—	—	—
			$\sum (y - \hat{y}) =$	$SSE = \sum (y - \hat{y})^2 =$

  a. Complete the table.
  b. Plot the least squares line on a scattergram of the data. Plot the following line on the same graph:

$$\hat{y} = 14 - 2.5x$$

  c. Show that SSE is larger for the line in part **b** than it is for the least squares line.

9.12 Construct a scattergram for the data in the following table.

x	.5	1	1.5
y	2	1	3

  a. Plot the following two lines on your scattergram:

$$y = 3 - x \quad \text{and} \quad y = 1 + x$$

  b. Which of these lines would you choose to characterize the relationship between x and y? Explain.
  c. Show that the sum of errors for both of these lines equals 0.
  d. Which of these lines has the smaller SSE?
  e. Find the least squares line for the data and compare it to the two lines described in part **a**.

9.13 Consider the following pairs of measurements:

x	5	3	−1	2	7	6	4
y	4	3	0	1	8	5	3

  a. Construct a scattergram for these data.
  b. What does the scattergram suggest about the relationship between x and y?
  c. Given that $SS_{xx} = 43.4286$, $SS_{xy} = 39.8571$, $\bar{y} = 3.4286$, and $\bar{x} = 3.7143$, calculate the least squares estimates of β_0 and β_1.

d. Plot the least squares line on your scattergram. Does the line appear to fit the data well? Explain.

e. Interpret the *y*-intercept and slope of the least squares line. Over what range of *x* are these interpretations meaningful?

Applying the Concepts—Basic

9.14 Real estate investors, home buyers, and home owners often use the appraised value of a property as a basis for predicting sale price. Data on sale prices and total appraised values of 92 residential properties sold in 1999 in an upscale Tampa, Florida, neighborhood named Tampa Palms are saved in the TAMPALMS file. The first five and last five observations of the data set are listed in the accompanying table.

⊘ TAMPALMS

Property	Appraised Value	Sale Price
1	$ 170,432	$ 180,000
2	212,827	245,100
3	68,130	85,400
4	65,505	87,900
5	68,655	84,200
⋮	⋮	⋮
88	195,862	244,000
89	176,850	219,000
90	95,718	132,000
91	137,108	156,900
92	183,704	263,000

Source: Hillsborough County (Florida) Property Appraiser's Office.

MINITAB Scatterplot for Exercise 9.14

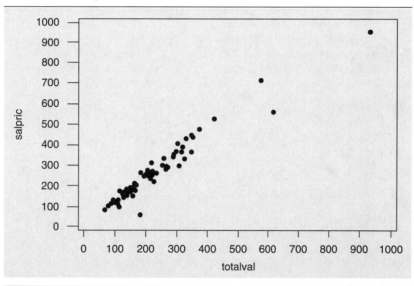

MINITAB Simple Linear Regression Output for Exercise 9.14

```
The regression equation is
salepric = 20.9 + 1.07 totalval

Predictor      Coef      SE Coef        T          P
Constant      20.942       6.446      3.25      0.002
totalval     1.06873     0.02709     39.45      0.000

S = 32.79      R-Sq = 94.5%      R-Sq(adj) = 94.5%

Analysis of Variance

Source           DF        SS          MS          F          P
Regression        1    1673142     1673142    1556.48      0.000
Residual Error   90      96746        1075
Total            91    1769888
```

a. Propose a straight-line model to relate the appraised property value x to the sale price y for residential properties in this neighborhood.

b. A MINITAB scatterplot of the data is shown on p. 453. [*Note*: Both sale price and total appraised value are shown in thousands of dollars.] Does it appear that a straight-line model will be an appropriate fit to the data? Yes

c. A MINITAB simple linear regression printout is also shown on p. 453. Find the equation of the best-fitting line through the data on the printout.

d. Interpret the y-intercept of the least squares line. Does it have a practical meaning for this application? Explain.

e. Interpret the slope of the least squares line. Over what range of x is the interpretation meaningful?

f. Use the least squares model to estimate the mean sale price of a property appraised at $300,000.

9.15 In Denver, Colorado, environmentalists have discovered a link between high arsenic levels in soil and a crabgrass killer used in the 1950s and 1960s (*Environmental Science & Technology*, Sept. 1, 2000). The recent discovery was based, in part, on the scattergrams shown below. The graphs plot the level of the metals cadmium and arsenic, respectively, against the distance from a former smelter plant for samples of soil taken from Denver residential properties.

a. Normally, the metal level in soil decreases as distance from the source (e.g., a smelter plant) increases. Propose a straight-line model relating metal level y to distance from the plant x. Based on the theory, would you expect the slope of the line to be positive or negative? $y = \beta_0 + \beta_1 x + \varepsilon$

b. Examine the scatterplot for cadmium. Does the plot support the theory, part **a**? Yes

c. Examine the scatterplot for arsenic. Does the plot support the theory, part **a**? (*Note:* This finding led investigators to discover the link between high arsenic levels and the use of the crabgrass killer.) No

9.16 In *Chance* (Winter, 2000), statistician Howard Wainer and two students compared men's and women's winning times in the Boston Marathon. One of the graphs used to illustrate gender differences is reproduced on p. 455. The scattergram plots the winning times (in minutes) against year in which the race was run. Men's times are represented by solid dots and women's times by open circles.

a. Consider only the winning times for men. Is there evidence of a linear trend? If so, propose a straight-line model for predicting winning time (y) based on

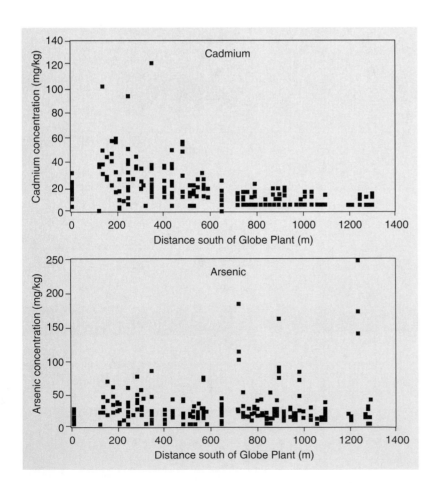

year (x). Would you expect the slope of this line to be positive or negative?

b. Repeat part **b** for women's times.

c. Which slope, men's or women's, will be greater in absolute value?   Women's

d. Would you recommend using the straight-line models to predict the winning time in the 2020 Boston Marathon? Why or why not?   No

9.17 Refer to the *Brain and Behavior Evolution* (Apr. 2000) study of the feeding behavior of blackbream fish, Exercise 2.110 (p. 89). Recall that the zoologists recorded the number of aggressive strikes of two blackbream fish feeding at the bottom of an aquarium in the 10-minute period following the addition of food. The table listing the weekly number of strikes and age of the fish (in days) is reproduced here.

⊘ BLACKBREAM

Week	Number of Strikes	Age of Fish (days)
1	85	120
2	63	136
3	34	150
4	39	155
5	58	162
6	35	169
7	57	178
8	12	184
9	15	190

Source: Shand, J., *et al.* "Variability in the location of the retinal ganglion cell area centralis is correlated with ontogenetic changes in feeding behavior in the Blackbream, Acanthopagrus 'butcher'." *Brain and Behavior*, Vol. 55, No. 4, Apr. 2000 (Figure H).

a. Write the equation of a straight-line model relating number of strikes (y) to age of fish (x).

b. Fit the model to the data using the method of least squares and give the least squares prediction equation. $\hat{y} = 175.70 - .8195x$

c. Give a practical interpretation of the value of $\hat{\beta}_0$, if possible.

d. Give a practical interpretation of the value of $\hat{\beta}_1$, if possible.

9.18 Is the number of games won by a major league baseball team in a season related to the team's batting average? The table below, reproduced from Exercise 2.111 (p. 90), shows the number of games won and the batting averages for the 14 teams in the American League for the 2000 Major League Baseball season.

⊘ ALWINS

Team	Games Won	Batting Ave.
New York	87	.277
Toronto	83	.275
Baltimore	74	.272
Boston	85	.267
Tampa Bay	69	.257
Cleveland	90	.288
Detroit	79	.275
Chicago	95	.286
Kansas City	77	.288
Minnesota	69	.270
Anaheim	82	.280
Texas	71	.283
Seattle	91	.269
Oakland	91	.270

Source: Major League Baseball, 2001.

SPSS Output for Exercise 9.18

Coefficients[a]

Model		Unstandardized Coefficients B	Std. Error	Standardized Coefficients Beta	t	Sig.
1	(Constant)	-.219	74.082		-.003	.998
	BATAVG	297.140	268.768	.304	1.106	.291

a. Dependent Variable: WINS

a. If you were to model the relationship between the mean (or expected) number of games won by a major league team and the team's batting average x, using a straight line, would you expect the slope of the line to be positive or negative? Explain.

b. In Exercise 2.111 you constructed a scattergram of the data. Does the pattern revealed by the scattergram agree with your answer to part **a**?

c. An SPSS printout of the simple linear regression is provided above. Find the estimates of the β's on the printout and write the equation of the least squares line. $\hat{y} = -.219 + 297.140x$

d. Graph the least squares line on your scattergram. Does your least squares line seem to fit the points on your scattergram?

e. Does the mean (or expected) number of games won appear to be strongly related to a team's batting average? Explain.   No

f. Interpret the values of $\hat{\beta}_0$ and $\hat{\beta}_1$ in the words of the problem.

Applying the Concepts—Intermediate

9.19 The quality of the orange juice produced by a manufacturer (e.g., Minute Maid, Tropicana) is constantly monitored. There are numerous sensory and chemical components that combine to make the best tasting orange juice. For example, one manufacturer has developed a quantitative index of the "sweetness" of orange juice. (The higher the index, the sweeter the juice.) Is there a relationship between the sweetness index and a chemical measure such as the amount of water soluble pectin (parts per million) in the orange juice? Data collected on these two variables for 24 production runs at a juice manufacturing plant are shown in the next table. Suppose a manufacturer wants to use simple linear regression to predict the sweetness (y) from the amount of pectin (x).

a. Find the least squares line for the data.

b. Interpret $\hat{\beta}_0$ and $\hat{\beta}_1$ in the words of the problem.

c. Predict the sweetness index if amount of pectin in the orange juice is 300 ppm. [*Note:* A measure of reliability of such a prediction is discussed in Section 9.8.]

9.20 Chemists at Kyushu University (Japan) examined the linear relationship between the maximum absorption rate

⊙ OJUICE

Run	Sweetness Index	Pectin (ppm)
1	5.2	220
2	5.5	227
3	6.0	259
4	5.9	210
5	5.8	224
6	6.0	215
7	5.8	231
8	5.6	268
9	5.6	239
10	5.9	212
11	5.4	410
12	5.6	256
13	5.8	306
14	5.5	259
15	5.3	284
16	5.3	383
17	5.7	271
18	5.5	264
19	5.7	227
20	5.3	263
21	5.9	232
22	5.8	220
23	5.8	246
24	5.9	241

Note: The data in the table are authentic. For confidentiality reasons, the manufacturer cannot be disclosed.

y (in nanomoles) and the Hammett substituent constant x for metacyclophane compounds (*Journal of Organic Chemistry*, July 1995). The data for variants of two compounds are given in the table on p. 457. The variants of compound 1 are labeled 1a, 1b, 1d, 1e, 1f, 1g, and 1h; the variants of compound 2 are 2a, 2b, 2c, and 2d.

a. Plot the data in a scattergram. Use two different plotting symbols for the two compounds. What do you observe?

b. Using only the data for compound 1, find $\hat{\beta}_0$ and $\hat{\beta}_1$. Interpret the results.   308.137; 41.713

c. Using only the data for compound 2, find $\hat{\beta}_0$ and $\hat{\beta}_1$. Interpret the results.   302.588; 64.053

⊘ ORGCHEM

Compound	Maximum Absorption y	Hammett Constant x
1a	298	0.00
1b	346	.75
1d	303	.06
1e	314	−.26
1f	302	.18
1g	332	.42
1h	302	−.19
2a	343	.52
2b	367	1.01
2c	325	.37
2d	331	.53

Source: Adapted from Tsuge, A., *et al.* "Preparation and spectral properties of disubstituted [2-2] metacyclophanes." *Journal of Organic Chemistry*, Vol. 60, No. 15, July 1995, pp. 4390–4391 (Table 1 and Figure 1).

9.21 A study was conducted to investigate how people with a hearing impairment communicate with their conversational partners (*Journal of the Academy of Rehabilitative Audiology*, Vol. 27, 1994). Each of thirteen hearing-impaired subjects, all fitted with a cochlear implant, participated in a structured communication interaction with a familiar conversational partner (a family member) and with an unfamiliar conversational partner (who was instructed not to take the initiative to repair breakdowns in communication). The total number of words used by the subject in each of the two conversations is given in the next table.
   a. Plot the data in a scattergram. Is there visual evidence of a linear relationship between x and y? If so, is it positive or negative?   Positive

b. Propose a straight-line model relating y to x.
c. A STATISTIX printout of the simple linear regression analysis is provided at the bottom of the page. Find the estimate of β_0 and β_1 on the printout. 20.128; .624
d. Interpret the values of $\hat{\beta}_0$ and $\hat{\beta}_1$.

⊘ HEARAID

Subject	Words with Familiar Partner x	Words with Unfamiliar Partner y
1	65	47
2	160	78
3	55	90
4	83	75
5	0	6
6	140	101
7	49	40
8	164	215
9	62	29
10	56	75
11	207	121
12	207	139
13	93	83

Source: Tye-Murray, N., *et al.* "Communication breakdowns: Partner contingencies and partner reactions." *Journal of the Academy of Rehabilitative Audiology*, Vol. 27, 1994, pp. 116–117 (Tables 6 & 7).

9.22 Two species of predatory birds, collard flycatchers and tits, compete for nest holes during breeding season on the island of Gotland, Sweden. Frequently, dead flycatchers are found in nest boxes occupied by tits. A field study examined whether the risk of mortality to flycatchers is related to the degree of competition between the two bird species for nest sites (*The Condor*, May 1995). The table (p. 458) gives data on the number y of flycatchers killed at each of 14 discrete locations (plots)

```
STATISTIX Output for Exercise 9.21

UNWEIGHTED LEAST SQUARES LINEAR REGRESSION OF UNFWORDS

PREDICTOR
VARIABLES     COEFFICIENT     STD ERROR      STUDENT'S T         P
---------     -----------     ---------      -------------      ------
CONSTANT         20.1275        19.2194           1.05          0.3174
FAMWORDS         0.62442        0.15899           3.93          0.0024

R-SQUARED                 0.5837     RESID. MEAN SQUARE (MSE)      1305.22
ADJUSTED R-SQUARED        0.5459     STANDARD DEVIATION            36.1279

SOURCE        DF      SS            MS            F         P
---------     ---     ----------    ----------    -----     ------
REGRESSION     1      20131.8       20131.8       15.42     0.0024
RESIDUAL      11      14357.5       1305.22
TOTAL         12      34489.2

CASES INCLUDED 13     MISSING CASES 0
```

on the island as well as the nest box tit occupancy x (that is, the percentage of nest boxes occupied by tits) at each plot. SAS was used to conduct a simple linear regression analysis for the model, $E(y) = \beta_0 + \beta_1 x$. The printout is shown below.

⊙CONDOR2

Plot	Number of Flycatchers Killed y	Nest Box Tit Occupancy x (%)
1	0	24
2	0	33
3	0	34
4	0	43
5	0	50
6	1	35
7	1	35
8	1	38
9	1	40
10	2	31
11	2	43
12	3	55
13	4	57
14	5	64

Source: Merila, J., and Wiggins, D. A. "Interspecific competition for nest holes causes adult mortality in the collard flycatcher." *The Condor,* Vol. 97, No. 2, May 1995, p. 449 (Figure 2), Cooper Ornithological Society.

a. Plot the data in a scattergram. Does the frequency of flycatcher casualties per plot appear to increase linearly with increasing proportion of nest boxes occupied by tits?   Yes
b. Find the estimates of β_0 and β_1 in the SAS printout. Interpret their values. $-3.047; 0.108$

SAS Output for Exercise 9.22

The REG Procedure
Dependent Variable: NOKILLED

Analysis of Variance

Source	DF	Sum of Squares	Mean Square	F Value	Pr > F
Model	1	19.11669	19.11669	16.03	0.0018
Error	12	14.31188	1.19266		
Corrected Total	13	33.42857			

Root MSE	1.09209	R-Square	0.5719	
Dependent Mean	1.42857	Adj R-Sq	0.5362	
Coeff Var	76.44618			

Parameter Estimates

Variable	DF	Parameter Estimate	Standard Error	t Value	Pr > |t|
Intercept	1	-3.04686	1.15533	-2.64	0.0217
TITPCT	1	0.10766	0.02689	4.00	0.0018

9.23 Refer to the *Journal of Experimental Psychology-Applied* (June 2000) study in which the "name game" was used to help groups of students learn the names of other students in the group, Exercise 7.84 (p. 385). Recall that the "name game" requires the first student in the group to state his/her full name, the second student to say his/her name and the name of the first student, the third student to say his/her name and the names of the first two students, etc. After making their introductions, the students listened to a seminar speaker for 30 minutes. At the end of the seminar, all students were asked to remember the full name of each of the other students in their group and the researchers measured the proportion of names recalled for each. One goal of the study was to investigate the linear trend between y = recall proportion and x = position (order) of the student during the game. The data (simulated based on summary statistics provided in the research article) for 144 students in the first eight positions are saved in the NAMEGAME2 file. The first five and last five observations in the data set are listed below. [*Note:* Since the student in position 1 actually must recall the names of all the other students, he or she is assigned the position number 9 in the data set.] Use the method of least squares to estimate the line, $E(y) = \beta_0 + \beta_1 x$. Interpret the β estimates in the words of the problem.

⊙NAMEGAME2

Position	Recall
2	0.04
2	0.37
2	1.00
2	0.99
2	0.79
⋮	⋮
9	0.72
9	0.88
9	0.46
9	0.54
9	0.99

Source: Morris, P.E., and Fritz, C.O. "The name game: Using retrieval practice to improve the learning of names." *Journal of Experimental Psychology-Applied,* Vol. 6, No. 2, June 2000 (data simulated from Figure 2).

Applying the Concepts—Advanced

9.24 The long jump is a track and field event in which a competitor attempts to jump a maximum distance into a sand pit after a running start. At the edge of the sand pit is a takeoff board. Jumpers usually try to plant their toes at the front edge of this board to maximize jumping distance. The absolute distance between the front edge of the takeoff board and the spot where the toe actually lands on the board prior to jumping is called "takeoff error." Is takeoff error in the long jump linearly related to best jumping distance? To answer this question, kinesiology researchers videotaped the performances of 18 novice long jumpers at a high school track meet (*Journal of Applied Biomechanics*, May 1995). The average takeoff error, x, and best jumping distance (out of three jumps), y, for each jumper are recorded in the table. If a jumper can reduce his/her average takeoff error by .1 meter, how much would you estimate his/her best jumping distance to change? Based on your answer, comment on the usefulness of the model for predicting best jumping distance.

LONGJUMP

Jumper	Best Jumping Distance y (meters)	Average Takeoff Error x (meters)
1	5.30	.09
2	5.55	.17
3	5.47	.19
4	5.45	.24
5	5.07	.16
6	5.32	.22
7	6.15	.09
8	4.70	.12
9	5.22	.09
10	5.77	.09
11	5.12	.13
12	5.77	.16
13	6.22	.03
14	5.82	.50
15	5.15	.13
16	4.92	.04
17	5.20	.07
18	5.42	.04

Source: Berg, W. P., and Greer, N. L. "A kinematic profile of the approach run of novice long jumpers." *Journal of Applied Biomechanics*, Vol. 11, No. 2, May 1995, p. 147 (Table 1).

9.3 MODEL ASSUMPTIONS

In Section 9.2 we assumed that the probabilistic model relating drug reaction time y to the percentage of drug x in the bloodstream is

$$y = \beta_0 + \beta_1 x + \varepsilon$$

We also recall that the least squares estimate of the deterministic component of the model, $\beta_0 + \beta_1 x$, is

$$\hat{y} = \hat{\beta}_0 + \hat{\beta}_1 x = -.1 + .7x$$

Now we turn our attention to the random component ε of the probabilistic model and its relation to the errors in estimating β_0 and β_1. We will use a probability distribution to characterize the behavior of ε. We will see how the probability distribution of ε determines how well the model describes the relationship between the dependent variable y and the independent variable x.

Step 3 in a regression analysis requires us to specify the probability distribution of the random error ε. We will make four basic assumptions about the general form of this probability distribution:

Assumption I The mean of the probability distribution of ε is 0. That is, the average of the values of ε over an infinitely long series of experiments is 0 for each setting of the independent variable x. This assumption implies that the mean value of y, $E(y)$, for a given value of x is $E(y) = \beta_0 + \beta_1 x$.

Assumption 2 The variance of the probability distribution of ε is constant for all settings of the independent variable x. For our straight-line model, this assumption means that the variance of ε is equal to a constant, say σ^2, for all values of x.

FIGURE 9.7

The Probability Distribution of ε

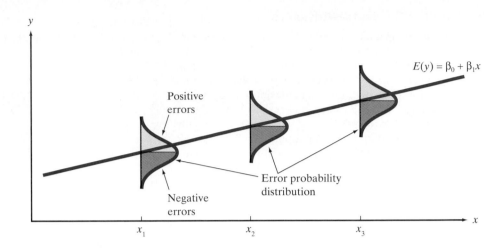

Assumption 3 The probability distribution of ε is normal.

Assumption 4 The values of ε associated with any two observed values of y are independent. That is, the value of ε associated with one value of y has no effect on the values of ε associated with other y values.

The implications of the first three assumptions can be seen in Figure 9.7, which shows distributions of errors for three values of x, namely, x_1, x_2, and x_3. Note that the relative frequency distributions of the errors are normal with a mean of 0 and a constant variance σ^2. (All the distributions shown have the same amount of spread or variability.) The straight line shown in Figure 9.7 is the line of means. It indicates the mean value of y for a given value of x. We denote this mean value as $E(y)$. Then, the line of means is given by the equation

$$E(y) = \beta_0 + \beta_1 x$$

These assumptions make it possible for us to develop measures of reliability for the least squares estimators and to develop hypothesis tests for examining the usefulness of the least squares line. We have various techniques for checking the validity of these assumptions, and we have remedies to apply when they appear to be invalid. These topics are beyond the scope of this text, but they are discussed in some of the chapter references. Fortunately, the assumptions need not hold exactly in order for least squares estimators to be useful. The assumptions will be satisfied adequately for many applications encountered in practice.

9.4 AN ESTIMATOR OF σ^2

It seems reasonable to assume that the greater the variability of the random error ε (which is measured by its variance σ^2), the greater will be the errors in the estimation of the model parameters β_0 and β_1 and in the error of prediction when $\hat{y}$ is used to predict y for some value of x. Consequently, you should not be surprised, as we proceed through this chapter, to find that σ^2 appears in the formulas for all confidence intervals and test statistics that we will be using.

In most practical situations, σ^2 is unknown and we must use our data to estimate its value. The best estimate of σ^2, denoted by s^2, is obtained by dividing the sum of squares of the deviations of the y values from the prediction line,

$$\text{SSE} = \sum (y_i - \hat{y}_i)^2$$

Suggested Exercise 9.28

by the number of degrees of freedom associated with this quantity. We use 2 df to estimate the two parameters β_0 and β_1 in the straight-line model, leaving $(n - 2)$ df for the error variance estimation.

TEACHING TIP
The value of the standard deviation will be calculated by all statistical software packages. The name that appears on the printout will vary widely. Use several examples of software output to illustrate this point.

Estimation of σ^2 for a (First-Order) Straight-Line Model

$$s^2 = \frac{\text{SSE}}{\text{Degrees of freedom for error}} = \frac{\text{SSE}}{n - 2}$$

where $\text{SSE} = \sum (y_i - \hat{y}_i)^2 = \text{SS}_{yy} - \hat{\beta}_1 \text{SS}_{xy}$

$$\text{SS}_{yy} = \sum (y_i - \bar{y})^2 = \sum y_i^2 - \frac{\left(\sum y_i\right)^2}{n}$$

To estimate the standard deviation σ of ε, we calculate

$$s = \sqrt{s^2} = \sqrt{\frac{\text{SSE}}{n - 2}}$$

We will refer to s as the **estimated standard error of the regression model**.

Warning: When performing these calculations, you may be tempted to round the calculated values of SS_{yy}, $\hat{\beta}_1$, and SS_{xy}. Be certain to carry at least six significant figures for each of these quantities to avoid substantial errors in calculation of the SSE.

In the drug reaction example, we previously calculated SSE $= 1.10$ for the least squares line $\hat{y} = -.1 + .7x$. Recalling that there were $n = 5$ data points, we have $n - 2 = 5 - 2 = 3$ df for estimating σ^2. Thus,

$$s^2 = \frac{\text{SSE}}{n - 2} = \frac{1.10}{3} = .367$$

is the estimated variance, and

$$s = \sqrt{.367} = .61$$

is the standard error of the regression model.

The values of s^2 and s can also be obtained from a simple linear regression printout. The SAS printout for the drug reaction example is reproduced in Figure 9.8. The value of s^2 is highlighted on the printout (in the **Mean Square** column in the row labeled **Error**). The value, $s^2 = .36667$, rounded to three decimal places, agrees with the one calculated by hand. The value of s is also highlighted in Figure 9.8 (next to the heading **Root MSE**). This value, $s = .60553$, agrees (except for rounding) with our hand-calculated value.

You may be able to grasp s intuitively by recalling the interpretation of a standard deviation given in Chapter 2 and remembering that the least squares line estimates the mean value of y for a given value of x. Since s measures the spread of the distribution of y values about the least squares line and these errors of prediction are assumed to be normally distributed, we should not be surprised to find that most (about 95%) of the observations lie within $2s$, or $2(.61) = 1.22$, of the least squares line. For this simple example (only five data points), all five data points fall within $2s$ of the least squares line. In Section 9.8, we use s to evaluate the error of prediction when the least squares line is used to predict a value of y to be observed for a given value of x.

FIGURE 9.8

SAS Printout for Time–Drug Regression

The REG Procedure
Dependent Variable: TIME_Y

Analysis of Variance

Source	DF	Sum of Squares	Mean Square	F Value	Pr > F
Model	1	4.90000	4.90000	13.36	0.0354
Error	3	1.10000	0.36667		
Corrected Total	4	6.00000			

Root MSE	0.60553	R-Square	0.8167
Dependent Mean	2.00000	Adj R-Sq	0.7556
Coeff Var	30.27650		

Parameter Estimates

Variable	DF	Parameter Estimate	Standard Error	t Value	Pr > |t|
Intercept	1	-0.10000	0.63509	-0.16	0.8849
DRUG_X	1	0.70000	0.19149	3.66	0.0354

Interpretation of s, the Estimated Standard Deviation of ε

We expect most ($\approx 95\%$) of the observed y values to lie within $2s$ of their respective least squares predicted values, $\hat{y}$.

EXERCISES 9.25–9.33

Learning the Mechanics

9.25 Calculate SSE and s^2 for each of the following cases:

  a. $n = 20, SS_{yy} = 95, SS_{xy} = 50, \hat{\beta}_1 = .75$

  b. $n = 40, \sum y^2 = 860, \sum y = 50,$
    $SS_{xy} = 2{,}700, \hat{\beta}_1 = .2$ 257.5; 6.7763

  c. $n = 10, \sum (y_i - \bar{y})^2 = 58,$
    $SS_{xy} = 91, SS_{xx} = 170$ 9.288; 1.161

9.26 Suppose you fit a least squares line to 12 data points and the calculated value of SSE is .429.

  a. Find s^2, the estimator of σ^2 (the variance of the random error term ε).   .0429

  b. What is the largest deviation that you might expect between any one of the 12 points and the least squares line?   0.414

9.27 Visually compare the scattergrams shown at right. If a least squares line were determined for each data set, which do you think would have the smallest variance, s^2? Explain.   Graph b

9.28 Refer to Exercises 9.10 and 9.13 (p. 452). Calculate SSE, s^2, and s for the least squares lines obtained in those exercises. Interpret the standard error of the regression model, s, for each.

Applying the Concepts—Basic

9.29 Refer to the simple linear regression relating number of agressive strikes y to age x for blackbream fish, Exercise 9.17 (p. 455).

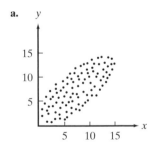
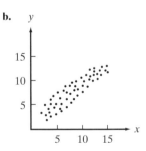
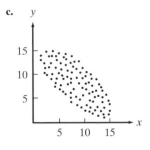

  a. Find SSE, s^2, and s.   1, 667.65; 238.24; 15.43
  b. Interpret the value of s.

9.30 *Statistical Bulletin* (Oct.–Dec. 1999) reported the average hospital charge and the average length of hospital stay for patients undergoing radical prostatectomies in a sample of 12 states. The data are listed in the accompanying table.

⬤ HOSPITAL

State	Average Hospital Charge ($)	Average Length of Stay (days)
Massachusetts	11,680	3.64
New Jersey	11,630	4.20
Pennsylvania	9,850	3.84
Minnesota	9,950	3.11
Indiana	8,490	3.86
Michigan	9,020	3.54
Florida	13,820	4.08
Georgia	8,440	3.57
Tennessee	8,790	3.80
Texas	10,400	3.52
Arizona	12,860	3.77
California	16,740	3.78

Source: Statistical Bulletin, Vol. 80, No. 4, Oct.–Dec. 1999, p. 13.

  a. Plot the data on a scattergram.
  b. Use the method of least squares to model the relationship between average hospital charge (y) and length of hospital stay (x).
  c. Find the estimated standard error of the regression model and interpret its value in the context of the problem. $s = 2,496.7$
  d. For a hospital stay of length $x = 4$ days, find $\hat{y} \pm 2s$.
  e. What fraction of the states in the sample have average hospital charges within $\pm 2s$ of the least squares line?   11/12

9.31 Refer to the simple linear regression relating total number of words in a conversation with an unfamiliar partner y to total number of words with a familiar partner x for hearing-impaired subjects, Exercise 9.21 (p. 457).
  a. Find SSE, s^2, and s on the STATISTIX printout (p. 457).   14, 357.5; 1, 305.2; 36.13
  b. Interpret the value of s.

9.32 Refer to the simple linear regression relating number of flycatchers killed y to nest box tit occupancy x for nest sites, Exercise 9.22 (p. 457).

  a. Find SSE, s^2, and s on the SAS printout (p. 458).
  b. Interpret the value of s.

Applying the Concepts—Advanced

9.33 To improve the quality of the output of any production process, it is necessary first to understand the capabilities of the process (*Out of the Crisis*, Deming, 1982). In a particular manufacturing process, the useful life of a cutting tool is related to the speed at which the tool is operated. The data in the accompanying table were derived from life tests for the two different brands of cutting tools currently used in the production process.
  a. Construct a scattergram for each brand of cutting tool.
  b. For each brand, use the method of least squares to model the relationship between useful life and cutting speed.
  c. Find SSE, s^2, and s for each least squares line.
  d. For a cutting speed of 70 meters per minute, find $\hat{y} \pm 2s$ for each least squares line.
  e. For which brand would you feel more confident in using the least squares line to predict useful life for a given cutting speed? Explain.   Brand B

⬤ CUTTOOL

Cutting Speed (meters per minute)	Useful Life (hours) Brand A	Brand B
30	4.5	6.0
30	3.5	6.5
30	5.2	5.0
40	5.2	6.0
40	4.0	4.5
40	2.5	5.0
50	4.4	4.5
50	2.8	4.0
50	1.0	3.7
60	4.0	3.8
60	2.0	3.0
60	1.1	2.4
70	1.1	1.5
70	.5	2.0
70	3.0	1.0

9.5 ASSESSING THE UTILITY OF THE MODEL: MAKING INFERENCES ABOUT THE SLOPE β_1

Now that we have specified the probability distribution of ε and found an estimate of the variance σ^2, we are ready to make statistical inferences about the model's usefulness for predicting the response y. This is step 4 in our regression modeling procedure.

FIGURE 9.9

Graphing the Model with $\beta_1 = 0$: $y = \beta_0 + \varepsilon$

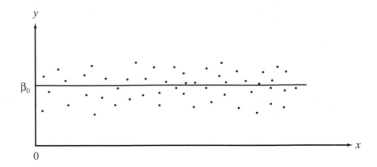

Refer again to the data of Table 9.1 and suppose the reaction times are *completely unrelated* to the percentage of drug in the bloodstream. What could be said about the values of β_0 and β_1 in the hypothesized probabilistic model

$$y = \beta_0 + \beta_1 x + \varepsilon$$

if x contributes no information for the prediction of y? The implication is that the mean of y—that is, the deterministic part of the model $E(y) = \beta_0 + \beta_1 x$—does not change as x changes. In the straight-line model, this means that the true slope, β_1, is equal to 0 (see Figure 9.9). Therefore, to test the null hypothesis that the linear model contributes no information for the prediction of y against the alternative hypothesis that the linear model is useful for predicting y, we test

$$H_0: \beta_1 = 0$$
$$H_a: \beta_1 \neq 0$$

If the data support the alternative hypothesis, we will conclude that x does contribute information for the prediction of y using the straight-line model [although the true relationship between $E(y)$ and x could be more complex than a straight line]. Thus, in effect, this is a test of the usefulness of the hypothesized model.

The appropriate test statistic is found by considering the sampling distribution of $\hat{\beta}_1$, the least squares estimator of the slope β_1, as shown in the following box.

Sampling Distribution of $\hat{\beta}_1$

If we make the four assumptions about ε (see Section 9.3), the sampling distribution of the least squares estimator $\hat{\beta}_1$ of the slope will be normal with mean β_1 (the true slope) and standard deviation

$$\sigma_{\hat{\beta}_1} = \frac{\sigma}{\sqrt{SS_{xx}}} \quad \text{(see Figure 9.10)}$$

We estimate $\sigma_{\hat{\beta}_1}$ by $s_{\hat{\beta}_1} = \dfrac{s}{\sqrt{SS_{xx}}}$ and refer to this quantity as the estimated standard error of the least squares slope $\hat{\beta}_1$.

FIGURE 9.10

Sampling Distribution of $\hat{\beta}_1$

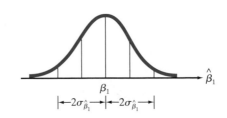

Since σ is usually unknown, the appropriate test statistic is a t statistic, formed as follows:

$$t = \frac{\hat{\beta}_1 - \text{Hypothesized value of } \beta_1}{s_{\hat{\beta}_1}} \qquad \text{where} \qquad s_{\hat{\beta}_1} = \frac{s}{\sqrt{SS_{xx}}}$$

Thus,

$$t = \frac{\hat{\beta}_1 - 0}{s/\sqrt{SS_{xx}}}$$

Note that we have substituted the estimator s for σ and then formed the estimated standard error $s_{\hat{\beta}_1}$ by dividing s by $\sqrt{SS_{xx}}$. The number of degrees of freedom associated with this t statistic is the same as the number of degrees of freedom associated with s. Recall that this number is $(n - 2)$ df when the hypothesized model is a straight line (see Section 9.4). The setup of our test of the usefulness of the straight-line model is summarized in the next box.

A Test of Model Utility: Simple Linear Regression

ONE-TAILED TEST	TWO-TAILED TEST
$H_0: \beta_1 = 0$	$H_0: \beta_1 = 0$
$H_a: \beta_1 < 0$ (or $H_a: \beta_1 > 10$)	$H_a: \beta_1 \neq 0$

Test statistic: $t = \dfrac{\hat{\beta}_1}{s_{\hat{\beta}_1}} = \dfrac{\hat{\beta}_1}{s/\sqrt{SS_{xx}}}$

Rejection region: $t < -t_\alpha$ *Rejection region:* $|t| > t_{\alpha/2}$

(or $t > t_\alpha$ when $H_a: \beta_1 > 0$)

where t_α and $t_{\alpha/2}$ are based on $(n - 2)$ degrees of freedom
Assumptions: The four assumptions about ε listed in Section 9.3.

Suggested Exercise 9.43

For the drug reaction example, we will choose $\alpha = .05$ and, since $n = 5$, t will be based on $n - 2 = 3$ df and the rejection region will be

$$|t| > t_{.025} = 3.182$$

We previously calculated $\hat{\beta}_1 = .7$, $s = .61$, and $SS_{xx} = 10$. Thus,

$$t = \frac{\hat{\beta}_1}{s/\sqrt{SS_{xx}}} = \frac{.7}{.61\sqrt{10}} = \frac{.7}{.19} = 3.7$$

Since this calculated t value falls in the upper-tail rejection region (see Figure 9.11), we reject the null hypothesis and conclude that the slope β_1 is not 0. The sample evidence indicates that amount of drug x in the blood stream contributes information for the prediction of reaction time y when a linear model is used.

We can reach the same conclusion by using the observed significance level (p-value) of the test from a computer printout. The MINITAB printout for the drug reaction example is reproduced in Figure 9.12. The test statistic and two-tailed p-value are highlighted on the printout. Since the p-value $= .035$ is smaller than $\alpha = .05$, we will reject H_0.

What conclusion can be drawn if the calculated t value does not fall in the rejection region or if the observed significance level of the test exceeds α? We know

FIGURE 9.11

Rejection Region and Calculated t Value for Testing H_0: $\beta_1 = 0$ Versus H_a: $\beta_1 \neq 0$

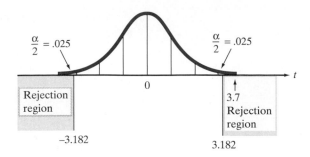

FIGURE 9.12

MINITAB Printout for Time–Drug Regression

```
The regression equation is
TIME_Y = - 0.100 + 0.700 DRUG_X


Predictor        Coef      SE Coef         T       P
Constant      -0.1000       0.6351     -0.16   0.885
DRUG_X         0.7000       0.1915      3.66   0.035


S = 0.6055       R-Sq = 81.7%     R-Sq(adj) = 75.6%


Analysis of Variance

Source           DF           SS        MS       F       P
Regression        1       4.9000    4.9000   13.36   0.035
Residual Error    3       1.1000    0.3667
Total             4       6.0000
```

from previous discussions of the philosophy of hypothesis testing that such a *t* value does *not* lead us to accept the null hypothesis. That is, we do not conclude that $\beta_1 = 0$. Additional data might indicate that β_1 differs from 0, or a more complex relationship may exist between *x* and *y*, requiring the fitting of a model other than the straight-line model.

Another way to make inferences about the slope β_1 is to estimate it using a confidence interval. This interval is formed as shown in the next box.

A $100(1 - \alpha)\%$ Confidence Interval for the Simple Linear Regression Slope β_1

$$\hat{\beta}_1 \pm t_{\alpha/2}s_{\hat{\beta}_1}$$

where the estimated standard error of $\hat{\beta}_1$ is calculated by

$$s_{\hat{\beta}_1} = \frac{s}{\sqrt{SS_{xx}}}$$

and $t_{\alpha/2}$ is based on $(n - 2)$ degrees of freedom.

Assumptions: The four assumptions about ε listed in Section 9.3.

For the drug reaction example, $t_{\alpha/2}$ is based on $(n - 2) = 3$ degrees of freedom. Therefore, a 95% confidence interval for the slope β_1, the expected change in reaction time for a 1% increase in the amount of drug in the bloodstream, is

$$\hat{\beta}_1 \pm t_{.025}s_{\hat{\beta}_1} = .7 \pm 3.182\left(\frac{s}{\sqrt{SS_{xx}}}\right) = .7 \pm 3.182\left(\frac{.61}{\sqrt{10}}\right) = .7 \pm .61$$

Thus, the interval estimate of the slope parameter β_1 is .09 to 1.31. In terms of this example, the implication is that we can be 95% confident that the *true* mean increase in reaction time per additional 1% of the drug is between .09 and 1.31 seconds. This inference is meaningful only over the sampled range of x—that is, from 1% to 5% of the drug in the bloodstream. Since all the values in this interval are positive, it appears that β_1 is positive and that the mean of y, $E(y)$, increases as x increases. However, the rather large width of the confidence interval reflects the small number of data points (and, consequently, a lack of information) in the experiment. We would expect a narrower interval if the sample size were increased.

EXERCISES 9.34–9.47

Learning the Mechanics

9.34 Construct both a 95% and a 90% confidence interval for β_1 for each of the following cases:
  a. $\hat{\beta}_1 = 31, s = 3, SS_{xx} = 35, n = 12$
  b. $\hat{\beta}_1 = 64, SSE = 1{,}960, SS_{xx} = 30, n = 18$
  c. $\hat{\beta}_1 = -8.4, SSE = 146, SS_{xx} = 64, n = 24$

9.35 Consider the following pairs of observations:

x	1	5	3	2	6	6	0
y	1	3	3	1	4	5	1

  a. Construct a scattergram for the data.
  b. Use the method of least squares to fit a straight line to the seven data points in the table.
  c. Plot the least squares line on your scattergram of part **a**.
  d. Specify the null and alternative hypotheses you would use to test whether the data provide sufficient evidence to indicate that x contributes information for the (linear) prediction of y.

  e. What is the test statistic that should be used in conducting the hypothesis test of part **d**? Specify the degrees of freedom associated with the test statistic.
  f. Conduct the hypothesis test of part **d** using $\alpha = .05$.

9.36 Refer to Exercise 9.35. Construct an 80% and a 98% confidence interval for β_1. $.617 \pm .166; .617 \pm .378$

9.37 Do the accompanying data provide sufficient evidence to conclude that a straight line is useful for characterizing the relationship between x and y?

y	4	2	5	3	2	4
x	1	4	5	3	2	4

Applying the Concepts—Basic

9.38 Refer to the data on sale prices and total appraised values of 92 residential properties in an upscale Tampa, Florida, neighborhood, Exercise 9.14 (p. 453). An SAS simple linear regression printout for the analysis is given here.

SAS Simple Linear Regression Output for Exercise 9.38

The REG Procedure
Dependent Variable: SALEPRIC

Analysis of Variance

Source	DF	Sum of Squares	Mean Square	F Value	Pr > F
Model	1	1673142	1673142	1556.48	<.0001
Error	90	96746	1074.95355		
Corrected Total	91	1769888			

Root MSE	32.78648	R-Square	0.9453
Dependent Mean	236.55761	Adj R-Sq	0.9447
Coeff Var	13.85983		

Parameter Estimates

Variable	DF	Parameter Estimate	Standard Error	t Value	Pr > |t|	95% Confidence Limits	
Intercept	1	20.94193	6.44617	3.25	0.0016	8.13550	33.74837
TOTALVAL	1	1.06873	0.02709	39.45	<.0001	1.01491	1.12254

a. Use the printout to determine whether there is a positive linear relationship between appraised property value x and sale price y for residential properties sold in this neighborhood. That is, determine if there is sufficient evidence (at $\alpha = .01$) to indicate that β_1, the slope of the straight-line model, is positive.

b. Find a 95% confidence interval for the slope, β_1, on the printout. Interpret the result practically.

c. What can be done to obtain a narrower confidence interval in part **b**?

9.39 Refer to Exercise 9.19 (p. 456) and the simple linear regression relating the sweetness index (y) of an orange juice sample to the amount of water soluble pectin (x) in the juice. Find a 90% confidence interval for the true slope of the line. Interpret the result.

9.40 Refer to Exercises 9.22 and 9.32 (pp. 457, 463) and the simple linear regression relating number of flycatchers killed y to nest box tit occupancy x.

a. Refer to the SAS printout (p. 458) and test whether y is positively linearly related to x. Use $\alpha = .01$.

b. Construct a 99% confidence interval for β_1. Practically interpret the result. $.1077 \pm .082$

9.41 The abundance of tropical seagrasses in Australia was the subject of research published in *Aquatic Botany* (Mar. 1995). Simple linear regression was used to relate the standing crop (y) of a seagrass species (measured in grams per meter squared) to the percentage (x) of land covered by the plants. Data collected for $n = 12$ sites at Rowes Bay, Australia, yielded the following results:

$$\hat{y} = .031 + .089x \qquad t = 2.34 \qquad p\text{-value} = .042$$

a. Give the practical interpretation, if possible, of the estimated y-intercept of the line.

b. Give the practical interpretation, if possible, of the estimated slope of the line.

c. At what α-level is there sufficient evidence of a linear relationship between standing crop and percentage cover of a seagrass species? Explain.

Applying the Concepts—Intermediate

9.42 The *British Journal of Sports Medicine* (Apr. 2000) published a study of the effect of massage on boxing performance. Two variables measured on amateur boxers were blood lactate concentration (mM) and the boxer's perceived recovery (28-point scale). Based on information provided in the article, the following data were obtained for 16 five-round boxing performances, where a massage was given to the boxer between rounds. Conduct a test to determine whether blood lactate level (y) is linearly related to perceived recovery (x). Use $\alpha = .10$.

⊘ BOXING2

Blood Lactate Level	Perceived Recovery
3.8	7
4.2	7
4.8	11
4.1	12
5.0	12
5.3	12
4.2	13
2.4	17
3.7	17
5.3	17
5.8	18
6.0	18
5.9	21
6.3	21
5.5	20
6.5	24

Source: Hemmings, B., Smith, M., Graydon, J., and Dyson, R. "Effects of massage on physiological restoration, perceived recovery, and repeated sports performance." *British Journal of Sports Medicine*, Vol. 34, No. 2, Apr. 2000 (data adapted from Figure 3).

9.43 How do eye and head movements relate to body movements when reacting to a visual stimulus? Scientists at the California Institute of Technology designed an experiment to answer this question and reported their results in *Nature* (Aug. 1998). Adult male rhesus monkeys were exposed to a visual stimulus (i.e., a panel of light-emitting diodes) and their eye, head, and body movements were electronically recorded. In one variation of the experiment, two variables were measured: active head movement (x, percent per degree) and body plus head rotation (y, percent per degree). The data for $n = 39$ trials were subjected to a simple linear regression analysis, with the following results: $\hat{\beta}_1 = .88, s_{\hat{\beta}_1} = .14$

a. Conduct a test to determine whether the two variables, active head movement x and body plus head rotation y are positively linearly related. Use $\alpha = .05$.

b. Construct and interpret a 90% confidence interval for β_1. $.88 \pm .24$

c. The scientists want to know if the true slope of the line differs significantly from 1. Based on your answer to part **b**, make the appropriate inference.

9.44 The U.S. Department of Agriculture has developed and adopted the Universal Soil Loss Equation (USLE) for predicting water erosion of soils. In geographic areas where runoff from melting snow is common, calculating the USLE requires an accurate estimate of snowmelt runoff erosion. An article in the *Journal of Soil and Water Conservation* (Mar.–Apr. 1995) used simple linear regression to develop a snowmelt erosion index. Data for 54 climatological stations in Canada were used to model the McCool winter-adjusted rainfall erosivity index, y, as a straight-line function of the once-in-5-year snowmelt runoff amount, x (measured in millimeters).

a. The data points are plotted in the scattergram shown on p. 469. Is there visual evidence of a linear trend?   Yes

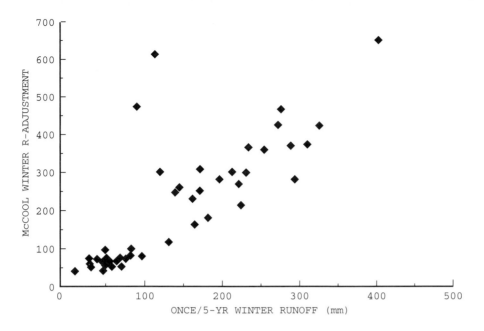

b. The data for seven stations were removed from the analysis due to lack of snowfall during the study period. Why is this strategy advisable?

c. The simple linear regression on the remaining $n = 47$ data points yielded the following results: $\hat{y} = -6.72 + 1.39x$, $s_{\hat{\beta}_1} = .06$. Use this information to construct a 90% confidence interval for β_1.

d. Interpret the interval, part **c**.

⊘NAMEGAME2

9.45 Refer to the *Journal of Experimental Psychology-Applied* (June 2000) name retrieval study, Exercise 9.23 (p. 458). Recall that the goal of the study was to investigate the linear trend between proportion of names recalled (y) and position (order) of the student (x) during the "name game." Is there sufficient evidence (at $\alpha = .01$) of a linear trend? Answer the question by analyzing the data for 144 students saved in the NAMEGAME2 file. $t = 2.858$

Applying the Concepts—Advanced

9.46 One of the most difficult tasks of developing and managing a global portfolio is assessing the risks of potential foreign investments. Duke University researcher C. R. Henry collaborated with two First Chicago Investment Management Company directors to examine the use of country credit ratings as a means of evaluating foreign investments (*Journal of Portfolio Management*, Winter 1995). To be effective, such a measure should help explain and predict the volatility of the foreign market in question. The researchers analyzed data on annualized risk (y) and average credit rating (x) for the 40 countries shown in the table on p. 470. A MINITAB printout for a simple linear regression analysis conducted on the data is shown below.

a. Locate the least squares estimates of β_0 and β_1 on the printout.   57.76, −.3996

MINITAB Output for Exercise 9.46

```
The regression equation is
RISK = 57.8 - 0.400 RATING

Predictor        Coef     SE Coef        T        P
Constant       57.755       6.128     9.43    0.000
RATING        -0.39961     0.09152    -4.37    0.000

S = 12.68      R-Sq = 33.4%     R-Sq(adj) = 31.7%

Analysis of Variance

Source           DF          SS         MS        F        P
Regression        1      3064.4     3064.4    19.07    0.000
Residual Error   38      6107.5      160.7
Total            39      9171.9
```

GLOBRISK

Country	Annualized Risk (%)	Average Credit Rating
Argentina	87.0	31.8
Australia	26.9	78.2
Austria	26.3	83.8
Belgium	22.0	78.4
Brazil	64.8	36.2
Canada	19.2	87.1
Chile	31.6	38.6
Colombia	31.5	44.4
Denmark	20.6	72.6
Finland	26.1	76.0
France	23.8	85.3
Germany	23.0	93.4
Greece	39.6	51.9
Hong Kong	34.3	69.6
India	30.0	46.6
Ireland	23.4	66.4
Italy	28.0	75.5
Japan	25.7	94.5
Jordan	17.6	33.6
Korea	30.7	62.2
Malaysia	26.7	64.4
Mexico	46.3	43.3
Netherlands	18.5	87.6
New Zealand	26.3	68.9
Nigeria	41.4	30.6
Norway	28.3	83.0
Pakistan	24.4	26.4
Philippines	38.4	29.6
Portugal	47.5	56.7
Singapore	26.4	77.6
Spain	24.8	70.8
Sweden	24.5	79.5
Switzerland	19.6	94.7
Taiwan	53.7	72.9
Thailand	27.0	55.8
Turkey	74.1	32.6
United Kingdom	21.8	87.6
United States	15.4	93.4
Venezuela	46.0	45.0
Zimbabwe	35.6	24.5

Source: Erb, C. B., Harvey, C. R., and Viskanta, T. E. "Country risk and global equity selection." *Journal of Portfolio Management*, Vol. 21, No. 2, Winter 1995, p. 76.

    b. Plot the data in a scattergram, then sketch the least squares line on the graph.

    c. Do the data provide sufficient evidence to conclude that credit rating (x) contributes information for the prediction of annualized risk (y)?

    d. Use the plot, part **b**, to locate any unusual data points (outliers).

    e. Eliminate the outlier(s), part **d**, from the data set and rerun the simple linear regression analysis. Note any dramatic changes in the results.

9.47 A recent civil suit revolved around a 5-building brick apartment complex located in the Bronx, New York, which began to suffer *spalling* damage (i.e., a separation of some portion of the face of a brick from its body). The owner of the complex alleged that the bricks were defectively manufactured. The brick manufacturer countered that poor design and shoddy management led to the damage. To settle the suit, an estimate of the rate of damage per 1,000 bricks, called the spall rate, was required (*Chance*, Summer 1994). The owner estimated the spall rate using several *scaffold-drop* surveys. (With this method, an engineer lowers a scaffold down at selected places on building walls and counts the number of visible spalls for every 1,000 bricks in the observation area.) The brick manufacturer conducted its own survey by dividing the walls of the complex into 83 wall segments and taking a photograph of each wall segment. (The number of spalled bricks that could be made out from each photo was recorded and the sum over all 83 wall segments used as an estimate of total spall damage.) In this court case, the jury was faced with the following dilemma: The scaffold-drop survey provided the most accurate estimate of spall rates in a given wall segment. Unfortunately, the drop areas were not selected at random from the entire complex; rather, drops were made at areas with high spall concentrations, leading to an overestimate of the total damage. On the other hand, the photo survey was complete in that all 83 wall segments in the complex were checked for spall damage. But the spall rate estimated by the photos, at least in areas of high spall concentration, was biased low (spalling damage cannot always be seen from a photo), leading to an underestimate of the total damage.

    The data in the table are the spall rates obtained using the two methods at 11 drop locations. Use the data, as did expert statisticians who testified in the case, to help the jury estimate the true spall rate at a given wall segment. Then explain how this information, coupled with the data (not given here) on all 83 wall segments, can provide a reasonable estimate of the total spall damage (i.e., total number of damaged bricks).

BRICKS

Drop Location	Drop Spall Rate (per 1,000 bricks)	Photo Spall Rate (per 1,000 bricks)
1	0	0
2	5.1	0
3	6.6	0
4	1.1	.8
5	1.8	1.0
6	3.9	1.0
7	11.5	1.9
8	22.1	7.7
9	39.3	14.9
10	39.9	13.9
11	43.0	11.8

Source: Fairley, W. B., *et al.* "Bricks, buildings, and the Bronx: Estimating masonry deterioration." *Chance*, Vol. 7. No. 3, Summer 1994, p. 36 (Figure 3). [*Note:* The data points are estimated from the points shown on a scatterplot.]

9.6   THE COEFFICIENT OF CORRELATION

TEACHING TIP
When working with simple linear regression, the correlation coefficient can adequately substitute for the β_1 parameter of the preceding sections.

Recall (from optional Section 2.9) that a **bivariate relationship** describes a relationship—or correlation—between two variables, x and y. Scattergrams are used to graphically describe a bivariate relationship. In this section we will discuss the concept of **correlation** and how it can be used to measure the linear relationship between two variables, x and y. A numerical descriptive measure of correlation is provided by the *Pearson product moment coefficient of correlation*, r.

DEFINITION 9.2

The **Pearson product moment coefficient of correlation,** r, is a measure of the strength of the *linear* relationship between two variables x and y. It is computed (for a sample of n measurements on x and y) as follows:

$$r = \frac{SS_{xy}}{\sqrt{SS_{xx}SS_{yy}}}$$

Note that the computational formula for the correlation coefficient r given in Definition 9.2 involves the same quantities that were used in computing the least squares prediction equation. In fact, since the numerators of the expressions for $\hat{\beta}_1$ and r are identical, you can see that $r = 0$ when $\hat{\beta}_1 = 0$ (the case where x contributes no information for the prediction of y) and that r is positive when the slope is positive and negative when the slope is negative. Unlike $\hat{\beta}_1$, the correlation coefficient r is *scaleless* and assumes a value between -1 and $+1$, regardless of the units of x and y.

A value of r near or equal to 0 implies little or no linear relationship between y and x. In contrast, the closer r comes to 1 or -1, the stronger the linear relationship between y and x. And if $r = 1$ or $r = -1$, all the sample points fall exactly on the least squares line. Positive values of r imply a positive linear relationship between y and x; that is, y increases as x increases. Negative values of r imply a negative linear relationship between y and x; that is, y decreases as x increases. Each of these situations is portrayed in Figure 9.13.

We demonstrate how to calculate the coefficient of correlation r using the data in Table 9.1 for the drug reaction example. The quantities needed to calculate r are SS_{xy}, SS_{xx}, and SS_{yy}. The first two quantities have been calculated previously and are repeated here for convenience:

$$SS_{xy} = 7, \quad SS_{xx} = 10, \quad SS_{yy} = \sum y^2 - \frac{\left(\sum y\right)^2}{n}$$

$$= 26 - \frac{(10)^2}{5} = 26 - 20 = 6$$

We now find the coefficient of correlation:

$$r = \frac{SS_{xy}}{\sqrt{SS_{xx}SS_{yy}}} = \frac{7}{\sqrt{(10)(6)}} = \frac{7}{\sqrt{60}} = .904$$

The fact that r is positive and near 1 in value indicates that the reaction time tends to increase as the amount of drug in the bloodstream increases—*for this sample of five subjects*. This is the same conclusion we reached when we found the calculated value of the least squares slope to be positive.

FIGURE 9.13

Values of r and Their Implications

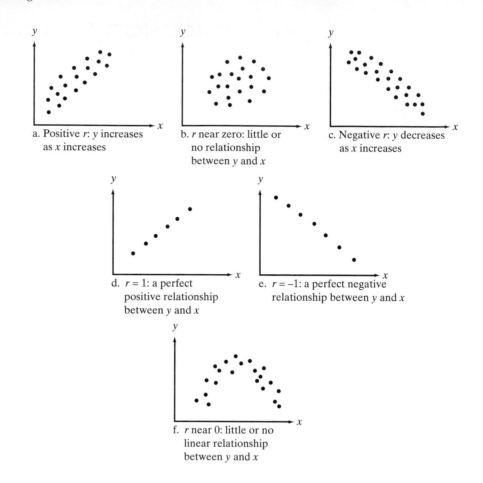

a. Positive *r*: *y* increases as *x* increases

b. *r* near zero: little or no relationship between *y* and *x*

c. Negative *r*: *y* decreases as *x* increases

d. *r* = 1: a perfect positive relationship between *y* and *x*

e. *r* = –1: a perfect negative relationship between *y* and *x*

f. *r* near 0: little or no linear relationship between *y* and *x*

E X A M P L E   9 . 1

r = .987

Legalized gambling is available on several riverboat casinos operated by a city in Mississippi. The mayor of the city wants to know the correlation between the number of casino employees and yearly crime rate. The records for the past 10 years are examined, and the results listed in Table 9.5 are obtained. Calculate the coefficient of correlation *r* for the data.

⊘ CASINO

TABLE 9.5   Data on Casino Employees and Crime Rate, Example 9.1

Year	Number of Casino Employees *x* (thousands)	Crime Rate *y* (number of crimes per 1,000 population)
1993	15	1.35
1994	18	1.63
1995	24	2.33
1996	22	2.41
1997	25	2.63
1998	29	2.93
1999	30	3.41
2000	32	3.26
2001	35	3.63
2002	38	4.15

FIGURE 9.14

MINITAB Correlation Printout for Example 9.1

Pearson correlation of EMPLOY and CRIME = 0.987
P-Value = 0.000

FIGURE 9.15

MINITAB Scattergram for Example 9.1

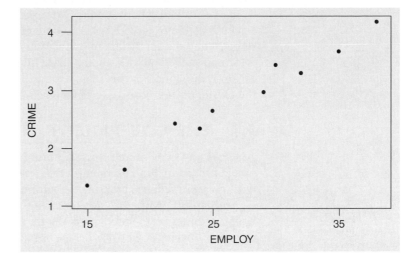

Suggested Exercise 9.48

Solution

Rather than use the computing formula given in Definition 9.2, we resort to a statistical software package. The data of Table 9.5 were entered into a computer and MINITAB was used to compute r. The MINITAB printout is shown in Figure 9.14.

The coefficient of correlation, highlighted on the printout, is $r = .987$. Thus, the size of the casino workforce and crime rate in this city are very highly correlated—at least over the past 10 years. The implication is that a strong positive linear relationship exists between these variables (see Figure 9.15). We must be careful, however, not to jump to any unwarranted conclusions. For instance, the mayor may be tempted to conclude that hiring more casino workers next year will increase the crime rate—that is, that there is a *causal relationship* between the two variables. However, high correlation does not imply causality. The fact is, many things have probably contributed both to the increase in the casino workforce and to the increase in crime rate. The city's tourist trade has undoubtedly grown since legalizing riverboat casinos and it is likely that the casinos have expanded both in services offered and in number. *We cannot infer a causal relationship on the basis of high sample correlation. When a high correlation is observed in the sample data, the only safe conclusion is that a linear trend may exist between x and y.* Another variable, such as the increase in tourism, may be the underlying cause of the high correlation between x and y.   ■

Keep in mind that the correlation coefficient r measures the linear correlation between x values and y values in the sample, and a similar linear coefficient of correlation exists for the population from which the data points were selected. The **population correlation coefficient** is denoted by the symbol ρ (rho). As you might expect, ρ is estimated by the corresponding sample statistic, r. Or, instead of estimating ρ, we might want to test the null hypothesis $H_0: \rho = 0$ against $H_a: \rho \neq 0$—that is, we can test the hypothesis that x contributes no information for the prediction of y by using the straight-line model against the alternative that the two variables are at least linearly related.

However, we already performed this *identical* test in Section 9.5 when we tested $H_0: \beta_1 = 0$ against $H_a: \beta_1 \neq 0$. That is, the null hypothesis $H_0: \rho = 0$ is equivalent to the hypothesis $H_0: \beta_1 = 0$.* When we tested the null hypothesis $H_0: \beta_1 = 0$ in connection with the drug reaction example, the data led to a rejection of the null hypothesis at the $\alpha = .05$ level. This rejection implies that the null hypothesis of a 0 linear correlation between the two variables (drug and reaction time) can also be rejected at the $\alpha = .05$ level. The only real difference between the least squares slope $\hat{\beta}_1$ and the coefficient of correlation r is the measurement scale. Therefore, the information they provide about the usefulness of the least squares model is to some extent redundant. For this reason, we will use the slope to make inferences about the existence of a positive or negative linear relationship between two variables.

9.7   THE COEFFICIENT OF DETERMINATION

Another way to measure the usefulness of the model is to measure the contribution of x in predicting y. To accomplish this, we calculate how much the errors of prediction of y were reduced by using the information provided by x. To illustrate, consider the sample shown in the scattergram of Figure 9.16a. If we assume that x contributes no information for the prediction of y, the best prediction for a value of y is the sample mean $\bar{y}$, which is shown as the horizontal line in Figure 9.16b. The vertical line segments in Figure 9.16b are the deviations of the points about the mean $\bar{y}$. Note that the sum of squares of deviations for the prediction equation $\hat{y} = \bar{y}$ is

$$SS_{yy} = \sum (y_i - \bar{y})^2$$

FIGURE 9.16

A Comparison of the Sum of Squares of Deviations for Two Models

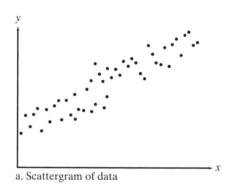

a. Scattergram of data

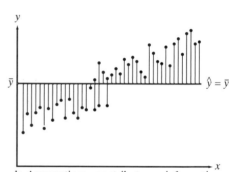

b. Assumption: x contributes no information for predicting y, $\hat{y} = \bar{y}$

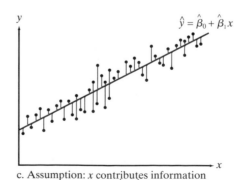

c. Assumption: x contributes information for predicting y, $\hat{y} = \hat{\beta}_0 + \hat{\beta}_1 x$

---

*The correlation test statistic that is equivalent to $t = \hat{\beta}_1/s_{\hat{\beta}_1}$ is $t = \dfrac{r}{\sqrt{(1 - r^2)/(n - 2)}}$

Now suppose you fit a least squares line to the same set of data and locate the deviations of the points about the line as shown in Figure 9.16c. Compare the deviations about the prediction lines in Figures 9.16b and 9.16c. You can see that

1. If x contributes little or no information for the prediction of y, the sums of squares of deviations for the two lines,

$$SS_{yy} = \sum (y_i - \bar{y})^2 \quad \text{and} \quad SSE = \sum (y_i - \hat{y}_i)^2$$

will be nearly equal.

2. If x does contribute information for the prediction of y, the SSE will be smaller than SS_{yy}. In fact, if all the points fall on the least squares line, then $SSE = 0$.

Then, the reduction in the sum of squares of deviations that can be attributed to x, expressed as a proportion of SS_{yy}, is

$$\frac{SS_{yy} - SSE}{SS_{yy}}$$

Note that SS_{yy} is the "total sample variation" of the observations around the mean $\bar{y}$ and that SSE is the remaining "unexplained sample variability" after fitting the line $\hat{y}$. Thus, the difference $(SS_{yy} - SSE)$ is the "explained sample variability" attributable to the linear relationship with x. Then a verbal description of the proportion is

$$\frac{SS_{yy} - SSE}{SS_{yy}} = \frac{\text{Explained sample variability}}{\text{Total sample variability}}$$
$$= \text{Proportion of total sample variability explained by the linear relationship}$$

In simple linear regression, it can be shown that this proportion—called the *coefficient of determination*—is equal to the square of the simple linear coefficient of correlation r (the Pearson product moment coefficient of correlation).

DEFINITION 9.3

The **coefficient of determination** is

$$r^2 = \frac{SS_{yy} - SSE}{SS_{yy}} = 1 - \frac{SSE}{SS_{yy}}$$

It represents the proportion of the total sample variability around $\bar{y}$ that is explained by the linear relationship between y and x. (In simple linear regression, it may also be computed as the square of the coefficient of correlation r.)

Note that r^2 is always between 0 and 1, because r is between -1 and $+1$. Thus, an r^2 of .60 means that the sum of squares of deviations of the y values about their predicted values has been reduced 60% by the use of the least squares equation $\hat{y}$, instead of $\bar{y}$, to predict y.

EXAMPLE 9.2

$r^2 = .82$

Calculate the coefficient of determination for the drug reaction example. The data are repeated in Table 9.6 for convenience. Interpret the result.

●STIMULUS
TABLE 9.6

Amount of Drug x (%)	Reaction Time y (seconds)
1	1
2	1
3	2
4	2
5	4

Solution

From previous calculations,

$$SS_{yy} = 6 \quad \text{and} \quad SSE = \sum (y - \hat{y})^2 = 1.10$$

Suggested Exercise 9.55

Then, from Definition 9.3, the coefficient of determination is given by

$$r^2 = \frac{SS_{yy} - SSE}{SS_{yy}} = \frac{6.0 - 1.1}{6.0} = \frac{4.9}{6.0} = .817$$

Another way to compute r^2 is to recall (Section 9.6) that $r = .904$. Then we have $r^2 = (.904)^2 = .817$. A third way to obtain r^2 is from a computer printout. This value is highlighted on the SPSS printout in Figure 9.17. Our interpretation is as follows: We know that using the amount of drug in the blood, x, to predict y with the least squares line

$$\hat{y} = -.1 + .7x$$

accounts for nearly 82% of the total sum of squares of deviations of the five sample y values about their mean. Or, stated another way, 82% of the sample variation in reaction time (y) can be "explained" by using amount (x) of drug in a straight-line model.

FIGURE 9.17

Portion of SPSS Printout for Time–Drug Regression

Model Summary

Model	R	R Square	Adjusted R Square	Std. Error of the Estimate
1	.904[a]	.817	.756	.61

a. Predictors: (Constant), DRUG_X

Practical Interpretation of the Coefficient of Determination, r^2

$100(r^2)$% of the sample variation in y (measured by the total sum of squares of deviations of the sample y values about their mean $\bar{y}$) can be explained by (or attributed to) using x to predict y in the straight-line model.

EXERCISES 9.48–9.61

Learning the Mechanics

9.48 Describe the slope of the least squares line if
   a. $r = .7$
   b. $r = -.7$
   c. $r = 0$
   d. $r^2 = .64$

9.49 Explain what each of the following sample correlation coefficients tells you about the relationship between the x and y values in the sample:
   a. $r = 1$
   b. $r = -1$
   c. $r = 0$
   d. $r = .90$
   e. $r = .10$
   f. $r = -.88$

9.50 Calculate r^2 for the least squares line in each of the following exercises. Interpret their values.
   a. Exercise 9.10 (p. 452)   .9438
   b. Exercise 9.13 (p. 453)   .8769

9.51 Construct a scattergram for each data set. Then calculate r and r^2 for each data set. Interpret their values.

   a.
x	-2	-1	0	1	2
y	-2	1	2	5	6

   $r = .9853$
   $r^2 = .9709$

   b.
x	-2	-1	0	1	2
y	6	5	3	2	0

   $r = -.9934$
   $r^2 = .9868$

   c.
x	1	2	2	3	3	3	4
y	2	1	3	1	2	3	2

   $r = 0$
   $r^2 = 0$

   d.
x	0	1	3	5	6
y	0	1	2	1	0

   $r = 0$
   $r^2 = 0$

Applying the Concepts—Basic

⊙ TAMPALMS

9.52 Refer to the data on sale prices and total appraised values of 92 residential properties recently sold in an upscale Tampa, Florida, neighborhood, Exercise 9.14 (p. 453). The MINITAB simple linear regression printout relating sale price (y) to appraised property value (x) is reproduced below, followed by a MINITAB correlation printout.
   a. Find the coefficient of correlation between appraised property value and sale price on the printout. Interpret this value.
   b. Find the coefficient of determination between appraised property value and sale price on the printout. Interpret this value.

9.53 Is there a relationship between leisure activities and high school performance? This question was investigated in the *Journal of Leisure Research* (Vol. 24, 1992). A list of 43 leisure activities was presented to 159 high school students and the students were asked to identify the activities they participated in each week. The article reports that the correlation between high school GPA and the number of leisure activities is $r = .13$, which has a two-tailed p-value of .0512.
   a. What are the appropriate null and alternative hypotheses to test whether the number of leisure activities and GPA are linearly related? $H_a: \rho \neq 0$
   b. Interpret the p-value in terms of this test.

MINITAB Output for Exercise 9.52

```
The regression equation is
SALEPRIC = 20.9 + 1.07 TOTALVAL

Predictor        Coef      SE Coef        T        P
Constant       20.942        6.446     3.25    0.002
TOTALVAL      1.06873      0.02709    39.45    0.000

S = 32.79        R-Sq = 94.5%      R-Sq(adj) = 94.5%

Analysis of Variance

Source            DF           SS          MS        F        P
Regression         1      1673142     1673142  1556.48    0.000
Residual Error    90        96746        1075
Total             91      1769888

Pearson correlation of SALEPRIC and TOTALVAL = 0.972
P-Value = 0.000
```

9.54 Many high school students experience "math anxiety." Does such an attitude carry over to learning computer skills? A researcher at Duquesne University investigated this question and published her results in *Educational Technology* (May–June 1995). A sample of high school students—902 boys and 828 girls—from public schools in Pittsburgh, Pennsylvania, participated in the study. Using 5-point Likert scales, where 1 = "strongly disagree" and 5 = "strongly agree," the researcher measured the students' interest and confidence in both mathematics and computers.

a. For boys, math confidence and computer interest were correlated at $r = .14$. Fully interpret this result.

b. For girls, math confidence and computer interest were correlated at $r = .33$. Fully interpret this result.

Applying the Concepts—Intermediate

9.55 *Perception & Psychophysics* (July 1998) reported on a study of how people view the 3-dimensional objects projected onto a rotating 2-dimensional image. Each in a sample of 25 university students viewed various depth-rotated objects (e.g., hairbrush, duck, shoe) until they recognized the object. The recognition exposure time—that is, the minimum time (in milliseconds) required for the subject to recognize the object—was recorded for each. In addition, each subject rated the "goodness of view" of the object on a numerical scale, where lower scale values correspond to better views. The table gives the correlation coefficient, r, between recognition exposure time and goodness of view for several different rotated objects.

Object	r	t
Piano	.447	2.40
Bench	−.057	.27
Motorbike	.619	3.78
Armchair	.294	1.47
Teapot	.949	14.50

a. Interpret the value of r for each object.

b. Calculate and interpret the value of r^2 for each object.

c. The table also includes the t-value for testing the null hypothesis of no correlation (i.e., for testing $H_0: \beta_1 = 0$). Interpret these results.

9.56 The fertility rate of a country is defined as the number of children a woman citizen bears, on average, in her lifetime. *Scientific American* (Dec. 1993) reported on the declining fertility rate in developing countries. The researchers found that family planning can have a great effect on fertility rate. The table on p. 479 gives the fertility rate, y, and contraceptive prevalence, x, (measured as the percentage of married women who use contraception) for each of 27 developing countries. An SAS printout of the simple linear regression analysis is provided below.

a. According to the researchers, "the data reveal that differences in contraceptive prevalence explain about 90% of the variation in fertility rates." Do you concur?

SAS Output for Exercise 9.56

The REG Procedure
Dependent Variable: FERTRATE

Analysis of Variance

Source	DF	Sum of Squares	Mean Square	F Value	Pr > F
Model	1	35.96633	35.96633	74.31	<.0001
Error	25	12.10033	0.48401		
Corrected Total	26	48.06667			

Root MSE	0.69571	R-Square	0.7483
Dependent Mean	4.51111	Adj R-Sq	0.7382
Coeff Var	15.42216		

Parameter Estimates

Variable	DF	Parameter Estimate	Standard Error	t Value	Pr > |t|
Intercept	1	6.73193	0.29034	23.19	<.0001
CONTPREV	1	-0.05461	0.00634	-8.62	<.0001

FERTRATE

Country	Contraceptive Prevalence x	Fertility Rate y
Mauritius	76	2.2
Thailand	69	2.3
Colombia	66	2.9
Costa Rica	71	3.5
Sri Lanka	63	2.7
Turkey	62	3.4
Peru	60	3.5
Mexico	55	4.0
Jamaica	55	2.9
Indonesia	50	3.1
Tunisia	51	4.3
El Salvador	48	4.5
Morocco	42	4.0
Zimbabwe	46	5.4
Egypt	40	4.5
Bangladesh	40	5.5
Botswana	35	4.8
Jordan	35	5.5
Kenya	28	6.5
Guatemala	24	5.5
Cameroon	16	5.8
Ghana	14	6.0
Pakistan	13	5.0
Senegal	13	6.5
Sudan	10	4.8
Yemen	9	7.0
Nigeria	7	5.7

Source: Robey, B., *et al.* "The fertility decline in developing countries." *Scientific American*, Dec. 1993, p. 62. [*Note:* The data values are estimated from a scatterplot.]

b. The researchers also concluded that "if contraceptive use increases by 18 percent, women bear, on average, one fewer child." Is this statement supported by the data? Explain.   Yes

9.57 In cotherapy two or more therapists lead a group. An article in the *American Journal of Dance Therapy* (Spring/Summer 1995) examined the use of cotherapy in dance/movement therapy. Two of several variables measured on each of a sample of 136 professional dance/movement therapists were years of formal training x and reported success rate y (measured as a percentage) of coleading dance/movement therapy groups.

a. Propose a linear model relating y to x.

b. The researcher hypothesized that dance/movement therapists with more years in formal dance training will report higher perceived success rates in cother-

apy relationships. State the hypothesis in terms of the parameter of the model, part **a**. $H_0: \beta_1 > 0$

c. The correlation coefficient for the sample data was reported as $r = -.26$. Interpret this result.

d. Does the value of r in part **c** support the hypothesis in part **b**? Test using $\alpha = .05$. [*Hint:* See the last paragraph and accompanying footnote of Section 9.6] No; $t = -3.12$

NAMEGAME2

9.58 Refer to the *Journal of Experimental Psychology-Applied* (June 2000) name retrieval study, Exercises 9.23 (p. 458) and 9.45 (p. 469). Find and interpret the values of r and r^2 for the simple linear regression relating the proportion of names recalled (y) and position (order) of the student (x) during the "name game."

BOXING2

9.59 Refer to the *British Journal of Sports Medicine* (April 2000) study of the effect of massage on boxing performance, Exercise 9.42 (p. 468). Find and interpret the values of r and r^2 for the simple linear regression relating the blood lactate concentration and the boxer's perceived recovery. $r = .5702; r^2 = .3257$

Applying the Concepts—Advanced

9.60 Botanists at the University of Toronto conducted a series of experiments to investigate the feeding habits of baby snow geese (*Journal of Applied Ecology*, Vol. 32, 1995). Goslings were deprived of food until their guts were empty, then were allowed to feed for 6 hours on a diet of plants or Purina Duck Chow. For each feeding trial, the change in the weight of the gosling after 2.5 hours was recorded as a percentage of initial weight. Two other variables recorded were digestion efficiency (measured as a percentage) and amount of acid-detergent fibre in the digestive tract (also measured as a percentage). The data for 42 feeding trials are listed in the table on p. 480.

a. The botanists were interested in the correlation between weight change (y) and digestion efficiency (x). Plot the data for these two variables in a scattergram. Do you observe a trend?

b. Find the coefficient of correlation relating weight change y to digestion efficiency x. Interpret this value. $r = .612$

c. Conduct a test to determine whether weight change y is correlated with a digestion efficiency x. Use $\alpha = .01$. $t = 4.89$

d. Repeat parts **b** and **c**, but exclude the data for trials that used duck chow from the analysis. What do you conclude? $r = .309, t = 1.81$

e. The botanists were also interested in the correlation between digestion efficiency y and acid-detergent fibre x. Repeat parts **a**–**d** for these two variables.

⬤ SNOWGEESE

Feeding Trial	Diet	Weight Change (%)	Digestion Efficiency (%)	Acid-Detergent Fibre (%)
1	Plants	−6	0	28.5
2	Plants	−5	2.5	27.5
3	Plants	−4.5	5	27.5
4	Plants	0	0	32.5
5	Plants	2	0	32
6	Plants	3.5	1	30
7	Plants	−2	2.5	34
8	Plants	−2.5	10	36.5
9	Plants	−3.5	20	28.5
10	Plants	−2.5	12.5	29
11	Plants	−3	28	28
12	Plants	−8.5	30	28
13	Plants	−3.5	18	30
14	Plants	−3	15	31
15	Plants	−2.5	17.5	30
16	Plants	−.5	18	22
17	Plants	0	23	22.5
18	Plants	1	20	24
19	Plants	2	15	23
20	Plants	6	31	21
21	Plants	2	15	24
22	Plants	2	21	23
23	Plants	2.5	30	22.5
24	Plants	2.5	33	23
25	Plants	0	27.5	30.5
26	Plants	.5	29	31
27	Plants	−1	32.5	30
28	Plants	−3	42	24
29	Plants	−2.5	39	25
30	Plants	−2	35.5	25
31	Plants	.5	39	20
32	Plants	5.5	39	18.5
33	Plants	7.5	50	15
34	Duck Chow	0	62.5	8
35	Duck Chow	0	63	8
36	Duck Chow	2	69	7
37	Duck Chow	8	42.5	7.5
38	Duck Chow	9	59	8.5
39	Duck Chow	12	52.5	8
40	Duck Chow	8.5	75	6
41	Duck Chow	10.5	72.5	6.5
42	Duck Chow	14	69	7

Source: Gadallah, F. L., and Jefferies, R. L. "Forage quality in brood rearing areas of the lesser snow goose and the growth of captive goslings." *Journal of Applied Biology,* Vol. 32, No. 2, 1995, pp. 281–282 (adapted from Figures 2 and 3).

9.61 A study published in *Psychosomatic Medicine* (Mar./Apr. 2001) explored the relationship between reported severity of pain and actual pain tolerance in 337 patients who suffer from chronic pain. Each patient reported his/her severity of chronic pain on a 7-point scale (1 = no pain, 7 = extreme pain). To obtain a pain tolerance level, a tourniquet was applied to the arm of each patient and twisted. The maximum pain level tolerated was measured on a quantitative scale.

a. According to the researchers, "correlational analysis revealed a small but significant inverse relationship between [actual] pain tolerance and the reported severity of chronic pain." Based on this statement, is the value of r for the 337 patients positive or negative?

b. Suppose that the result reported in part **a** is significant at $\alpha = .05$. Find the approximate value of r for the sample of 337 patients. [*Hint:* Use the formula $t = r\sqrt{(n - 2)}/\sqrt{(1 - r^2)}$.] $r = -.0895$

9.8 USING THE MODEL FOR ESTIMATION AND PREDICTION

If we are satisfied that a useful model has been found to describe the relationship between reaction time and amount of drug in the bloodstream, we are ready for step 5 in our regression modeling procedure: using the model for estimation and prediction.

The most common uses of a probabilistic model for making inferences can be divided into two categories. The first is the use of the model for estimating the mean value of y, $E(y)$, for a specific value of x.

For our drug reaction example, we may want to estimate the mean response time for all people whose blood contains 4% of the drug.

The second use of the model entails predicting a new individual y value for a given x.

That is, we may want to predict the reaction time for a specific person who possesses 4% of the drug in the bloodstream.

In the first case, we are attempting to estimate the mean value of y for a very large number of experiments at the given x value. In the second case, we are trying to predict the outcome of a single experiment at the given x value. Which of these model uses—estimating the mean value of y or predicting an individual new value of y (for the same value of x)—can be accomplished with the greater accuracy?

Before answering this question, we first consider the problem of choosing an estimator (or predictor) of the mean (or a new individual) y value. We will use the least squares prediction equation

$$\hat{y} = \hat{\beta}_0 + \hat{\beta}_1 x$$

both to estimate the mean value of y and to predict a specific new value of y for a given value of x. For our example, we found

$$\hat{y} = -.1 + .7x$$

so that the estimated mean reaction time for all people when $x = 4$ (drug is 4% of blood content) is

$$\hat{y} = -.1 + .7(4) = 2.7 \text{ seconds}$$

The same value is used to predict a new y value when $x = 4$. That is, both the estimated mean and the predicted value of y are $\hat{y} = 2.7$ when $x = 4$, as shown in Figure 9.18.

The difference between these two model uses lies in the relative accuracy of the estimate and the prediction. These accuracies are best measured by using the sampling errors of the least squares line when it is used as an estimator and as a predictor, respectively. These errors are reflected in the standard deviations given in the next box.

FIGURE 9.18

Estimated Mean Value and Predicted Individual Value of Reaction Time y for $x = 4$

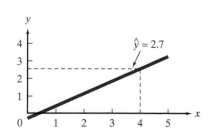

TEACHING TIP
Use the formulas to illustrate why the prediction interval for *y* is always wider than the confidence interval for $E(y)$.

Sampling Errors for the Estimator of the Mean of *y* and the Predictor of an Individual New Value of *y*

1. The standard deviation of the sampling distribution of the estimator $\hat{y}$ of the mean value of *y* at a specific value of *x*, say x_p, is

$$\sigma_{\hat{y}} = \sigma \sqrt{\frac{1}{n} + \frac{(x_p - \bar{x})^2}{SS_{xx}}}$$

where σ is the standard deviation of the random error ε. We refer to $\sigma_{\hat{y}}$ as the standard error of $\hat{y}$.

2. The standard deviation of the prediction error for the predictor $\hat{y}$ of an individual new *y* value at a specific value of *x* is

$$\sigma_{(y-\hat{y})} = \sigma \sqrt{1 + \frac{1}{n} + \frac{(x_p - \bar{x})^2}{SS_{xx}}}$$

where σ is the standard deviation of the random error ε. We refer to $\sigma_{(y-\hat{y})}$ as the standard error of prediction.

The true value of σ is rarely known, so we estimate σ by *s* and calculate the estimation and prediction intervals as shown in the next two boxes.

TEACHING TIP
Use a graph of the confidence and prediction bands around the regression line to illustrate the fact that the confidence interval is always wider than the prediction interval.

A $100(1 - \alpha)\%$ Confidence Interval for the Mean Value of *y* at $x = x_p$

$$\hat{y} + t_{\alpha/2}(\text{Estimated standard error of } \hat{y})$$

or

$$\hat{y} \pm t_{\alpha/2}s \sqrt{\frac{1}{n} + \frac{(x_p - \bar{x})^2}{SS_{xx}}}$$

where $t_{\alpha/2}$ is based on $(n - 2)$ degrees of freedom.

A $100(1 - \alpha)\%$ Prediction Interval* for an Individual New Value of *y* at $x = x_p$

$$\hat{y} \pm t_{\alpha/2}(\text{Estimated standard error of prediction})$$

or

$$\hat{y} \pm t_{\alpha/2}s \sqrt{1 + \frac{1}{n} + \frac{(x_p - \bar{x})^2}{SS_{xx}}}$$

where $t_{\alpha/2}$ is based on $(n - 2)$ degrees of freedom.

---

*The term *prediction interval* is used when the interval formed is intended to enclose the value of a random variable. The term *confidence interval* is reserved for estimation of population parameters (such as the mean).

EXAMPLE 9.3

2.7 ± 1.1

Find a 95% confidence interval for the mean reaction time when the concentration of the drug in the bloodstream is 4%.

Solution

For a 4% concentration, $x = 4$ and the confidence interval for the mean value of y is

$$\hat{y} \pm t_{\alpha/2}s\sqrt{\frac{1}{n} + \frac{(x_\mathrm{p} - \bar{x})^2}{\mathrm{SS}_{xx}}} = \hat{y} \pm t_{.025}s\sqrt{\frac{1}{5} + \frac{(4 - \bar{x})^2}{\mathrm{SS}_{xx}}}$$

where $t_{.025}$ is based on $n - 2 = 5 - 2 = 3$ degrees of freedom. Recall that $\hat{y} = 2.7$, $s = .61$, $\bar{x} = 3$, and $\mathrm{SS}_{xx} = 10$. From Table IV in Appendix A, $t_{.025} = 3.182$. Thus, we have

$$2.7 \pm (3.182)(.61)\sqrt{\frac{1}{5} + \frac{(4 - 3)^2}{10}} = 2.7 \pm (3.182)(.61)(.55)$$
$$= 2.7 \pm (3.182)(.34)$$
$$= 2.7 \pm 1.1$$

Therefore, when the percentage of drug in the bloodstream is 4%, the standard error of $\hat{y}$ is .34 and the corresponding 95% confidence interval for the mean reaction time for all possible subjects is 1.6 to 3.8 seconds. Note that we used a small amount of data (small sample size) for purposes of illustration in fitting the least squares line. The interval would probably be narrower if more information had been obtained from a larger sample.

EXAMPLE 9.4

2.7 ± 2.2

Predict the reaction time for the next performance of the experiment for a subject with a drug concentration of 4%. Use a 95% prediction interval.

Solution

To predict the response time for an individual new subject for whom $x = 4$, we calculate the 95% prediction interval as

$$\hat{y} \pm t_{\alpha/2}s\sqrt{1 + \frac{1}{n} + \frac{(x_\mathrm{p} - \bar{x})^2}{\mathrm{SS}_{xx}}} = 2.7 \pm (3.182)(.61)\sqrt{1 + \frac{1}{5} + \frac{(4 - 3)^2}{10}}$$
$$= 2.7 \pm (3.182)(.61)(1.14)$$
$$= 2.7 \pm (3.182)(.70)$$
$$= 2.7 \pm 2.2$$

Suggested Exercise 9.67

Therefore, the standard error of prediction when the drug concentration is 4% is .70, and we predict with 95% confidence that the reaction time for this new individual will fall in the interval from .5 to 4.9 seconds. Like the confidence interval for the mean value of y, the prediction interval for y is quite large. This is because we have chosen a simple example (only five data points) to fit the least squares line. The width of the prediction interval could be reduced by using a larger number of data points.

Both the confidence interval for $E(y)$ and the prediction interval for y can be obtained using a statistical software package. Figures 9.19 and 9.20 are SAS printouts showing confidence intervals and prediction intervals, respectively, for the data in the drug example.

FIGURE 9.19

SAS Printout Showing 95% Confidence Intervals for E(y)

Dependent Variable: TIME_Y

Output Statistics

Obs	DRUG_X	Dep Var TIME_Y	Predicted Value	Std Error Mean Predict	95% CL Mean		Residual
1	1	1.0000	0.6000	0.4690	-0.8927	2.0927	0.4000
2	2	1.0000	1.3000	0.3317	0.2445	2.3555	-0.3000
3	3	2.0000	2.0000	0.2708	1.1382	2.8618	0
4	4	2.0000	2.7000	0.3317	1.6445	3.7555	-0.7000
5	5	4.0000	3.4000	0.4690	1.9073	4.8927	0.6000

FIGURE 9.20

SAS Printout Showing 95% Prediction Intervals for y

Dependent Variable: TIME_Y

Output Statistics

Obs	DRUG_X	Dep Var TIME_Y	Predicted Value	Std Error Mean Predict	95% CL Predict		Residual
1	1	1.0000	0.6000	0.4690	-1.8376	3.0376	0.4000
2	2	1.0000	1.3000	0.3317	-0.8972	3.4972	-0.3000
3	3	2.0000	2.0000	0.2708	-0.1110	4.1110	0
4	4	2.0000	2.7000	0.3317	0.5028	4.8972	-0.7000
5	5	4.0000	3.4000	0.4690	0.9624	5.8376	0.6000

TEACHING TIP
Point out that the confidence and prediction intervals can be found on most computer output. The interpretation of these intervals is the key. Now is a good time to review the two symbols $E(y)$ and y.

The 95% confidence interval for $E(y)$ when $x = 4$ is highlighted in Figure 9.19. The interval shown on the printout, (1.6445, 3.7555), agrees (except for rounding) with the interval calculated in Example 9.3. The 95% prediction interval for y when $x = 4$ is highlighted in Figure 9.20. Again, except for rounding, the SAS interval (.5028, 4.8972) agrees with the one computed in Example 9.4.

A comparison of the confidence interval for the mean value of y and the prediction interval for a new value of y for 4% drug concentration ($x = 4$) is shown in Figure 9.21. Note that the prediction interval for an individual new value of y is *always* wider than the corresponding confidence interval for the mean value of y. You can see this by examining the formulas for the two intervals and by studying Figure 9.21.

FIGURE 9.21

A 95% Confidence Interval and Prediction Interval for Response Time when x = 4

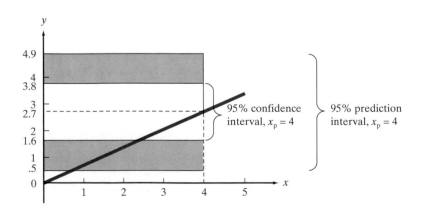

FIGURE 9.22

*Error of Estimating the Mean
Value of y for a Given Value of x*

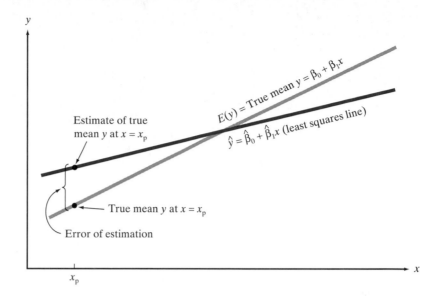

The error in estimating the mean value of y, $E(y)$, for a given value of x, say x_p, is the distance between the least squares line and the true line of means, $E(y) = \beta_0 + \beta_1 x$. This error, $[\hat{y} - E(y)]$, is shown in Figure 9.22. In contrast, *the error* $(y_p - \hat{y})$ *in predicting some future value of y is the sum of two errors*—the error of estimating the mean of y, $E(y)$, shown in Figure 9.22, plus the random error that is a component of the value of y to be predicted (see Figure 9.23). Consequently, the error of predicting a particular value of y will be larger than the error of estimating the mean value of y for a particular value of x. Note from their formulas that both the error of estimation and the error of prediction take their smallest values when $x_p = \bar{x}$. The farther x_p lies from $\bar{x}$, the larger will be the errors of estimation and prediction. You can see why this is true by noting the deviations for different values of x_p between the line of means $E(y) = \beta_0 + \beta_1 x$ and the predicted line of means $\hat{y} = \hat{\beta}_0 + \hat{\beta}_1 x$ shown in Figure 9.23. The deviation is larger at the extremes of the interval where the largest and smallest values of x in the data set occur.

FIGURE 9.23

*Error of Predicting a Future
Value of y for a Given Value of x*

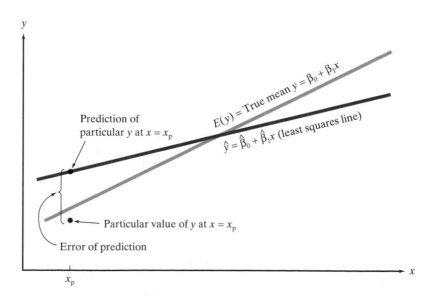

FIGURE 9.24

Confidence Intervals for Mean Values and Prediction Intervals for New Values

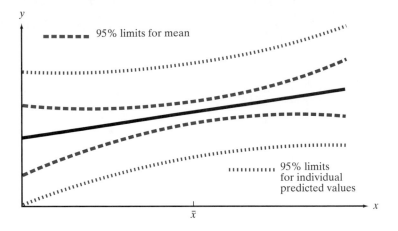

TEACHING TIP
The confidence interval formula is an easy place to illustrate why the intervals are narrowest at $x_p = \bar{x}$.

Both the confidence intervals for mean values and the prediction intervals for new values are depicted over the entire range of the regression line in Figure 9.24. You can see that the confidence interval is always narrower than the prediction interval, and that they are both narrowest at the mean $\bar{x}$, increasing steadily as the distance $|x - \bar{x}|$ increases. In fact, when x is selected far enough away from $\bar{x}$ so that it falls outside the range of the sample data, it is dangerous to make any inferences about $E(y)$ or y.

Caution

Using the least squares prediction equation to estimate the mean value of y or to predict a particular value of y for values of x that fall outside the range of the values of x contained in your sample data may lead to errors of estimation or prediction that are much larger than expected. Although the least squares model may provide a very good fit to the data over the range of x values contained in the sample, it could give a poor representation of the true model for values of x outside this region.

The confidence interval width grows smaller as n is increased; thus, in theory, you can obtain as precise an estimate of the mean value of y as desired (at any given x) by selecting a large enough sample. The prediction interval for a new value of y also grows smaller as n increases, but there is a lower limit on its width. If you examine the formula for the prediction interval, you will see that the interval can get no smaller than $\hat{y} \pm z_{\alpha/2}\sigma$.* Thus, the only way to obtain more accurate predictions for new values of y is to reduce the standard deviation of the regression model, σ. This can be accomplished only by improving the model, either by using a curvilinear (rather than linear) relationship with x or by adding new independent variables to the model, or both. Consult the references to learn more about these methods of improving the model.

---

*The result follows from the facts that, for large n, $t_{\alpha/2} \approx z_{\alpha/2}$, $s \approx \sigma$, and the last two terms under the radical in the standard error of the predictor are approximately 0.

EXERCISES 9.62–9.73

Learning the Mechanics

9.62 Consider the following pairs of measurements:

x	1	2	3	4	5	6	7
y	3	5	4	6	7	7	10

    a. Construct a scattergram for these data.
    b. Find the least squares line, and plot it on your scattergram.
    c. Find s^2.   .8
    d. Find a 90% confidence interval for the mean value of y when $x = 4$. Plot the upper and lower bounds of the confidence interval on your scattergram.
    e. Find a 90% prediction interval for a new value of y when $x = 4$. Plot the upper and lower bounds of the prediction interval on your scattergram. 6 ± 1.927
    f. Compare the widths of the intervals you constructed in parts **d** and **e**. Which is wider and why?

9.63 Consider the following pairs of measurements:

x	y	x	y
1	−1	−2	−6
−1	−5	3	4
2	1	−1	−4
0	−3	5	4
4	7	1	0

    Given that $SS_{xx} = 47.6$, $SS_{yy} = 168.1$, $SS_{xy} = 85.6$, and $\hat{y} = -2.458 + 1.7983x$:
    a. Construct a scattergram for the data.
    b. Plot the least squares line on your scattergram.
    c. Use a 95% confidence interval to estimate the mean value of y when $x = 5$. Plot the upper and lower bounds of the interval on your scattergram.
    d. Repeat part **c** for $x = 1.2$ and $x = -2$.
    e. Compare the widths of the three confidence intervals you constructed in parts **c** and **d** and explain why they differ.

9.64 Refer to Exercise 9.63.
    a. Using no information about x, estimate and calculate a 95% confidence interval for the mean value of y. [*Hint:* Use the one-sample t methodology of Section 7.3.] $-.3 \pm 3.091$
    b. Plot the estimated mean value and the confidence interval as horizontal lines on your scattergram.
    c. Compare the confidence intervals you calculated in parts **c** and **d** of Exercise 9.63 with the one you calculated in part **a** of this exercise. Does x appear to contribute information about the mean value of y?

    d. Check the answer you gave in part **c** with a statistical test of the null hypothesis $H_0: \beta_1 = 0$ against $H_a: \beta_1 \neq 0$. Use $\alpha = .05$. $t = 9.32$

9.65 Consider the pairs of measurements shown in the following table.

x	4	6	0	5	2	3	2	6	2	1
y	3	5	−1	4	3	2	0	4	1	1

    For these data, $SS_{xx} = 38.900$, $SS_{yy} = 33.600$, $SS_{xy} = 32.8$, and $\hat{y} = -4.14 + .843x$.
    a. Construct a scattergram for these data.
    b. Plot the least squares line on your scattergram.
    c. Use a 95% confidence interval to estimate the mean value of y when $x_p = 6$. Plot the upper and lower bounds of the interval on your scattergram.
    d. Repeat part **c** for $x_p = 3.2$ and $x_p = 0$.
    e. Compare the widths of the three confidence intervals you constructed in parts **c** and **d** and explain why they differ.

9.66 In fitting a least squares line to $n = 10$ data points, the following quantities were computed:

$$SS_{xx} = 32, \bar{x} = 3, SS_{yy} = 26, \bar{y} = 4, SS_{xy} = 28$$

    a. Find the least squares line. $\hat{y} = 1.375 + .875x$
    b. Graph the least squares line.
    c. Calculate SSE.   1.5
    d. Calculate s^2.   .1875
    e. Find a 95% confidence interval for the mean value of y when $x_p = 2.5$. $3.5625 \pm .3279$
    f. Find a 95% prediction interval for y when $x_p = 4$.

Applying the Concepts—Basic

9.67 Refer to the data on sale prices and total appraised values of 92 residential properties in an upscale Tampa, Florida, neighborhood, Exercises 9.14 and 9.38 (pp. 453, 467).
    a. In Exercise 9.38, you determined that appraised property value x and sale price y are positively related for homes sold in the Tampa Palms subdivision. Does this result guarantee that appraised value will yield accurate predictions of sale price? Explain.   No
    b. MINITAB was used to predict the sale price of a residential property in the subdivision with a total appraised value of $300,000. Locate a 95% prediction interval for the sale price of this property on the printout (p. 488) and interpret the result.
    c. Locate a 95% confidence interval for $E(y)$ on the printout and interpret the result.   (332.95, 350.17)

MINITAB Output for Exercise 9.67

```
Predicted Values for New Observations

New Obs      Fit     SE Fit        95.0% CI            95.0% PI
1         341.56       4.33    ( 332.95, 350.17) (  275.86, 407.26)


Values of Predictors for New Observations

New Obs    TOTALVAL
1               300
```

9.68 Refer to the simple linear regression of sweetness index y and amount of pectin x for $n = 24$ orange juice samples, Exercise 9.19 (p. 456). The SPSS printout of the analysis is shown on p. 489. A 90% confidence interval for the mean sweetness index, $E(y)$, for each value of x is shown on the SPSS spreadsheet. Select an observation and interpret this interval.

9.69 Refer to the simple linear regression of number of flycatchers killed y and nest box tit occupancy x for $n = 14$ nest sites, Exercises 9.22 and 9.32 (pp. 457, 463).

    a. A 95% prediction interval for y when $x = 64$ is shown at the bottom of the SAS printout shown below. Interpret this interval.

    b. How would the width of a 95% confidence interval for $E(y)$ when $x = 64$ compare to the interval, part **a**?   Narrower

    c. Would you recommend using the model to predict the number of flycatchers killed at a site with a nest box tit occupancy of 15%? Explain.   No

Applying the Concepts—Intermediate

🔵 NAMEGAME2

9.70 Refer to the *Journal of Experimental Psychology-Applied* (June 2000) name retrieval study, Exercises 9.23, 9.45, and 9.58 (pp. 458, 469, 479).

    a. Find a 99% confidence interval for the mean recall proportion for students in the fifth position during the "name game." Interpret the result.

    b. Find a 99% prediction interval for the recall proportion of a particular student in the fifth position during the "name game." Interpret the result.

    c. Compare the two intervals, parts **a** and **b**. Which interval is wider? Will this always be the case? Explain.   Prediction; yes

9.71 The Sasakawa Sports Foundation conducted a national survey to assess the physical activity patterns of Japanese adults. The table below lists the frequency (average number of days in the past year) and duration of time (average number of minutes per single activity) Japanese adults spent participating in a sample of 11 sports activities.

🔵 JAPANSPORTS

Activity	Frequency x (days/year)	Duration y (minutes)
Jogging	135	43
Cycling	68	99
Aerobics	44	61
Swimming	39	60
Volleyball	30	80
Tennis	21	100
Softball	16	91
Baseball	19	127
Skating	7	115
Skiing	10	249
Golf	5	262

Source: J. Bennett, ed. *Statistics in Sport.* London: Arnold, 1998 (adapted from Figure 9.6).

SAS Output for Exercise 9.69

		Output Statistics				
Obs	titpct	Dep Var nokilled	Predicted Value	Std Error Mean Predict	95% CL Predict	
1	24	0	-0.4631	0.5554	-3.1326	2.2064
2	33	0	0.5058	0.3719	-2.0078	3.0194
3	34	0	0.6135	0.3559	-1.8891	3.1161
4	43	0	1.5824	0.2944	-0.8820	4.0468
5	50	0	2.3360	0.3695	-0.1760	4.8479
6	35	1.0000	0.7211	0.3412	-1.7718	3.2140
7	35	1.0000	0.7211	0.3412	-1.7718	3.2140
8	38	1.0000	1.0441	0.3073	-1.4278	3.5159
9	40	1.0000	1.2594	0.2949	-1.2053	3.7241
10	31	2.0000	0.2905	0.4074	-2.2492	2.8301
11	43	2.0000	1.5824	0.2944	-0.8820	4.0468
12	55	3.0000	2.8742	0.4643	0.2887	5.4598
13	57	4.0000	3.0896	0.5073	0.4659	5.7132
14	64	5.0000	3.8431	0.6700	1.0516	6.6347

SPSS Output for Exercise 9.68

Model Summary[b]

Model	R	R Square	Adjusted R Square	Std. Error of the Estimate
1	.478[a]	.229	.194	.215

a. Predictors: (Constant), PECTIN
b. Dependent Variable: SWEET

ANOVA[b]

Model		Sum of Squares	df	Mean Square	F	Sig.
1	Regression	.301	1	.301	6.520	.018[a]
	Residual	1.017	22	4.622E-02		
	Total	1.318	23			

a. Predictors: (Constant), PECTIN
b. Dependent Variable: SWEET

Coefficients[a]

Model		Unstandardized Coefficients		Standardized Coefficients		
		B	Std. Error	Beta	t	Sig.
1	(Constant)	6.252	.237		26.422	.000
	PECTIN	-2.31E-03	.001	-.478	-2.554	.018

a. Dependent Variable: SWEET

	run	sweet	pectin	lower90m	upper90m
1	1	5.2	220	5.64898	5.83848
2	2	5.5	227	5.63898	5.81613
3	3	6.0	259	5.57819	5.72904
4	4	5.9	210	5.66194	5.87173
5	5	5.8	224	5.64337	5.82560
6	6	6.0	215	5.65564	5.85493
7	7	5.8	231	5.63284	5.80379
8	8	5.6	268	5.55553	5.71011
9	9	5.6	239	5.61947	5.78019
10	10	5.9	212	5.65946	5.86497
11	11	5.4	410	5.05526	5.55416
12	12	5.6	256	5.58517	5.73592
13	13	5.8	306	5.43785	5.65219
14	14	5.5	259	5.57819	5.72904
15	15	5.3	284	5.50957	5.68213
16	16	5.3	383	5.15725	5.57694
17	17	5.7	271	5.54743	5.70434
18	18	5.5	264	5.56591	5.71821
19	19	5.7	227	5.63898	5.81613
20	20	5.3	263	5.56843	5.72031
21	21	5.9	232	5.63125	5.80075
22	22	5.8	220	5.64898	5.83848
23	23	5.8	246	5.60640	5.76091
24	24	5.9	241	5.61587	5.77454

a. Write the equation of a straight-line model relating duration (y) to frequency (x). $y = \beta_0 + \beta_1 x + \varepsilon$
b. Find the least squares prediction equation.
c. Is there evidence of a linear relationship between y and x? Test using $\alpha = .05$. $t = -2.051$
d. Use the least squares line to predict the duration of time Japanese adults participate in a sport that they play 25 times a year. Form a 95% confidence interval around the prediction and interpret the result.

⬤SNOWGEESE

9.72 Refer to the *Journal of Applied Ecology* feeding study of the relationship between weight change y of baby snow geese and digestion efficiency x, Exercise 9.60 (p. 479).
a. Fit the simple linear regression model to the data.
b. Do you recommend using the model to predict weight change y? Explain. $t = 4.90$; $p < .001$
c. Use the model to form a 95% confidence interval for the mean weight change of all baby snow geese with a digestion efficiency of $x = 15\%$. Interpret the interval. $(-2.579, 0.481)$

Applying the Concepts—Advanced

9.73 Refer to Exercise 9.33 (p. 463). The data are reproduced in the table.
a. Use a 90% confidence interval to estimate the mean useful life of a brand A cutting tool when the cutting speed is 45 meters per minute. Repeat for brand B. Compare the widths of the two intervals and comment on the reasons for any difference.
b. Use a 90% prediction interval to predict the useful life of a brand A cutting tool when the cutting speed is 45 meters per minute. Repeat for brand B. Compare the widths of the two intervals to each other and to the two intervals you calculated in part **a**. Comment on the reasons for any differences.

c. Note that the estimation and prediction you performed in parts **a** and **b** were for a value of x that was not included in the original sample. That is, the value $x = 45$ was not part of the sample. However, the value is within the range of x values in the sample, so that the regression model spans the x value for which the estimation and prediction were made. In such situations, estimation and prediction represent **interpolations**.

Suppose you were asked to predict the useful life of a brand A cutting tool for a cutting speed of $x = 100$ meters per minute. Since the given value of x is outside the range of the sample x values, the prediction is an example of **extrapolation**. Predict the useful life of a brand A cutting tool that is operated at 100 meters per minute, and construct a 95% confidence interval for the actual useful life of the tool. What additional assumption do you have to make in order to ensure the validity of an extrapolation?

⬤CUTTOOL

Cutting Speed (meters per minute)	Useful Life (Hours)	
	Brand A	Brand B
30	4.5	6.0
30	3.5	6.5
30	5.2	5.0
40	5.2	6.0
40	4.0	4.5
40	2.5	5.0
50	4.4	4.5
50	2.8	4.0
50	1.0	3.7
60	4.0	3.8
60	2.0	3.0
60	1.1	2.4
70	1.1	1.5
70	.5	2.0
70	3.0	1.0

9.9   A COMPLETE EXAMPLE

In the preceding sections we presented the basic elements necessary to fit and use a straight-line regression model. In this section we will assemble these elements by applying them in an example with the aid of computer software.

Suppose a fire insurance company wants to relate the amount of fire damage in major residential fires to the distance between the burning house and the nearest fire station. The study is to be conducted in a large suburb of a major city; a sample of 15 recent fires in this suburb is selected. The amount of damage, y, and the distance between the fire and the nearest fire station, x, are recorded for each fire. The results are given in Table 9.7.

⊘ FIREDAM
TABLE 9.7   Fire Damage Data

Distance from Fire Station *x* (miles)	Fire Damage *y* (thousands of dollars)
3.4	26.2
1.8	17.8
4.6	31.3
2.3	23.1
3.1	27.5
5.5	36.0
.7	14.1
3.0	22.3
2.6	19.6
4.3	31.3
2.1	24.0
1.1	17.3
6.1	43.2
4.8	36.4
3.8	26.1

Step 1 First, we hypothesize a model to relate fire damage, *y*, to the distance from the nearest fire station, *x*. We hypothesize a straight-line probabilistic model:

$$y = \beta_0 + \beta_1 x + \varepsilon$$

Step 2 Next, we enter the data of Table 9.7 into a computer and use statistical software to estimate the unknown parameters in the deterministic component of the hypothesized model. The STATISTIX printout for the simple linear regression analysis is shown in Figure 9.25. The least squares estimate of the slope β_1 and intercept β_0, highlighted on the printout, are

$$\hat{\beta}_1 = 4.91933$$

$$\hat{\beta}_0 = 10.2779$$

and the least squares equation is (rounded)

$$\hat{y} = 10.28 + 4.92x$$

This prediction equation is graphed in Figure 9.26 along with a plot of the data points.

The least squares estimate of the slope, $\hat{\beta}_1 = 4.92$, implies that the estimated mean damage increases by \$4,920 for each additional mile from the fire station. This interpretation is valid over the range of *x*, or from .7 to 6.1 miles from the station. The estimated *y*-intercept, $\hat{\beta}_0 = 10.28$, has the interpretation that a fire 0 miles from the fire station has an estimated mean damage of \$10,280. Although this would seem to apply to the fire station itself, remember that the *y*-intercept is meaningfully interpretable only if *x* = 0 is within the sampled range of the independent variable. Since *x* = 0 is outside the range, $\hat{\beta}_0$ has no practical interpretation.

FIGURE 9.25

STATISTIX Printout for Fire Damage Regression

```
UNWEIGHTED LEAST SQUARES LINEAR REGRESSION OF DAMAGE

PREDICTOR
VARIABLES      COEFFICIENT    STD ERROR      STUDENT'S T       P
---------      -----------    ---------      -----------     ------
CONSTANT         10.2779       1.42028          7.24         0.0000
DISTANCE         4.91933       0.39275         12.53         0.0000


R-SQUARED              0.9235    RESID. MEAN SQUARE (MSE)    5.36546
ADJUSTED R-SQUARED     0.9176    STANDARD DEVIATION          2.31635

SOURCE          DF       SS          MS          F          P
---------       ---    --------    --------     -----     ------
REGRESSION       1     841.766     841.766     156.89     0.0000
RESIDUAL        13     69.7510     5.36546
TOTAL           14     911.517

PREDICTED/FITTED VALUES OF DAMAGE

LOWER PREDICTED BOUND    22.324      LOWER FITTED BOUND    26.190
PREDICTED VALUE          27.496      FITTED VALUE          27.496
UPPER PREDICTED BOUND    32.667      UPPER FITTED BOUND    28.801
SE (PREDICTED VALUE)     2.3939      SE (FITTED VALUE)     0.6043


UNUSUALNESS (LEVERAGE)   0.0681
PERCENT COVERAGE         95.0
CORRESPONDING T          2.16

PREDICTOR VALUES: DISTANCE = 3.5000
```

FIGURE 9.26

STATISTIX Plot of Least Squares Line for Fire Damage Regression

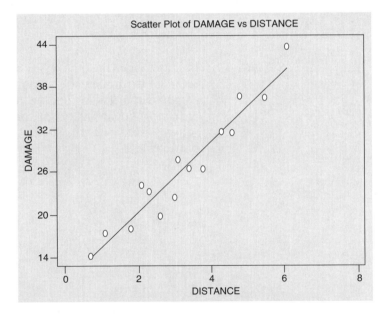

Step 3   Now we specify the probability distribution of the random error component ε. The assumptions about the distribution are identical to those listed in Section 9.3. Although we know that these assumptions are not completely satisfied (they rarely are for practical problems), we are willing to assume they are approximately satisfied for this example. The estimate of the standard deviation σ of ε, highlighted on the printout, is

$$s = 2.31635$$

This implies that most of the observed fire damage (y) values will fall within approximately $2s = 4.64$ thousand dollars of their respective predicted values when using the least squares line.

Step 4 We can now check the usefulness of the hypothesized model—that is, whether x really contributes information for the prediction of y using the straight-line model. First, test the null hypothesis that the slope β_1 is 0, that is, that there is no linear relationship between fire damage and the distance from the nearest fire station, against the alternative hypothesis that fire damage increases as the distance increases. We test

$$H_0: \beta_1 = 0$$
$$H_a: \beta_1 > 0$$

The two-tailed observed significance level for testing $H_a: \beta_1 \neq 0$, highlighted on the printout, is .0000. Thus, the p-value for our one-tailed test is half of this value. This small p-value leaves little doubt that mean fire damage and distance between the fire and station are at least linearly related, with mean fire damage increasing as the distance increases.

We gain additional information about the relationship by forming a confidence interval for the slope β_1. A 95% confidence interval is

$$\hat{\beta}_1 \pm t_{.025}s_{\hat{\beta}_1}$$

where $\hat{\beta}_1 = 4.919$ and its standard error, $s_{\hat{\beta}_1} = .393$, are both obtained from the printout. The value $t_{.025}$, based on $n - 2 = 13$ df, is 2.160. Therefore, the 95% confidence interval is

$$\hat{\beta}_1 \pm t_{.025}s_{\hat{\beta}_1} = 4.919 \pm (2.160)(.393) = 4.919 \pm .849 = (4.070, 5.768)$$

We estimate that the interval from \$4,070 to \$5,768 encloses the mean increase (β_1) in fire damage per additional mile distance from the fire station.

Another measure of the utility of the model is the coefficient of determination, r^2. The value (highlighted on the printout) is $r^2 = .9235$, which implies that about 92% of the sample variation in fire damage (y) is explained by the distance (x) between the fire and the fire station.

The coefficient of correlation, r, that measures the strength of the linear relationship between y and x is not shown on the STATISTIX printout and must be calculated. Using the facts that $r = \sqrt{r^2}$ in simple linear regression and that r and $\hat{\beta}_1$ have the same sign, we find

$$r = +\sqrt{r^2} = \sqrt{.9235} = .96$$

The high correlation confirms our conclusion that β_1 is greater than 0; it appears that fire damage and distance from the fire station are positively correlated. All signs point to a strong linear relationship between y and x.

Step 5 We are now prepared to use the least squares model. Suppose the insurance company wants to predict the fire damage if a major residential fire were to occur 3.5 miles from the nearest fire station. The predicted value (highlighted at the bottom of the printout) is $\hat{y} = 27.496$, while the 95% prediction interval (also highlighted) is $(22.324, 32.667)$. Therefore, with 95% confidence we predict fire damage in a major residential fire 3.5 miles from the nearest station to be between \$22,324 and \$32,667.

One caution before closing: We would not use this prediction model to make predictions for homes less than .7 mile or more than 6.1 miles from the nearest fire station. A look at the data in Table 9.7 reveals that all the x values fall between .7 and 6.1. It is dangerous to use the model to make predictions outside the region in which the sample data fall. A straight line might not provide a good model for the relationship between the mean value of y and the value of x when stretched over a wider range of x values.

9.10 A NONPARAMETRIC TEST FOR CORRELATION (OPTIONAL)

When the simple linear regression assumptions (Section 9.3) are violated (e.g., the random error ϵ has a highly skewed distribution), an alternative method of analysis may be required. One technique is to apply a nonparametric test for correlation based on ranks.

To illustrate, suppose 10 new paintings are shown to two art critics and each critic ranks the paintings from 1 (best) to 10 (worst). We want to determine whether the critics' ranks are related. Does a correspondence exist between their ratings? If a painting is ranked high by critic 1, is it likely to be ranked high by critic 2? Or do high rankings by one critic correspond to low rankings by the other? That is, are the rankings of the critics *correlated*?

If the rankings are as shown in the "Perfect Agreement" columns of Table 9.8, we immediately notice that the critics agree on the rank of every painting. High ranks correspond to high ranks and low ranks to low ranks. This is an example of *perfect positive correlation* between the ranks. In contrast, if the rankings appear as shown in the "Perfect Disagreement" columns of Table 9.8, high ranks for one critic correspond to low ranks for the other. This is an example of *perfect negative correlation*.

In practice, you will rarely see perfect positive or negative correlation between the ranks. In fact, it is quite possible for the critics' ranks to appear as shown in Table 9.9.

TABLE 9.8   Rankings of 10 Paintings by Two Critics

	Perfect Agreement		Perfect Disagreement	
Painting	**Critic 1**	**Critic 2**	**Critic 1**	**Critic 2**
1	4	4	9	2
2	1	1	3	8
3	7	7	5	6
4	5	5	1	10
5	2	2	2	9
6	6	6	10	1
7	8	8	6	5
8	3	3	4	7
9	10	10	8	3
10	9	9	7	4

TABLE 9.9   Rankings of Paintings: Less Than Perfect Agreement

	Critic		Difference Between Rank 1 and Rank 2	
Painting	**1**	**2**	**d**	**d^2**
1	4	5	−1	1
2	1	2	−1	1
3	9	10	−1	1
4	5	6	−1	1
5	2	1	1	1
6	10	9	1	1
7	7	7	0	0
8	3	3	0	0
9	6	4	2	4
10	8	8	0	0
				$\Sigma d^2 = 10$

You will note that these rankings indicate some agreement between the critics, but not perfect agreement, thus indicating a need for a measure of rank correlation.

Spearman's rank correlation coefficient, r_s, provides a measure of correlation between ranks. The formula for this measure of correlation is given in the next box. We also give a formula that is identical to r_s when there are no ties in rankings; this provides a good approximation to r_s when the number of ties is small relative to the number of pairs.

Note that if the ranks for the two critics are identical, as in the second and third columns of Table 9.8, the differences between the ranks, d, will all be 0. Thus,

$$r_s = 1 - \frac{6 \sum d^2}{n(n^2 - 1)} = 1 - \frac{6(0)}{10(99)} = 1$$

That is, *perfect positive correlation* between the pairs of ranks is characterized by a Spearman correlation coefficient of $r_s = 1$. When the ranks indicate perfect disagreement, as in the fourth and fifth columns of Table 9.8, $\Sigma d_i^2 = 330$ and

$$r_s = 1 - \frac{6(330)}{10(99)} = -1.$$

Thus, *perfect negative correlation* is indicated by $r_s = -1$.

TEACHING TIP
A perfect correlation will exist when all the differences are 0. This occurs when both u and v are ranked exactly the same for every observation.

Spearman's Rank Correlation Coefficient

$$r_s = \frac{SS_{uv}}{\sqrt{SS_{uu}SS_{vv}}}$$

where

$$SS_{uv} = \sum (u_i - \bar{u})(v_i - \bar{v}) = \sum u_i v_i - \frac{\left(\sum u_i\right)\left(\sum v_i\right)}{n}$$

$$SS_{uu} = \sum (u_i - \bar{u})^2 = \sum u_i^2 - \frac{\left(\sum u_i\right)^2}{n}$$

$$SS_{vv} = \sum (v_i - \bar{v})^2 = \sum v_i^2 - \frac{\left(\sum v_i\right)^2}{n}$$

u_i = Rank of the ith observation in sample 1
v_i = Rank of the ith observation in sample 2
n = Number of pairs of observations (number of observations in each sample)

Shortcut Formula for r_s*

$$r_s = 1 - \frac{6 \sum d_i^2}{n(n^2 - 1)}$$

where

$$d_i = u_i - v_i \quad \text{(difference in the ranks of the ith observations for sample 1 and sample 2)}$$

---

*The shortcut formula is not exact when there are tied measurements, but it is a good approximation when the total number of ties is not large relative to n.

TEACHING TIP
The interpretation of the correlation is exactly the same here as it is for the parametric correlation studied in Section 9.6.

For the data of Table 9.9,

$$r_s = 1 - \frac{6\sum d^2}{n(n^2 - 1)} = 1 - \frac{6(10)}{10(99)} = 1 - \frac{6}{99} = .94$$

The fact that r_s is *close* to 1 indicates that the critics tend to agree, but the agreement is not perfect.

The value of r_s always falls between −1 and +1, with +1 indicating perfect positive correlation and −1 indicating perfect negative correlation. The closer r_s falls to +1 or −1, the greater the correlation between the ranks. Conversely, the nearer r_s is to 0, the less the correlation.

Note that the concept of correlation implies that two responses are obtained for each experimental unit. In the art critics example, each painting received two ranks (one for each critic) and the objective of the study was to determine the degree of positive correlation between the two rankings. Rank correlation methods can be used to measure the correlation between any pair of variables. If two variables are measured on each of *n* experimental units, we rank the measurements associated with each variable separately. Ties receive the average of the ranks of the tied observations. Then we calculate the value of r_s for the two rankings. This value measures the rank correlation between the two variables. We illustrate the procedure in Example 9.5.

EXAMPLE 9.5

A study is conducted to investigate the relationship between cigarette smoking during pregnancy and the weights of newborn infants. A sample of 15 women smokers kept accurate records of the number of cigarettes smoked during their pregnancies, and the weights of their children were recorded at birth. The data are given in Table 9.10. Calculate and interpret Spearman's rank correlation coefficient for the data.

⊘ NEWBORN
TABLE 9.10   Data and Calculations for Example 9.5

Woman	Cigarettes per Day	Rank	Baby's Weight (pounds)	Rank	d	d^2
1	12	1	7.7	5	−4	16
2	15	2	8.1	9	−7	49
3	35	13	6.9	4	9	81
4	21	7	8.2	10	−3	9
5	20	5.5	8.6	13.5	−8	64
6	17	3	8.3	11.5	−8.5	72.25
7	19	4	9.4	15	−11	121
8	46	15	7.8	6	9	81
9	20	5.5	8.3	11.5	−6	36
10	25	8.5	5.2	1	7.5	56.25
11	39	14	6.4	3	11	121
12	25	8.5	7.9	7	1.5	2.25
13	30	12	8.0	8	4	16
14	27	10	6.1	2	8	64
15	29	11	8.6	13.5	−2.5	6.25
					Total =	795

Solution

We first rank the number of cigarettes smoked per day, assigning a 1 to the smallest number (12) and a 15 to the largest (46). Note that the two ties receive the averages of their respective ranks. Similarly, we assign ranks to the 15 babies' weights. Since the number of ties is relatively small, we will use the shortcut formula to calculate r_s. The differences d between the ranks of the babies' weights and the ranks of the number of cigarettes smoked per day are shown in Table 9.10. The squares of the differences, d^2, are also given. Thus,

$$r_s = 1 - \frac{6 \sum d_i^2}{n(n^2 - 1)} = 1 - \frac{6(795)}{15(15^2 - 1)} = 1 - 1.42 = -.42$$

The value of r_s can also be obtained using a computer. A SAS printout of the analysis is shown in Figure 9.27. The value of r_s, highlighted on the printout, agrees (except for rounding) with our hand-calculated value of $-.42$.

FIGURE 9.27

SAS Printout for Example 9.5

```
              The CORR Procedure

     2  Variables:   CIGARETS WEIGHT

     Spearman Correlation Coefficients,  N = 15
            Prob > |r| under HO:  Rho=0

                        CIGARETS      WEIGHT

     CIGARETS           1.00000      -0.42473
                                      0.1145


     WEIGHT            -0.42473       1.00000
                        0.1145
```

This negative correlation coefficient indicates that in this sample an increase in the number of cigarettes smoked per day is *associated with* (but is not necessarily the *cause of*) a decrease in the weight of the newborn infant. Can this conclusion be generalized from the sample to the population? That is, can we conclude that weights of newborns and the number of cigarettes smoked per day are negatively correlated for the populations of observations for *all* smoking mothers?

If we define ρ as the **population rank correlation coefficient** [i.e., the rank correlation coefficient that could be calculated from all (x, y) values in the population], this question can be answered by conducting the test

H_0: $\rho = 0$ (no population correlation between ranks)

H_a: $\rho < 0$ (negative population correlation between ranks)

Test statistic: r_s (the *sample* Spearman rank correlation coefficient)

To determine a rejection region, we consult Table XII in Appendix A, which is partially reproduced in Table 9.11. Note that the left-hand column gives values of n, the number of pairs of observations. The entries in the table are values for an upper-tail rejection region, since only positive values are given. Thus, for $n = 15$ and $\alpha = .05$, the value .441 is the boundary of the upper-tailed rejection region, so that $P(r_s > .441) = .05$ if H_0: $\rho = 0$ is true. Similarly, for negative values of r_s, we have $P(r_s < -.441) = .05$ if $\rho = 0$. That is, we expect to see $r_s < -.441$ only 5% of the time if there is really no relationship between the ranks of the variables.

TABLE 9.11   Reproduction of Part of Table XII in Appendix A: Critical Values of Spearman's Rank Correlation Coefficient

n	$\alpha = .05$	$\alpha = .025$	$\alpha = .01$	$\alpha = .005$
5	.900	—	—	—
6	.829	.886	.943	—
7	.714	.786	.893	—
8	.643	.738	.833	.881
9	.600	.683	.783	.833
10	.564	.648	.745	.794
11	.523	.623	.736	.818
12	.497	.591	.703	.780
13	.475	.566	.673	.745
14	.457	.545	.646	.716
15	.441	.525	.623	.689
16	.425	.507	.601	.666
17	.412	.490	.582	.645
18	.399	.476	.564	.625
19	.388	.462	.549	.608
20	.377	.450	.534	.591

The lower-tailed rejection region is therefore

$$\textit{Rejection region }(\alpha = .05): \quad r_s < -.441$$

Since the calculated $r_s = -.42$ is not less than $-.441$, we cannot reject H_0 at the $\alpha = .05$ level of significance. That is, this sample of 15 smoking mothers provides insufficient evidence to conclude that a negative correlation exists between number of cigarettes smoked and the weight of newborns for the populations of measurements corresponding to all smoking mothers. (Note that the p-value of .1145 shown on the SAS printout, Figure 9.27, confirms this result.) This does not, of course, mean that no relationship exists. A study using a larger sample of smokers and taking other factors into account (father's weight, sex of newborn child, etc.) would be more likely to reveal whether smoking and the weight of a newborn child are related. ■

A summary of Spearman's nonparametric test for correlation is given in the next box.

Spearman's Nonparametric Test for Rank Correlation

ONE-TAILED TEST

$H_0: \rho = 0$

$H_a: \rho > 0$
[or $H_a: \rho < 0$]

TWO-TAILED TEST

$H_0: \rho = 0$

$H_a: \rho \neq 0$

Test statistic: r_s, the sample rank correlation (see the formulas for calculating r_s)

Rejection region: $r_s > r_{s,\alpha}$
[or $r_s < -r_{s,\alpha}$ when $H_a: \rho_s < 0$]

where $r_{s,\alpha}$ is the value from Table XII corresponding to the upper-tail area α and n pairs of observations

Rejection region: $|r_s| > r_{s,\alpha/2}$

where $r_{s,\alpha/2}$ is the value from Table XII corresponding to the upper-tail area $\alpha/2$ and n pairs of observations

> ## Spearman's Nonparametric Test for Rank Correlation (*continued*)
>
> *Assumptions:* 1. The sample of experimental units on which the two variables are measured is randomly selected.
> 2. The probability distributions of the two variables are continuous.
>
> *Ties:* Assign tied measurements the average of the ranks they would receive if they were unequal but occurred in successive order. For example, if the third-ranked and fourth-ranked measurements are tied, assign each a rank of $(3 + 4)/2 = 3.5$. The number of ties should be small relative to the total number of observations.

✓ EXERCISES 9.74–9.84

Learning the Mechanics

9.74 Use Table XII of Appendix A to find the rejection region for Spearman's nonparametric test for rank correlation in each of the following situations:

a. $H_0: \rho = 0, H_a: \rho \neq 0, n = 10, \alpha = .05$
b. $H_0: \rho = 0, H_a: \rho > 0, n = 20, \alpha = .025$ $r_s > .450$
c. $H_0: \rho = 0, H_a: \rho < 0, n = 30, \alpha = .01$ $r_s < -.432$

9.75 Compute Spearman's rank correlation coefficient for each of the following pairs of sample observations:

a.
x	33	61	20	19	40
y	26	36	65	25	35

b.
x	89	102	120	137	41
y	81	94	75	52	136

c.
x	2	15	4	10
y	11	2	15	21

d.
x	5	20	15	10	3
y	80	83	91	82	87

9.76 The following sample data were collected on variables *x* and *y:*

 LM9_76

x	0	3	0	−4	3	0	4
y	0	2	2	0	3	1	2

a. Specify the null and alternative hypotheses that should be used in conducting a hypothesis test to determine whether the variables *x* and *y* are correlated.

b. Conduct the test of part **a** using $\alpha = .05$. $r_s = .745$
c. What is the approximate *p*-value of the test of part **b**?
d. What assumptions are necessary to ensure the validity of the test of part **b**?

Applying the Concepts—Basic

9.77 Refer to the *British Journal of Sports Medicine* (Apr. 2000) study of the effect of massaging boxers between rounds, Exercise 9.42 (p. 468). Two variables measured on the boxers were blood lactate level (*y*) and the boxer's perceived recovery (*x*). The data for 16 five-round boxing performances are reproduced here.

 BOXING2

Blood Lactate Level	Perceived Recovery
3.8	7
4.2	7
4.8	11
4.1	12
5.0	12
5.3	12
4.2	13
2.4	17
3.7	17
5.3	17
5.8	18
6.0	18
5.9	21
6.3	21
5.5	20
6.5	24

Source: Hemmings, B., Smith, M., Graydon, J., and Dyson, R. "Effects of massage on physiological restoration, perceived recovery, and repeated sports performance." *British Journal of Sports Medicine*, Vol. 34, No. 2, Apr. 2000 (data adapted from Figure 3).

a. Rank the values of the 16 blood lactate levels.

b. Rank the values of the 16 perceived recovery values.

c. Use the ranks, parts **a** and **b**, to compute Spearman's rank correlation coefficient. Give a practical interpretation of the result.

d. Find the rejection region for a test to determine whether y and x are rank correlated. Use $\alpha = .10$.

e. What is the conclusion of the test, part **d**? State your answer in the words of the problem.

9.78 The Milan Overall Dementia Assessment (MODA) is a neuropsychologically oriented test that provides a measure of an individual's cognitive deterioration. MODA scores range from 0 to 100, with higher scores indicating a greater degree of deterioration. A team of psychologists and neurologists administered the MODA to a sample of 30 patients with Alzheimer's disease (*Neuropsychologia*, June 1995). In addition, all patients were given a "face matching" test in which they were requested to match photographs of unknown faces. Scores on this test were recorded as the number of matching errors, with a maximum score of 27.

a. The researchers reported Spearman's rank correlation between the MODA and face-matching scores as $r_s = .48$. Interpret this result.

b. Test the hypothesis of a positive correlation between MODA score and face-matching score in Alzheimer's patients. Use $\alpha = .05$.   Reject H_0

9.79 Two expert wine tasters were asked to rank six brands of wine. Their rankings are shown below.

◉ WINETASTE

Brand	Expert 1	Expert 2
A	6	5
B	5	6
C	1	2
D	3	1
E	2	4
F	4	3

a. Calculate Spearman's rank correlation between the rankings of the two experts. $r_s = .657$

b. Do the data present sufficient evidence to indicate a positive correlation in the rankings of the two experts? Test Using $\alpha = .10$.

9.80 The metabolic and cardiopulmonary responses during maximal effort exercise in persons with multiple sclerosis (MS) was studied (*Clinical Kinesiology*, Spring 1995). The following variables were measured for each of 10 MS patients:

1. Expanded Disability Status Scale (EDSS)—Ratings range from 1 to 4.5

2. Peak oxygen uptake (liters per minute) during an arm cranking exercise (ARM)

3. Peak oxygen uptake (liters per minute) during a leg cycling exercise (LEG)

4. Peak oxygen uptake (liters per minute) during a combined leg cycling and arm cranking exercise (LEG/ARM)

Spearman's rank correlation was calculated for each pair of variables. The accompanying table gives the correlation r_s between the variables in the corresponding row and column.

	EDSS	ARM	LEG/ARM	LEG
EDSS	1.000	−.439	−.655	−.512
ARM		1.000	.634	.722
LEG/ARM			1.000	.890
LEG				1.000

Source: Ponichtera-Mulcare, J. A., *et al.* "Maximal aerobic exercise of individuals with multiple sclerosis using three modes of energy." *Clinical Kinesiology*, Vol. 49, No. 1, Spring 1995, p. 10 (Table 2).

a. Explain why the table shows rank correlations of 1.000 in the diagonal.

b. Practically interpret each of the other rank correlations in the table.

c. Is there sufficient evidence (at $\alpha = .01$) to conclude that EDSS and peak oxygen uptake are negatively correlated? Perform the test for each of the three exercises. $r_s = -.439$

9.81 Refer to the *Brain and Behavior Evolution* (Apr. 2000) study of the feeding behavior of blackbream fish, Exercises 2.110 (p. 89) and 9.17 (p. 455). Recall that the zoologists recorded the number of aggressive strikes of two blackbream fish feeding at the bottom of an aquarium in the 10-minute period following the addition of food. The table listing the weekly number of strikes and age of the fish (in days) is reproduced here.

◉ BLACKBREAM

Week	Number of Strikes	Age of Fish (days)
1	85	120
2	63	136
3	34	150
4	39	155
5	58	162
6	35	169
7	57	178
8	12	184
9	15	190

Source: Shand, J., *et al.* "Variability in the location of the retinal ganglion cell area centralis is correlated with ontogenetic changes in feeding behavior in the Blackbream, Acanthopagrus 'butcher'." *Brain and Behavior*, Vol. 55, No. 4, Apr. 2000 (Figure H).

a. Find Spearman's correlation coefficient relating number of strikes (y) to age of fish (x).

b. Conduct a nonparametric test to determine whether number of strikes (y) and age (x) are negatively correlated. Test using $\alpha = .01$.

Applying the Concepts—Intermediate

9.82 Refer to the *Journal of Experimental Psychology-Applied* (June 2000) study in which the "name game" was used to help groups of students learn the names of other students in the group, Exercise 9.23 (p. 458). Recall that one goal of the study was to investigate the relationship between proportion of names recalled (y) by a student and position (order) of the student during the game (x). The data for 144 students in the first eight positions are saved in the NAMEGAME2 file. (The first five and last five observations in the data set are listed here.)

⊚ NAMEGAME2

Position	Recall
2	0.04
2	0.37
2	1.00
2	0.99
2	0.79
⋮	⋮
9	0.72
9	0.88
9	0.46
9	0.54
9	0.99

Source: Morris, P.E., and Fritz, C.O. "The name game: Using retrieval practice to improve the learning of names." *Journal of Experimental Psychology-Applied*, Vol. 6, No. 2, June 2000 (data simulated from Figure 2).

a. To properly apply the parametric test for correlation based on the Pearson coefficient of correlation, r (Section 9.6), both the x and y variables must be normally distributed. Demonstrate that this assumption is violated for this data. What are the consequences of this violation?

b. Find Spearman's rank correlation coefficient on the SAS printout provided below and interpret its value.

c. Find the observed significance level for testing for zero rank correlation on the SAS printout and interpret its value. $p = .0130$

d. At $\alpha = .05$, is there sufficient evidence of rank correlation between proportion of names recalled (y) by a student and position (order) of the student during the game (x)?

SAS Output for Exercise 9.82

```
         The CORR Procedure

   2  Variables:   POSITION RECALL

Spearman Correlation Coefficients, N = 144
     Prob > |r| under HO: Rho=0.

                  POSITION    RECALL

   POSITION       1.00000     0.20652
                               0.0130

   RECALL         0.20652     1.00000
                  0.0130
```

9.83 Psychologists theorize that certain types of recreation have a tranquilizing effect on highly excitable mentally retarded patients. In one experiment, nine mentally retarded patients were monitored for 1 week, and the total number of hours each spent in a specially designed recreation room was recorded. Also recorded was the number of times during the week that some type of tranquilizing medication had to be given to a patient in a highly excited state. Using the data in the table, determine whether there is sufficient evidence to indicate that recreation is negatively rank correlated with the number of times a tranquilizer has to be given. Test using $\alpha = .05$. $r_s = -.812$

⊚ RECTIME

Patient	Recreation Time (hours)	Tranquilizers
1	16	3
2	22	1
3	10	4
4	8	9
5	14	5
6	34	2
7	26	0
8	13	10
9	5	9

9.84 *The Wall Street Journal* (Feb. 7, 2001) reported on a Harris Interactive, Inc. survey of consumers to rate the reputations of America's most visible companies. The 1999 and 2000 ranks of 15 randomly selected companies are listed in the accompanying table. Conduct a test to determine if the 1999 and 2000 reputation ranks are positively correlated. Use $\alpha = .05$.

⊚ COREP

Company	1999 Rank	2000 Rank
Johnson & Johnson	1	1
Anheuser-Busch	14	6
Disney	10	8
Microsoft	15	9
FedEx	21	13
Wal-Mart	6	14
Coca-Cola	2	16
McDonalds	24	24
Yahoo!	19	27
Sears	29	35
America Online	26	39
Kmart	37	41
Toyota	28	19
Home Depot	8	4
IBM	17	7

Source: Harris Interactive, the Reputation Institute.

STATISTICS IN ACTION
Can "Dowsers" Really Detect Water?

The act of searching for and finding underground supplies of water using nothing more than a divining rod is commonly known as "dowsing." Although widely regarded among scientists as no more than a superstitious relic from medieval times, dowsing remains popular in folklore and, to this day, there are individuals who claim to have this mysterious skill.

Many dowsers in Germany claim that they respond to "earthrays" that emanate from the water source. These earthrays, say the dowsers, are a subtle form of radiation potentially hazardous to human health. As a result of these claims, the German government in the mid-1980s conducted a 2-year experiment to investigate the possibility that dowsing is a genuine skill. If such a skill could be demonstrated, reasoned government officials, then dangerous levels of radiation in Germany could be detected, avoided, and disposed of.

A group of university physicists in Munich, Germany, were provided a grant of 400,000 marks ($\approx$$250,000) to conduct the study. Approximately 500 candidate dowsers were recruited to participate in preliminary tests of their skill. To avoid fraudulent claims, the 43 individuals who seemed to be the most successful in the preliminary tests were selected for the final, carefully controlled, experiment.

The researchers set up a 10-meter long line on the ground floor of a vacant barn, along which a small wagon could be moved. Attached to the wagon was a short length of pipe, perpendicular to the test line, that was connected by hoses to a pump with running water. The location of the pipe along the line for each trial of the experiment was assigned using a computer-generated random number. On the upper floor of the barn, directly above the experimental line, a 10-meter test line was painted. In each trial, a dowser was admitted to this upper level and required, with his or her rod, stick, or other tool of choice, to ascertain where the pipe with running water on the ground floor was located.

Each dowser participated in at least one test series, that is, a sequence of from 5 to 15 trials (typically 10), with the pipe randomly repositioned after each trial. (Some dowsers undertook only one test series, selected others underwent more than 10 test series.) Over the 2-year experimental period, the 43 dowsers participated in a total of 843 tests. The experiment was "double-blind" in that neither the observer (researcher) on the top floor nor the dowser knew the pipe's location, even after a guess was made. [*Note:* Before the experiment began, a professional magician inspected the entire arrangement for potential deception or cheating by the dowsers.]

For each trial, two variables were recorded: the actual pipe location (in decimeters from the beginning of the line) and the dowser's guess (also measured in decimeters). Based on an examination of these data, the German physicists concluded in their final report that although most dowsers did not do particularly well in the experiments, "some few dowsers, in particular tests, showed an extraordinarily high rate of success, which can scarcely if at all be explained as due to chance . . . a real core of dowser-phenomena can be regarded as empirically proven . . ." (Wagner, Betz, and König, 1990).

This conclusion was critically assessed by J. T. Enright, a professor of behavioral physiology at the University of California–San Diego (*Skeptical Inquirer*, Jan./Feb. 1999). Using scatterplots and the notion of correlation, Enright concluded exactly the opposite of the German physicists. According to Enright, "the Munich experiments constitute as decisive and complete a failure as can be imagined of dowsers to do what they claim they can."

Focus

a. Using scatterplots, Enright provided several hypothetical examples of outcomes that might be expected from the dowsing experiments, assuming various arbitrary categories of dowser skill. Let x = dowser's guess and y = pipe location. Construct a scatterplot of hypothetical data that would reflect perfect prediction by the dowsers.

b. Repeat part **a** for dowsers that have good (but not perfect) skill.

c. Repeat part **a** for dowsers that have no skill (i.e., that are making random guesses).

d. Enright presented a scatterplot of the data for all 843 tests performed in the Munich barn. A reproduction of this plot is displayed in Figure 9.28. Based on this graph, what inference would you make about the overall ability of the dowsers? Explain. [*Note:* In this plot, dowser's guess is shown on the vertical axis to make the graph comparable to the scatterplots you drew in parts **a–c.**]

e. Recall that the certain German physicists found that "some few dowsers . . . showed an extraordinarily high rate of success." Consequently, they might argue that the scatterplot in Figure 9.28 obscures these outstanding performances since it lumps all 43 dowsers (the majority of which are unskilled) together. In the researchers' final report, they identified three dowsers (those numbered 99, 18, and 108) as having particularly impressive results. All three of these "best" dowsers performed the experiment multiple times. The best test series (sequence of trials) for

STATISTICS IN ACTION *(continued)*

each of these three dowsers was identified; these data are listed in Table 9.12 and stored in the DOWSING file. Conduct a complete simple linear regression analysis of the data in order to make an inference about the overall performance of each of the three best dowsers.

f. The data in Table 9.12 represent the outcome of the dowsing experiment in its most favorable light. Can these results be reproduced by the best dowsers in comparable tests? Remember, each of these three dowsers did participate in other test series. Enright plotted the data for these other series in which the three best dowsers performed; this scatterplot is reproduced in Figure 9.29. Comment on the performance of the three best dowsers during these "rest of the best" trials.

g. According to Enright, "there is another way of evaluating the results from those dowsers who produced the best test series.... Suppose that they had always left their dowsing equipment at home in the closet, and had simply, in each and every test, just guessed that the pipe was located exactly at the middle of the test line." Replace the dowsers' guesses in Table 9.12 with the midpoint of the test line, 50 decimeters, and repeat the analysis of part **e**. What do you conclude from the analysis?

h. Give a critical assessment of the Munich dowsing experiments. With whom do you tend to agree, the German physicists or J. T. Enright?

⊘ DOWSING

TABLE 9.12   Dowsing Trial Results: Best Series for the Three Best Dowsers

Trial	Dowser Number	Pipe Location	Dowser's Guess
1	99	4	4
2	99	5	87
3	99	30	95
4	99	35	74
5	99	36	78
6	99	58	65
7	99	40	39
8	99	70	75
9	99	74	32
10	99	98	100
11	18	7	10
12	18	38	40
13	18	40	30
14	18	49	47
15	18	75	9
16	18	82	95
17	108	5	52
18	108	18	16
19	108	33	37
20	108	45	40
21	108	38	66
22	108	50	58
23	108	52	74
24	108	63	65
25	108	72	60
26	108	95	49

Source: Enright, J. T. "Testing dowsing: The failure of the Munich experiments." *Skeptical Inquirer*, Jan./Feb. 1999, p. 45 (Figure 6a).

(continues on next page)

STATISTICS IN ACTION *(continued)*

FIGURE 9.28

Results from All 843
Dowsing Trials

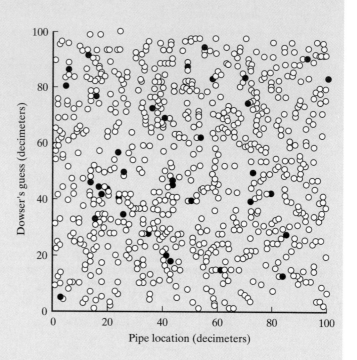

FIGURE 9.29

Results for Other Test Series that
Three Best Dowsers Participated In

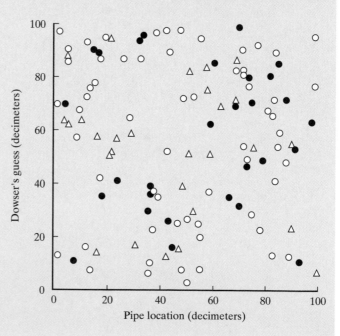

Suggestions for class discussion of the above case can be found in the *Instructor's Notes*.

QUICK REVIEW

Key Terms

Note: Starred (*) items are from the optional section in this chapter.

Bivariate relationship   471
Coefficient of correlation   471
Coefficient of determination   475
Confidence interval for mean
   of *y*   482
Dependent variable   442
Deterministic model   441
Deterministic relationship   441
Errors of prediction   445
Estimated standard error of the
   regression model   461
Independent variable   442

Least squares estimates   446
Least squares line (or prediction
   equation)   445, 446
Line of means   443
Method of least squares   445
Pearson product moment coefficient
   of correlation   471
Population correlation coefficient   473
*Population rank correlation
   coefficient   497
Prediction interval for *y*   482
Predictor variable   442

Probabilistic model   441
Probabilistic relationship   441
Random error   441
Regression analysis   442
Regression line   445
Response variable   442
Scattergram   444
Slope   442
*Spearman's rank correlation
   coefficient   495
Straight-line (first-order) model   442
y-intercept   442

Key Formulas

$$\hat{\beta}_1 = \frac{SS_{xy}}{SS_{xx}}, \qquad \hat{\beta}_0 = \bar{y} - \hat{\beta}_1 \bar{x}$$

Least squares estimates of β's   446

$$\text{where} \quad SS_{xy} = \sum xy - \frac{(\sum x)(\sum y)}{n}$$

$$SS_{xx} = \sum x^2 - \frac{(\sum x)^2}{n}$$

$$\hat{y} = \hat{\beta}_0 + \hat{\beta}_1 x$$

Least squares line   446

$$SSE = \sum (y_i - \hat{y}_i)^2 = SS_{yy} - \hat{\beta}_1 SS_{xy}$$

Sum of squared errors   461

$$\text{where} \quad SS_{yy} = \sum y^2 - \frac{(\sum y)^2}{n}$$

$$s^2 = \frac{SSE}{n-2}$$

Estimated variance of σ^2 of ε 461

$$s_{\hat{\beta}_1} = \frac{s}{\sqrt{SS_{xx}}}$$

Estimated standard error of $\hat{\beta}_1$ 465

$$t = \frac{\hat{\beta}_1}{s_{\hat{\beta}_1}}$$

Test statistic for $H_0: \beta_1 = 0$ 465

$$\hat{\beta}_1 \pm (t_{\alpha/2}) s_{\hat{\beta}_1}$$

$(1 - \alpha)100\%$ confidence interval for β_1 466

$$r = \frac{SS_{xy}}{\sqrt{SS_{xx}SS_{yy}}} = \pm\sqrt{r^2} \text{ (same sign as } \hat{\beta}_1)$$

Coefficient of correlation   471

$$r^2 = \frac{SS_{yy} - SSE}{SS_{yy}}$$

Coefficient of determination   475

$$\hat{y} \pm (t_{\alpha/2})s\sqrt{\frac{1}{n} + \frac{(x_p - \bar{x})^2}{SS_{xx}}}$$

$(1 - \alpha)100\%$ confidence interval for $E(y)$ when $x = x_p$ 482

$$\hat{y} \pm (t_{\alpha/2})s\sqrt{1 + \frac{1}{n} + \frac{(x_p - \bar{x})^2}{SS_{xx}}}$$

$(1 - \alpha)100\%$ prediction interval for y when $x = x_p$ 482

$$*r_s = 1 - \frac{6\Sigma d^2}{n(n^2 - 1)}$$

Spearman's rank correlation coefficient 495

where d_i = difference between ranks of ith observations for sample 1 and sample 2

Language Lab

Symbol	Pronunciation	Description
y		Dependent variable (variable to be predicted or modeled)
x		Independent (predictor) variable
$E(y)$		Expected (mean) value of y
β_0	beta-zero	y-intercept of true line
β_1	beta-one	Slope of true line
$\hat{\beta}_0$	beta-zero hat	Least squares estimate of y-intercept
$\hat{\beta}_1$	beta-one hat	Least squares estimate of slope
ε	epsilon	Random error
$\hat{y}$	y-hat	Predicted value of y
$(y - \hat{y})$		Error of prediction
SE		Sum of errors (will equal zero with least squares line)
SSE		Sum of squared errors (will be smallest for least squares line)
SS_{xx}		Sum of squares of x-values
SS_{yy}		Sum of squares of y-values
SS_{xy}		Sum of squares of cross-products, $x \cdot y$
r		Coefficient of correlation
r^2	R-squared	Coefficient of determination
x_p		Value of x used to predict y
$*r_s$	r-s	Spearman's rank correlation coefficient

 SUPPLEMENTARY EXERCISES 9.85–9.100

Learning the Mechanics

9.85 Consider the following sample data:

y	5	1	3
x	5	1	3

a. Construct a scattergram for the data.
b. It is possible to find many lines for which $\Sigma(y - \hat{y}) = 0$. For this reason, the criterion $\Sigma(y - \hat{y}) = 0$ is not used for identifying the "best-fitting" straight line. Find two lines that have $\Sigma(y - \hat{y}) = 0$.
c. Find the least squares line. $\hat{y} = x$
d. Compare the value of SSE for the least squares line to that of the two lines you found in part **b.** What principle of least squares is demonstrated by this comparison?

9.86 In fitting a least squares line to $n = 15$ data points, the following quantities were computed: $SS_{xx} = 55$, $SS_{yy} = 198$, $SS_{xy} = -88$, $\bar{x} = 1.3$, and $\bar{y} = 35$.

a. Find the least squares line. $\hat{y} = 37.08 - 1.6x$
b. Graph the least squares line.
c. Calculate SSE. 57.2
d. Calculate s^2. 4.4
e. Find a 90% confidence interval for β_1. Interpret this estimate. $-1.6 \pm .501$
f. Find a 90% confidence interval for the mean value of y when $x = 15$. 13.08 ± 6.929
g. Find a 90% prediction interval for y when $x = 15$.

9.87 Consider the following 10 data points:

🔵 LM9_87

x	3	5	6	4	3	7	6	5	4	7
y	4	3	2	1	2	3	3	5	4	2

a. Plot the data on a scattergram.
b. Calculate the values of r and r^2. $-.1245; .0155$
c. Is there sufficient evidence to indicate that x and y are linearly correlated? Test at the $\alpha = .10$ level of significance. $t = -.35$

*9.88 Refer to Exercise 9.87.

 a. Calculate Spearman's rank correlation coefficient, r_s, for the data in the table.
 b. Conduct a nonparametric test for correlation. Use $\alpha = .10$.

Applying the Concepts—Basic

9.89 A study was conducted to determine whether a student's final grade in an introductory sociology course is linearly related to his or her performance on the verbal ability test administered before college entrance. The verbal test scores and final grades for a random sample of 10 students are shown in the table at right. A STATISTIX printout of the simple linear regression is also provided.

 a. Find the least squares line relating y to x.
 b. Plot the data points and graph the least squares line.
 c. Do the data provide sufficient evidence to indicate that a positive correlation exists between verbal score and final grade? Use $\alpha = .01$. $t = 4.38$
 d. Find a 95% confidence interval for the slope β_1.
 e. Predict a student's final grade in the introductory course when his or her verbal test score is 50. Use a 95% prediction interval. (57.95, 100.17)
 f. Find a 95% confidence interval for the mean final grade for all students scoring 50 on the college entrance verbal exam. (72.51, 85.61)

◉ SOCIOL

Student	Verbal Ability Test Score x	Final Sociology Grade y
1	39	65
2	43	78
3	21	52
4	64	82
5	57	92
6	47	89
7	28	73
8	75	98
9	34	56
10	52	75

9.90 A breeder of thoroughbred horses wishes to model the relationship between the gestation period and the life span of a horse. The breeder believes that the two variables may follow a linear trend. The information in the table on p. 508 was supplied to the breeder from various thoroughbred stables across the state. (Note that the horse has the greatest variation of gestation period of any species owing to seasonal and nutritional factors.) A MINITAB printout of the simple linear regression analysis is also shown on p. 508.

STATISTIX Output for Exercise 9.89

```
UNWEIGHTED LEAST SQUARES LINEAR REGRESSION OF FINGRADE


PREDICTOR
VARIABLES      COEFFICIENT    STD ERROR    STUDENT'S T       P
---------      -----------    ---------    -------------   ------

CONSTANT         40.7842       8.50686        4.79         0.0014
VERBAL           0.76556       0.17498        4.38         0.0024


R-SQUARED            0.7052     RESID. MEAN SQUARE (MSE)     75.7532
ADJUSTED R-SQUARED   0.6684     STANDARD DEVIATION           8.70363


SOURCE        DF       SS          MS          F       P
----------    ---   ----------  ----------   -----   ------

REGRESSION     1     1449.97     1449.97      19.14   0.0024
RESIDUAL       8     606.026     75.7532
TOTAL          9     2056.00


PREDICTED/FITTED VALUES OF FINGRADE


LOWER PREDICTED BOUND    57.950     LOWER FITTED BOUND    72.513
PREDICTED VALUE          79.062     FITTED VALUE          79.062
UPPER PREDICTED BOUND    100.17     UPPER FITTED BOUND    85.611
SE (PREDICTED VALUE)     9.1552     SE (FITTED VALUE)     2.8399


UNUSUALNESS (LEVERAGE)   0.1065
PERCENT COVERAGE         95.0
CORRESPONDING T          2.31


PREDICTOR VALUES: VERBAL = 50.000
```

MINITAB Output for Exercise 9.90

```
The regression equation is
LIFESPAN = 18.9 + 0.0109 GESTATE


Predictor          Coef      SE Coef        T        P
Constant         18.890        4.499     4.20    0.009
GESTATE         0.01087      0.01337     0.81    0.453


S = 1.971       R-Sq = 11.7%    R-Sq(adj) = 0.0%


Analysis of Variance


Source          DF           SS          MS        F        P
Regression       1        2.571       2.571     0.66    0.453
Residual Error   5       19.429       3.886
Total            6       22.000
```

HORSES

Horse	Gestation Period x (days)	Life Span y (years)
1	416	24
2	279	25.5
3	298	20
4	307	21.5
5	356	22
6	403	23.5
7	265	21

    a. Do the data provide sufficient evidence to support the breeder's hypothesis? Test using $\alpha = .05$.

    b. Find a 90% confidence interval for β_1. Interpret this interval. Does the interval support part **a**?

*9.91 A psychologist ranked a random sample of 10 children on two subjective scales according to the amount of paranoid behavior and the amount of aggressiveness they exhibit. The rankings according to these two criteria are given in the table. Compute Spearman's rank correlation coefficient. Is there evidence of a relationship between aggression and paranoia in children as judged by this psychologist? Test using $\alpha = .05$.

PARANOIA

Child	Paranoia	Aggression
1	7	5
2	3	1
3	6	4
4	1	2
5	2	8
6	4	7
7	10	9
8	8	3
9	5	6
10	9	10

9.92 Refer to the *American Scientist* (July–Aug. 1998) study of the relationship between self-avoiding and unrooted walks, Exercise 2.113 (p. 90). Recall that in a self-avoiding walk you never retrace or cross your own path; while an unrooted walk is a path in which the starting and ending points are impossible to distinguish. The possible number of walks of each type of various lengths are reproduced in the table. Consider the straight-line model $y = \beta_0 + \beta_1 x + \varepsilon$, where x is walk length (number of steps).

WALKS

Walk Length (number of steps)	Unrooted Walks	Self-Avoiding Walks
1	1	4
2	2	12
3	4	36
4	9	100
5	22	284
6	56	780
7	147	2,172
8	388	5,916

Source: Hayes, B. "How to avoid yourself." *American Scientist*, Vol. 86, No. 4, July–Aug. 1988, p. 317 (Figure 5).

    a. Use the method of least squares to fit the model to the data if y is the possible number of unrooted walks.

    b. Interpret $\hat{\beta}_0$ and $\hat{\beta}_1$ in the estimated model, part **a**.

    c. Repeat parts **a** and **b** if y is the possible number of self-avoiding walks. $\hat{y} = -1762 + 650x$

    d. Find a 99% confidence interval for the number of unrooted walks possible when walk length is four steps.

    e. Would you recommend using simple linear regression to predict the number of walks possible when walk length is 15 steps? Explain.   No

Applying the Concepts—Intermediate

9.93 Refer to Exercise 2.127 (p. 98). The data on the value of 50 Beanie Babies collector's items, published in *Beanie World Magazine*, are reproduced here. Can age of a Beanie Baby be used to accurately predict its market value? Answer this question by conducting a complete simple linear regression analysis on the data.

⊘ BEANIE

Name	Age (months) as of Sept. 1998	Retired (R)/ Current (C)	Value ($)
1. Ally the Alligator	52	R	55.00
2. Batty the Bat	12	C	12.00
3. Bongo the Brown Monkey	28	R	40.00
4. Blackie the Bear	52	C	10.00
5. Bucky the Beaver	40	R	45.00
6. Bumble the Bee	28	R	600.00
7. Crunch the Shark	21	C	10.00
8. Congo the Gorilla	28	C	10.00
9. Derby the Coarse Mane Horse	28	R	30.00
10. Digger the Red Crab	40	R	150.00
11. Echo the Dolphin	17	R	20.00
12. Fetch the Golden Retriever	5	C	15.00
13. Early the Robin	5	C	20.00
14. Flip the White Cat	28	R	40.00
15. Garcia the Teddy	28	R	200.00
16. Happy the Hippo	52	R	20.00
17. Grunt the Razorback	28	R	175.00
18. Gigi the Poodle	5	C	15.00
19. Goldie the Goldfish	52	R	45.00
20. Iggy the Iguana	10	C	10.00
21. Inch the Inchworm	28	R	20.00
22. Jake the Mallard Duck	5	C	20.00
23. Kiwi the Toucan	40	R	165.00
24. Kuku the Cockatoo	5	C	20.00
25. Mystic the Unicorn	11	R	45.00
26. Mel the Koala Bear	21	C	10.00
27. Nanook the Husky	17	C	15.00
28. Nuts the Squirrel	21	C	10.00
29. Peace the Tie Dyed Teddy	17	C	25.00
30. Patty the Platypus	64	R	800.00
31. Quacker the Duck	40	R	15.00
32. Puffer the Penguin	10	C	15.00
33. Princess the Bear	12	C	65.00
34. Scottie the Scottie	28	R	28.00
35. Rover the Dog	28	R	15.00
36. Rex the Tyrannosaurus	40	R	825.00
37. Sly the Fox	28	C	10.00
38. Slither the Snake	52	R	1,900.00
39. Skip the Siamese Cat	21	C	10.00
40. Splash the Orca Whale	52	R	150.00
41. Spooky the Ghost	28	R	40.00
42. Snowball the Snowman	12	R	40.00
43. Stinger the Scorpion	5	C	15.00
44. Spot the Dog	52	R	65.00
45. Tank the Armadillo	28	R	85.00
46. Stripes the Tiger (Gold/Black)	40	R	400.00
47. Teddy the 1997 Holiday Bear	12	R	50.00
48. Tuffy the Terrier	17	C	10.00
49. Tracker the Basset Hound	5	C	15.00
50. Zip the Black Cat	28	R	40.00

Source: Beanie World Magazine, Sept. 1998.

9.94 Common maize rust is a serious disease of sweet corn. Researchers in New York state have developed an action threshold for initiation of fungicide applications based on a regression equation relating maize rust incidence to severity of the disease (*Phytopathology*, Vol. 80, 1990). In one particular field, data were collected on more than 100 plants of the sweet corn hybrid Jubilee. For each plant, incidence was measured as the percentage of leaves infected (*x*) and severity was calculated as the log (base 10) of the average number of infections per leaf (*y*). A simple linear regression analysis of the data produced the following results:

$$\hat{y} = -.939 + .020x$$
$$r^2 = .816$$
$$s = .288$$

a. Interpret the value of $\hat{\beta}_1$.
b. Interpret the value of r^2.
c. Interpret the value of *s*.
d. Calculate the value of *r* and interpret it.   .903
e. Use the result, part **d**, to test the utility of the model. Use $\alpha = .05$. (Assume $n = 100$.) $t = 20.83$
f. Predict the severity of the disease when the incidence of maize rust for a plant is 80%. [*Note:* Take the antilog (base 10) of $\hat{y}$ to obtain the predicted average number of infections per leaf.   4.58

9.95 At temperatures approaching absolute zero ($-273°$C), helium exhibits traits that seem to defy many laws of Newtonian physics. An experiment has been conducted with helium in solid form at various temperatures near absolute zero. The solid helium is placed in a dilution refrigerator along with a solid impure substance, and the fraction (in weight) of the impurity passing through the solid helium is recorded. (This phenomenon of solids passing directly through solids is known as *quantum tunneling*.) The data are given in the table.

⬤HELIUM

Temperature x (°C)	Proportion of Impurity
−262.0	.315
−265.0	.202
−256.0	.204
−267.0	.620
−270.0	.715
−272.0	.935
−272.4	.957
−272.7	.906
−272.8	.985
−272.9	.987

a. Find the least squares estimates of the intercept and slope. Interpret them.   −13.49; −.0528
b. Use a 95% confidence interval to estimate the slope β_1. Interpret the interval in terms of this application. Does the interval support the hypothesis that tem-

perature contributes information about the proportion of impurity passing through helium?
c. Interpret the coefficient of determination for this model. $r^2 = .8538$
d. Find a 95% prediction interval for the percentage of impurity passing through solid helium at $-273°$C. Interpret the result.   (.5987, 1.2653)
e. Note that the value of *x* in part **d** is outside the experimental region. Why might this lead to an unreliable prediction?
*f. Use a nonparametric method to test for a correlation between temperature (*x*) and impurity proportion (*y*). Use $\alpha = .05$.

9.96 Neurologists have found that the hippocampus, a structure found in the brain, plays an important role in short-term memory. Refer to the *American Journal of Psychiatry* (July 1995) study of the relationship between hippocampal volume and short-term verbal memory of 21 Vietnam vets with combat related post-traumatic stress disorder (PTSD), Exercise 2.114 (p. 90). Recall that magnetic resonance imaging was used to measure the volume, *x*, of the right hippocampus (in cubic millimeters) of each subject while verbal memory retention, *y*, of each subject was measured by the percent retention subscale of the Wechsler Memory Scale. The scattergram for the data is reproduced here.

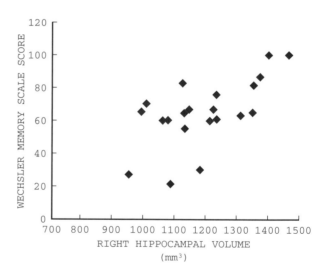

a. Propose a straight-line model relating verbal memory *y* with right hippocampal volume *x* to support the theory that smaller hippocampal volume would be associated with deficits in short-term verbal memory in patients with PTSD.
b. Based on the theory, would you expect the slope of the line to be positive or negative? Explain.
c. The coefficient of correlation between right hippocampal volume and verbal retention was $r = .64$. Interpret this value.

d. A statistical test of $H_0: \beta_1 = 0$ versus $H_a: \beta_1 > 0$ resulted in a *p*-value smaller than .05. Interpret this result.

9.97 Researchers at the University of North Carolina-Greensboro investigated a model for the rate of seed germination (*Journal of Experimental Botany*, Jan. 1993). In one experiment, alfalfa seeds were placed in a specially constructed germination chamber. Eleven hours later, the seeds were examined and the change in free energy (a measure of germination rate) was recorded. The results for seeds germinated at seven different temperatures are given in the table. The data were used to fit a simple linear regression model, with y = change in free energy and x = temperature.

⊘ SEEDGERM

Change in Free

Energy (kJ/mol) y	Temperature (K) x
7	295
6.2	297.5
9	291
9.5	289
8.5	301
7.8	293
11.2	286.5

Source: Hageseth, G. T., and Cody, A. L. "Energy-level model for isothermal seed germination." *Journal of Experimental Botany*, Vol. 44, No. 258, Jan. 1993, p. 123 (Figure 9).

a. Plot the points in a scattergram.

b. Find the least squares prediction equation.

c. Plot the least squares line, part **b** on the scattergram of part **a**.

d. Conduct a test of model adequacy. Use $\alpha = .01$.

e. Use the plot, part **c**, to locate any unusual data points (outliers).

f. Eliminate the outlier, part **e**, from the data set and repeat parts **a–d**.

9.98 One of the most common types of "information retrieval" processes is document-database searching. An experiment was conducted to investigate the variables that influence search performance in the Medline database and retrieval system (*Journal of Information Science*, Vol. 21, 1995). Simple linear regression was used to model the fraction y of the set of potentially informative documents that are retrieved using Medline as a function of the number x of terms in the search query, based on a sample of $n = 124$ queries. The results are summarized below:

$$\hat{y} = .202 + .135x;$$
$$t(\text{for testing } H_0: \beta_1 = 0) = 4.98$$
$$\text{Two-tailed } p\text{-value} = .001$$

a. Is there sufficient evidence to indicate that x and y are linearly related? Test using $\alpha = .01$.

b. If appropriate, use the model to predict the fraction of documents retrieved for a search query with $x = 3$ terms.   .607

9.99 Is there a link between the loneliness of parents and that of their offspring? This question was examined in the *Journal of Marriage and Family* (Aug. 1986). The participants in the study were 130 female college undergraduates and their parents. Each triad of daughter, mother, and father completed the UCLA Loneliness Scale, a 20-item questionnaire designed to assess loneliness and several variables theoretically related to loneliness, such as social accessibility to others, difficulty in making friends, and depression.

a. The correlation between daughter's loneliness score y and mother's loneliness score x was determined to be $r = .26$. Interpret this value.

b. The correlation between daughter's loneliness score y and father's loneliness score x was determined to be $r = .19$. Interpret this value.

c. The correlation between daughter's loneliness score y and mother's self-esteem score x was determined to be $r = .14$. Interpret this value.

d. The correlation between daughter's loneliness score y and father's assertiveness score x was determined to be $r = .01$. Interpret this value.

e. Calculate the coefficient of determination r^2, for parts **a–d**. Interpret the results.

Applying the Concepts—Advanced

9.100 Sometimes it is known from theoretical considerations that the straight-line relationship between two variables, x and y, passes through the origin of the xy-plane. Consider the relationship between the total weight of a shipment of 50-pound bags of flour, y, and the number of bags in the shipment, x. Since a shipment containing $x = 0$ bags (i.e., no shipment at all) has a total weight of $y = 0$, a straight-line model of the relationship between x and y should pass through the point $x = 0, y = 0$. In such a case you could assume $\beta_0 = 0$ and characterize the relationship between x and y with the following model:

$$y = \beta_1 x + \varepsilon$$

The least squares estimate of β_1 for this model is

$$\hat{\beta}_1 = \frac{\sum x_i y_i}{\sum x_i^2}$$

From the records of past flour shipments, 15 shipments were randomly chosen and the data shown in the table on p. 512 were recorded.

a. Find the least squares line for the given data under the assumption that $\beta_0 = 0$. Plot the least squares line on a scattergram of the data. $\hat{y} = 46.3992x$

⊕FLOUR

Weight of Shipment	Number of 50-Pound Bags in Shipment
5,050	100
10,249	205
20,000	450
7,420	150
24,685	500
10,206	200
7,325	150
4,958	100
7,162	150
24,000	500
4,900	100
14,501	300
28,000	600
17,002	400
16,100	400

b. Find the least squares line for the given data using the model

$$y = \beta_0 + \beta_1 x + \varepsilon$$

(i.e., do not restrict β_0 to equal 0). Plot this line on the same scatterplot you constructed in part **a**.

c. Refer to part **b**. Why might $\hat{\beta}_0$ be different from 0 even though the true value of β_0 is known to be 0?

d. The estimated standard error of $\hat{\beta}_0$ is equal to

$$s\sqrt{\frac{1}{n} + \frac{\bar{x}^2}{SS_{xx}}}$$

Use the t statistic

$$t = \frac{\hat{\beta}_0 - 0}{s\sqrt{(1/n) + (\bar{x}^2/SS_{xx})}}$$

to test the null hypothesis $H_0: \beta_0 = 0$ against the alternative $H_a: \beta_0 \neq 0$. Use $\alpha = .10$. Should you include β_0 in your model? $t = .906$

STUDENT PROJECTS

Many dependent variables in all areas of research serve as the subjects of regression modeling efforts. We list five such variables here:

1. Crime rate in various communities
2. Daily maximum temperature in your town
3. Grade point average of students who have completed one academic year at your college
4. Gross Domestic Product of the United States
5. Points scored by your favorite football team in a single game

Choose one of these dependent variables or choose some other dependent variable for which you want to construct a prediction model. There may be a large number of independent variables that should be included in a prediction equation for the dependent variable you choose. List three potentially important independent variables, x_1, x_2, and x_3,

that you think might be (individually) strongly related to your dependent variable. Next, obtain 10 data values, each of which consists of a measure of your dependent variable y and the corresponding values of x_1, x_2, and x_3.

a. Use the least squares formulas given in this chapter to fit three straight-line models—one for each independent variable—for predicting y.

b. Interpret the sign of the estimated slope coefficient $\hat{\beta}_1$ in each case, and test the utility of each model by testing $H_0: \beta_1 = 0$ against $H_a: \beta_1 \neq 0$. What assumptions must be satisfied to ensure the validity of these tests?

c. Calculate the coefficient of determination r^2 for each model. Which of the independent variables predicts y best for the 10 sampled sets of data? Is this variable necessarily best in general (that is, for the entire population)? Explain.

REFERENCES

Chatterjee, S., and Price, B. *Regression Analysis by Example*, 2nd ed. New York: Wiley, 1991.

Draper, N., and Smith, H. *Applied Regression Analysis*, 2nd ed. New York: Wiley, 1981.

Mendenhall, W. *Introduction to Linear Models and the Design and Analysis of Experiments*. Belmont, Ca.: Wadsworth, 1968.

Mendenhall, W., and Sincich, T. *A Second Course in Statistics: Regression Analysis*, 5th ed. Upper Saddle River, N. J.: Prentice Hall, 1996.

Mosteller, F., and Tukey, J. W. *Data Analysis and Regression: A Second Course in Statistics*. Reading, Mass.: Addison-Wesley, 1977.

Neter, J., Kutner, M., Cjnachtsheim, C., and Wasserman, W. *Applied Linear Statistical Models*, 4th ed. Homewood, Ill.: Richard Irwin, 1996.

Rousseeuw, P. J., and Leroy, A. M. *Robust Regression and Outlier Detection*. New York: Wiley, 1987.

A P P E N D I X   A

Tables

TABLE I Random Numbers

Row	1	2	3	4	5	6	7	8	9	10	11	12	13	14
1	10480	15011	01536	02011	81647	91646	69179	14194	62590	36207	20969	99570	91291	90700
2	22368	46573	25595	85393	30995	89198	27982	53402	93965	34095	52666	19174	39615	99505
3	24130	48360	22527	97265	76393	64809	15179	24830	49340	32081	30680	19655	63348	58629
4	42167	93093	06243	61680	07856	16376	39440	53537	71341	57004	00849	74917	97758	16379
5	37570	39975	81837	16656	06121	91782	60468	81305	49684	60672	14110	06927	01263	54613
6	77921	06907	11008	42751	27756	53498	18602	70659	90655	15053	21916	81825	44394	42880
7	99562	72905	56420	69994	98872	31016	71194	18738	44013	48840	63213	21069	10634	12952
8	96301	91977	05463	07972	18876	20922	94595	56869	69014	60045	18425	84903	42508	32307
9	89579	14342	63661	10281	17453	18103	57740	84378	25331	12566	58678	44947	05585	56941
10	85475	36857	53342	53988	53060	59533	38867	62300	08158	17983	16439	11458	18593	64952
11	28918	69578	88231	33276	70997	79936	56865	05859	90106	31595	01547	85590	91610	78188
12	63553	40961	48235	03427	49626	69445	18663	72695	52180	20847	12234	90511	33703	90322
13	09429	93969	52636	92737	88974	33488	36320	17617	30015	08272	84115	27156	30613	74952
14	10365	61129	87529	85689	48237	52267	67689	93394	01511	26358	85104	20285	29975	89868
15	07119	97336	71048	08178	77233	13916	47564	81056	97735	85977	29372	74461	28551	90707
16	51085	12765	51821	51259	77452	16308	60756	92144	49442	53900	70960	63990	75601	40719
17	02368	21382	52404	60268	89368	19885	55322	44819	01188	65255	64835	44919	05944	55157
18	01011	54092	33362	94904	31273	04146	18594	29852	71585	85030	51132	01915	92747	64951
19	52162	53916	46369	58586	23216	14513	83149	98736	23495	64350	94738	17752	35156	35749
20	07056	97628	33787	09998	42698	06691	76988	13602	51851	46104	88916	19509	25625	58104
21	48663	91245	85828	14346	09172	30168	90229	04734	59193	22178	30421	61666	99904	32812
22	54164	58492	22421	74103	47070	25306	76468	26384	58151	06646	21524	15227	96909	44592
23	32639	32363	05597	24200	13363	38005	94342	28728	35806	06912	17012	64161	18296	22851
24	29334	27001	87637	87308	58731	00256	45834	15398	46557	41135	10367	07684	36188	18510
25	02488	33062	28834	07351	19731	92420	60952	61280	50001	67658	32586	86679	50720	94953
26	81525	72295	04839	96423	24878	82651	66566	14778	76797	14780	13300	87074	79666	95725
27	29676	20591	68086	26432	46901	20849	89768	81536	86645	12659	92259	57102	80428	25280
28	00742	57392	39064	66432	84673	40027	32832	61362	98947	96067	64760	64584	96096	98253
29	05366	04213	25669	26422	44407	44048	37937	63904	45766	66134	75470	66520	34693	90449
30	91921	26418	64117	94305	26766	25940	39972	22209	71500	64568	91402	42416	07844	69618
31	00582	04711	87917	77341	42206	35126	74087	99547	81817	42607	43808	76655	62028	76630
32	00725	69884	62797	56170	86324	88072	76222	36086	84637	93161	76038	65855	77919	88006
33	69011	65795	95876	55293	18988	27354	26575	08625	40801	59920	29841	80150	12777	48501
34	25976	57948	29888	88604	67917	48708	18912	82271	65424	69774	33611	54262	85963	03547
35	09763	83473	73577	12908	30883	18317	28290	35797	05998	41688	34952	37888	38917	88050

(continues on next page)

TABLE I Random Numbers (continued)

Row	1	2	3	4	5	6	7	8	9	10	11	12	13	14
36	91576	42595	27958	30134	04024	86385	29880	99730	55536	84855	29080	09250	79656	73211
37	17955	56349	90999	49127	20044	59931	06115	20542	18059	02008	73708	83517	36103	42791
38	46503	18584	18845	49618	02304	51038	20655	58727	28168	15475	56942	53389	20562	87338
39	92157	89634	94824	78171	84610	82834	09922	25417	44137	48413	25555	21246	35509	20468
40	14577	62765	35605	81263	39667	47358	56873	56307	61607	49518	89656	20103	77490	18062
41	98427	07523	33362	64270	01638	92477	66969	98420	04880	45585	46565	04102	46880	45709
42	34914	63976	88720	82765	34476	17032	87589	40836	32427	70002	70663	88863	77775	69348
43	70060	28277	39475	46473	23219	53416	94970	25832	69975	94884	19661	72828	00102	66794
44	53976	54914	06990	67245	68350	82948	11398	42878	80287	88267	47363	46634	06541	97809
45	76072	29515	40980	07391	58745	25774	22987	80059	39911	96189	41151	14222	60697	59583
46	90725	52210	83974	29992	65831	38857	50490	83765	55657	14361	31720	57375	56228	41546
47	64364	67412	33339	31926	14883	24413	59744	92351	97473	89286	35931	04110	23726	51900
48	08962	00358	31662	25388	61642	34072	81249	35648	56891	69352	48373	45578	78547	81788
49	95012	68379	93526	70765	10592	04542	76463	54328	02349	17247	28865	14777	62730	92277
50	15664	10493	20492	38391	91132	21999	59516	81652	27195	48223	46751	22923	32261	85653
51	16408	81899	04153	53381	79401	21438	83035	92350	36693	31238	59649	91754	72772	02338
52	18629	81953	05520	91962	04739	13092	97662	24822	94730	06496	35090	04822	86774	98289
53	73115	35101	47498	87637	99016	71060	88824	71013	18735	20286	23153	72924	35165	43040
54	57491	16703	23167	49323	45021	33132	12544	41035	80780	45393	44812	12512	98931	91202
55	30405	83946	23792	14422	15059	45799	22716	19792	09983	74353	68668	30429	70735	25499
56	16631	35006	85900	98275	32388	52390	16815	69290	82732	38480	73817	32523	41961	44437
57	96773	20206	42559	78985	05300	22164	24369	54224	35083	19687	11052	91491	60383	19746
58	38935	64202	14349	82674	66523	44133	00697	35552	35970	19124	63318	29686	03387	59846
59	31624	76384	17403	53363	44167	64486	64758	75366	76554	31601	12614	33072	60332	92325
60	78919	19474	23632	27889	47914	02584	37680	20801	72152	39339	34806	08930	85001	87820
61	03931	33309	57047	74211	63445	17361	62825	39908	05607	91284	68833	25570	38818	46920
62	74426	33278	43972	10110	89917	15665	52872	73823	73144	88662	88970	74492	51805	99378
63	09066	00903	20795	95452	92648	45454	09552	88815	16553	51125	79375	97596	16296	66092
64	42238	12426	87025	14267	20979	04508	64535	31355	86064	29472	47689	05974	52468	16834
65	16153	08002	26504	41744	81959	65642	74240	56302	00033	67107	77510	70625	28725	34191
66	21457	40742	29820	96783	29400	21840	15035	34537	33310	06116	95240	15957	16572	06004
67	21581	57802	02050	89728	17937	37621	47075	42080	97403	48626	68995	43805	33386	21597
68	55612	78095	83197	33732	05810	24813	86902	60397	16489	03264	88525	42786	05269	92532
69	44657	66999	99324	51281	84463	60563	79312	93454	68876	25471	93911	25650	12682	73572
70	91340	84979	46949	81973	37949	61023	43997	15263	80644	43942	89203	71795	99533	50501

TABLE I Random Numbers (continued)

Row \ Column	1	2	3	4	5	6	7	8	9	10	11	12	13	14
71	91227	21199	31935	27022	84067	05462	35216	14486	29891	68607	41867	14951	91696	85065
72	50001	38140	66321	19924	72163	09538	12151	06878	91903	18749	34405	56087	82790	70925
73	65390	05224	72958	28609	81406	39147	25549	48542	42627	45233	57202	94617	23772	07896
74	27504	96131	83944	41575	10573	08619	64482	73923	36152	05184	94142	25299	84387	34925
75	37169	94851	39117	89632	00959	16487	65536	49071	39782	17095	02330	74301.	00275	48280
76	11508	70225	51111	38351	19444	66499	71945	05422	13442	78675	84081	66938	93654	59894
77	37449	30362	06694	54690	04052	53115	62757	95348	78662	11163	81651	50245	34971	52924
78	46515	70331	85922	38329	57015	15765	97161	17869	45349	61796	66345	81073	49106	79860
79	30986	81223	42416	58353	21532	30502	32305	86482	05174	07901	54339	58861	74818	46942
80	63798	64995	46583	09785	44160	78128	83991	42865	92520	83531	80377	35909	81250	54238
81	82486	84846	99254	67632	43218	50076	21361	64816	51202	88124	41870	52689	51275	83556
82	21885	32906	92431	09060	64297	51674	64126	62570	26123	05155	59194	52799	28225	85762
83	60336	98782	07408	53458	13564	59089	26445	29789	85205	41001	12535	12133	14645	23541
84	43937	46891	24010	25560	86355	33941	25786	54990	71899	15475	95434	98227	21824	19585
85	97656	63175	89303	16275	07100	92063	21942	18611	47348	20203	18534	03862	78095	50136
86	03299	01221	05418	38982	55758	92237	26759	86367	21216	98442	08303	56613	91511	75928
87	79626	06486	03574	17668	07785	76020	79924	25651	83325	88428	85076	72811	22717	50585
88	85636	68335	47539	03129	65651	11977	02510	26113	99447	68645	34327	15152	55230	93448
89	18039	14367	64337	06177	12143	46609	32989	74014	64708	00533	35398	58408	13261	47908
90	08362	15656	60627	36478	65648	16764	53412	09013	07832	41574	17639	82163	60859	75567
91	79556	29068	04142	16268	15387	12856	66227	38358	22478	73373	88732	09443	82558	05250
92	92608	82674	27072	32534	17075	27698	98204	63863	11951	34648	88022	56148	34925	57031
93	23982	25835	40055	67006	12293	02753	14827	23235	35071	99704	37543	11601	35503	85171
94	09915	96306	05908	97901	28395	14186	00821	80703	70426	75647	76310	88717	37890	40129
95	59037	33300	26695	62247	69927	76123	50842	43834	86654	70959	79725	93872	28117	19233
96	42488	78077	69882	61657	34136	79180	97526	43092	04098	73571	80799	76536	71255	64239
97	46764	86273	63003	93017	31204	36692	40202	35275	57306	55543	53203	18098	47625	88684
98	03237	45430	55417	63282	90816	17349	88298	90183	36600	78406	06216	95787	42579	90730
99	86591	81482	52667	61582	14972	90053	89534	76036	49199	43716	97548	04379	46370	28672
100	38534	01715	94964	87288	65680	43772	39560	12918	86537	62738	19636	51132	25739	56947

Source: Abridged from W. H. Beyer (ed.), *CRC Standard Mathematical Tables*, 24th edition. (Cleveland: The Chemical Rubber Company), 1976. Reproduced by permission of the publisher.

TABLE II   Binomial Probabilities

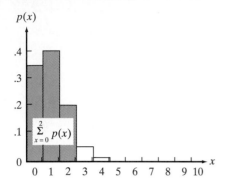

Tabulated values are $\sum_{x=0}^{k} p(x)$. *(Computations are rounded at the third decimal place.)*

a. $n = 5$

k	.01	.05	.10	.20	.30	.40	.50	.60	.70	.80	.90	.95	.99
0	.951	.774	.590	.328	.168	.078	.031	.010	.002	.000	.000	.000	.000
1	.999	.977	.919	.737	.528	.337	.188	.087	.031	.007	.000	.000	.000
2	1.000	.999	.991	.942	.837	.683	.500	.317	.163	.058	.009	.001	.000
3	1.000	1.000	1.000	.993	.969	.913	.812	.663	.472	.263	.081	.023	.001
4	1.000	1.000	1.000	1.000	.998	.990	.969	.922	.832	.672	.410	.226	.049

b. $n = 6$

k	.01	.05	.10	.20	.30	.40	.50	.60	.70	.80	.90	.95	.99
0	.941	.735	.531	.262	.118	.047	.016	.004	.001	.000	.000	.000	.000
1	.999	.967	.886	.655	.420	.233	.109	.041	.011	.002	.000	.000	.000
2	1.000	.998	.984	.901	.744	.544	.344	.179	.070	.017	.001	.000	.000
3	1.000	1.000	.999	.983	.930	.821	.656	.456	.256	.099	.016	.002	.000
4	1.000	1.000	1.000	.998	.989	.959	.891	.767	.580	.345	.114	.033	.001
5	1.000	1.000	1.000	1.000	.999	.996	.984	.953	.882	.738	.469	.265	.059

c. $n = 7$

k	.01	.05	.10	.20	.30	.40	.50	.60	.70	.80	.90	.95	.99
0	.932	.698	.478	.210	.082	.028	.008	.002	.000	.000	.000	.000	.000
1	.998	.956	.850	.577	.329	.159	.063	.019	.004	.000	.000	.000	.000
2	1.000	.996	.974	.852	.647	.420	.227	.096	.029	.005	.000	.000	.000
3	1.000	1.000	.997	.967	.874	.710	.500	.290	.126	.033	.003	.000	.000
4	1.000	1.000	1.000	.995	.971	.904	.773	.580	.353	.148	.026	.004	.000
5	1.000	1.000	1.000	1.000	.996	.981	.937	.841	.671	.423	.150	.044	.002
6	1.000	1.000	1.000	1.000	1.000	.998	.992	.972	.918	.790	.522	.302	.068

TABLE II *(continued)*

d. *n* = 8

k	.01	.05	.10	.20	.30	.40	.50	.60	.70	.80	.90	.95	.99
0	.923	.663	.430	.168	.058	.017	.004	.001	.000	.000	.000	.000	.000
1	.997	.943	.813	.503	.255	.106	.035	.009	.001	.000	.000	.000	.000
2	1.000	.994	.962	.797	.552	.315	.145	.050	.011	.001	.000	.000	.000
3	1.000	1.000	.995	.944	.806	.594	.363	.174	.058	.010	.000	.000	.000
4	1.000	1.000	1.000	.990	.942	.826	.637	.406	.194	.056	.005	.000	.000
5	1.000	1.000	1.000	.999	.989	.950	.855	.685	.448	.203	.038	.006	.000
6	1.000	1.000	1.000	1.000	.999	.991	.965	.894	.745	.497	.187	.057	.003
7	1.000	1.000	1.000	1.000	1.000	.999	.996	.983	.942	.832	.570	.337	.077

e. *n* = 9

k	.01	.05	.10	.20	.30	.40	.50	.60	.70	.80	.90	.95	.99
0	.914	.630	.387	.134	.040	.010	.002	.000	.000	.000	.000	.000	.000
1	.997	.929	.775	.436	.196	.071	.020	.004	.000	.000	.000	.000	.000
2	1.000	.992	.947	.738	.463	.232	.090	.025	.004	.000	.000	.000	.000
3	1.000	.999	.992	.914	.730	.483	.254	.099	.025	.003	.000	.000	.000
4	1.000	1.000	.999	.980	.901	.733	.500	.267	.099	.020	.001	.000	.000
5	1.000	1.000	1.000	.997	.975	.901	.746	.517	.270	.086	.008	.001	.000
6	1.000	1.000	1.000	1.000	.996	.975	.910	.768	.537	.262	.053	.008	.000
7	1.000	1.000	1.000	1.000	1.000	.996	.980	.929	.804	.564	.225	.071	.003
8	1.000	1.000	1.000	1.000	1.000	1.000	.998	.990	.960	.866	.613	.370	.086

f. *n* = 10

k	.01	.05	.10	.20	.30	.40	.50	.60	.70	.80	.90	.95	.99
0	.904	.599	.349	.107	.028	.006	.001	.000	.000	.000	.000	.000	.000
1	.996	.914	.376	.376	.149	.046	.011	.002	.000	.000	.000	.000	.000
2	1.000	.988	.930	.678	.383	.167	.055	.012	.002	.000	.000	.000	.000
3	1.000	.999	.987	.879	.650	.382	.172	.055	.011	.001	.000	.000	.000
4	1.000	1.000	.998	.967	.850	.633	.377	.166	.047	.006	.000	.000	.000
5	1.000	1.000	1.000	.999	.953	.834	.623	.367	.150	.033	.002	.000	.000
6	1.000	1.000	1.000	.999	.989	.945	.828	.618	.350	.121	.013	.001	.000
7	1.000	1.000	1.000	1.000	.998	.988	.945	.833	.617	.322	.070	.012	.000
8	1.000	1.000	1.000	1.000	1.000	.998	.989	.954	.851	.624	.264	.086	.004
9	1.000	1.000	1.000	1.000	1.000	1.000	.999	.994	.972	.893	.651	.401	.096

(continues on next page)

TABLE II   (continued)

g. $n = 15$

k	.01	.05	.10	.20	.30	.40	.50	.60	.70	.80	.90	.95	.99
0	.860	.463	.206	.035	.005	.000	.000	.000	.000	.000	.000	.000	.000
1	.990	.829	.549	.167	.035	.005	.000	.000	.000	.000	.000	.000	.000
2	1.000	.964	.816	.398	.127	.027	.004	.000	.000	.000	.000	.000	.000
3	1.000	.995	.944	.648	.297	.091	.018	.002	.000	.000	.000	.000	.000
4	1.000	.999	.987	.838	.515	.217	.059	.009	.001	.000	.000	.000	.000
5	1.000	1.000	.998	.939	.722	.403	.151	.034	.004	.000	.000	.000	.000
6	1.000	1.000	1.000	.982	.869	.610	.304	.095	.015	.001	.000	.000	.000
7	1.000	1.000	1.000	.996	.950	.787	.500	.213	.050	.004	.000	.000	.000
8	1.000	1.000	1.000	.999	.985	.905	.696	.390	.131	.018	.000	.000	.000
9	1.000	1.000	1.000	1.000	.996	.966	.849	.597	.278	.061	.002	.000	.000
10	1.000	1.000	1.000	1.000	.999	.991	.941	.783	.485	.164	.013	.001	.000
11	1.000	1.000	1.000	1.000	1.000	.998	.982	.909	.703	.352	.056	.005	.000
12	1.000	1.000	1.000	1.000	1.000	1.000	.996	.973	.873	.602	.184	.036	.000
13	1.000	1.000	1.000	1.000	1.000	1.000	1.000	.995	.965	.833	.451	.171	.010
14	1.000	1.000	1.000	1.000	1.000	1.000	1.000	1.000	.995	.965	.794	.537	.140

h. $n = 20$

k	.01	.05	.10	.20	.30	.40	.50	.60	.70	.80	.90	.95	.99
0	.818	.358	.122	.012	.001	.000	.000	.000	.000	.000	.000	.000	.000
1	.983	.736	.392	.069	.008	.001	.000	.000	.000	.000	.000	.000	.000
2	.999	.925	.677	.206	.035	.004	.000	.000	.000	.000	.000	.000	.000
3	1.000	.984	.867	.411	.107	.016	.001	.000	.000	.000	.000	.000	.000
4	1.000	.997	.957	.630	.238	.051	.006	.000	.000	.000	.000	.000	.000
5	1.000	1.000	.989	.804	.416	.126	.021	.002	.000	.000	.000	.000	.000
6	1.000	1.000	.998	.913	.608	.250	.058	.006	.000	.000	.000	.000	.000
7	1.000	1.000	1.000	.968	.772	.416	.132	.021	.001	.000	.000	.000	.000
8	1.000	1.000	1.000	.990	.887	.596	.252	.057	.005	.000	.000	.000	.000
9	1.000	1.000	1.000	.997	.952	.755	.412	.128	.017	.001	.000	.000	.000
10	1.000	1.000	1.000	.999	.983	.872	.588	.245	.048	.003	.000	.000	.000
11	1.000	1.000	1.000	1.000	.995	.943	.748	.404	.113	.010	.000	.000	.000
12	1.000	1.000	1.000	1.000	.999	.979	.868	.584	.228	.032	.000	.000	.000
13	1.000	1.000	1.000	1.000	1.000	.994	.942	.750	.392	.087	.002	.000	.000
14	1.000	1.000	1.000	1.000	1.000	.998	.979	.874	.584	.196	.011	.000	.000
15	1.000	1.000	1.000	1.000	1.000	1.000	.994	.949	.762	.370	.043	.003	.000
16	1.000	1.000	1.000	1.000	1.000	1.000	.999	.984	.893	.589	.133	.016	.000
17	1.000	1.000	1.000	1.000	1.000	1.000	1.000	.996	.965	.794	.323	.075	.001
18	1.000	1.000	1.000	1.000	1.000	1.000	1.000	.999	.992	.931	.608	.264	.017
19	1.000	1.000	1.000	1.000	1.000	1.000	1.000	1.000	.999	.988	.878	.642	.182

TABLE II *(continued)*

i. *n* = 25

k \ p	.01	.05	.10	.20	.30	.40	.50	.60	.70	.80	.90	.95	.99
0	.778	.277	.072	.004	.000	.000	.000	.000	.000	.000	.000	.000	.000
1	.974	.642	.271	.027	.002	.000	.000	.000	.000	.000	.000	.000	.000
2	.998	.873	.537	.098	.009	.000	.000	.000	.000	.000	.000	.000	.000
3	1.000	.966	.764	.234	.033	.002	.000	.000	.000	.000	.000	.000	.000
4	1.000	.993	.902	.421	.090	.009	.000	.000	.000	.000	.000	.000	.000
5	1.000	.999	.967	.617	.193	.029	.002	.000	.000	.000	.000	.000	.000
6	1.000	1.000	.991	.780	.341	.074	.007	.000	.000	.000	.000	.000	.000
7	1.000	1.000	.998	.891	.512	.154	.022	.001	.000	.000	.000	.000	.000
8	1.000	1.000	1.000	.953	.677	.274	.054	.004	.000	.000	.000	.000	.000
9	1.000	1.000	1.000	.983	.811	.425	.115	.013	.000	.000	.000	.000	.000
10	1.000	1.000	1.000	.994	.902	.586	.212	.034	.002	.000	.000	.000	.000
11	1.000	1.000	1.000	.998	.956	.732	.345	.078	.006	.000	.000	.000	.000
12	1.000	1.000	1.000	1.000	.983	.846	.500	.154	.017	.000	.000	.000	.000
13	1.000	1.000	1.000	1.000	.994	.922	.655	.268	.044	.002	.000	.000	.000
14	1.000	1.000	1.000	1.000	.998	.966	.788	.414	.098	.006	.000	.000	.000
15	1.000	1.000	1.000	1.000	1.000	.987	.885	.575	.189	.017	.000	.000	.000
16	1.000	1.000	1.000	1.000	1.000	.996	.946	.726	.323	.047	.000	.000	.000
17	1.000	1.000	1.000	1.000	1.000	.999	.978	.846	.488	.109	.002	.000	.000
18	1.000	1.000	1.000	1.000	1.000	1.000	.993	.926	.659	.220	.009	.000	.000
19	1.000	1.000	1.000	1.000	1.000	1.000	.998	.971	.807	.383	.033	.001	.000
20	1.000	1.000	1.000	1.000	1.000	1.000	1.000	.991	.910	.579	.098	.007	.000
21	1.000	1.000	1.000	1.000	1.000	1.000	1.000	.998	.967	.766	.236	.034	.000
22	1.000	1.000	1.000	1.000	1.000	1.000	1.000	1.000	.991	.902	.463	.127	.002
23	1.000	1.000	1.000	1.000	1.000	1.000	1.000	1.000	.998	.973	.729	.358	.026
24	1.000	1.000	1.000	1.000	1.000	1.000	1.000	1.000	1.000	.996	.928	.723	.222

TABLE III   Normal Curve Areas

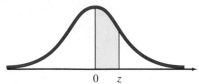

z	.00	.01	.02	.03	.04	.05	.06	.07	.08	.09
.0	.0000	.0040	.0080	.0120	.0160	.0199	.0239	.0279	.0319	.0359
.1	.0398	.0438	.0478	.0517	.0557	.0596	.0636	.0675	.0714	.0753
.2	.0793	.0832	.0871	.0910	.0948	.0987	.1026	.1064	.1103	.1141
.3	.1179	.1217	.1255	.1293	.1331	.1368	.1406	.1443	.1480	.1517
.4	.1554	.1591	.1628	.1664	.1700	.1736	.1772	.1808	.1844	.1879
.5	.1915	.1950	.1985	.2019	.2054	.2088	.2123	.2157	.2190	.2224
.6	.2257	.2291	.2324	.2357	.2389	.2422	.2454	.2486	.2517	.2549
.7	.2580	.2611	.2642	.2673	.2704	.2734	.2764	.2794	.2823	.2852
.8	.2881	.2910	.2939	.2967	.2995	.3023	.3051	.3078	.3106	.3133
.9	.3159	.3186	.3212	.3238	.3264	.3289	.3315	.3340	.3365	.3389
1.0	.3413	.3438	.3461	.3485	.3508	.3531	.3554	.3577	.3599	.3621
1.1	.3643	.3665	.3686	.3708	.3729	.3749	.3770	.3790	.3810	.3830
1.2	.3849	.3869	.3888	.3907	.3925	.3944	.3962	.3980	.3997	.4015
1.3	.4032	.4049	.4066	.4082	.4099	.4115	.4131	.4147	.4162	.4177
1.4	.4192	.4207	.4222	.4236	.4251	.4265	.4279	.4292	.4306	.4319
1.5	.4332	.4345	.4357	.4370	.4382	.4394	.4406	.4418	.4429	.4441
1.6	.4452	.4463	.4474	.4484	.4495	.4505	.4515	.4525	.4535	.4545
1.7	.4554	.4564	.4573	.4582	.4591	.4599	.4608	.4616	.4625	.4633
1.8	.4641	.4649	.4656	.4664	.4671	.4678	.4686	.4693	.4699	.4706
1.9	.4713	.4719	.4726	.4732	.4738	.4744	.4750	.4756	.4761	.4767
2.0	.4772	.4778	.4783	.4788	.4793	.4798	.4803	.4808	.4812	.4817
2.1	.4821	.4826	.4830	.4834	.4838	.4842	.4846	.4850	.4854	.4857
2.2	.4861	.4864	.4868	.4871	.4875	.4878	.4881	.4884	.4887	.4890
2.3	.4893	.4896	.4898	.4901	.4904	.4906	.4909	.4911	.4913	.4916
2.4	.4918	.4920	.4922	.4925	.4927	.4929	.4931	.4932	.4934	.4936
2.5	.4938	.4940	.4941	.4943	.4945	.4946	.4948	.4949	.4951	.4952
2.6	.4953	.4955	.4956	.4957	.4959	.4960	.4961	.4962	.4963	.4964
2.7	.4965	.4966	.4967	.4968	.4969	.4970	.4971	.4972	.4973	.4974
2.8	.4974	.4975	.4976	.4977	.4977	.4978	.4979	.4979	.4980	.4981
2.9	.4981	.4982	.4982	.4983	.4984	.4984	.4985	.4985	.4986	.4986
3.0	.4987	.4987	.4987	.4988	.4988	.4989	.4989	.4989	.4990	.4990

Source: Abridged from Table I of A. Hald, *Statistical Tables and Formulas* (New York: Wiley), 1952. Reproduced by permission of A. Hald.

TABLE IV **Critical Values of *t***

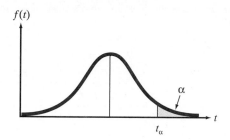

$f(t)$

α

t_α

ν	$t_{.100}$	$t_{.050}$	$t_{.025}$	$t_{.010}$	$t_{.005}$	$t_{.001}$	$t_{.0005}$
1	3.078	6.314	12.706	31.821	63.657	318.31	636.62
2	1.886	2.920	4.303	6.965	9.925	22.326	31.598
3	1.638	2.353	3.182	4.541	5.841	10.213	12.924
4	1.533	2.132	2.776	3.747	4.604	7.173	8.610
5	1.476	2.015	2.571	3.365	4.032	5.893	6.869
6	1.440	1.943	2.447	3.143	3.707	5.208	5.959
7	1.415	1.895	2.365	2.998	3.499	4.785	5.408
8	1.397	1.860	2.306	2.896	3.355	4.501	5.041
9	1.383	1.833	2.262	2.821	3.250	4.297	4.781
10	1.372	1.812	2.228	2.764	3.169	4.144	4.587
11	1.363	1.796	2.201	2.718	3.106	4.025	4.437
12	1.356	1.782	2.179	2.681	3.055	3.930	4.318
13	1.350	1.771	2.160	2.650	3.012	3.852	4.221
14	1.345	1.761	2.145	2.624	2.977	3.787	4.140
15	1.341	1.753	2.131	2.602	2.947	3.733	4.073
16	1.337	1.746	2.120	2.583	2.921	3.686	4.015
17	1.333	1.740	2.110	2.567	2.898	3.646	3.965
18	1.330	1.734	2.101	2.552	2.878	3.610	3.922
19	1.328	1.729	2.093	2.539	2.861	3.579	3.883
20	1.325	1.725	2.086	2.528	2.845	3.552	3.850
21	1.323	1.721	2.080	2.518	2.831	3.527	3.819
22	1.321	1.717	2.074	2.508	2.819	3.505	3.792
23	1.319	1.714	2.069	2.500	2.807	3.485	3.767
24	1.318	1.711	2.064	2.492	2.797	3.467	3.745
25	1.316	1.708	2.060	2.485	2.787	3.450	3.725
26	1.315	1.706	2.056	2.479	2.779	3.435	3.707
27	1.314	1.703	2.052	2.473	2.771	3.421	3.690
28	1.313	1.701	2.048	2.467	2.763	3.408	3.674
29	1.311	1.699	2.045	2.462	2.756	3.396	3.659
30	1.310	1.697	2.042	2.457	2.750	3.385	3.646
40	1.303	1.684	2.021	2.423	2.704	3.307	3.551
60	1.296	1.671	2.000	2.390	2.660	3.232	3.460
120	1.289	1.658	1.980	2.358	2.617	3.160	3.373
∞	1.282	1.645	1.960	2.326	2.576	3.090	3.291

Source: This table is reproduced with the kind permission of the Trustees of Biometrika from E. S. Pearson and H.O. Hartley (eds.), *The Biometrika Tables for Statisticians*, Vol. 1, 3d ed., Biometrika, 1966.

TABLE V   Critical Values of T_L and T_U for the Wilcoxon Rank Sum Test: Independent Samples

Test statistic is the rank sum associated with the smaller sample (if equal sample sizes, either rank sum can be used).

a. $\alpha = .025$ one-tailed; $\alpha = .05$ two-tailed

n_2 \ n_1	3 T_L	3 T_U	4 T_L	4 T_U	5 T_L	5 T_U	6 T_L	6 T_U	7 T_L	7 T_U	8 T_L	8 T_U	9 T_L	9 T_U	10 T_L	10 T_U
3	5	16	6	18	6	21	7	23	7	26	8	28	8	31	9	33
4	6	18	11	25	12	28	12	32	13	35	14	38	15	41	16	44
5	6	21	12	28	18	37	19	41	20	45	21	49	22	53	24	56
6	7	23	12	32	19	41	26	52	28	56	29	61	31	65	32	70
7	7	26	13	35	20	45	28	56	37	68	39	73	41	78	43	83
8	8	28	14	38	21	49	29	61	39	73	49	87	51	93	54	98
9	8	31	15	41	22	53	31	65	41	78	51	93	63	108	66	114
10	9	33	16	44	24	56	32	70	43	83	54	98	66	114	79	131

b. $\alpha = .05$ one-tailed; $\alpha = .10$ two-tailed

n_2 \ n_1	3 T_L	3 T_U	4 T_L	4 T_U	5 T_L	5 T_U	6 T_L	6 T_U	7 T_L	7 T_U	8 T_L	8 T_U	9 T_L	9 T_U	10 T_L	10 T_U
3	6	15	7	17	7	20	8	22	9	24	9	27	10	29	11	31
4	7	17	12	24	13	27	14	30	15	33	16	36	17	39	18	42
5	7	20	13	27	19	36	20	40	22	43	24	46	25	50	26	54
6	8	22	14	30	20	40	28	50	30	54	32	58	33	63	35	67
7	9	24	15	33	22	43	30	54	39	66	41	71	43	76	46	80
8	9	27	16	36	24	46	32	58	41	71	52	84	54	90	57	95
9	10	29	17	39	25	50	33	63	43	76	54	90	66	105	69	111
10	11	31	18	42	26	54	35	67	46	80	57	95	69	111	83	127

Source: From F. Wilcoxon and R. A. Wilcox, "Some Rapid Approximate Statistical Procedures," 1964, 20–23.

TABLE VI **Critical Values of T_0 in the Wilcoxon Paired Difference Signed Ranks Test**

One-Tailed	Two-Tailed	$n = 5$	$n = 6$	$n = 7$	$n = 8$	$n = 9$	$n = 10$
$\alpha = .05$	$\alpha = .10$	1	2	4	6	8	11
$\alpha = .025$	$\alpha = .05$		1	2	4	6	8
$\alpha = .01$	$\alpha = .02$			0	2	3	5
$\alpha = .005$	$\alpha = .01$				0	2	3
		$n = 11$	$n = 12$	$n = 13$	$n = 14$	$n = 15$	$n = 16$
$\alpha = .05$	$\alpha = .10$	14	17	21	26	30	36
$\alpha = .025$	$\alpha = .05$	11	14	17	21	25	30
$\alpha = .01$	$\alpha = .02$	7	10	13	16	20	24
$\alpha = .005$	$\alpha = .01$	5	7	10	13	16	19
		$n = 17$	$n = 18$	$n = 19$	$n = 20$	$n = 21$	$n = 22$
$\alpha = .05$	$\alpha = .10$	41	47	54	60	68	75
$\alpha = .025$	$\alpha = .05$	35	40	46	52	59	66
$\alpha = .01$	$\alpha = .02$	28	33	38	43	49	56
$\alpha = .005$	$\alpha = .01$	23	28	32	37	43	49
		$n = 23$	$n = 24$	$n = 25$	$n = 26$	$n = 27$	$n = 28$
$\alpha = .05$	$\alpha = .10$	83	92	101	110	120	130
$\alpha = .025$	$\alpha = .05$	73	81	90	98	107	117
$\alpha = .01$	$\alpha = .02$	62	69	77	85	93	102
$\alpha = .005$	$\alpha = .01$	55	61	68	76	84	92
		$n = 29$	$n = 30$	$n = 31$	$n = 32$	$n = 33$	$n = 34$
$\alpha = .05$	$\alpha = .10$	141	152	163	175	188	201
$\alpha = .025$	$\alpha = .05$	127	137	148	159	171	183
$\alpha = .01$	$\alpha = .02$	111	120	130	141	151	162
$\alpha = .005$	$\alpha = .01$	100	109	118	128	138	149
		$n = 35$	$n = 36$	$n = 37$	$n = 38$	$n = 39$	
$\alpha = .05$	$\alpha = .10$	214	228	242	256	271	
$\alpha = .025$	$\alpha = .05$	195	208	222	235	250	
$\alpha = .01$	$\alpha = .02$	174	186	198	211	224	
$\alpha = .005$	$\alpha = .01$	160	171	183	195	208	
		$n = 40$	$n = 41$	$n = 42$	$n = 43$	$n = 44$	$n = 45$
$\alpha = .05$	$\alpha = .10$	287	303	319	336	353	371
$\alpha = .025$	$\alpha = .05$	264	279	295	311	327	344
$\alpha = .01$	$\alpha = .02$	238	252	267	281	297	313
$\alpha = .005$	$\alpha = .01$	221	234	248	262	277	292
		$n = 46$	$n = 47$	$n = 48$	$n = 49$	$n = 50$	
$\alpha = .05$	$\alpha = .10$	389	408	427	446	466	
$\alpha = .025$	$\alpha = .05$	361	379	397	415	434	
$\alpha = .01$	$\alpha = .02$	329	345	362	380	398	
$\alpha = .005$	$\alpha = .01$	307	323	339	356	373	

Source: From F. Wilcoxon and R. A. Wilcox, "Some Rapid Approximate Statistical Procedures," 1964, p. 28.

TABLE VII   Percentage Points of the *F*-Distribution, $\alpha = .10$

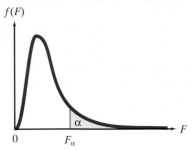

ν_2 \ ν_1	**NUMERATOR DEGREES OF FREEDOM**								
	1	**2**	**3**	**4**	**5**	**6**	**7**	**8**	**9**
1	39.86	49.50	53.59	55.83	57.24	58.20	58.91	59.44	59.86
2	8.53	9.00	9.16	9.24	9.29	9.33	9.35	9.37	9.38
3	5.54	5.46	5.39	5.34	5.31	5.28	5.27	5.25	5.24
4	4.54	4.32	4.19	4.11	4.05	4.01	3.98	3.95	3.94
5	4.06	3.78	3.62	3.52	3.45	3.40	3.37	3.34	3.32
6	3.78	3.46	3.29	3.18	3.11	3.05	3.01	2.98	2.96
7	3.59	3.26	3.07	2.96	2.88	2.83	2.78	2.75	2.72
8	3.46	3.11	2.92	2.81	2.73	2.67	2.62	2.59	2.56
9	3.36	3.01	2.81	2.69	2.61	2.55	2.51	2.47	2.44
10	3.29	2.92	2.73	2.61	2.52	2.46	2.41	2.38	2.35
11	3.23	2.86	2.66	2.54	2.45	2.39	2.34	2.30	2.27
12	3.18	2.81	2.61	2.48	2.39	2.33	2.28	2.24	2.21
13	3.14	2.76	2.56	2.43	2.35	2.28	2.23	2.20	2.16
14	3.10	2.73	2.52	2.39	2.31	2.24	2.19	2.15	2.12
15	3.07	2.70	2.49	2.36	2.27	2.21	2.16	2.12	2.09
16	3.05	2.67	2.46	2.33	2.24	2.18	2.13	2.09	2.06
17	3.03	2.64	2.44	2.31	2.22	2.15	2.10	2.06	2.03
18	3.01	2.62	2.42	2.29	2.20	2.13	2.08	2.04	2.00
19	2.99	2.61	2.40	2.27	2.18	2.11	2.06	2.02	1.98
20	2.97	2.59	2.38	2.25	2.16	2.09	2.04	2.00	1.96
21	2.96	2.57	2.36	2.23	2.14	2.08	2.02	1.98	1.95
22	2.95	2.56	2.35	2.22	2.13	2.06	2.01	1.97	1.93
23	2.94	2.55	2.34	2.21	2.11	2.05	1.99	1.95	1.92
24	2.93	2.54	2.33	2.19	2.10	2.04	1.98	1.94	1.91
25	2.92	2.53	2.32	2.18	2.09	2.02	1.97	1.93	1.89
26	2.91	2.52	2.31	2.17	2.08	2.01	1.96	1.92	1.88
27	2.90	2.51	2.30	2.17	2.07	2.00	1.95	1.91	1.87
28	2.89	2.50	2.29	2.16	2.06	2.00	1.94	1.90	1.87
29	2.89	2.50	2.28	2.15	2.06	1.99	1.93	1.89	1.86
30	2.88	2.49	2.28	2.14	2.05	1.98	1.93	1.88	1.85
40	2.84	2.44	2.23	2.09	2.00	1.93	1.87	1.83	1.79
60	2.79	2.39	2.18	2.04	1.95	1.87	1.82	1.77	1.74
120	2.75	2.35	2.13	1.99	1.90	1.82	1.77	1.72	1.68
∞	2.71	2.30	2.08	1.94	1.85	1.77	1.72	1.67	1.63

Denominator Degrees of Freedom

Source: From M. Merrington and C. M. Thompson, "Tables of Percentage Points of the Inverted Beta (*F*)-Distribution," *Biometrika*, 1943, 33, 73–88.

TABLE VII *(continued)*

ν_2 \ ν_1	NUMERATOR DEGREES OF FREEDOM									
	10	12	15	20	24	30	40	60	120	∞
1	60.19	60.71	61.22	61.74	62.00	62.26	62.53	62.79	63.06	63.33
2	9.39	9.41	9.42	9.44	9.45	9.46	9.47	9.47	9.48	9.49
3	5.23	5.22	5.20	5.18	5.18	5.17	5.16	5.15	5.14	5.13
4	3.92	3.90	3.87	3.84	3.83	3.82	3.80	3.79	3.78	3.76
5	3.30	3.27	3.24	3.21	3.19	3.17	3.16	3.14	3.12	3.10
6	2.94	2.90	2.87	2.84	2.82	2.80	2.78	2.76	2.74	2.72
7	2.70	2.67	2.63	2.59	2.58	2.56	2.54	2.51	2.49	2.47
8	2.54	2.50	2.46	2.42	2.40	2.38	2.36	2.34	2.32	2.29
9	2.42	2.38	2.34	2.30	2.28	2.25	2.23	2.21	2.18	2.16
10	2.32	2.28	2.24	2.20	2.18	2.16	2.13	2.11	2.08	2.06
11	2.25	2.21	2.17	2.12	2.10	2.08	2.05	2.03	2.00	1.97
12	2.19	2.15	2.10	2.06	2.04	2.01	1.99	1.96	1.93	1.90
13	2.14	2.10	2.05	2.01	1.98	1.96	1.93	1.90	1.88	1.85
14	2.10	2.05	2.01	1.96	1.94	1.91	1.89	1.86	1.83	1.80
15	2.06	2.02	1.97	1.92	1.90	1.87	1.85	1.82	1.79	1.76
16	2.03	1.99	1.94	1.89	1.87	1.84	1.81	1.78	1.75	1.72
17	2.00	1.96	1.91	1.86	1.84	1.81	1.78	1.75	1.72	1.69
18	1.98	1.93	1.89	1.84	1.81	1.78	1.75	1.72	1.69	1.66
19	1.96	1.91	1.86	1.81	1.79	1.76	1.73	1.70	1.67	1.63
20	1.94	1.89	1.84	1.79	1.77	1.74	1.71	1.68	1.64	1.61
21	1.92	1.87	1.83	1.78	1.75	1.72	1.69	1.66	1.62	1.59
22	1.90	1.86	1.81	1.76	1.73	1.70	1.67	1.64	1.60	1.57
23	1.89	1.84	1.80	1.74	1.72	1.69	1.66	1.62	1.59	1.55
24	1.88	1.83	1.78	1.73	1.70	1.67	1.64	1.61	1.57	1.53
25	1.87	1.82	1.77	1.72	1.69	1.66	1.63	1.59	1.56	1.52
26	1.86	1.81	1.76	1.71	1.68	1.65	1.61	1.58	1.54	1.50
27	1.85	1.80	1.75	1.70	1.67	1.64	1.60	1.57	1.53	1.49
28	1.84	1.79	1.74	1.69	1.66	1.63	1.59	1.56	1.52	1.48
29	1.83	1.78	1.73	1.68	1.65	1.62	1.58	1.55	1.51	1.47
30	1.82	1.77	1.72	1.67	1.64	1.61	1.57	1.54	1.50	1.46
40	1.76	1.71	1.66	1.61	1.57	1.54	1.51	1.47	1.42	1.38
60	1.71	1.66	1.60	1.54	1.51	1.48	1.44	1.40	1.35	1.29
120	1.65	1.60	1.55	1.48	1.45	1.41	1.37	1.32	1.26	1.19
∞	1.60	1.55	1.49	1.42	1.38	1.34	1.30	1.24	1.17	1.00

Denominator Degrees of Freedom

TABLE VIII   Percentage Points of the *F*-Distribution, $\alpha = .05$

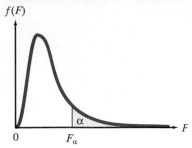

ν_2 \ ν_1	NUMERATOR DEGREES OF FREEDOM								
	1	**2**	**3**	**4**	**5**	**6**	**7**	**8**	**9**
1	161.4	199.5	215.7	224.6	230.2	234.0	236.8	238.9	240.5
2	18.51	19.00	19.16	19.25	19.30	19.33	19.35	19.37	19.38
3	10.13	9.55	9.28	9.12	9.01	8.94	8.89	8.85	8.81
4	7.71	6.94	6.59	6.39	6.26	6.16	6.09	6.04	6.00
5	6.61	5.79	5.41	5.19	5.05	4.95	4.88	4.82	4.77
6	5.99	5.14	4.76	4.53	4.39	4.28	4.21	4.15	4.10
7	5.59	4.74	4.35	4.12	3.97	3.87	3.79	3.73	3.68
8	5.32	4.46	4.07	3.84	3.69	3.58	3.50	3.44	3.39
9	5.12	4.26	3.86	3.63	3.48	3.37	3.29	3.23	3.18
10	4.96	4.10	3.71	3.48	3.33	3.22	3.14	3.07	3.02
11	4.84	3.98	3.59	3.36	3.20	3.09	3.01	2.95	2.90
12	4.75	3.89	3.49	3.26	3.11	3.00	2.91	2.85	2.80
13	4.67	3.81	3.41	3.18	3.03	2.92	2.83	2.77	2.71
14	4.60	3.74	3.34	3.11	2.96	2.85	2.76	2.70	2.65
15	4.54	3.68	3.29	3.06	2.90	2.79	2.71	2.64	2.59
16	4.49	3.63	3.24	3.01	2.85	2.74	2.66	2.59	2.54
17	4.45	3.59	3.20	2.96	2.81	2.70	2.61	2.55	2.49
18	4.41	3.55	3.16	2.93	2.77	2.66	2.58	2.51	2.46
19	4.38	3.52	3.13	2.90	2.74	2.63	2.54	2.48	2.42
20	4.35	3.49	3.10	2.87	2.71	2.60	2.51	2.45	2.39
21	4.32	3.47	3.07	2.84	2.68	2.57	2.49	2.42	2.37
22	4.30	3.44	3.05	2.82	2.66	2.55	2.46	2.40	2.34
23	4.28	3.42	3.03	2.80	2.64	2.53	2.44	2.37	2.32
24	4.26	3.40	3.01	2.78	2.62	2.51	2.42	2.36	2.30
25	4.24	3.39	2.99	2.76	2.60	2.49	2.40	2.34	2.28
26	4.23	3.37	2.98	2.74	2.59	2.47	2.39	2.32	2.77
27	4.21	3.35	2.96	2.73	2.57	2.46	2.37	2.31	2.25
28	4.20	3.34	2.95	2.71	2.56	2.45	2.36	2.29	2.24
29	4.18	3.33	2.93	2.70	2.55	2.43	2.35	2.28	2.22
30	4.17	3.32	2.92	2.69	2.53	2.42	2.33	2.27	2.21
40	4.08	3.23	2.84	2.61	2.45	2.34	2.25	2.18	2.12
60	4.00	3.15	2.76	2.53	2.37	2.25	2.17	2.10	2.04
120	3.92	3.07	2.68	2.45	2.29	2.17	2.09	2.02	1.96
∞	3.84	3.00	2.60	2.37	2.21	2.10	2.01	1.94	1.88

Denominator Degrees of Freedom

Source: From M. Merrington and C. M. Thompson, "Tables of Percentage Points of the Inverted Beta (*F*)-Distribution." *Biometrika*, 1943, 33, 73–88.

TABLE VIII *(continued)*

ν_2 \ ν_1	NUMERATOR DEGREES OF FREEDOM									
	10	**12**	**15**	**20**	**24**	**30**	**40**	**60**	**120**	**∞**
1	241.9	243.9	245.9	248.0	249.1	250.1	251.1	252.2	253.3	254.3
2	19.40	19.41	19.43	19.45	19.45	19.46	19.47	19.48	19.49	19.50
3	8.79	8.74	8.70	8.66	8.64	8.62	8.59	8.57	8.55	8.53
4	5.96	5.91	5.86	5.80	5.77	5.75	5.72	5.69	5.66	5.63
5	4.74	4.68	4.62	4.56	4.53	4.50	4.46	4.43	4.40	4.36
6	4.06	4.00	3.94	3.87	3.84	3.81	3.77	3.74	3.70	3.67
7	3.64	3.57	3.51	3.44	3.41	3.38	3.34	3.30	3.27	3.23
8	3.35	3.28	3.22	3.15	3.12	3.08	3.04	3.01	2.97	2.93
9	3.14	3.07	3.01	2.94	2.90	2.86	2.83	2.79	2.75	2.71
10	2.98	2.91	2.85	2.77	2.74	2.70	2.66	2.62	2.58	2.54
11	2.85	2.79	2.72	2.65	2.61	2.57	2.53	2.49	2.45	2.40
12	2.75	2.69	2.62	2.54	2.51	2.47	2.43	2.38	2.34	2.30
13	2.67	2.60	2.53	2.46	2.42	2.38	2.34	2.30	2.25	2.21
14	2.60	2.53	2.46	2.39	2.35	2.31	2.27	2.22	2.18	2.13
15	2.54	2.48	2.40	2.33	2.29	2.25	2.20	2.16	2.11	2.07
16	2.49	2.42	2.35	2.28	2.24	2.19	2.15	2.11	2.06	2.01
17	2.45	2.38	2.31	2.23	2.19	2.15	2.10	2.06	2.01	1.96
18	2.41	2.34	2.27	2.19	2.15	2.11	2.06	2.02	1.97	1.92
19	2.38	2.31	2.23	2.16	2.11	2.07	2.03	1.98	1.93	1.88
20	2.35	2.28	2.20	2.12	2.08	2.04	1.99	1.95	1.90	1.84
21	2.32	2.25	2.18	2.10	2.05	2.01	1.96	1.92	1.87	1.81
22	2.30	2.23	2.15	2.07	2.03	1.98	1.94	1.89	1.84	1.78
23	2.27	2.20	2.13	2.05	2.01	1.96	1.91	1.86	1.81	1.76
24	2.25	2.18	2.11	2.03	1.98	1.94	1.89	1.84	1.79	1.73
25	2.24	2.16	2.09	2.01	1.96	1.92	1.87	1.82	1.77	1.71
26	2.22	2.15	2.07	1.99	1.95	1.90	1.85	1.80	1.75	1.69
27	2.20	2.13	2.06	1.97	1.93	1.88	1.84	1.79	1.73	1.67
28	2.19	2.12	2.04	1.96	1.91	1.87	1.82	1.77	1.71	1.65
29	2.18	2.10	2.03	1.94	1.90	1.85	1.81	1.75	1.70	1.64
30	2.16	2.09	2.01	1.93	1.89	1.84	1.79	1.74	1.68	1.62
40	2.08	2.00	1.92	1.84	1.79	1.74	1.69	1.64	1.58	1.51
60	1.99	1.92	1.84	1.75	1.70	1.65	1.59	1.53	1.47	1.39
120	1.91	1.83	1.75	1.66	1.61	1.55	1.50	1.43	1.35	1.25
∞	1.83	1.75	1.67	1.57	1.52	1.46	1.39	1.32	1.22	1.00

Denominator Degrees of Freedom

TABLE IX   Percentage Points of the *F*-Distribution, α = .025

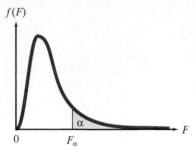

ν_2	ν_1 NUMERATOR DEGREES OF FREEDOM								
	1	2	3	4	5	6	7	8	9
1	647.8	799.5	864.2	899.6	921.8	937.1	948.2	956.7	963.3
2	38.51	39.00	39.17	39.25	39.30	39.33	39.36	39.37	39.39
3	17.44	16.04	15.44	15.10	14.88	14.73	14.62	14.54	14.47
4	12.22	10.65	9.98	9.60	9.36	9.20	9.07	8.98	8.90
5	10.01	8.43	7.76	7.39	7.15	6.98	6.85	6.76	6.68
6	8.81	7.26	6.60	6.23	5.99	5.82	5.70	5.60	5.52
7	8.07	6.54	5.89	5.52	5.29	5.12	4.99	4.90	4.82
8	7.57	6.06	5.42	5.05	4.82	4.65	4.53	4.43	4.36
9	7.21	5.71	5.08	4.72	4.48	4.32	4.20	4.10	4.03
10	6.94	5.46	4.83	4.47	4.24	4.07	3.95	3.85	3.78
11	6.72	5.26	4.63	4.28	4.04	3.88	3.76	3.66	3.59
12	6.55	5.10	4.47	4.12	3.89	3.73	3.61	3.51	3.44
13	6.41	4.97	4.35	4.00	3.77	3.60	3.48	3.39	3.31
14	6.30	4.86	4.24	3.89	3.66	3.50	3.38	3.29	3.21
15	6.20	4.77	4.15	3.80	3.58	3.41	3.29	3.20	3.12
16	6.12	4.69	4.08	3.73	3.50	3.34	3.22	3.12	3.05
17	6.04	4.62	4.01	3.66	3.44	3.28	3.16	3.06	2.98
18	5.98	4.56	3.95	3.61	3.38	3.22	3.10	3.01	2.93
19	5.92	4.51	3.90	3.56	3.33	3.17	3.05	2.96	2.88
20	5.87	4.46	3.86	3.51	3.29	3.13	3.01	2.91	2.84
21	5.83	4.42	3.82	3.48	3.25	3.09	2.97	2.87	2.80
22	5.79	4.38	3.78	3.44	3.22	3.05	2.93	2.84	2.76
23	5.75	4.35	3.75	3.41	3.18	3.02	2.90	2.81	2.73
24	5.72	4.32	3.72	3.38	3.15	2.99	2.87	2.78	2.70
25	5.69	4.29	3.69	3.35	3.13	2.97	2.85	2.75	2.68
26	5.66	4.27	3.67	3.33	3.10	2.94	2.82	2.73	2.65
27	5.63	4.24	3.65	3.31	3.08	2.92	2.80	2.71	2.63
28	5.61	4.22	3.63	3.29	3.06	2.90	2.78	2.69	2.61
29	5.59	4.20	3.61	3.27	3.04	2.88	2.76	2.67	2.59
30	5.57	4.18	3.59	3.25	3.03	2.87	2.75	2.65	2.57
40	5.42	4.05	3.46	3.13	2.90	2.74	2.62	2.53	2.45
60	5.29	3.93	3.34	3.01	2.79	2.63	2.51	2.41	2.33
120	5.15	3.80	3.23	2.89	2.67	2.52	2.39	2.30	2.22
∞	5.02	3.69	3.12	2.79	2.57	2.41	2.29	2.19	2.11

Denominator Degrees of Freedom

Source: From M. Merrington and C. M. Thompson, "Tables of Percentage Points of the Inverted Beta (*F*)-Distribution," *Biometrika*, 1943, 33, 73–88.

TABLE IX *(continued)*

ν_2 \ ν_1	NUMERATOR DEGREES OF FREEDOM									
	10	**12**	**15**	**20**	**24**	**30**	**40**	**60**	**120**	**∞**
1	968.6	976.7	984.9	993.1	997.2	1,001	1,006	1,010	1,014	1,018
2	39.40	39.41	39.43	39.45	39.46	39.46	39.47	39.48	39.49	39.50
3	14.42	14.34	14.25	14.17	14.12	14.08	14.04	13.99	13.95	13.90
4	8.84	8.75	8.66	8.56	8.51	8.46	8.41	8.36	8.31	8.26
5	6.62	6.52	6.43	6.33	6.28	6.23	6.18	6.12	6.07	6.02
6	5.46	5.37	5.27	5.17	5.12	5.07	5.01	4.96	4.90	4.85
7	4.76	4.67	4.57	4.47	4.42	4.36	4.31	4.25	4.20	4.14
8	4.30	4.20	4.10	4.00	3.95	3.89	3.84	3.78	3.73	3.67
9	3.96	3.87	3.77	3.67	3.61	3.56	3.51	3.45	3.39	3.33
10	3.72	3.62	3.52	3.42	3.37	3.31	3.26	3.20	3.14	3.08
11	3.53	3.43	3.33	3.23	3.17	3.12	3.06	3.00	2.94	2.88
12	3.37	3.28	3.18	3.07	3.02	2.96	2.91	2.85	2.79	2.72
13	3.25	3.15	3.05	2.95	2.89	2.84	2.78	2.72	2.66	2.60
14	3.15	3.05	2.95	2.84	2.79	2.73	2.67	2.61	2.55	2.49
15	3.06	2.96	2.86	2.76	2.70	2.64	2.59	2.52	2.46	2.40
16	2.99	2.89	2.79	2.68	2.63	2.57	2.51	2.45	2.38	2.32
17	2.92	2.82	2.72	2.62	2.56	2.50	2.44	2.38	2.32	2.25
18	2.87	2.77	2.67	2.56	2.50	2.44	2.38	2.32	2.26	2.19
19	2.82	2.72	2.62	2.51	2.45	2.39	2.33	2.27	2.20	2.13
20	2.77	2.68	2.57	2.46	2.41	2.35	2.29	2.22	2.16	2.09
21	2.73	2.64	2.53	2.42	2.37	2.31	2.25	2.18	2.11	2.04
22	2.70	2.60	2.50	2.39	2.33	2.27	2.21	2.14	2.08	2.00
23	2.67	2.57	2.47	2.36	2.30	2.24	2.18	2.11	2.04	1.97
24	2.64	2.54	2.44	2.33	2.27	2.21	2.15	2.08	2.01	1.94
25	2.61	2.51	2.41	2.30	2.24	2.18	2.12	2.05	1.98	1.91
26	2.59	2.49	2.39	2.28	2.22	2.16	2.09	2.03	1.95	1.88
27	2.57	2.47	2.36	2.25	2.19	2.13	2.07	2.00	1.93	1.85
28	2.55	2.45	2.34	2.23	2.17	2.11	2.05	1.98	1.91	1.83
29	2.53	2.43	2.32	2.21	2.15	2.09	2.03	1.96	1.89	1.81
30	2.51	2.41	2.31	2.20	2.14	2.07	2.01	1.94	1.87	1.79
40	2.39	2.29	2.18	2.07	2.01	1.94	1.88	1.80	1.72	1.64
60	2.27	2.17	2.06	1.94	1.88	1.82	1.74	1.67	1.58	1.48
120	2.16	2.05	1.94	1.82	1.76	1.69	1.61	1.53	1.43	1.31
∞	2.05	1.94	1.83	1.71	1.64	1.57	1.48	1.39	1.27	1.00

Denominator Degrees of Freedom

TABLE X   Percentage Points of the *F*-Distribution, $\alpha = .01$

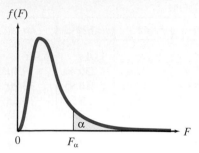

ν_2 \ ν_1	NUMERATOR DEGREES OF FREEDOM									
	1	**2**	**3**	**4**	**5**	**6**	**7**	**8**	**9**
1	4,052	4,999.5	5,403	5,625	5,764	5,859	5,928	5,982	6,022
2	98.50	99.00	99.17	99.25	99.30	99.33	99.36	99.37	99.39
3	34.12	30.82	29.46	28.71	28.24	27.91	27.67	27.49	27.35
4	21.20	18.00	16.69	15.98	15.52	15.21	14.98	14.80	14.66
5	16.26	13.27	12.06	11.39	10.97	10.67	10.46	10.29	10.16
6	13.75	10.92	9.78	9.15	8.75	8.47	8.26	8.10	7.98
7	12.25	9.55	8.45	7.85	7.46	7.19	6.99	6.84	6.72
8	11.26	8.65	7.59	7.01	6.63	6.37	6.18	6.03	5.91
9	10.56	8.02	6.99	6.42	6.06	5.80	5.61	5.47	5.35
10	10.04	7.56	6.55	5.99	5.64	5.39	5.20	5.06	4.94
11	9.65	7.21	6.22	5.67	5.32	5.07	4.89	4.74	4.63
12	9.33	6.93	5.95	5.41	5.06	4.82	4.64	4.50	4.39
13	9.07	6.70	5.74	5.21	4.86	4.62	4.44	4.30	4.19
14	8.86	6.51	5.56	5.04	4.69	4.46	4.28	4.14	4.03
15	8.68	6.36	5.42	4.89	4.56	4.32	4.14	4.00	3.89
16	8.53	6.23	5.29	4.77	4.44	4.20	4.03	3.89	3.78
17	8.40	6.11	5.18	4.67	4.34	4.10	3.93	3.79	3.68
18	8.29	6.01	5.09	4.58	4.25	4.01	3.84	3.71	3.60
19	8.18	5.93	5.01	4.50	4.17	3.94	3.77	3.63	3.52
20	8.10	5.85	4.94	4.43	4.10	3.87	3.70	3.56	3.46
21	8.02	5.78	4.87	4.37	4.04	3.81	3.64	3.51	3.40
22	7.95	5.72	4.82	4.31	3.99	3.76	3.59	3.45	3.35
23	7.88	5.66	4.76	4.26	3.94	3.71	3.54	3.41	3.30
24	7.82	5.61	4.72	4.22	3.90	3.67	3.50	3.36	3.26
25	7.77	5.57	4.68	4.18	3.85	3.63	3.46	3.32	3.22
26	7.72	5.53	4.64	4.14	3.82	3.59	3.42	3.29	3.18
27	7.68	5.49	4.60	4.11	3.78	3.56	3.39	3.26	3.15
28	7.64	5.45	4.57	4.07	3.75	3.53	3.36	3.23	3.12
29	7.60	5.42	4.54	4.04	3.73	3.50	3.33	3.20	3.09
30	7.56	5.39	4.51	4.02	3.70	3.47	3.30	3.17	3.07
40	7.31	5.18	4.31	3.83	3.51	3.29	3.12	2.99	2.89
60	7.08	4.98	4.13	3.65	3.34	3.12	2.95	2.82	2.72
120	6.85	4.79	3.95	3.48	3.17	2.96	2.79	2.66	2.56
∞	6.63	4.61	3.78	3.32	3.02	2.80	2.64	2.51	2.41

Denominator Degrees of Freedom

Source: From M. Merrington and C. M. Thompson, "Tables of Percentage Points of the Inverted Beta (*F*)-Distribution," *Biometrika*, 1943, 33, 73–88.

TABLE X   *(continued)*

ν_2 \ ν_1	10	12	15	20	24	30	40	60	120	∞
1	6,056	6,106	6,157	6,209	6,235	6,261	6,287	6,313	6,339	6,366
2	99.40	99.42	99.43	99.45	99.46	99.47	99.47	99.48	99.49	99.50
3	27.23	27.05	26.87	26.69	26.60	26.50	26.41	26.32	26.22	26.13
4	14.55	14.37	14.20	14.02	13.93	13.84	13.75	13.65	13.56	13.46
5	10.05	9.89	9.72	9.55	9.47	9.38	9.29	9.20	9.11	9.02
6	7.87	7.72	7.56	7.40	7.31	7.23	7.14	7.06	6.97	6.88
7	6.62	6.47	6.31	6.16	6.07	5.99	5.91	5.82	5.74	5.65
8	5.81	5.67	5.52	5.36	5.28	5.20	5.12	5.03	4.95	4.86
9	5.26	5.11	4.96	4.81	4.73	4.65	4.57	4.48	4.40	4.31
10	4.85	4.71	4.56	4.41	4.33	4.25	4.17	4.08	4.00	3.91
11	4.54	4.40	4.25	4.10	4.02	3.94	3.86	3.78	3.69	3.60
12	4.30	4.16	4.01	3.86	3.78	3.70	3.62	3.54	3.45	3.36
13	4.10	3.96	3.82	3.66	3.59	3.51	3.43	3.34	3.25	3.17
14	3.94	3.80	3.66	3.51	3.43	3.35	3.27	3.18	3.09	3.00
15	3.80	3.67	3.52	3.37	3.29	3.21	3.13	3.05	2.96	2.87
16	3.69	3.55	3.41	3.26	3.18	3.10	3.02	2.93	2.84	2.75
17	3.59	3.46	3.31	3.16	3.08	3.00	2.92	2.83	2.75	2.65
18	3.51	3.37	3.23	3.08	3.00	2.92	2.84	2.75	2.66	2.57
19	3.43	3.30	3.15	3.00	2.92	2.84	2.76	2.67	2.58	2.49
20	3.37	3.23	3.09	2.94	2.86	2.78	2.69	2.61	2.52	2.42
21	3.31	3.17	3.03	2.88	2.80	2.72	2.64	2.55	2.46	2.36
22	3.26	3.12	2.98	2.83	2.75	2.67	2.58	2.50	2.40	2.31
23	3.21	3.07	2.93	2.78	2.70	2.62	2.54	2.45	2.35	2.26
24	3.17	3.03	2.89	2.74	2.66	2.58	2.49	2.40	2.31	2.21
25	3.13	2.99	2.85	2.70	2.62	2.54	2.45	2.36	2.27	2.17
26	3.09	2.96	2.81	2.66	2.58	2.50	2.42	2.33	2.23	2.13
27	3.06	2.93	2.78	2.63	2.55	2.47	2.38	2.29	2.20	2.10
28	3.03	2.90	2.75	2.60	2.52	2.44	2.35	2.26	2.17	2.06
29	3.00	2.87	2.73	2.57	2.49	2.41	2.33	2.23	2.14	2.03
30	2.98	2.84	2.70	2.55	2.47	2.39	2.30	2.21	2.11	2.01
40	2.80	2.66	2.52	2.37	2.29	2.20	2.11	2.02	1.92	1.80
60	2.63	2.50	2.35	2.20	2.12	2.03	1.94	1.84	1.73	1.60
120	2.47	2.34	2.19	2.03	1.95	1.86	1.76	1.66	1.53	1.38
∞	2.32	2.18	2.04	1.88	1.79	1.70	1.59	1.47	1.32	1.00

TABLE XI   Critical Values of χ^2

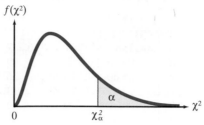

Degrees of Freedom	$\chi^2_{.995}$	$\chi^2_{.990}$	$\chi^2_{.975}$	$\chi^2_{.950}$	$\chi^2_{.900}$
1	.0000393	.0001571	.0009821	.0039321	.0157908
2	.0100251	.0201007	.0506356	.102587	.210720
3	.0717212	.114832	.215795	.351846	.584375
4	.206990	.297110	.484419	.710721	1.063623
5	.411740	.554300	.831211	1.145476	1.61031
6	.675727	.872085	1.237347	1.63539	2.20413
7	.989265	1.239043	1.68987	2.16735	2.83311
8	1.344419	1.646482	2.17973	2.73264	3.48954
9	1.734926	2.087912	2.70039	3.32511	4.16816
10	2.15585	2.55821	3.24697	3.94030	4.86518
11	2.60321	3.05347	3.81575	4.57481	5.57779
12	3.07382	3.57056	4.40379	5.22603	6.30380
13	3.56503	4.10691	5.00874	5.89186	7.04150
14	4.07468	4.66043	5.62872	6.57063	7.78953
15	4.60094	5.22935	6.26214	7.26094	8.54675
16	5.14224	5.81221	6.90766	7.96164	9.31223
17	5.69724	6.40776	7.56418	8.67176	10.0852
18	6.26481	7.01491	8.23075	9.39046	10.8649
19	6.84398	7.63273	8.90655	10.1170	11.6509
20	7.43386	8.26040	9.59083	10.8508	12.4426
21	8.03366	8.89720	10.28293	11.5913	13.2396
22	8.64272	9.54249	10.9823	12.3380	14.0415
23	9.26042	10.19567	11.6885	13.0905	14.8479
24	9.88623	10.8564	12.4011	13.8484	15.6587
25	10.5197	11.5240	13.1197	14.6114	16.4734
26	11.1603	12.1981	13.8439	15.3791	17.2919
27	11.8076	12.8786	14.5733	16.1513	18.1138
28	12.4613	13.5648	15.3079	16.9279	18.9392
29	13.1211	14.2565	16.0471	17.7083	19.7677
30	13.7867	14.9535	16.7908	18.4926	20.5992
40	20.7065	22.1643	24.4331	26.5093	29.0505
50	27.9907	29.7067	32.3574	34.7642	37.6886
60	35.5346	37.4848	40.4817	43.1879	46.4589
70	43.2752	45.4418	48.7576	51.7393	55.3290
80	51.1720	53.5400	57.1532	60.3915	64.2778
90	59.1963	61.7541	65.6466	69.1260	73.2912
100	67.3276	70.0648	74.2219	77.9295	82.3581

Source: From C. M. Thompson, "Tables of the Percentage Points of the χ^2-Distribution," *Biometrika*, 1941, 32, 188–189. Reproduced by permission of the *Biometrika* Trustees.

TABLE XI *(continued)*

Degrees of Freedom	$\chi^2_{.100}$	$\chi^2_{.050}$	$\chi^2_{.025}$	$\chi^2_{.010}$	$\chi^2_{.005}$
1	2.70554	3.84146	5.02389	6.63490	7.87944
2	4.60517	5.99147	7.37776	9.21034	10.5966
3	6.25139	7.81473	9.34840	11.3449	12.8381
4	7.77944	9.48773	11.1433	13.2767	14.8602
5	9.23635	11.0705	12.8325	15.0863	16.7496
6	10.6446	12.5916	14.4494	16.8119	18.5476
7	12.0170	14.0671	16.0128	18.4753	20.2777
8	13.3616	15.5073	17.5346	20.0902	21.9550
9	14.6837	16.9190	19.0228	21.6660	23.5893
10	15.9871	18.3070	20.4831	23.2093	25.1882
11	17.2750	19.6751	21.9200	24.7250	26.7569
12	18.5494	21.0261	23.3367	26.2170	28.2995
13	19.8119	22.3621	24.7356	27.6883	29.8194
14	21.0642	23.6848	26.1190	29.1413	31.3193
15	22.3072	24.9958	27.4884	30.5779	32.8013
16	23.5418	26.2962	28.8454	31.9999	34.2672
17	24.7690	27.5871	30.1910	33.4087	35.7185
18	25.9894	28.8693	31.5264	34.8053	37.1564
19	27.2036	30.1435	32.8523	36.1908	38.5822
20	28.4120	31.4104	34.1696	37.5662	39.9968
21	29.6151	32.6705	35.4789	38.9321	41.4010
22	30.8133	33.9244	36.7807	40.2894	42.7956
23	32.0069	35.1725	38.0757	41.6384	44.1813
24	33.1963	36.4151	39.3641	42.9798	45.5585
25	34.3816	37.6525	40.6465	44.3141	46.9278
26	35.5631	38.8852	41.9232	45.6417	48.2899
27	36.7412	40.1133	43.1944	46.9630	49.6449
28	37.9159	41.3372	44.4607	48.2782	50.9933
29	39.0875	42.5569	45.7222	49.5879	52.3356
30	40.2560	43.7729	46.9792	50.8922	53.6720
40	51.8050	55.7585	59.3417	63.6907	66.7659
50	63.1671	67.5048	71.4202	76.1539	79.4900
60	74.3970	79.0819	83.2976	88.3794	91.9517
70	85.5271	90.5312	95.0231	100.425	104.215
80	96.5782	101.879	106.629	112.329	116.321
90	107.565	113.145	118.136	124.116	128.299
100	118.498	124.342	129.561	135.807	140.169

TABLE XII   Critical Values of Spearman's Rank Correlation Coefficient

The α values correspond to a one-tailed test of H_0: $\rho = 0$. The value should be doubled for two-tailed tests.

n	α = .05	α = .025	α = .01	α = .005	n	α = .05	α = .025	α = .01	α = .005
5	.900		—	—	18	.399	.476	.564	.625
6	.829	.886	.943	—	19	.388	.462	.549	.608
7	.714	.786	.893	—	20	.377	.450	.534	.591
8	.643	.738	.833	.881	21	.368	.438	.521	.576
9	.600	.683	.783	.833	22	.359	.428	.508	.562
10	.564	.648	.745	.794	23	.351	.418	.496	.549
11	.523	.623	.736	.818	24	.343	.409	.485	.537
12	.497	.591	.703	.780	25	.336	.400	.475	.526
13	.475	.566	.673	.745	26	.329	.392	.465	.515
14	.457	.545	.646	.716	27	.323	.385	.456	.505
15	.441	.525	.623	.689	28	.317	.377	.448	.496
16	.425	.507	.601	.666	29	.311	.370	.440	.487
17	.412	.490	.582	.645	30	.305	.364	.432	.478

Source: From E. G. Olds, "Distribution of Sums of Squares of Rank Differences for Small Samples," *Annals of Mathematical Statistics*, 1938, 9.

Data Sets on CD-ROM That Accompanies Text

File Name	Location	Number of Variables	Description of Variables
• APHASIA	Table 2.1	1	Type of aphasia
• BLOODLOSS	Example 2.1	3	Patient, drug type & complication type
• CEODEGREES	Ex. 2.9	1	Highest degree obtained
• HEARLOSS	Ex. 2.10	1	Type of hearing loss (coded)
• DOLPHIN	Ex. 2.13	2	Whistle category & number
• DIGITS	Ex. 2.14	2	1st digit & number of occurrences
• OILSPILL	Ex. 2.15	2	Spillage amount & cause of spill
• EPAGAS	Table 2.3	1	Gas mileage rating
• LOANDEFAULT	Table 2.5	1	Percent of loans in default
• SPIDERS	Ex. 2.24	1	Fluid loss
• BRAINPMI	Ex. 2.25	1	Postmortem interval
• BULIMIA	Ex. 2.27	2	Group & FNE score
• NEUROSPECT	Ex. 2.28	2	Density ratio & patient type
• SATURN	Ex. 2.29	2	Satellite event & percent of light lost
• SAT	Ex. 2.30	3	State, 1990 SAT & 2000 SAT
• LISTEN	Ex. 2.31	3	Group, response rate & percent correct
• STLRUNS	Ex. 2.32	1	Runs scored by St. Louis
• CUBSRUNS	Ex. 2.32	1	Runs scored by Chicago
• WOMENPOWER	Ex. 2.44	4	Name, age, company & title
• AMMONIA	Ex. 2.45	1	Ammonia concentration
• CONDOR	Ex. 2.46	3	Plot, number killed & breeders
• PAF	Ex. 2.48	1	PAF inhibition percentage
• SLI	Ex. 2.51	5	Subject, gender, group, DIQ & percent pronoun errors
• LEFTEYE	Ex. 2.52	1	Cylinder power measurement
• RATMAZE	Table 2.9	1	Maze run time
• SHIPSANIT	Ex. 2.72	2	Ship name & sanitation score
• REACTION	Table 2.10	2	Stimulus type & reaction time
• LM2_98	Ex. 2.98	2	Sample & value
• MBASAL	Ex. 2.100	1	Top salary offer
• DOWNTIME	Ex. 2.103	2	Customer & downtime
• MEDFACTORS	Table 2.11	2	Number of factors & length of stay
• BLACKBREAM	Ex. 2.110	3	Week, number of strikes & age
• ALWINS	Ex. 2.111	3	Team, wins & batting average
• EYECUE	SIA Chap. 2	4	Gender, group, deviation & judgment
• WALKS	Ex. 2.113	3	Steps, unrooted walks & self-avoiding walks
• CRASH	Ex. 2.115	2	Driver star rating & head injury rating
• BEANIE	Ex. 2.127	5	Number, name, age retired, status & value
• TORNADO	Ex. 2.130	3	State, number of tornados & peak month
• TRAPS	Ex. 2.134	2	Species & number killed
• EVOS	Ex. 2.135	4	Transect, number of sea birds, length & oil status
• GALAXY2	Ex. 2.137	1	Galaxy velocity
• SUICIDE	Ex. 2.139	7	Victim, jail days, marital status, race, charge, suicide time & year
• LM4_72	Ex. 4.72	1	Quantitative value
• DARTS	Ex. 4.75	2	Distance from target & frequency
• ALBATAVG	Ex. 4.76	2	AL player & batting average

File Name	Location	Number of Variables	Description of Variables
• NLBATAVG	Ex. 4.76	2	NL player & batting average
• RUBIDIUM	Ex. 4.139	1	Rubidium amount
• INSOMNIA	Table 4.10	1	Time to fall asleep
• HOSPLOS	Table 5.1	1	Length of stay
• AIRNOSHOWS	Example 5.1	1	Number of unoccupied seats
• CHARITY	Ex. 5.14	1	Charitable amount
• CORR34	Ex. 5.17	1	Study correlation
• ATTIMES	Ex. 5.18	1	Twins attention time
• PRINTHEAD	Table 5.4	1	Number of characters printed
• LEADCOPP	Ex. 5.27	2	Lead level & copper level
• DECAY	Ex. 5.28	1	Daily decay rate
• PIGEONS	Ex. 5.29	1	Pigeon crop weight
• SOUND	Ex. 5.30	1	Sound pressure level
• TRAFFIC	Ex. 5.31	4	Station, route type, 30th hour volume & 100th hour volume
• SNACK	Ex. 5.40	1	Taste rating (coded)
• SCALLOPS	Table 5.6	1	Scallop weight measurement
• MATHCPU	Ex. 5.81	4	Solution time
• PCB	Ex. 6.26	1	Number of joints inspected
• BONES	Ex. 6.27	1	Length-to-width ratio
• EMISSIONS	Example 6.5	1	Engine emission level
• IRONTEMP	Ex. 6.54	1	Crankshaft pouring temperature
• SUBABUSE	Table 6.4	1	Substance abuse score
• LM6_78	Ex. 6.78	1	Value
• LYMPHO	Ex. 6.82	2	Case & lymphocyte cell count
• GUPPY	Ex. 6.84	1	Number of fish
• REMISSION	Ex. 6.96	1	Cancer remission time
• DIETSTUDY	Table 7.1	2	Diet & weight loss
• READING	Table 7.2	2	Method & reading score
• LM7_5	Ex. 7.5	2	Sample & value
• HWSTUDY	Ex. 7.15	4	Home work condition, level of involvement in science, math & language
• BILLBOARD	Ex. 7.17	3	Pre/Post SoundScan, week & number of new artists
• PAIREDSCORES	Table 7.3	3	Pair, new method score & standard method score
• PAIREDGRADS	Table 7.5	3	Pair, male salary & female salary
• LM7_27	Ex. 7.27	3	Pair, value 1 & value 2
• LM7_28	Ex. 7.28	3	Pair, value 1 & value 2
• LM7_29	Ex. 7.29	3	Pair, x-value & y-value
• DEPOPROV	Ex. 7.31	3	Patient, pre-treatment level & post-treatment level
• FMATTITUDES	Ex. 7.32	3	Student, father attitude & mother attitude
• MSSTUDY	Ex. 7.35	8	Pair, gender, MS age, MS height, MS weight, nonMS age, nonMS height & nonMS weight
• TWINSIQ	Ex. 7.36	3	Pair, twin A IQ & twin B IQ
• RATPUPS	Ex. 7.38	3	Litter, male swims & female swims
• HOMOPHONE	Ex. 7.39	3	Patient, time 1 confusion error & time 2 confusion error
• INOSITOL	Ex. 7.40	3	Patient, placebo panic attacks & inositol panic attacks
• REACTION	Table 7.6	2	Drug & reaction time
• LM7_51	Ex. 7.51	2	Sample & value
• LM7_52	Ex. 7.52	2	Sample & value
• LM7_53	Ex. 7.53	2	Sample & value
• EYEMOVE	Ex. 7.57	2	Deaf status & eye movement rate
• COLORAIN	Ex. 7.58	2	Station & precipitation
• INFOSYS	Ex. 7.59	2	Sector & IS expenditures
• COWBIRD	Ex. 7.60	2	Flycatcher group & vocalization rate
• LM7_63	Ex. 7.63	3	Pair, A value & B value
• SOFTPAPER	Table 7.8	3	Judge, A rating & B rating

File Name	Location	Number of Variables	Description of Variables
• CRIMEPLAN	Table 7.10	3	Expert, A rating & B rating
• FACEVT	Ex. 7.68	3	Group, face score & video score
• AGGTWINS	Ex. 7.69	3	Twin set, 1st born score & 2nd born score
• ATLAS	Ex. 7.70	3	Theme, teacher ranking & alumni ranking
• HYPOGLY	Ex. 7.71	3	Mouse, X blood sugar & Y blood sugar
• POWVEP	Ex. 7.72	3	POW, VEP 157 days & VEP 379 days
• GOLFCRD	Table 7.12	2	Golf ball brand & driving distance
• SORPRATE	Ex. 7.82	2	Organic solvent & sorption rate
• HAIRPAIN	Ex. 7.83	2	Hair color & pain threshold
• NAMEGAME	Ex. 7.84	2	Group & percent of names recalled
• FACES	Ex. 7.85	2	Dominance rating & facial expression rating
• ROACH	Table 7.15	3	Trail deviation, roach group & extract type
• LM7_90	Ex. 7.90	2	Treatment & value
• LM7_91	Ex. 7.91	3	Pair, value 1 & value 2
• LM7_94	Ex. 7.94	2	Sample & value
• LM7_95	Ex. 7.95	2	x-value & y-value
• BACTERIA	Ex. 7.101	2	Location & bacteria count
• PUPILL	Ex. 7.102	3	Consumer, reading 1 & reading 2
• BIOFEED	Ex. 7.104	3	Subject, before blood pressure & after blood pressure
• RAZORS	Ex. 7.109	2	Blade type & number of shaves
• MICE	Ex. 7.112	2	Group & maze test score
• AMI	Ex. 8.42	3	Heart status, alcohol consumption & number
• TRAUMA	Ex. 8.44	3	Seatbelt status, race & number
• ETHNIC	Ex. 8.45	3	Region, political strategy & number
• PIGFARM	Ex. 8.46	3	Size, education level & number
• MOTH	Ex. 8.47	3	Food choice, egg fed & number
• SIMKILLS	Ex. 8.48	3	Stop status, number of kills & number
• HYPNOSIS	Ex. 8.49	3	SHSS, CAHS & number
• GANGS	Ex. 8.50	3	Gang score, weapon & number
• HIVVAC1	Ex. 8.51	3	Group, MN strain & number
• HIVVAC2	Ex. 8.51	3	Group, MN strain & number
• HIVVAC3	Ex. 8.51	3	Group, MN strain & number
• TUMORS	Ex. 8.55	3	Cancer, diet & number
• LIFESTYLE	Ex. 8.68	3	Husband lifestyle, wife lifestyle & number
• SISEBERT	SIA Chap. 8	3	Movie, Siskel rating & Ebert rating
• MEDLYONS	SIA Chap. 8	3	Movie, Medved rating & Lyons rating
• HEARIMP	Ex. 8.57	2	Hearing loss & number of patients
• FATHSON	Ex. 8.59	3	Father's job, son's job & number
• QUAKE	Ex. 8.60	3	Insurance status, county & number
• SCANNER	Ex. 8.64	3	Brand, number of purchases & market share
• DISRUPT	Ex. 8.66	3	Delay status, behavior & number
• STIMULUS	Table 9.1	2	Drug amount & reaction time
• TAMPALMS	Ex. 9.14	2	Appraisal value & sale price
• OJUICE	Ex. 9.19	3	Run, sweetness index & pectin amount
• ORGCHEM	Ex. 9.20	3	Compound, maximum absorption & Hammett constant
• HEARAID	Ex. 9.21	3	Subject, familiar words & unfamiliar words
• CONDOR2	Ex. 9.22	3	Plot, number killed & tit occupancy
• NAMEGAME2	Ex. 9.23	2	Position & name recall percentage
• LONGJUMP	Ex. 9.24	3	Jumper, best distance & takeoff error
• HOSPITAL	Ex. 9.30	3	State, average change & average length of stay
• CUTTOOL	Ex. 9.33	3	Speed, brand A life & brand B life
• BOXING2	Ex. 9.42	2	Blood lactate level & perceived recovery
• GLOBRISK	Ex. 9.46	3	Country, risk & credit rating
• BRICKS	Ex. 9.47	3	Location, drop spall rate & photo spall rate

File Name	Location	Number of Variables	Description of Variables
• CASINO	Table 9.5	3	Year, number of employees & crime rate
• FERTRATE	Ex. 9.56	2	Contraceptive prevalence & fertility rate
• SNOWGEESE	Ex. 9.60	5	Trial, diet, weight change, digest efficiency & acid fibre
• JAPANSPORTS	Ex. 9.71	3	Activity, frequency & duration
• FIREDAM	Table 9.7	2	Distance from station & fire damage
• DOWSING	Table 9.12	4	Trial, dowser, pipe location & guess
• NEWBORN	Table 9.10	2	Number of cigarettes & baby weight
• LM9_76	Ex. 9.76	2	x-value & y-value
• WINETASTE	Ex. 9.79	3	Brand, expert 1 ranking & expert 2 ranking
• RECTIME	Ex. 9.83	2	Recreation time & number of tranquilizers
• COREP	Ex. 9.84	3	Company, 1999 rank & 2000 rank
• LM9_87	Ex. 9.87	2	x-value & y-value
• SOCIOL	Ex. 9.89	2	Verbal score & final grade
• HORSES	Ex. 9.90	2	Gestation period & life span
• PARANOIA	Ex. 9.91	3	Child, paranoia ranking & aggression ranking
• HELIUM	Ex. 9.95	2	Temperature & impurity proportion
• SEEDGERM	Ex. 9.97	2	Energy change & temperature
• FLOUR	Ex. 9.100	2	Shipment weight & number of bags

Calculation Formulas for Analysis of Variance
(Independent Sampling)

$$CM = \text{Correction for mean}$$

$$= \frac{(\text{Total of all observations})^2}{\text{Total number of observations}} = \frac{(\sum y_i)^2}{n}$$

$$SS(\text{Total}) = \text{Total sum of squares}$$

$$= (\text{Sum of squares of all observations}) - CM = \sum y_i^2 - CM$$

$$SST = \text{Sum of squares for treatments}$$

$$= \left(\begin{array}{c} \text{Sum of squares of treatment totals with} \\ \text{each square divided by the number of} \\ \text{observations for that treatment} \end{array} \right) - CM$$

$$= \frac{T_1^2}{n_1} + \frac{T_2^2}{n_2} + \cdots + \frac{T_p^2}{n_p} - CM$$

$$SSE = \text{Sum of squares for error} = SS(\text{Total}) - SST$$

$$MST = \text{Mean square for treatments} = \frac{SST}{p-1}$$

$$MSE = \text{Mean square for error} = \frac{SSE}{n-p}$$

$$F = \text{Test statistic} = \frac{MST}{MSE}$$

where

$$n = \text{Total number of observations}$$
$$p = \text{Number of treatments}$$
$$T_i = \text{Total for treatment } i \ (i = 1, 2, \ldots, p)$$

CHAPTER 1

1.11 qualitative; qualitative **1.13a.** quantitative **b.** qualitative **c.** qualitative **d.** quantitative **1.15b.** qualitative **e.** survey data **1.17a.** qualitative; qualitative; qualitative; quantitative **b.** sample **1.19a.** quantitative **b.** quantitative **c.** qualitative **d.** quantitative **e.** qualitative **f.** quantitative **g.** qualitative **1.21a.** designed experiment **b.** amateur boxers **c.** heart rate (quantitative); blood lactate level (quantitative) **d.** no difference between the two groups of boxers **e.** no **1.23a.** designed experiment **b.** inferential statistics **c.** all possible burn patients **1.25b.** number of headers per game and IQ **c.** both quantitative **1.27b.** answer to question posed; qualitative **d.** survey data

CHAPTER 2

2.1 relative frequencies: A, .08; B, .18; C, .45; D, .15; F, .14 **2.3a.** relative frequencies: Black, .191; White, .623; Sumatran, .029; Javan, .005; Indian, .151 **c.** .814; .185 **2.5a.** qualitative; infant, child, medical, infant & medical, child & medical, infant & child, infant & child & medical **b.** infant, .061; child, .565; medical, .276; infant & medical, .001; child & medical, .030; infant & child, .062; infant & child & medical, .004 **d.** .312 **2.7a.** pie chart **b.** breast cancer (19%) **c.** 19% **2.9** 68% of sampled CEOs had advanced degrees **2.11a.** "favorable/recommended"; .635 **b.** yes **2.13** Over half of the whistle types were "Type a."

2.15a.

Cause of Spillage	Frequency
Collision	1
Grounding	1
Fire/explosion	1
Hull failure	1
Unknown	
Total	5

b. It does not appear that one cause is more likely to occur than any other. **2.17** frequencies: 50, 75, 125, 100, 25, 50, 50, 25 **2.19a.** frequency histogram **b.** 14 **c.** 49 **2.23a.** 28.5% **b.** 71% **c.** slightly skewed to the right **d.** skewed to the left **2.25** most of the PMI's range from 3 to 7.5

2.27a.

Stem	Leaf
0	6 7 8
1	0 0 1 3 3 3 3 4 5 6 6 8 9 9 9
2	0 0 1 1 3 4 5
3	

b. yes

2.29a.

Stem	Leaf
0	1 1 2 3 4 5 5 9
1	1 2 3
2	0 0
3	9
4	6
5	6
6	1 5
7	
8	
9	
10	0

c. eclipses **2.31a.** response rate of the familiarity group is higher **b.** familiarity group had the highest accuracy rate; control group had the lowest accuracy rate **2.33a.** 12 **b.** 40 **c.** 7 **d.** 21 **e.** 144 **2.35a.** 11.2 **b.** 12 **c.** 30 **2.37** mode = 15; mean = 14.545; median = 15 **2.41a.** 8.5 **b.** 25 **c.** .78 **d.** 13.44 **2.43a.** mean < median **b.** mean > median **c.** mean = median **2.45a.** 1.47 **b.** 1.49 **2.49b.** probably none **2.51a.** qualitative, qualitative, quantitative, quantitative **c.** mean = 93.6; median = 91.5; mode = 86 **d.** mean = 95.3; median = 92; mode = 92 **e.** mean = 101.9; median = 103; mode = 113 **g.** SLI: mean = 30.17, median = 32.46; YND: mean = 46.88, median = 43.72; OND: mean = 0, median = 0 **h.** three centers **2.53a.** median **b.** mean **2.55a.** $s^2 = 4.8889$; $s = 2.211$ **b.** $s^2 = 3.3333$; $s = 1.826$ **c.** $s^2 = .1868$; $s = .432$ **2.57a.** 5, 3.7, 1.92 **b.** 99, 1,949.25, 44.15 **c.** 98, 1,307.84, 36.16 **2.59** data set 1: 1, 1, 2, 2, 3, 3, 4, 4, 5, 5; data set 2: 1, 1, 1, 1, 1, 5, 5, 5, 5, 5 **2.61a.** 3, 1.3, 1.1402 **b.** 3, 1.3, 1.1402 **c.** 3, 1.3, 1.1402 **2.63a.** .18 **b.** .0041 **c.** .064 **d.** morning **2.65a.** $s^2 = 36.178$, $s = 6.1048$ **b.** $s = 8$ **c.** $s = 5$ **d.** decrease **2.67a.** dollars; quantitative **b.** at least 3/4; at least 8/9; nothing; nothing **2.69a.** $\approx 68\%$ **b.** $\approx 95\%$ **c.** $\approx$ all **2.71** range/6 = 104.17, range/4 = 156.25; no **2.73b.** 59.82 ± 3(53.362) **2.75a.** at most 25% **b.** $\approx 2.5\%$ **2.77a.** at least 8/9 of the velocities will fall within 936 ± 3(10) **b.** no **2.79a.** adults: 6.45 ± 2.89, 6.45 ± 2(2.89); adolescents: 10.89 ± 2.48, 10.89 ± 2(2.48) **b.** adults: 31, 44; adolescents: 13, 18 **2.81a.** 19 ± 195 **b.** 7 ± 147 **c.** SAT-Math **2.83a.** 25%; 75% **b.** 50%; 50% **c.** 80%; 20% **d.** 16%; 84% **2.85a.** $z = 2$ **b.** $z = -3$ **c.** $z = -2$ **d.** $z = 1.67$ **2.87a.** skewed right **b.** 90% of the scores are below 660 **c.** 6% exceed your score **2.89a.** $z = -2.92$ **b.** $z = .75$ **2.91a.** -0.2 **b.** -0.06 **c.** $z = -4.65$ **2.93a.** $z = 2.0: 3.7$; $z = -1.0: 2.2$; $z = .5: 2.95$; $z = -2.5: 1.45$ **b.** 1.9 **c.** mound-shaped; symmetric **2.97a.** 4 **b.** 3, 5.75 **c.** 2.75 **d.** skewed to the right **e.** 50%; 75% **f.** potential outlier: 14; two outliers: 18.5 and 19.5 **2.99a.** -1.26 **b.** no **2.101a.** median for 1990 is smaller than median for 2000 **b.** IQR for 1990 is smaller than IQR for 2000 **c.** no **2.103b.** customers 238, 268, 269, and 264 **c.** $z = 1.92, 2.06, 2.13, 3.14$ **2.105** group 1: $M = 48$, IQR = 12; group 2: $M = 47.5$, IQR = 12; group 3: $M = 46$, IQR = 4 **2.107** too little; $z = -2.5$ **2.109** slight positive linear trend **2.111** very slight increasing trend **2.113a.** increasing **b.** increasing **2.115a.** no trend **b.** no trend **2.117a.** $-1, 1, 2$ **b.** $-2, 2, 4$ **c.** 1, 3, 4 **d.** .1, .3, .4 **2.119a.** 3.1234 **b.** 9.0233 **c.** 9.7857 **2.121a.** $\bar{x} = 5.67, s^2 = 1.0667, s = 1.03$ **b.** $\bar{x} = -\$1.50, s^2 = 11.5, s = \$3.39$ **c.** $\bar{x} = .4125\%, s^2 = .0883, s = .30\%$ **d.** 3; $10; .7375% **2.123** yes, positive **2.125** More than half (60.2%) of the cars received 4 star ratings. **2.127a.** current: 42.0%; retired: 58.0% **b.** Most (40 of 50) Beanie babies have values less than $100. **c.** yes, positive **2.129c.** 26.4 ± 21.2 **d.** 24.5 ± 22.4 **2.131a.** yes, Texas **b.** $\bar{x} = 16.2046$, $M = 10$, $s^2 = 476.0939, s = 21.8196$ **c.** skewed to the right **d.** at least 0; at least 3/4; at least 8/9 **e.** $\bar{x} \pm s: 47/50 = .94$; $\bar{x} \pm 2s: 49/50 = .98$; $\bar{x} \pm 3s: 49/50 = .98$ **2.133a.** frequency histogram **b.** drop in frequency in years following disaster **2.135a.** seabirds: quantitative; length: quantitative; oil: qualitative **b.** transect **c.** oiled: 38%; unoiled: 62% **e.** distributions are similar **f.** 3.27 ± 13.4 **g.** 3.495 ± 11.936 **h.** unoiled **2.137b.** yes **c.** A1775A: $\bar{x} = 19,462.2$, $s = 532.29$; A1775B: $\bar{x} = 22,838.5$, $s = 560.98$ **d.** cluster A1775A **2.139a.** quantitative: days and year **b.** lesser crimes **c.** yes **d.** mean = 41.4 days; median = 15 days **e.** no; $z = 2.38$ **f.** no

CHAPTER 3

3.1a. .5 **b.** .3 **c.** .6 **3.3** $P(A) = .55; P(B) = .50; P(C) = .70$ **3.5a.** $P(A) = 4/9$
b. $P(B) = 1/3$ **c.** 0 **3.7** $P(A) = 1/12; P(B) = 1/4; P(C) = 1/2; P(D) = 1/2$
3.9 $P(A) = 1/10; P(B) = 6/10; P(C) = 3/10$ **3.11** .05 **3.13a.** .01 **b.** yes **3.15a.** 1,
2, 3, 4, 5, 6, 7, 8, 9 **c.** .147, .101, .104, .133, .097, .157, .120, .083, .058 **d.** .248 **e.** .418 **3.17a.**
Total population, Agricultural change, Presence of industry, Growth, Population concentration **b.** .18, .05, .27, .05, .45 **c.** .63 **3.19a.** .385 **b.** .378 **c.** .184 **3.21** probabilities
are the same **3.23a.** 1 to 2 **b.** .5 **c.** .4 **3.25b.** $P(A) = 7/8$; $P(B) = 1/2$;
$P(A \cup B) = 7/8; P(A^c) = 1/8; P(A \cap B) = 1/2$ **c.** 7/8 **d.** no **3.27a.** 3/4 **b.** 13/20
c. 1 **d.** 2/5 **e.** 1/4 **f.** 7/20 **g.** 1 **h.** 1/4 **3.29a.** .65 **b.** .72 **c.** .25 **d.** .35 **e.** .72
f. .65 **g.** A and C, B and C **3.31** $A = \{$eighth-grader scores above 653 on mathematics
assessment test$\}; P(A^c) = .95$ **3.33a.** $\{11, 13, 15, 17, 29, 31, 33, 35\}$ **b.** $\{2, 4, 6, 8, 10, 11,$
13, 15, 17, 20, 22, 24, 26, 28, 29, 31, 33, 35, 1, 3, 5, 7, 9, 19, 21, 23, 25, 27$\}$ **c.** $P(A) = 9/19$;
$P(B) = 9/19; P(A \cap B) = 4/19$; $P(A \cup B) = 14/19; P(C) = 9/19$ **d.** $\{11, 13, 15, 17\}$
e. 14/19, no **f.** 2/19 **g.** $\{1, 2, 3, \ldots, 29, 31, 33, 35\}$ **h.** 16/19 **3.35a.** .15, no **b.** .36, no
c. 0, yes **d.** .82, no **e.** .46, no **3.37a.** .216 **b.** .151 **c.** .547 **d.** .475 **3.39a.** 1/2
b. 5/6 **3.41** $P(A|B) = .25, P(B|A) = .5$ **3.43b.** $P(A) = 3/4$; $P(B) = 1/2$;
$P(A \cap B) = 1/2$ **c.** $P(A|B) = 1; P(B|A) = 2/3$ **3.45** .364 **3.47a.** .857 **b.** .5
c. .333 **3.49a.** .28 **b.** .143 **3.51a.** .333 **b.** .120 **c.** no **d.** probability of playing
Center depends on race **3.53a.** .557 **b.** .075 **c.** .028 **d.** .504 **3.55a.** .553
b. $P(W|CA) = 1$; $P(W|CB) = .873$; $P(W|CC) = .559$; $P(W|BA) = .741$;
$P(W|BB) = .547$; $P(W|BC) = .342$; $P(W|AA) = .536$; $P(W|AB) = .216$;
$P(W|AC) = .077$ **c.** .856 **d.** .552 **3.57a.** $P(A) = 7/8; P(B) = 3/8; P(C) = 3/8$;
$P(D) = 1/2$; $P(A \cap B) = 3/8$; $P(A \cap D) = 3/8$; $P(B \cap C) = 0$; $P(B \cap D) = 3/8$
b. $P(B|A) = 3/7; P(A|D) = 3/4; P(C|B) = 0$ **c.** No pairs are independent. **3.59** no
3.61a. false **b.** true **c.** false **3.63** .095 **3.65a.** .1 **b.** .522 **c.** The person is guessing. **3.67a.** .02 **b.** .08 **3.69** **b.** .147 **c.** .85 **d.** .65 **e.** yes **f.** African Americans
are stopped for speeding more often than expected. **3.71a.** .116 **b.** .728 **3.73a.** .25
b. .0156 **c.** .4219 **d.** .001, .000000001, .9970 **3.75a.** probability of any sequence is .00098
b. .00196 **c.** .99804 **3.77a.** 35,820,200 **b.** 1/35,820,200 **c.** highly unlikely
3.81a. .000186 **c.** no **3.83.** $P(A) = 1/4; P(B) = 1/2$ **e.** $P(A^c) = 3/4; P(B^c) = 1/2$;
$P(A \cap B) = 1/4; P(A \cup B) = 1/2; P(A|B) = 1/2; P(B|A) = 1$ **f.** no, no **3.85a.** no
b. yes **c.** no **3.87a.** no **b.** .3, .1 **c.** .37 **3.89a.** false **b.** true **c.** true **d.** false
3.91a. .25 **b.** .13 **c.** .75 **d.** .0325 **3.93** **c.** .6637, .1815, .1263, .0214, .0071 **d.** .2029
3.95a. .03125 **b.** .15625 **c.** Pheromone has no effect. **3.97a.** .64, .32, .04 **b.** .72, .22, .06
c. dependent **3.99a.** .4096 **b.** .0016 **c.** .9984 **3.101a.** .9639 **b.** .3078 **c.** .0361 **d.** 3
3.103a. .3886 **b.** .3049 **c.** .1720 **d.** no **3.105** .4167, .2431, .1418, $(21/36)^{n-1}(15/36)$
3.107 .993 **3.109a.** .738; .262; .028; .205 **b.** .0207; .0537 **c.** .0744 **3.111a.** 2/9 **b.** 1/12
c. 7/18

CHAPTER 4

4.3a. discrete **b.** continuous **c.** continuous **d.** discrete **e.** continuous **f.** continuous
4.5a. continuous **b.** discrete **c.** discrete **d.** discrete **e.** discrete **f.** continuous **4.7a.** .25
b. .40 **c.** .75 **4.9a.** .7 **b.** .3 **c.** 1 **d.** .2 **e.** .8 **4.11a.** 3.8 **b.** 10.56 **c.** 3.2496 **e.** no
f. yes **4.13a.** $\mu_x = 1, \mu_y = 1$ **b.** distribution of x **c.** $\mu_x = 1, \sigma = .6; \mu_y = 1, \sigma = .2$
4.15a. 1 **b.** .24 **c.** .39 **d.** 1.8 **e.** .9899 **f.** .96 **4.17a.** .9985, .0015, 0, 0 **c.** 1 **4.19a.** .15
b. .15 **4.21** $\mu = -\$.50$ **4.23** \$.25 **4.25** 7/8 **4.27a.** discrete **b.** binomial **d.** $\mu = 3.5$,
$\sigma = 1.025$ **4.29a.** $\mu = 12.5, \sigma^2 = 6.25, \sigma = 2.5$ **b.** $\mu = 16$, $\sigma^2 = 12.8$, $\sigma = 3.578$
c. $\mu = 60, \sigma^2 = 24, \sigma = 4.899$ **d.** $\mu = 63, \sigma^2 = 6.3, \sigma = 2.510$ **e.** $\mu = 48$, $\sigma^2 = 9.6$,
$\sigma = 3.098$ **f.** $\mu = 40, \sigma^2 = 38.4, \sigma = 6.197$ **4.31b.** .03125, .15625, .3125, .3125,
.15625, .03125 **4.33b.** .05 **c.** 10 **4.35a.** .001 **b.** .322 **c.** .994 **4.37** **b.** 1/30 **c.** .1455
d. .1559 **4.39a.** .383 **b.** .983 **c.** 20 **4.41a.** .7908 **b.** .0555 **4.43b.** $\mu = 2.4, \sigma = 1.47$
c. $p = .9$, $q = .1, n = 24, \mu = 21.6, \sigma = 1.47$ **4.45** no **4.47a.** .4772 **b.** .3413 **c.** .4987
d. .2190 **e.** .4772 **f.** .3413 **g.** .4545 **h.** .2190 **4.49a.** 1.645 **b.** 1.96 **c.** -1.96
d. 1.28 **e.** 1.28 **4.51a.** $z = 1$ **b.** $z = -1$ **c.** $z = 0$ **d.** $z = -2.5$ **e.** $z = 3$

4.53a. 30 **b.** 14.32 **c.** 40.24 **d.** 16.84 **e.** 19.76 **f.** 36.72 **g.** 48.64 **4.55a.** .9406 **b.** .9406 **c.** .1140 **4.57a.** .1515 **b.** .1660 **4.59a.** .5124 **b.** .2912 **c.** .0027 **4.61a.** .0735 **b.** .3651 **c.** 7.29 **4.63** $d = 56.24$ **4.65a.** .3745 **b.** .7553 **4.67a.** $z_L = -.675, z_U = .675$ **b.** $-2.68, 2.68$ **c.** $-4.69, 4.69$ **d.** .0074, 0 **4.69a.** .68 **b.** .95 **c.** .997 **4.71** plot c **4.73a.** not normal **4.75a.** no **4.77a.** skewed to the right **b.** approximately normal **4.79** The lowest score of 0 is less than one standard deviation below the mean. **4.81a.** yes **b.** $\mu = 10, \sigma^2 = 6$ **c.** .726 **d.** .7291 **4.83a.** .1788 **b.** .5236 **c.** .6950 **4.85a.** 10 **b.** 3.082 **c.** .16 **d.** .4364 **4.87a.** $\mu = 6,000, \sigma^2 = 2,400$ **b.** .0202 **c.** no **4.89a.** no **b.** yes **c.** yes **4.91a.** 0 **b.** .4641 **c.** no **4.93a.** 300 **b.** 800 **c.** 1 **4.95c.** 1/16 **4.97c.** .05 **d.** no **4.99a.** 5 **b.** $E(\bar{x}) = 5$ **c.** $E(M) = 4.778$ **d.** $\bar{x}$ **4.103a.** $\mu = 2.9, \sigma^2 = 3.29, \sigma = 1.814$ **c.** $\mu_{\bar{x}} = 2.9, \sigma_{\bar{x}} = 1.283$ **4.105a.** $\mu = 20$, $\sigma = 2$ **b.** approximately normal, yes **c.** $z = -2.25$ **d.** $z = 1.50$ **4.107a.** .8944 **b.** .0228 **c.** .1292 **d.** .9699 **4.109a.** $\mu_{\bar{x}} = 89.34, \sigma_{\bar{x}} = 1.3083$ **c.** .8461 **d.** .0367 **4.111** .0838 **4.113a.** $\mu_{\bar{x}} = 5.1, \sigma_{\bar{x}} = .4981$ **b.** yes **c.** .2119 **d.** .4071 **4.115a.** .0013 **b.** program did decrease mean number of sick days **4.117a.** .0031 **b.** more likely if $\mu = 156$; less likely if $\mu = 158$ **c.** less likely if $\sigma = 2$; more likely if $\sigma = 6$ **4.119a.** .243 **b.** .131 **c.** .36 **d.** .157 **e.** .128 **f.** .121 **4.121a.** .192 **b.** .228 **c.** .772 **d.** .987 **e.** .960 **f.** 14; 4.2; 2.05 **g.** .975 **4.123a.** .9821 **b.** .0179 **c.** .9505 **d.** .3243 **e.** .9107 **f.** 0764 **4.125a.** .6915 **b.** .0228 **c.** .5328 **d.** .3085 **e.** 0 **f.** .9938 **4.127a.** .5 **b.** .0606 **c.** .0985 **d.** .8436 **4.131a.** yes **b.** .051 **c.** .757 **d.** .192 **e.** 3.678 **4.133** .8315 **4.135a.** .008 **b.** n large **4.137a.** 520; 13.5 **b.** no **4.139** approx. normal **4.141a.** approximately normal with $\mu_{\bar{x}} = 107, \sigma_{\bar{x}} = 1.637$ **b.** .0336 **c.** no **4.143** 0; unlikely

CHAPTER 5

5.1a. 1.645 **b.** 2.58 **c.** 1.96 **d.** 1.28 **5.3a.** 10.9375 ± 2.0094 **5.5a.** 83.2 ± 1.25 **c.** 83.2 ± 1.65 **d.** increases **e.** yes **5.9** yes **5.11b.** no assumptions needed **c.** 95% confident claim is false **5.13a.** 19 ± 7.826 **b.** 7 ± 5.90 **c.** SAT-Math **5.15a.** $1.13 \pm .672$ **b.** yes **5.17** (.3526, .4921) **5.19a.** $\mu_Y: 4.17 \pm .095$; μ_{MA}: $4.04 \pm .057$; $\mu_O: 4.31 \pm .062$ **b.** more likely **5.23a.** 2.228 **b.** 2.567 **c.** -3.707 **d.** -1.771 **5.25a.** 5 ± 1.876 **b.** 5 ± 2.394 **c.** 5 ± 3.754 **d.** (a) $5 \pm .780$, (b) $5 \pm .941$, (c) 5 ± 1.276; width decreases **5.27a.** 2.8856 ± 4.034 **b.** $.4083 \pm .2564$ **5.29** (0.61, 2.13) **5.31c.** (1,633.05, 2,778.85) **e.** (1,532.39, 2,658.81) **5.33a.** 17.82 ± 3.31 **b.** 14.14 ± 3.05 **c.** possibly **5.35** approximately normal with $\mu_{\hat{p}} = p$ and $\sigma_{\hat{p}} = \sqrt{\dfrac{pq}{n}}$.

5.37a. yes **b.** $.64 \pm .067$ **5.39a.** yes **b.** no **c.** yes **d.** no **5.41** $.556 \pm .044$ **5.43** $.219 \pm .024$ **5.45a.** $.368 \pm .129$ **b.** $.605 \pm .130$ **5.47a.** .636 **b.** $.636 \pm .167$ **5.49a.** $.524 \pm .095$ **5.51** 519 **5.53a.** 482 **b.** 214 **5.55** 34 **5.57a.** $n = 16: W = .98$; $n = 25: W = .784; n = 49: W = .56; n = 100: W = .392; n = 400: W = .196$ **5.59** 984 **5.61** 14,785 **5.63** 129 **5.65** 271 **5.67a.** 1,083 **b.** wider **c.** 38.3% **5.69a.** $t = 2.086$ **b.** $z = 1.96$ **c.** $z = 1.96$ **d.** $z = 1.96$ **e.** neither t nor z **5.71** (1) p; (2) μ; (3) μ; (4) μ **5.73a.** $.257 \pm .003$ **b.** $.241 \pm .003$ **5.75** $2 \pm .47$ **5.77a.** $.660 \pm .029$ **b.** $.301 \pm .035$ **5.79a.** $.044 \pm .162$ **b.** no evidence to indicate that this species has tendency to inbreed **5.81** (.39, 1.23) **5.83** $49.70 \pm .1498$ **5.85a.** $.378 \pm .156$ **b.** $.703 \pm .147$ **c.** 41.405 ± 21.493 **d.** $.378 \pm .156$ **5.87a.** .094 **b.** yes **c.** $.094 \pm .037$ **5.89a.** 49.3 ± 8.60 **c.** Population is normal. **d.** 35

CHAPTER 6

6.1 null hypothesis **6.3** α **6.7** no **6.9a.** $H_0: p = .45$ **b.** $H_0: \mu = 2.5$ **6.11a.** $H_0: p = .75, H_a: p > .75$ **6.13a.** $H_0: \mu = 15, H_a: \mu < 15$ **b.** conclude that mean level of mercury in 2000 is less than 15 ppm when mean is equal to 15 ppm **c.** conclude that mean level of mercury in 2000 is equal to 15 ppm when mean is less than 15 ppm **6.15b.** .37, .24 **6.17g.** (a) .025, (b) .05, (c) $\approx$ .005, (d) $\approx$ .10, (e) .10, (f) $\approx$ .01 **6.19a.** $z = 1.67$, reject H_0 **b.** $z = 1.67$, fail to reject H_0 **6.21** yes, $z = 12.36$ **6.23a.** $H_0: \mu = 10, H_a: \mu > 10$; one-tailed **b.** $z > 1.645$ **c.** $z = 1.29$, fail to reject H_0 **6.25a.** $H_0: \mu = 16, H_a: \mu < 16$ **b.** $z = -4.31$, reject H_0 **6.27a.** $z = 4.03$, reject H_0 **6.29a.** $z = -3.72$, reject H_0

6.31a. fail to reject H_0 **b.** reject H_0 **c.** reject H_0 **d.** fail to reject H_0 **e.** fail to reject H_0
6.33 .0150   **6.35** p-value = .9279, fail to reject H_0 **6.37a.** fail to reject H_0 **b.** fail to reject H_0 **c.** reject H_0 **d.** fail to reject H_0 **6.39a.** .0985; fail to reject H_0 at $\alpha = .05$
b. smaller   **6.41** .0002; reject H_0 at $\alpha = .05$ **6.43b.** reject H_0 at $\alpha = .05$ **c.** reject H_0 at $\alpha = .05$ **6.45** small n, normal population   **6.49a.** $t < -2.160$ or $t > 2.160$ **b.** $t > 2.500$
c. $t > 1.397$ **d.** $t < -2.718$ **e.** $t < -1.729$ or $t > 1.729$ **f.** $t < -2.353$
6.51a. $t = -2.064$; fail to reject H_0 **b.** $t = -2.064$; fail to reject H_0 **c.** (a) $.05 < p$-value $< .10$; (b) $.10 < p$-value $< .20$ **6.53** $t = -.43$; fail to reject H_0 **6.55a.** H_a:
$\mu \neq 1.0$ **b.** $t = 1.44$ **c.** p-value = .169, fail to reject H_0 **6.57** $t = 2.97$, fail to reject H_0: $\mu = .100$ **6.59** yes, $t = 1.95$ **6.61a.** yes   **b.** no   **c.** yes   **d.** no   **e.** no
6.63a. $z = -2.00$ **c.** reject H_0 **d.** .0228   **6.65a.** $z = 1.13$, fail to reject H_0 **b.** .1292
6.67 yes, $z = 1.74$ **6.69** no, $z = -2.37$ **6.71a.** H_0: $p = .5$, H_a: $p \neq .5$ **b.** $z = 5.99$
c. yes   **d.** reject H_0 **6.73** $z = 33.47$, reject H_0: $p = .5$ **6.75a.** yes, $z = 1.85$ **b.** yes,
$z = 3.09$ **6.77a.** .063   **b.** .500   **c.** .004   **d.** .151; .1515   **e.** .212; .2119   **6.79** $S = 16$,
p-value = .115, do not reject H_0 **6.81a.** H_0: $\eta = 1.5$, H_a: $\eta > 1.5$ **b.** 3   **c.** .855   **d.** do not reject H_0 **6.83a.** H_0: $\eta = 5$ **b.** $z = 6.07$, p-value = 0   **c.** reject H_0 **6.85a.** yes,
$z = 2.12$ **b.** no   **6.87** H_0, H_a, α **6.89** alternative   **6.91a.** $z = -1.78$, fail to reject H_0
b. $z = -1.78$, fail to reject H_0 **c.** $.29 \pm .063$ **d.** $.29 \pm .083$ **e.** 549   **6.93a.** .1288
b. normal population   **c.** .2576   **6.95a.** H_0: Drug is unsafe, H_a: Drug is safe   **c.** α
6.97a. $t = -3.46$, reject H_0 **b.** normal population   **6.99a.** $t = 3.512$; $.025 < p < .05$
b. .0362   **c.** normal population   **d.** Type I   **6.101a.** $z = .70$, fail to reject H_0 **b.** yes
6.103a. no, $z = 1.41$ **b.** small α **c.** power increases   **6.105** $z = 21.66$, reject H_0
6.107a. .1949; Type II error   **b.** .05; Type I error   **c.** .8051

CHAPTER 7

7.1a. 35 ± 24.5 **b.** p-value = .0052, reject H_0 **c.** p-value = .0026   **d.** p-value = .4238,
fail to reject H_0 **e.** independent random samples   **7.3a.** no   **b.** no   **c.** no   **d.** yes
e. no   **7.5a.** .5989   **b.** $t = -2.39$, reject H_0 **c.** $-1.24 \pm .98$ **7.7a.** p-value = .1150, fail to reject H_0 at $\alpha = .10$ **b.** p-value = .0575, reject H_0 at $\alpha = .10$ **7.9a.** $z = -2.64$, reject H_0 **b.** $z = .27$, fail to reject H_0 **7.11a.** $(-.60, 7.96)$ **b.** independent random samples from normal populations with equal variances   **7.13a.** no, $z = 1.19$ **b.** $.5 \pm .69$; yes
c. independent random samples   **d.** .2340   **7.15a.** $t = -7.24$, p-value = 0; reject H_0
b. $t = -.50$, p-value = .62; fail to reject H_0 **c.** $t = -1.25$, p-value = .21; fail to reject H_0
d. independent random samples   **7.17a.** $(.078, 2.082)$ **b.** .077; yes   **7.19a.** $-.52 \pm 3.546$
b. both populations normal and $\sigma_1^2 = \sigma_2^2$ **c.** 8.59 ± 3.152 **7.21** $t = -.46$, fail to reject H_0
7.23a. no standard deviations reported   **b.** $s_1 = s_2 = 5$ **c.** $s_1 = s_2 = 6$ **7.25a.** $t > 1.833$
b. $t > 1.328$ **c.** $t > 2.776$ **d.** $t > 2.896$ **7.27a.** $\bar{x}_D = 2$, $s_D^2 = 2$ **b.** $\mu_D = \mu_1 - \mu_2$
c. 2 ± 1.484 **d.** $t = 3.46$, reject H_0 **7.29a.** $t = .81$, fail to reject H_0 **b.** p-value $> .20$
7.31a. 749.5 ± 278.1 **7.33a.** H_0: $\mu_D = 0$, H_a: $\mu_D \neq 0$ **b.** $z = 2.08$, p-value = .0376
c. reject H_0 **7.35b.** $t = .66$, fail to reject H_0 **c.** height: p-value = .210, fail to reject H_0;
weight: p-value = .149, fail to reject H_0 **7.37a.** H_0: $\mu_D = 0$, H_a: $\mu_D > 0$ **b.** paired difference test   **d.** leadership: fail to reject H_0 at $\alpha = .05$; popularity: reject H_0 at $\alpha = .05$; intellectual self-confidence: reject H_0 at $\alpha = .05$ **7.39** $t = -2.306$, p-value = .0163, reject
H_0 at $\alpha = .05$ **7.41** $n_1 = n_2 = 24$ **7.43** yes   **7.45** $n_1 = n_2 = 49$ **7.47** $n_1 = n_2 = 1{,}729$
7.49 $n_1 = n_2 = 136$ **7.51b.** $T_B = 42.5$, reject H_0 **7.53a.** $T_1 = 62.5$ **b.** yes,
$T_1 = 62.5$ **7.55a.** H_0: $D_{female} = D_{normal}$, H_a: $D_{female} > D_{normal}$ **c.** 174.5   **d.** 150.5
e. $z > 1.28$ **f.** $z = 1.72$, reject H_0 **7.57a.** $T_1 = 150.5$, reject H_0 **b.** $z = 3.44$, reject H_0
7.59a. yes, $T_2 = 105$ **b.** less than   **7.61a.** data not normal   **b.** $z = -6.47$, reject H_0
c. $z = -.39$, do not reject H_0 **d.** $z = -1.35$, do not reject H_0 **7.63a.** H_a: $D_A > D_B$
b. $T_- = 3.5$, reject H_0 **7.65a.** H_a: $D_A > D_B$ **b.** $z = 2.499$, reject H_0 **c.** .0062   **7.67a.** data not normal   **b.** H_0: $D_{father} = D_{mother}$, H_a: $D_{father} \neq D_{mother}$ **d.** $T_+ = 22$; $T_- = 44$ **e.** $T_+ \leq 11$
f. do not reject H_0 **7.69** no, p-value = .449   **7.71** no, $T_- = 13$ **7.73a.** 6.59   **b.** 16.69
c. 1.61   **d.** 3.87   **7.75e.** 37.5; 5.21   **7.77a.** 4; 38   **b.** $F = 14.80$; reject H_0 **c.** sample means
7.79a. independent samples   **b.** treatments: 3, 6, 9, 12 robots; dep. var.: energy expanded
c. H_0: $\mu_3 = \mu_6 = \mu_9 = \mu_{12}$, H_a: At least two μ's differ   **d.** reject H_0 **7.81a.** H_0:
$\mu_{young} = \mu_{middle} = \mu_{old}$ **b.** reject H_0 **c.** H_0: $\mu_{young} = \mu_{middle} = \mu_{old}$; fail to reject H_0

7.83a. independant samples   **b.** $F = 6.79$; reject H_0 **c.** .0041   **7.85** $F = 3.96$; do not reject H_0 **7.87a.** 52.32   **b.** 77.24

c.

Source	df	SS	MS	F
Treatment	2	52.32	26.16	12.89
Error	38	77.24	2.03	
Total	40	129.56		

d. yes, $F = 12.89$ **f.** Subjects in the first two groups were not randomly and independently assigned to their groups.   **7.89a.** $3.9 \pm .31$ **b.** $z = 20.60$, reject H_0 **c.** $n_1 = n_2 = 346$ **7.91a.** $t = 5.73$, reject H_0 **b.** 3.8 ± 1.84 **7.93a.** reject H_0 **b.** fail to reject H_0 **c.** reject H_0 **d.** reject H_0 **e.** fail to reject H_0 **f.** fail to reject H_0 **7.95a.** no, $r_s = .40$ **b.** yes, $T_- = 1.5$ **7.97a.** $H_0: \mu_1 - \mu_2 = 0, H_a: \mu_1 - \mu_2 \neq 0$ **b.** $z = -7.69$, reject H_0 **7.99a.** yes   **b.** yes   **7.101b.** $H_0: \mu_1 - \mu_2 = 0, H_a: \mu_1 - \mu_2 > 0$ **c.** p-value $= .148$, fail to reject H_0 at $\alpha = .05$ **d.** both populations normal and $\sigma_{plant}^2 = \sigma_{upstream}^2$ **7.103a.** t-test   **b.** yes   **c.** no   **d.** fail to reject H_0 **e.** reject H_0 at $\alpha = .05$ **7.105a.** Infrequency: $F = 155.8$, reject H_0; Obvious: $F = 49.7$, reject H_0; Subtle: $F = 10.3$, reject H_0; Obvious-Subtle: $F = 45.4$, reject H_0; Dissimulation: $F = 39.1$, reject H_0 **b.** no   **c.** Infrequency: $\mu_{CSH} < (\mu_{NFP}, \mu_{FP}) < \mu_{CSFB}$; Obvious: $\mu_{CSH} < (\mu_{NFP}, \mu_{FP}) < \mu_{CSFB}$; Subtle: no significant differences; Obvious-Subtle: $(\mu_{CSH}, \mu_{NFP}, \mu_{FP}) < \mu_{CSFB}$; Dissimulation: $(\mu_{CSH}, \mu_{NFP}, \mu_{FP}) < \mu_{CSFB}$ **7.107** $-.33 \pm .22$ **7.109a.** $T_1 = 82$ **b.** no, paired design   **7.111** yes, $t = -2.19$

CHAPTER 8

8.3a. $z < -2.33$ **b.** $z < -1.96$ **c.** $z < -1.645$ **d.** $z < -1.28$ **8.5a.** $.07 \pm .067$ **b.** $.06 \pm .086$ **c.** $-.15 \pm .131$ **8.7** $z = 1.16$, fail to reject H_0 **8.9a.** .143   **b.** .049   **c.** $z = 2.27$, fail to reject H_0 **d.** reject H_0 **8.11** $z = -1.201$, fail to reject H_0 **8.13** yes; p-value $= .033$, reject H_0 at $\alpha = .05$ **8.15** $z = 1.21$, fail to reject H_0 at $\alpha = .01$ **8.17a.** yes, $z = 13.02$ **8.19a.** $n_1 = n_2 = 29,954$ **8.21** $n_1 = n_2 = 130$ **8.23** $n_1 = 520$, $n_2 = 260$ **8.25a.** 18.3070   **b.** 29.7067   **c.** 23.5418   **d.** 79.4900   **8.27a.** $\chi^2 > 5.99147$ **b.** $\chi^2 > 7.77944$ **c.** $\chi^2 > 11.3449$ **8.29a.** no, $\chi^2 = 3.293$ **8.31a.** Pottery type; burnished, monochrome, painted, other   **b.** $p_1 = p_2 = p_3 = p_4 = .25$ **c.** $H_0: p_1 = p_2 = p_3 = p_4 = .25$ **d.** $\chi^2 = 436.59$ **e.** p-value $= .0000$ **8.33** $\chi^2 = 16$, reject H_0 **8.35a.** $\chi^2 = 17.16$, reject H_0 **b.** $.376 \pm .103$ **c.** yes   **8.37a.** $\chi^2 > 26.2962$ **b.** $\chi^2 > 15.9871$ **c.** $\chi^2 > 9.21034$ **8.39a.** column 1: 36%, 64%; column 2: 53.1%, 46.9%; column 3: 67.9%, 32.1%; totals: 57.5%, 42.5%   **8.41a.** B_1: 29.9%; B_2: 44.2%; B_3: 29.6%   **b.** B_1: 47.0%; B_2: 32.5%; B_3: 49.3%   **c.** B_1: 23.1%; B_2: 23.3%; B_3: 21.1%

8.43a.

		LLD		
		Yes	No	Total
Genetic Trait	Yes	21	15	36
	No	150	306	456
	Total	171	321	492

b. $\chi^2 = 9.52$, reject H_0 **8.45** $\chi^2 = 35.41$, reject H_0 **8.47** $\chi^2 = 6.171$, reject H_0 **8.49a.** no

b.

		CAHS		
		Low	Medium	High
	Low	32	14	2
SHSS:C	Medium	11	14	6
	High	6	16	29

c. yes   **d.** $\chi^2 = 46.70$, reject H_0 **8.51a.** $\chi^2 = 4.407$, reject H_0 **b.** no   **c.** .0438   **d.** .0057; .0003   **e.** p-value $= .0498$, reject H_0 **8.53a.** no, $\chi^2 = 2.133$ **b.** $.233 \pm .057$ **8.55a.** possibly not   **c.** $\chi^2 = 12.9$ **d.** yes, reject H_0 **e.** $.233 \pm .200$ **8.57a.** $\chi^2 = 1.143$, do not reject H_0 **b.** $.179 \pm .119$ **8.59** yes, $\chi^2 = 180.874$ **8.61** $z = 2.55$, reject H_0 **8.63** yes, $\chi^2 = 9.35$ **8.65** $z = 2.45$, reject H_0 **8.67** $\chi^2 = 39.27$, reject H_0 **8.69a.** $\chi^2 = 9.647$ **b.** 11.0705   **c.** no   **d.** $.05 < p\text{-value} < .10$

CHAPTER 9

9.3 $\beta_1 = 1/3, \beta_0 = 14/3, y = 14/3 + (1/3)x$ **9.9** no   **9.13c.** $\hat{\beta}_1 = .198, \hat{\beta}_0 = .020$ **e.** -1 to 7   **9.15a.** $y = \beta_0 + \beta_1 x + \varepsilon$; negative   **b.** yes   **c.** no   **9.17a.** $y = \beta_0 + \beta_1 x + \varepsilon$ **b.** $\hat{y} = 175.7033 - .8195x$ **9.19a.** $\hat{y} = 6.25 - .0023x$ **c.** 5.56   **9.21a.** yes; positive   **b.** $y = \beta_0 + \beta_1 x + \varepsilon$ **c.** $\hat{\beta}_0 = 20.1275, \hat{\beta}_1 = .62442$ **9.23** $\hat{y} = .5704 + .0264x$ **9.25a.** 57.5; 3.19444   **b.** 257.5; 6.7763   **c.** 9.288; 1.1610   **9.27** graph   **b**   **9.29a.** SSE $= 1,667.65, s^2 = 238.24, s = 15.43$ **9.31a.** SSE $= 14,357.5, s^2 = 1305.22, s = 36.1279$ **9.33b.** Brand A: $\hat{y} = 6.62 - .0727x$; Brand B: $\hat{y} = 9.31 - .1077x$ **c.** Brand A: SSE $= 19.056, s^2 = 1.466, s = 1.211$;   Brand   B: SSE $= 4.833, s^2 = .372, s = .610$ **d.** Brand   A: $(-.891, 3.593)$;   Brand   B: $(.551, 2.991)$ **e.** Brand   B   **9.35b.** $\hat{y} = .544 + .617x$ **d.** $H_0: \beta_1 = 0, H_a: \beta_1 \neq 0$ **e.** $t = 5.50$, df $= 5$ **f.** reject H_0 **9.37** no, $t = .627$ **9.39** $(-.0039, -.0007)$ **9.41c.** $\alpha > .042$ **9.43a.** $t = 6.286$, reject H_0 **b.** $.88 \pm .24$ **c.** no evidence that the true slope differs from 1   **9.45** yes, $t = 2.858$ **9.47** $\hat{y} = 2.55 + 2.76x$ **9.49a.** perfect positive linear   **b.** perfect negative linear   **c.** strong positive linear   **d.** weak positive linear   **e.** strong negative linear   **9.51a.** $r = .9853, r^2 = .9709$ **b.** $r = -.9934, r^2 = .9868$ **c.** $r = 0, r^2 = 0$ **d.** $r = 0, r^2 = 0$ **9.53a.** $H_0: \rho = 0, H_a: \rho \neq 0$ **b.** do not reject H_0 at $\alpha = .05$ **9.55b.** piano: $r^2 = .1998$; bench: $r^2 = .0032$; motorbike: $r^2 = .3832$, armchair: $r^2 = .0864$; teapot: $r^2 = .9006$ **c.** Reject H_0 for all objects except bench and armchair.   **9.57a.** $y = \beta_0 + \beta_1 x + \varepsilon$ **b.** $\beta_1 > 0$ **d.** no, $t = -3.12$ **9.59** $r = .5702, r^2 = .3251$ **9.61a.** negative   **b.** $r < -.0895$ **9.63c.** 6.5335 ± 1.9487 **d.** -6.0546 ± 1.7225 **9.65c.** 4.644 ± 1.118 **d.** $-.414 \pm 1.717$ **9.67a.** no   **b.** $(275.86, 407.26)$ **c.** $(332.95, 350.17)$ **9.69a.** $(1.0516, 6.6347)$ **b.** narrower   **c.** no   **9.71a.** $y = \beta_0 + \beta_1 x + \varepsilon$ **b.** $\hat{y} = 155.912 - 1.086x$ **c.** no, $t = -2.05$ **d.** $(-21.56, 279.09)$ **9.73a.** Brand A: $3.349 \pm .587$; Brand B: $4.464 \pm .296$ **b.** Brand A: 3.349 ± 2.224;   Brand B: 4.464 ± 1.120 **c.** $-.65 \pm 3.606$ **9.75a.** .4   **b.** $-.9$ **c.** $-.2$ **d.** .2   **9.77c.** .713   **d.** $|r_s| > .425$ **e.** reject H_0 **9.79** no, $r_s = .657$ **9.81a.** $-.733$ **b.** do not reject H_0 **9.83** $r_s = -.812$, reject H_0 **9.85b.** $\hat{y} = x; \hat{y} = 3$ **c.** $\hat{y} = x$ **d.** least squares line has the smallest SSE   **9.87b.** $-.1245; .0155$ **c.** no, $t = -.35$ **9.89a.** $\hat{y} = 40.7842 + .76556x$ **c.** yes, $t = 4.38$ **d.** $.76556 \pm .4035$ **e.** $(72.513, 85.611)$ **f.** $(57.95, 100.17)$ **9.91** no, $r_s = .479$ **9.93** $\hat{y} = -92.458 + 8.347x; t = 3.248$, reject H_0; $r = .42$ **9.95a.** $\hat{\beta}_0 = -13.49, \hat{\beta}_1 = -.0528$ **b.** $-.0528 \pm .0178$; yes   **c.** $r^2 = .8538$ **d.** $(.5987, 1.2653)$ **9.97b.** $\hat{y} = 78.516 - .2389x$ **d.** $t = -2.31$, do not reject H_0 **e.** $(301, 8.5)$ **f.** $\hat{y} = 139.759 - .4497x; t = -15.35$, reject H_0 **9.99e.** (a): $r^2 = .0676$; (b): $r^2 = .0361$; (c): $r^2 = .0196$; (d): $r^2 = .0001$

INDEX

LICENSE AGREEMENT

YOU SHOULD CAREFULLY READ THE FOLLOWING TERMS AND CONDITIONS BEFORE BREAKING THE SEAL ON THE PACKAGE. AMONG OTHER THINGS, THIS AGREEMENT LICENSES THE ENCLOSED SOFTWARE TO YOU AND CONTAINS WARRANTY AND LIABILITY DIS-CLAIMERS. BY BREAKING THE SEAL ON THE PACKAGE, YOU ARE ACCEPTING AND AGREEING TO THE TERMS AND CONDITIONS OF THIS AGREEMENT. IF YOU DO NOT AGREE TO THE TERMS OF THIS AGREEMENT, DO NOT BREAK THE SEAL. YOU SHOULD PROMPTLY RE-TURN THE PACKAGE UNOPENED.

LICENSE.

Subject to the provisions contained herein, Prentice-Hall, Inc. ("PH") hereby grants to you a non-exclusive, non-transferable license to use the object code version of the computer software product ("Software") contained in the package on a single computer of the type identified on the package.

SOFTWARE AND DOCUMENTATION.

PH shall furnish the Software to you on media in machine-readable object code form and may also provide the standard documentation ("Documentation") containing instructions for operation and use of the Software.

LICENSE TERM AND CHARGES.

The term of this license commences upon delivery of the Software to you and is perpetual unless earlier terminated upon default or as otherwise set forth herein.

TITLE.

Title, and ownership right, and intellectual property rights in and to the Software and Documentation shall remain in PH and/or in suppliers to PH of programs contained in the Software. The Software is provided for your own internal use under this license. This license does not include the right to sublicense and is personal to you and therefore may not be assigned (by operation of law or otherwise) or transferred without the prior written consent of PH. You acknowledge that the Software in source code form remains a confidential trade secret of PH and/or its suppliers and therefore you agree not to attempt to decipher or decompile, modify, disassemble, reverse engineer or prepare derivative works of the Software or develop source code for the Software or knowingly allow others to do so. Further, you may not copy the Documentation or other written materials accompanying the Software.

UPDATES.

This license does not grant you any right, license, or interest in and to any improvements, modifications, enhancements, or updates to the Software and Documentation. Updates, if available, may be obtained by you at PH's then current standard pricing, terms, and conditions.

LIMITED WARRANTY AND DISCLAIMER.

PH warrants that the media containing the Software, if provided by PH, is free from defects in material and workmanship under normal use for a period of sixty (60) days from the date you purchased a license to it.

THIS IS A LIMITED WARRANTY AND IT IS THE ONLY WARRANTY MADE BY PH. THE SOFTWARE IS PROVIDED 'AS IS' AND PH SPECIFI-CALLY DISCLAIMS ALL WARRANTIES OF ANY KIND, EITHER EXPRESS OR IMPLIED, INCLUDING, BUT NOT LIMITED TO, THE IMPLIED WAR-RANTY OF MERCHANTABILITY AND FITNESS FOR A PARTICULAR PURPOSE. FURTHER, COMPANY DOES NOT WARRANT, GUARANTY OR MAKE ANY REPRESENTATIONS REGARDING THE USE, OR THE RESULTS OF THE USE, OF THE SOFTWARE IN TERMS OF CORRECT-NESS, ACCURACY, RELIABILITY, CURRENTNESS, OR OTHERWISE AND DOES NOT WARRANT THAT THE OPERATION OF ANY SOFTWARE WILL BE UNINTERRUPTED OR ERROR FREE. COMPANY EXPRESSLY DISCLAIMS ANY WARRANTIES NOT STATED HEREIN. NO ORAL OR WRITTEN INFORMATION OR ADVICE GIVEN BY PH, OR ANY PH DEALER, AGENT, EMPLOYEE OR OTHERS SHALL CREATE, MODIFY OR EXTEND A WARRANTY OR IN ANY WAY INCREASE THE SCOPE OF THE FOREGOING WARRANTY, AND NEITHER SUBLICENSEE OR PUR-CHASER MAY RELY ON ANY SUCH INFORMATION OR ADVICE. If the media is subjected to accident, abuse, or improper use; or if you violate the terms of this Agreement, then this warranty shall immediately be terminated. This warranty shall not apply if the Software is used on or in conjunction with hardware or programs other than the unmodified version of hardware and programs with which the Software was designed to be used as described in the Documentation.

LIMITATION OF LIABILITY.

Your sole and exclusive remedies for any damage or loss in any way connected with the Software are set forth below. UNDER NO CIRCUMSTANCES AND UNDER NO LEGAL THEORY, TORT, CONTRACT, OR OTHERWISE, SHALL PH BE LIABLE TO YOU OR ANY OTHER PERSON FOR ANY INDI-RECT, SPECIAL, INCIDENTAL, OR CONSEQUENTIAL DAMAGES OF ANY CHARACTER INCLUDING, WITHOUT LIMITATION, DAMAGES FOR LOSS OF GOODWILL, LOSS OF PROFIT, WORK STOPPAGE, COMPUTER FAILURE OR MALFUNCTION, OR ANY AND ALL OTHER COMMER-CIAL DAMAGES OR LOSSES, OR FOR ANY OTHER DAMAGES EVEN IF PH SHALL HAVE BEEN INFORMED OF THE POSSIBILITY OF SUCH DAMAGES, OR FOR ANY CLAIM BY ANY OTHER PARTY. PH'S THIRD PARTY PROGRAM SUPPLIERS MAKE NO WARRANTY, AND HAVE NO LIABILITY WHATSOEVER, TO YOU. PH's sole and exclusive obligation and liability and your exclusive remedy shall be: upon PH's election, (i) the replace-ment of your defective media; or (ii) the repair or correction of your defective media if PH is able, so that it will conform to the above warranty; or (iii) if PH is un-able to replace or repair, you may terminate this license by returning the Software. Only if you inform PH of your problem during the applicable warranty period will PH be obligated to honor this warranty. You may contact PH to inform PH of the problem as follows:

SOME STATES OR JURISDICTIONS DO NOT ALLOW THE EXCLUSION OF IMPLIED WARRANTIES OR LIMITATION OR EXCLUSION OF CON-SEQUENTIAL DAMAGES, SO THE ABOVE LIMITATIONS OR EXCLUSIONS MAY NOT APPLY TO YOU. THIS WARRANTY GIVES YOU SPE-CIFIC LEGAL RIGHTS AND YOU MAY ALSO HAVE OTHER RIGHTS WHICH VARY BY STATE OR JURISDICTION.

MISCELLANEOUS.

If any provision of this Agreement is held to be ineffective, unenforceable, or illegal under certain circumstances for any reason, such decision shall not affect the validity or enforceability (i) of such provision under other circumstances or (ii) of the remaining provisions hereof under all circumstances and such provision shall be reformed to and only to the extent necessary to make it effective, enforceable, and legal under such circumstances. All headings are solely for convenience and shall not be considered in interpreting this Agreement. This Agreement shall be governed by and construed under New York law as such law applies to agreements between New York residents entered into and to be performed entirely within New York, except as required by U.S. Government rules and regulations to be governed by Federal law.

YOU ACKNOWLEDGE THAT YOU HAVE READ THIS AGREEMENT, UNDERSTAND IT, AND AGREE TO BE BOUND BY ITS TERMS AND CON-DITIONS. YOU FURTHER AGREE THAT IT IS THE COMPLETE AND EXCLUSIVE STATEMENT OF THE AGREEMENT BETWEEN US THAT SU-PERSEDES ANY PROPOSAL OR PRIOR AGREEMENT, ORAL OR WRITTEN, AND ANY OTHER COMMUNICATIONS BETWEEN US RELATING TO THE SUBJECT MATTER OF THIS AGREEMENT.

U.S. GOVERNMENT RESTRICTED RIGHTS.

Use, duplication or disclosure by the Government is subject to restrictions set forth in subparagraphs (a) through (d) of the Commercial Computer-Restricted Rights clause at FAR 52.227-19 when applicable, or in subparagraph (c) (1) (ii) of the Rights in Technical Data and Computer Software clause at DFARS 252.227-7013, and in similar clauses in the NASA FAR Supplement.